THE DEVELOPMENT OF
HUMAN GENE THERAPY

**COLD SPRING HARBOR
MONOGRAPH SERIES**

The Lactose Operon
The Bacteriophage Lambda
The Molecular Biology of Tumour Viruses
Ribosomes
RNA Phages
RNA Polymerase
The Operon
The Single-Stranded DNA Phages
Transfer RNA:
    Structure, Properties, and Recognition
    Biological Aspects
Molecular Biology of Tumor Viruses, Second Edition:
    DNA Tumor Viruses
    RNA Tumor Viruses
The Molecular Biology of the Yeast *Saccharomyces:*
    Life Cycle and Inheritance
    Metabolism and Gene Expression
Mitochondrial Genes
Lambda II
Nucleases
Gene Function in Prokaryotes
Microbial Development
The Nematode *Caenorhabditis elegans*
Oncogenes and the Molecular Origins of Cancer
Stress Proteins in Biology and Medicine
DNA Topology and Its Biological Implications
The Molecular and Cellular Biology of the Yeast *Saccharomyces:*
    Genome Dynamics, Protein Synthesis, and Energetics
    Gene Expression
    Cell Cycle and Cell Biology
Transcriptional Regulation
Reverse Transcriptase
The RNA World
Nucleases, Second Edition
The Biology of Heat Shock Proteins and Molecular Chaperones
Arabidopsis
Cellular Receptors for Animal Viruses
Telomeres
Translational Control
DNA Replication in Eukaryotic Cells
Epigenetic Mechanisms of Gene Regulation
C. elegans II
Oxidative Stress and the Molecular Biology of Antioxidant Defenses
RNA Structure and Function
The Development of Human Gene Therapy

THE DEVELOPMENT OF HUMAN GENE THERAPY

Edited by

Theodore Friedmann
University of California, San Diego

COLD SPRING HARBOR LABORATORY PRESS
Cold Spring Harbor, New York

**THE DEVELOPMENT OF
HUMAN GENE THERAPY**

Monograph 36
© 1999 by Cold Spring Harbor Laboratory Press,
   Cold Spring Harbor, New York
All rights reserved
Printed in the United States of America

Project Coordinator: Liz Powers
Production Editor: Christopher Bianco
Desktop Editor: Susan Schaefer, Danny deBruin
Interior Book Designer: Emily Harste

ISBN 0-87969-528-5
ISSN 0270-1847

Library of Congress Cataloging-in-Publication Date

The development of human gene therapy / edited by Theodore Friedmann.
        p. cm. -- (Cold Spring Harbor monograph series, ISSN
   0270-1847 ; monograph 36)
      Includes bibliographical references and index.
      ISBN 0-87969-528-5
      1. Gene therapy. 2. Human genetics. I. Friedmann, Theodore,
   1935–   . II. Series: Cold Spring Harbor monograph series ; 36.
      RB155.8.D48 1998
      616'.042--dc21   98-25205
                                                                    CIP

All Cold Spring Harbor Laboratory Press publications may be ordered directly from Cold
Spring Harbor Laboratory Press, 10 Skyline Drive, Plainview, New York 11803-2500. Phone:
1-800-843-4388 in Continental U.S. and Canada. All other locations: (516) 349-1930. FAX:
(516) 349-1946. E-mail: cshpress@cshl.org. For a complete catalog of Cold Spring Harbor
Laboratory Press publications, visit our World Wide Web Site http://www.cshl.org

*The editor wishes to dedicate this book to
teachers and colleagues who inspired,
and to family who tolerated.
Special thanks to
Giulio Barbero, Fred Sanger,
Christian Anfinsen, J. Edwin Seegmiller,
Renato Dulbecco, Richard Roblin,
and Ingrid.*

Contents

Preface

The development of the concepts and tools of human gene therapy are something of an anomaly in the history of medicine. In most instances, general acceptance of important new directions in medicine by the medical profession and by society in general has in the past been relatively slow and occurred only after demonstrations of the true therapeutic efficacy of new therapeutic tools. Of course, the first appearance of epochal new directions in medicine has often come to medical and public attention explosively and controversially, as in the case of the discovery of anesthesia and antibiotics, the first kidney transplants, and even to greater extent, the early stages of heart transplantation. However, it has usually been only after clinical successes have accumulated that the importance of these new fields of medicine have come to dominate medical thought.

Gene therapy has been different—It has had a somewhat upside-down history, and it has moved more quickly than most new areas of medicine to become a central driving force, long before clinical successes have occurred. First, it must be remembered that gene therapy is really two things: It is the concept itself of attacking human disease at the site of the causative genetic defect, and it is also the implementation of that concept into clinical and therapeutic reality. Although it was in the earliest days usually an embarrassment to mention "gene therapy" in polite scientific circles, a rapid and successful evolution of the concept from strange new idea to broad medical and societal acceptance emerged, fueled by a pervasive public curiosity and interest in all things genetic and a persistent and, at least in the early days, generally responsible coverage by the media. Despite a period in the early-mid 1980s during which some in the public policy and theological arenas fomented mistrust and misunderstanding of the goals and techniques of gene therapy, the obvious correctness and inevitability of the concept, together with the receptiveness of the public and its public policy officials, produced an astounding medical and societal acceptance of the concept of human gene therapy long before it was able to provide truly believable clinical benefits. That acceptance has not been seriously eroded by the recent retreat by the general gene therapy community from the period of overexpectation and overstatement, as reviewed in chapters of this volume. The success of the concept of gene therapy has been phenomenal and represents a truly epochal new direction for medicine.

The universal acceptance of that concept indicates that the conceptual revolution is over. There are few if any who doubt that gene therapy will soon be a clinical reality. The road to convincing clinical application is slow, incremental but inexorable, thereby adding clinical reality to the conceptual victory. By the time this volume appears in print, a handful of reports will almost certainly begin convincingly to show clinical improvement in patients with several genetic diseases and even with some kinds of cancer. The imminent success of the field owes an enormous debt to those who illuminated the mechanisms of pathogenesis of human disease, those who provided the initial concept of genetic correction of human disease, those who developed the genetic and cellular tools for effective gene transfer, those who called for ethical examination and justification of the field, and still others who developed the public policy and regulatory mechanisms for implementing human studies, and of course to clinicians who pursued early clinical studies even with imperfect tools.

Why a Cold Spring Harbor Laboratory book on gene therapy and why now? As it does for many other areas of biomedicine, the Laboratory serves to keep its fingers on the pulse of much of modern biomedicine and to serve as one of the most important avenues for the communication of scientific advances through its meetings, symposia, and publications. Very early in the evolution of the field, the Laboratory convened an influential conference at its Banbury Center in 1981 on the prospects for human gene therapy. This book continues the Laboratory's commitment to this new area of medicine as a vital and important one. The book is not meant to serve as a source of detailed technical information for any of the current approaches to human gene therapy or as a primer on "how to do it." That kind of analysis is doomed to become dated very quickly in any rapidly evolving field of biomedical science and especially in human gene therapy, which is so early in its evolution. Most investigators in the field believe that the tools which will eventually be used for wide-scale delivery of really effective gene therapy in a clinical setting will be different from the tools that we currently have available. Rather, the book is meant to chronicle some of the major developments in the evolution of this new field of medicine, to illustrate how concepts have shaped the development of technology, and to emphasize the still immature nature of the field. The book therefore is meant as a mixture of history and technology. We have included a chapter describing very recently developed technology that, if confirmed, will lead eventually to one of the Holy Grails of human gene therapy—clinically useful correction of a mutation. This new sequence mismatch correction technology is as yet unconfirmed, but it does represent one of the first and most convincing reports of a relatively efficient correction of a mutant sequence in a way that can prevent, and possibly reverse, a disease phenotype. Even in the event that this specific approach does not produce an effective clinical application, another and probably similar approach will eventually do so. We therefore thought that this volume would be seriously incomplete without a preliminary discussion of sequence correction as an approach to gene therapy.

It is now approximately 30 years since the concept of human gene therapy began seriously to take shape. We have witnessed the birth and firm establishment of a concept for a completely new paradigm for the control of human disease. That concept is about to prove itself clinically. The immediate future will be even more heady than the recent past, but it will not be free of difficulties. The amazing speed of gene discovery spurred by the human genome project and the inevitable extension of these new tools of human genetics to potentially contentious applications including genetic enhancement of nondisease traits, germ line modification, and the still unresolved inequities and injustices associated with the current upheaval in health care delivery will require that we stay on our ethical toes in delivering this powerful new technology to saving life, easing suffering, and improving the quality of life. We hope that this book will serve as an anchor point for part of the early history of this important field of medicine.

I am very grateful to John Inglis and his staff at the Cold Spring Harbor Laboratory Press for their devoted and hard work in developing and producing this book. Very special thanks go especially to the project coordinator Inez Sialiano for coordinating the occasionally difficult logistical issues regarding authors and manuscripts, and to the production editors Christopher Bianco and Dorothy Brown and the Desktop editor Danny deBruin for turning all the disparate pieces into a whole.

May, 1998   **T. Friedmann**

1

The Origins, Evolution, and Directions of Human Gene Therapy

Theodore Friedmann

Center for Molecular Genetics
University of California, San Diego
La Jolla, California, 92093–0634

Medicine is on the brink of a new era—that of molecular genetic medicine. As in the case of previous conceptual and technical revolutions, we are witnessing the early stages of a quantum change in the way in which we understand and confront human disease. Like previous revolutions associated with the development of the sciences of human anatomy, pathology, medical pathology, microbiology, and chemical pathology, the revolution of molecular genetics is opening doors to new and definitive approaches to therapy that were previously only the stuff of dreams and scientific fantasy. This approach has come to be called "gene therapy" and aspects of its inception and development have been reviewed elsewhere (Friedmann 1992, 1996, 1997; Weatherall 1995; Wilson 1995; Brenner 1996; Curiel et al. 1996; Wagner 1996; Blaese 1997; Felgner 1997). Interestingly, the birth and evolution of gene therapy have been, if not unique, certainly one of the more unusual scientific developments in modern biomedicine. It has become a dominant concept in the treatment of disease long before it has given its first compelling evidence of true clinical effectiveness. It is, nevertheless, an established, powerful, and inexorable driving force in modern medicine.

THE EARLY RATIONALE FOR GENE THERAPY

During the late 1960s and early 1970s, the explosion in biochemical, human, and molecular genetics led to the beginnings of a completely unprecedented new understanding of human biology: normal and abnormal. The surge of understanding of molecular genetic mechanisms defining the flow of genetic information, as well as the beginnings of an understanding of the human genome and the role of genetic mechanisms in human traits, made it clear that molecular genetics would have an enormous impact on understanding the pathogenesis and detection of human disease. But what about therapy? With the exception of the discovery of anesthetics and modern surgical anesthetics and the discovery of antibiotic therapy, modern approaches to disease treatment stem largely from the development of chemical pathology by Archibald Garrod shortly after the beginning of the 20th century. As seen from this vantage point, human genetic disease and many other human disorders and defects result from the disruption of the normal metabolic pathways by inherited mistakes in the factors that regulate those pathways. Inspired by the rediscovery of the work of Gregor Mendel, Garrod termed the disorders that result from these mistakes "inborn errors of metabolism" and the factors responsible for them later came to be called "genes." This epochal view of the chemical pathology of human disease not only illuminated the mechanisms of genetic disease, but also suggested approaches to therapy based on manipulations of metabolic pathways—providing missing metabolic products, inhibiting enzyme activity, removing toxic metabolites from cells, etc. But although such approaches to therapy may have been, and continue to be, elegant in principle, they have usually proven to be inadequate in a clinical setting. Even the simplest single gene defects have pleiomorphic metabolic manifestations, and the biochemical manipulation of only one of the multiple metabolic disturbances has only limited effects on the complex and interacting sets of metabolic steps. This fact, together with the emerging understanding of human genetic defects seen in the light of the molecular genetics of the 1960s and 1970s, made it obvious that targeting therapies everywhere in the scheme of pathogenesis except at the site of the underlying defect itself—the mutant gene itself—is destined in many cases to be inadequate. The molecular genetic advances in the 1960s and 1970s were to change forever the simple Garrodian metabolic view of metabolic defects as the principal basis for therapeutic intervention.

By the mid 1960s, the notion was emerging that a continued therapeutic reliance on the purely metabolic approach to the treatment of human genetic disease was likely to remain, at best, only partially effective and made it vital to imagine therapy aimed directly at the underlying causative genetic defects, i.e., the defective genes. The route to this goal was the use of gene transfer agents able to introduce and express normal copies of disease-causing genes into genetically defective human cells. Viruses seemed to be ideal agents to carry out such a therapeutic gene transfer since they had evolved to perform precisely this task with great efficiency and specificity. The disease targets that were the most obvious and accessible at the time were the simple Mendelian inborn errors of metabolism in which correction of a single gene defect could be expected to bring about a complete correction of the disease phenotype.

At the same time, work with mammalian viruses was also advancing very quickly. In the mid 1960s, Renato Dulbecco demonstrated that a family of tumor viruses, represented by the papovaviruses SV40 and polyomavirus, transformed cells from a normal to a tumorigenic phenotype by introducing viral genetic material into the cells through the highly efficient route of virus transduction, by integrating their DNA into the infected cell genome in a stable and heritable manner and by expressing at least some of the viral genetic information, the so-called transforming genes, as stable and heritable new cellular functions. Although the effect of this gene transfer is obviously detrimental to an animal harboring such transformed cells, the phenomenon of efficient gene transfer by these naturally occurring viruses suggested to us the possibility that these or other viruses might somehow be suitably modified genetically to persuade them to transfer not deleterious genes into cells but genetic material that could complement genetic defects and thereby correct some simple, single-gene human disease phenotypes, i.e., "swords into plowshares" in the best medical tradition.

A rudimentary approach to virus-mediated transfer of foreign genetic material had been suggested by Stanfield Rogers who reported in 1968 (Rogers and Pfuderer 1968) that polylysine could be found in plants infected with poly(A)-modified tobacco mosaic virus (TMV), leading these authors to conclude that "the next step is to build a meaningful sequence on a virus RNA or DNA, replicate the modified virus to the amount needed, and use the virus to transmit the information." It was irrelevant whether or not the specific approach taken by Rogers toward the use of a genetically engineered virus as a potentially therapeutic transducing vector was truly feasible at the time. It obviously was not. The importance of this report was that it suggested that efficient virus-mediated transduction with a more suitably engineered virus might indeed serve to introduce wild-type alleles of deleterious mutant genes into defective human cells as an approach to therapy of genetic disease.

The extension of that suggestion to development of gene transfer vectors from animal viruses came some years later through suggestions that the known naturally occurring versions of the papovavirus, called "pseudovirions," containing random fragments of cellular DNA might be useful as potential gene transfer agents (Friedmann 1971; Aposhian et al. 1972) if it were possible to introduce potentially therapeutic genes into such particles. The process seemed to be doomed by the probable inefficiency of introducing the appropriate genetic segment into such particles and the impossibility of propagating and expanding them to usefully high concentrations. As an alternative, my colleagues and I in the Dulbecco Laboratory began to envision that the insertion of foreign genes into mammalian cells via engineered recombinant viruses would become a preferred avenue for introducing therapeutic genes into defective cells in a way that would permit integration of the foreign genes into the host genome and thereby would stabilize the expression of its therapeutic gene. Obviously, this concept was not ready for testing at that time, since truly effective methods for isolating normal copies of the disease-related genes and of producing recombinant DNA molecules were to become possible only later that year, and the use of recombinant DNA methods to produce recombinant virus vectors did not become possible until almost a decade later (Shimotohno and Temin 1981; Wei et al. 1981; Tabin et al. 1982).

Despite these technical and conceptual gaps, it seemed clear to several of us at the time that such a technology was inevitable and that what might be possible with TMV, as proposed by Rogers, would certainly become possible with mammalian viruses. To the extent that was possible at the time, we offered a view of the needs for a new, gene-based approach to therapy, the technical approaches to such treatment, and some of the important ethical and policy problems that were sure to emerge from such a direction (Friedmann and Roblin 1972). At the practical level, we imagined two methods of clinical application—direct in vivo application of vector by an approach now known as the in vivo model for gene therapy and also genetic modification of cells in vitro followed by grafting the normalized cells back into the patient (now called the ex vivo model of gene therapy). In addition to the technical issues of therapeutic gene transfer by transduction, we pointed to a number of the public policy, ethics and regulatory components and identified ethical and public policy issues that would surely emerge as a result of the development of such a capability and still represent some of the most important problems facing human gene therapy. We concluded that the scientific basis for therapeutically useful gene transfer was not solid enough yet to clinical application, although we recognized the urgent medical need in some cases of serious and otherwise untreatable diseases for early application of even inadequate knowledge for the treatment of desperate diseases. To ensure that such early studies be carried out wisely, we urged that a review mechanism, in the form of hospital review committees, be established to protect patients from premature clinical application of incomplete science, especially in the case of gene therapy where we saw the manipulation to be irreversible. Furthermore, we envisioned the emergence of some of the same kinds of regulatory issues that face any new medical treatments, drugs, and devices and the involvement of federal funding agencies (as well as the Food and Drug Administration [FDA]) in the quality control review of gene therapy reagents and in the approval of human clinical studies.

At around the same time, Rogers and his colleagues were already carrying out studies with the delivery of a viral agent to treat a human inborn error of metabolism, i.e., hyperargininemia. Rogers had concluded that some scientists who worked with the Shope papillomavirus (SPV) in the laboratory had lower than normal serum levels of serum arginine. He surmised that SPV carried a gene for a viral arginase into infected cells and attributed the lowered arginine level in the plasma of investigators to a subclinical infection with SPV that he assumed stably introduced a viral arginase gene into the workers. Shortcutting his own previously published engineered virus approach, Rogers proceeded in the early 1970s with a human clinical study of a virus-mediated "gene transfer," using not a modified virus but rather a naturally occurring fully wild-type Shope virus. In this study, Rogers' clinical colleagues treated two young German arginase-deficient girls with injections of SPV in an attempt to introduce the putative viral arginase gene and thereby to reconstitute the arginase deficiency in the patients' cells (Rogers et al. 1973; Terheggen et al. 1975). To this day, there has been little or no rigorous biochemical, metabolic, or clinical follow up of the study and no convincing indication that the girls were helped or harmed by the experiment. To the extent that all experiments should be designed and followed up to provide information, the study was not a success.

Other than for this use of naturally occurring, unmodified wild-type viruses as gene transfer agents, the notion of using recombinant viruses for human gene therapy was to remain merely theoretical until several major technical obstacles could be overcome, especially those related to the isolation of suitable wild-type genes and their introduction and expression in human cells. The advent of recombinant DNA technology in 1972 finally provided methods both for isolating mutant and normal copies of disease-related genes and for creating recombinant virus genomes. The potential for transferring such molecules with acceptable efficiency into human and other mammalian cells also received a boost by the development in 1973 of the improved calcium phosphate transfection method by Graham and van der Eb (1973).

THE GLOBIN GENE TRANSFER EXPERIMENT OF MARTIN CLINE

The tools of molecular genetics provided the means for isolating, cloning, and characterizing disease-related genes, and one of the earliest such genes was the human β-globin gene, defects of which cause important human diseases such as sickle-cell anemia and forms of thalassemia. The calcium phosphate transfection technique of Graham and van der Eb was quickly applied to the human clinical problem of β-thalassemia by the clinician Martin Cline at the University of California, Los Angeles. His team had reported that foreign genes could be introduced by the calcium phosphate method into bone marrow cells of mice and that such cells could be selected for with sufficient efficiency to allow partial repopulation of the marrow by genetically modified cells (Cline et al. 1980; Mercola et al. 1980). On the basis of these studies, bone marrow cells from two patients with β-thalassemia were transfected with plasmids containing the normal human β-globin gene and were then re-transfused into the patients by procedures similar to those that seemed to have succeeded in mice. We know now, and some critics of the studies at the time also commented, that the efficiency of gene transfer into the kinds of bone marrow cells (stem cells) that must be genetically modified if this kind of study were to succeed is exceedingly poor, making it highly unlikely that such an approach could have succeeded. Unfortunately, the studies, carried out in Italy and Israel, had not received the necessary approval from the University Human Subjects Committee. The studies were terminated and Cline was severely punished. To this day, there has been no definitive follow-up study of those patients, no conclusive demonstration of the fate of the introduced normal globin genes, and no thorough clinical studies of the patients. Largely because of the controversy surrounding this experiment, the mandate of the DNA Advisory Committee at the National Institutes of Health (NIH), established in 1976 to regulate recombinant DNA research, was expanded to include human gene therapy studies.

THE APPEARANCE OF EFFICIENT GENE TRANSFER TECHNOLOGY

In 1981, several groups independently reported the development of the first truly efficient methods (the replication-defective and highly engineered recombinant murine-based retroviral vectors retroviruses) for delivery of potentially therapeutic forms of wild-type genes to mammalian cells to correct the defects caused by mutant genes (Shimotohno and Temin 1981; Wei et al. 1981; Tabin et al. 1982). Finally, with this technical tool in hand, it became possible to test the principle of stable correction of genetic defects through the introduction of foreign genetic material in the form of a transducing viral vector. Such a demonstration came in 1983 with the correction by a retroviral vector not only of the enzymic defect, but also of the associated purine metabolic aberrations in cells derived from patients with Lesch Nyhan syndrome (Willis et al. 1984).

The retroviral vectors, of course, were not the solution to all gene transfer problems because of their relatively low titers, in vivo instability, and inability to infect nonreplicating cells. For that reason, a number of other powerful viral and nonviral systems soon were added to the gene transfer armamentarium, including vectors derived from human adenovirus, herpes simplex virus, adeno-associated virus (Mulligan 1993; Berns and Giraud 1995; Smith 1995), and others. At the same time, very important nonviral gene transfer systems such as those represented by liposomes and by conjugates of DNA with agents designed to facilitate recognition of specific cell surface receptors and protect the newly introduced intracellular DNA from degradation were beginning to appear (Wu and Wu 1987; Curiel et al. 1991; Wagner et al. 1992; Zatloukal et al. 1993; Douglas et al. 1996; Zeigler et al. 1996; Felgner 1997). Very rapidly, these gene transfer tools were used to demonstrate successful transfer of functional genes into almost every conceivable human target cell and to demonstrate expression either of a reporter gene or of disease-related genes. At last, the door had been opened for definitive model studies to test the potential for correction of clinical disease in real patients.

These technical advances quickly led to an explosion of interest that has continued to grow unabated in gene transfer studies at the basic and molecular levels as well as in preclinical animal studies. Figure 1 indicates the yearly number of citations for the phrase "gene therapy" in the published biomedical literature. The dip at the end of 1997 may reflect either slow appearance in the databases of papers published toward the end of the year, some slowdown in the rate of growth of the gene therapy literature, or both.

With such a rapid growth of information on gene transfer into virtually all tissues and cells relevant to human disease, little seemed to stand in the way of relatively easy application of these gene transfer methodologies and in vitro results to the clinic. Reports in the lay and scientific press made much of initial clinical proposals, all but promising quick cures for some otherwise dire diseases such as cancer, cystic fibrosis, muscular dystrophy, and others. Books appeared representing themselves as practical primers for clinicians (Culver 1994). It all seemed so simple and the level of optimism was very high for quick clinical success. Clinical human studies began.

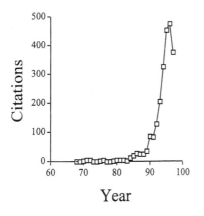

Figure 1 Number of citation to "gene therapy" contained in the Medline data base, beginning in 1968. Other references to "genetic engineering," "genetic therapy," or related phrases have not been included.

HUMAN CLINICAL STUDIES

Clinical trials on the transfer of foreign genes into human beings began in 1989 and 1990. Since the technologies were so new and untested, all initial studies were formally phase I studies, ostensibly designed to test safety rather than efficacy of gene transfer procedures. The first human clinical application was a cell marking study designed to determine by a typical ex vivo approach only if a foreign gene, (the one encoding the neomycin phosphotransferase gene) could be transferred into human tumor-infiltrating lymphocytes with a retroviral vector, if such genetically modified cells could be returned safely to the patient and if they would target in vivo back to the tumor of their origin (Rosenberg et al. 1990). Quickly thereafter came the ex vivo experiment with retroviral vector transfer of the normal gene for adenosine deaminase (ADA) into T cells of patients with severe combined immunodeficiency disorder (SCID) resulting from ADA deficiency (Cline 1986; Grompe et al. 1991; van Beusechem et al. 1992; Hughes et al. 1994). Over the ensuing several years, many other human gene transfer studies followed, utilizing all the viral and nonviral gene transfer tools to study the possibility of gene therapy for many forms of cancer, metabolic diseases such as cystic fibrosis (Marx 1992; Coutelle et al. 1993; Zabner et al. 1993; Crystal et al. 1994, 1995a,b; Sorscher et al. 1994) familial hypercholesterolemia (Miyanohara et al. 1988; Grossman and Wilson 1992; Kozarsky et al. 1993; Grossman et al. 1995), hemophilia, Gaucher's disease, neuromuscular and degenerative neurological disorders, and even infectious diseases such as AIDS (Marasco et al. 1993; Sarver and Rossi 1993; Yu et al. 1994). By 1995, more than 200 approved gene therapy studies were under way, with more than 1000 patients enrolled, largely in the United States and with growing research and clinical gene therapy activity in Italy, England, France, Germany, Japan, and other countries. Informal and word of mouth reports from some of these studies at scientific and medical meetings were extremely optimistic, with the expected highly rosy reports appearing in the lay press.

THE DIFFICULTIES EMERGE

By the fall of 1995, it was becoming clear that therapeutically useful clinical application in humans was going to be far more complicated than had been hoped. The field had gone through 5 years or more of highly visible and highly advertised studies, implied effectiveness, and an enormously high level of anticipation of clinical efficacy. Many investigators and observers had become concerned about the apparent gulf between promises and accomplishments in the field of gene therapy (Friedmann 1994, 1996; Verma 1994; Leiden 1995). Responding to that concern, the director of the NIH, Dr. Harold Varmus, convened two advisory committees to advise him on the questions of whether the review and evaluation processes for clinical studies by the Recombinant DNA Advisory Committee (RAC) of the NIH and the FDA were appropriate and effective and whether the scientific basis of this field was solid (Varmus 1995). To add further urgency to the need for a comprehensive re-evaluation of the ongoing approaches to gene transfer and gene therapy in humans, several publications appeared in the biomedical literature in the fall of 1995, reporting some of the first rigorous results from several of the clinical studies that had seemed so promising in earlier informal reports, particularly those dealing with in vivo adenovirus-mediated transfer of the cystic fibrosis transmembrane conductance regulator (CFTR) into the airway of patients with cystic fibrosis (Knowles et al. 1995), the retrovirus vector-mediated ex vivo correction of the ADA deficiency (Blaese et al. 1995; Bordignon et al. 1995), and of the low density lipoprotein (LDL) receptor defect in hepatocytes of patients with familial hypercholesterolemia (Grossman et al. 1995). Although these studies were indeed able to prove a degree of stable gene transfer and even prolonged gene expression of a foreign gene into human patients, they were all somewhat disappointing because there was far from convincing evidence for clinical efficacy than was expected, or at least was hoped for. It came as a somewhat rude surprise to some that clinical human gene therapy was not going to be easy.

The disappointment among many investigators to these publications was palpable and the reaction in the scientific and lay press to the report of one of the two NIH advisory committees was pointed and harsh. The NIH committee chaired by Drs. Stuart Orkin of the Boston Children's Hospital and Arno Motulsky of the University of Washington supported several previously published critical commentaries (Friedmann 1994, 1996; Verma 1994; Leiden 1995) by stating clearly that the potential for clinically helpful gene therapy with current tools had been greatly overstated to the public and by criticizing aspects of the scientific underpinnings for some of the clinical studies (see http://www.nih.gov/news/panelrep.html). The committee recommended a renewed emphasis on basic issues of vector design, delivery, and regulated gene expression. The second committee, chaired by Dr. Inder Verma, recommended changes to the review process for human clinical gene therapy studies (see http://www.nih.gov/news/panelrep.html). The press took the opportunity of the advisory committee's reports, particularly that of the Orkin–Motulsky committee, and of the publications of the disappointing clinical results to remind the gene therapy community that it had inappropriately been promising, or at least heavily implying, quick cures to the general public and that it had failed to deliver.

FEDERAL REGULATORY OVERSIGHT AND THE EMERGENCE OF THE BIOETHICS COMMUNITY

With respect to the social, ethical, and regulatory issues, progress has been comprehensive, steady, and relatively smooth. As a consequence of the growth in federal government funding of medical research in the United States, federal government oversight of much of human experimentation through the establishment of local institutional review boards had become a universally accepted oversight mechanism in human medical experimentation. Starting from the Asilomar Conference in 1975, federal government oversight of recombinant DNA research per se became institutionalized through local Institutional Biosafety Committees and the national Recombinant DNA Advisory Committee (RAC). Since recombinant DNA research is one of the essential enabling technologies for gene therapy, the transition of federal government oversight of gene therapy efforts in the United States following the Cline experiments constituted a fairly seamless extension of existing regulatory operations in the United States and, very quickly afterward, was emulated in other countries. As a consequence, it became impossible for scientists working with federal support to carry out human gene therapy without extensive prior review by their peers and government groups such as the RAC and the FDA.

GENE THERAPY AND THE BIOTECHNOLOGY INDUSTRY

One area that we did not fully appreciate in 1972 was the extremely powerful role that the commercial world would come to have in the development of the gene therapy technology. The modern biotechnology industry, which existed only in the most rudimentary form in 1972, is now playing as powerful a part in the development of new gene therapy technology as is the academic world. This development might have been anticipated from a realization that gene therapy is an area of medicine that, more than many others, requires not only novel concepts and demonstration of proofs of principle, but also a great deal of expensive implementation, scale-up, and the manufacture of clinical-grade gene transfer reagents. Neither academia nor the commercial world alone is able, by virtue of its ethos, strength, or technical capabilities, to carry out all of these activities effectively. In many cases, some aspects of gene therapy that cannot and probably should not be carried out in academia, e.g., large-scale development, are rapidly being moved, often by academic centers themselves, into the commercial arena. Conversely, although it is disingenuous to claim that novelty and innovation are exclusive activities of academia; it is true that most adventuresome, high-risk and high-payoff studies that depart from conventional therapies are still more likely to arise in the somewhat less fettered world of academia and research institutes than of industry. In the general area of biotechnology and particularly in gene therapy, eventual broad-scale clinical success will require a heavy collaboration between these two disparate but mutually dependent and more and more interactive worlds.

CURRENT GENE THERAPY STUDIES AND DIRECTIONS

The complexities inherent in therapeutic gene transfer into human beings have now been taken to heart by many more basic and clinical investigators than was the case only several years ago, most of whom have become somewhat more circumspect and even more modest in their expectations of immediate clinical efficacy. That is not to say that the urgency or the long-range optimism to institute clinical studies have weakened, but rather that there is an acknowledged recognition of the need for increased scientific and clinical rigor in both the basic molecular and cell biology as well as in the design and implementation of clinical studies. If anything, the advance of the technology has continued and now, more than ever, makes it clear that clinical success, at least at the research level and the proof of principle clinical level, is not far off. In fact, it seems very likely that convincing clinical therapeutic results will have been demonstrated by the time this volume appears in print.

Target Diseases

There are at the present time more than 200 approved clinical studies with more than 2500 enrolled patients in clinical gene therapy trials worldwide. The target diseases span the entire spectrum of human disease—simple Mendelian genetic disorders; complex disorders such as atherosclerosis, diabetes, obesity, and others; degenerative disorders such as arthritis; neuromuscular and degenerative neurological disorders such as Parkinson's and Alzheimer's diseases; and even infectious diseases such as AIDS. Many of these clinical approaches are described in other chapters of this book. Because of the clinical urgency of cancer, the magnitude of the problem, the ineffectiveness of most current treatments, the desperation of the public, the opportunity for large-scale biotechnical and pharmaceutical involvement, and the realization that most or all cancer is the manifestation of genetically defined mechanisms, most current clinical gene therapy studies involve cancer as the target disease. Because of this enormous effort, it seems likely that one or another form of cancer will provide some of the most convincing early clinical applications, probably through the induction of an augmented immune response to the tumor cell or modification of cellular programs that determine cell senescence and death. Of course, the classic genetic diseases, such as the inborn errors of metabolism, remain important disease targets and the single gene defects, particularly those that involve the absence of humoral factors such as the hemophilias, seem especially well positioned to fall to a genetic onslaught. Genetic approaches to diseases that are not primarily genetic, for example, degenerative disorders such as arthritis and Parkinson's disease and infectious diseases such as AIDS and malaria, are also sure to be developed.

Emerging Technologies

It is commonly stated that today's tools and techniques are likely not to be those that will eventually provide truly effective clinical application. That may be a bit simplistic, and it seems very probable that even the imperfect gene transfer vectors already available will within the next 2–3 years be used in achieving initial true clinical benefit for patients. But the methods are improving quickly, and new techniques will increase markedly the rapid and economical application of large-scale therapeutic programs. Viral vectors are being made increasingly simple by removing all but the most essential elements required for infection, integration, or stabilization of the new genetic information and transgene persistence (Bramson et al. 1995; Kay 1997; Parks and Graham 1997). At the same time, modifications are being made to the cell and tissue-recognition mechanisms in order to redirect target cell recognition and entry to tissue-specific sites and sites of integration of foreign genes into exposed cells (Neda et al. 1991; Russell et al. 1993; Chu and Dornburg 1995; Cosset et al. 1995; Kasahara et al. 1995; Somia et al. 1995; Wickham et al. 1995, 1996; Douglas et al. 1996; Krasnykh et al. 1996; Hall et al. 1997). Conversely, nonviral systems at the same time are being made more and more virus-like. They are being augmented with elements for compaction and more effective encapsidation of the incoming genes, ligands for efficient and possibly cell-specific gene delivery are being introduced, and additional modifications are being made to ensure in vivo stability of the agents (Bowenkamp et al. 1996; Kikuchi et al. 1996; Wu and Zern 1996). Semisynthetic and synthetic gene complexes are being developed that have shown great promise in cultured cells and will no doubt be extended eventually to in vivo delivery. Autonomously replicating artificial chromosomes and other stable, episomal elements with very high capacity for foreign DNA are emerging with the potential advantages of stability and very large capacity for foreign DNA (Sun and Vos 1996; Harrington et al. 1997; Nelson 1997). We are also seeing the beginnings of technology aimed at true sequence modification through recombination between a corrective foreign genetic sequence and a mutant endogenous gene (Kunzelmann et al. 1996; Chan and Glazer 1997). Although all these approaches are in their infancy, it is self-evident that they will ripen into the next generation of gene transfer technologies.

Genetic Enhancement?

One of the strong recommendations of the NIH advisory committees in 1995 was that the NIH, through the Office of Recombinant DNA Activities and the RAC convene Gene Therapy Policy Workshops to anticipate and evaluate quickly any conceptual and technical developments that may portend unusually promising advances or pose particularly contentious problems. The first of these meetings dealing with the potential and advisability of genetic manipulation for nondisease characteristics such as physical, cognitive, and personality traits was held in September 1997 in Bethesda. It is probably fair to say that the general consensus of that conference was that enhancement manipulation, for better or for worse, is likely to become feasible, that it has already achieved a certain level of public acceptance (Walters and Palmer 1997), but that more enlightened public and policy discourse is urgently needed to prepare the public, policy makers, and the scientists themselves for the inevitable enhancement pressures and inevitable applications. A simple denial of the possibilities and refusal to prepare our scientific, philosophical, theological, political, and policy communities would be too facile and ill-advised. Our societal experiences in this area have not been entirely constructive and the penchant that human beings have for using powerful scientific knowledge and techniques for ill as well as for good must convince us that we need broad and effective new community mechanisms to regulate and guide applications in science in general and genetics in particular. It is not so long ago that our society tolerated, and in fact grew up around, notions of using genetic knowledge to achieve truly malevolent ends (Kevles 1985). Remember the genetic principles espoused by Charles Davenport, Director of the Cold Spring Harbor Genetics Station and Eugenics Record Office from 1904 to 1926, regarding the genetic underpinnings of racial characteristics, principles that were very instrumental in the development of discriminatory and murderous racial policies of the subsequent decades:

- that race is a prime determinant of behavior. For instance, Poles tend to be "independent, self-reliant, clannish"; Italians tend to "crimes of personal violence"; and Hebrews are "intermediate between slovenly Greeks and tidy Swedes, Germans and Bohemians"
- that the infusion of "new blood" into a population makes the population "darker, smaller, more mercurial, prone to crimes of larceny, kidnapping, assault, murder, rape, sex immorality
- that such foreign (defective) germ plasm would persist
- that undesirable traits such as "nomadism," "shiftlessness," and "thalassophilia" are racially determined

So seriously did the Cold Spring Harbor group take their long-range eugenics mission that they even apparently had a communal laboratory song, *"Serious we have to be, working for posterity."*

Unfortunately, as is often the case, our scientific establishments find themselves reacting to events rather than establishing and formulating principles in advance. Additional technical and policy conferences on technical and policy questions are being scheduled.

Germ-line Modification?

The impressive successes during the past several decades with transgenic animals should convince even the most cautious skeptic that modifications of these methods to allow genetic modification of the human germ line will eventually become possible. This technology, together with the controversy surrounding the recent reports of the first apparent successes in mammalian cloning (Campbell et al. 1996; Kass 1997; Mirsky and Rennie 1997; Schnieke et al. 1997), has given new urgency to the necessity of understanding the pressures for and problems associated with human germ-line modification. Many of the same technical, ethical, metaphysical, theological, and policy issues raised by genetic enhancement are certain to be recapitulated and magnified by the questions surrounding germ-line genetic modification. Although any potential clinical indications for carrying out germ-line genetic modification are still quite unconvincing, the technology of gene transfer into germ cells will not remain out of reach for long, and human applications are probably inevitable and the confluence of technologies of germ-line modification and gene therapy is inexorable. A symposium, also in September 1997, was organized by the American Association for the Advancement of Science around the question of germ-line genetic modification for therapy or for enhancement purposes.

SUMMARY AND CONCLUSION: A MORE CIRCUMSPECT AND FORWARD-LOOKING GENE THERAPY FIELD MUST EMERGE

Has gene therapy arrived as an effective therapeutic strategy? The honest answer is yes and no. The past 25 years have seen the almost universal acceptance of the concept of gene therapy as a central force in medicine. Few, if any, can doubt that much of human disease will be treated through an attack on genetic targets or modification of gene expression in the diseased organism. That conceptual part of the revolution of human gene therapy is essentially over—gene therapy is an established concept in medicine. The remainder of the revolution now lies in the hard work of implementation. Part of this practical task is "merely" technical, and that ensures success. The obvious appropriateness of the concept of gene therapy, the volume of work, the exploding number of investigators, the maturation of many of the current gene transfer and disease model systems, the discovery of an ever-growing number of disease-related genes, the vast improvements in the preparation of faithful animal models for human disease, and the immense impetus supplied by the energy, get-it-done approach and, yes, even the profit motive of industry, all combine to indicate strongly that clinically effective gene therapy will become a reality. It will come, and we think that it will come soon. Much of the medical and scientific revolution of human gene therapy has already occurred and the field is established. It cannot be unborn.

REFERENCES

Aposhian H.V., Qasba P.K., Osterman J.V., and Waddell A. 1972. Polyoma pseudovirions: An experimental model for the development of DNA for gene therapy. *Fed. Proc.* **31:** 1310–1314.

Berns K.I. and Giraud C. 1995. Adenovirus and adeno-associated virus as vectors for gene therapy. *Ann. N.Y. Acad. Sci.* **772:** 95–104.

Blaese R.M. 1997. Gene therapy for cancer. *Sci. Am.* **276:** 111–115.

Blaese R.M., Culver K.W., Miller A.D., Carter C.S., Fleisher T., Clerici M., Shearer G., Chang I., Chiang Y., Tolstoshev P., et al. 1995. T lymphocyte-directed gene therapy for ADA-SCID: Initial trial results after 4 years. *Science* **270:** 475–480.

Bordignon C., Notarangelo L.D., Nobili N., Ferrari G., Casorati G., Panina P., Mazzolari E., Maggioni D., Rossi C., Servida P., et al. 1995. Gene therapy in peripheral blood lymphocytes and bone marrow for ADA-immunodeficient patients. *Science* **270:** 470–475.

Bowenkamp K.E., David D., Lapchak P.L., Henry M.A., Granholm A.C., Hoffer B.J., and Mahalik T.J. 1996. 6-Hydroxydopamine induces the loss of the dopaminergic phenotype in substantia nigra neurons of the rat —A possible mechanism for restoration of the nigrostriatal circuit mediated by glial cell line-derived neurotrophic factor. *Exp. Brain Res.* **111:** 1–7.

Bramson J.L., Graham F.L., and Gauldie J. 1995. The use of adenoviral vectors for gene therapy and gene transfer *in vivo*. *Curr. Opin. Biotechnol.* **6:** 590–595.

Brenner M.K. 1996. Gene transfer to hematopoietic cells. *N. Engl. J. Med.* **335:** 337–339.

Campbell K.H., McWhir J., Ritchie W.A., and Wilmut I. 1996. Sheep cloned by nuclear transfer from a cultured cell line. *Nature* **380:** 64–66.

Chan P.P. and Glazer P.M. 1997. Triplex DNA: Fundamentals, advances, and potential applications for gene therapy. *J. Mol. Med.* **75:** 267–282.

Chu T.-H.T. and Dornburg R. 1995. Retroviral vector particles displaying the antigen-binding site of an antibody enable cell-type-specific gene transfer. *J. Virol.* **69:** 2659–2663.

Cline M. 1986. Gene therapy: Current status and guture directions. *Schweiz. Med. Wochenschr.* **116:** 1459–1464.

Cline M.J., Stang H.D., Mercola K.E., Morse L., Ruprecht R., Brown J., and Salser W. 1980. Gene transfer in intact animals. *Nature* **284:** 422–425.

Cosset F.-L., Morling F.J., Takeuchi Y., Weiss R.A., Collins M.K.L., and Russell S.J. 1995. Retroviral retargeting by envelopes expressing an N-terminal binding domain. *J. Virol.* **69:** 6314–6322.

Coutelle C., Caplen N., Hart S., Huxley C., and Williamson R. 1993. Gene therapy for cystic fibrosis. *Arch. Dis. Child.* **68:** 437–440.

Crystal R.G., McElvaney N.G., Rosenfeld M.A.,Chu C.S., Mastrangeli A., Hay J.G., Brody S.L., Jaffe H.A., Eissa N.I., and Danel C. 1994. Administration of an adenovirus containing the human *CFTR* cDNA to the respiratory tract of individuals with cystic fibrosis. *Nat. Genet.* **8:** 42–51.

Crystal R.G., Jaffe A., Brody S., Mastrangeli A., McElvaney N.G., Rosenfeld M., Chu C.S., Danel C., Hay J., and Eissa T. 1995a. A phase 1 study, in cystic fibrosis patients, of the safety, toxicity, and biological efficacy of a single administration of a replication deficient, recombinant adenovirus carrying the cDNA of the normal cystic fibrosis transmembrane conductance regulator gene in the lung. *Hum. Gene Ther.* **6:** 643–666.

Crystal R.G., Mastrangeli A., Sanders A., Cooke J., King T., Gilbert F., Henschke C.,

Pascal W., Herena J., Harvey B.G., et al.. 1995b. Evaluation of repeat administration of a replication deficient, recombinant adenovirus containing the normal cystic fibrosis transmembrane conductance regulator cDNA to the airways of individuals with cystic fibrosis. *Hum. Gene Ther.* **6:** 667–703.

Culver K.W. 1994. *Gene therapy: A handbook for physicians.* Mary Ann Liebert, New York.

Curiel D.T., Pilewski J.M., and Albelda S.M. 1996. Gene therapy approaches for inherited and acquired lung diseases. *Am. J. Respir. Cell Mol. Biol.* **14:** 1–18.

Curiel D., Agarwal S., Wagner E., and Cotten M. 1991. Adenovirus enhancement of transferrin-polylysine-mediated gene delivery. *Proc. Natl. Acad. Sci.* **88:** 8850–8854.

Douglas J.T., Rogers B.E., Rosenfeld M.E., Michael S.I., Feng M.Z., and Curiel D.T. 1996. Targeted gene delivery by tropism-modified adenoviral vectors. *Bio/Technology.* **14:** 1574–1578.

Felgner P.L. 1997. Nonviral strategies for gene therapy. *Sci. Am.* **276:** 102–106.

Friedmann T. 1971. In vitro reassembly of shell-like particles from disrupted polyoma virus. *Proc. Natl. Acad. Sci.* **68:** 2574–2578.

——. 1992. A brief history of gene therapy. *Nat. Genet.* **2:** 93–98.

——. 1994. The promise and overpromise of human gene therapy. *Gene Ther.* **1:** 217–218.

——. 1996. Gene therapy—An immature genie but certainly out of the bottle. *Nat. Med.* **2:** 144–147.

——. 1997. Overcoming the obstacles to gene therapy. *Sci. Am.* **276:** 96–101.

Friedmann T. and Roblin R. 1972. Gene therapy for human genetic disease? *Science* **175:** 949–955.

Graham F.L. and van der Eb A.J. 1973. A new technique for the assay of infectivity of human adenovirus 5 DNA. *Virology* **52:** 456–467.

Grompe M., Mitani K., Lee C.-C., Jones S.N., and Caskey C.T. 1991. Gene therapy in man and mice: Adenosine deaminase deficiency, ornithine transcarbamylase deficiency, and Duchenne muscular dystrophy. *Adv. Exp. Med. Biol.* **309B:** 51–56.

Grossman M. and Wilson J.M. 1992. Frontiers in gene therapy: LDL receptor replacement for hypercholesterolemia. *J. Lab. Clin. Med.* **119:** 457–460.

Grossman M., Rader D.J., Muller D.W.M., Kolansky D.M., Kozarsky K., Clark B.J. III, Stein E.A., Lupien P.J., Brewer H.B. Jr., Raper S.E., et al. 1995. A pilot study of ex vivo gene therapy for homozygous familial hypercholesterolemia. *Nat. Med.* **1:** 1148–1154.

Hall F.L., Gordon E.M., Wu L., Zhu N.L., Skotzko M.J., Starnes V.A., and Anderson W.F. 1997. Targeting retroviral vectors to vascular lesions by genetic engineering of the MoMLV gp70 envelope protein. *Hum. Gene Therapy* **8:** 2183–2192.

Harrington J.J., Van Bokkelen G., Mays R.W., Gustashaw K., and Willard H.F. 1997. Formation of de novo centromeres and construction of first-generation human artificial minichromosome. *Nat. Genet.* **15:** 345–355.

Hughes M., Vassilakos A., Andrews D.W., Hortelano G., Belmont J.W., and Chang P.L. 1994. Delivery of a secretable adenosine deaminase through microcapsules—A novel approach to somatic gene therapy. *Hum. Gene Ther.* **5:** 1445–1455.

Kasahara N., Dozy A., and Kan Y.W. 1995. Targeting retroviral vectors to specific cells. *Science* **269:** 417.

Kass L.R. 1997. The wisdom of repugnance. *The New Republic,* June 2: 17–26.

Kay M.A. 1997. Adenoviral vectors for hepatic gene transfer in animals. *Chest* (suppl.) **111:** 138S–142S.

Kevles D.J. 1985. *In the name of eugenics.* Alfred A. Knopf, New York.

Kikuchi A., Sugaya S., Ueda H., Tanaka K., Aramaki Y., Hara T., Arima H., Tsuchiya S. and Fuwa T. 1996. Efficient gene transfer to EGF receptor overexpressing cancer cells by means of EGF-labeled cationic liposomes. *Biochem. Biophys. Res. Commun.* **227:** 666–671.

Knowles M.R., Hohneker K.W., Zhou Z., Knowles M.R., Hohneker K.W., Zhou Z., Olsen J.C., Noah T.L., Hu P.C., Leigh M.W., Engelhardt J.F., Edwards L.J., Jones K.R., et al. 1995. A controlled study of adenoviral-mediated gene transfer in the nasal epithelium of patients with cystic fibrosis. *N. Engl. J. Med.* **333:** 823–831.

Kozarsky K., Grossman M., and Wilson J.M. 1993. Adenovirus-mediated correction of the genetic defect in hepatocytes from patients with familial hypercholesterolemia. *Somat. Cell Mol. Genet.* **19:** 449–458.

Krasnykh V.N., Mikheeva G.V., Douglas J.T., and Curiel D.T. 1996. Generation of recombinant adenovirus vectors with modified fibers for altering viral tropism. *J. Virol.* **70:** 6839–6846.

Kunzelmann K., Legendre J.Y., Knoell D.L., Escobar L.C., Xu Z., and Gruenert D.C. 1996. Gene targeting of CFTR DNA in CF epithelial cells. *Gene Ther.* **3:** 859–867.

Leiden J.M. 1995. Gene therapy–promise, pitfalls and prognosis. *N. Engl. J. Med.* **333:** 871–872.

Marasco W.A., Haseltine W.A., and Chen S.-Y. 1993. Design, intracellular expression and activity of a human anti-human immunodeficiency virus type 1 gp120 single chain antibody. *Proc. Natl. Acad. Sci.* **90:** 7889–7893.

Marx J. 1992. Gene therapy for CF advances. *Science* **255:** 289.

Mercola K.E., Stang H.D., Browne J., Salser W., and Cline M.J. 1980. Insertion of a new gene of viral origin into bone marrow cells of mice. *Science* **208:** 1033–1035.

Mirsky S. and Rennie J. 1997. What cloning means for gene therapy. *Sci. Am.* **276:** 122–123.

Miyanohara A., Sharkey M.F., Witztum J.L., Steinberg D., and Friedmann T. 1988. Efficient expression of retroviral vector-transduced human low density lipoprotein (LDL) receptor in LDL receptor-deficient rabbit fibroblasts in vitro. *Proc. Natl. Acad. Sci.* **85:** 6538–6542.

Mulligan R.C. 1993. The basic science of gene therapy. *Science* **260:** 926–932.

Neda H., Wu C.H., and Wu G.Y. 1991. Chemical modification of an ecotropic murine leukemia virus results in redirection of its target cell specificity. *J. Biol. Chem.* **266:** 14143–14146.

Nelson N.J. 1997. Faux chromosomes hit the streets, hold real promise for gene therapy. *J. Natl. Cancer Inst.* **89:** 908–910.

Parks R.J. and Graham F.L. 1997. A helper-dependent system for adenovirus vector production helps define a lower limit for efficient DNA packaging. *J. Virol.* **71:** 3293–3298.

Rogers S. and Pfuderer P. 1968. Use of viruses as carriers of added genetic information. *Nature* **219:** 749–751.

Rogers S., Lowenthal A., Terheggen H.G., and Columbo J.P. 1973. Induction of arginase activity with the Shope papilloma virus in tissue culture cells from an argininemic patient. *J. Exp. Med.* **137:**·1091–1096.

Rosenberg S.A., Aebersold P., Cornetta K., Rosenberg S.A., Aebersold P., Cornetta K., Kasid A., Morgan R.A., Moen R., Karson E.M., Lotze M.T., Yang J.C., Topalian S.L., et al. 1990. Gene transfer into humans—Immunotherapy of patients with advanced melanoma, using tumor-infiltrating lymphocytes modified by retroviral gene transduction. *N. Engl. J. Med.* **323:** 570–578.

Russell S.J., Hawkins R.E., and Winter G. 1993. Retroviral vectors displaying functional antibody fragments. *Nucleic Acids Res.* **21:** 1081–1085.

Sarver N. and Rossi J. 1993. Gene therapy: A bold direction for HIV-1 treatment. *AIDS Res. Hum. Retroviruses* **9:** 483–487.

Schnieke A.E., Kind A.J., Ritchie W.A., Mycock K., Scott A.R., Ritchie M., Wilmut I., Colman A., and Campbell K.H. 1997. Human factor IX transgenic sheep produced by transfer of nuclei from transfected fetal fibroblasts. *Science* **278:** 2130–2133.

Shimotohno K. and Temin H.M. 1981. Formation of infectious progeny virus after insertion of herpes simplex thymidine kinase gene into DNA of an avian retrovirus. *Cell* **26:** 67–77.

Smith A.E. 1995. Viral vectors in gene therapy. *Annu. Rev. Microbiol.* **49:** 807–838.

Somia N.V., Zoppé M., and Verma I.M. 1995. Generation of targeted retroviral vectors by using single-chain variable fragment: An approach to in vivo gene delivery. *Proc. Natl. Acad. Sci.* **92:** 7570–7574.

Sorscher E.J., Logan J.J., Frizzell R.A., Lyrene R.K., Bebok Z., Dong J.Y., Duvall M.D., Felgner P.L., Matalon S., Walker L., et al. 1994. Gene therapy for cystic fibrosis using cationic liposome-mediate gene transfer: A phase I trial of safety and efficacy in the nasal airway. *Hum. Gene Ther.* **5:** 1259–1277.

Sun T.-Q. and Vos J.-M.H. 1996. Engineering 100- to 300kb DNA as persisting extrachromosomal elements in human cells using human artificial episomal chromosome system. *Methods Mol. Genet.* **8:** 167–188.

Tabin C.J., Hoffmann J.W., Goff S.P., and Weinberg R.A. 1982. Adaptation of a retrovirus as a eucaryotic vector transmitting the herpes simplex thymidine kinase gene. *Mol. Cell. Biol.* **2:** 426–436.

Terheggen H.G., Lowenthal A., Lavinha F., Columbo J.P., and Rogers S. 1975. Unsuccessful trial of gene replacement in arginase deficiency. *Z. Kinderheilkd.* **119:** 1–3.

Van Beusechem V.W., Kukler A., Heidt P.J., and Valerio D. 1992. Long-term expression of human adenosine deaminase in rhesus monkeys transplanted with retrovirus-infected bone-marrow cells. *Proc. Natl. Acad. Sci.* **89:** 7640–7644.

Varmus H. 1995. NIH review of gene therapy protocols. *Science* **267:** 1889.

Verma I. 1994. Gene therapy; hopes, hypes and hurdles. *Mol. Med.* **1:** 2–3.

Wagner E., Zatloukal K., Cotten M., Kirlappos H., Mechtler K., Curiel D.T., and Birnstiel M.L. 1992. Coupling of adenovirus to transferrin-polylysine/DNA complexes greatly enhances receptor-mediated gene delivery and expression of transfected genes. *Proc. Natl. Acad. Sci.* **89:** 6099–6103.

Wagner J.A. 1996. Gene therapy. *N. Engl. J. Med.* **334:** 332–333.

Walters L. and Palmer J.G. 1997. *The ethics of human gene therapy.* Oxford University Press, Oxford.

Weatherall D.J. 1995. Scope and limitations of gene therapy. *Br. Med. Bull.* **51:** 1–11.

Wei C., Gibson M., Spear P.G., and Scolnick E.M. 1981. Construction and isolation of a transmissible retrovirus containing the src gene from Harvey murine sarcoma virus and the thymidine kinase gene from herpes simplex virus type 1. *J. Virol.* **39:** 935–944.

Wickham T.J., Carrion M.E., and Kovesdi I. 1995. Targeting of adenovirus penton base to new receptors through replacement of its RGD motif with other receptor-specific peptide motifs. *Gene Ther.* **2:** 750–756.

Wickham T.J., Segal D.M., Roelvink P.W., Carrion M.E., Lizonova A., Lee G.M., and

Kovesdi I. 1996. Targeted adenovirus gene transfer to endothelial and smooth muscle cells by using bispecific antibodies. *J. Virol.* **70:** 6831–6838.

Willis R.C..W, Jolly D.J., Miller A.D., Plent M.M., Esty A.C., Anderson P.J., Chang H.C., Jones O.W., Seegmiller J.E., and Friedmann T. 1984. Partial phenotypic correction of human Lesch-Nyhan (hypoxanthine-guanine phosphoribosyltransferase-deficient) lymphoblasts with a transmissible retroviral vector. *J. Biol. Chem.* **259:** 7842–7849.

Wilson J.M. 1995. Gene therapy for cystic fibrosis: Challenges and future directions. *J. Clin. Invest.* **96:** 2547–2554.

Wu G.Y. and Wu C.H. 1987. Receptor-mediated in vitro gene transformation by a soluble DNA carrier system. *J. Biol. Chem.* **262:** 4429–4432.

Wu J. and Zern M.A. 1996. Modification of liposomes for liver targeting. *J. Hepatol.* **24:** 757–763.

Yu M., Poeschla E., and Wong-Staal F. 1994. Progress towards gene therapy for HIV infection. *Gene Ther.* **1:** 13–26.

Zabner J., Coutre L.A., Gregory R.J., Graham S.M., Smith A.E., and Welsh M.J. 1993. Adenovirus-mediated gene transfer gene transfer transiently corrects the chloride transport defect in nasal epithelia of patients with cystic fibrosis. *Cell* **75:** 207–216.

Zatloukal K., Schmidt W., Cotten M., Wagner E., Stingl G., and Birnstiel M.L. 1993. Somatic gene therapy for cancer: The utility of transferrinfection in generating `tumor vaccines'. *Gene* **135:** 199–207.

Zeigler S.T., Kerby J.D., Curiel D.T., Diethelm A.G., and Thompson J.A. 1996. Molecular conjugate-mediated gene transfer into isolated human kidneys. *Transplantation* **61:** 812–817.

WWW RESOURCE

http://www.nih.gov/news/panelrep.html—Orkin S.H. and Motulsky A.G. 1995. Report and recommendations of the panel to assess the NIH investment in research on gene therapy. Submitted to the meeting of the Advisory Committee to the NIH Director, December 7, 1995.

2
Retroviral Vectors

Jiing-Kuan Yee
Chiron Technologies, Center for Gene Therapy
San Diego, California 92121

The combination of the advancements in the knowledge of gene regulation and improvements in gene delivery into cells has resulted in significant progress in the field of human gene therapy during the past 15years. In the United States alone, there were more than 50 gene therapy protocols approved by the National Institutes of Health (NIH) during 1997. The gene transfer vectors involved in these clinical trials include vectors derived from retroviruses, plasmids, adenoviruses, pox-viruses, and adeno-associated viruses (AAV). Because of their many unique properties, vectors derived from retroviruses, especially from the Moloney murine leukemia virus (Mo-MLV), have been used extensively in a majority of clinical trials for somatic gene therapy (Miller 1992; Mulligan 1993; Crystal 1995). Each retroviral particle consists of an envelope on the outside, which is derived from the host plasma membrane during virus budding, and two single-stranded RNAs encapsidated in a nucleocapsid structure. Interaction between the virus-encoded envelope protein, which is embedded in the virion envelope, with cell surface receptors allows the virus to enter into the infected cells. On infection, both RNAs serve as templates to generate one copy of a linear double-stranded DNA in the cytoplasm via reverse transcription using virion-associated reverse transcriptase. The viral genomic DNA migrates into the nucleus and integrates into the host chromosome. The integration step is mediated by the virion-associated integrase and is a highly efficient and specific reaction that results in the generation of a proviral DNA that is colinear with its unintegrated form. The unique properties that make retroviruses ideal gene transfer vectors include: (1) the receptor for the envelope protein is ubiquitously expressed in mammalian cells; (2) infection by retroviruses such as Mo-MLV is nonpathogenic; (3) integration is highly specific, and DNA rearrangement that is frequently observed in other types of DNA transfer methods does not occur; (4) up to 8-kb pairs of foreign DNA fragments may be inserted into a retroviral vector; and (5) the long terminal repeat (LTR) of retroviruses contains an efficient promoter and enhancer that function in various cell types and can direct the expression of the gene inserted in a retroviral vector.

GENERATION OF RETROVIRAL VECTORS

To test the feasibility of using a retroviral vector for gene delivery, Wei et al. (1981) constructed a bacteriophage λ clone containing the LTR of a Harvey murine sarcoma virus followed by the *src* gene and the thymidine kinase (*tk*) gene of a herpes simplex virus (HSV). Transformed foci derived from the transfection of this construct into NIH-3T3 cells were isolated. To rescue the recombinant virus, the foci were superinfected with wild-type Mo-MLV. The rescued virus was found to induce focus formation in NIH-3T3 cells and convert NIH-3T3(TK⁻) cells into TK⁺ transformants. Shimotohno and Temin (1981) inserted the HSV *tk* gene into a cloned DNA genome of a spleen necrosis virus (SNV) and demonstrated that infectious recombinant vector containing the *tk* gene could be rescued from chicken cells cotransfected with the vector DNA and a clone of the SNV genome. Infection of chicken or rat TK⁻ cells with the recombinant virus transformed them to the TK⁺ phenotype. These two cases were the first to demonstrate that retroviruses could be used as vectors for efficient gene delivery into mammalian cells. Tabin et al. (1982) took a similar approach, except that in their Mo-MLV-based vector, only the HSV *tk* gene was present. Again, they demonstrated that the TK⁺ phenotype was transmitted to NIH-3T3(TK⁻) cells with the infectious tk vector rescued by the wild-type Mo-MLV. Despite the efficient transduction of the *tk* gene with these retroviral vectors, the presence of a replication-competent helper virus in the vector preparations can lead to multiple integration events that can result in the activation of potentially harmful genes such as oncogenes in the transduced cells. Significant effort was therefore contributed to overcome the problem of helper virus contamination.

PACKAGING CELL LINES

The approaches to solving the problem of helper virus contamination are based on several earlier observations. First, because retroviral genomic RNA is strongly preferred over the spliced subgenomic RNA and cellular mRNAs to be packaged into virus particles, the viral genomic RNA must contain a specific sequence that can be recognized by viral packaging machinery. The potential site for the packaging signal was identified from a study by Linial et al. (1978) by obtaining a mutant of the Rous sarcoma virus (RSV), SE21Q1b. Although virus particles were generated from the infected cells, they contained mostly cellular RNAs and only 1% of the RSV genomic RNA. Analysis of the mutant RNA demonstrated a 150-bp deletion somewhere between 300 and 600 bp from the left end of the provirus, suggesting that this region contained a site that is necessary for the recognition of the genomic RNA by viral-encoded proteins during the assembly process. On the basis of this study, Mann et al. (1983) defined a site, the ψ sequence, that is required for the efficient packaging of Mo-MLV genomic RNA into virus particles. Deletion of about 350 bp from this putative packaging signal resulted in little release of the infectious virus after transfecting the defective viral DNA into cells. Stable NIH-3T3-derived cell lines harboring this defective genome were established. In one such packaging cell line, ψ-2, the efficiency of packaging the mutated genomic RNA into virus particles was estimated to be less than 1% of the wild-type genomic RNA. The transient transfection of a retroviral vector construct containing the *Escherichia coli* gene encoding xanthine-guanine phosphoribosyltransferase (gpt) under the control of the 5´LTR resulted in the generation of an infectious gpt vector that gave gpt[+] colonies in transduced NIH-3T3 cells. No reverse transcriptase activity was detected in these gpt[+] cells, suggesting that the replication-competent helper virus was absent from these cells. Watanabe and Temin (1983) separated the cloned DNA genome of avian reticuloendotheliosis virus into *gag/pol* and *env* genes, respectively, under the control of the 5´LTR. The ψ sequence was removed from both expression constructs. Stable cell lines derived from the cotransfection of these two constructs and a selectable marker into canine cell line D17 were established. These packaging cell lines supported the production of a replication-defective, *tk* gene-containing retroviral vector derived from the SNV without helper virus contamination.

Because only murine cells express the Mo-MLV ecotropic receptor (Albritton et al. 1989), which is the cellular amino acid transporter, the host range for the retroviral vector generated from the ψ-2 cells is quite limited. To expand the host range, Cone and Mulligan (1984) constructed a hybrid MLV genome containing the *gag* and *pol* genes from the Mo-MLV and an envelope (*env*) gene from amphotropic virus 4070A. The receptor for the amphotropic envelope protein is the membrane phosphate transporter that is ubiquitously expressed in mammalian cells (Kavanaugh et al. 1994). Retroviral vectors carrying the amphotropic envelope protein can therefore infect most mammalian cells, including human cells. The ψ sequence in the hybrid genome was deleted, and the resulting plasmid was transfected into NIH-3T3 cells along with a selectable marker. Stable packaging cell lines were established, and retroviral producer clones capable of producing vector titers of more than 10^5 per milliliter were generated (Cone and Mulligan 1984). As predicted, these vectors not only transduced murine cells, they also transduced all human cells tested. Culture supernatants from the transduced cells showed no reverse transcriptase activity and contained no infectious vector as assayed by the inability to retransduce new NIH-3T3 cells, suggesting that the vectors generated from such packaging cell lines were free of helper virus contamination.

Deletion of the ψ sequence from the Mo-MLV packaging construct significantly reduces the possibility of generating a helper virus but does not completely eliminate it. The deleted ψ sequence in the packaging construct can, in principle, be corrected through homologous recombination between the retroviral vector and the packaging construct. Indeed, low-level contamination of a replication-competent helper virus in the vector preparations has been detected periodically from such packaging cell lines (Mann et al. 1983; Cone and Mulligan 1984). To further minimize the occurrence of such an event, Miller and Buttimore constructed an Mo-MLV packaging construct, pPAM3, with the ψ sequence removed and the Mo-MLV *env* gene replaced with the 4070A *env* gene (Miller and Buttimore 1986). In addition, the 3´LTR of pPAM3 was deleted and replaced with the simian virus 40 (SV40) polyadenylation signal. This deletion removes the site for initiation of second-strand Mo-MLV DNA synthesis during reverse transcription. The 5´ end of the 5´LTR in pPAM3 was also removed, thus eliminating the integration signal and preventing virus integration. An NIH-3T3-based packaging cell line, PA317, was established by stable transfection of the pPAM3. Producer clones derived from PA317 with a titer of 10^7 per milliliter were established, and no helper virus was detected in these vector preparations (Miller and Buttimore 1986).

These modifications of the Mo-MLV DNA genome significantly reduce the possibility of generating a helper virus but do not completely eliminate the regions of homology between the packaging construct and the retroviral vector. To further reduce the possibility of generating a helper virus, Markowitz et al. (1988), as well as Danos and Mulligan (1988), constructed two packaging constructs with one containing the *gag* and *pol* genes and the other containing the *env* gene. Gene expressions from both constructs were driven by the Mo-MLV LTR, and an SV40 sequence served as the polyadenylation signal. The ψ sequence in both constructs was removed. These two packaging constructs were introduced into NIH-3T3 cells by two consecutive transfection steps rather than by cotransfection, and the packaging cell lines isolated were shown to generate vector titers between 10^5 and 10^6 per milliliter. Various assays for helper virus contamination in vectors prepared from such packaging cell lines failed to detect the presence of a helper virus, suggesting that the *gag/pol* and *env* genes, when present in two different plasmids, may provide an efficient and safer packaging line for use in human gene therapy.

The packaging cell lines described so far were all derived from NIH-3T3 cells. Retroviral vectors generated from such packaging cell lines suffer several disadvantages:

1. NIH-3T3 cells express endogenous MLV-like viral sequences that can be copackaged with the vector RNA genome into the virus particle and participate in recombination events to generate a helper virus (Scadden et al. 1990; Ronfort et al. 1995), especially during the scale-up production of a clinical-grade vector.
2. The presence of such endogenous viral sequences can compete with the recombinant retroviral vector genome for packaging in the producer cells, resulting in lower vector titers.
3. Vectors generated from murine-based cell lines are rapidly inactivated by the complements in human serum (Takeuchi et al. 1994), making them unsuitable for in vivo gene delivery in humans.

In contrast, cell lines derived from primates contain little endogenous MLV-like sequences, and retroviruses generated from such cell lines were shown to be complement-resistant (Takeuchi et al. 1994). To test whether primate cell lines are more suitable for Mo-MLV vector production than murine cell lines, Cosset et al. (1995) generated packaging cell lines based on human HT1080 fibrosarcoma cells. Retroviral vectors generated from such packaging cell lines were resistant to human serum. The vector titers were generally 10–100-fold higher than those from NIH-3T3-based packaging cell lines, a result that could be due to the absence of endogenous viruses to compete for packaging. Rigg et al. (1996) used a similar approach to establish human 293-based packaging cell lines and demonstrated that vectors generated from such packaging cell lines were resistant to human serum inactivation and had higher vector titers. These results suggest that packaging cells derived from primate cell lines may be more suitable for retroviral vector generation than those derived from murine cell lines.

HOST RANGE

The host range of a retroviral vector is mostly determined by the interaction between the envelope protein and the corresponding cell surface receptor. The receptor for the 4070A envelope protein is Ram-1, a sodium-dependent phosphate transporter (Kavanaugh et al. 1994). The cell-surface receptor for the Gibbon ape leukemia virus (GaLV) is Glvr-1, another sodium-dependent phosphate transporter (Kavanaugh et al. 1994). Other C-type retroviruses, such as xenotropic MLV and feline endogenous virus RD114, have been shown to infect human cells via different receptors from those for amphotropic MLV or GaLV (Sommerfelt and Weiss 1990; Miller et al. 1991). However, the receptors for xenotropic MLV or RD114 remain to be identified. Like the amphotropic envelope protein, these other envelope proteins can be used to pseudotype MLV particles quite efficiently (Miller et al. 1991; Battini et al. 1995; Takeuchi et al. 1994; Cosset et al. 1995). Because receptor expression may vary among different cell types (Kavanaugh et al. 1994), MLV particles pseudotyped with different envelope proteins may show preferential transduction of particular cell types. Bunnell et al. (1995) have shown that primary human peripheral blood lymphocytes (PBLs) can be transduced more efficiently with an MLV-based vector pseudotyped with the GaLV envelope protein than with one pseudotyped with the amphotropic envelope protein. Porter et al. (1996) confirmed this result and showed that a vector pseudotyped with the RD114 envelope protein also demonstrated a higher transduction efficiency of human PBLs than one pseudotyped with the amphotropic envelope protein. von Kalle et al. (1994) demonstrated a more efficient transduction of primary human CD34$^+$ cells with a vector pseudotyped with the GaLV envelope protein than one pseudotyped with the amphotropic envelope protein, although the advantage was only twofold. These studies demonstrate the possibility of enhancing retrovirus transduction efficiency of particular cell types via the use of different viral envelope proteins.

The G protein derived from the vesicular stomatitis virus (VSV), a member of the rhabdovirus virus family, was shown to be incorporated efficiently into MLV particles (Emi et al. 1991). Although the mechanism for such phenomenon, called "pseudotype formation," remains unclear, the VSV-G pseudotyped vector has the following potential advantages:

1. These vectors should have an expanded host range, because VSV is able to infect mammalian as well as nonmammalian cells (Weiss 1979–1980).
2. VSV can be concentrated by methods such as ultracentrifugation or precipitation by polyethylene glycol (PEG) to high titers without sig-

nificant loss of infectivity, whereas similar procedures fail to concentrate MLV.

Burns et al. (1993) confirmed these potential advantages and demonstrated that VSV-G pseudotyped retroviral vectors could deliver genes into cell lines derived from hamsters and fish that are refractory to infection by amphotropic vectors. In addition, the pseudotyped vector could be concentrated by ultracentrifugation to titers greater than 10^9 per milliliter (Burns et al. 1993). These unique properties of VSV-G pseudotyped vectors not only extend the use of retroviral vectors for genetic studies in previously inaccessible species (Lin et al. 1994), but they may also facilitate direct gene transfer in human gene therapy trials. Because the stable expression of VSV-G is toxic to most mammalian cells, Chen et al. (1996) and Ory et al. (1996) used the tetracycline-inducible system described by Gossen and Bujard (1992) to regulate VSV-G expression in stable packaging cell lines. These cell lines generated VSV-G pseudotyped retroviral vectors with titers ranging from 10^5 to 10^7 per milliliter on induction of VSV-G expression. Such cell lines should facilitate the large-scale production of VSV-G pseudotyped vectors for preclinical human gene therapy studies.

Most of the virus-encoded envelope proteins possess no selectivity for target cells. As a first step toward developing retroviral vectors that are capable of targeting specific cell types, Valsesia-Wittmann et al. (1994) modified the envelope protein of the avian leukosis virus (ALV) by inserting a 16-amino acid RGD-containing peptide known to be the target for several cellular integrin receptors. ALV-derived vectors containing this mutant envelope protein were able to transduce the avian cells through the subgroup A receptor at levels similar to those of the wild type, suggesting that the insertion did not abolish the function of the envelope protein. Importantly, the vector was able to transduce mammalian cells, whereas the vector containing the wild-type ALV envelope protein failed to transduce mammalian cells. Preincubation of the target cells with a blocking peptide containing the RGD sequence completely blocked the transduction of the vector with the mutant envelope protein (Valsesia-Wittmann et al. 1994). Although the transduction efficiency of this vector in mammalian cells remained rather low and the envelope was not tissue-specific, this study represents the first successful attempt to modify a retroviral envelope protein to redirect tissue tropism. Kasahara et al. (1994) generated a chimeric protein containing the polypeptide hormone erythropoietin (EPO) inserted into the ecotropic Mo-MLV envelope protein. The chimeric protein, together with the wild-type envelope protein, was incorporated into viral particles, and the vector was able to specifically transduce human erythroid cell lines expressing the EPO receptor, but not human lines without the receptor. Somia et al. (1995) generated a chimeric protein with a single-chain antibody specific for the low-density lipoprotein receptor (LDLR) fused with the ecotropic envelope protein. MLV particles containing this chimeric protein failed to transduce any human or murine cell line, probably due to the fact that some envelope protein functions, such as trimer formation for efficient intracellular transportation of the envelope protein to the cell surface, were affected by the insertion. This problem was overcome by the coexpression of both the wild-type ecotropic envelope protein and the chimeric protein in the same producer cells, resulting in viral particles that are capable of transducing both murine and human cells. A quail cell line expressing no LDLR could not be transduced by this vector, whereas the same cell line transfected with an LDLR-expressing plasmid could be transduced by the vector. Chu and Dornburg (1995) took a similar approach to produce a chimeric protein with a single-chain antibody fused with the SNV envelope protein. This single-chain antibody was specific for an antigen expressing in human breast tumor and colon carcinoma cells. Vectors derived from cells expressing both the wild-type and the chimeric proteins specifically transduced a human carcinoma cell line, whereas vectors containing only the chimeric protein failed to transduce these cells. Addition of the antibody to the vector preparation effectively blocked transduction, suggesting that transduction was mediated by the antibody portion of the chimeric protein. For targeting human T cells, Mammano et al. (1997) tested pseudotype formation between an MLV-based vector and the HIV envelope protein. No infectious particles were detected. However, truncation of the cytoplasmic domain of the HIV envelope protein from 150 amino acids to only 7 amino acids resulted in the efficient incorporation of the mutated envelope protein into MLV particles. These particles were able to efficiently transduce $CD4^+$ HeLa cells, and pretreatment of the vector particles with soluble CD4 abolished the transduction, demonstrating that the specific interaction between the HIV envelope protein and the cell surface CD4 molecule allowed virus entry. Thus, the incorporation of heterologous cell surface proteins, including chimeric proteins containing foreign ligands fused with the Mo-MLV envelope protein, can either expand or limit the host range of MLV-based vectors. These proteins can potentially increase the transduction efficiency of retroviral vectors for specific target cells as well as increase the safety of using such vectors for human gene therapy trials in vivo.

TARGETED INTEGRATION

One drawback of using retroviral vectors is that the integration event shows little target specificity. In most cases, the sites of integration are random, although a minor fraction of the integration events are mapped to preferred "hot spots" or to "open" chromatin regions detected as DNase hypersensitive sites (Vijaya et al. 1986; Rohdewohld et al. 1987; Shih et al. 1989). Random insertion can lead to the activation of proto-oncogenes or the inactivation of genes that are important for normal cell functions (Hayward et al. 1981; Jenkins et al. 1981; Nusse and Varmus 1982). These are the undesirable outcomes of using retroviral vectors for gene therapy. Whether strategies can be designed for retroviral vectors to integrate into specific sites on chromosomes remains to be seen. However, recent studies suggest some possible approaches to investigate site-specific integration. Ji et al. (1993) reported that the integration sites for the yeast retrovirus-like element Ty1 on chromosome III were not mapped in a cellular gene, but were clustered upstream of tRNA genes. For Ty3, efficient integrations were mapped within 4 bp of the transcription initiation site of the tRNA genes. Chalker and Sandmeyer (1992) showed that the selective integration of Ty3 required a functional promoter element for the RNA polymerase III factors TFIIIB and TFIIIC. Using an in vitro system, Kirchner et al. (1995) demonstrated that TFIIIB and TFIIIC, but not RNA polymerase III, were required for site-specific integration. Bushman (1994) constructed a fusion protein between the human immunodeficiency virus 1 (HIV-1) integrase and the λ repressor. In in vitro integration reactions containing several target DNAs, the fusion protein was able to selectively direct the integration of substrate DNA to the λ operator region. Addition of the wild-type λ repressor in the reactions abolished the selective integration, suggesting that binding of the λ repressor to the operator region was important. Thus, modification of the integrase may, in the future, increase the selectivity of the integration sites for retroviral vectors.

GENE EXPRESSION IN HEMATOPOIETIC CELLS

Because many genetic disorders can potentially be treated with bone marrow transplantation (BMT), gene transfer into pluripotent hematopoietic stem cells (HSCs) with MLV-based vectors has been an area of intensive studies. Joyner et al. (1983) first demonstrated that mouse hematopoietic progenitor cells could be transduced with a retroviral vector containing the bacterial gene for neomycin resistance (*neo*). The transduced cells, when plated in G418-containing methylcellulose, gave rise to granulo-cyte-macrophage colony-forming units (GM-CFU), pure erythroid colonies (E-CFU), and colonies containing erythroid cells plus cells from one of the other lineages (MIX/E-CFU). Miller et al. (1984) demonstrated that mouse bone marrow cells transduced with a retroviral vector containing the cDNA-encoding human hypoxanthine phosphoribosyltransferase (HPRT) could reconstitute the hematopoietic system of lethally irradiated recipient mice, and human HPRT protein could be detected in these animals. Williams et al. (1984) took a similar approach and demonstrated that lethally irradiated mice could be engrafted with donor hematopoietic cells transduced with a *neo*-containing retroviral vector. Using the random chromosomal integration site of a retroviral vector as an unique marker in the reconstituted mice, Dick et al. (1985) demonstrated the transduction of primitive pluripotent stem cells that are capable of producing both myeloid and lymphoid progeny as well as more committed progenitor cells that are capable of producing either the myeloid or the lymphoid progeny. Using the same approach, Keller et al. (1985) demonstrated that pluripotent stem cells capable of producing both myeloid and lymphoid progeny were transduced with a retroviral vector. In addition, they showed the self-renewal capacity of these cells by serial BMT and reconstitution of the lethally irradiated mice. These studies clearly indicate that it may be possible to introduce appropriate genes into HSCs with the aim of correcting certain genetic defects.

Two major obstacles, however, impede the progress of retrovirus-mediated gene transfer into HSCs. First, the transduction efficiency of primate pluripotent progenitor cells with MLV-based retroviral vectors is quite low. van Beusechem et al. (1993) cocultivated rhesus monkey bone-marrow cells with retroviral vector producer lines, followed by autologous BMT. One year after BMT, several monkeys showed a persisting presence of the provirus in approximately 0.1% of the peripheral blood mononuclear cells (PBMCs) and granulocytes. Using producer cells with significantly higher vector titers in cocultivation failed to significantly increase the transduction efficiency, suggesting that, under the culture conditions used, only a minor fraction of the HSCs were cycling and thus were susceptible to transduction by the retroviral vector. Dunbar et al. (1995) reported a preliminary human autologous transplantation study in myeloma and breast cancer patients of retroviral gene transfer to bone marrow and peripheral-blood-derived CD34-enriched cells. They detected only 0.1–0.01% bone marrow or peripheral blood cells positive for the retroviral DNA sequence up to 24 months after BMT. Again, the overall efficiency of retroviral gene transfer in this clinical trial was too low to be considered useful for therapeutic applications. To overcome this problem, culture conditions that stimulate quiescent HSCs into proliferation and simultaneously maintain the pluripotent properties of the stem cells need to be vigorously pursued.

The second problem is that the Mo-MLV LTR does not function efficiently as a promoter in HSCs. Williams et al. (1986) reconstituted lethally irradiated mice with bone marrow cells transduced with a retroviral vector containing both the Mo-MLV LTR and an internal promoter derived from the early promoter of SV40. S1 nuclease analysis of spleen RNAs isolated from the reconstituted mice showed that transcription from the SV40 internal promoter was undetectable, whereas the transcription from the 5′LTR was variable but much lower than that in NIH-3T3 cells transduced with the same vector. This study suggests that the promoter activities of both the LTR and the SV40 promoter are down-regulated in hematopoietic cells relative to that in NIH-3T3 cells. With serial BMT and reconstitution, Challita and Kohn (1994) observed gene expression in a high percentage of spleen colony forming units (CFU-S) from primary bone marrow recipients. The percentage decreased to 10% in secondary recipients and became undetectable in tertiary recipients despite the presence of the retroviral DNA sequence in most CFU-S of the secondary and tertiary bone marrow recipients. They also demonstrated that the lack of gene expression could be correlated, for the most part, with an increase in the methylation of the proviral DNA sequence. By following the vector integration pattern with Southern blot analysis, they were able to show that the cells capable of forming CFU-S in the secondary recipients had the characteristics of long-lived, pluripotent HSCs. It is in these cells that the LTR promoter activity is frequently inactive. The high percentage of CFU-S expressing the introduced gene in the primary recipients therefore may be derived from committed progenitor cells instead of the "true" HSCs.

Similar to HSCs, the Mo-MLV LTR also functions poorly in pluripotent embryonic carcinoma (EC) and embryonic stem (ES) cells, despite the efficient infection of these two cell types by Mo-MLV (Peries et al. 1977; Gautsch 1980; Speers et al. 1980). ES cells have been used as models of hematopoietic stem cells. Under permissive conditions, ES cells can be induced to differentiate into hematopoietic cells in culture or after introduction into mice in vivo (Keller et al. 1993; Wiles 1993). The mechanisms for the suppression of the LTR promoter function in EC and ES cells include:

1. The enhancer interacts with cellular factors that down-regulate the enhancer activity in undifferentiated EC cells (Speck and Baltimore 1987).
2. The 5´-noncoding region, including the primer-binding site in the viral genome, functions as a negative element for transcription in EC cells (Niwa et al. 1983; Gorman et al. 1985; Weiher et al. 1987; Akgun et al. 1991; Tsukiyama et al. 1992).
3. A negative control region (NCR) upstream of the enhancer binds an EC cell-specific protein that suppresses the LTR promoter activity (Flanagan et al. 1989, 1992).

Whether the same mechanisms operating in EC and ES cells also have a role in down-regulating the Mo-MLV LTR in HSCs remains to be determined.

Several new MLV-based vectors have been designed to overcome the inactivity of the Mo-MLV LTR in stem cells. Grez et al. (1990) developed a retroviral vector called the murine embryonic stem cell virus (MESV) whose LTRs were derived from a PCC4-cell-passaged myeloproliferative sarcoma virus (PCMV). PCMV grows efficiently in NIH-3T3 and EC cells such as F9 and PCC4. Sequence analysis of the PCMV genome demonstrates multiple mutations in the LTR relative to that of the Mo-MLV, and these mutations have been shown to account for the efficient replication of PCMV in EC cells (Stocking et al. 1985; Hilberg et al. 1987). Besides the PCMV LTR, MESV also contains an altered primer binding site in the 5´-untranslated region that inactivates the negative element for transcription in this region (Weiher et al. 1987; Grez et al. 1991). Unlike Mo-MLV-derived vectors, transduction of ES cells with MESV allows efficient gene expression (Grez et al. 1990). Robbins et al. (1997) also developed a retroviral vector that functions efficiently in EC cells. This vector contains the enhancer derived from the myeloproliferative sarcoma virus (MPSV) that was shown to function more efficiently than the Mo-MLV enhancer in EC cells and hematopoietic cells. A functional SP1 transcription factor binding site within the MPSV enhancer has been attributed to its increased activity in EC cells (Prince and Rigby 1991). In addition, this new vector contains an altered primer binding site in the 5´-untranslated region that is similar to the MESV vector. The NCR upstream of the Mo-MLV enhancer was also removed from this vector. Transcription from this vector was significantly increased compared with an Mo-MLV LTR-containing vector in infected ES and EC cells, and the increased expression could be correlated with a decrease in the methylation state of the proviral 5′ LTR (Robbins et al. 1997). Whether these modified vectors can ultimately give rise to long-term gene expression in HSCs remains to be determined.

Riviere et al. (1995) compared gene expressions from several retroviral vectors in the hematopoietic cells of mice following BMT. These vectors contained either the Mo-MLV LTR, the MPSV LTR, a hybrid between the Mo-MLV LTR and the MPSV enhancer, or a hybrid between the Mo-MLV LTR and the Friend virus enhancer, respectively. In addition, these vectors retained both the λ sequence and the Mo-MLV splice donor and acceptor sequences that are necessary for the generation of the subgenomic viral *env* mRNA. The gene of interest was inserted downstream from the splice acceptor site. Stable expression of the introduced gene was observed in cells derived from both the lymphoid and the myeloid lineages 12–14 months following BMT. These results strongly suggest that these LTRs, including the wild-type Mo-MLV LTR, can function efficiently in hematopoietic cells. Because these vectors, called MFG, contain the authentic splice sites of Mo-MLV and were shown to give more efficient splicing and gene expression than the vectors containing no authentic Mo-MLV splice acceptor site in transduced cells (Drall et al. 1996), the gene expression shutdown from the Mo-MLV LTR in hematopoietic cells reported in previous studies may simply be due to the specific vector design rather than the LTR function. Whether the MFG vectors can transduce the "true" HSCs and remain active in transcription following serial BMT remains to be resolved.

GENE EXPRESSION IN FIBROBLASTS

As in hematopoietic cells, the expression of retroviral vector-delivered genes in fibroblasts is also not stable. Li et al. (1989) examined the long-term functional and structural stabilities of retroviral vectors in transduced murine cell lines. They concluded that a number of interacting variables could exert important influences on proviral stability. These variables include the nature of the transduced cell, the sequence of the inserted gene, the relative position of the inserted gene in the retroviral vector, the integration site, and the presence or absence of selection pressure. Reporter gene expression from an internal promoter was likely to be more unstable than that from the 5′LTR. The loss of proviral gene expression was often accompanied by deletions of the proviral DNA sequence; however, in some cases the inactive clones retained an apparently intact provirus, and the provirus could be rescued by superinfection with a helper virus, suggesting that chromosomal instability in established cell lines cannot completely account for the loss of the marker gene expression. Hoeben et al. (1991) monitored the expression of the gene encoding *E. coli* β-galactosidase (β-gal) from the 5´LTR in two clones derived from transduced murine B77 fibroblast cell lines. They showed that in one clone, the β-gal gene was only expressed in 8% of its cell population at any given time. Treatment of 5-azacytidine, a DNA demethylating agent, increased the β-gal-positive cell population to over 60%, and removal of the 5-azacytidine resulted in a gradual decline of the fraction of β-gal-positive cells. This epigenetic shutdown of gene expression was probably caused by position effects of the integration site, because the same phenomenon was not observed in the other clone. After observing that a large portion of isolated B77 colonies transduced with this vector contained considerable numbers of β-gal-negative cells, they concluded that inactivation of the Mo-MLV LTR was a frequent event in murine fibroblast cell lines. Palmer et al. (1991) transduced primary rat skin fibroblasts with retroviral vectors carrying genes encoding human adenosine deaminase (ADA) and *neo*. The transduced skin was grafted onto rats and then removed at various times after transplantation for ADA and vector DNA analyses. One month after transplantation, vector-encoded ADA expression was decreased more than 1500-fold, whereas the suppression was not seen during long-term culture of the transduced cells in vitro. Cellular or antibody-mediated immune responses to the grafted tissues were not detected in the transplanted animals, suggesting that expression of the ADA gene from the 5´LTR was shut down. Because the grafted cells were polyclonal, epigenetic events due to a specific integration site cannot account for the gene expression shutdown. These studies conclusively demonstrate that the expression of retrovirus-delivered genes in fibroblasts, like in HSCs, is not always stable and alternative vector designs with appropriate *cis*-regulatory sequences may be necessary to direct proper gene expression in fibroblasts.

REGULATION OF GENE EXPRESSION BY RETROVIRAL VECTORS

The promoter and enhancer in the Mo-MLV LTR function in most cell types. For tissue-specific gene expression, it may be necessary to use an internal, tissue-specific promoter to drive the expression of the inserted gene. However, transcription from the 5´LTR has been shown, under some conditions, to interfere with the regulation of gene expression from downstream promoters. To obtain the proper regulation of gene expression from an internal promoter, it may be necessary to eliminate the promoter and enhancer function from the Mo-MLV LTR. The possibility of aberrant activation of flanking cellular genes by the Mo-MLV enhancer due to retrovirus integration should also be minimized by the removal of the enhancer from both LTRs. Yu et al. (1986) constructed a retroviral vector called a self-inactivating (SIN) vector that contained a deletion of the Mo-MLV enhancer and part of the promoter in the 3´LTR. When vectors derived from this construct were used to transduce cells, the deletion was transferred to the 5´LTR, resulting in the transcriptional inactivation of the provirus in the transduced cell. The inserted gene was shown to express almost entirely from the internal promoter, although small amounts of mRNA initiated from the inactivated 5´LTR were still detectable. Yee ct al. (1987) inserted the human HPRT cDNA under the control of a human metallothionein (MT-IIA) promoter into a SIN vector from which the enhancer, the "CAAT" box, and the "TATA" box in the LTRs had been removed. Expression of the HPRT cDNA was shown to initiate from the internal MTIIA promoter, and incubation of the transduced cells with cadmium sulfate induced HPRT expression threefold from the MTIIA promoter. To ascertain that the promoter function of the 5´LTR was completely inactivated by the deletions, a replication-competent helper virus was applied to the transduced cells, and no SIN vector could be rescued from the transduced cells. One drawback of SIN vectors is that their titers were at least 10–100-fold lower than the parental retroviral vector containing functional LTRs (Yu et al. 1986; Yee et al. 1987). This problem hampers the general use of such vectors in human gene therapy trials.

To test targeted expression with a retroviral vector, Cone et al. (1987a) inserted a functionally rearranged murine λ1 immunoglobulin gene into a retroviral vector in the opposite orientation relative the 5´LTR. While the transduction of NIH-3T3 cells or cell lines derived from T-cell lineage showed little expression of the λ1 mRNA, the transduction of cell lines derived from B cell lineage with this vector resulted in the detection of λ1 mRNA, although the absolute levels of mRNA expressed from the retroviral vector were much lower than those from a control hybridoma cell line. This study demonstrated that retroviral vectors can direct the expression of an inserted gene in a tissue-specific manner and that the presence of the LTR enhancer does not interfere with the tissue specificity of the internal promoter. By using a similar approach, Cone et al. (1987b) introduced the human β-globin gene under the control of its own promoter into murine erythroleukemia (MEL) cells by retroviral transduction. The induction of MEL differentiation with dimethyl sulfoxide resulted in the up-regulation of β-globin mRNA 5–30-fold in either the absence or the presence of the Mo-MLV enhancer. To test the tissue-specific expression of the human β-globin gene in vivo, Dzierzak et al. (1988) used the same vectors described above to transduce mouse bone-marrow cells followed by the BMT of lethally irradiated mice. RNA samples isolated from various tissues including blood, spleen, and bone marrow demonstrated that the human β-globin gene was specifically expressed in cells derived from erythroid lineage. Only small amounts of human β-globin mRNA were found in B and T lymphocytes or in macrophages. The level of the human β-globin mRNA was between 0.4% and 4% of the endogenous mouse β-globin mRNA levels. Removal of the Mo-MLV enhancer from the retroviral vector enhanced the degree of tissue specificity of the β-globin expression.

Peng et al. (1988) constructed a retroviral vector containing an internal promoter of the human α1-antitrypsin (αAT) gene driving the transcription of the phenylalanine hydroxylase (PAH) cDNA. While the transcript initiated from the 5′LTR was observed in PA317-derived producer cells, the transcript initiated from the αAT promoter was undetectable. In contrast, both transcripts were readily detectable in transduced primary mouse hepatocytes in culture, suggesting that the αAT promoter continued to exhibit tissue-specific activity even when inserted in a retroviral vector containing the LTR enhancer. Significantly, the levels of PAH mRNA generated from the internal αAT promoter were comparable to the constitutive levels of PAH mRNA in human liver. These studies demonstrate the feasibility of using tissue-specific *cis*-regulatory sequences in a retroviral vector for targeted gene expression.

An alternative way to regulate the expression of the introduced gene in transduced cells is to insert inducible promoters into a retroviral vector. The MTIIA promoter in the SIN vector mentioned above is one such example because this promoter can be activated by either cadmium sulfate or dexamethasone. However, such promoters in general have some basal level of activity, even in the absence of the inducers, making them unsuitable for studying gene functions in cells or for expressing gene products whose intracellular levels need to be strictly regulated. The tetracycline-inducible system reported by Gossen and Bujard (1992) represents a significant advancement in alleviating the background expression problem. This system consists of two components: a fusion gene encoding tTA, which is a fusion protein between the tetracycline repressor and VP16, the strong transactivator derived from HSV, and tetO, a minimum promoter containing only the TATA box linked to seven copies of the operator region of the tetracycline operon. The association of tetracycline with tTA prevents the binding of this transactivator with the tetO promoter. Removal of the tetracycline allows the binding of tTA to the tetO promoter, and VP16 in tTA then activates transcription from this promoter. Because even low concentrations of tetracycline are sufficient to block tTA function, and most mammalian cells can tolerate tetracycline to a certain extent, this system provides a tightly regulated on/off switch for gene expression in many cell types. To test inducible gene expression, the complete tetracycline-inducible system was inserted into a retroviral vector by four different groups (Hofmann et al. 1996; Hwang et al. 1996; Iida et al. 1996; Paulus et al. 1996). In these vectors, the gene of interest was placed under the control of the inducible promoter and the tTA gene was controlled either by the 5´LTR or by another internal promoter. Inducible gene expression from the tetO promoter in transduced cells was demonstrated when the tetracycline was removed from the culture medium and very low background activity of the reporter gene was detected in the presence of tetracycline. One concern of this system is that the Mo-MLV enhancer in the LTRs may fortuitously activate the tetO promoter present in the same vector, thereby increasing the basal level expression of the reporter gene. The data from these studies, however, do not support this hypothesis, because the reporter gene expression from the tetO promoter remained low or undetectable with or without the Mo-MLV enhancer and was readily induced upon removal of the tetracycline. Using a combination of retroviral vectors for efficient gene delivery and a powerful inducible system for the regulation of gene expression not only can facilitate the study of gene function, but may also be applied in the future to techniques aimed at the clinical application in human gene therapy trials.

LENTIVIRAL VECTORS

One major disadvantage of using Mo-MLV-derived vectors is the inability to transduce nonproliferating cells because the migration of the Mo-MLV preintegration complex into the nucleus depends on the breakdown of the nuclear membrane during mitosis (Roe et al. 1993). This disadvantage limits the use of Mo-MLV-based vectors for in vivo gene transfer into nondividing cells such as hepatocytes, myofibers, HSCs, and neurons. In contrast, lentiviruses are capable of infecting nondividing cells due to the presence of three virus-encoded proteins: the matrix protein (MA), Vpr, and integrase (Bukrinsky et al. 1993; Heinzinger et al. 1994; Gallay et al. 1997). Both MA and integrase contain a nuclear localization signal (NLS) and were shown to be recognized by proteins in the importin/karyopherin pathway. The NLS in Vpr has not been identified, and the nature of its cellular receptor remains to be defined. The presence of these three proteins in lentiviral particles allows the transportation of the preintegration complex into the nucleus through the nuclear pores. To test whether lentivirus-derived vectors can indeed deliver genes into non-dividing cells, Naldini et al. (1996) generated HIV-1-derived vectors by using a transient transfection scheme. Three plasmids were used in the transfection: (1) a packaging plasmid containing all HIV genes, except the *env* gene and the *vpu* gene, under the control of the CMV promoter; (2) an expression plasmid for VSV-G; and (3) a vector construct with both HIV-1 LTRs and the putative packaging signal linked with the gene of interest controlled by an internal CMV promoter. Cotransfection of these three plasmids into human 293T cells resulted in the generation of HIV vectors with titers ranging between 10^5 and 10^6 per milliliter. Such vectors were shown to transduce growth-arrest fibroblast lines or primary macrophages in culture, whereas Mo-MLV-derived vectors failed to transduce these cells. Direct injection of concentrated HIV vectors into the brain, retina, liver, and muscle resulted in the highly efficient transduction of neurons, photoreceptor cells, hepatocytes, and muscle cells, respectively (Naldini et al. 1996; Kafri et al. 1997; Miyoshi et al. 1997). Thus, lentivirus-based vectors may offer substantial promise for stable in vivo gene delivery to treat human diseases. For details of lentiviral vectors, please review the chapter by Naldini and Verma (this volume).

FUTURE PROSPECTS

As retroviral vectors containing therapeutic genes were used in a majority of human gene therapy trials, clinical efficacy has not been definitively demonstrated at this time in any of these protocols. Insufficient understanding of the disease biology, the interaction between retroviral vectors and the host, and the host immune response toward the gene product probably account for the inability to translate preclinical studies of gene therapy into effective clinical treatment. A better understanding of the biology of the target cell, for example, the culture conditions to stimulate HSCs to proliferate without affecting their pluripotent properties, may allow more efficient gene transfer into HSCs. Proper vector design, together with *cis*-regulatory elements that function in stem cells, may overcome the difficulty of achieving persistent gene expression in HSCs or their differentiated progeny cells. Improvement of the retroviral packaging cell lines to increase vector titers may allow direct gene delivery in clinical trials, and this approach will significantly reduce the cost of culturing primary cells in ex vivo gene therapy trials. An understanding of the immune response toward retrovirus-encoded proteins such as the envelope protein or the protein expressed from the transduced gene may allow the development of methods to control the immune response. Finally, the development of stable packaging cell lines for lentiviral vector production and of sensitive assays for helper-virus detection will undoubtedly have a major impact in the field of human gene therapy in the future.

ACKNOWLEDGMENT

The author thanks Dr. Julie Johnston for her critical reading of this manuscript.

REFERENCES

Akgun E., Ziegler M., and Grez M. 1991. Determinants of retrovirus gene expression in embryonal carcinoma cells. *J. Virol.* **65:** 382–388.

Albritton L.M., Tseng L., Scadden D., and Cunningham J.M. 1989. A putative murine ecotropic retrovirus receptor gene encodes a multiple membrane-spanning protein and confers susceptibility to virus infection. *Cell* **57:** 659–666.

Battini J.L., Danos O., and Heard J.M. 1995. Receptor-binding domain of murine leukemia virus envelope glycoproteins. *J. Virol.* **69:** 713–719.

Bukrinsky M.I., Haggerty S., Dempsey M.P., Sharova N., Adzhubel A., Spitz L., Lewis P., Goldfarb D., Emerman M., and Stevenson M. 1993. A nuclear localization signal within HIV-1 matrix protein that governs infection of non-dividing cells. *Nature* **365:** 666–669.

Bunnell B.A., Muul L.M., Donahue R.E., Blaese R.M., and Morgan R.A. 1995. High efficiency retroviral-mediated gene transfer into human and nonhuman primate peripheral blood lymphocytes. *Proc. Natl. Acad. Sci.* **92:** 7739–7743.

Burns J.C., Friedmann T., Driever W., Burrascano M., and Yee J.K. 1993. Vesicular stomatitis virus G glycoprotein pseudotyped retroviral vectors: Concentration to very high titer and efficient gene transfer into mammalian and nonmammalian cells. *Proc. Natl. Acad. Sci.* **90:** 8033–8037.

Bushman F. 1994. Tethering human immunodeficiency virus 1 integrase to a DNA site directs integration to nearby sequences. *Proc. Natl. Acad. Sci.* **91:** 9233–9237.

Chalker D.L. and Sandmeyer S.B. 1992. Ty3 integrates within the region of RNA polymerase III transcription initiation. *Genes Dev.* **6:** 117–128.

Challita P.M. and Kohn D.B. 1994. Lack of expression from a retroviral vector after transduction of murine hematopoietic stem cells is associated with methylation in vivo. *Proc. Natl. Acad. Sci.* **91:** 2567–2571.

Chen S.T., Iida A., Guo L., Friedmann T., and Yee J.K. 1996. Generation of packaging cell lines for VSV-G pseudotyped retroviral vectors using a modified tetracycline-inducible system. *Proc. Natl. Acad. Sci.* **93:** 10057–10062.

Chu T.T. and Dornburg R. 1995. Retroviral vector particles displaying the antigen-binding site of an antibody enable cell-type-specific gene transfer. *J. Virol.* **69:** 2659–2663.

Cone R.D. and Mulligan R.C. 1984. High-efficiency gene transfer into mammalian cells: Generation of helper-free recombinant retrovirus with broad mammalian host range.

Proc. Natl. Acad. Sci. **81:** 6349–6353.

Cone R.D., Reilly E.B., Eisen H.N., and Mulligan R.C. 1987a. Tissue-specific expression of functionally rearranged λ1 Ig gene through a retrovirus vector. *Science* **236:** 954–957.

Cone R.D., Weber-Benarous A., Baorto D., and Mulligan R.C. 1987b. Regulated expression of a complete human β-globin gene encoded by a transmissible retrovirus vector. *Mol. Cell. Biol.* **7:** 887–897.

Cosset F., Takeuchi Y., Battini J., Weiss R.A., and Collins M.K.L. 1995. High-titer packaging cells producing recombinant retroviruses resistant to human serum. *J. Virol.* **69:** 7430–7436.

Crystal R.G. 1995. Transfer of genes to humans: Early lessons and obstacles to success. *Science* **270:** 404–410.

Danos O. and Mulligan R.C. 1988. Safe and efficient generation of recombinant retroviruses with amphotropic and ecotropic host ranges. *Proc. Natl. Acad. Sci.* **85:** 6460–6464.

Dick J.E., Magli M.C., Huszar D., Philips R.A., and Bernstein A. 1985. Introduction of a selectable gene into primitive stem cells capable of long-term reconstitution of the hemopoietic system of WWᵛ mice. *Cell* **42:** 71–79.

Drall W.J., Skelton D.C., Yu X.J., Riviere I., Lehn P., Mulligan R.C., and Kohn D.B. 1996. Increased levels of spliced RNA account for augmented expression from the MFG retroviral vector in hematopoietic cells. *Gene Ther.* **3:** 37–48.

Dunbar C.E., Cottler-Fox M., O'Shaughnessy J.A., Doren S., Carter C., Berenson R., Brown S., Moen R., Greenblatt J., Stewart F.M., Leitman S.F., Wilson W.H., Cowan K., Young N.S., and Nienhuis A.W. 1995. Retrovirally marked CD34-enriched peripheral blood and bone marrow cells contribute to long-term engraftment after autologous transplantation. *Blood* **85:** 3048–3057.

Dzierzak E.A., Papayannopoulou T., and Mulligan R.C. 1988. Lineage-specific expression of a human β-globin gene in murine bone marrow transplant recipients reconstituted with retrovirus-transduced stem cells. *Nature* **331:** 35–41.

Emi N., Friedmann T., and Yee J.K. 1991. Pseudotype formation of murine leukemia virus with the G protein of vesicular stomatitis virus. *J. Virol.* **65:** 1202–1207.

Flanagan J.R., Krieg A.M., Max E.E., and Khan A.S. 1989. Negative control region at the 5′ end of murine leukemia virus long terminal repeats. *Mol. Cell. Biol.* **9:** 739–746.

Flanagan J.R., Becker K.G., Ennist D.L., Gleason S.L., Driggers P.H., Levi B.Z., Appella E., and Ozato K. 1992. Cloning of a negative transcription factor that binds to the upstream conserved region of Moloney murine leukemia virus. *Mol. Cell. Biol.* **12:** 38–44.

Gallay P., Hope T., Chin D., and Trono D. 1997. HIV-1 infection of nondividing cells through the recognition of integrase by the importin/karyopherin pathway. *Proc. Natl. Acad. Sci.* **94:** 9825–9830.

Gautsch J.W. 1980. Embryonal carcinoma stem cells lack a function required for virus replication. *Nature* **285:** 110–112.

Gorman C.M., Rigby P.W.J., and Lane D.P. 1985. Negative regulation of viral enhancers in undifferentiated embryonic stem cells. *Cell* **42:** 519–526.

Gossen M. and Bujard H. 1992. Tight control of gene expression in mammalian cells by tetracycline-responsive promoters. *Proc. Natl. Acad. Sci.* **89:** 5547–5551.

Grez M., Akgun E., Hilberg F., and Ostertag W. 1990. Embryonic stem cell virus, a recombinant murine retrovirus with expression in embryonic stem cells. *Proc. Natl. Acad.*

Sci. **87:** 9202–9206.

Grez M., Zornig M., Nowock J., and Ziegler M. 1991. A single point mutation activates the Moloney murine leukemia virus long terminal repeat in embryonal stem cells. *J. Virol.* **65:** 4691–4698.

Hayward W.S., Neel B.G., and Astrin S.M. 1981. Activation of a cellular onc gene by promoter insertion in ALV-induced lymphoid leukosis. *Nature* **290:** 475–480.

Heinzinger N.K., Bukrinsky M.I., Haggerty S.A., Ragland A.M., Kewalramani V., Lee M., Gendelman H.E., Ratner L., Stevenson M., and Emerman M. 1994. The Vpr protein of human immunodeficiency virus type 1 influences nuclear localization of viral nucleic acids in nondividing host cells. *Proc. Natl. Acad. Sci.* **91:** 7311–7315.

Hilberg F., Stocking C., Ostertag W., and Grez M. 1987. Functional analysis of a retroviral host-range mutant: Altered long terminal repeat sequences allow expression in embryonal carcinoma cells. *Proc. Natl. Acad. Sci.* **84:** 5232–5236.

Hoeben R.C., Migchielsen A.A.J., van der Jagt M., van Ormondt H., and van der Eb A. 1991. Inactivation of the Moloney murine leukemia virus long terminal repeat in murine fibroblast cell lines is associated with methylation and dependent on its chromosomal position. *J. Virol.* **65:** 904–912.

Hofmann A., Nolan G.P., and Blau H.M. 1996. Rapid retroviral delivery of tetracycline-inducible genes in a single autoregulatory cassette. *Proc. Natl. Acad. Sci.* **93:** 5185–5190.

Hwang J., Scuric Z., and Anderson W.F. 1996. Novel retroviral vector transferring a suicide gene and a selectable marker gene with enhanced gene expression by using a tetracycline-responsive expression system. *J. Virol.* **70:** 8138–8141.

Iida A., Chen S.T., Friedmann T., and Yee J.K. 1996. Inducible gene expression by retrovirus-mediated transfer of a modified tetracycline-regulated system. *J. Virol.* **70:** 6054–6059.

Jenkins N.A., Copeland N.G., Taylor B.A., and Lee B.K. 1981. Dilute (d) coat color mutation of DBA/2J mice is associated with the site of integration of an ecotropic MuLV genome. *Nature* **293:** 370–374.

Ji H., Moore D.P., Blomberg M.A., Braiterman L.T., Voytas D.F., Natsoulis G., and Boeke J.D. 1993. Hotspots for unselected Ty1 transposition events on yeast chromosome III are near tRNA genes and LTR sequences. *Cell* **73:** 1007–1018.

Joyner A., Keller G., Phillips R.A., and Bernstein A. 1983. Retrovirus transfer of a bacterial gene into mouse haematopoietic progenitor cells. *Nature* **305:** 206–208.

Kafri T., Blomer U., Peterson D.A., Gage F.H., and Verma I.M. 1997. Sustained expression of genes delivered directly into liver and muscle by lentiviral vectors. *Nat. Genet.* **17:** 314–317.

Kasahara N., Dozy A.M., and Kan Y.W. 1994. Tissue-specific targeting of retroviral vectors through ligand-receptor interactions. *Science* **266:** 1373–1376.

Kavanaugh M.P., Miller D.G., Zhang W., Law W., Kozak S.L., Kabat D., and Miller A.D. 1994. Cell-surface receptors for gibbon ape leukemia virus and amphotropic murine retrovirus are inducible sodium-dependent phosphate symporters. *Proc. Natl. Acad. Sci.* **91:** 7071–7075.

Keller G., Kennedy M., Papayannopoulou T., and Wiles M.V. 1993. Hematopoietic commitment during embryonic stem cell differentiation in culture. *Mol. Cell. Biol.* **13:** 473–486.

Keller G., Paige C., Gilboa E., and Wagner E.F. 1985. Expression of a foreign gene in myeloid and lymphoid cells derived from multipotent haematopoietic precursors.

Nature **318:** 149–154.

Kirchner J., Connolly C.M., and Sandmeyer S.B. 1995. Requirement of RNA polymerase III transcription factors for in vitro position-specific integration of a retroviruslike element. *Science* **267:** 1488–1491.

Li X., Yee J.K., Wolff J.A., and Friedmann T. 1989. Factors affecting long-term stability of Moloney murine leukemia virus-based vectors. *Virology* **171:** 331–341.

Lin S., Gaiano N., Culp P., Burns J.C., Friedmann T., Yee J.K., and Hopkins N. 1994. Integration and germ line transmission of a pseudotyped retroviral vector in zebrafish. *Science* **265:** 666–669.

Linial M., Medeiros E., and Hayward W.S. 1978. An avian oncovirus mutant (SE21Q1b) deficient in genomic RNA: Biological and biochemical characterization. *Cell* **15:** 1371–1381.

Mammano F., Salvatori F., Indraccolo S., de Rossi A., Chieco-Bianchi L., and Gottlinger H.G. 1997. Truncation of the human immunodeficiency virus type 1 envelope glycoprotein allows efficient pseudotyping of Moloney murine leukemia virus particles and gene transfer into CD4+ cells. *J. Virol.* **71:** 3341–3345.

Mann R., Mulligan R.C., and Baltimore D. 1983. Construction of a retrovirus packaging mutant and its use to produce helper-free defective retrovirus. *Cell* **33:** 153–159.

Markowitz D., Goff S., and Bank A. 1988. A safe packaging line for gene transfer: Separating viral genes on two different plasmids. *J. Virol.* **62:** 1120–1124.

Miller A.D. 1992. Human gene therapy comes of age. *Nature* **357:** 455–460.

Miller A.D. and Buttimore C. 1986. Redesign of retrovirus packaging lines to avoid recombination leading to helper virus production. *Mol. Cell. Biol.* **6:** 2895–2902.

Miller A.D., Eckner R.J., Jolly D.J., Friedmann T., and Verma I.M. 1984. Expression of a retrovirus encoding human HPRT in mice. *Science* **225:** 630–632.

Miller A.D., Garcia J.V., von Suhr N., Lynch C.M., Wilson C., and Eiden M.V. 1991. Construction and properties of retrovirus packaging cells based on gibbon ape leukemia virus. *J. Virol.* **65:** 2220–2224.

Miyoshi H., Takahashi M., Gage F.H., and Verma I.M. 1997. Stable and efficient gene transfer into the retina using an HIV-based lentiviral vector. *Proc. Natl. Acad. Sci.* **94:** 10319–10323.

Mulligan R.C. 1993. The basic science of gene therapy. *Science* **260:** 926–932.

Naldini L., Blomer U., Gallay P., Ory D., Mulligan R., Gage F.H., Verma I.M., and Trono D. 1996. In vivo gene delivery and stable transduction of nondividing cells by a lentiviral vector. *Science* **272:** 263–267.

Niwa O., Yokota Y., Ishida H., and Sugahara T. 1983. Independent mechanisms involved in suppression of Moloney murine leukemia virus genome during differentiation of murine teratocarcinoma cells. *Cell* **32:** 1105–1113.

Nusse R. and Varmus H.E. 1982. Many tumors induced by the mouse mammary tumor virus contain a provirus integrated in the same region of the host genome. *Cell* **31:** 99–109.

Ory D.S., Neugeboren B.A., and Mulligan R.C. 1996. A stable human-derived packaging cell line for production of high titer retrovirus/vesicular stomatitis virus G pseudotypes. *Proc. Natl. Acad. Sci.* **93:** 11400–11406.

Palmer T.D., Rosman G.J., Osborne R.A., and Miller A.D. 1991. Genetically modified skin fibroblasts persist long after transplantation but gradually inactivate introduced genes. *Proc. Natl. Acad. Sci.* **88:** 1330–1334.

Paulus W., Baur I., Boyce F.M., Breakefield X.O., and Reeves S.A. 1996. Self-contained, tetracycline-regulated retroviral vector system for gene delivery to mammalian cells. *J. Virol.* **70:** 62–67.

Peng H., Armentano D., MacKenzie-Graham L., Shen R., Darlington G., Ledley F., and Woo S.L.C. 1988. Retroviral-mediated gene transfer and expression of human phenylalanine hydroxylase in primary mouse hepatocytes. *Proc. Natl. Acad. Sci.* **85:** 8146–8150.

Peries J., Alves-Cardoso E., Canivet M., Devons-Guillemin M.C., and Lasneret J. 1977. Lack of multiplication of ectropic murine C-type viruses in mouse teratocarcinoma primitive cells. *J. Natl. Cancer Inst.* **59:** 463–465.

Porter C.D., Parkar M.H., Collins M.K.L., Levinsky R.J., and Kinnon C. 1996. Efficient retroviral transduction of human bone marrow progenitor and long-term culture-initiating cells: Partial reconstitution of cells from patients with X-linked chronic granulomatous disease by gp91-phox expression. *Blood* **87:** 3722–3730.

Prince V.E. and Rigby P.W.J. 1991. Derivatives of Moloney murine sarcoma virus capable of being transcribed in embryonal carcinoma stem cells have gained a functional Sp1 site. *J. Virol.* **65:** 1803–1811.

Rigg R.J., Chen J., Dando J.S., Forestell S.P., Plavec I., and Bohnlein E. 1996. A novel human amphotropic packaging cell line: High titer, complement resistance, and improved safety. *Virology* **218:** 290–295.

Riviere I., Brose K., and Mulligan R.C. 1995. Effects of retroviral vector design on expression of human adenosine deaminase in murine bone marrow transplant recipients engrafted with genetically modified cells. *Proc. Natl. Acad. Sci.* **92:** 6733–6737.

Robbins P.B., Yu X., Skelton D.M., Pepper K.A., Wasserman R.M., Zhu L., and Kohn D.B. 1997. Increased probability of expression from modified retroviral vectors in embryonal stem cells and embryonal carcinoma cells. *J. Virol.* **71:** 9466–9474.

Roe T., Reynolds T.C., Yu G., and Brown P.O. 1993. Integration of murine leukemia virus DNA depends on mitosis. *EMBO J.* **12:** 2099–2108.

Rohdewohld H., Weiher H., Reik W., Jaenisch R., and Breindl M. 1987. Retrovirus integration and chromatin structure: Moloney murine leukemia proviral integration sites map near DNAseI-hypersensitive sites. *J. Virol.* **61:** 336–343.

Ronfort C., Girod A., Cosset F.L., Legras C., Nigon V.M., Chebloune Y., and Verdier G. 1995. Defective retroviral endogenous RNA is efficiently transmitted by infectious particles produced on an avian retroviral vector packaging cell line. *Virology* **207:** 271–275.

Scadden D.T., Fuller B., and Cunningham J.M. 1990. Human cells infected with retrovirus vectors acquire an endogenous murine provirus. *J. Virol.* **64:** 424–427.

Shih C.-C., Stoye J.P., and Coffin J.M. 1989. Highly preferred targets for retroviral integration. *Cell* **53:** 531–537.

Shimotohno K. and Temin H.M. 1981. Formation of infectious progeny virus after insertion of herpes simplex thymidine kinase gene into DNA of an avian retrovirus. *Cell* **26:** 67–77.

Somia N.V., Zoppe M., and Verma I.M. 1995. Generation of targeted retroviral vectors by using single-chain variable fragment: An approach to in vivo gene delivery. *Proc. Natl. Acad. Sci.* **92:** 7570–7574.

Sommerfelt M.A. and Weiss R.A. 1990. Receptor interference groups of 20 retroviruses plating on human cells. *Virology* **176:** 58–69.

Speck N.A. and Baltimore D. 1987. Six distinct nuclear factors interact with the 75-base-pair repeat of the Moloney murine leukemia virus enhancer. *Mol. Cell. Biol.* **7:**

1101–1110.

Speers W.C., Gautsch J.W., and Dixon F.J. 1980. Silent infection of murine embryonal carcinoma cells by Moloney murine leukemia virus. *Virology* **105:** 241–244.

Stocking C., Kollek R., Bergholz U., and Ostertag W. 1985. Long terminal repeat sequences impart hematopoietic transformation properties to the myeloproliferative sarcoma virus. *Proc. Natl. Acad. Sci.* **82:** 5746–5750.

Tabin C.J., Hoffmann J.W., Goff S.P., and Weinberg R.A. 1982. Adaptation of a retrovirus as a eucaryotic vector transmitting the herpes simplex virus thymidine kinase gene. *Mol. Cell. Biol.* **2:** 426–436.

Takeuchi Y., Cosset F.L., Lachmann P.J., Okada H., Weiss R.A., and Collins M.K.L. 1994. Type C retrovirus inactivation by human complement is determined by both the viral genome and producer cell. *J. Virol.* **68:** 8001–8007.

Tsukiyama T., Ueda H., Hirose S., and Niwa O. 1992. Embryonal long terminal repeat-binding protein is a murine homolog of FTZ-F1, a member of the steroid receptor superfamily. *Mol. Cell. Biol.* **12:** 1286–1291.

Valsesia-Wittmann S., Drynda A., Deleage G., Aumailley M., Heard J., Danos O., Verdier G., and Cosset F. 1994. Modifications in the binding domain of avian retrovirus envelope protein to redirect the host range of retroviral vectors. *J. Virol.* **68:** 4609–4619.

van Beusechem V.W., Bakx T.A., Kaptein L.C.M., Bart-Baumeister J.A.K., Kukler A., Braakman E., and Valerio D. 1993. Retrovirus-mediated gene transfer into Rhesus monkey hematopoietic stem cells: The effect of viral titers on transduction efficiency. *Hum. Gene Ther.* **4:** 239–247.

Vijaya S., Steffen D.L., and Robinson H.L. 1986. Acceptor sites for retroviral integrations map near DNAse I-hypersensitive sites in chromatin. *J. Virol.* **60:** 683–692.

von Kalle C., Kiem H., Goehle S., Darovsky B., Heimfeld S., Torok-Storb B., Storb R., and Schuening F.G. 1994. Increased gene transfer into human hematopoietic progenitor cells by extended in vitro exposure to a pseudotyped retroviral vector. *Blood* **84:** 2890–2897.

Watanabe S. and Temin H.M. 1983. Construction of a helper cell line for avian reticuloendotheliosis virus cloning vectors. *Mol. Cell. Biol.* **3:** 2241–2249.

Wei C., Gibson M., Spear P.G., and Scolnick E.M. 1981. Construction and isolation of a transmissible retrovirus containing the src gene of Harvey murine sarcoma virus and the thymidine kinase gene of herpes simplex virus type 1. *J. Virol.* **39:** 935–944.

Weiher H., Barklis E., Ostertag W. and Jaenisch R. 1987. Two distinct sequence elements mediate retroviral gene expression in embryonal carcinoma cells. *J. Virol.* **61:** 2742–2746.

Weiss, R.A. 1979–1980. Rhabdovirus pseudotypes. In *Rhabdoviruses* (ed. D.H.L. Bishop), pp. 52–65. CRC Press, Boca Raton, Florida.

Wiles M.V. 1993. Embryonic stem cell differentiation in vitro. *Methods Enzymol.* **225:** 900–918.

Williams D.A., Lemischka I.R., Nathan D.G., and Mulligan R.C. 1984. Introduction of new genetic material into pluripotent haematopoietic stem cells of the mouse. *Nature* **310:** 476–480.

Williams D.A., Orkin S.H., and Mulligan R.C. 1986. Retrovirus-mediated transfer of human adenosine deaminase gene sequences into cells in culture and into murine hematopoietic cells in vivo. *Proc. Natl. Acad. Sci.* **83:** 2566–2570.

Yee J.K., Moores J.C., Jolly D.J., Wolff J.A., Respess J.G., and Friedmann T. 1987. Gene expression from transcriptionally disabled retroviral vectors. *Proc. Natl. Acad. Sci.* **84:**

5197–5201.

Yu S., von Ruden T., Kantoff P.W., Garber C., Seiberg M., Ruther U., Anderson W.F., Wagner E.F., and Gilboa E. 1986. Self-inactivating retroviral vectors designed for transfer of whole genes into mammalian cells. *Proc. Natl. Acad. Sci.* **83:** 3194–3198.

3

Lentiviral Vectors

Luigi Naldini

Cell Genesys, Inc.
Foster City, California 94404

Inder M. Verma

The Salk Institute
La Jolla, California 92037

Somatic gene therapy requires the efficient delivery and sustained expression of a therapeutic gene into the tissues of a human body. Such an approach has tremendous therapeutic potential for several inherited and acquired diseases. However, major obstacles must be overcome to fulfill these high expectations. Among them, the development of more effective gene delivery systems is widely viewed as a critical challenge to gene therapy investigators (Verma and Somia 1997).

Retroviral vectors have long been favored as a gene transfer tool for several reasons. First, they integrate efficiently into the genome of the target cell. Second, they do not transfer any viral gene, thus alleviating the risk of immune response against the transduced cells. Both of these properties are likely to be crucial for achieving sustained expression of the transgene. Third, the genome and the life cycle of retroviruses are relatively simple and well studied, which has allowed a continuous improvement in the vector design and the generation of stable producer systems amenable to characterization and scaleup. Fourth, retroviral vectors can transfer a sizeable amount of DNA, up to 7.0 kb of foreign genetic material (Miller et al. 1993).

Until now, the retroviral vectors used in clinical trials have been derived from onco-retroviruses such as the Moloney murine leukemia virus (MLV). A major drawback of these vectors is that they can only transduce cells that divide shortly after infection (Miller et al. 1990). While actively dividing cells are found in several body tissues, they are short-lived and continuously replaced. The long-lived specialized cells of the liver, heart, skeletal muscles, and brain do not proliferate in the adult individual and thus severely limit the utility of retroviral vectors for important targets of gene therapy. Where applicable, alternative approaches have been attempted such as culturing in vitro the tissue precursor cells, triggering their proliferation to allow infection by the retroviral vectors, followed by reimplantation of the transduced cells in the host (Dai et al. 1992; Grossman et al. 1995). While this ex vivo approach had a limited success with peripheral blood lymphocytes (Blaese et al. 1995; Bordignon et al. 1995; Kohn et al. 1995), the poor in vivo engraftment of the transduced cells, and the transcriptional shut-off of the transgene have been major handicaps in most other tissues (Palmer et al. 1991; Scharfmann et al. 1991).

The block to oncoretroviral infection in nondividing cells has been characterized at the molecular level (Roe et al. 1993; Lewis and Emerman 1994). In the course of the infection, the virus delivers its nucleoprotein core in the cytoplasm of the target cells. Here uncoating and reverse transcription take place, but the newly synthesized DNA must reach the nucleus for integration to occur. Given the large molecular size of the preintegration complex, only the disruption of the nuclear membrane occurring during mitotic prophase would allow it access to the chromatin.

A possible solution to this problem was offered by the study of lentiviruses, a family of complex retroviruses typically associated with infection of macrophages and lymphocytes. As most tissue macrophages are terminally differentiated cells, and thus do not divide, lentiviruses must have evolved a strategy to circumvent the block to infection inherent to other prototypic retroviruses. Several researchers have shown that, at least in the case of the human lentivirus HIV, the viral preintegration complex interacts with the nuclear import machinery of the infected cell, and it is actively transported to the nucleus through the nucleopores (Bukrinsky et al. 1993). The complex mimics other nuclear bound cellular proteins by displaying nuclear localization signals on its components, thus exploiting a cellular pathway for its own sake. Several candidate nuclear localization signals have been identified in at least three components of the HIV preintegration complex, the Gag matrix, Vpr, and integrase proteins (Heinzinger et al. 1994; Gallay et al. 1995a,b, 1997). Although the relative importance of each of them is still under investigation, the emerging picture is that of a redundant apparatus ensuring a vital function in the virus life cycle.

We therefore developed vectors from HIV-1, with the goal of maintaining the ability of the virus to infect nondividing cells while completely disabling its replicative and pathogenetic potential. Early studies (Naldini et al. 1996a,b) proved the ability of HIV-derived vectors to infect human cells growth-arrested in vitro and, most importantly, to transduce neurons after direct injection into the brain of adult rats. The vector was remarkably efficient at transferring marker genes in vivo into the brain and achieved long-term expression in the absence of detectable pathology. In this chapter, we review the design and function of lentiviral vectors, and discuss recent advances that improve their biosafety and demonstrate their potential for gene delivery into important target tissues of gene therapy (Blömer et al. 1997; Kafri et al. 1997; Miyosh et al. 1997; Zufferey et al. 1997).

DESIGN AND FUNCTION OF LENTIVIRAL VECTORS

The general design of viral vectors is based on the strategy of segregation of *cis*-acting sequences involved in the transfer of the viral genome to target cells from the sequences encoding the *trans*-acting viral proteins. The prototype vector particle is assembled by viral proteins expressed in the producer cell from construct(s) stripped of all viral *cis*-acting sequences. The viral *cis*-acting sequences are linked to the transgene and are introduced in the same cell. As the vector particle can only transfer the latter construct, the infection process is limited to a single round without spreading. The efficiency and biosafety of an actual vector system depend on the extent to which the ideal situation of complete segregation of *cis*- and *trans*-acting functions of the viral genome is achieved.

The lentiviral vector that we developed is a replication-defective, hybrid viral particle made by core proteins derived from HIV-1 and the envelope of an unrelated virus, either the vesicular stomatitis virus (VSV; Burns et al. 1993) or the amphotropic MLV. The vector transfers an expression cassette for the transgene flanked by *cis*-acting sequence of HIV-1 and without any accompanying viral gene (See Fig.1).

Some features of the lentiviral vector design are similar to those recently utilized for generating MLV-based retroviral vectors from improved, split-genome packaging constructs. The packaging functions for the lentiviral vector are provided by two separate expression plasmids that use transcriptional signals unrelated to those of the virus. A packaging construct, derived from the HIV-1 proviral DNA, expresses the viral Gag, Pol, Tat, and Rev proteins from heterologous transcription signals. Most of the *env* gene has been deleted. When introduced alone in cells, this construct generates noninfectious particles, as they lack both an envelope and a packaged genome that can be transferred. A separate construct expresses a heterologous envelope that is incorporated into the vector particles (pseudotyping) and allows entry into the target cells. Only a third construct, the transfer vector, expresses RNA that contains the viral *cis*-acting sequences required for efficient packaging by the vector particles, reverse transcription and integration in the target cells. When

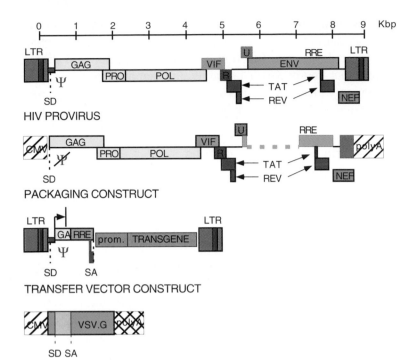

Figure 1 Schematic drawing of the HIV provirus and the three constructs used to make lentiviral vector. The viral LTRs, the reading frames of the viral genes, the major 5´ splice donor site (SD), the packaging sequence (ψ), and the Rev response element (RRE) are indicated. The packaging construct expresses the viral genes from heterologous transcriptional signals. Most of the envelope gene and the packaging sequence have been deleted. The transfer vector construct contains an expression cassette for the transgene flanked by the HIV LTRs. Downstream from the 5´ LTR, the vector contains the HIV leader sequence, the first 360 bp of the *gag* gene, the RRE element, and the splice acceptor sites of the third exon of *tat* and *rev*. A separate expression construct encodes an heterologous envelope to pseudotype the vector, here shown coding for the protein G of the (VSV-G). Only the relevant part of the constructs is shown.

expressed together with the other two constructs, it transfers an expression cassette for the transgene into target cells.

The use of a heterologous envelope is a key feature in the vector design. It broadens the tissue tropism of the vector, allowing the infection of several more cell types than allowed by the strict macrophage–lymphocyte tropism of lentiviruses. Most importantly, it makes impossible the generation of wild-type virus of either parental type during vector production by any mechanism. In fact, neither the HIV envelope gene nor VSV genes other than the envelope are present in any of the components used to generate the vector.

The packaging construct was derived from the HIV-1 proviral clones HXB2 and NL43. It expresses the viral Gag, Pol, Tat, and Rev proteins from the immediate early human cytomegalovirus (CMV) enhancer/promoter substituted for the viral 5′ long terminal repeat (LTR), and the rat insulin polyadenylation site substituted for the viral 3′ LTR. It does not express the viral envelope, as most of its sequence (1402 bp) were deleted except for the fragment containing the Rev response element (RRE), the splice acceptor sites, and the second exons of Tat and Rev. Earlier versions of this packaging construct varied in the complement of viral accessory genes expressed in addition to the structural *gag* and *pol* genes. In a latest, "minimal" version, the sequences encoding the four HIV accessory genes, *vif, vpr, vpu* and *nef,* have been entirely deleted. The CMV promoter sequence was joined to the HIV genome 32 bp upstream of the major 5′ splice donor. Downstream from the splice donor, the construct has a 39 bp deletion in the ψ packaging sequence.

Several features of the core packaging construct prevent its transfer to the target cells. As discussed below, the new 5′ end removes or disrupts all the structural motifs associated with dimerization and encapsidation of the viral RNA with the possible exception of a stem loop encoded by the 5′ part of the *gag* gene (SL4, see following paragraph). Furthermore, the deletion of both LTRs and of the primer binding site will prevent reverse transcription and integration of any encapsidated transcript.

The ψ sequence, a *cis*-acting region near the 5′ end of the HIV-1 genome, both upstream (Kim et al. 1994; Vicenzi et al. 1994; Geigenmuller and Linial 1996; McBride and Panganiban 1996; Paillart et al. 1996) and downstream from the major splice donor site (Lever et al. 1989; Aldovini et al. 1990; Luban and Goff 1994) appears to be essential for genome dimerization and encapsidation (Lever et al. 1989; Aldovini and Young 1990; Kim et al. 1994; Luban and Goff 1994; Vicenzi et al. 1994; Kaye et al. 1995; Geigenmuller and Linial 1996; McBride and Panganiban 1996; Paillart et al. 1996). According to the latest analyses, this region is organized into four putative RNA stem loops (SL1-4). SL2 contains the splice donor site. The other three stem loops (SL1, SL3, SL4) contain binding sites for the Gag protein nucleocapsid. At least two of them (SL1 and SL3) contribute to RNA encapsidation in vivo, and one (SL1) is implicated in dimer initiation through a kissing-loop mechanism. Mutations or deletions affecting one or more of these stem loops dramatically decrease the viral infectivity (Lever et al. 1989; Aldovini and Young 1990; Luban and Goff

1994; Kaye et al. 1995; Clever and Parslow 1997; Laughrea et al. 1997; McBride and Panganiban 1997). Additional sequences upstream of SL1, and including the TAR loop, are also required for encapsidation (McBride et al. 1997). The changes at the 5′ end of the core packaging construct delete all HIV 5′ sequences upstream of SL1, the putative left stem, and half of the loop of SL1; the 39-bp deletion in the ψ sequence deletes the putative right stem of SL2, and the whole of SL3. Recently, McBride and Panganiban (1996) reported the encapsidation efficiency of HIV-1 transcripts carrying 5′ deletions. For the transcripts derived from a construct comparable to that discussed here, the encapsidation efficiency relative to the wild-type counterpart was reduced to less than 0.1, and to less than 0.02 (the background signal) in the presence of competing wild-type RNA (McBride and Panganiban 1996).

The transfer vector contains viral sequences derived from the HIV-1 proviral clone HXB2 and an expression cassette for the transgene driven by an internal promoter, typically that of the housekeeping phosphoglycerokinase gene (PGK) or the immediate early enhancer/promoter of the human CMV. Alternatively, several others promoters, including tissue-specific and regulatable ones, can be used. The vector contains the complete HIV leader sequence and the 5′ splice donor site, followed by approximately 360 bp of the *gag* gene (encoding SL4; with the *gag* reading frame closed by a synthetic stop codon) as well as *env* sequences (~700 bp) containing the RRE and the splice acceptor sites of the third exon of the *tat* and *rev* genes. For the requirement of HIV-derived *cis*-acting sequences in the transfer vector see also the work of other investigators (Poznansky et al. 1991; Shimada et al. 1991; Buchschacher and Panganiban 1992; Carroll et al. 1994; Parolin et al. 1994; Berkowitz et al. 1995; Kaye et al. 1995; McBride et al. 1997).

The design of the transfer vector reflects and exploits one aspect of the complexity of lentiviruses, namely, that they encode their own regulators of transcription. In the presence of the Tat and Rev regulatory proteins, expressed by the packaging construct and thus only in producer cells, vector transcription from the HIV LTR is strongly stimulated and the same would be expected for the nuclear export of unspliced RNA containing the RRE motif. As this full-length "viral" RNA has a full complement of packaging signals it is efficiently encapsidated by the vector particles. On the other hand, in target cells, where the *trans*-activators are no more available, transcription from the LTR is suppressed and any residual transcript undergoes splicing of the region containing the ψ sequence. Transcription of the transgene is now driven almost entirely by the internal promoter. It should be noted that such a tight regulation of the transcription from the lentiviral LTR offers a distinctive advantage over conventional MLV-derived retroviral vectors. In the latter case, the constitutive enhancers in the viral LTRs often override the transcriptional control of an internal promoter, causing accumulation of "viral" transcripts in producer cells that may be the substrate of recombination and augmenting the risk of insertional mutagenesis.

IMPROVEMENTS IN VECTOR BIOSAFETY

To be considered for clinical applications, lentiviral vectors must comply to the strictest safety standards, as dictated by the choice of parental virus. The generation of an infectious retroviral vector requires all the *trans*-acting functions encoded by the *gag*, *pol*, and *env* genes (the packaging components). Thus, it is theoretically possible that these genes could be rejoined to the viral *cis*-acting sequences present in the transfer vector to generate a replication competent recombinant (RCR). The likelihood of this event is hard to predict and is dependent on the presence of residual *cis*-acting sequences in the packaging construct that allow some level of encapsidation and on the extent of homology between packaging and transfer vector. Formation of heterozygous vector particles and homologous recombination during reverse transcription is the pathway most often incriminated for generating recombinants. Vectors generated from split-genome packaging constructs require multiple recombination events and thus reduce the likelihood of an RCR outbreak. In the case of the lentiviral vector, the structural functions are not only separate, but they are contributed by two different viruses, namely, HIV-1 and VSV (a rhabdovirus). The RNA expressing the VSV envelope has no sequence homology, and presumably, no residual *cis*-acting function that would promote its recombination with the other two components derived from the HIV. Thus, it can be predicted that generation of RCR during vector production will be an unlikely event.

Recently, we have begun to identify the minimal set of HIV functions necessary for transduction of nondividing cells both in vitro and in vivo. We showed that a vector made from a packaging construct in which the accessory genes *vif, vpr, vpu,* and *nef* had been deleted was as efficient as one made from a wild-type construct at transducing transgenes into growth-arrested cells and monocyte-derived macrophages in culture, and into adult neurons in vivo (Fig. 2) (Zufferey et al. 1997). The removal of five genes essential for HIV pathogenesis from the packaging construct eliminates the possibility that even the unlikely RCR arising during vector production would have the pathogenetic features of HIV. The four

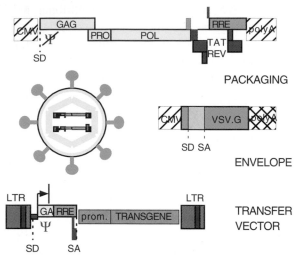

Figure 2 Schematic drawing of a second generation lentiviral vector particle, and of the three constructs used to generate it. The contribution of each construct is indicated by the different colors. The packaging construct provides the protein core of the vector, derived from the *gag* and *pol* genes of the HIV-1. The envelope construct provides the outer envelope proteins, derived from the unrelated VSV. The transfer vector construct provides the genome of the particle, contains HIV-1 *cis*-acting sequences and an expression cassette for the transgene. It is the only portion transferred to the target cells. Only a minimal set of HIV genes are used to generate the vector, as the viral genes *env, vif, vpr, vpu*, and *nef* have been deleted from the packaging construct.

HIV accessory genes have a crucial role in the exceedingly high rate of replication of the virus in vivo and consequently drive the viral burden on the host. In at least a subset of long-term survivors of HIV infection, the absence of disease was associated with infection by a *nef*-defective strain (Deacon et al. 1995; Kirchhoff et al. 1995). Similarly, adult rhesus monkeys inoculated with a *nef*-deleted SIV exhibit only low levels of viremia and not only remain asymptomatic (Kestler 1991), but also become resistant to subsequent challenges with wild-type SIV, leading to the suggestion that *nef*-mutated viruses could be the basis for live-attenuated AIDS vaccines (Daniel et al. 1992; Wyand et al. 1997). A similar case can be made for *vpr*, that also promotes viral replication in vivo, as indicated by the rapid emergence of revertants in monkeys inoculated with *vpr*-defective SIV mutants (Lang et al. 1993). Correspondingly, HIV-1 strains with mutations in *vpr* were isolated from a long-term nonprogressing mother and her infected child (Wang et al. 1996). The *vif* gene of both HIV and

SIV is essential for growth in lymphocytes and macrophages, the natural targets of these viruses. Finally, the *vpu* gene is likely to be equally crucial in view of its high degree of conservation among HIV-1 isolates. HIV-1 strains defective in *nef, vif*, or *vpu* displayed attenuated phenotypes in the SCID-hu mouse model of HIV-1 infection (Aldrovandi and Zack 1996).

In conclusion, we can predict that the only features of the parental HIV shared by any of the unlikely RCR originating during the production of the lentiviral vector would be those dependent on the *gag* and *pol* genes. This implies that RCR monitoring during vector manufacturing and in the transduced host may be performed by the well validated and sensitive assays based on HIV *gag* detection, such as the p24 immuno-capture assay and the *gag* RNA-PCR assay. Most importantly, any RCR would be sensitive to the variety of anti-HIV drugs that target reverse transcriptase or the viral protease, and are currently used in the clinic with satisfactory efficacy.

Another important advantage of the new minimal packaging construct is that it can be stably introduced into a cell line. The production of lentiviral vectors poses serious challenges because of the toxic nature of some of the components used in the system. In particular the HIV accessory gene product Vpr causes cell cycle arrest in G_2 (Levy et al. 1993; Rogel et al. 1995) and is thus incompatible with the generation of stable expressor clones. The VSV-G protein is also highly toxic when expressed to signifi-cant levels, mostly due to its efficient fusogenic activity. Accordingly, the first approach adopted to generate vectors was the transient cotransfection of the required constructs into human kidney 293T cells. The availability of a stable producer system amenable to better characterization and scale-up will considerably increase the potential for the use of lentiviral vectors in a clinical setting. A number of approaches are being taken toward real-izing this goal, which include the elimination of viral components and the use of inducible expression systems (Carroll et al. 1994; Corbeau et al. 1996; Yu et al. 1996; Srinivasakumar et al. 1997).

Potential for Direct In Vivo Gene Delivery

The promise of lentiviral vectors for direct gene delivery was demon-strated by introducing into the central nervous system genes expressing β-galactosidase and green fluorescent protein (GFP). After injection into the corpus striatum or the hippocampus of the brain of adult rats, highly efficient transduction and stable expression of the transgene was observed in neurons, without any detectable pathology. Triple labeling confirmed that almost 90% of the striatal cells transduced by the lentiviral vector were terminally differentiated neurons (Naldini et al. 1996a,b; Blömer et al. 1997). Animals analyzed 9 months after a single injection of the vector, the longest time tested so far, showed no decrease in the average level of transgene expression and no sign of tissue pathology or immune reaction (Blömer et al. 1997). We propose that lifelong expression of an exogenous gene can be achieved in normal animals by a single injection of the lentiviral vector. Stable and efficient gene delivery was also shown in the retina. Lentiviral vector carrying the GFP gene was injected into the subretinal space of rat eyes. The GFP gene under the control of the CMV promoter was efficiently expressed in both photoreceptor cells and retinal pigment epithelium. However, the use of the rhodopsin promoter resulted in expression predominantly in photoreceptor cells. Up to 80% of the area of whole retina expressed the GFP, and the expression persisted for at least 12 weeks with no apparent decrease (Miyoshi et al. 1997).

We have further reported the ability of lentiviral vectors to introduce genes directly into the liver and muscle of rats. Sustained expression of GFP could be observed for more than 22 weeks in the liver, the longest time tested. Similar sustained expression was observed in transduced skeletal muscles. At a minimum, 3–4% of the total liver tissue was transduced by a single injection of vector. Furthermore, no inflammation or recruitment of lymphocytes could be detected at the site of injection (Kafri et al. 1997). The vector could also be efficiently readministered.

Other groups have also used VSV-G-pseudotyped replication-defective HIV vectors and reported significantly improved transduction of growth-arrested human primary fibroblasts, macrophages, and CD34[+] hematopoietic progenitor cells in vitro over MLV-based vectors (Poeschla et al. 1996; Reiser et al. 1996).

An ideal vector should have the following three properties: (1) ability to integrate and sustain expression of the transgene; (2) no immunological consequences attributed to the vector; and (3) no adverse pathological consequences to the recipient. We believe the available data to date points to lentiviral vectors fulfilling these desired features. Further modifications of "minimal" lentiviral vectors to allow regulation of the transgene will be the next step toward making them a versatile delivery system for human gene therapy.

ACKNOWLEDGMENTS

This work was supported by funds from the National Institutes of Health, March of Dimes, American Cancer Society, the H.N. and Frances C. Berger Foundation, Pardee Foundation, Valley Foundation, and the Fritz

Burns Foundation. Dr. Verma is an American Cancer Society Professor of Molecular Biology.

REFERENCES

Aldovini A. and Young R.A. 1990. Mutations of RNA and protein sequences involved in human immunodeficiency virus type 1 packaging result in production of noninfectious virus. *J. Virol..* **64:** 1920–1926.

Aldrovandi G.M. and Zack J.A. 1996. Replication and pathogenicity of HIV type 1 accessory gene mutants in SCID-hu mice. *J. Virol.* **70:** 1505–1507.

Berkowitz R.D., Hammarskjöld M.-L., Helga-Maria C., Rekosh D., and Goff S.P. 1995. 5′ regions of HIV-1 RNAs are not sufficient for encapsidation: Implications for the HIV-1 packaging signal. *Virology* **212:** 718–723.

Blaese R.M., Culver K.W., Miller A.D., Carter C.S., Fleisher T., Clerici M., Shearer G., Chang L., Chiang Y., Tolstoshev P., et al. 1995. T lymphocyte-directed gene therapy for ADA-SCID: Initial trial results after 4 years. *Science* **270:** 475–480.

Blömer U., Naldini L., Kafri T., Trono D., Verma I.M., and Gage F.H. 1997. Highly efficient and sustained gene transfer in adult neurons with a lentivirus vector. *J. Virol.* **71:** 6641–6649.

Bordignon C., Notarangelo L.D., Nobili N., Ferrari G., Casorati G., Panina P., Mazzolari E., Maggioni D., Rossi C., Servida P., et al. 1995. Gene therapy in peripheral blood lymphocytes and bone marrow for ADA-immunodeficient patients. *Science* **270:** 470–475.

Buchschacher G.L.J. and Panganiban A.T. 1992. Human immunodeficiency virus vectors for inducible expression of foreign genes. *J. Virol.* **66:** 2731–2739.

Bukrinsky M.I., Haggerty S., Dempsey M.P., Sharova N., Adzhubel A., Spitz L., Lewis P., Goldfarb D., Emerman M., and Stevenson M. 1993. A nuclear localization signal within HIV-1 matrix protein that governs infection of non-dividing cells. *Nature* **365:** 666–669.

Burns J.C., Friedmann T., Driever W., Burrascano M., and Yee J.-K. 1993. Vesicular Stomatitis virus G glycoprotein pseudotyped retroviral vectors: Concentration to very high titer and efficient gene transfer into mammalian and non-mammalian cells. *Proc. Natl. Acad. Sci.* **90:** 8033–8037.

Carroll R., Lin J.T., Dacquel E.J., Mosca J.D., Burke D.S., and St. Louis DC. 1994. A human immunodeficiency virus type 1 (HIV-1)-based retroviral vector system utilizing stable HIV-1 packaging cell lines. *J. Virol.* **68:**6047–6051.

Clever J.L. and Parslow T.G. 1997. Mutant human immunodeficiency virus type 1 genomes with defects in RNA dimerization or encapsidation. *J. Virol.* **71:** 3407–3414.

Corbeau P., Kraus G., and Wong-Staal F. 1996. Efficient gene transfer by a human immunodeficiency virus type 1 (HIV-1)-derived vector utilizing a stable HIV packaging cell line. *Proc. Natl. Acad. Sci.* **93:** 14070–14075.

Dai Y., Roman M., Naviaux R.K., and Verma I.M. 1992. Gene therapy via primary myoblasts: Long-term expression of factor IX protein following transplantation in vivo. *Proc. Natl. Acad. Sci.* **89:** 10892–10895.

Daniel M.D., Kirchhoff F., Czajak S.C., Sehgal P.K., and Desrosiers R.C. 1992. Protective effect of a live-attenuated SIV vaccine with a deletion in the *nef* gene. *Science* **258:** 1938–1941.

Deacon N.J., Tsykin A., Solomon A., Smith K., Ludford-Menting M., and Hooker D.J., McPhee D.A., Greenway A.L., Ellett A., Chatfield C., et al. 1995. Genomic structure of an attenuated quasi species of HIV-1 from a blood transfusion donor and recipients. *Science* **270:** 988–991.

Gallay P., Chin D., Hope T.J., and Trono D. 1997. HIV-1 infection of nondividing cells mediated through the recognition of integrase by the import/karyopherin pathway. *Proc. Natl. Acad. Sci.* **94:** 9825–9830.

Gallay P., Swingler S., Aiken C., and Trono D. 1995a. HIV-1 infection of nondividing cells: C-terminal tyrosine phosphorylation of the viral matrix protein is a key regulator. *Cell* **80:** 379–388.

Gallay P., Swingler S., Song J., Bushman F., and Trono D. 1995b. HIV nuclear import is governed by the phosphotyrosine-mediated binding of matrix to the core domain of integrase. *Cell* **83:** 569–576.

Geigenmuller U. and Linial M.L. 1996. Specific binding of human immunodeficiency virus type 1 (HIV-1) Gag-derived proteins to a 5′ HIV-1 genomic RNA sequence. *J. Virol.* **70:** 667–671.

Grossman M., Radar D.J., Muller D.W.M., Kolansky D.M., Kozarsky K., Clark III B.J., Stein E.A., Lupien P.J., Brewer H.B. Jr., Raper S.E.,. et al. 1995. A pilot study of ex vivo gene therapy for homozygous familial hypercholesterolaemia. *Nat. Med.* **1:** 1148–1154.

Heinzinger N.K., Bukinsky M.I., Haggerty S.A., Ragland A.M., Kewalramani V., Lee M.A., Gendelman H.E., Ratner L., Stevenson M., and Emerman M. 1994. The Vpr protein of human immunodeficiency virus type 1 influences nuclear localization of viral nucleic acids in nondividing host cells. *Proc. Natl. Acad. Sci.* **91:** 7311–7315.

Kafri T., Blömer U., Peterson D.A., Gage F.H., and Verma I.M. 1997. Sustained expression of genes delivered directly into liver and muscle by lentiviral vectors. *Nat. Genet.* **17:** 314–317.

Kaye J.F., Richardson J.H., and Lever A.M.L. 1995. *cis*-acting sequences involved in human immunodeficiency virus type 1 RNA packaging. *J. Virol.* **69:** 6588–6592.

Kestler III H.W., Ringler D.J., Mori K., Panicali D.L., Sehgal P.K., Daniel M.D., and Desrosiers R.C. 1991. Importance of the *nef* gene for maintenance of high virus loads and for development of AIDS. *Cell* **65:** 651–662.

Kim H.J., Lee K., and O'Rear J.J. 1994. A short sequence upstream of the 5′ major splice site is important for encapsidation of HIV-1 genomic RNA. *Virology* **198:** 336–340.

Kirchhoff F., Greenough T.C., Brettler D.B., Sullivan J.L., and Desrosiers R.C. 1995. Absence of intact *nef* sequences in a long-term survivor with nonprogressive HIV-1 infection. *N. Engl. J. Med.* **332:** 228–232.

Kohn D.B., Weinberg K.I., Nolta J.A., Heiss L.N., Lenarsky C., Crooks G.M., Hanley M.E., Annett G., Brooks J.S., and el-Khoureiy A. 1995. Engraftment of gene-modified umbilical cord blood cells in neonates with adenosine deaminase deficiency. *Nat. Med.* **1:** 1017–1023.

Lang S.M., Strahl-Henning C., Coulibaly C., Hunsmann G., Muller J., Muller-Hermalink H., Fuchs D., Wachter H., Daniell M.M., et al. 1993. Importance of vpr for infection of rhesus monkey with simian immunodeficiency virus. *J. Virol.* **67:** 902–912.

Laughrea M., Jette L., Mak J., Kleiman L., Liang C., and Wainberg M.A. 1997. Mutations in the kissing-loop hairpin of human immunodeficiency virus type 1 reduce viral infectivity as well as genomic RNA packaging and dimerization. *J. Virol.* **71:** 3397–3406.

Lever A., Gottlinger H., Haseltine W., and Sodroski J. 1989. Identification of a sequence required for efficient packaging of human immunodeficiency virus type 1 RNA into virions. *J. Virol.* **63:** 4085–4087.

Levy D.N., Fernandes L.S., Williams W.V., and Weiner D.B. 1993. Induction of cell differentiation by human immunodeficiency virus 1 vpr. *Cell* **72:** 541–550.

Lewis P.F. and Emerman M. 1994. Passage through mitosis is required for oncoretroviruses but not for the human immunodeficiency virus. *J. Virol.* **68:** 510.

Luban J. and Goff SP. 1994. Mutational analysis of *cis*-acting packaging signals in human immunodeficiency virus type 1 RNA. *J. Virol.* **68:** 3784–3793.

McBride M.S. and Panganiban A. 1996. The human immunodeficiency virus type 1 encapsidation site is a multipartite RNA element composed of functional hairpin structures. *J. Virol.* **70:** 2963–2973.

——. 1997. Position dependence of functional hairpins important for human immunodeficiency virus type 1 RNA encapsidation in vivo. *J. Virol.* **71:** 20050–20058.

McBride M.S., Schwartz M.D., and Panganiban A. 1997. Efficient encapsidation of human immunodeficiency virus type 1 vectors and further characterization of *cis* elements required for encapsidation. *Proc. Natl. Acad. Sci. J. Virol.* **71:** 4544–4554.

Miller A.D., Miller D.G., Garcia J.V., and Lynch C.M. 1993. Use of retroviral vectors for gene transfer and expression. *Methods Enzymol.* **217:** 581–599.

Miller D.G., Adam M.A., and Miller A.D. 1990. Gene transfer by retrovirus vectors occurs only in cells that are actively replicating at the time of infection. *Mol. Cell. Biol.* **10:** 4239–4242.

Miyoshi H., Takahashi M., Gage F.H., and Verma I.M. 1997. Stable and efficient gene transfer into the retina using an HIV-based lentiviral vector. *Proc. Natl. Acad. Sci.* **94:** 10319–10323.

Naldini L., Blömer U., Gage F.H., Trono D., and Verma I.M. 1996a. Efficient transfer, integration, and sustained long-term expression of the transgene in adult rat brains injected with a lentiviral vector. *Proc. Natl. Acad. Sci.* **93:** 11382–11388.

Naldini L., Blömer U., Gallay P., Ory D., Mulligan R., Gage F.H., Verma I.M., and Trono D. 1996b. In vivo gene delivery and stable transduction of nondividing cells by a lentiviral vector. *Science* **272:** 263–267.

Paillart J.C., Skripkin E., Ehresmann B., Ehresmann C., and Marquet R. 1996. A loop-loop "kissing" complex is the essential part of the dimer linkage of genomic HIV-1 RNA. *Proc. Natl. Acad. Sci.* **93:** 5572–5577.

Palmer T.D., Rosman G.J., Osborne W.R.A., and Miller A.D. 1991. Genetically modified skin fibroblasts persist long after transplantation but gradually inactivate introduced genes. *Proc. Natl. Acad. Sci.* **88:** 1330–1334.

Parolin C., Dorfman T., Palu G., Gottlinger H., and Sodroski J. 1994. Analysis in human immunodeficiency virus type 1 vectors of *cis*-acting sequences that affect gene transfer into human lymphocytes. *J. Virol.* **68:** 3888–3895.

Poeschla E., Corbeau P., and Wong-Staal F. 1996. Development of HIV vectors for anti-HIV gene therapy. *Proc. Natl. Acad. Sci.* **93:** 11395–11399.

Poznansky M., Lever A., Bergeron L., Haseltine W., and Sodroski J. 1991. Gene transfer into human lymphocytes by a defective human immunodeficiency virus type 1 vector. *J. Virol.* **65:** 532–536.

Reiser J., Harmison G., Kluepfel-Stahl S., Brady R.O., Karlsson S., and Schubert M. 1996. Transduction of nondividing cells pseudotyped defective high-titer HIV type 1 particles. *Proc. Natl. Acad. Sci.* **93:** 15266–15271.

Roe T., Reynolds T.C., Yu G., and Brown P.O. 1993. Integration of murine leukemia virus

DNA depends on mitosis. *EMBO J.* **12:** 2099–2108.

Rogel M.E., Wu L.I., and Emerman M. 1995.The human immunodeficiency virus type 1 vpr gene prevents cell proliferation during chronic infection. *J. Virol.* **69:** 882–888.

Scharfmann R., Axelrod J.H., and Verma I.M. 1991. Long-term in vivo expression of retrovirus-mediated gene transfer in mouse fibroblast implants. *Proc. Natl. Acad. Sci.* **88:** 4626–4630.

Shimada T., Fujii H., Mitsuya A., and Nienhuis W. 1991. Targeted and highly efficient gene transfer into CD4+ cells by a recombinant human immunodeficiency virus retroviral vector. *J. Clin. Invest.* **88:** 1043–1047.

Srinivasakumar N., Chazal N., Helga-Maria C., Prasad S., Hammarskjöld M., and Rekosh D. 1997. The effect of viral regulatory protein expression on gene delivery by human immunodeficiency virus type 1 vectors produced in stable packaging cell lines. *J. Virol.* **71:** 5841–5848.

Verma I.M. and Somia N. 1997. Gene therapy promises, problems and prospects. *Nature* **389:** 239–242.

Vicenzi E., Dimitrov D.S., Engelman A., Migone T.S., Purcell D.F.J., Leonard J., Englund G., and Martin M.A. 1994. An integration-defective U5 deletion mutant of human immunodeficiency virus type 1 reverts by eliminating additional long terminal repeat sequences. *J. Virol.* **68:** 7879–7890.

Wang B., Ge Y.C., Palasanthiran P., Xiang S., Ziegler J., Dwyer D.E., Randle C., Dowton D., Cunningham A., and Saksena N.K. 1996. Gene defects clustered at the C-terminus of the vpr gene of HIV-1 in long-term nonprogressing mother and child pair: In vivo evolution of vpr quasispecies in blood and plasma. *Virology* **223:** 224–232.

Wyand M.S., Manson K.H., Lackner A.A., and Desrosiers R.C. 1997. Resistance of neonatal monkeys to live attenuated vaccine strains of simian immunodeficiency virus. *Nature* **3:** 32–36.

Yu H., Rabson A.B., Kaul M., Ron Y., and Dougherty J.P. 1996. Inducible human immunodeficiency virus type 1 packaging cell lines. *J. Virol.* **70:** 4530–4537.

Zufferey R., Nagy D., Mandel R.J., Naldini L., and Trono D. 1997. Multiply attenuated lentiviral vector achieves efficient gene delivery in vivo. *Nat. Biotechnol.* **15:** 871–875.

4

Structure and Genetic Organization of Adenovirus Vectors

Mary M. Hitt and Robin J. Parks

Department of Biology
McMaster University
Hamilton, Ontario
Canada L8S 4K1

Frank L. Graham

Departments of Biology and Pathology
McMaster University
Hamilton, Ontario
Canada L8S 4K1

Adenovirus (Ad) was first isolated in 1953 (Rowe et al. 1953) and not long after was recognized as an invaluable tool for the investigation of various aspects of mammalian molecular biology from oncogenesis to DNA replication, transcriptional regulation, and protein synthesis. Some of the features that made Ad preferable over other viruses for such studies include the ease with which the Ad genome can be experimentally manipulated, the ability to propagate Ad to high titer, and the efficiency with which Ad infects a wide variety of both quiescent and proliferating cells of various species and cell types. These properties have also made Ad an attractive vehicle for gene transfer and transgene expression in mammalian cells with numerous recent applications of Ad as a vector for human gene therapy (for review, see Bramson et al. 1995; Hitt et al. 1996; and this volume). This chapter overviews the structure and lytic cycle of wild-type Ad as well as the structure of first generation, second generation, and helper-dependent Ad vectors designed for gene therapy.

GENOMIC ORGANIZATION AND VIRION STRUCTURE

Adenoviruses have been isolated from many different avian and mammalian species, although the human Ads have been the most extensively studied. The approximately 50 known serotypes of human Ad have been divided into 6 subgroups based on immunological, biological, and DNA sequence similarities. The best characterized serotypes are adenovirus type 2 (Ad2) and Ad5 of subgroup C, Ad12 of subgroup A and Ad7 of subgroup B. However, all human Ads examined thus far share many similarities in terms of genomic organization and biological properties (for review, see Shenk 1996). The 30–40 kbp double-stranded linear DNA genome is flanked by short (about 100–140 bp) inverted terminal repeat (ITR) sequences, each containing an origin of DNA replication. A *cis*-acting packaging domain, required for encapsidation of the genome, is located near the ITR at the "left" (relative to the conventional map of Ad; Fig. 1). The genome can be divided into regions that are expressed, for the most part, either early or late after infection of the host cell, with the onset of DNA replication delineating the boundary of the two phases. Nearly all of the genome is transcribed by the host RNA polymerase II. The early region transcription units are early region 1A (E1A), E1B, E2, E3, and E4. The gene-encoding protein IX (pIX), which is colinear with E1B but utilizes a different promoter, is expressed at an intermediate time, as is the pIVa2 gene. Most late transcripts initiate from the major late promoter

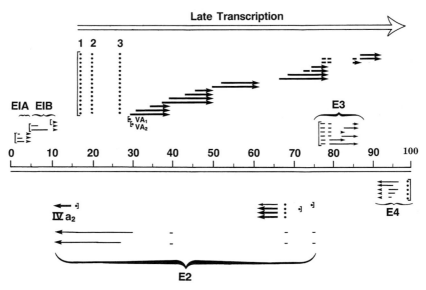

Figure 1 Transcription map of Ad5. Early Ad transcripts are shown as *light arrows*, the primary major late transcript is indicated by the *open arrow*, and processed late mRNAs are shown as *bold arrows*, with the arrows indicating the direction of transcription. The tripartite leader of the late mRNAs is indicated by 1, 2, and 3. The genome length is given in map units with each map unit corresponding to 360 bp.

(MLP) and are subsequently processed to generate five families of 3´-coterminal transcripts (L1 to L5), each containing the well characterized tripartite leader sequence at the 5´ end. Other late transcripts include those from the late E2 promoter as well as the RNA polymerase III transcribed VA RNAs.

The Ad genome is packaged in a nonenveloped icosahedral virion about 140 nm in diameter. The 12 penton subunits of the viral capsid are found at the vertices of the icosahedron, each containing a penton base from which a fiber protein protrudes. The facets of the virion are composed of hexon subunit arrays, with minor capsid proteins functioning to stabilize the virion (Stewart et al. 1991, 1993 and references therein).

LYTIC CYCLE OF ADENOVIRUS

Viral entry into the host cell occurs in two steps: attachment and internalization. Attachment to most cell types is mediated by high-affinity binding of fiber to host cell receptors, which appear to have some degree of subgroup specificity (Defer et al. 1990; Stevenson et al. 1995; Hong et al. 1997). Although the identities of the fiber receptors have not yet been definitively demonstrated, potential candidates include the $\alpha 2$ domain of the heavy chain of human major histocompatibility complex (MHC) class I molecules (Hong et al. 1997) and a protein that also acts as a receptor for Coxsackie B viruses (Bergelson et al. 1997) encoded on human chromosome 21 that was previously implicated as carrying an Ad fiber receptor locus (Mayr and Freimuth 1997). Internalization of the virus is promoted by binding of the penton base of Ad to α_v integrins on the cell surface (Wickham et al. 1993). For attachment to hematopoietic cells, which lack a fiber receptor, Ad may utilize penton base binding to a class of integrins distinct from that required for internalization (Huang et al. 1996). Poor transduction by Ad has been correlated with a lack of target cell receptors for either attachment or internalization or both (Acsadi et al. 1994; Freimuth 1996; Hashimoto et al. 1997; Zabner et al. 1997).

After binding to host cell integrins, the virus is internalized by receptor-mediated endocytosis. Once inside the cell, penton base mediates lysis of the endosomal membrane releasing the contents of the endosome into the cytosol (Seth et al. 1984; Curiel et al. 1991). The process of Ad receptor-mediated endocytosis has been exploited for efficient nonviral gene transfer by cointernalization of foreign DNA with the Ad particles (Yoshimura et al. 1993; Michael and Curiel 1994; Cotten 1995). The virion is disassembled stepwise throughout the process of internalization and nuclear import (Greber et al. 1993, 1997).

Ad transcription, replication, and virus packaging take place within the nucleus of the infected cell. The E1A transcription unit is the first Ad sequence to be expressed during viral infection (for review, see Shenk and Flint 1991; Bayley and Mymryk 1994). Alternative splicing of the primary E1A transcript results in expression of two major E1A proteins that are involved in transcriptional regulation of the virus and are required for Ad replication under normal circumstances. In addition, E1A stimulates the host cell to enter S phase by binding to cellular factors that block cell cycle progression. Like E1A and many other Ad transcripts, E1B is alternatively spliced, yielding two major proteins which are necessary for blocking host mRNA transport, stimulating viral mRNA transport (for review, see Shenk 1996), and blocking E1A-induced apoptosis (White et al. 1991). The E1 region of most, if not all, subgroups is capable of transforming rodent cells in tissue culture, although only subgroup A and (to a lesser extent) B are capable of inducing tumors in animals (for review, see Bayley and Mymryk 1994).

The E2 region, which encodes proteins required for viral DNA replication, can be divided into two subregions, E2a and E2b, which share the same promoters but have transcripts that are polyadenylated at different sites. E2a encodes the 72-kD DNA-binding protein, and E2b encodes the DNA polymerase and the terminal protein precursor (pTP) .

At least seven proteins are produced from the E3 region, many of which are involved in evasion of the host immune system (for review, see Wold and Gooding 1989). The E3 gp19 kD protein binds to MHC class I molecules, causing their retention in the endoplasmic reticulum, thus reducing recognition of the infected host cell by cytotoxic T cells. The E3 14.7 kD protein and E3 10.4 kD/14.5 kD complex may have an even more dramatic effect on the host immune system by inhibiting cytolysis and inflammation induced by tumor necrosis factor (Krajcsi et al. 1996; Sparer et al. 1996). Although E3 proteins do not appear to be required for virus growth in tissue culture, the E3 11.6 kD "death" protein has recently been reported to accelerate lysis and to induce release of virus from the host cell (Tollefson et al. 1996).

The E4 region (for review, see Leppard 1997) encodes at least six proteins, some of which function in DNA replication, late gene expression, and host protein shut off. E4 open reading frame 3 (ORF3) and ORF6 products have redundant functions in facilitating virus growth, both increasing the stability of late viral transcripts in the nucleus. In addition, the E4 ORF6 protein complexes with the E1B 55 kD protein to facilitate viral RNA transport. Other E4 gene products are not critical for virus production in tissue culture (Bridge and Ketner 1989). The E4

ORF4 product indirectly down-regulates E1A expression, resulting in a drastic reduction in viral DNA accumulation unless E4 ORF3 or ORF6 is present to counteract the effects of ORF4 (Bridge et al. 1993; Medghalchi et al. 1997). The E4 ORF6/7 product binds to the cellular transcription factor E2F, which in turn activates the Ad E2 region, increasing expression of the replication proteins. Recent evidence suggests that the E4 ORF6 product may also have some oncogenic potential (Moore et al. 1996; Nevels et al. 1997).

Viral DNA replication begins within 8 hours postinfection (for review, see Ginsberg 1984). Replication initiates at either ITR with pTP serving as a primer for DNA synthesis following covalent linkage to dCMP catalyzed by Ad polymerase. Synthesis continues for the length of the genome by a strand displacement mechanism involving the E2-encoded DNA binding protein and Ad polymerase as well as cellular factors. The ITRs of the displaced viral strand can anneal to form a panhandle structure which also acts as a template for initiation of DNA synthesis (Daniell 1976; Lippe and Graham 1989).

Late gene expression occurs after the onset of viral DNA replication. Nearly all of the protein expressed at this time is viral, due to the activities of the E1 and E4 gene products described previously. Transcripts from the MLP encode most of the virion structural proteins, including penton, hexon, and fiber (encoded by L2, L3, and L5, respectively). VA RNAs, which are required for translation of late viral transcripts, are also expressed late (Thimmapaya et al. 1982). Protein IX contributes to the heat stability of Ad as well as to the efficiency of packaging full-length genomes (Colby and Shenk 1981; Ghosh-Choudhury et al. 1987). Virus production continues for about 2 days postinfection at which time lysis occurs, releasing up to 1000 infectious viral particles per cell.

FIRST-GENERATION ADENOVIRUS VECTORS

Most human Ad vectors have been constructed using derivatives of Ad2 or Ad5, which are 95% homologous and have been completely sequenced (Roberts et al. 1984; Chroboczek et al. 1992), although Ad7- and Ad4-based vectors have also been reported (Lubeck et al. 1989; Lindley et al. 1994; Abrahamsen et al. 1997; Chengalvala et al. 1997). First-generation vectors for gene therapy typically have deletions in the E1 region to render the vector replication-defective, thus preventing virus production and lysis of the target cell. These vectors can be propagated in the E1-complementing human embryonic kidney cell-derived 293 line (Graham et al. 1977). The largest viable E1 deletion reported for first-generation vectors extends from nucleotide 342 to 3523 (Bett et al. 1994), removing nearly all the E1 sequences between the viral packaging domain and the coding sequence for pIX.

The E3 region, which is dispensable for growth in culture, is also deleted in many Ad vectors to allow a greater cloning capacity. The largest E3 deletion reported to date is 3.1 kbp (from nucleotide 27865 to 30995 of Ad5), which prevents expression of all E3 genes (Bett et al. 1994). Because Ad is capable of packaging a genome that is 5% larger than the wild-type genome, combining the largest E1/E3 deletions provides a total cloning capacity of 8.3 kbp. However, there may be cases in which expression of the E3 region is desirable, particularly in instances where evasion of the host immune system is an advantage, and vector systems have been designed that maintain or restore at least part of the E3 region (Bett et al. 1994; Lee et al. 1995; Ginsberg 1996; Bruder et al. 1997). Nevertheless, expression of the E3 19 kD protein does not guarantee prolonged transgene expression in vivo (Schowalter et al. 1997) probably in part because the level of E3 expression required is quite high. It is noteworthy that many of the Ad vectors in use today were derived from the Ad5 strain *dl*309 or its derivative plasmid, pJM17, which carry a deletion/substitution in E3, but can still express the E3 19 kD and 11.6 kD proteins (Bett et al. 1995; Gingras et al. 1996).

There are several strategies commonly used for the construction of first-generation Ad vectors (Gerard and Meidell 1995; Hitt et al. 1995; Spector and Samaniego 1995). To insert a foreign gene or cDNA in place of the E1 region, the desired gene (usually together with exogenous promoter and polyadenylation sequences) is first inserted into a shuttle plasmid in which the insertion site replacing E1 is flanked by the remaining Ad sequences from the left end of the genome. The sequence is then rescued into virus either by in vitro ligation of the shuttle plasmid with DNA derived from the right end of the genome followed by transfection into 293 cells or more usually by in vivo homologous recombination following cotransfection of the shuttle plasmid with restricted viral DNA or with a second plasmid carrying Ad genomic sequences (Fig. 2). A number of Ad genomic plasmids have been engineered to minimize the rescue of virus that does not contain the insert. This can be accomplished by increasing the size of the parental genomic plasmid such that only recombination with the shuttle will generate a virus small enough to be packaged (McGrory et al. 1988), or by removing the packaging signal from the parental genomic plasmid such that only recombination with the shuttle plasmid, which carries the packaging signal, will yield a packageable virus (Bett et al. 1994). Alternatively, replacing the E1 region of the

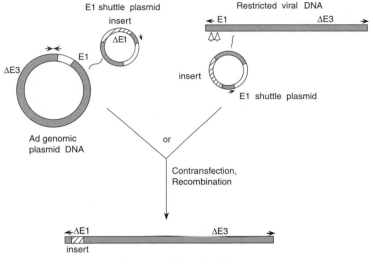

Figure 2 Construction of first-generation recombinant adenovirus vectors by in vivo homologous recombination. First-generation Ad vectors can be rescued following cotransfection of an E1 shuttle plasmid carrying the foreign DNA insert with either a circular Ad genome plasmid (*top left*) or purified Ad DNA that has been cleaved by restriction enzymes to reduce the recovery of parental plaques (*top right*). Homologous recombination in 293 cells between the transfected DNAs results in the production of recombinant virus containing the foreign DNA insert (*bottom*). Ad viral DNA sequences are indicated by *shaded bars*, the foreign DNA insert is indicated by a *hatched bar*, plasmid DNA sequences are indicated by *white bars*, and the ITRs are indicated by *black arrows*. Restriction sites suitable for cleavage of the linear viral DNA are shown as *open arrows*.

parental genomic DNA with markers such as thymidine kinase (Imler et al. 1995) or β-galactosidase (Schaack et al. 1995a) allows for selection or screening to eliminate unrecombined parental plaques. These strategies have been successful for the rescue of probably thousands of Ad vectors to date. However, several alternative strategies have been proposed recently to facilitate rescue of the insert into Ad genomic sequences prior to transfection of 293 cells, for situations in which homologous recombination in 293 cells is deemed too inefficient. Such strategies include homologous recombination in *Escherichia coli* between the insert and genomic sequences maintained as a plasmid (Chartier et al. 1996), homologous recombination in yeast between the insert and Ad genomic sequences maintained as a yeast artificial chromosome (Ketner et al.

1994), and direct cloning of the insert into plasmids carrying a full-length Ad genome (Ghosh-Choudhury et al. 1986; Ghosh-Choudhury and Graham 1987; Miyake et al. 1996).

First-generation Ad vectors are extremely useful for gene transfer in mammalian systems and potentially excellent as recombinant viral vaccines (for review, see Graham and Prevec 1992). However, for many gene therapy applications, there are certain problems with these vectors that must first be solved. Of primary importance is the need to reduce the host immune response to the vector (see Wivel et al., this volume). Even E1-deleted viruses express low levels of viral genes in vitro and in vivo (Nevins 1981; Spergel et al. 1992; Engelhardt et al. 1993; Rich et al. 1993), which are sufficient to induce an immune response against Ad (Yang et al. 1994). This immune response can eliminate transgene expression due to recognition and clearance of the vector-infected cells (Yang et al. 1994, 1995, 1996; Yang and Wilson 1995) and prevent transgene expression following repeat administration of the vector as well (Dai et al. 1995; Gilgenkrantz et al. 1995; Yang et al. 1995). Ad virion proteins in the original innoculum can also elicit an anti-Ad immune response (McCoy et al. 1995). Furthermore, the response against the transgene product can limit the duration of expression (Tripathy et al. 1996; Christ et al. 1997). For persistent transgene expression, either the Ad vectors must be modified to reduce expression of the immunogenic Ad proteins or the immune system of the host must be modulated to prevent an anti-Ad and/or antitransgene product response. Attempts to modify the vectors have led to the development of second generation and helper-dependent Ad vectors discussed in the following sections.

Another problem with the standard strategies for Ad vector rescue and large-scale vector production is the inadvertent generation of E1-positive replication-competent Ad (RCA). RCA can also be a problem when vectors are administered in vivo to the airway, a primary site of infection for wild-type adenovirus. RCA contamination is deleterious for several reasons: RCA can direct high levels of expression of late viral genes that are sufficient to induce an immune response when administered in vivo. In addition, RCA can act as helper for replication of E1-deleted vectors, potentially increasing the effective dose of the vector, as well as its mobilization throughout the recipient and, in theory at least, to bystanders. In culture, contaminating RCA can be generated by recombination between the vector and the E1 sequences, extending from nucleotide 1 to 4344 (Louis et al. 1997), carried by the complementing 293 cell line. The amount of RCA in a virus preparation can be amplified during propagation (Lochmuller et al. 1994), particularly if the RCA has a significant growth advantage over the recombinant vector. In an attempt to reduce RCA generation, alternative E1-complementing cell lines have been derived by transforming human A549 cells with either inducible or constitutive expression cassettes carrying the Ad sequence from nucleotide 505 to 4034 (Imler et al. 1996). These cell lines express E1A proteins and the E1B 19 kD protein, but cannot generate RCA by homologous recombination and can be used to propagate E1-deleted viruses, although with a lower yield of virus than with 293 cells. In another approach to reduce the occurrence of RCA, the vector backbone has been modified by deleting or rearranging the pIX gene, which is colinear with E1B, thereby reducing the frequency of recombination between E1 sequences in 293 cells and those in the Ad vector (Krougliak and Graham 1995; Hehir et al. 1996). Alternatively, insertion of a DNA "stuffer" sequence into the E3 region of the vector can reduce the generation of RCA, since recombination between the "stuffed" vector and the E1 sequences in 293 cells would yield a virus too large to be packaged (Parks et al. 1996).

SECOND-GENERATION ADENOVIRUS VECTORS

To combat the immune response against Ad gene therapy vectors, modifications have been made in the E1-deleted vector backbone, principally in the E2 and the E4 regions, to further attenuate viral gene expression in infected cells (for review, see Yeh and Perricaudet 1997). The E2 region encodes the 72-kD DNA-binding protein (DBP), the DNA polymerase, and the pTP, all required for replication of the genome. All three of these coding sequences have been targeted in the development of second-generation Ad vectors. Engelhardt et al. (1994) first demonstrated that transgene expression could be prolonged in treated animals if expression of DBP were blocked, in this case by using a β-galactosidase-encoding, E1-deleted, temperature-sensitive DBP mutant. However, experiments using different animal models and a vector carrying the same temperature-sensitive mutation but a different transgene demonstrated minimal improvement in persistent expression compared to first-generation vectors (Fang et al. 1996). Cell lines have been constructed which complement DBP deletion mutants, one derived from 293 cells transformed with a fragment of the Ad genome carrying the E2a (DBP) gene under control of the E2 promoter (Zhou et al. 1996), and a second cell line derived from human A549 cells transformed with the E1A and E2a region genes under control of separate inducible promoters (Gorziglia et al. 1996). Both cell lines support growth of E1/E2a-deleted Ad vectors, which will be useful for examining transgene persistence in vivo. Cell lines have also been constructed that complement Ad DNA polymerase mutants (Amalfitano et al. 1996), pTP mutants (Schaack et al. 1995b; Langer and Schaack 1996), or both (Amalfitano and Chamberlain 1997), and a pTP-deleted vector is now available (Schaack et al. 1996) for in vivo expression analysis.

In addition, a number of modifications have been made in the E4 region, a complex regulatory region that is required indirectly for viral DNA replication. The ability of E4 ORF6 to substitute for the entire E4 region in facilitating virus replication in culture has allowed Armentano and colleagues (1995) to construct E1-deleted vectors, containing only ORF6 in E4, that can be propagated in 293 cells. Producing cell lines capable of fully complementing both E1 and E4 has been more of a challenge because coexpression of E4 ORF6 with E1B blocks host protein synthesis, which one would expect to be lethal to the cell. Recently, E1/E4-complementing cell lines have been generated by transforming 293 cells with all or part of the E4 region under control of inducible promoters. 293 cells have been transformed with the entire E4 region under control of the mouse α inhibin promoter (Wang et al. 1995) or the mouse mammary tumor virus (MMTV) promoter (Krougliak and Graham 1995). 293 cell derivatives have also been generated that express E4 ORF6 or ORF6 together with ORF7 under control of the MMTV promoter (Gao et al. 1996; Yeh et al. 1996) or the metallothionein promoter (Brough et al. 1996; Gao et al. 1996). Although these complementing lines should theoretically support growth of vectors with the entire E4 region deleted, in practice such deletions in the vector backbone have resulted in reduced expression of fiber (encoded by a viral DNA segment adjacent to the E4 region) and consequently reduced vector yield (Brough et al. 1996). In contrast, E1-deleted vectors containing either a fragment of E4 (Gao et al. 1996; Yeh et al. 1996; Wang et al. 1997) or spacer DNA (Brough et al. 1996) in place of the entire E4 region have been successfully propagated using these complementing cell lines. Such vectors are generally referred to as E1/E4-deleted vectors even though a portion of the E4 region may be expressed upon infection at high multiplicities with some of these vectors. Cells transduced with E1/E4-deleted vectors yield high levels of transgene expression, and reduced levels of viral late gene expression in vitro have been reported (Dedieu et al. 1997). Although prolonged transgene expression has been observed following in vivo administration of E1/E4-deleted vectors compared to E1-deleted vectors (Wang et al. 1997), an intact E4 region appears to be required for persistent transgene expression in vivo in at least some cases (Armentano et al. 1997; Brough et al. 1997).

In a separate approach to disabling the E4 region of the vector, Fang et al. (1997) have replaced the E4 promoter in an E1-deleted vector with a synthetic promoter containing GAL4-binding sites. This vector can be propagated in a 293-derived cell line engineered to express a GAL4-VP16 fusion protein that activates the synthetic promoter, inducing expression of E4 from the vector backbone. Viral gene expression is greatly reduced in noncomplementing cell lines infected with the E4 mutant vector compared to first generation vectors, but persistence in vivo has not yet been reported.

In summary, deletions in the E2 and E4 regions of Ad vectors greatly reduce viral late gene expression compared to first generation vectors, which should result in a reduced immune response against the vector and vector-infected cells in vivo. However, further investigation is required to ascertain which mutations achieve the highest level of prolonged transgene expression in vivo.

HELPER-DEPENDENT ADENOVIRUS VECTORS

Helper-dependent Ad vectors (HDAd) have the potential to allow for elimination of all viral coding sequences, conceptually at least, the simplest method of avoiding all viral gene expression while at the same time enormously increasing the cloning capacity of the vectors. Consequently, there has been renewed interest in the development of HDAd for gene delivery. Since HDAds are deleted for large regions of Ad protein coding sequences, a second "helper" virus is required within the infected cell to provide replicative functions in *trans* during vector propagation (Fig. 3). HDAds retain many of the advantages of first-generation Ad vectors, mainly high transduction efficiency of both replicating and nonreplicating cells, but also have the added advantages of increased cloning capacity (up to approximately 37 kb), increased safety, and the potential for reduced immune responses due to the elimination of all viral coding sequences. Unfortunately, there exist certain technical problems that have prevented the development of HDAd for gene therapy. These are mainly the difficulty in generating relatively pure, high-titer stocks of the HDAd (typically the helper virus is present in vast excess over the HDAd), and the DNA instability of the HDAd viruses during propagation in the presence of the helper.

The first suggestion that Ad vectors could be generated that were deleted for large regions of the genome came after the characterization by cesium chloride buoyant density gradient of virus harvested from multiple high multiplicity of infection (moi) passages. Burlingham et al.

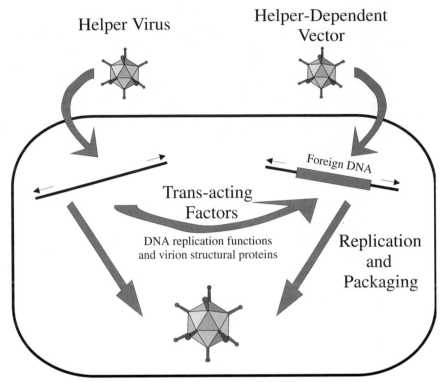

Figure 3 General scheme for the production of HDAd. The HDAd, which is deleted for most, if not all, Ad-coding sequences must be propagated in the presence of a helper virus that provides in *trans* all of the factors required for DNA replication and packaging.

(1974) found that there was a virus subpopulation that was of the same approximate density as wild-type Ad, but which contained DNA that hybridized strongly to cellular DNA sequences. It was subsequently determined that these viruses contained only approximately 1 kb of Ad DNA, derived from the left end of the Ad genome, with the majority of Ad-coding sequences replaced by a large inverted repeat derived from cellular DNA. These viruses were given the designation symmetric recombinants or SYREC (Deuring et al. 1981). That SYREC were only observed after high moi passages was almost certainly due to the requirement of these viruses for helper functions provided during coinfection of cells with wild-type virus.

In the late 1970s, spontaneous recombinants between Ads and SV40 were identified (Hassell et al. 1978; Tjian 1978), in which large regions of the Ad genome, mainly the Ad E3 and E4 transcriptional units, were replaced by SV40 sequences. One of these recombinants, Ad2⁺D2, produced large quantities of T antigen, fused to an Ad structural protein, from the Ad major late promoter and, for the first time, allowed for the purification of large quantities of a T-antigen-related protein that retained biological activity. Because of the loss of essential E4 functions, Ad2⁺D2 could only be propagated in the presence of a wild-type helper virus, usually present in at least a tenfold excess. To ensure that Ad2⁺D2 was maintained in the mixed virus culture, the viruses were propagated in COS cells, which are only permissive for Ad replication in the presence of T antigen, thus providing a strong selection for maintenance of Ad2⁺D2. Later, T-antigen-expressing Ad–SV40 recombinants with specific structures were constructed in vitro, again under the control of either the E2 or major late promoters, and were maintained by growth selection in COS cells (Solnick 1981; Thummel et al. 1981). Because of the difficulty in maintaining stable vector stocks in the absence of effective selection, and difficulty in purifying vectors away from the helper virus, use of HDAd was limited and, once 293 cells became widely available, many researchers developed and used expression systems based on more versatile E1-deleted Ad vectors (see above).

The prospect of using HDAd for transgene transfer for gene therapy was first proposed by Mitani et al. (1995), who produced a vector that was deleted for essential regions of the Ad genome (L1, L2, VAI, VAII, pTP), and that encoded a β-galactosidase/neomycin fusion gene. The vector DNA and wild-type Ad2 helper DNA were cotransfected into 293 cells, and the resulting blue "plaques," presumably containing both the lacZ–HDAd and Ad2 helper virus, were isolated and amplified by serial passage on 293 cells. The vector could be partially purified from helper virus by cesium chloride buoyant-density centrifugation, thanks to differences in virion density resulting from genome size differences. The resulting vector stocks still contained an approximately 200-fold excess of helper virus over vector, but could transduce lacZ into recipient COS cells. Unfortunately, analysis of the vector DNA showed that it had undergone rearrangement, presumably by recombination between the vector genome and the Ad2 helper virus or the E1-complementing Ad5 sequences encoded within the 293 cells, resulting in a loss of some vector sequences.

Strategies similar to those employed by Mitani et al. (1995) were used to generate "fully deleted" vectors encoding cDNAs for CFTR (Fisher et al. 1996), or the Becker or full-length Duchenne muscular dystrophy (DMD) alleles (Haecker et al. 1996; Kumar-Singh and Chamberlain 1996), although helper functions were provided by a first-generation Ad5 vector, encoding human secreted alkaline phosphatase (hSEAP), instead of wild-type Ad. The inclusion of hSEAP permitted use of a simple assay for the presence of helper virus. The vector titer was increased by serial passage of crude vector extracts on 293 cells. Once again, the vector could be partially purified from the helper virus by centrifugation, and the helper content of the final "purified" vector stocks ranged from approximately a 1000-fold excess (Fisher et al. 1996) to 4% (Kumar-Singh and Chamberlain 1996) of the vector titer. Interestingly, vectors that were constructed significantly below the size of the Ad genome tended to be unstable and underwent DNA rearrangement, primarily multimerization events, resulting in an increase in the final size of the vector to approximately the size of wild-type Ad (Fisher et al. 1996; Haecker et al. 1996); however, all of the above vectors were able to transduce cells and allowed for expression of the transgene.

In an attempt to reduce the quantity of helper virus present in the final vector preparations, Kochanek et al. (1996) used a derivative of a first-generation Ad5 helper virus, SV5, that was deleted for 91 bp of the Ad packaging signal, including three of the five elements believed to be essential for Ad DNA packaging (Grable and Hearing 1990). This mutation resulted in a 90-fold reduction in the packaging efficiency of the helper virus compared to wild-type Ad. Thus, during serial amplification of an HDAd encoding a wild-type packaging signal in cells coinfected with SV5, the HDAd was preferentially packaged, resulting in a considerable improvement in the purity of the HDAd after cesium chloride centrifugation (~1% helper virus contamination). In this manner, an HDAd was produced that contained the full-length DMD cDNA under the regulation of the murine creatine kinase promoter and a HCMV-β-galactosidase expression cassette, designated AdDYSβ-gal. This vector could induce expression of both β-gal and DMD, and treatment of *mdx* mice resulted in partial, albeit transient, phenotypic correction (Clemens et al. 1996). In normal mice, transduction with AdDYSβ-gal resulted in inflammation, and histologic examination of transduced tissues showed an infiltration by CD4[+] and CD8[+] cells (Chen et al. 1997). Transgene expression in wild-type mice was transient (less than 42 days); however, in β-gal-tolerized mice (i.e., *lacZ* transgenic animals), transgene expression was substantially prolonged (greater than 82 days). These results were interpreted to mean that, in the absence of an immune response to the transgene, an HDAd might provide long-term expression in vivo.

Recently, we and others have employed bacteriophage recombinase systems to prevent packaging of the helper virus, or to remove large regions of coding sequence from first generation vectors (Lieber et al.

1996; Parks et al. 1996; Hardy et al. 1997). Lieber et al. (1996) used the Cre recombinase to excise most Ad-coding sequences from a first-generation vector containing *lox*P sites flanking the majority of the Ad genome, one *lox*P site located just to the right of the transgene expression cassette that replaced the E1 region and the second located within the E3 region, generating, after Cre-mediated recombination, an HDAd of approximately 9 kb (Fig. 4). In this system, the unrecombined vector itself acts as the complementing helper virus and the E4 transcription unit is retained in the final HDAd. Unfortunately, these vectors were unstable in vivo and provided only limited transgene expression. Preliminary results suggested that these vectors could be stabilized in vivo by the coadministration of a first-generation Ad vector, which apparently provided low-level replicative functions that were necessary for vector persistence and transgene expression (Lieber et al. 1996). It has subsequently been shown that first-generation Ad vectors do not undergo detectable replication in vivo, at least in the liver (Nelson and Kay 1997). Thus, the reason for the poor transgene expression from the "mini"-HDAd remains unclear.

A different Cre/*lox*P-based system for the generation of HDAd involves the use of a first-generation helper virus that has the Ad packaging signal flanked by *lox*P recognition sites (Fig. 5; Parks et al. 1996; Hardy et al. 1997). This virus is easily propagated in normal 293 cells; however, upon infection of a 293-derived cell line that stably expresses the Cre recombinase, the packaging signal is excised, rendering the helper virus genome unpackageable. Without the packaging signal, the

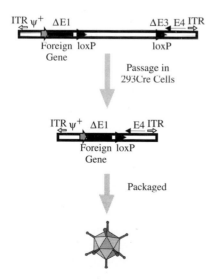

Figure 4 A Cre/*lox*P system for the generation of HDAd. A first generation Ad vector that contains *loxP* sites flanking the majority of the Ad genome generates, after passage through Cre-expressing 293 cells, an HDAd of approximately 9 kb. In this system, the unrecombined vector itself acts as the complementing helper virus.

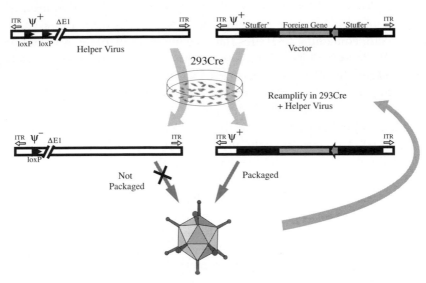

Figure 5 An alternative Cre/*lox*P system for the generation of HDAd. The helper virus has a packaging signal (ψ) flanked by *loxP* sites such that upon infection of 293Cre cells, ψ is excised rendering the helper virus unpackageable. The helper virus is still able to replicate and provides in *trans* all of the functions required for replication and packaging of the HDAd. The titer of the HDAd is increased by serial passage through helper-virus-infected 293Cre cells. (Adapted from Parks et al. 1996.)

helper virus DNA retains the ability to replicate and expresses all of the functions required in *trans* for the replication and packaging of an HDAd. The titer of the vector is increased by serial passage in 293Cre cells, and new helper virus must be added at each passage. This system facilitates the generation of high-titer HDAd preparations with substantially reduced quantities of contaminating helper virus, typically ranging from 0.1 to 0.01% of the HDAd titer.

Using the Cre/*lox*P HDAd system, it has recently been determined that Ad virions have a lower limit for efficient DNA packaging of approximately 75% of the wild-type genome length (Parks and Graham 1997). HDAd vectors were constructed that ranged in size from 15.1 kb to 33.6 kb, and used in the Cre/*lox*P helper-dependent system (Parks et al. 1996). Vectors greater than or equal to 27.7 kb were packaged with equal efficiencies, whereas vectors with smaller genomes were inefficiently packaged. Similar trends were observed during subsequent rounds of vector amplification, with smaller vectors continuing to be recovered at reduced efficiencies. Analysis of vector DNA after amplification showed that vectors below 75% of the Ad genome length had undergone DNA rearrangements, resulting in a final vector size of greater than approximately 27 kb, whereas larger vectors were unaltered. Interestingly, a vector of approximately 15 kb, about half the size of the Ad genome, was recovered as a mix of head-to-tail and tail-to-tail covalent dimers. This observation raises the possibility of constructing vectors of fractional size of the Ad genome (i.e., 1/2, 1/3, etc.) that would be reasonably expected to multimerize spontaneously during amplification, and therefore encode multiple copies of a transgene within a single virion. The tendency of small vectors to rearrange explains the observation of Fisher et al. (1996) and Haecker et al. (1996) who observed multimerization of vectors that were less than the subsequently determined 75% lower limit for efficient DNA packaging. This may also partially explain the results of Lieber et al. (1996) if their 9 kb vector was somehow destabilized by its small size.

Schiedner et al. (1998) examined transgene expression from an HDAd deleted of all viral coding sequences and encoding a genomic copy of the human α_1-antitrypsin gene (approximately 19 kb), including the endogenous promoter, generated using the Cre/*lox*P system (Parks et al. 1996). A 9 kb "stuffer" DNA segment, consisting of an intronic fragment of the hypoxanthine–guanine phosphoribosyltransferase gene, was also included in the vector in order to increase the size above the lower limit for efficient DNA packaging. This vector showed "tissue-specific" gene expression in cell culture, and high-level, stable transgene expression in C57BL/6J mice for greater than 10 months after liver transduction. Moreover, histopathologic examination revealed that those animals treated with the HDAd showed essentially normal liver morphology, whereas control animals treated with a first generation Ad vector had significant acute and chronic liver injury. Taken together, these data suggest that HDAd vectors have very significant advantages over first generation Ad vectors for prolonged transgene expression and decreased host immune and inflammatory responses.

SUMMARY AND FUTURE DIRECTIONS

Adenovirus vectors provide the most efficient system available today for gene transfer into a wide variety of mammalian cell types. First-generation E1-deleted Ad vectors induce high-level transgene expression, the duration of which is limited in vivo to a large extent by a host anti-Ad immune response targeting the vector-infected cells. For some applications, such as cancer gene therapy where the goal of the therapy is clearance of the target tissue, such an immune response may be highly desirable. However, for gene replacement therapy, or other therapies requiring persistent expression, an immune response against first-generation vectors must be avoided. Second-generation vectors have been designed with additional regions of the genome deleted to reduce the level of expression of viral antigens to which the recipient is exposed. Many of these second-generation vectors have yet to be tested for their ability to prolong transgene expression in vivo. Investigators are well on their way to producing helper-dependent systems to generate what may prove to be the ultimate Ad vector, one that contains no Ad protein-coding sequences at all. Such vectors should induce the minimum anti-Ad response possible without immunosuppression of the host. Apart from minimizing the host immune response, the next important milestones will likely be the development of tissue-specific Ad vectors and the development of Ad vectors that direct either integration into the host chromosome or episomal maintenance of the transgene to allow permanent correction of disease.

ACKNOWLEDGMENTS

The authors are supported by grants from the National Institutes of Health (USA), Natural Sciences and Engineering Research Council (NSERC), Medical Research Council, and the National Cancer Institute of Canada. R.J.P. is supported by a NSERC Postdoctoral Fellowship. F.L.G. is a Terry Fox Research Scientist of the National Cancer Institute.

REFERENCES

Abrahamsen L., Kong H.L., Mastrangeli A., Brough D., Lizonova A., Crystal R.G., and Falck-Pedersen E. 1997. Construction of an adenovirus type 7a E1A-vector. *J. Virol.* **71:** 8946–8951.

Acsadi G., Jani A., Huard J., Blaschuk K., Massie B., Holland P., Lochmuller H., and Karpati G. 1994. Cultured human myoblasts and myotubes show markedly different transducibility by replication-defective adenovirus recombinants. *Gene Ther.* **1:** 338–340.

Amalfitano A. and Chamberlain J.S. 1997. Isolation and characterization of packaging cell lines that coexpress the adenovirus E1, DNA polymerase, and preterminal proteins: Implications for gene therapy. *Gene Ther.* **4:** 258–263.

Amalfitano A., Begy C.R., and Chamberlain J.S. 1996. Improved adenovirus packaging cell lines to support the growth of replication-defective gene-delivery vectors. *Proc. Natl. Acad. Sci.* **93:** 3352–3356.

Armentano D., Sookdeo C.C., Hehir K.M., Gregory R.J., St. George J.A., Prince G.A., Wadsworth S.C., and Smith A.E. 1995. Characterization of an adenovirus gene transfer vector containing an E4 deletion. *Hum. Gene Ther.* **6:** 1343–1353.

Armentano D., Zabner J., Sacks C., Sookdeo C.C., Smith M.P., St. George J.A.,

Wadsworth S.C., Smith A.E., and Gregory R.J. 1997. Effect of the E4 region on the persistence of transgene expression from adenovirus vectors. *J Virol.* **71:** 2408–2416.

Bayley S.T. and Mymryk J.S. 1994. Adenovirus E1A proteins and transformation: A review. *Int. J. Oncol.* **5:** 425–444.

Bergelson J.M., Cunningham J.A., Droguett G., Kurt-Jones E.A., Krithivas A., Hong J.S., Horwitz M.S., Crowell R.L., and Finberg R.W. 1997. Isolation of a common receptor for Coxsackie B viruses and adenoviruses 2 and 5. *Science* **275:** 1320–1323.

Bett A.J., Krougliak V., and Graham F.L. 1995. DNA sequence of the deletion/insertion in early region 3 of Ad5 *dl*309. *Virus Res.* **39:** 75–82.

Bett A.J., Haddara W., Prevec L., and Graham F.L. 1994. An efficient and flexible system for construction of adenovirus vectors with inserts or deletion in early regions 1 and 3. *Proc. Natl. Acad. Sci.* **91:** 8802–8806.

Bramson J.L., Graham F.L., and Gauldie J. 1995. The use of adenoviral vectors for gene therapy and gene transfer *in vivo. Curr. Opin. Biotechnol.* **6:** 590–595.

Bridge E. and Ketner G. 1989. Redundant control of adenovirus late gene expression by early region 4. *J. Virol.* **63:** 631–638.

Bridge E., Medghalchi S., Ubol S., Leesong M., and Ketner G. 1993. Adenovirus early region 4 and viral DNA synthesis. *Virology* **193:** 794–801.

Brough D.E., Lizonova A., Hsu C., Kulesa V.A., and Kovesdi I. 1996. A gene transfer vector-cell line system for complete functional complementation of adenovirus early regions E1 and E4. *J. Virol.* **70:** 6497–6501.

Brough D.E., Hsu C.H., Kulesa V.A., Lee G.M., Cantolupa L.J., Lizonova A., and Kovesdi I. 1997. Activation of transgene expression by early region 4 is responsible for a high level of persistent transgene expression from adenovirus vectors *in vivo. J. Virol.* **71:** 9206–9213.

Bruder J.T., Jie T., McVey D.L., and Kovesdi I. 1997. Expression of gp19K increases the persistence of transgene expression from an adenovirus vector in the mouse lung and liver. *J. Virol.* **71:** 7623–7628.

Burlingham B.T., Brown D.T., and Doerfler W. 1974. Incomplete particles of adenovirus. I. Characteristics of the DNA associated with incomplete adenovirions of types 2 and 12. *Virology* **60:** 419–430.

Chartier C., Degryse E., Gantzer M., Dieterle A., Pavirani A., and Mehtali M. 1996. Efficient generation of recombinant adenovirus vectors by homologous recombination in *Escherichia coli. J. Virol.* **70:** 4805–4810.

Chen H.-H., Mack L.M., Kelly R., Ontell M., Kochanek S., and Clemens P.R. 1997. Persistence in muscle of an adenoviral vector that lacks all viral genes. *Proc. Natl. Acad. Sci.* **94:** 1645–1650.

Chengalvala M.V., Bhat B.M., Bhat R.A., Dheer S.K., Lubeck M.D., Purcell R.H., and Murthy K.K. 1997. Replication and immunogenicity of Ad7-, Ad4-, and Ad5-hepatitis B virus surface antigen recombinants, with or without a portion of E3 region, in chimpanzees. *Vaccine* **15:** 335–339.

Christ M., Lusky M., Stoeckel F., Dreyer D., Dieterle A., Michou A.I., Pavirani A., and Mehtali M. 1997. Gene therapy with recombinant adenovirus vectors: Evaluation of the host immune response. *Immunol. Lett.* **57:** 19–25.

Chroboczek J., Bieber F., and Jacrot B. 1992. The sequence of the genome of adenovirus type 5 and its comparison with the genome of adenovirus type 2. *Virology* **186:** 280–285.

Clemens P.R., Kochanek S., Sunada Y., Chan S., Chen H.-H., Campbell K.P., and Caskey

C.T. 1996. *In vivo* muscle gene transfer of full-length dystrophin with an adenoviral vector that lacks all viral genes. *Gene Ther.* **3:**965–972.

Colby W.W. and Shenk T. 1981. Adenovirus type 5 virions can be assembled *in vivo* in the absence of detectable polypeptide IX. *J. Virol.* **39:** 977–980.

Cotten M. 1995. Adenovirus-augmented, receptor-mediated gene delivery and some solutions to the common toxicity problems. In *The molecular repertoire of adenoviruses III: Biology and pathogenesis* (ed. W. Doerfler and P. Boehm), pp. 283–295. Springer, Berlin.

Curiel D.T., Agarwal S., Wagner E., and Cotten M. 1991. Adenovirus enhancement of transferrin-polylysine-mediated gene delivery. *Proc. Natl. Acad. Sci.* **88:** 8850–8854.

Dai Y., Schwarz E.M., Gu D., Zhang W.W., Sarvetnick N., and Verma I.M. 1995. Cellular and humoral immune responses to adenoviral vectors containing factor IX gene: Tolerization of factor IX and vector antigens allows for long-term expression. *Proc. Natl. Acad. Sci.* **92:** 1401–1405.

Daniell E. 1976. Genome structure of incomplete particles of adenovirus *J. Virol.* **19:** 685–708.

Dedieu J.F., Vigne E., Torrent C., Jullien C., Mahfouz I., Caillaud J.M., Aubailly N., Orsini C., Guillaume J.M., Opolon P., Delaere P., Perricaudet M., and Yeh P. 1997. Long term gene delivery into the liver of immunocompetent mice with E1/E4-defective adenoviruses. *J. Virol.* **71:** 4626–4637.

Defer C., Belin M.-T., Caillet-Boudin M.-L., and Boulanger P. 1990. Human adenovirus-host cell interactions: Comparative study with members of subgroups B and C. *J. Virol.* **64:** 3661–3673.

Deuring R., Klotz G., and Doerfler W. 1981. An unusual symmetric recombinant between adenovirus type 12 DNA and human cell DNA. *Proc. Natl. Acad. Sci.* **78:** 3142–3146.

Engelhardt J.F., Ye X., Doranz B., and Wilson J.M. 1994. Ablation of E2A in recombinant adenoviruses improves transgene persistence and decreases inflammatory response in mouse liver. *Proc. Natl. Acad. Sci.* **91:** 6196–6200.

Engelhardt J.F., Yang Y., Stratford-Perricaudet L.D., Allen E.D., Kozarsky K., Perricaudet M., Yankaskas J.R., and Wilson J.M. 1993. Direct gene transfer of human CFTR into human bronchial epithelia of xenografts with E1-deleted adenoviruses. *Nat. Genet.* **4:** 27–34.

Fang B., Koch P., and Roth J.A. 1997. Diminishing adenovirus gene expression and viral replication by promoter replacement. *J. Virol.* **71:** 4798–4803.

Fang B., Wang H., Gordon G., Bellinger D.S., Read M.S., Brinklaus K.M., Woo S.L.C., and Eisensmith R.C. 1996. Lack of persistence of E1- recombinant adenoviral vectors containing a temperature sensitive E2a mutation in immunocompetent mice and hemophilia B dogs. *Gene Ther.* **3:** 217–222.

Fisher K.J., Choi H., Burda J., Chen S.-J., and Wilson, J.M. 1996. Recombinant adenovirus deleted of all viral genes for gene therapy of cystic fibrosis. *Virology* **217:**11–22.

Freimuth P. 1996. A human cell line selected for resistance to adenovirus infection has reduced levels of the virus receptor. *J. Virol.* **70:** 4081–4085.

Gao G.P., Yang Y., and Wilson J.M. 1996. Biology of adenovirus vectors with E1 and E4 deletions for liver-directed gene therapy. *J. Virol.* **70:** 8934–8943.

Gerard R.D. and Meidell R.S. 1995. Adenovirus vectors. In *DNA cloning: A practical approach* (ed. B.D. Hames and D. Glover), pp. 285–306. Oxford University Press, Oxford, United Kingdom.

Ghosh-Choudhury G. and Graham, F.L. 1987. Stable transfer of a mouse dihydrofolate reductase gene into a deficient cell line using a human adenovirus vector. *Biochem. Biophys. Res. Commun.* **147:** 964–973.

Ghosh-Choudhury G., Haj-Ahmad Y., and Graham F.L. 1987. Protein IX, a minor component of the human adenovirus capsid, is essential for the packaging of full length genomes. *EMBO J.* **6:** 1733–1739.

Ghosh-Choudhury G., Haj-Ahmad Y., Brinkley P., Rudy J., and Graham, F.L. 1986. Human adenovirus cloning vectors based on infectious bacterial plasmids. *Gene* **50:** 161–171.

Gilgenkrantz H., Duboc D., Juillard V., Couton D., Pavirani A., Guillet J.G., Briand P., and Kahn A. 1995. Transient expression of genes transferred *in vivo* into heart using first-generation adenoviral vectors: Role of the immune response. *Hum. Gene Ther.* **6:** 1265–1274.

Gingras M.C., Arevalo P., and Aguilar-Cordova E. 1996. Potential salmon sperm origin of the E3 region insert of the adenovirus 5 *dl*309 mutant. *Cancer Gene Ther.* **3:** 151–154.

Ginsberg H.S., ed. 1984. *The adenoviruses*, Plenum Press, New York.

—— 1996. The ups and downs of adenovirus vectors. *Bull. N.Y. Acad. Med.* **73:** 53–58.

Gorziglia M.I., Kadan M.J., Yei S., Lim J., Lee G.M., Luthra R., and Trapnell B.C. 1996. Elimination of both E1 and E2a from adenovirus vectors further improves prospects for *in vivo* human gene therapy. *J. Virol.* **70:** 4173–4178.

Grable M. and Hearing P. 1990. Adenovirus type 5 packaging domain is composed of a repeated element that is functionally redundant. *J. Virol.* **64:** 2047–2056.

Graham F.L. and Prevec L. 1992. Adenovirus-based expression vectors and recombinant vaccines. In *Vaccines: New approaches to immunological problems* (ed. R.W. Ellis), pp. 363–389. Butterworth-Heinemann, Boston, Massachusetts.

Graham F.L., Smiley J., Russell W.C., and Nairn R. 1977. Characteristics of a human cell line transformed by DNA from human adenovirus 5. *J. Gen. Virol.* **36:** 59–72.

Greber U.F., Willetts M., Webster P., and Helenius A. 1993. Stepwise dismantling of adenovirus 2 during entry into cells. *Cell* **75:** 477–486.

Greber U.F., Suomalainen M., Stidwell R.P., Boucke K., Ebersold M.W., and Helenius A. 1997. The role of the nuclear pore complex in adenovirus DNA entry. *EMBO J.* **16:** 5998–6007.

Haecker S.E., Stedman H.H., Balice-Gordon R.J., Smith D.B.J., Greenish J.P., Mitchell M.M., Wells A., Sweeney H.L., and Wilson J.M. 1996. *In vivo* expression of full-length human dystrophin from adenoviral vectors deleted of all viral genes. *Hum. Gene Ther.* **7:** 1907–1914.

Hardy S., Kitamura M., Harris-Stansil T., Dai Y., and Phipps M.L. 1997. Construction of adenovirus vectors through Cre-*lox* recombination. *J. Virol.* **71:** 1842–1849.

Hashimoto Y., Kohri K., Akita H., Mitani K., Ikeda K., and Nakanishi M. 1997. Efficient transfer of genes into senescent cells by adenovirus vectors via highly expressed $\alpha_v \beta_5$ integrin. *Biochem. Biophys. Res. Commun.* **240:** 88–92.

Hassell J.A., Lukanidin E., Fey G., and Sambrook J. 1978. The structure and expression of two defective adenovirus 2/simian virus 40 hybrids. *J. Mol. Biol.* **120:** 209–247.

Hehir K.M., Armentano D., Cardoza L.M., Choquette T.L., Berthelette P.B., White G.A., Couture L.A., Everton M.B., Keegan J., Martin J.M., Pratt D.A., Smith M.P., Smith A.E., and Wadsworth S.C. 1996. Molecular characterization of replication-competent variants of adenovirus vectors and genome modifications to prevent their occurrence. *J. Virol.* **70:** 8459–8467.

Hitt M.M., Addison C.L., and Graham F.L. 1996. Human adenovirus vectors for gene transfer into mammalian cells. *Adv. Pharmacol.* **40:** 137–206.

Hitt M., Bett A.J., Addison C.L., Prevec L., and Graham F.L. 1995. Techniques for human adenovirus vector construction and characterization. *Methods Mol. Genet.* **7:** 13–30.

Hong S.S., Karayan L., Tournier J., Curiel D.T., and Boulanger P.A. 1997. Adenovirus type 5 fiber knob binds to MHC class I α2 domain at the surface of human epithelial and B lymphoblastoid cells. *EMBO J.* **16:** 2294–2306.

Huang S., Kamata T., Takada Y., Ruggeri Z.M., and Nemerow G.R. 1996. Adenovirus interaction with distinct integrins mediates separate events in cell entry and gene delivery to hematopoietic cells. *J. Virol.* **70:** 4502–4508.

Imler J.-L., Chartier C., Dieterle A., Dreyer D., Mehtali M., and Pavirani A. 1995. An efficient procedure to select and recover recombinant adenovirus vectors. *Gene Ther.* **2:** 263–268.

Imler J.-L., Chartier C., Dreyer D., Dieterle A., Sainte-Marie M., Faure T., Pavirani A., and Mehtali M. 1996. Novel complementation cell lines derived from human lung carcinoma A549 cells support the growth of E1-deleted adenovirus vectors. *Gene Ther.* **3:** 75–84.

Ketner G., Spencer F., Tugendreich S., Connelly C., and Hieter P. 1994. Efficient manipulation of the human adenovirus genome as an infectious yeast artificial chromosome clone. *Proc. Natl. Acad. Sci.* **91:** 6186–6190.

Kochanek S., Clemens P.R., Mitani K., Chen H.-H., Chan S., and Caskey C.T. 1996. A new adenoviral vector: Replacement of all viral coding sequences with 28 kb of DNA independently expressing both full-length dystrophin and β-galactosidase. *Proc. Natl. Acad. Sci.* **93:** 5731–5736.

Krajcsi P., Dimitrov T., Hermiston T.W., Tollefson A.E., Ranheim T.S., Vande Pol S.B., Stephenson A.H., and Wold W.S. 1996. The adenovirus E3-14.7K protein and the E3-10.4K/14.5K complex of proteins, which independently inhibit tumor necrosis factor (TNF)-induced apoptosis, also independently inhibit TNF-induced release of arachidonic acid. *J. Virol.* **70:** 4904–4913.

Krougliak V. and Graham F.L. 1995. Development of cell lines capable of complementing E1, E4, and protein IX defective adenovirus type 5 mutants. *Hum. Gene Ther.* **6:** 1575–1586.

Kumar-Singh R. and Chamberlain J.S. 1996. Encapsidated adenovirus minichromosomes allow delivery and expression of a 14 kb dystrophin cDNA to muscle cells. *Hum. Mol. Genet.* **5:** 913–921.

Langer S.J. and Schaack J. 1996. 293 cell lines that inducibly express high levels of adenovirus type 5 precursor terminal protein. *Virology* **221:** 172–179.

Lee M.G., Abina M.A., Haddada H., and Perricaudet M. 1995. The constitutive expression of the immunomodulatory gp19k protein in E1−, E3− adenoviral vectors strongly reduces the host cytotoxic T cell response against the vector. *Gene Ther.* **2:** 256–262.

Leppard K.N. 1997. E4 gene function in adenovirus, adenovirus vector and adeno-associated virus infections. *J. Gen. Virol.* **78:** 2131–2138.

Lieber A., He C.-Y., Kirillova I., and Kay M.A. 1996. Recombinant adenoviruses with large deletions generated by Cre-mediated excision exhibit different biological properties compared with first-generation vectors *in vitro* and *in vivo*. *J. Virol.* **70:** 8944–8960.

Lindley T., Virk K.P., Ronchetti-Blume M., Goldberg K., Lee S.G., Eichberg J.W., Hung P.P., and Cheng S.M. 1994. Construction and characterization of adenovirus co-expressing hepatitis B virus surface antigen and interleukin-6. *Gene* **138:** 165–170.

Lippe R. and Graham F.L. 1989. Adenoviruses with nonidentical terminal sequences are viable. *J. Virol.* **63:** 5133–5141.

Lochmuller H., Jani A., Huard J., Prescott S., Simoneau M., Massie B., Karpati G., and Ascadi G. 1994. Emergence of early region 1-containing replication-competent adenovirus in stocks of replication-defective adenovirus recombinants (ΔE1+ΔE3) during multiple passages in 293 cells. *Hum. Gene Ther.* **5:** 1485–1491.

Louis N., Evelegh C., and Graham F.L. 1997. Cloning and sequencing of the cellular-viral junctions from the human adenovirus type 5 transformed 293 cell line. *Virology* **233:** 423–429.

Lubeck M.D., Davis A.R., Chengalvala M., Natuk R.J., Morin J.E., Molnar-Kimber K., Mason B.B., Bhat B.M., Mizutani S., Hung P.P., and Purcell R.H. 1989. Immunogenicity and efficacy testing in chimpanzees of an oral hepatitis B vaccine based on live recombinant adenovirus. *Proc. Natl. Acad. Sci.* **86:** 6763–6767.

Mayr G.A. and Freimuth P. 1997. A single locus on human chromosome 21 directs the expression of a receptor for adenovirus type 2 in mouse A9 cells. *J. Virol.* **71:** 412–418.

McCoy R.D., Davidson B.L., Roessler B.J., Huffnagle G.B., Janich S.L., Laing T.J., and Simon R.H. 1995. Pulmonary inflammation induced by incomplete or inactivated adenoviral particles. *Hum. Gene Ther.* **6:** 1553–1560.

McGrory J., Bautista D., and Graham F.L. 1988. A simple technique for the rescue of early region I mutations into infectious human adenovirus type 5. *Virology* **163:** 614–617.

Medghalchi S., Padmanabhan R., and Ketner G. 1997. Early region 4 modulates adenovirus DNA replication by two genetically separable mechanisms. *Virology* **236:** 8–17.

Michael S.I. and Curiel D.T. 1994. Strategies to achieve targeted gene delivery via the receptor-mediated endocytosis pathway. *Gene Ther.* **1:** 223–232.

Mitani K., Graham F.L., Caskey C.T., and Kochanek S. 1995. Rescue, propagation, and partial purification of a helper virus-dependent adenovirus vector. *Proc. Natl. Acad. Sci.* **92:** 3854–3858.

Miyake S., Makimura M., Kanegae Y., Harada S., Sato Y., Takamori K., Tokuda C., and Saito I. 1996. Efficient generation of recombinant adenoviruses using adenovirus NA-terminal protein complex and a cosmid bearing the full-length virus genome. *Proc. Natl. Acad. Sci.* **93:** 1320–1324.

Moore M., Horikoshi N., and Shenk T. 1996. Oncogenic potential of the adenovirus E4orf6 protein. *Proc. Natl. Acad. Sci.* **93:** 11295–11301.

Nelson J.E. and Kay M.A. 1997. Persistence of recombinant adenovirus is not dependent on vector DNA replication. *J. Virol.* **71:** 8902–8907.

Nevels M., Rubenwolf S., Spruss T., Wolf H., and Dobner T. 1997. The adenovirus E4orf6 protein can promote E1A/E1B–induced focus formation by interfering with p53 tumor suppressor function. *Proc. Natl. Acad. Sci.* **94:** 1206–1211.

Nevins J.R. 1981. Mechanism of activation of early viral transcription by the adenovirus E1A gene product. *Cell* **26:** 213–220.

Parks R.J. and Graham F.L. 1997. A helper-dependent system for adenovirus vector production helps define a lower limit for efficient DNA packaging. *J. Virol.* **71:** 3293–3298.

Parks R.J., Chen L., Anton M., Sankar U., Rudnicki M.A., and Graham F.L. 1996. A helper-dependent adenovirus vector system: Removal of helper virus by Cre-mediated excision of the viral packaging signal. *Proc. Natl. Acad. Sci.* **93:** 13565–13570.

Rich D.P., Couture L.A., Cardoza L.M., Guiggio V.M., Armentano D., Espino P.C., Hehir K., Welsh M.J., Smith A.E., and Gregory R.J. 1993. Development and analysis of recombinant adenoviruses for gene therapy of cystic fibrosis. *Hum. Gene Ther.* **4:**

461–476.

Roberts R.J., O'Neill K.E., and Yen C.I. 1984. DNA sequences from the adenovirus 2 genome. *J. Biol. Chem.* **259:** 13968–13985.

Rowe W.P., Huebner R.J., Gilmore L.K., Parrott R.H., and Ward T.G. 1953. Isolation of a cytopathic agent from human adenoids undergoing spontaneous degeneration in tissue culture. *Proc. Soc. Exp. Biol. Med.* **84:** 570–573.

Schaack J., Guo X., and Langer S.J. 1996. Characterization of a replication-incompetent adenovirus type 5 mutant deleted for the preterminal protein gene. *Proc. Natl. Acad. Sci.* **93:** 14686–14691.

Schaack J., Langer S., and Guo X. 1995a. Efficient selection of recombinant adenoviruses by vectors that express β-galactosidase. *J. Virol.* **69:** 3920–3923.

Schaack J., Guo X., Ho W.Y.-W., Karlok M., Chen C., and Ornelles D. 1995b. Adenovirus type 5 precursor terminal protein-expressing 293 and HeLa cell lines. *J. Virol.* **69:** 4079–4085.

Schiedner G., Morral N., Parks R.J., Wu Y, Koopmans S.C., Langston C., Graham F.L., Beaudet A.L., and Kochanek S. 1998. Genomic DNA transfer with a high capacity adenovirus vector that has all viral genes deleted results in improved *in vivo* gene expression and decreased toxicity. *Nat.Genet.* **18:** 180–183.

Schowalter D.B., Tubb J.C., Liu M., Wilson C.B., and Kay M.A. 1997. Heterologous expression of adenovirus E3-gp19K in an E1a-deleted adenovirus vector inhibits MHC I expression *in vitro*, but does not prolong transgene expression *in vivo*. *Gene Ther.* **4:** 351–360.

Seth P., Fitzgerald D., Ginsberg H., Willingham M., and Pastan I. 1984. Evidence that the penton base of adenovirus is involved in potentiation of toxicity of Pseudomonas exotoxin conjugated to epidermal growth factor. *Mol. Cell. Biol.* **4:** 1528–1533.

Shenk T. 1996. *Adenoviridiae*: The viruses and their replication. In *Fields virology, 3rd edition* (ed. B.N. Fields et al.), pp. 2111–2171. Lippincott-Raven, Philadelphia, Pennsylvania.

Shenk T. and Flint J. 1991. Transcriptional and transforming activities of the adenovirus E1A proteins. *Adv. Cancer Res.* **57:** 47–85.

Solnick D. 1981. Construction of an adenovirus-SV40 recombinant producing SV40 T antigen from an adenovirus late promoter. *Cell* **24:** 135–143.

Sparer T.E., Tripp R.A., Dillehay D.L., Hermiston T.W., Wold W.S., and Gooding L.R. 1996. The role of human adenovirus early region 3 proteins (gp19K, 10.4K, 14.5K, and 14.7K) in a murine pneumonia model. *J. Virol.* **70:** 2431–2439.

Spector D.J. and Samaniego L.A. 1995. Construction and isolation of recombinant adenoviruses with gene replacements. In *Methods Mol. Genet.* **7:** 31–44.

Spergel J.M., Hsu W., Akira S., Thimmappaya B., Kishimoto T., and Chen-Kiang S. 1992. NF-IL6, a member of the C/EBP family, regulates E1A-responsive promoters in the absence of E1A. *J. Virol.* **66:** 1021–1030.

Stevenson S.C., Rollence M., White B., Weaver L., and McClelland A. 1995. Human adenovirus serotypes 3 and 5 bind to two different cellular receptors via the fiber head domain. *J. Virol.* **69:** 2850–2857.

Stewart P.L., Fuller S.D., and Burnett R.M. 1993. Difference imaging of adenovirus: Bridging the resolution gap between X-ray crystallography and electron microscopy. *EMBO J.* **12:** 2589–2599.

Stewart P.L., Burnett R.M., Cyrklaff M., and Fuller S.D. 1991. Image reconstruction reveals the complex molecular organization of adenovirus. *Cell* **67:** 145–154.

Thimmapaya B., Weinberger C., Sheider R.J., and Shenk T. 1982. Adenovirus VAI RNA is required for efficient translation of viral mRNAs at late times after infection. *Cell* **31:** 543–551.

Thummel C., Tjian R, and Grodzicker T. 1981. Expression of SV40 T antigen under control of adenovirus promoters. *Cell* **23:** 825–836.

Tjian R. 1978. The binding site on SV40 DNA for a T antigen-related protein. *Cell* **13:** 165–179.

Tollefson A.E., Scaria A., Hermiston T.W., Ryerse J.S., Wold L.J., and Wold W.S. 1996. The adenovirus death protein (E3-11.6K) is required at very late stages of infection for efficient cell lysis and release of adenovirus from infected cells. *J. Virol.* **70:** 2296–2306.

Tripathy S.K., Black H.B., Goldwasser E., and Leiden J.M. 1996. Immune responses to transgene-encoded proteins limit the stability of gene expression after injection of replication-defective adenovirus vectors. *Nat. Med.* **2:** 545–550.

Wang Q., Jia X.C., and Finer M.H. 1995. A packaging cell line for propagation of recombinant adenovirus vectors containing two lethal gene region deletions. *Gene Ther.* **2,** 775–783.

Wang Q., Greenburg G., Bunch D., Farson D., and Finer M.H. 1997. Persistent transgene expression in mouse liver following *in vivo* gene transfer with a ΔE1/ΔE4 adenovirus vector. *Gene Ther.* **4:** 393–400.

White E., Cipriani R., Sabbatini P., and Denton A. 1991. Adenovirus E1B 19-kilodalton protein overcomes the cytotoxicity of E1A proteins. *J. Virol.* **65:** 2968–2978.

Wickham T.J., Mathias P., Cheresh D.A., and Nemerow G.R. 1993. Integrins $\alpha_V\beta_3$ and $\alpha_V\beta_5$ promote adenovirus internalization but not virus attachment. *Cell* **73:** 309–319.

Wold W.S. and Gooding L.R. 1989. Adenovirus region E3 proteins that prevent cytolysis by cytotoxic T cells and tumor necrosis factor. *Mol. Biol. Med.* **6:** 433–452.

Yang Y. and Wilson J.M. 1995. Clearance of adenovirus-infected hepatocytes by MHC class I-restricted CD4+ CTLs *in vivo*. *J. Immunol.* **155:** 2564–2570.

Yang Y., Li Q., Ertl H.C.J., and Wilson J.M. 1995. Cellular and humoral immune responses to viral antigens create barriers to lung-directed gene therapy with recombinant adenoviruses. *J. Virol.* **69:** 2004–2015.

Yang Y., Jooss K.U., Su Q., Ertl H.C.J., and Wilson J.M. 1996. Immune responses to viral antigens vs. transgene product in the elimination of recombinant adenovirus infected hepatocytes *in vivo*. *Gene Ther.* **3:** 137–144.

Yang Y., Nunes F.A., Berencsi K., Furth E.F., Gonczol E., and Wilson J.M. 1994. Cellular immunity to viral antigens limits E1-deleted adenoviruses for gene therapy. *Proc. Natl. Acad. Sci.* **91:** 4407–4411.

Yeh P. and Perricaudet M. 1997. Advances in adenoviral vectors: From genetic engineering to their biology. *FASEB J.* **11:** 615–623.

Yeh P., Dedieu J.F., Orsini C., Vigne E., Denefle P., and Perricaudet M. 1996. Efficient dual transcomplementation of adenovirus E1 and E4 regions from a 293-derived cell line expressing a minimal E4 functional unit. *J. Virol.* **70:** 559–565.

Yoshimura K., Rosenfeld M.A., Seth P., and Crystal R.G. 1993. Adenovirus-mediated augmentation of cell transfection with unmodified plasmid vectors. *J. Biol. Chem.* **268:** 2300–2303.

Zabner J., Freimuth P., Puga A., Fabrega A., and Welsh M.J. 1997. Lack of high affinity fiber receptor activity explains the resistance of ciliated airway epithelia to adenovirus infection. *J. Clin. Invest.* **100:** 1144–1149.

Zhou H., O'Neal W., Morral N., and Beaudet A.L. 1996. Development of a complementing cell line and a system for construction of adenovirus vectors with E1 and E2a deleted. *J. Virol.* **70:** 7030–7038.

5

Adenovirus Vectors

Nelson A. Wivel, Guang–Ping Gao, and James M. Wilson
The Wistar Institute and The Institute for Human Gene Therapy and
Department of Molecular and Cellular Engineering
University of Pennsylvania
Philadelphia, Pennsylvania 19104–4268

From a historical perspective, the murine retroviruses were the first recombinant viruses that were used in human gene transfer studies. As to be expected with any developing system, there were significant positive and negative aspects. Transduction of the target cells was done in an ex vivo manner that accommodated both biological and safety issues. In the ex vivo setting, it was possible to control the target cell population being transduced by methods of selection such as fluorescence-activated cell sorting, and expanding the cells in culture assured the requisite dividing cells necessary for retrovirus-mediated gene transfer. The in vitro environment excluded the presence of human complement that was known to inactivate retroviruses. While retroviral vectors integrate into the chromosomes of the targeted cells, thus developing a potential for long-term gene expression, the pattern of strictly random insertion raises the issue of insertional mutagenesis with its predisposition to oncogenicity.

To develop in vivo strategies for gene transfer that could target terminally differentiated cells, such as those in the respiratory tract or liver, it was necessary to consider the use of other viruses. It soon became obvious that adenoviruses had properties that were advantageous for in vivo gene delivery because they could transfer recombinant genes into a wide variety of dividing and nondividing cells. Because of their natural affinity for respiratory epithelium, adenoviral vectors were first used to study cystic fibrosis transmembrane conductance regulator (CFTR) gene transfer in patients with cystic fibrosis. Adenoviral vectors can be grown to much higher titers (10^{11} pfu/ml) as compared with the retrovirus vectors (10^6 to 10^7 ffu/ml); because they are nonenveloped viruses, they are more stable during the process of concentration and purification, and they are much less sensitive to complement-mediated inactivation. Because adenoviruses do not integrate into the host-cell chromosome, gene expression is most likely to be transient and there is a need for repeat administration if prolonged gene expression is necessary. Most often the second administration of adenoviral vector is inefficient or impossible because of the cellular and humoral immune responses that mimic the immune responses to any viral infection. Vectors, like the parent viruses, are internalized by macrophages and the vector genome expresses viral proteins that are presented by major histocompatibility (MHC) class I molecules to CD8$^+$ T cells that are cytotoxic for vector-infected cells. Viral capsid proteins from the input vector are presented by MHC class II molecules to CD4$^+$ T helper cells that stimulate cytotoxic T lymphocytes (CTLs) that will destroy transduced target cells and provoke inflammation. In addition, there is activation of B cells and the necessary CD4$^+$ T-cell subsets, leading to a humoral response and the production of neutralizing antibodies that block adenovirus receptor sites on target cells and thus abrogate the intended effects of the second vector administration (Wilson 1996). Clearly, one of the principal challenges in the continued development of adenoviral vectors for use in gene therapy is the need to create immunomodulatory schema that will permit effective gene expression after repeat vector administration. The other challenge is to minimize the number of viral genes in the recombinant vector genome and, thus, reduce toxicities related to the vector itself or toxicities mediated through the immune response.

E1-DELETED RECOMBINANT ADENOVIRAL VECTORS

The first-generation recombinant, replication-defective adenoviruses that were developed for gene therapy contained deletions of the entire E1a and part of the E1b portions of the genome (Kozarsky and Wilson 1993; Krougliak and Graham 1995). This E1 region is essential for the regulation of adenoviral transcription and is necessary for viral replication. Using an adenovirus-transformed human embryonic kidney cell line (293) that provides the E1a and E1b region genes in *trans*, it is possible to grow the replication-defective virus (Graham et al. 1977). The virus vector produced by the 293 cell line is capable of infecting cells and can express the transgene, but it cannot replicate in a cell that does not carry the E1 region DNA unless there is a very high multiplicity of infection.

Although it was initially assumed that deletion of E1 sequences in the first-generation recombinant adenovirus might be sufficient to essentially ablate expression of other early and late viral genes, this has not proven to be the case. Several studies have shown that deletion of E1 can be overcome in vitro with either high multiplicities of infection or through the action of cellular *trans*-activators with E1-like activity that could override the absence of transcription by the recombinant vector (Imperiale et al. 1984; Spergel et al. 1992). Clearly, the presence of viral gene expression can lead to the production of foreign proteins that will stimulate destructive cellular immune responses and the concomitant loss of transgene expression.

Much of the experimental work in animal model systems has focused on the liver and the lung. The data derived from studying these selected organ sites have defined the principal immune barriers to adenovirus-mediated gene therapy. One of the earlier studies, using an E1-deleted vector containing *lacZ* as the transgene, established some important para-meters with respect to transient gene expression in the liver. The role of the immune system in influencing the stability of transgene expression was demonstrated by comparing immunocompetent and genetically athymic (*nu/nu*) strains of mice. *lacZ* expression was demonstrated by X-Gal histochemistry of hepatocytes following retrograde instillation of the vector into the biliary tract. More than 80% of the hepatocytes were pos-itive 2 days after gene transfer in both strains, but expression was extin-guished by day 21 in the immunocompetent strain (CBA mice); there was no diminution of expression in the athymic mice after 60 days (Yang et al. 1994b). CTLs specific for adenoviral proteins were detected in the CBA mice and there was substantial liver pathology in CBA mice given the vector as opposed to naive mice and *nu/nu* mice. These results were consistent with the hypothesis that recombinant adenovirus vectors express low levels of viral protein, that destructive cellular immune responses develop, that vector-infected cells are eliminated in the pres-ence of an inflammatory reaction, and that the liver is repopulated with hepatocytes that do not contain a transgene.

A number of experiments have been conducted to more precisely define the cellular immune mechanisms responsible for target-cell destruction following exposure to adenoviral vectors. Several antigen-specific immune responses were studied using adoptive transfer and genetic "knock-out" mice. It was established by in vitro assays that expo-sure to the virus produced antigen-specific CTLs and that T_{H1} cells were activated, as seen in proliferation assays. Two experimental mouse mod-els were created to define the functional significance of these cellular immune responses. E1-deleted vectors were administered to β_2-microglobulin knock-out mice that are deficient in both MHC class I expression and CD8[+] cells, and to another strain of mice in which the

MHC II heavy chain was interrupted, causing a deficiency of MHC class II expression and an absence of CD4+ cells. The MHC-class-II-deficient animals were fully competent in eliminating targeted hepatocytes when compared to immunocompetent mice, but the MHC-class-I-deficient animals exhibited a vigorous T_{H1} response and were able to mobilize CD4+ cells, but not to eliminate hepatocytes containing the transgene (Yang et al. 1994a). Additional data suggesting that CTLs are primary effectors were developed through adoptive transfer experiments in which virus-activated CD4+ and CD8+ cells from primed C57BL/6 mice were selectively introduced into RAG2− mice stably expressing adenoviral *lacZ*; following transfer, there was a subsequent loss of transgene expression and a lymphocyte-dominated hepatitis developed. The lack of detectable immune responses in RAG2− and *nu/nu* mice, both of which contain normal natural killer (NK) cells, is consistent with the observation that CTLs are the primary effectors of the immune reaction (Yang et al. 1994a; Yang and Wilson 1995).

The role of T_{H1} activation in response to adenovirus exposure has been more clearly delineated as a result of studies in genetically deficient mice. T-helper cells can differentiate along one of two pathways, the T_{H1} pathway that enhances cellular responses or the T_{H2} pathway that is the axis for activation of B cells and generation of antibody. Using mice that were genetically deficient in CD4+ or mice made deficient in CD4+ cells as a result of monoclonal antibody depletion, it was possible to demonstrate stable transgene expression and thus confirm the importance of T-helper function (Yang et al. 1995b). These results do conflict with experiments in the mouse deficient in CD4+ cells due to MHC class II germ line interruption; these animals develop compensatory T-helper activity. As mechanisms of supporting the enhancement of the antigen-specific CTL response, T_{H1} cells secrete interleukin (IL-2) and interferon-γ (IFN-γ) which upregulates MHC class I presentation of viral antigens in target cells. By using mice genetically deficient in IFN-γ and mice deficient in IFN-γ as a result of monoclonal antibody depletion, it was possible to demonstrate stable transgene expression. Although there was a marked CTL response in these animals, there was a major loss of T-helper activity; these results support the hypothesis that a IFN-γ-enhanced presentation of MHC-class-I-associated viral antigens is essential to the development of an effective CTL response (Yang et al. 1995b).

Related studies have been done using the lung as the target organ instead of the liver and the results are confirmatory in that the instillation of a *lacZ*-expressing adenovirus vector into the lungs of C57BL/6 mice elicited CTL responses to both viral proteins and the transgene product, β-galactosidase, with the subsequent elimination of transgene-containing cells (Yang et al. 1996b). In two experimental models in which the mice should be tolerant to the transgene, there were CTL responses to viral proteins in the absence of transgene-specific CTLs. Thus, it seems clear that CTLs specific for adenoviral antigens have a major contributory role in limiting transgene expression.

One of the other major factors impinging on the success of gene transfer is the capacity for a CTL response to the transgene that has the potential to function as a neoantigen in and of itself. Recent experiments have defined some of the parameters by which a transgene can act to induce a CTL-directed immune response. Several different strains of mice were given intramuscular injections consisting of an E1-deleted adenovirus vector containing either human or mouse erythropoietin (EPO). When the transgene was human EPO (foreign protein), the peak effect of gene transfer was seen at 14 days, whereas mouse EPO (self protein) produced elevated hematocrits for several months (Tripathy et al. 1996). Although there was no evidence for autoimmunity in this particular study, the fact remains that a foreign protein could break tolerance and lead to an autoimmune response against the homologous self-protein.

Another significant limitation in the use of adenoviral vectors is the difficulty in obtaining successful gene transfer after a second administration of the virus (Smith et al. 1993; Kozarsky et al. 1994; Yei et al. 1994). Primary exposure to the virus has the potential not only to induce a cellular immune response, but also to direct the formation of neutralizing antibodies that will bind the virus at cell receptor sites following a second administration. The humoral immune response is mediated by MHC class II presentation of viral proteins plus activation of B cells and T_H cells. In mouse models where gene transfer is liver-directed, initial exposure of the virus produces a mixed T_{H1} (IL-2 and IFN-γ) and T_{H2} (IL-4) response with the formation of neutralizing antibodies of IgG2a (T_{H1}-derived), IgA, and IgGl (T_{H2}-derived) isotypes (Wilson 1995). This presence of neutralizing antibodies directly correlates with a proportional decrease in gene transfer when the vector is given for the second time. It was shown that MHC class II mice failed to produce antibodies and demonstrated undiminished levels of gene transfer subsequent to a second exposure to adenoviral antigens.

In another approach, investigators took advantage of the observation that the adenovirus early region (E3) gene encodes for several proteins that modulate the immune response of the host to adenovirus-infected cells. By using a recombinant E1-deleted vector containing the E3 gene driven by a cytomegalovirus promoter, it was possible to carry out repeated gene transfer to the liver in the presence of a marked inhibition of antiviral antibody and adenovirus-specific CTL responses (Ilan et al. 1997c).

Although considerable data have accrued concerning the antigen-specific immunity that develops as a result of exposure to E1-deleted adenoviral vectors, less is known about nonantigen-specific immunity or innate immunity. Innate immune mechanisms have been defined as the first-line host defense against potential pathogens, including barrier functions, humoral functions, and cell-mediated functions (Fearon and Locksley 1996). In an attempt to evaluate the contribution of the innate immune component of host defense to adenoviral vectors, El-deleted vectors expressing the β-galactosidase transgene or expressing no gene were administered to immunocompetent or immunodeficient mice and the amount of vector genome was quantified in the liver. Approximately 90% of vector DNA was eliminated within 24 hours, whereas only 9% of the initial amount of vector DNA was cleared over the subsequent 3 weeks (Worgall et al. 1997). This early-stage elimination of DNA was independent of the transgene and observed in both immunocompetent and immunodeficient mice. It is not known if the elimination of adenovirus DNA represents loss of functionally active virus vector or whether it reflects the degradation of functionally inactive virus; however, the rate of clearance apparently was identical even when different particle-to-pfu ratios were used. These results suggest that gene expression is derived from a very small part of the total vector that is administered, at least when the intravenous route is used.

Other possible elements of an antiviral innate immune response could involve the activation of macrophages, NK cells, complement, and cytokine release. Recent experiments have attempted to define the role of Kupffer cell activation in the response to administration of adenoviral proteins. Activation of cytokines, particularly tumor necrosis factor (TNF) and IL-6, is part of the innate immune response to viruses; TNF is primarily the product of activated Kupffer cells, whereas IL-6 is produced by a variety of cell types including splenic macrophages. NF-κB is an ubiquitous transcription factor that can be activated by viruses. It can implement the cellular immune response against viruses by *trans*-activating the transcription of inflammatory cytokines, including TNF and IL-6. When mice were given gadolinium chloride to temporarily deplete Kupffer cells, and then given intravenously, first-generation adenovirus or adenovirus vectors lacking most early and late gene expression, NF-κB activation was observed within minutes, suggesting that the vectors had a direct effect on the hepatocyte (Lieber et al. 1997b). Kupffer cell depletion almost eliminated TNF release after vector administration, but there was a very robust IL-6 release. Depletion of Kupffer cells had little effect on adenovirus-mediated hepatocyte apoptosis seen on day 3 after vector infusion, but did increase hepatocellular DNA synthesis. Clearly, there are a number of complex interactions between adenovirus vectors and the liver, and eliminating the early toxicity induced by these gene delivery vehicles will be a challenging task.

Some discussion has been devoted to the cytotoxic effects associated with the immune responses to adenoviral proteins, but the relationship of adenoviruses to host-cell apoptosis is influenced by factors other than the CTL reaction. It has been established that wild-type adenoviruses contain several early region genes that modulate apoptotic cell death. It is known that TNF-α can trigger apoptosis, but the E1B 19 kD and E3 14.7 kD and an E3 10.4 kD/E3 14.5 kD protein complex can protect adenovirus-infected cells from lysis by TNF-α (Shen and Shenk 1995). Whereas the adenovirus E1A proteins induce apoptosis, this can be inhibited by either the E1B 19 kD protein or by Bcl-2 (Rao et al. 1992). Since some of the current vectors are deleted in the E1 and E3 regions, it could be postulated that there might be a diminished effect on the lysis of cells. However, there is in vitro data from two separate cell systems to indicate that E1-deleted adenovirus vectors can exert effects on both cell proliferation and apoptosis. In a study of primary human airway epithelial cells, there was a dose-dependent relationship between the vector multiplicity of infection (moi) and the efficiency of gene transfer; at the same time, there was a concomitant slowing of cell proliferation that was the result of increased apoptotic cell death and lower recruitment of cells into S phase (Teramoto et al. 1995). Ultraviolet inactivation of the vector genes abolished the effects on cell proliferation. When isolated rat pancreatic islets were incubated in the presence of an E1-deleted adenovirus vector containing the β-galactosidase transgene at different mois, there was a significant increase in apoptosis that was dose-dependent (Weber et al. 1997). Thus, in two types of terminally differentiated cells, it has been demonstrated that deletion of early region genes yields vectors that retain the property of inducing apoptosis in a targeted cell population despite the fact that these early region genes encode proteins regulatory for apoptosis. Recent studies involving the construction of adenovirus vectors that are deleted in both the E1 and E4 regions have the potential to alter the phenotype for apoptosis. It has been demonstrated in vitro that the E4orf6 protein can bind to the p53 cellular tumor suppressor protein, and block its ability to activate transcription, with the result that the induction of p53-mediated apoptosis is prevented (Dobner et al. 1996; Moore et al. 1996).

There is a considerable literature that documents the relationship of the immune response to adenovirus vectors and its effect on transgene expression, but there is a direct toxicity seen in the lungs and the liver that also is the result of host immunity. Prior to the time that adenoviruses became candidates for gene delivery, there were extensive studies of the intrapulmonary pathology associated with wild-type adenovirus infection, using the mouse and the cotton rat as models (Ginsberg et al. 1991; Prince et al. 1993). In these animal models, the histopathology was characterized by grading the mononuclear cell (lymphocytes, macrophages, and plasma cells) infiltration in the peribronchial, perivascular, interstitial, and alveolar areas on a scale of 1 to 4 (where 1 represents a very mild infiltration and 4 represents a severe pneumonitis that is extensive in the particular histopathological section that is being viewed). There was a very high correlation between the severity of the inflammatory reaction and the input dose of the virus. When pathological analyses were done in the mouse with E1-deleted vectors expressing *lacZ*, the tissue changes were similar to that observed with wild-type adenovirus and the severity was dose-dependent. When longitudinal studies were done using a single administration of vector, peak inflammation was observed at day 10 with a readily detectable degree of regression by day 28 (Yang et al. 1996c). One of the first nonhuman primate toxicity studies was done using an E1-deleted vector containing the CFTR transgene and the observed inflammatory changes were dose-dependent and largely localized to the delivery site in the right lower lobe of the lung (Simon et al. 1993). Nonhuman primates have been used to study the acute responses to airway delivery of recombinant E1,E3-deleted vectors expressing the CFTR transgene, and microscopic changes in the lungs revealed a dose-dependent increase in inflammatory cells that was primarily localized to the site of vector delivery (right lower lobe of the lung). The areas of inflammation tended to be patchy but were perivascular, peribronchial, and interstitial in location (Brody et al. 1994).

Similar to the lung, there is a demonstrable inflammatory response in the liver when that organ is the site of vector delivery; a mild to moderate portal and lobular hepatitis is common and associated with transient gene expression (Li et al. 1993). Recent studies on liver-directed gene transfer in nonhuman primates, using E1-deleted vectors expressing the *lacZ* gene, utilized three routes of administration: common bile duct, portal vein, and saphenous vein. Gene transfer via the common bile duct was associated with a very low level of transduction efficiency and produced portal inflammation, whereas β-galactosidase expression was essentially equivalent whether by portal or saphenous vein, but each mode of delivery was associated with mild to moderate portal and lobular hepatitis (Sullivan et al. 1997). Generalized immunosuppression with cyclophosphamide/prednisone allowed for sustained expression of β-galactosidase for up to 35 days and there was greatly diminished liver inflammation.

Another challenge in the use of E1-deleted adenovirus vectors involves the search for promoters that can contribute to optimum gene expression. As would be expected, expression vectors have been constructed to utilize very strong promoters that are most often of viral origin, such as the Rous sarcoma virus long terminal repeat, the SV40 early promoter, and the human cytomegalovirus immediate–early gene promoter. There is an inherent disadvantage in the use of these viral promoters and it is their lack of tissue specificity, thus raising certain safety issues such as transgene expression in nontargeted cells. One of the most problematic outcomes would be the transduction of gonadal tissues with resulting germ-line gene modification. To restrict expression to the desired target cells, there are two principal levels at which adenovirus vectors could be controlled, the level of cell entry and the level of gene expression. To establish specificity at the cell level, one could attempt alteration of virus capsid proteins so that they would bind to cell-specific receptors (Douglas and Curiel 1997). However, the most common approach has been to use tissue-specific promoters and this has met with mixed success.

In some of the earlier in vitro studies with adenoviruses, it was shown that high levels of E1B expression were seen in hepatocytes, but not in myeloma cells, when the E1B promoter was replaced by the rat albumin promoter (Babiss et al. 1986). To determine if a tissue-specific promoter inserted into an adenoviral vector expressing the *lacZ* gene would function, investigators used the human ventricular/slow muscle myosin light chain 1 promoter. When the vector was tested both in vitro and in vivo, there was no muscle-specific gene expression, but rather there was reporter gene expression in muscle fascia, and it has been postulated that there were adenovirus *cis* elements, such as the E1A enhancer, that directed this connective tissue expression (Shi et al. 1997). Similar results were seen when a recombinant adenovirus expressing the *lacZ* gene was placed under the control of 5′-untranslated DNA sequences of the CFTR gene. Results seen in cultured cells from the respiratory tract and from lung sections taken after intratracheal instillation in the mouse revealed β-galactosidase expression that was not restricted to epithelial cells known to express CFTR (Imler et al. 1996).

Some model systems have yielded data that documents tissue-specific targeting when appropriate promoters were employed. In vitro and in vivo studies using an adenoviral vector with a *lacZ* transgene and a smooth muscle cell (SMC)-specific promoter, SM22α, revealed expression only in primary rat aortic SMC or in the SMCs of the tunica media and neointima. No vascular endothelial cells or adventitial cells had β-galactosidase activity, and intravenous injection of the vector showed no reporter gene expression in the lungs or liver (Kim et al. 1997).

OTHER MODIFICATIONS OF ADENOVIRUS VECTORS

E2 Mutations

With the accrual of data from several laboratories, the shortcomings of the first-generation E1-deleted recombinant adenoviruses became rather well characterized. Following the expression of viral proteins, the stimulation of cellular immune responses and destruction of infected cells, there was a consistent pattern of transient gene expression. Further crippling of the virus had the potential to decrease the expression of viral proteins. Since the transgenes may have varying degrees of immunogenicity, this confounds the interpretation of vector performance, and renders negative studies difficult to interpret. A second generation of adenovirus vectors was created by introducing a temperature-sensitive mutation into the E2A gene of an E1-deleted recombinant. In the initial studies, the transgene was *lacZ* and both the mouse liver and lungs were the target organs. Mice receiving the second-generation vector intravenously expressed the transgene in the liver for at least 70 days, whereas those receiving the first-generation vector were limited to 14 days of expression. In addition, the inflammatory response, as measured by the presence of CD8$^+$ T cells, was reduced measurably (Engelhardt et al. 1994b). When mice received an intratracheal installation of the E2a-mutated vector, there was a marked reduction in infiltration of CD8$^+$ cells and a dramatic reduction in lung inflammation on day 7. By substituting the CFTR transgene for *lacZ*, it was possible to demonstrate CFTR protein expression at 21 days, while a first-generation vector produced a negative result at this time point (Yang et al. 1994c). These studies were extended using the cotton rat as a model, a species that is known to be permissive for replication of human adenoviruses (Prince et al. 1993). Intratracheal instillation of the E2a-defective adenoviral vector containing the CFTR transgene yielded expression that was more stable than could be obtained with a first-generation vector, and there was a diminished CD8$^+$ T-cell response

(Engelhardt et al. 1994a). However, the overall histopathology seen in the lungs was indistinguishable when compared to first-generation vectors and suggested either a direct toxicity of the virus or a nonantigen-specific immune response mediated by NK cells. Furthermore, when a xenograft model was devised in which cotton rat trachea was grown in *nu/nu* mice, there was no difference between first- and second-generation vectors in duration of transgene expression, suggesting that the vector genome in each may be relatively stable as long as there are no destructive host immune responses. Primate toxicology studies have been done using a second-generation E1-deleted, E2a-defective adenovirus containing the CFTR transgene. When compared directly with first-generation vectors, the inflammatory responses were markedly diminished but were prolonged and still quite detectable at day 21 (Goldman et al. 1995). Another group of investigators reported a lack of persistence of gene expression in immunocompetent mice and dogs when an E1-deleted vector containing a temperature-sensitive mutation in E2A was used (Fang et al. 1996).

While the initial reports described the results of temperature-sensitive mutations in the E2a gene, the data were at variance with other models. More recently, vectors have been created that involve a deletion of the E2a gene. A complementing cell line was developed that permitted construction of an E1- and E2a-deleted vector. Although the vector titers were 10–30-fold less than recombinants with E2a wild-type regions, such vectors were capable of expressing either β-galactosidase or human α-1-antitrypsin (hAAT) following intravenous injection in mice (Zhou et al. 1996). When these vectors were tested in several strains of mice to determine the duration of gene expression, there did not appear to be any advantage in using an E2a-deleted vector containing β-galactosidase, while much more prolonged expression was achieved with hAAT. These results suggested that the host response to reporter genes, rather than the host response to adenoviral proteins, was of primary importance (Morral et al. 1997). Although the varying experimental conditions in this model preclude a direct comparison with the initial studies with E2a-deleted vectors, it stands to reason that the E2a-ts mutant vector is not equivalent to an E2a-deleted vector.

Another group of investigators constructed an adenovirus vector deleted in both E1 and E2a and characterized it in vitro. Metabolic DNA-labeling studies revealed no detectable de novo vector DNA synthesis and metabolic protein labeling failed to detect de novo hexon protein synthesis, although gene transfer and expression were comparable to that seen with first generation E1-deleted vectors (Gorziglia et al. 1996).

E4 Mutations

The adenovirus E4 gene encodes multiple functions that are essential to the virus life cycle. Primary transcripts from the E4 promoter are subject to alternative splicing events that produce at least 18 distinct mRNAs and these would be predicted to encode seven different polypeptides, most of which have been detected in infected cells (Leppard 1997). In addition, E4 encodes functions required for viral DNA replication and inhibition of host cell protein synthesis. Clearly, the deletion of the E1 and E4 regions would render a vector more replication-defective in that two recombinational events would be required to produce a replication-competent virus, and it would further reduce the potential for viral gene expression. Recently, investigators created new transcomplementing cell lines (27-18 and 10-3) that only accommodate open reading frame 6 (ORF 6) of the adenovirus but can provide necessary E4 functions for viral replication and packaging. An E1,E4-deleted, *lacZ* containing vector was produced and injected intravenously into C57BL/6 and C3H mice. Although there was less expression of late viral proteins, less toxicity, and less apoptosis in the liver, there was little difference in CTL activity to target cells with an identical histocompatibility complex. Transgene expression was no more stable with the E1,E4-deleted vector than with the E1-deleted vector (Gao et al. 1996). When the experiments were repeated in *nu/nu* mice, both generations of adenoviral vectors were stably expressed, indicating that the E4 deletion did not destabilize the genome. To determine if the results were due to an immune response to the transgene, the experiments were repeated in ROSA-26 mice that have a β-galactosidase gene incorporated into the germ line. It turned out that the adenovirus-vector-encoded β-galactosidase gene was expressed at significantly greater levels than the endogenous transgene. Expression of *lacZ* from the E1-deleted vector was reduced by at least tenfold over 60 days while there was a less than twofold reduction when the E1,E4-deleted vector was used (Gao et al. 1996). Thus, expression of the vector-encoded transgene is enhanced, by deleting both E1 and E4, when its product is eliminated as an immunologic target.

In a separate study, E1-deleted and E1,E4-deleted vectors were compared both in vitro and in vivo, and the results in tumor-derived cell lines revealed reduced levels of E2A-specific RNA, a notable shutoff of late gene and protein expression, and no apparent cytotoxicity. Since the transgene was *lacZ*, a *lacZ* transgenic mouse model was used to compare transgene expression in the liver. Apparently, there was no immune response to the β-galactosidase protein, as long-term expression with E1-deleted vectors mimicked that seen in immunodeficient mice. The E1,E4-

deleted vector promoted the extrachromosomal persistence of the viral DNA for at least 3 months, but there was not a similarly sustained expression of β-galactosidase, a finding that was attributed to the use of an RSV promoter (Dedieu et al. 1997).

Another group of investigators demonstrated positive results when comparing E1-deleted and E1,E4-deleted vectors in immunocompetent CBA mice. Analysis of liver sections revealed that β-galactosidase expression following transduction with the E1 vector peaked at day 7 and was 2% of the maximum at day 77. When the E1,E4-deleted vector was used, maximum expression occurred at day 14 and was about 40% of the maximum at day 77 (Wang et al. 1997).

Although E4 constitutes only 10% of the adenoviral genome, its products not only are essential to viral replication, but are also responsible for a complex set of phenotypes ranging from inhibition of the tumor suppressor gene p53 to providing helper functions for the adeno-associated virus. To construct E1,E4-deleted vectors that could be grown to the same titer as first-generation vectors, one group of investigators developed a vector-cell line system for complete functional complementation of both E1 and E4 by transforming 293 cells with an E4-ORF6 expression cassette containing an inducible metallothionein promoter (Brough et al. 1996). Initial results with a construct containing the *lacZ* transgene were disappointing in that the virus yield was 30-fold less when compared with E1-deleted vectors. This particular vector product had reduced fiber protein and there was an accumulation of mRNA. When the vector was modified by using a spacer sequence placed between late region 5 and the right inverted terminal repeat, there was a more efficient expression of fiber protein and the yields were comparable to first-generation vectors. There are two important characteristics of this system: without overlapping sequences between the cell lines and the vector genome, no generation of replication-competent virus can occur as a result of homologous recombination; and the deletion of E1, E3, and E4 creates a transgene capacity of 10 kb.

As an affirmation of the complexities inherent in this vector system, recent data suggest that persistence of transgene expression, instead of persistence of the gene itself, can be affected by the E4 status of the vector backbone with the cytomegalovirus promoter being specifically down-regulated in the absence of a wild-type E4 gene. Vectors expressing either CFTR or β-galactosidase were constructed with several variations in the E4 region, one with wild-type E4, one with E4-ORF6, and one with a complete deletion of E4. These vectors were delivered to the lungs in mice and the results were consistent with the hypothesis that nonimmunological factors significantly influence the period of gene expression (Armentano et al. 1997). All vectors containing the E1a promoter showed transient expression, but persistence of expression with the cytomegalovirus promoter was dependent on the presence of wild-type E4; using E4-ORF6 alone was not sufficient to achieve the effect seen with the wild-type gene. Since there was no correlation between transient expression and the disappearance of vector DNA, the concept of promoter down-regulation was advanced.

RECOMBINANT ADENOVIRUSES DELETED OF ALL VIRAL GENES

Recombinant retroviruses and adeno-associated viruses have been successfully used in gene transfer studies and both have been deleted of all their structural genes, leaving only *cis*-acting sequences necessary for packaging and integration. Producing analogous types of recombinant adenoviruses has proven to be challenging. One of the first attempts to do this involved production of viral proteins necessary for viral replication and assembly by coinfecting an E1-deleted helper virus together with a plasmid (ΔrAd) containing a minimal virus genome consisting only of inverted terminal repeats (ITRs) and adjacent packaging sequences. A principal challenge in this strategy is to maximize the titers of recombinant while minimizing the titers of helper virus, a considerable problem since the helper virus has a selective growth advantage. To discriminate between these two populations of virus, the *lacZ* gene was incorporated into the ΔrAd genome and the alkaline phosphatase gene was inserted in place of the E1 gene in the helper virus. Fortunately, the virions containing the ΔrAd genome sedimented at a lower density than the helper virus, and therefore, cesium gradients could be used to separate the two populations. Coinfection of 293 cells yielded approximately 3×10^6 *lacZ* transducing particles, a much lower titer than could be achieved with first- and third-generation vectors (Fisher et al. 1996).

Another strategy was used to create recombinant adenoviruses with large (25 kb) deletions and that were substantially void of helper virus. This required the use of a heterologous site-specific recombination system, the P1 bacteriophage Cre-*lox* system, containing a 38-kD recombinase (Cre) and a 34-bp *lox*P target sequence (Sternberg and Hamilton 1981). Recombinant adenovirus with two parallel *lox* sites flanking the pIX expression unit and the entire E2 region were generated and amplified to high titers. This virus was used to infect 293 cells stably expressing Cre recombinase that generated an efficient excision of the intervening 25-kb region joining the left genome end with the ITR, packaging signal, and hAAT expression cassette and joining the right genome end with the E4 region and right ITR together (Lieber et al. 1996). This 25-kb deletion yielded a virus deficient in E1, E2, E3, and late gene expression. In vitro gene transfer was successful in several cell lines, but when in vivo gene transfer to the mouse liver was attempted, deleted vector DNA rapidly declined within 30 minutes, the virus genome was cleared 7 hours postinfusion, and serum hAAT was detectable only on days 1 and 2. There was no evidence of in vivo hepatocellular toxicity; thus, it was speculated that early viral DNA synthesis was necessary to maintain the vector genome and allow for persistence. Later experiments by this same group of investigators established that coexpression of the adenoviral E2-preterminal protein from the vector (*cis*) or in *trans* stabilizes the minigenome in vitro and in vivo without evidence of cellular toxicity (Lieber et al. 1997a).

In a relevant and somewhat related study, some of the parameters that determine stability of the adenovirus vectors have been identified. There is apparently a direct relationship between genome size and the efficiency of DNA packaging. Adenoviral vectors with genomes greater than or equal to about 27 kb are packaged with equal efficiencies, whereas those with smaller genomes ae much less efficient. By amplifying vector DNA in helper-virus-infected cells, it was possible to demonstrate that vectors containing less than 75% of the wild-type genome had DNA rearrangements, whereas larger vectors were unaltered (Parks and Graham 1997).

One of the principal challenges to the effective use of gene therapy is the size constraints imposed on the insertion of various transgenes. One of the appealing features of deleted vectors is the capacity to accommodate large inserts. Successful genetic correction of inherited muscle diseases such as Duchenne muscular dystrophy will require long-term expression of full-length dystrophin. Several different groups of researchers have been able to successfully generate adenoviral vectors deleted of all viral open reading frames that contain a full-length dystrophin minigene. There is evidence from in vitro (primary *mdx* muscle cells that are dystrophin-deficient) and in vivo (muscle fibers in *mdx* mice) systems that successful transduction takes place (Clemens et al. 1995, 1996; Haecker et al. 1996; Kochanek et al. 1996; Kumar-Singh and Chamberlain 1996; Chen et al. 1997). One of the more encouraging results is that several of the groups reported expression of recombinant dystrophin in the skeletal muscle of *mdx* mice at 28 days after injection. In some of those constructs that contained the *lacZ* transgene in addition to full-length dystrophin, there was evidence of an immune response to β-galactosidase expression, and this was abrogated in *lacZ* transgenic mice (Chen et al. 1997).

A very recent report presents some extremely interesting and significant findings with regard to the in vivo use of fully deleted adenovirus vectors in several strains of mice. It has been possible to construct an adenovirus vector with all the viral coding sequences deleted and to insert the complete hAAT locus (28 kb). This vector differs substantively from other vector systems that use heterologous eukaryotic or viral promoters. By using a genomic DNA that contains its own endogenous regulatory regions, it was possible to observe tissue-specific transcriptional regulation and to obtain high levels of very stable gene expression that persisted for more than 8 months. When liver specimens were sampled periodically to determine the degree of target organ injury, there were no significant lymphocytic infiltrations and only sporadic CD4$^+$ or CD8$^+$ lymphocytes were detectable by immunostaining (Schiedner et al. 1998). This contrasts with previous work using an adenoviral vector deleted of all viral open reading frames, except E4, where expression was transient in the liver (Lieber et al. 1996).

ADJUNCTS TO ENHANCE ADENOVIRUS VECTOR GENE THERAPY

Despite the fact that adenovirus vectors can be used to transduce nondividing cells in vivo with some degree of efficiency, the well-demonstrated host immune responses have continued to present a major barrier, particularly since a principal alternative to long-term gene expression is readministration of the vector-gene construct. Vector modification will primarily impact on CTL responses, but it will have little or no effect on development of neutralizing antibodies. Thus, it is not surprising that there have been a number of experimental attempts to induce host immunosuppression. Some of the experimental models would most probably not be suitable for application in humans, but they are illustrative of what can be achieved by modulation of the immune response.

It has been possible to demonstrate tolerance to adenoviral antigens by using an oral regimen. Mutant Gunn rats that lack a bilirubin transferase enzyme and have life-long hyperbilirubinemia were given adenovirus extracts every other day for 21 days via gastroduodenostomy. Subsequent administration of a recombinant adenovirus containing the bilirubin-UDP-glucuronosyltransferase-1 (BUGT) transgene resulted in reduction of bilirubin levels by 70%. After bilirubin levels began to rise, a second dose of vector (given on day 98), markedly reduced serum bilirubin again. Neutralizing antibodies to adenovirus and CTLs were markedly inhibited, and transplantation of splenocytes to naïve animals adoptively transferred the tolerance (Ilan et al. 1997b).

At least three different immunosuppressive agents have been shown to prolong transgene expression following delivery of adenoviral vectors to rodent liver and lung. Cyclophosphamide, an alkylating agent used frequently in the treatment of various cancers, was administered simultaneously with an adenovirus vector containing the *lacZ* transgene into C57BL/6 mice, either intravenously or intratracheally. The rationale for this approach was to inactivate $CD4^+$ cells and thus prevent the consequences of a CTL response and B-cell function. Since $CD4^+$ T-cell activation is MHC-class-II-dependent and related to input viral capsid proteins, the immune modulation only needs to be transient. Cyclophosphamide blocked the activation and mobilization of $CD4^+$ and $CD8^+$ T cells and, at higher doses, no neutralizing antibodies developed. As a result, transgene expression was prolonged and local liver and lung inflammation was reduced (Jooss et al. 1996). In a related study, it was demonstrated that immunosuppression of mice with either cyclosporine A or cyclophosphamide allowed these animals to produce and secrete canine Factor IX for more than 5 months as compared to untreated, immunocompetent mice where expression was limited to 7–10 days (Dai et al. 1995).

FK506 is an immunosuppressant that shares many of the properties of cyclosporine A, and it has been shown to inhibit naïve and primed CTLs and to inhibit IL-2 secretion by these cells. When FK506 was administered to Gunn rats with inherited hyperbilirubinemia for 3 days beginning 24 hours before the adenoviral vector delivery, there were low or undetectable antibody titers to the recombinant adenovirus and minimal CTL responses to adenovirus-infected cells. Repeat administration of the vector containing the BUGT transgene, at days 42 and 98, reduced the bilirubin levels as effectively as the first injection (Ilan et al. 1997a).

Deoxyspergualin is an immunosuppressant currently being tested for use in organ transplantation. It appears to exhibit a potent, long-term effect on antigen-specific B cells and has been shown to effectively prevent production of a specific antibody when coadministered with protein antigens (Tepper 1993). In studies designed to utilize transient immunosuppression, deoxyspergualin could be used interchangeably with cyclophosphamide at the time of administration of an adenovirus vector containing Factor IX as the transgene. In C57BL/6 mice, the formation of neutralizing antibodies to adenovirus was prevented and a second administration of the Factor IX vector was successful (Smith et al. 1996).

Another important type of transient immune blockade was designed to take advantage of the observation that neutralizing antibodies to adenoviral proteins are generated in the context of class II MHC and dependent on CD4[+] cells. Using C57BL/6 mice with the liver as the target organ and adenovirus vectors containing either alkaline phosphatase or *lacZ* as the reporter genes, it was possible to acutely deplete CD4[+] cells with a monoclonal antibody at the time of vector administration. High-level transgene expression was stable for 1 month. This transient CD4[+] cell depletion inhibited the formation of both neutralizing antibody and virus specific IgA antibody (Yang et al. 1996a). In related experiments, IL-12 was coadministered with the virus vector and there was a selective block in secretion of antigen-specific IgA that was concurrent with a 20-fold reduction in neutralizing antibody; a high level of gene transfer to airway epithelial cells occurred after a second administration of vector. Apparently, this result was related to the selective inactivation of the T_{H2} subset of T-helper cells. Similar results were seen in C57BL/6 mice when the liver was the target organ. Again, the transient suppression of CD4 function was successful in preventing the generation of neutralizing antibody, and allowed an effective repeat administration of the vector (Yang et al. 1995a).

Although the previous studies were done in syngeneic mice, analogous results were achieved in outbred rats when a nondepleting anti-CD4 antibody was given eight times intraperitoneally over a 21-day period. There was prolonged reporter gene expression in the lung along with greatly reduced inflammation, and the absence of neutralizing antibodies permitted a successful administration of the vector (Lei et al. 1996).

Another interesting and reasonably successful approach to preventing CD4[+] T-cell activation by adenovirus vectors is predicated on direct interference with T-cell priming via the CD40 ligand-CD40 interactions. Initially, it was demonstrated that adenovirus-mediated transgene expression was stable in mice genetically deficient in CD40 ligand; such mice did not develop neutralizing antibodies and repeat vector administration was effective. By using an antibody to CD40 ligand that was administered along with the adenovirus vector, it was possible to achieve stabilization of transgene expression, diminished production of neutralizing antibody, and secondary transgene expression after vector readministration to the lung (Yang et al. 1996b; Scaria et al. 1997). The efficiency of this approach was improved by the coadministration of two reagents, muCTLA4Ig which blocks the CD28/B7 costimulatory pathway, and a CD40 ligand antibody. Adenovirus-mediated gene expression was observed for up to 1 year in more than 90% of the treated mice, and secondary gene transfer was seen in more than 50% of the mice even when the immunosuppressive effects of the two agents were no longer detectable. However, neither of these agents alone was effective in permitting readministration of the vector in the liver (Kay et al. 1997).

Another strategy has been pursued to evaluate the role of liver macrophages or Kupffer cells on adenovirus-vector-mediated gene transfer. Using an E1, partially E3-deleted vector containing chloroamphenical acetyltransferase (CAT) as the transgene, C57BL/6 and BALB/c mice were chosen for these experiments. Depletion of Kupffer cells was achieved by the intravenous administration of a multilamellar liposome containing dichloromethylene-bisphosphonate. Evaluation of the amount of CAT expression in the liver in control and liposome-treated mice demonstrated a 27-fold increase in the livers of the treated animals. A similar pattern was observed in immunodeficient BALB/c *nu/nu* mice. Related analyses revealed a delayed clearance of vector DNA and prolonged transgene expression. However, this particular protocol did not prevent the development of CTLs or neutralizing immunity directed against the vector. These results suggest that this particular cellular element of the innate immune system has an important role in host defense against adenoviral vectors (Wolff et al. 1997).

SUMMARY

The last two decades of progress in recombinant DNA technology have made the transfer of normal genes into genetically deficient cells a common reality. Both viral and nonviral delivery systems have been under recent development, but it still remains that the viral vectors possess a number of natural properties that facilitate efficient gene transfer. The focus of attention on adenoviruses is rationally based since recombinant vectors can transduce nondividing cells and the absence of integration bypasses the risk of insertional mutagenesis. Because of the natural tropism for pulmonary epithelial cells, these vectors are legitimate candidates to provide gene therapy for diseases such as cystic fibrosis. Intravenous delivery of adenovirus vectors results in transduction of sufficient numbers of hepatocytes to postulate their use in a variety of liver-based diseases. A principal challenge to the effective use of adenovirus vectors is posed by the immune system and its response to the expression of viral proteins, but recent mutations in vectors and the use of transient immune blockades of various types suggest that these problems are amenable to experimental manipulation. It is reasonable to assume that in the relatively near future, there will be significant refinements in the use of adenovirus-mediated in vivo gene transfer that will improve the potential for successful therapeutic applications in humans.

REFERENCES

Armentano D.A., Zabner J., Sachs C., Sookdeo C.C., Smith M.P., St. George J.A., Wadsworth S.C., Smith A.E., and Gregory R.J. 1997. Effect of the E4 region on the persistence of transgene expression from adenovirus vectors. *J. Virol.* **71:** 2408–2416.

Babiss L.E., Friedman J.M., and Darnell J.E. 1986. Cellular promoters incorporated into the adenovirus genome: Effects of viral regulatory elements on transcription rates and cell specificity of albumin and β-globin promoters. *Mol. Cell. Biol.* **6:** 3798–3806.

Brody S.L., Metzger M., Danel C., Rosenfeld M.A., and Crystal R.G. 1994. Acute responses of non-human primates to airway delivery of an adenovirus vector containing the human cystic fibrosis transmembrane conductance regulator cDNA. *Hum. Gene Ther.* **5:** 821–836.

Brough D.E., Lizonova A., Hsu C., Kulera V.A., and Kovesdi I. 1996. A gene transfer vector-cell line system for complete functional complementation of adenovirus early regions E1 and E4. *J. Virol.* **70:** 6497–6501.

Chen H.-H., Mack L.M., Kelly R., Ontell M., Kochanek S., and Clemens P.R. 1997. Persistence in muscle of an adenoviral vector that lacks all viral genes. *Proc. Natl. Acad. Sci.* **94:** 1645–1650.

Clemens P.R., Krause T.L., Chan S., Korb K.E., Graham F.L., and Caskey C.T. 1995. Recombinant truncated dystrophin minigenes: Construction, expression and adenoviral delivery. *Hum. Gene Ther.* **6:** 1477–1485.

Clemens P.R., Kochanek S., Sunada Y., Chan S., Chen H-H., Campbell K.P., and Caskey C.T. 1996. In vivo muscle gene transfer of full-length dystrophin with an adenoviral vector that lacks all viral genes. *Gene Ther.* **3:** 965–972.

Dai Y., Schwarz E.M., Gu D., Zhang W.-W., Sarvetnick N., and Verma I.M. 1995. Cellular and humoral immune responses to adenoviral vectors containing factor IX gene: Tolerization of factor IX and vector antigens allows for long-term expression. *Proc. Natl. Acad. Sci.* **92:** 1401–1405.

Dedieu J.-F., Vigne E., Torrent C., Jullien C., Mahfouz I., Caillaud J.-M., Aubailly N., Orsini C., Guillaume J.-M., Opolon P., Delaere P., Perricaudet M., and Yeh P. 1997. Long-term gene delivery into the livers of immunocompetent mice with E1/E4-defective adenoviruses. *J. Virol.* **71:** 4626–4637.

Dobner T., Horikoshi N., Rubenwolf S., and Shenk T. 1996. Blockage by adenovirus E4orf6 of transcriptional activation by the p53 tumor suppressor. *Science* **272:** 1470–1473.

Douglas J.T. and Curiel D.T. 1997. Strategies to accomplish targeted gene delivery to muscle cells employing tropism-modified adenoviral vectors. *Neuromusc. Disord.* **7:** 284–298.

Engelhardt J.F., Litzky L., and Wilson J.M. 1994a. Prolonged transgene expression in cotton rat lung with recombinant adenoviruses defective in E2a. *Hum. Gene Ther.* **5:** 1217–1229.

Engelhardt J.F., Ye X., Doranz B., and Wilson J.M. 1994b. Ablation of E2a in recombinant adenoviruses improves transgene persistence and decreases inflammatory response in mouse liver. *Proc. Natl. Acad. Sci.* **91:** 6196–6200.

Fang B., Wang H., Gordon G., Bellinger D.A., Read M.S., Brinkhous K.M., Woo S.L.C., and Eisensmith R.C. 1996. Lack of persistence of E1⁻ recombinant adenoviral vectors containing a temperature-sensitive E2A mutation in immunocompetent mice and dogs. *Gene Ther.* **3:** 217–222.

Fearon D.T. and Locksley R.M. 1996. The instructive role of innate immunity in the acquired immune response. *Science* **272:** 50–54.

Fisher K.J., Choi H., Burda J., Chen S-J., and Wilson J.M. 1996. Recombinant adenovirus deleted of all viral genes for gene therapy of cystic fibrosis. *Virology* **217:** 11–22.

Gao G.R., Yang Y., and Wilson J.M. 1996. Biology of adenovirus vectors with E1 and E4 deletions for liver-directed gene therapy. *J. Virol.* **70:** 8934–8943.

Ginsberg H.S., Moldawer L.L., Sehgal P.B., Redington M., Kilian P.L., Chanock R.M., and Prince G.A. 1991. A mouse model for investigating the molecular pathogenesis of adenovirus pneumonia. *Proc. Natl. Acad. Sci.* **88:** 1651–1655.

Goldman M.J., Litzky L.A., Engelhardt J.F., and Wilson J.M. 1995. Transfer of the CFTR gene to the lung of nonhuman primates with E1-deleted, E2a-defective recombinant adenoviruses: A preclinical toxicology study. *Hum. Gene Ther.* **6:** 839–851.

Gorziglia M.I., Kadan M.J., Yei S., Lim J., Lee G.M., Luthra R., and Trapnell B. 1996. Elimination of both E1 and E2a from adenovirus vectors further improves prospects for in vivo human gene therapy. *J. Virol.* **70:** 4173–4178.

Graham F.L., Smiley J., Russell W.C., and Nairn R. 1977. Characteristics of a human cell line transformed by DNA from human adenovirus type 5. *J. Gen. Virol.* **36:** 59–74.

Haecker S.E., Stedman H.H., Balice-Gordon R.J., Smith D.B.J., Greelish J.P., Mitchell M.A., Wells A., Sweeney H.L., and Wilson J.M. 1996. In vivo expression of full-length human dystrophin from adenoviral vectors deleted of all viral genes. *Hum. Gene Ther.* **7:** 1907–1914.

Ilan Y., Jona V.K., Sengupta K., Davidson A., Horwitz M.S., Chowdhury N.R., and Chowdhury J.R. 1997a. Transient immunosuppression with FK 506 permits long-term expression of therapeutic genes into the liver using recombinant adenoviruses in the rat. *Hepatology* **26:** 949–956.

Ilan Y., Prakash R., Davidson A., Jona V., Droguett G., Horwitz M.S., Chowdhury N.R., and Chowdhury J.R. 1997b. Oral tolerization to adenoviral antigens permits long-term gene expression using recombinant adenoviral vectors. *J. Clin. Invest.* **99:** 1098–1106.

Ilan Y., Droguett G., Chowdhury N.R., Li Y., Sengupta K., Thummala N.R., Davidson A., Chowdhury J.R., and Horwitz M.S. 1997c. Insertion of the adenoviral E3 region into a recombinant viral vector prevents antiviral humoral and cellular responses and permits long-term gene expression. *Proc. Natl. Acad. Sci.* **94:** 2587–2592.

Imler J.L., Dupuit F., Chartier C., Accart N., Dieterle A., Schultz H., Puchelle E., and Pavirani A. 1996. Targeting cell-specific gene expression with an adenovirus vector containing the *lac*Z gene under the control of the CFTR promoter. *Gene Ther.* **3:** 49–58.

Imperiale M.J., Kao H.T., Feldman L.T., Nevins J.R., and Strickland S. 1984. Common control of the heat shock gene and early adenovirus genes: Evidence for a cellular E1A-like activity. *Mol. Cell. Biol.* **4:** 867–874.

Jooss K., Yang Y., and Wilson J.M. 1996. Cyclophosphamide diminishes inflammation and prolongs transgene expression following delivery of adenoviral vectors to mouse lung and liver. *Hum. Gene Ther.* **7:** 1555–1566.

Kay M.A., Meuse L., Gown A.M., Linsley P., Hollenbaugh D., Aruffo A., Ochs H.D., and Wilson C.B. 1997. Transient immunomodulation with anti-CD40 ligand antibody and CTLA4Ig enhances persistence and secondary adenovirus-mediated gene transfer into mouse liver. *Proc. Natl. Acad. Sci.* **94:** 4686–4691.

Kim S., Lin H., Barr E., Chu L., Leiden J.M., and Parmacek M.S. 1997. Transcriptional targeting of replication-defective adenovirus transgene expression to smooth muscle cells in vivo. *J. Clin. Invest.* **100:** 1006–1014.

Kochanek S., Clemens P.R., Mitani K., Chen H.-H., Chan S., and Caskey C.T. 1996. A new adenoviral vector: Replacement of all viral coding sequences with 28 kb of DNA independently expressing both full-length dystrophin and β-galactosidase. *Proc. Natl. Acad. Sci.* **93:** 5731–5736.

Kozarsky K.F. and Wilson J.M. 1993. Gene therapy: Adenoviral vectors. *Curr. Opin. Genet. Dev.* **3:** 499–503.

Kozarsky K.F., McKinley D.R., Austin L.L., Raper S.E., Stratford-Perricaudet L.D., and Wilson J.M. 1994. In vivo correction of low density lipoprotein receptor deficiency in the Watanabe heritable hyperlipidemic rabbit with recombinant adenoviruses. *J. Biol. Chem.* **269:** 13695–13702.

Krougliak V. and Graham F.L. 1995. Development of cell lines capable of complementing E1, E4, and protein IX defective adenovirus type 5 mutants. *Hum. Gene Ther.* **6:** 1575–1586.

Kumar-Singh R. and Chamberlain J.S. 1996. Encapsidated adenovirus minichromosomes allow delivery and expression of a 14 kb dystrophin cDNA to muscle cells. *Hum. Mol. Genet.* **5:** 913–921.

Lei D., Lehmann M., Shellito J.E., Nelson S., Siegling A., Volk H-D., and Kolls J.K. 1996. Nondepleting anti-CD4 antibody treatment prolongs lung-directed E1-deleted adenovirus-mediated gene expression in rats. 1996. *Hum. Gene Ther.* **7:** 2273–2279.

Leppard K.N. 1997. E4 gene function in adenovirus, adenovirus vector and adeno-associated virus infections. *J. Gen. Virol.* **78:** 2131–2138.

Li Q., Kay M.A., Finegold M., Stratford-Perricaudet L.D., and Woo S.L.C. 1993. Assessment of recombinant adenoviral vectors for hepatic gene therapy. *Hum. Gene Ther.* **4:** 403–409.

Lieber A., He C.-Y., and Kay M.A. 1997a. Adenoviral preterminal protein stabilizes miniadenoviral genomes in vitro and in vivo. *Nat. Biotechnol.* **15:** 1383–1387.

Lieber A., He C.-Y., Kirillova I., and Kay M.A. 1996. Recombinant adenoviruses with large deletions generated by cre-mediated excision exhibit different biological properties compared with first-generation vectors in vitro and in vivo. *J. Virol.* **70:** 8944–8960.

Lieber A., He C.-Y., Meuse L., Schowalter D., Kirillova I., Winther B., and Kay M.A. 1997b. The role of Kupffer cell activation and viral gene expression in early liver toxicity after infusion of recombinant adenovirus vectors. *J. Virol.* **71:** 8798–8807.

Moore M., Horikoshi N., and Shenk T. 1996. Oncogenic potential of the adenovirus E4orf6 protein. *Proc. Natl. Acad. Sci.* **93:** 11295–11301.

Morral N., O'Neal W., Zhou H., Langston C., and Beaudet A. 1997. Immune responses to reporter proteins and high viral dose limit duration of expression with adenoviral vectors: Comparison of E2a wild type and E2a deleted vectors. *Hum. Gene Ther.* **8:** 1275–1286.

Parks R.J. and Graham F.L. 1997. A helper-dependent system for adenovirus vector production helps define a lower limit for efficient DNA packaging. *J. Virol.* **71:** 3293–3298.

Prince G.A., Porter D.D., Jenson A.B., Horswood R.C., Chanock R.M., and Ginsberg H.S. 1993. Pathogenesis of adenovirus type 5 pneumonia in cotton rats (Sigmodon hespidus). *J. Virol.* **67:** 101–111.

Rao L., Debbas M., Sabbatini P., Hochenbery D., Korsmeyer S., and White E. 1992. The adenovirus E1A proteins induce apoptosis which is inhibited by E1B19-kDA and Bcl-2 proteins. *Proc. Natl. Acad. Sci.* **89:** 7742–7746.

Scaria A., St. George J.A., Gregory R.J., Noelle R.J., Wadsworth S.C., Smith A.E., and Kaplan J.M. 1997. Antibody to CD40 ligand inhibits both humoral and cellular immune responses to adenoviral vectors and facilitates repeated administration to mouse airway. *Gene Ther.* **4:** 611–617.

Schiedner G., Morral N., Parks R.J., Wu Y., Koopmans S.C., Langston C., Graham F.L., Beaudet A.L., and Kochanek S. 1998. Genomic DNA Transfer with a high-capacity adenovirus vector results in improved in vivo gene expression and decreased toxicity. *Nat. Genet.* **18:** 180–183.

Shen Y. and Shenk T.E. 1995. Viruses and apoptosis. *Curr. Opin. Genet. Dev.* **5:** 105–111.

Shi Q., Wang Y., and Worton R. 1997. Modulation of the specificity and activity of a cellular promoter in an adenoviral vector. *Hum. Gene Ther.* **8:** 403–410.

Simon R.H., Engelhardt J.F., Yang Y., Zepeda M., Weber-Pendleton S., Grossman M., and Wilson J.M. 1993. Adenovirus-mediated transfer of the CFTR gene to lung of nonhuman primates: Toxicity study. *Hum. Gene Ther.* **4:** 771–780.

Smith T.A.G., White B.D., Gardner J.M., Kaleko M., and McClelland T. 1996. Transient immunosuppression permits successful repetitive administration of an adenovirus vector. *Gene Ther.* **3:** 496–502.

Smith T.A., Mehaffey M.G., Kayda D.B., Saunders J.M., Yei S., Trapnell B.C., McClemmand A., and Kaleko M. 1993. Adenovirus mediated expression of therapeutic plasma levels of human factor IX in mice. *Nat. Genet.* **5:** 397–402.

Spergel J.M., Hsu W., Akira S., Thimmappaya B., Kishimoto T., and Chen-Kiang S. 1992. NF-IL6, a member of the C/EBP family, regulates E1A-responsive promoters in the absence of E1A. *J. Virol.* **66:** 1021–1030.

Sternberg N. and Hamilton D. 1981. Bacteriophage P1 site-specific recombination. I. Recombination between lox P sites. *J. Mol. Biol.* **150:** 467–486.

Sullivan D.E., Srikanta D., Du H., Hiramatsu N., Aydin F., Kolls J., Blanchard J., Baskin G., and Gerber M.A. 1997. Liver-directed gene transfer in non-human primates. *Hum. Gene Ther.* **8:** 1195–1206.

Tepper M.A. 1993. Deoxyspergualin: Mechanism of action studies of a novel immunosuppressive drug. *Ann. N.Y. Acad. Sci.* **696:** 123–132.

Teramoto S., Johnson L.G., Huang W., Leigh M., and Boucher R. 1995. Effect of adenoviral vector infection on cell proliferation in cultured primary human airway epithelial cells. *Hum. Gene Ther.* **6:** 1045–1053.

Tripathy S.K., Black H.B., Goldwasser E., and Leiden J.M. 1996. Immune responses to transgene-encoded proteins limit the stability of gene expression after injection of replication-defective adenovirus vectors. *Nat. Med.* **2:** 545–550.

Wang Q., Greenburg G., Bunch D., Farson D., and Finer M.H. 1997. Persistent transgene expression in mouse liver following in vivo gene transfer with a ΔE1/ΔE4 adenovirus vector. *Gene Ther.* **4:** 393–400.

Weber M., Deng S., Kucher T., Shaked A., Ketchum R.J., and Brayman K.L. 1997. Adenoviral transfection of isolated pancreatic islets: A study of programmed cell death (apoptosis) and islet function. *J. Surg. Res.* **69:** 23–32.

Wilson J.M. 1995. Adenovirus-mediated gene transfer to liver. *Adv. Drug Delivery Rev.* **17:** 303–307.

———. 1996. Adenoviruses as gene-delivery vehicles. *N. Engl. J. Med.* **334:** 1185–1187.

Wolff G., Worgall S., van Rooijen N., Song W.-R., Harvey B.-G., and Crystal R.G. 1997. Enhancement of in vivo adenovirus-mediated gene transfer and expression by prior depletion of tissue macrophages in the target organ. *J. Virol.* **71:** 624–629.

Worgall S., Wolff G., Falck-Pedersen E., and Crystal R.G. 1997. Innate immune mechanisms dominate elimination of adenoviral vectors following in vivo administration. *Hum. Gene Ther.* **8:** 37–44.

Yang Y. and Wilson J.M. 1995. Clearance of adenovirus-infected hepatocytes by MHC class I-restricted CD4+ CTLs in vivo. *J. Immunol.* **155:** 2564–2570.

Yang Y., Ertl H.C.J., and Wilson J.M. 1994a. MHC class I-restricted cytotoxic T lymphocytes to viral antigens destroy hepatocytes in mice infected with E1-deleted recombinant adenoviruses. *Immunity* **1:** 433–442.

Yang Y., Greenough K., and Wilson J.M. 1996a. Transient immune blockade prevents formation of neutralizing antibody to recombinant adenovirus and allows repeated gene transfer to mouse liver. *Gene Ther.* **3:** 412–420.

Yang Y., Su Q., and Wilson J.M. 1996b. Role of viral antigens in destructive cellular immune responses to adenovirus vector-transduced cells in mouse lungs. *J. Virol.* **70:** 7209–7212.

Yang Y., Trinchiere G., and Wilson J.M. 1995a. Recombinant IL-12 prevents formation of blocking IgA antibodies to recombinant adenovirus and allows repeated gene therapy to mouse lung. *Nat. Med.* **1:** 890–893.

Yang Y., Xiang Z., Ertl H.C.J., and Wilson J.M. 1995b. Upregulation of class I major histocompatibility complex antigens by interferon γ is necessary for elimination of recombinant adenovirus-infected hepatocytes in vivo. *Proc. Natl. Acad. Sci.* **92:** 7257–7261.

Yang Y., Nunes F.A., Berencsi K., Furth E.E., Gonczol E., and Wilson J.M. 1994b. Cellular immunity to viral antigens limits E1-deleted adenoviruses for gene therapy. *Proc. Natl. Acad. Sci.* **91:** 4407–4411.

Yang Y., Nunes F.A., Berencsi K., Gonczol E., Engelhardt J.E., and Wilson J.M. 1994c. Inactivation of E2a in recombinant adenoviruses improves the prospect for gene therapy for cystic fibrosis. *Nat. Genet.* **7:** 362–369.

Yang Y., Su Q., Greval I.S., Schilz R., Flavell R.A., and Wilson J.M. 1996c. Transient subversion of CD40 ligand function diminishes immune responses to adenovirus vectors in mouse liver and lung tissues. *J. Virol.* **70:** 6370–6377.

Yei S., Mittereder N., Tang K., O'Sullivan C., and Trapnell B.C. 1994. Adenovirus-mediated gene expression for cystic fibrosis: Quantitative evaluation of repeated in vivo vector administration to the lung. *Gene Ther.* **1:** 192–200.

Zhou H., O'Neal W., Morral N., and Beaudet A.L. 1996. Development of a complementing cell line and a system for construction of adenovirus vectors with E1 and E2a deleted. *J. Virol.* **70:** 7030–7038.

6

Strategies to Adapt Adenoviral Vectors for Gene Therapy Applications: Targeting and Integration

Paul N. Reynolds and David T. Curiel

Gene Therapy Program, Comprehensive Cancer Center
University of Alabama at Birmingham
Birmingham, Alabama 35294-3300

For the effective application of gene therapy strategies to human disease, Anderson suggested that certain criteria should be met; namely, that vectors should deliver a therapeutic gene specifically to a target cell, that resultant gene expression should be at an appropriate level and for an appropriate period of time, and that delivery and expression of the therapeutic gene should be achieved within an acceptable safety margin (Anderson and Fletcher 1980). These criteria remain largely unmet. However, in recent years, disappointment in the results of clinical trials has forced a refocusing on the basics of vector design, resulting in steady advancements in vector technology that now show promise for more successful gene therapy.

Development of vectors that have in vivo efficacy is critical because many diseases for which gene therapy can be rationally considered require direct in situ gene delivery and cannot feasibly be addressed by an ex vivo approach. Replication-incompetent adenovirus is a potential candidate vector for clinical gene therapy based on several key attributes that include ease of production to high titer, infection of both dividing and nondividing cells, and systemic stability, which has allowed for efficient in vivo gene expression (Brody and Crystal 1994). However, the virus has several important limitations including its widespread tropism, stimulation of inflammatory and immune responses, and short-term transgene expression (Yang et al. 1995, 1996; Tomko et al. 1997). This chapter focuses chiefly on the issue of targeted gene delivery to address the limitations brought about by native viral tropism. To date, several groups have sought to exploit the fundamental advantages of adenovirus by using it in specific contexts where the recognized limitations were felt to be less important. For example, it was thought that the issue of the widespread tropism of the virus could be circumvented by administering the vector by direct injection, particularly in the context of tumors. However, in phase I human trials, dissemination beyond the injected site has been found (D.R. Wilson, unpubl.). Application to "compartmentalized" disease has also met with problems. For example, poor transduction efficiency has been noted following administration into the pleural space for therapy of mesothelioma (Esandi et al. 1997; S.M. Albelda, unpubl.), and in the peritoneum, effective use of antitumor gene therapy has been limited by concurrent transduction of the liver with subsequent toxicity (Yee et al. 1996). Further limitations have arisen in the application to pulmonary disease. Here, prior clinical experience had indicated that the virus had a natural tropism for the respiratory tract, thus direct administration of vector to the airways for cystic fibrosis therapy seemed a rational approach Crystal et al. 1994; Zabner et al. 1993, 1996; Knowles et al. 1995; Bellon et al. 1997). In reality, the levels of transduction achieved were lower than expected because differentiated airway epithelial cells lack sufficient adenoviral receptors and the integrins required for viral internalization (Grubb et al. 1994; Goldman and Wilson 1995; Goldman et al. 1996; Zabner et al. 1997). Thus, even in these apparently favorable anatomical locations, there is a strong case for developing a vector with cell-specific targeting properties. Despite the limitations of adenovirus, its basic advantages, in particular its in vivo efficacy, justify using this virus as a starting point in the development of improved vector systems.

ADENOVIRAL ENTRY PATHWAY

Strategies for the retargeting of viral vectors were first applied to retroviruses and were based on a sound understanding of viral entry mechanisms (Cosset and Russell 1996). The entry mechanisms of adenovirus, including the recent identification of primary adenoviral receptors, are now well understood and allow for a rational approach to the targeting of adenoviral vectors (Fig. 1).

The adenovirus is an unenveloped icosahedral particle with 12 fibers projecting from the surface (Shenk 1996). During the assembly phase of viral replication, fiber monomers trimerize in the cytoplasm, then bind to a viral penton base protein that is subsequently incorporated into the viral capsid. At the distal tip of each fiber monomer is a globular region referred to as the knob domain. It is this knob region which binds to cellular adenoviral receptors, initially anchoring the virus to the cells. Two

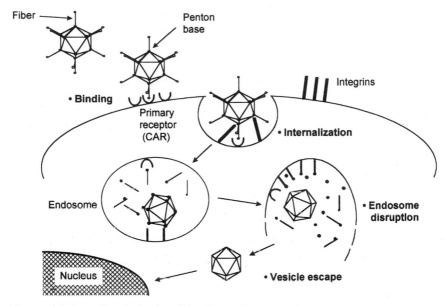

Figure 1 Adenovirus infection: Binding and entry pathway.

cellular receptors for adenovirus have recently been described. The cox-sackie/adenoviral receptor (CAR) (Bergelson et al. 1997; Tomko et al. 1997) binds both adenovirus and group-B coxsackie viruses. The murine and human receptors are made up of 365 and 352 amino acids, respectively, and are 91% identical. The extracellular region appears to contain two immunoglobulin-like domains. In a separate report, viral binding to the α2 domain of major histocompatability complex class I has also been shown (Hong et al. 1997). Following attachment, viral entry requires a second step, which involves the interaction between Arg-Gly-Asp (RGD) motifs in the penton base with cell surface integrins αvβ3 or αvβ5, which then leads to receptor-mediated endocytosis of the virion (Wickham et al. 1993). In the endosome, the virus undergoes a stepwise disassembly and endosomal lysis occurs (a process mediated by the penton base and low endosomal pH), followed by transport of the viral DNA to the cell nucleus. This endosomolysis step is critical for efficient gene delivery, and the ability of the adenovirus to effect endosomal escape is one of the key factors in its efficiency as a vector. Importantly, viral entry and endosomal escape are functionally uncoupled (Michael et al. 1993); thus, entry via a nonnative, cell-specific pathway does not appear to compromise downstream delivery of DNA to the nucleus. Based on the foregoing, a logical place to start in the development of a targeted adenoviral vector is manipulation of the knob domain.

Several groups are now developing strategies to impart targeting ability to adenoviral vectors. The strategies currently being used fall under two broad headings: immunological and genetic.

IMMUNOLOGICAL RETARGETING

Immunological retargeting strategies are based on the use of bispecific conjugates, typically a conjugate between an antibody directed against a component of the virus and a targeting antibody or ligand. True targeting requires a simultaneous abolition of native targeting and introduction of new tropism; thus, Douglas et al. (1996) developed a neutralizing monoclonal antibody against the knob region of adenovirus. This was achieved by immunizing mice with adenovirus and recombinant knob protein, then developing hybridomas that produced antibodies capable of neutralizing native adenoviral infection (as determined by a cytopathic effect assay using HeLa cells). The Fab fragment of an antibody generated in this way (1D6.14) was then conjugated to folate to effect targeting to the folate receptor. Folate was chosen because the folate receptor is up-regulated on several tumor types and the receptor internalizes after ligand binding (Weitman et al. 1992). Using this conjugate, adenoviral infection was redirected away from the native receptor to the folate receptor (Fig. 2). The gene transfer efficiency of this approach was approximately 70% of that seen with the native virus, which contrasted to the experience with retroviral retargeting where redirection had generally resulted in a dramatic fall in infectivity (Cosset and Russell 1996). Although the binding of the complex to cells was clearly mediated by folate, the mechanism of viral internalization was not established. In this regard, folate is normally internalized by potocytosis, and enters the cell via a caveolus (Anderson et al. 1992). Normally, the size of this caveolus would be too small to encompass the folate-virus complex, but whether it could enlarge under these circumstances or whether viral entry was effected by the usual integrin-mediated pathway is unknown. Using a slightly different approach, Wickham et al. (1996) developed an adenovirus that contained a FLAG epitope, DYKDDDDK, introduced into the penton base region, and then used a bispecific antibody directed against FLAG and αv integrins to direct binding to integrins on endothelial and smooth muscle cells. Because the retargeting bispecific conjugate was larger than the viral fiber, it was hypothesized that the conjugate would be functionally available for binding by extending outward past the knob domain. In this way, by using integrins as the attachment target on these cells that express low levels of CAR, gene delivery was enhanced.

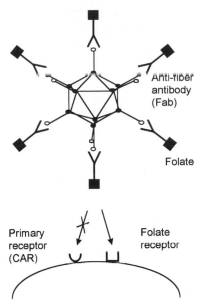

Figure 2 Schema for altering adenoviral tropism by an immunological targeting approach. A retargeting moiety formed by conjugating folate to the Fab fragment of an antiknob antibody is used to direct adenoviral binding to the folate receptor.

A further development in the immunological retargeting approach has been reported by Watkins et al. (1997). This group generated a bacteriophage library displaying single chain antibodies (scFvs) derived from the spleen of a mouse immunized against adenoviral knob. From this library a suitable neutralizing antiknob scFv was isolated, then a fusion protein between this and epidermal growth factor (EGF) was produced. This "adenobody" was then successfully used to retarget adenoviral gene delivery, resulting in enhanced transduction of EGF receptor (EGFR) expressing cells. Interestingly, this study showed that retargeting in this way appeared to bypass the need for the penton base–cell integrin interaction for internalization of the virus. This was shown by using an excess of an RGD-containing peptide that competes with penton for binding to integrins. This peptide reduced native viral transduction but had no effect on the transduction levels achieved with the retargeted vector complex, thus implying that in the latter case, viral entry was achieved by EGF receptor-mediated internalization. Taking a different approach to EGF receptor targeting, we have used a conjugate between 1D6.14 Fab and a monoclonal antibody that binds to EGFR (mab 425) (Murthy et al. 1987; Wersall et al. 1997). Using this approach, we have demonstrated retargeted delivery to two murine fibroblast cell lines, one stably expressing human EGFR, the other expressing a mutant of EGFR that does not internalize (C.R. Miller., unpubl.). Therefore, depending on the target selected, one may be able to exploit internalization mechanisms of either the targeted receptor itself or, if a noninternalizing target is selected, to use the native integrin pathway. The potential for overcoming a lack of integrins in some settings, or the ability to target to noninternalizing cell surface molecules if integrins are sufficient, implies a very broad potential applicability of retargeted vectors.

Following the initial demonstration that immunological retargeting of adenovirus could be achieved, further studies have explored the potential for therapeutic application of this approach. These applications might be considered in the context of several worthwhile goals of targeting with differing levels of stringency, including general nonspecific enhancement of delivery to a broad range of cell types, transduction of previously untransducible cells (relevant to both ex vivo and in vivo applications), targeting to enhance gene delivery and potentially reduce toxicity in a loco-regional or compartmental context, and cell-specific gene delivery following intravenous administration of vector.

At the lowest level of stringency, retargeting approaches to achieve enhanced gene delivery, even if in the absence of a clear specificity advantage, are worthwhile in compartmentalized disease contexts. Such approaches at the very least should allow for the use of lower doses of virus, thus potentially reducing direct viral toxic effects, dissemination from the administration site, and innate immune responses, which are clearly dose-dependent. In this regard, we have used basic fibroblast growth factor (FGF2) as a targeting ligand, based on the knowledge that FGF receptors are upregulated in a number of tumor types. A conjugate between 1D6.14 Fab and FGF2 was used to retarget adenoviral infection of a number of different tumor lines with varying baseline levels of susceptibility to adenoviral infection. Enhancements in transduction from two- to greater than tenfold were seen (Rogers et al. 1998).

With regard to the transduction of resistant cells, Kaposi's sarcoma (KS) is an example of a disease where gene therapy applications have been limited in part because of the poor transducibility of this tumor (J.A. Campain, unpubl.). Goldman et al. (1997) used the FGF2 retargeting approach to investigate retargeted gene delivery to previously untransfectable KS cell lines. The results demonstrated a dramatic increase in transduction. Furthermore, the potential therapeutic utility was shown by transfecting cells with a retargeted adenovirus carrying the gene for herpes simplex thymidine kinase (AdCMVHSV-TK). These cells were then tar more susceptible to killing by the subsequent administration of the prodrug ganciclovir than cells that had been infected with the untargeted virus. T lymphocytes are another cell type that is resistant to adenoviral infection, due to a lack of both CAR and αv integrins. Thus, Wickham et al. (1997a) successfully retargeted adenoviral vectors by using a conjugate between the anti-FLAG antibody and anti-CD3, thereby achieving significantly enhanced transduction. Thus, this study indicates the benefits of targeting can also be exploited in contexts relevant to ex vivo therapy.

In view of the direct cytotoxic effects of adenovirus, the ability to increase the number of transduced cells without resorting to an increase in viral dose is extremely important. For example, in the vasculature, where transduction is limited by a relatively low level of CAR expression (Wickham et al. 1996), transduction efficiency increases with escalating viral dose over a fairly narrow range, then dramatically falls as cytotoxicity supervenes and leads to a loss of infected cells. Schulick et al. (1995a) found a maximal transduction efficiency in vascular smooth-muscle cells of approximately 40% with 5×10^{10} pfu, which decreased to 0 at 10^{11} pfu. Similar results were also found in the endothelium (Schulick et al. 1995b). Using the bispecific anti–FLAG-anti-integrin approach, Wickham et al. (1996) achieved a seven- to ninefold enhancement of endothelial cell transduction. As FGF receptor expression is up-regulated in proliferating vasculature, and is thus relevant to a number of pathological processes including tumor angiogenesis, we examined the effect of FGF2 targeting of adenoviral gene delivery to proliferating endothelial cells. Here, an approximate 30-fold enhancement in luciferase expression was seen. Flow cytometry analysis of cells transfected with a β-galactosidase-encoding vector demonstrated that FGF2 retargeting led to both an increase in the number of transduced cells and an increase in the amount of gene expression per cell. In contrast, when FGF2 retargeting was used in the infection of quiescent, confluent cells, transgene expression was actually reduced, thus indicating a degree of cell-specific targeting based on the level of expression of the targeted receptor (P.N. Reynolds, unpubl.).

Toxicity at high viral doses has also been seen in murine carcinoma models, where escalating the dose of a herpes simplex thymidine kinase (HSV-TK)-encoding virus eventually led to deaths from toxicity before complete tumor eradication could be achieved (Yee et al. 1996). Also, in the context of HSV-TK, increasing the amount of transgene expression per cell (rather than the number of transduced cells) by manipulation of promoters did not increase therapeutic effect once a threshold level of expression was achieved (Elshami et al. 1997). Thus, Rancourt and colleagues investigated the use of FGF2 retargeting of AdCMVHSV-TK in a murine model of ovarian carcinoma (C. Rancourt, unpubl.). First, FGF receptor expression on the ovarian carcinoma line SKOV3.ip1 was confirmed by radiolabeled FGF binding. Then, enhancement of adenovirally mediated gene delivery to these cells using FGF2 retargeting in vitro using the luciferase reporter gene was demonstrated. Next, validation of the retargeting approach in vivo was obtained. Tumors were established in nude mice by intraperitoneal inoculation with SKOV3.ip1 cells, followed 5 days later with a peritoneal injection of either AdCMVLuc alone or with FGF2 retargeting. Mice were sacrificed and luciferase expression in the tumors was quantified. Tumors from the mice that had received the retargeted vector had tenfold greater level of luciferase expression. Thus, these results established that the immunological retargeting was efficacious in vivo. Tumors were then established in mice as before, followed by intraperitoneal injection of either placebo, AdCMVHSV-TK alone, or AdCMVHSV-TK with FGF2 retargeting (with viral doses of 10^8 and 10^9 pfu). Mice were then divided into two groups and received either ganciclovir or placebo for 14 days and were monitored for survival. The mice that did not receive ganciclovir had no survival advantage over the mice that received no gene therapy. When the survival curves for the mice who received ganciclovir were analyzed, two important results emerged (Fig. 3). First, there was a statistically significant increase in survival with FGF2 retargeting compared to the untargeted vector at each dose of virus. Second, the survival curve for the mice treated with 10^8 pfu of virus with FGF2 retargeting was the same as that for 10^9 pfu of untargeted vector. Thus, FGF2 retargeting enabled a tenfold reduction in viral dose for the same therapeutic outcome. These results indicate a potential for increasing the clinical utility and therapeutic index of adenoviral vectors by using a retargeting approach.

Successful targeting following intravenous administration of a targeted adenoviral vector has yet to be reported. In this context, one of the major hurdles to overcome is hepatic uptake of injected virus, which accounts for greater than 90% of the injected dose. While it was initially considered that this problem was predominantly due to nonspecific uptake related to hepatic reticulo-endothelial functions, more recent evidence suggests that there may be a receptor-mediated component and that this may be potentially addressed by modification of native viral tropism. To this end, Zinn et al. (1998) investigated the in vivo distribution of technetium-labeled adenovirus serotype 5 (Ad5) knob. They found that the majority of it localized to the liver within 10 minutes of injection. This

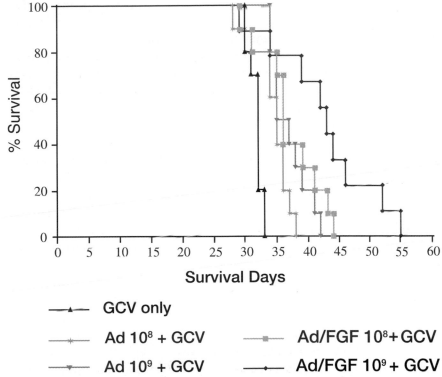

Figure 3 Retargeting adenovirus to enhance therapeutic gene delivery improves survival in a murine model of ovarian carcinoma. Mice with peritoneal ovarian tumors received intraperitoneal injections of placebo or adenovirus carrying the herpes simplex thymidine kinase gene (10^8 or 10^9 pfu)–either unmodified virus (Ad) or virus complexed to a retargeting moiety (formed by conjugating FGF2 to antiknob Fab) (Ad/FGF). Mice were then treated with ganciclovir (GCV) and survival monitored. Mice treated with retargeted adenovirus had a significant increase in median survival, Ad(10^8 pfu) vs. Ad/FGF(10^8 pfu), $p = 0.0025$; Ad(10^9 pfu) vs. Ad/FGF(10^9 pfu), $p = 0.007$.

localization could be inhibited by prior injection of an excess of unlabeled Ad5 knob, but not by an excess of serotype 3 (Ad3) knob (which binds to a different receptor), thus indicating the specificity of the hepatic uptake. When labeled Ad5 knob was complexed with 1D6.14 Fab prior to injection, hepatic uptake was markedly reduced, providing further evidence of the receptor-dependent nature of the uptake and indicating a degree of stability of the Fab to knob bond in the bloodstream.

The developments using immunological retargeting strategies have established a number of important principles. Modification of tropism has successfully been achieved, indicating that true cell-specific delivery is possible. The evidence to date suggests that limitations in transduction due to either a deficiency in CAR, αv integrins, or both, may be overcome by retargeting. Efficiency of gene delivery with retargeted complexes is not only comparable to wild-type vector, but in many cases has resulted in a substantial improvement in gene delivery, which is itself a worthwhile goal. Retargeted complexes are efficacious in vivo, at least in a compartmental context. The use of a retargeted vector has been shown to enhance a therapeutic endpoint and finally, the limitations of intravenous application of these vectors, imposed by hepatic uptake of virus, appears, at least in part, to be a receptor-mediated phenomenon and may therefore also be overcome by retargeting. The full potential of immunological retargeting, however, is yet to be defined and there are certain practical and theoretical limitations to this approach. Large-scale production of bispecific antibody conjugates of consistent configuration is difficult when using the hetero-bifunctional cross-linkers that have so far been reported. Also, clearance of retargeting complexes and activation of the complement system may limit in vivo application. Although further protein engineering refinements such as the fusion protein "adenobody" approach may address some of these issues, the stability of the targeting complex-virion bond following systemic delivery remains a concern, especially when attempting intravenous administration. Thus, development of another approach, genetic retargeting, is also being pursued.

GENETIC RETARGETING

In view of the practical and theoretical limitations of the immunological approach to retargeting mentioned previously, development of targeted vectors by genetic manipulation of the virus itself has proceeded alongside the immunological strategies. In addition, immunological targeting approach is not likely to be sufficient for application to the controlled replicating viral vector systems being developed to improve gene delivery to malignant tumors. In these systems, the need for precise targeting of both the initial viral dose and the progeny viruses will be particularly important and only achievable by genetic modifications.

Based on the knowledge of native viral binding, a rational place to begin in the development of genetically targeted vectors is with the knob domain, and most strategies have so far focussed on this region. The question of whether viral tropism could be modified genetically was initially addressed by Krasnykh et al. (1996). A chimeric adenovirus containing the serotype 3 knob on the Ad5 fiber shaft and capsid was produced by homologous recombination in 293 cells, using a modification of the shuttle and rescue plasmid technique developed by Frank Graham (Graham and Prevec 1991). Because Ad5 and Ad3 recognize different receptors, the tropism of the chimeric vector could be assessed by blocking infection of cells with an excess of free Ad5 or Ad3 recombinant knob. This confirmed that an "Ad5" vector possessing Ad3 tropism had been produced. A similar strategy was reported by Stevenson et al. (1997), who also demonstrated differences in transduction efficiency of various cell lines depending on whether wild-type or chimeric fiber vectors were used. Following the initial proof of principle, the incorporation of specific targeting ligands has been investigated.

When modifying the knob domain, there are important structural constraints that must be addressed. Because it is essential for fibers to trimerize to allow attachment to the penton base for subsequent capsid formation, any modification to the fiber must not perturb trimerization (Novelli and Boulanger 1991). Incorporation of a small ligand (gastrin releasing peptide) at the carboxy-terminal of fiber, and subsequent generation of trimers of this chimeric fiber, was initially reported by Michael et al. (1995). It has since been discovered that there are limits to the size of peptides that can be used in this way. Although the limits probably relate to the actual sequence used rather than to its length alone, it appears that trimerization is much less likely to occur with ligands longer than 25–30 amino acids. On the other hand, deletion of eight amino acids from the carboxyl terminus leads to failure of trimerization. Despite these limitations, the addition of small peptide ligands may have utility. For example, we have recently added a moiety containing six histidine residues (6-His) to the knob carboxy-terminal, successfully rescued the virus, and confirmed the binding availability of the 6-His moiety by using nickel column chromatography. This virus may be used to address the concerns regarding the stability of the bond between adenoviral vectors and immunological retargeting complexes. Theoretically, using an immunological retargeting complex containing an anti-6-His antibody, followed by the use of the nickel monoperoxyphthalic acid (MMPP) technology, might allow the formation of a stable covalent bond between the virus and the targeting complex (Fancy et al. 1996). Wickham et al. (1997b) recently reported a number of carboxy-terminal modifications, including one containing an RGD motif (21 amino acids) for the purpose of targeting to cell-surface integrins, and one containing seven lysine residues for the purpose of targeting to cell surface heparan sulfates. Production of a virus with longer carboxy-terminal additions was also attempted. A virus containing a 32-amino-acid peptide for targeting the laminin receptor was produced, but it failed to bind its target receptor, whereas a virus containing a 27-amino-acid E-selectin-binding peptide could not be produced. Neither of the successfully produced viruses had any modification to native tropism; thus, their potential application is in the context of enhancing delivery to otherwise poorly transfectable cells, rather than true cell-specific targeting. Nevertheless, enhanced vascular delivery using the polylysine virus was demonstrated in vivo in a vascular injury model and may have utility in the therapy of angioplasty restenosis. The RGD containing virus was shown to enhance gene delivery to endothelial and smooth muscle cells in vitro.

In view of the limitations involved in attaching ligands to the carboxyl terminus, other regions of the knob may ultimately prove to be better sites for ligand incorporation. In this regard, we have investigated the HI loop region of the knob. X-ray crystallographic modeling of the three-dimensional structure of trimeric knob indicates that the HI loop is located on the outer aspect in an area potentially available for interaction with receptors (Fig. 4) (Xia et al. 1994, 1995). Also, this region does not appear to be directly involved in trimerization, contains mostly hydrophilic amino acids, and is different lengths in different Ad serotypes, suggesting that there may be less rigid structural constraints than at the carboxyl terminus. As initial proof of concept, we have successfully incorporated a FLAG epitope into the HI loop (Krasnykh et al. 1998), using a technique of recombinant virus generation involving homologous recombination in *Escherichia coli* (Chartier et al. 1996). Affinity binding to an M2 matrix column confirmed that this epitope was available for binding in the context of the intact virion. Other ligands with more relevant targeting potential have now been incorporated, including a cyclic RGD peptide (which has affinity for tumor vasculature) (Pasqualini et al. 1997) and somatostatin. The size constraints of ligand incorporation at this site are yet to be determined, and thus the incorporation of large ligands such as EGF and scFvs is currently being investigated. However, it is likely that the sheer size of an scFv will require an alternate strategy such as complete replacement of the entire knob domain.

Ultimately, for true targeting to be achieved, modification to ablate native tropism will need to be addressed. It may be that incorporation of large ligands into the HI loop will simultaneously ablate native tropism by steric hindrance; however, if this is not the case then further modifications will be required. In this regard, receptor-binding epitopes within the knob that may be suitable for mutagenesis strategies have been identified (Hong and Boulanger 1995). Clearly, if complete replacement of

HI Loop

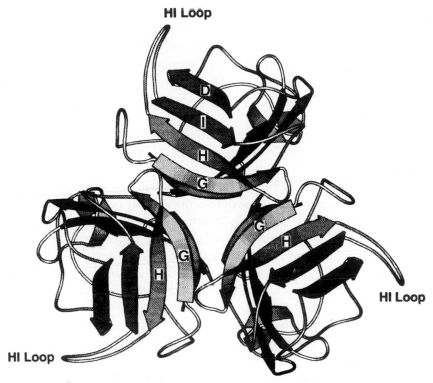

HI Loop

HI Loop

Figure 4 The knob trimer, showing the position of the HI loop. (Reprinted, with permission, from Xia et al. 1995 [copyright Springer-Verlag].)

the knob with a targeting and trimerization moiety could be achieved, this would simultaneously ablate native tropism. Of course, an integral part of any such strategy will be the use of permissive cell lines possessing the relevant target receptor to allow rescue and propagation of the virus.

Genetic modification strategies are not limited to the fiber. Wickham et al. (1995) introduced modifications into the penton base for targeting to cell-specific integrins. Hexon capsid proteins might also be exploited for targeting. An attractive aspect of this approach is the number of hexon proteins, 720, compared to the 36 knob regions; thus, hexon modification might have the potential for higher affinity binding. Such a strategy will still need to take into account the need to ablate native knob-dependent binding and address the stoichiometric issues relating to the fiber projecting out from the capsid, to ensure physical accessibility of the ligands introduced into the hexon.

As progress is being made in the development of retargeted vectors, the importance of identifying truly cell-specific ligands has been highlighted. Although there are many established ligands and antibodies that may be candidates in certain settings, in many cases, such as mature airway epithelium, truly specific targets have yet to be discovered, thus requiring further target definition. This is especially relevant in the context of those genetic retargeting strategies that attempt to target with small peptide ligands. In this regard, the use of bacteriophage panning techniques have shown potential utility for target definition (Barry et al. 1996; Pasqualini and Ruoslahti 1996; Pasqualini et al. 1997). Bacteriophage can be engineered to express peptide sequences of various lengths and configurations (e.g., linear or cyclic) on their surface. Libraries of phage can be generated that express all possible sequences of peptide of a defined length or configuration. Using this library, one can pan against target proteins, cells, or even organs and tissues in vivo. By isolating the phage from the library that have affinity for the target of interest, serial rounds of panning can ultimately identify peptide sequences that show particular affinity for the target. For example, using this approach in vivo, Pasqualini and colleagues (1997) have identified a double-cyclic RGD-containing peptide with particular affinity for tumor vasculature. Similar strategies are also being used to define scFvs with targeting potential (Neri et al. 1997).

It has clearly been shown that viral tropism can be modified by genetic strategies. However, concurrent ablation of native tropism has not yet been achieved. The targeting ligands that have been successfully introduced at this stage are limited to short peptides, and incorporation of larger ligands may require new approaches. Nevertheless, progress in this area has been rapid and the results to date, coupled with the immunological retargeting results, indicate that development of a systemically stable, cell-specific vector is a very realistic aim.

INTEGRATION

The use of adenovirus as a vector has other important limitations apart from issues of targeting. One of the problems is the transient nature of the gene expression achievable with these vectors. While this may not be a central issue in the context of cancer gene therapy, where transient expression of a toxin gene may in fact be advantageous, there are of course many situations of inherited genetic disorders where lifelong correction or compensation is necessary. Although issues relating to the immunogenicity of adenoviral vectors are important in this context, another significant issue is that transgenes delivered by adenoviral vectors are almost always not integrated into the host-cell DNA and are thus lost from subsequent generations of cells. In contrast, one of the chief advantages of retrovirus and adeno-associated virus (AAV) vectors is their ability to integrate transgene DNA, thus allowing for long-term gene expression. Use of retroviruses in vivo has been limited by issues of low titer production and complement-mediated inactivation of the vector. AAV vectors have also been limited by low titer production, the presence of contaminating helper adenovirus, and relatively small packaging capacity, although some of these problems have recently been addressed (Ferrari et al. 1997). When comparing the individual features of adenoviral and retroviral vectors, it becomes apparent that the relative strengths of one vector complements the weaknesses of the other. Thus, we developed a chimeric system to take advantage of the positive attributes of both, hypothesizing that such a system would potentially allow for high efficiency in vivo gene delivery coupled with integration of the delivered transgene and long-term gene expression (Feng et al. 1997). In addition to exploiting the individual attributes of the virus vectors, this approach seeks to capitalize on the knowledge that in vivo use of retroviral producer cells as vectors can achieve a level of gene delivery superior to administration of retroviruses themselves (Culver et al. 1992).

The strategy for the adenovirus/retrovirus chimera system is illustrated in Figure 5. Two adenoviral vectors were constructed, one carrying genes encoding retroviral packaging functions (*gag, pol*, and amphotropic *env*) (AdCMVAmpg), the other containing retroviral vector sequences including LTRs, packaging signal, and a green fluorescent protein (GFP) transgene (AdLNCMVGFP). Cells are coinfected with these two vectors, resulting in the transient in situ generation of retroviral producer cells. Progeny of replication-incompetent retroviruses are then released to infect neighbor cells with resultant transgene integration and long-term gene expression. For in vivo application, this system would take advantage of the initial systemic stability and highly efficient gene delivery properties of the adenovirus, followed by in situ generation of retroviruses to avoid the problems of low titer production and systemic instability of this vector, yet exploit its integration capabilities.

This approach was first validated in vitro by infecting cells with either one or both adenoviruses, then using the supernatants generated to infect naïve cells. The naïve cells incubated in the supernatant generated from the cells infected with the two viruses expressed GFP for at least 20 days, whereas those exposed to the supernatant from the cells infected with AdLNCMVGFP alone did not express any GFP. These results were

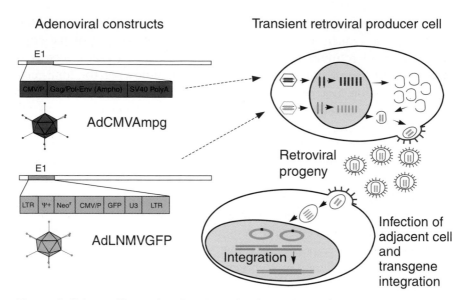

Figure 5. Schema illustrating the adenovirus/retrovirus chimera concept.

therefore consistent with retrovirus production from the initial coinfection with the adenoviral vectors. Subsequently, it was shown by Southern blot that those cells expressing GFP long-term had integration of proviral sequence into their chromosomal DNA.

To validate this approach in vivo, a murine ovarian carcinoma model was used. In the first instance, SKOV3.ip1 cells were infected in vitro with either AdLNCMVGFP alone (as a control) or AdCMVGFP plus AdCMVAmpg, then mixed 1:4 with naïve cells and implanted into nude mice. When sacrificed 20 days later, the tumors from the mice receiving cells infected with both vectors had large sheets of GFP-expressing cells comprising greater than 25% of the tumor mass and therefore consistent with lateral retroviral infection. In contrast, the tumors from the control mice had only rare isolated GFP-expressing cells. In a more stringent experiment, tumors were first established in nude mice by intraperitoneal injection of SKOV3.ip1 cells, and then the mice were then injected with either AdLNCMVGFP alone or AdCMVGFP plus AdCMVAmpg. Only those tumors from mice injected with both vectors expressed GFP when sacrificed. The adenovirus/retrovirus approach combines the favorable attributes of two separate vectors into the one vector system, thus illustrating the key concept of using chimerism to functionally combine the attributes of distinct vectors.

CONCLUSION

The studies discussed in this chapter illustrate the substantial progress being made toward one of the key criteria for successful gene therapy: specificity of delivery to appropriate target cells. Recent advances, such as those outlined in this chapter, reflect the commitment of the field to important basic vectorology issues, and should provide a secure foundation from which to extend the application of gene therapy strategies to human disease.

ACKNOWLEDGMENTS

Work cited in this manuscript that was conducted at the Gene Therapy Program, University of Alabama at Birmingham was supported by the following grants: NIH RO1CA74242 (DTC), NIH RO1CA68245 (DTC), NIH HL50255 (DTC), American Heart Association 965075-1W (DTC), a grant from the U.S. Department of Defense (DTC), a grant from the American Lung Association (DTC), and a grant from Allen and Hanbury's and the Thoracic Society of Australian and New Zealand (PR).

REFERENCES

Anderson R.G., Kamen B.A., Rothberg K.G., and Lacey S.W. 1992. Potocytosis: Sequestration and transport of small molecules by caveolae. *Science* **255:** 410–411.

Anderson W.F. and Fletcher J.C. 1980. Sounding boards. Gene therapy in human beings: When is it ethical to begin? *N. Engl. J. Med.* **303:** 1293–1297.

Barry M.A., Dower W.J., and Johnston S.A. 1996. Toward cell-targeting gene therapy vectors: Selection of cell-binding peptides from random peptide-presenting phage libraries. *Nat. Med.* **2:** 299–305.

Bellon G., Michel-Calemard L., Thouvenot D., Jagneaux V., Poitevin F., Malcus C., Accart N., Layani M.P., Aymard M., Bernon H., Bienvenu J., Courtney M., Doring G., Gilly B., Gilly R., Lamy D., Levrey H., Morel Y., Paulin C., Perraud F., Rodillon L., Sene C., So S., Touraine-Moulin F., Pavirani A., et al. 1997. Aerosol administration of a recombinant adenovirus expressing CFTR to cystic fibrosis patients: A phase I clinical trial. *Hum. Gene Ther.* **8:** 15–25.

Bergelson J.M., Cunningham J.A., Droguett G., Kurt-Jones E.A., Krithivas A., Hong J.S., Horwitz M.S., Crowell R.L., and Finberg R.W. 1997. Isolation of a common receptor for Coxsackie B viruses and adenoviruses 2 and 5. *Science* **275:** 1320–1323.

Brody S.L. and Crystal R.G. 1994. Adenovirus-mediated in vivo gene transfer. *Ann. N.Y. Acad. Sci.* **716:** 90–101.

Chartier C., Degryse E., Gantzier M., Dieterle A., Pavirani A., and Mehtali M. 1996. Efficient generation of recombinant adenovirus vectors by homologous recombination in *Escherichia coli. J. Virol.* **70:** 4805–4810.

Cosset F.L. and Russell S.J. 1996. Targeting retrovirus entry. *Gene Ther.* **3:** 946–956.

Crystal R.G., McElvaney N.G., Rosenfeld M.A., Chu C.S., Mastrangeli A., Hay J.G.,

Brody S.L., Jaffe H.A., Eissa N.T., and Danel C. 1994. Administration of an adenovirus containing the human CFTR cDNA to the respiratory tract of individuals with cystic fibrosis. *Nat. Genet.* **8:** 42–51.

Culver K.W., Ram Z., Wallbridge S., Ishii H., Oldfield E.H., and Blaese R.M. 1992. In vivo gene transfer with retroviral vector-producer cells for treatment of experimental brain tumors. *Science* **256:** 1550–1552.

Douglas J.T., Rogers B.E., Rosenfeld M.E., Michael S.I., Feng M., and Curiel D.T. 1996. Targeted gene delivery by tropism-modified adenoviral vectors. *Nat. Biotechnol.* **14:** 1574–1578.

Elshami A. A., Cook J.W., Amin K.M., Choi H., Park J.Y., Coonrod L., Sun J., Molnarkimber K., Wilson J.M., Kaiser L.R., and Albelda S.M. 1997. The effect of promoter strength in adenoviral vectors containing herpes simplex virus thymidine kinase on cancer gene therapy in vitro and in vivo. *Cancer Gene Ther.* **4:** 213–221.

Esandi M.C., van Someren G.D., Vincent A.J., van Bekkum D.W., Valerio D., Bout A., and Noteboom J.L. 1997. Gene therapy of experimental malignant mesothelioma using adenovirus vectors encoding the HSVtk gene. *Gene Ther.* **4:** 280–287.

Fancy D.A., Melcher K., Johnston S.A., and Kodadek T. 1996. New chemistry for the study of multiprotein complexes: The six-histidine tag as a receptor for a protein crosslinking reagent. *Chem. Biol.* **3:** 551–559.

Feng M.Z., Jackson W.H., Goldman C.K., Rancourt C., Wang M.H., Dusing S.K., Siegal G., and Curiel D.T. 1997. Stable in vivo gene transduction via a novel adenoviral/retroviral chimeric vector. *Nat. Biotechnol.* **15:** 866–870.

Ferrari F.K., Xiao X., McCarty D., and Samulski R.J. 1997. New developments in the generation of Ad-free, high-titer rAAV gene therapy vectors. *Nat. Med.* **3:** 1295–1297.

Goldman C.K., Rogers B.E., Douglas J.T., Sosnowski B.A., Ying W.B., Siegal G.P., Baird A., Campain J.A., and Curiel D.T. 1997. Targeted gene delivery to Kaposi's sarcoma cells via the fibroblast growth factor receptor. *Cancer Res.* **57:** 1447–1451.

Goldman M.J. and Wilson J.M. 1995. Expression of $\alpha_v \beta_5$ integrin is necessary for efficient adenovirus-mediated gene transfer in the human airway. *J. Virol.* **69:** 5951–5958.

Goldman M., Su Q., and Wilson J.M. 1996. Gradient of RGD-dependent entry of adenoviral vector in nasal and intrapulmonary epithelia: implications for gene therapy of cystic fibrosis. *Gene Ther.* **3:** 811–818.

Graham F. and Prevec L. 1991. Manipulation of adenovirus vectors. *Methods Mol. Biol.* **7:** *109–129.*

Grubb B.R., Pickles R.J., Ye H., Yankaskas J.R., Vick R.N., Engelhardt J.F., Wilson J.M., Johnson L.G., and Boucher R.C. 1994. Inefficient gene transfer by adenovirus vector to cystic fibrosis airway epithelia of mice and humans. *Nature* **371:** 802–806.

Hong S.S. and Boulanger P. 1995. Protein ligands of the human adenovirus type 2 outer capsid identified by biopanning of a phage-displayed peptide library on separate domains of wild-type and mutant penton capsomers. *EMBO J.* **14:** 4714–4727.

Hong S. S., Karayan L., Tournier J., Curiel D.T., and Boulanger P.A. 1997. Adenovirus type 5 fiber knob binds to MHC class I α2 domain at the surface of human epithelial and B lymphoblastoid cells. *EMBO J.* **6:** 2294–2306.

Knowles M.R., Hohneker K.W., Zhou Z., Olsen J.C., Noah T.L., Hu P.C., Leigh M.W., Engelhardt J.F., Edwards L.J., Jones K.R. and et al. 1995. A controlled study of adenoviral-vector-mediated gene transfer in the nasal epithelium of patients with cystic fibrosis [see comments]. *N. Engl. J. Med.* **333:** 823–831.

Krasnykh V.N., Mikheeva G.V., Douglas J.T., and Curiel D.T. 1996. Generation of recombinant adenovirus vectors with modified fibers for altering viral tropism. *J. Virol.* **70:** 6839–6846.

Krasnykh V., Dmitriev I., Mikheeva G., Miller C.R., Belousova N., and Curiel D.T. 1998. Characterization of an adenovirus vector containing a heterologous peptide epitope in the HI loop of the fiber knob. *J. Virol.* **72:** 1844–1852.

Michael S.I., Hong J.S., Curiel D.T., and Engler J.A. 1995. Addition of a short peptide ligand to the adenovirus fiber protein. *Gene Ther.* **2:** 660–668.

Michael S.I., Huang C.H., Romer M.U., Wagner E., Hu P.C., and Curiel D.T. 1993. Binding-incompetent adenovirus facilitates molecular conjugate-mediated gene transfer by the receptor-mediated endocytosis pathway. *J. Biol. Chem.* **268:** 6866–6869.

Murthy U., Basu A., Rodeck U., Herlyn M., Ross A.H., and Das M. 1987. Binding of an antagonistic monoclonal antibody to an intact and fragmented EGF-receptor polypeptide. *Arch. Biochem. Biophys.* **252:** 549–560.

Neri D., Carnemolla B., Nissim A., Leprini A., Querze G., Balza E., Pini A., Tarli L., Halin C., Neri P., Zardi L., and Winter G. 1997. Targeting by affinity matured recombinant antibody fragments of an angiogenesis associated fibronectin isoform. *Nat. Biotechnol.* **15:** 1271–1275.

Novelli A. and Boulanger P.A. 1991. Deletion analysis of functional domains in baculovirus-expressed adenovirus type 2 fiber. *Virology* **185:** 365–376.

Pasqualini R. and Ruoslahti E. 1996. Organ targeting in vivo using phage display peptide libraries. *Nature* **380:** 364–366.

Pasqualini R., Koivunen E., and Ruoslahti E. 1997. α v integrins as receptors for tumor targeting by circulating ligands. *Nat. Biotechnol.* **15:** 542–546.

Rogers B.E., Douglas J.T., Sosnowski B.A., Ying W., Pierce G., Buchsbaum D.J., Della Mana D., Baird A., and Curiel D.T. 1998. Enhanced in vivo gene delivery to human ovarian cancer xenografts utilizing a tropism-modified adenovirus vector. *Tumor Targetting* **3:** 25–31.

Schulick A.H., Newman K.D., Virmani R., and Dichek D.A. 1995a. In vivo gene transfer into injured carotid arteries. Optimization and evaluation of acute toxicity. *Circulation* **91:** 2407–2414.

Schulick A.H., Dong G., Newman K.D., Virmani R., and Dichek D.A. 1995b. Endothelium-specific in vivo gene transfer. *Circ. Res.* **77:** 475–485.

Shenk T. 1996. Adenoviridae: the viruses and their replication. In *Fields virology, 3rd edition* (ed. B.N Fields et al.), pp. 2111–2148. Lippincott-Raven, Philadelphia, Pennsylvania.

Stevenson S.C., Rollence M., Marshallneff J., and McClelland A. 1997. Selective targeting of human cells by a chimeric adenovirus vector containing a modified fiber protein. *J. Virol.* **71:** 4782–4790.

Tomko R.P., Xu R., and Philipson L. 1997. HCAR and MCAR: The human and mouse cellular receptors for subgroup C adenoviruses and group B coxsackieviruses. *Proc. Natl. Acad. Sci.* **94:** 3352–3356.

Watkins S.J., Mesyanzhinov V.V., Kurochkina L.P., and Hawkins R.E. 1997. The adenobody approach to viral targeting–Specific and enhanced adenoviral gene delivery. *Gene Ther.* **4:** 1004–1012.

Weitman S.D., Lark R.H., Coney L.R., Fort D.W., Frasca V., Zurawski V.R., Jr., and Kamen B.A. 1992. Distribution of the folate receptor GP38 in normal and malignant cell lines and tissues. *Cancer Res.* **52:** 3396–3401.

Wersall P., Ohlsson I., Biberfeld P., Collins V.P., von Krusenstjerna S., Larsson S.,

Mellstedt H., and Boethius J. 1997. Intratumoral infusion of the monoclonal antibody, mAb 425, against the epidermal-growth-factor receptor in patients with advanced malignant glioma. *Cancer Immunol. Immunother.* **44:** 157–164.

Wickham T.J., Carrion M.E., and Kovesdi I. 1995. Targeting of adenovirus penton base to new receptors through replacement of its RGD motif with other receptor-specific peptide motifs. *Gene Ther.* **2:** 750–756.

Wickham T.J., Mathias P., Cheresh D.A., and Nemerow G.R. 1993. Integrins alpha v beta 3 and alpha v beta 5 promote adenovirus internalization but not virus attachment. *Cell* **73:** 309–319.

Wickham T.J., Lee G.M., Titus J.A., Sconocchia G., Bakacs T., Kovesdi I., and Segal D.M. 1997a. Targeted adenovirus-mediated gene delivery to T cells via CD3. *J. Virol.* **71:** 7663–7669.

Wickham T.J., Segal D.M., Roelvink P.W., Carrion M.E., Lizonova A., Lee G.M., and Kovesdi I. 1996. Targeted adenovirus gene transfer to endothelial and smooth muscle cells by using bispecific antibodies. *J. Virol.* **70:** 6831–6838.

Wickham T.J., Tzeng E., Shears L.L., Roelvink P.W., Li Y., Lee G.M., Brough D.E., Lizonova A., and Kovesdi I. 1997b. Increased in vitro and in vivo gene transfer by adenovirus vectors containing chimeric fiber proteins. *J. Virol.* **71:** 8221–8229.

Xia D., Henry L.J., Gerard R.D., and Deisenhofer J. 1994. Crystal structure of the receptor-binding domain of adenovirus type 5 fiber protein at 1.7 Å resolution. *Structure* **2:** 1259–1270.

———. 1995. Structure of the receptor binding domain of adenovirus type 5 fiber protein. In *The molecular repertoire of adenoviruses 1* (ed. W. Doerfler and P. Bohm), pp. 39–46. Springer-Verlag, Berlin.

Yang Y., Li Q., Ertl H.C., and Wilson J.M. 1995. Cellular and humoral immune responses to viral antigens create barriers to lung-directed gene therapy with recombinant adenoviruses. *J. Virol.* **69:** 2004–2015.

Yang Y., Jooss K.U., Su Q., Ertl H.C., and Wilson J.M. 1996. Immune responses to viral antigens versus transgene product in the elimination of recombinant adenovirus-infected hepatocytes in vivo. *Gene Ther.* **3:** 137–144.

Yee D., McGuire S.E., Brunner N., Kozelsky T.W., Allred D.C., Chen S.H., and Woo S.L. 1996. Adenovirus-mediated gene transfer of herpes simplex virus thymidine kinase in an ascites model of human breast cancer. *Hum. Gene Ther.* **7:** 1251–1257.

Zabner J., Freimuth P., Puga A., Fabrega A., and Welsh M.J. 1997. Lack of high affinity fiber receptor activity explains the resistance of ciliated airway epithelia to adenovirus infection. *J. Clin. Invest.* **100:** 1144–1149.

Zabner J., Couture L.A., Gregory R.J., Graham S.M., Smith A.E., and Welsh M.J. 1993. Adenovirus-mediated gene transfer transiently corrects the chloride transport defect in nasal epithelia of patients with cystic fibrosis. *Cell* **75:** 207–216.

Zabner J., Ramsey B.W., Meeker D.P., Aitken M.L., Balfour R.P., Gibson R.L., Launspach J., Moscicki R.A., Richards S.M., Standaert T.A., et al. 1996. Repeat administration of an adenovirus vector encoding cystic fibrosis transmembrane conductance regulator to the nasal epithelium of patients with cystic fibrosis. *J. Clin. Invest.* **97:** 1504–1511.

Zinn K.R., Douglas J.J., Smyth C., Liu H.-G., Wu Q., Krasnykh V.N., Mountz J.D., Curiel D.T., and Mountz J.M. 1998. Imaging and tissue biodistribution of 99-m-Tc-labeled adenovirus knob (serotype 5). *Gene Ther.* **5:** (in press).

7

Adeno-associated Viral Vectors

Richard Jude Samulski

University of North Carolina Gene Therapy Center and
Department of Pharmacology
University of North Carolina at Chapel Hil
Chapel Hill, North Carolina

Mitch Sally

University of North Carolina Gene Therapy Center
Chapel Hill, North Carolina

Nicholas Muzyczka

University of Florida Gene Therapy Center and
Department of Microbiology
University of Florida at Gainesville
Gainesville, Florida

The discipline of gene therapy is relatively young when compared to other fields of study. However, from the start, there has been a clear vision of the expectations required to achieve successful clinical results. These have included (1) efficient transduction of the target cell, (2) long-term expression of the therapeutic gene, (3) lack of immune response to the vector or transduced cell, and (4) absence of toxicity to the patient following delivery. Until recently, this mandate has been met only in part by the numerous viral and nonviral delivery systems. Now, definitive in vivo results meeting all of the above criteria have been established by using the parvovirus adeno-associated virus (AAV) vector system supporting a thorough testing of this vector in the clinical arena. This chapter describes in detail the AAV vector system, production methods, successful in vivo models, and questions related to AAV biology that, when understood, may further enhance the existing vector system.

Research employing AAV as a vector for gene therapy has been driven by the desire to exploit the unique biology and life cycle of this virus. This human parvovirus exhibits many natural features that are absent from alternative vectors. AAV's most prominent feature, which suits it to applications for long-term gene therapy, is the tendency to establish latent infections through integration into the chromosomal DNA. The fact that AAV lytic cycle generally depends on the presence of a coinfecting helper virus means that, essentially, any uninfected helper cell can potentially serve as a host for latent infection. A second salient feature that distinguishes AAV from many other potential viral vectors is its lack of pathogenicity. Theoretically, a virus that has never been associated with any clinical disease poses relatively little risk when adapted as a vector. Other attractive characteristics are its stability in multiple cell types, genome simplicity, and ease of genetic manipulation. While these features are attractive, some obstacles to the application of recombinant AAV (rAAV) vectors have recently surfaced. Given the cryptic nature of this virus, it is not surprising that new insights into its biology have emerged only after the testing of recombinant vectors began.

RECOMBINANT AAV VECTORS

Plasmid Substrates

Genetic manipulation of AAV became possible when the viral genome containing intact palindromic terminal repeats (TR) was cloned into bacterial plasmids (Samulski et al. 1982; Laughlin et al. 1983). These TRs, which are the *cis*-elements required for AAV rescue, replication, and packaging, have made it possible to generate wild-type (wt) virus by transfecting the recombinant plasmid DNA into adenovirus-infected cells. These infectious plasmid clones have also provided a template for generating and studying mutant AAV (Hermonat and Muzyczka 1984; Labow et al. 1986; Tratschin et al. 1984, 1986). rAAV vectors are made in much the same way. The two AAV-specific genes required in *trans*, *rep,* and *cap*, are deleted by removing 96% of the genome, leaving only the 145-bp TR sequences of the parent virus (Samulski et al. 1989). A heterologous gene is cloned in place of the rep and cap genes between the TRs, generating the AAV vector substrate. To generate virus, this substrate is cotransfected with a second plasmid expressing the *rep* and *cap* genes in Ad-infected cells. After rescue and replication of the vector in permissive cells, the single-stranded progeny DNA molecules from the rAAV construct are then encapsidated into AAV virions. All genes within approximately 5% of the size of the wild-type AAV (wtAAV) genome (5 kb) will be efficiently packaged (Muzyczka 1992). While ostensibly simple to execute, this procedure was initially inefficient and difficult, stemming from the need to transfect vector, and AAV helper plasmids, followed by infection with the adenovirus.

rAAV Production

Under optimized conditions, the transient transfection procedure can yield virus titers up to 10^{10} infectious units of rAAV per milliliter. This represents approximately 50–150 infectious virus particles produced per transfected cell. In contrast, wtAAV generates 1000–5000 infectious units per cell, which is equivalent to what is seen in an adenovirus infection. Thus, the current procedure for making rAAV still has room for improvement. Although improvement in productions is expected, the quality of the virus currently produced is sufficient for preclinical and clinical experiments.

Two major concerns that must be addressed in all rAAV vector preparations relate to the presence of wtAAV and the Ad helper virus. The problem of wtAAV contamination was largely removed by eliminating homologous overlap between the vector plasmid and the AAV helper plasmid carrying the complementing genes. Because there is a lack of sequence homology between the AAV helper and the vector plasmid, wtAAV is not generated through homologous recombination. The absence of wtAAV in rAAV preparations greatly minimizes the introduction of wild-type-expressed gene products that could induce a host immune response, a problem that is often encountered with other viral vectors. Producer cell lines, which would eliminate the transfection step, are currently being investigated (Yang et al. 1993; Holscher et al. 1994; Clark et al. 1995). These hold promise and will be essential for the development of scaleable production methods. However, an evolution of new AAV helper constructs (Flotte et al. 1995; Li et al. 1997; Vincent et al. 1997; Xiao et al. 1998) that appear to be more effective than the original helper plasmids are also being developed and tested in the efficient transient transfection system. Additionally, incremental improvements in rAAV titers are being achieved through variations on the transfection procedure (Clark et al. 1995; Flotte et al. 1995; Mamounas et al. 1995; Tamoyose et al. 1996). Although it is generally believed that AAV packaging cell lines will surpass the current transient transfection procedure, many parameters involved with vector construction (e.g., testing promoter strength and therapeutic gene) can best be analyzed using transient production procedures before investing in the development of a stable producer cell line. We anticipate that the new helper constructs will eventually lead to the generation of stable AAV helper cell lines that will contain both the rAAV vector of choice and the AAV complementing genes. These cell lines would rely only on the infection of helper adenovirus to generate rAAV vectors.

To date, rAAV vector purification has relied on traditional $CsCl_2$ centrifugation. Again, while this has been adequate to generate a preclinical virus, more sophisticated methods, preferably high-throughput chromatograpy, will be required for purification of the vector for large-scale needs in clinical trials. Efforts in this direction are still in their infancy.

Ad-free AAV Vectors

Although the current procedure for rAAV production has essentially removed the generation of wtAAV recombinants, it relies on the presence of wild-type adenovirus (wtAd) to provide essential helper functions for rAAV replication. This results in the coproduction of wtAd along with rAAV, thus creating a problem of wtAd contamination in rAAV preparations. Typically, the issue of wtAd contamination has been addressed through the purification of rAAV on multiple $CsCl_2$ gradients, column chromatography, and/or heat inactivation of Ad through the incubation of rAAV preps at 56°C for 30 minutes (Snyder et al. 1996). Each of these procedures has been documented to remove infectious Ad contamination, yet complete removal of Ad capsid components is still not always possible. Because several Ad proteins (fiber, hexon, etc.) are known to produce a CTL immune response (Yang et al. 1994, 1995; Yang and Wilson 1995) this represents a significant drawback of these rAAV preparations (Monahan et al. 1998).

To overcome these difficulties, adenovirus mini-plasmids that carry all of the helper genes required for efficient AAV production have been generated (Ferrari et al. 1997; Xiao et al. 1998). The use of noninfectious Ad plasmids has had a significant impact on rAAV production (Fig. 1). The rAAV titers obtained with Ad mini-plasmids are 40–60-fold higher than those obtained with conventional methods of wtAd infection (Ferrari et al. 1997; Xiao et al. 1998). This increase (up to 1000 transducing units/cell) is probably due to the lack of competition between the AAV and adenovirus for Ad gene products in addition to removing the lytic pathway of this helper virus. More importantly, the bulk of the Ad structural genes have been deleted, thus eliminating the risk of contaminating Ad proteins that may copurify with rAAV vectors. Therefore, this method provides a way in which sufficient quantities of highly pure rAAV vectors can be generated for preclinical and phase I clinical studies. Using this approach, one is dependent on transient transfection to introduce the Ad mini-plasmids. Efforts to establish AAV packaging cell lines that do not require Ad infection or Ad mini-helper construct transfection are being initiated.

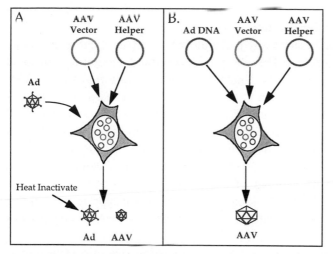

Figure 1 Diagram of rAAV production (*A*) Classic Method—the rAAV genome is introduced in *trans* with a plasmid expressing the AAV *rep* and *cap* genes. The Ad helper effect is supplied by Ad infection. (*B*) T.O.A.D. Method—differs from the clasic method in that the Ad helper functions are supplied by a nonreplicating Ad minigenomic plasmid.

Some of the in vivo variables seen with rAAV transduction can be attributed to the presence of Ad-contaminating components as well as wtAAV. Despite our best attempts, complete removal of Ad is not always possible. With this genetic approach of using mini-Ad plasmids for removing Ad infection and propagation, there is no risk of Ad contamination. The ability to generate high-titer pure rAAV has allowed considerable progress to be made in understanding transduction by these vectors. The use of AAV packaging strategies that eliminate Ad infection will now be essential when evaluating these vectors in vivo (see Monahan et al. 1998).

Titering rAAV

Despite this progress in growing rAAV, a lack of standard terminology to describe the infectious dose used in each experiment is still a persistent source of confusion in the field of rAAV vector research. This shortcoming arose in part because it is not clear how to quantify AAV, as is done with most viruses. The requirement for superinfection with a helper virus precludes the use of plaque assays to measure the infectious titer. It is therefore necessary to detect either the viral gene products or the replicating genome to infer the titer.

The replication center assay is the most broadly applicable technique for titering rAAV vectors. This assay relies on the detection of replicated vector DNA in single cells coinfected with adenovirus and wtAAV (Yakobson et al. 1987; Snyder et al. 1996). While this assay is technically demanding, it does measure two parameters that are common to all rAAV-based vectors: the presence of intact TRs and the encapsidation of vector DNA into an infectious particle. Because the TRs are the only *cis*-acting sequences required for integration, the replication center assay also serves to measure the number of potential long-term transducing units (McLaughlin et al. 1988). As long as a standard cell type is used in this assay, it will yield a value that can be compared to any rAAV construct or to the wild-type virus.

The replication center assay typically measures one infectious unit as corresponding to approximately 20–100 physical virus particles (de la Maza and Carter 1980). Of course, this ratio between infectious units and particles will vary among different cell types and tissues. We have adopted 293 cells as a standard for measuring infectious rAAV vectors, but similar titers are obtained from HeLa cells.

The infectious center assay (Clark et al. 1995; Snyder et al. 1996) also provides a consistent standard for the comparison of various vector titers. However, it does not necessarily predict a third value, the transducing unit titer. Again, this value is greatly influenced by cell type as well as promoter activity and specificity. In 293 cells infected with rAAV vectors expressing the bacterial *lacZ* gene under the control of the cytomegalovirus (CMV) immediate early promoter, the ratio of infectious titer to transducing units is approximately 10:1. Therefore, the ratio of physical virus particles to transducing units is approximately 200–1000:1 with this assay. For reasons that we will describe later in this chapter, the differences between the infectious center assay and the transduction titer can be eliminated.

For this review, we will attempt to frame our discussions using the convention of infectious rAAV particles per cell. It is arguably clearer to report rAAV titers in terms of physical virus particles, which do not change from one cell type or infection condition to another. However, with the aim of expanding on the mechanisms of rAAV transduction, we feel that focusing on the number of potentially active genomes entering a particular nucleus may be more meaningful. In some of the specific experiments discussed, we have estimated the ratios of infectious units to cells from reported values in transducing units or in physical particles.

BARRIERS TO rAAV TRANSDUCTION

The ability of rAAVs to efficiently transduce cells has been determined experimentally, largely through in vitro use of immortalized cell lines. However, the relevance of these results to the transduction of primary and nondividing cells in vitro and in vivo has come into question. The first observations raising these concerns stemmed from differences in the titer of rAAV virus stocks before and after purification on CsCl density gradients or after heating for 30 minutes at 56°C (Ferrari et al. 1996; Fisher et al. 1996b). Both of these procedures are employed to remove or inactivate contaminating adenovirus. It was soon realized that it was not the titer of the rAAV that was changing, but that the ability to detect rAAV infection as a transduction event was enhanced in the presence of Ad coinfection. Full restoration of the original rAAV titer could be achieved by adding back the adenovirus. The helper effects normally provided by Ad for wild-type AAVs were not expected to contribute to the expression of rAAV because the rAAV vectors were not capable of replication and the transgenes were not driven by either Ad or AAV promoters. This previously unrecognized helper function was observed with many heterologous promoters that were also not expected to respond to Ad-specific transcription factors.

Ad Enhancement of rAAV

Several potential steps existed where Ad could have exerted such an effect. The first step involved the potential participation in the uptake of the virus or the intracellular release of the rAAV genome from the capsid. The second step dealt with the conversion of the single-stranded rAAV virion DNA to the duplex form, which then could serve as a template for transcription. Finally, the stimulation of transcription from the transgene via an enhancer effect from the AAV TR was seen as a step at which the Ad could exert its effect. In general, most studies to date suggest that it is the synthesis of the second strand of the rAAV genome that is limiting for the transduction of normal cells.

Adenovirus infection is a complex process that involves several distinct steps prior to viral gene expression. Among these steps are adsorption and entry, endosomal disruption, and transport to the nucleus. The idea that Ad has a role in AAV adsorption, penetration, or translocation to the nucleus had been dismissed previously based on early efforts to define the Ad helper effect for wtAAV (Rose and Koczot 1972). Similarly, the novel helper effect provided for rAAV vector transduction failed to correlate with any of these steps. Induction was still observed when staggered Ad infection was used, making co-adsorption and penetration unlikely explanations (Ferrari et al. 1996). Additionally, there was no requirement for an intact Ad virion, because transfected Ad DNA was sufficient to induce high levels of transduction in rAAV-infected cells (Ferrari et al. 1996). This result suggested that an Ad-specific or Ad-induced gene product was mediating enhancement.

Adenovirus E4 ORF 6

A series of carefully characterized Ad mutants was used to discern which gene product or products were responsible for the helper effect (Ferrari et al. 1996; Fisher et al. 1996b). Ad with mutations in E1, VAI, E2, or E4 were coinfected with rAAV and transduction rates were compared. The E4 deletion mutants were the only group that failed to enhance transduction. Additional mutants were used to map this effect to E4 open reading frame 6 (ORF 6). Then, using a transfected plasmid expressing only the E4 ORF 6 gene product, the possibility that other interactions with the Ad genome were needed for the helper effect was eliminated. Interestingly, the E4 ORF 3 gene product, which is redundant with ORF 6 for some activities associated with lytic Ad infection (Ohman et al. 1993; Weiden and Ginsberg 1994) could not provide the helper effect for rAAV. This confirms previous studies that the E4 ORF 6 product was unique in its ability to help wtAAV infection.

Identification of the factor that mediated the enhancement of rAAV transduction allowed for partial characterization of the mechanism behind this effect. By comparing the DNase sensitivity of rAAV genomes in cells transfected with plasmids expressing E4 ORF 6 versus cells transfected with a control plasmid, the idea that E4 ORF 6 might promote uncoating of the virion was eliminated (C. Summerford and R. Samulski, unpubl.). These studies demonstrated that the virion DNA was DNase-sensitive at 24 hours postinfection, regardless of the expression of the Ad proteins, and no difference in DNA sensitivity was detected in E4-expressing cells versus controls.

The possibility remained that the E4 protein was modulating the transcription of the rAAV transgene, either directly or indirectly, through *cis* elements within the TR. The TRs have an enhancer-like activity that is dependent on the *trans* effects of the AAV Rep protein (Beaton et al. 1989; McCarty et al. 1991); however, it was difficult to reconcile this fact with the magnitude of the helper effect. Up to 1000-fold increases were observed in the number of cells transduced. Furthermore, the rAAV transgenes failed to show any such response to Ad infection when transfected as plasmid or existing in the proviral state (Ferrari et al. 1996). In the latter two situations, the transgene would be in an identical position relative to the TRs as in the infecting virion DNA. All of these data suggested that, whereas neither duplex plasmid nor integrated templates of identical sequences were sensitive to Ad enhancement, the virion form of DNA was, in fact, sensitive.

Second-strand Synthesis Is a Rate-limiting Step

The findings in the previous subsection were consistent with an induction of second-strand synthesis by Ad or the Ad E4 ORF 6 gene product. Experiments correlating the induction of transgene expression with the conversion of the single-stranded virion DNA to duplex offered evidence supporting this conclusion (Ferrari et al. 1996; Fisher et al. 1996b). When DNA from rAAV-infected cells was analyzed on denaturing agarose gels (Ferrari et al. 1996; Fisher et al. 1996a), conversion was observed as the appearance of dimer-sized single-stranded molecules. These dimer-length molecules arose by priming the DNA synthesis from the hairpinned TR of a single-stranded virion DNA. The mechanism for inducing synthesis of the rAAV second strand is not yet known. Several other functions that have been associated with E4 ORF 6 are (1) binding to the cellular tumor suppresser protein, p53, blocking its transcriptional *trans*-activation activity (Dobner et al. 1996); (2) enhancing the splicing of leader sequences of specific Ad late mRNAs (Ohman et al. 1993; Nordqvist et al. 1994); (3) prevention of concatemer formation during Ad DNA replication (Weiden and Ginsberg 1994); (4) binding to E1b to facilitate Ad mRNA transport (Samulski and Shenk 1988; Sarnow et al. 1984); and (5) inhibiting E4 ORF 4 in the regulation of Ad DNA replication (Bridge et al. 1993).

There is still some uncertainty as to whether the products of Ad E1 are also necessary for the enhancement of rAAV transduction. Fisher et al. (1996b) have suggested that E1 at least strengthens the effect based on two observations. First, Ad mutants that are deleted in E1, but express E4, enhance transduction in 293 cells which express E1. The same result is not found in HeLa cells, which do not express constitutive E1 product. However, this observation can be explained easily on the basis of the effect of E1 on the E4 promoter. Their second finding, in a 293 cell line expressing E4 under an inducible metallothionein promoter, involved rAAV transduction enhancement in a Zn^{++} concentration-dependent manner. In contrast, in a similar HeLa cell line, the enhancement does not reach proportionate levels. E1 is known to induce host-cell synthesis as well as many of the cellular enzymes required for DNA repair. In addition, E1 can augment the CMV promoter. For these various reasons, E1 still may play some role.

While the induction of rAAV second-strand synthesis by the E4 ORF 6 gene product may represent an adaptation evolved by wtAAV to further exploit its Ad helper, the phenomenon is clearly not unique to Ad-infected cells. A number of cytotoxic cell culture treatments have been found to facilitate the same step. These treatments include hydroxyurea, UV and X-ray irradiation, some topoisomerase inhibitors, and heat shock (Russell et al. 1995; Ferrari et al. 1996). Most of these treatments, with the possible exception of heat shock, result in DNA damage. It is reasonable to hypothesize that the induction of DNA damage repair synthesis is the mechanism that allows the conversion of a single-strand rAAV virion DNA to double-strand transcriptionally active templates (in a manner analogous to filling a gapped region of chromosomal DNA). Indeed, studies of AAV synthesis in vitro, using uninfected HeLa cell extracts, have identified several cellular enzymes that are required for and which may be involved in repair. These enzymes include RPA, PCNA, RFC, and either pol δ or pol ε (Ni et al. 1998).

Experiments with mutant cell lines lacking p53 and pRB suggest that these factors are not necessary for the induction of second-strand synthesis (F.K. Ferrari and R.J. Samulski, unpubl.). This result argues against the possibility that apoptotic pathways mediated by these factors lead to enhanced transduction. However, it should be noted that, while this cell line is deficient in p53 and Rb, many other genetic alterations have also occurred that may obscure the definitive answer to this question. Analysis of rAAV transduction in primary mouse embryo cells deficient in p53 or Rb would resolve this issue.

The observation that cytotoxic treatments could induce rAAV second-strand synthesis, and hence transduction, was similar to earlier experiments in which those types of treatments were shown to induce helper-virus-independent replication of wtAAV (Yakobson et al. 1987, 1989; Yalkinoglu et al. 1988). However, in these cases, UV or hydroxyurea treatments controlled replication of the viral DNA, whether the genome was presented as a single-stranded virion DNA or as a duplex plasmid. These findings indicate that the changes that take place in cells following DNA damage contribute to more than simply overcoming the barrier to duplex formation.

Yalkinoglu et al. (1991) made the more remarkable observation that plasmids containing a single copy of the TR were replicated in cells treated with genotoxic agents. The replication origin on these molecules was mapped to the TR using cell-free extracts from carcinogen-induced cells. This result suggests that the AAV TR can specifically recruit replication factors under conditions of cell stress. It then provides a plausible mechanism for the induction of second-strand synthesis when using rAAV vectors. The conditions that enhance rAAV transduction might induce the same factors to be recruited by the palindromic TR, which is already in a duplex hairpin conformation in the single-stranded virion DNA. Further, it suggests that AAV makes multiple adaptations to exploit cellular responses to environmental factors. This aspect of AAV biology has not been extensively explored as yet, but will certainly have a great impact on the future use of rAAV vectors in vivo for gene therapy.

Cellular Factors and AAV TRs

A host cellular protein that interacts with the single-stranded TR of the AAV genome has been identified (Qing et al. 1997). This protein has been designated the single-stranded D-sequence binding protein (ssD-BP) and, when phosphorylated at tyrosine residues, appears to inhibit second-strand synthesis of the rAAV genome. Furthermore, in the presence of Ad E4 ORF 6, hydroxyurea, and kinase inhibitors, the ssD-BP converts to a dephosphorylated state and rAAV second-strand synthesis (Qing et al. 1997). Whether the dephosphorylation process correlates with the induction of DNA damage by other agents remains to be determined, but it appears that the phosphorylation state of the ssD-BP may have a very important role in the life cycle of the AAV. Furthermore, it appears that the ratio of phosphorylated to unphosphorylated ssD-BP can differ between cell types. The amount of unphosphorylated ssD-BP in a panel of cell lines correlates with the efficiency by which AAV could transduce cells (Qing et al. 1998). Therefore, monitoring the fate of ssD-BP in various cell types may help us better identify appropriate tissues for gene therapy by AAV vectors.

What is the impact of these findings on the use of AAV vectors for gene therapy? Comprehensive analysis of rAAV's ability to enhance transduction in vivo remains to be fully explored, especially in inefficient cells. Clearly, the coinfection of Ad with rAAV in vivo can significantly enhance rAAV transduction. While the use of Ad coinfection is not plausible for clinical trials, it is easy to envision the use of cytotoxic agents such as X-ray or chemotherapy to enhance rAAV transduction in cancer gene therapy. Furthermore, hydroxyurea is already currently being used in the treatment of sickle cell anemia and may represent a reagent that can safely enhance rAAV transduction in vivo (Charache et al. 1995).

Although the use of genotoxic agents in vivo would have inherent risks, their significant effect on rAAV transduction observed in vitro warrants further study for their possible use in human gene therapy. The identification of a host cellular factor that may be a mediator for AAV second-strand synthesis is also a very important observation. Understanding the mechanism behind the induction of second-strand synthesis should lead to the finding of less toxic manners in which enhancement of rAAV transduction can be achieved. Mutant TRs that are not subject to the cellular factor may also provide a strategy to overcome this rate-limiting step.

Cellular Uptake of AAV Virions

The initial entry of AAV into the host cell serves as another potential barrier to rAAV transduction. Although AAV infects a very wide range of cell types, there are cell lines that do not support the initial binding of AAV. These lines include the human megacaryocytic cell lines, MB-02 and MO7e (Ponnazhagan et al. 1996), and the human leukemic cell line, UT7 EPO (Mizukami et al. 1996). The discovery of cell lines that apparently lack a cell surface receptor for AAV, combined with the observation that trypsin treatment of cells can abolish AAV binding to the cell surface, suggested that AAV uses a specific protein receptor to mediate infection (Mizukami et al. 1996). Such a receptor must be ubiquitously expressed based on AAV's extremely broad host range. In recent work, it was hypothesized that cell surface proteoglycans may serve as a receptor for the AAV (Summerford and Samulski 1998). This idea was based on an observation that AAV could be purified on a cellulose sulfate column. Cellulose sulfate columns are known only to purify viruses that interact with negatively charged surface molecules. Furthermore, cell surface proteoglycans are present on a wide variety of cell types and are expressed throughout the animal kingdom, making these membrane proteins reasonable candidate receptors for mediating AAV infection.

Heparan Sulfate as a Receptor for AAV

Proteoglycans are proteins that are modified by the posttranslational attachment of polysaccaharide glycosaminoglycan (GAG) moieties. Glycosaminoglycans are composed of repeating unbranched disaccharide units that undergo various degrees of sulfation. They differ by (1) the monosaccharides that comprise their chains, (2) the specific sites of sulfation, and (3) their susceptibility to bacterial enzymes known to cleave distinct monosaccharide linkages. The four most widely distributed GAGs found to be associated with plasma membrane proteins are heparan sulfate and the chondroitin sulfates A, B, and C. The functions of membrane-associated proteoglycans are diverse and include roles in cellular differentiation, adhesion, and growth. An interesting aspect of proteoglycans is that they are known to function as cellular receptors for some bacteria and several animal viruses, including herpes simplex virus types 1 and 2, cytomegleovirus, dengue virus, and the foot-and-mouth disease type-O virus. This fact led to the investigation of the role of proteoglycans in the AAV life cycle. Using defined mutant CHO cell lines with defects in the biosynthetic pathway of GAGs and standard biochemical assays, heparan sulfate proteoglycan has been identified as a cellular receptor for AAV type-2 virions (Summerford and Samulski 1998). Further analysis in various cell lines and in vivo suggests that heparan sulfate (HS) proteoglycan serves as a primary attachment receptor for AAV-2 and that this interaction is important for efficient infection by the AAV (C. Summerford and R.J. Samulski, unpubl.).

On the basis of AAV's broad host range and ability to infect such a wide variety of cells in tissue culture, it is tempting to speculate that AAV can independently use heparan sulfate proteoglycan to mediate productive viral entry. HS proteoglycans are known to undergo endocytosis and have been implicated in the direct internalization of other ligands. Alternatively, evidence exists for the possible use of a secondary receptor in AAV infection. When cells are treated with heparan sulfate digestive enzymes, AAV infection is more sensitive to the removal of HS than AAV binding. The possibility that the density of receptors may be an important factor influencing AAV entry can be postulated based on the absence of a one-to-one correlation between the amount of virus bound to viral infection ratio. In other words, the probability of AAV/HS complex finding another receptor molecule to interact with may be important for facilitating AAV entry. Furthermore, the CHO cell line that is deficient in HS synthesis still supports a low level of virus binding and infection, which suggests that there are potentially other low-copy-number receptors that may mediate infection, albeit inefficiently. Therefore, the idea that a secondary receptor could be involved in AAV uptake should not be ruled out. However, it is possible that, depending on the cell type and the amount of plasma membrane turnover, the presence of HS may be sufficient to mediate AAV entry. Further studies are required to address this issue.

TRANSDUCTION OF NONDIVIDING CELLS

Attempts to characterize the apparent discrepancy in transduction rates between transformed cells and nondividing cells have spawned several of the recent reports of enhancement of rAAV transduction. Russell et al. (1994) investigated the ability of rAAV vectors to transduce stationary versus dividing primary human fibroblasts. Little differences in transduction rates were noted between cells in stationary and rapidly dividing cultures when a *neo* reporter construct was used. However, a greater than 20-fold transduction rate difference resulted when an alkaline phosphatase (AP) reporter was used. These cultures did not require selection and cell division to measure rAAV gene expression. Furthermore, when transduced cells in a stationary culture were evaluated for incorporation of [^{3}H]thymidine as a marker for passage through the S phase, 90% of the cells had replicated nuclear DNA. This observation was interpreted to mean that, although rAAV vectors were capable of transducing cells in the absence of cell division, the efficiency was greater in cells that were undergoing DNA synthesis.

One explanation for the discrepancy between *neo* and AP transduction in nondividing cells has been postulated to be the persistence of single-stranded virion DNA in the nuclei of infected cells. After 12 days in culture, stationary human fibroblast cells retained approximately 10% of the original infectious dose of virion DNA. When the cells were passaged for G-418 selection and allowed to divide, these singled-stranded genomes were available to be converted into duplex DNA for expression of the transgene and the observed G-418 resistance.

Differences in susceptibility to rAAV transduction have also been recorded between primary and transformed cells. Halbert et al. (1995) compared primary human epithelial cells to the epithelial cell line, IB3, immortalized by an Ad/SV40 chimera. Even though both cultures were actively dividing, the transformed cells were transduced at a 60-fold higher frequency than the primary cells. This effect might have derived partially through the expression of the Ad E4 region in the transformed cell line. However, similar effects, approaching a 15-fold increase, were observed in cells transformed with SV40 large T antigen or the HPV E6, E7 genes. It must be noted that these effects did not manifest until the oncogentically transformed cells had been passaged more than five times. This finding demonstrated that it was not the oncogene that directly mediated the enhancement of transduction with rAAV vectors. Rather, it more likely can be attributed to the changes in endogenous gene expression, which are the hallmark of cellular transformation. In addition, with the identification of a primary attachment receptor for AAV, it is now possible to establish if the primary step of vector attachment was also contributing to these observations.

Efficient Transduction of Nondividing Cells

Although it is not yet clear as to why there are differences in transduction rates between primary and immortalized cells, the evidence for the occurrence of transduction in the absence of cell division is overwhelming. Recombinant AAV vectors have been used to transduce adult animal neural and muscle tissue that are clearly not dividing (McCown et al. 1996; Xiao et al. 1996). Furthermore, experiments designed to correlate rAAV transduction with any specific phase of the cell cycle have indicated that passage through the S phase is neither necessary for transduction nor sufficient to ensure that vector gene expression will take place in an rAAV-infected cell (F.K. Ferrari et al., in prep.). At present, rAAV has demonstrated efficient transduction in vivo in nondividing cells, but this feature is not common to all cell types. Therefore, it appears that several steps are required to obtain successful gene transduction in nondividing cells in vivo. This would include (1) the primary step of viral attachment through the AAV receptor, (2) the efficient conversion to a double-stranded template, and (3) the sufficient expression from the AAV vector whether episomal or integrated.

EPISOMAL EXPRESSION

As more attention is focused on the kinetics of AAV infection, it is becoming clear that the rate-limiting step for rAAV transduction in many cell types is conversion of the single-stranded virion DNA to duplex. The new double-stranded molecule then serves as either a substrate for gene expression or for integration into the host chromosome. Estimates of the integration efficiency of rAAV vectors vary from 0.4% to 80%, depending on such factors as the presence of the AAV Rep protein and the method of detecting transduction. In cases employing nonselectable markers, there was compelling evidence that a significant fraction, or even a preponderance, of the observed transgene expression was from episomal templates (Bachmann et al. 1979; Flotte et al. 1994; Goodman et al. 1994; Bertran et al. 1996). Further perspective on the nature of rAAV transduction may be gained from previous studies with wtAAV.

Dependent on the expression of the AAV Rep protein, the wtAAV displays a strong preference for integration into a specific region of chromosome 19 (Walsh et al. 1992; Kearns et al. 1996). The chromosome 19 integration site contains a sequence that is known to bind to Rep protein as well as a site where the single-strand endonucleolytic activity of Rep may occur (Kotin et al. 1992; Snyder et al. 1993; McCarty et al. 1994; Weitzman et al. 1994). Cleavage at this site might therefore be the initiating event for the integration of wtAAV. Whether or not this cleavage is the cause, the dependence of targeted integration on the Rep protein in *trans* suggests that, prior to integration, some virion DNA was converted to duplex. Based on these results, it is clear that an episomal wtAAV genome is capable of gene expression, and it is probable that this capacity is shared by the rAAV.

Several studies have evaluated the status of the AAV genome in infected cells. Flotte et al. (1994) sought rAAV-specific DNA in both the chromosomal and episomal fractions of cells transduced with a *lacZ* expressing vector. While Southern blotting readily revealed episomal copies of the genome, no rAAV-specific sequences were observed in association with the cellular genomic DNA. This result suggested the relative infrequency of integration for AAV molecules.

The episomal vector demonstrated electrophoretic mobility that is consistent with a linear molecule 8 kb in length, approximately two times the expected size of the double-stranded monomer vector genome. The authors proposed that this represented either a circular form of the vector DNA molecule or a duplex dimer. No clear explanation exists as to how such a molecule would be derived in the absence of the AAV *rep* gene products, because the duplex dimer form is observed during DNA replication of AAV. Synthesis of the complementary strand of single-stranded virion AAV begins at one of the hairpinned terminal repeat sequences (Straus et al. 1976; Hauswirth and Berns 1977, 1979; Lusby et al. 1980; Muzyczka 1992), and two different fates can ensue. Termination of DNA synthesis can occur at the 5′ end of the opposite TR, or this 5′ end may be displaced, an act that continues synthesis, extending and copying the TR. Either termination will result in a monomer duplex product that would not be expected to reinitiate DNA replication without the presence of the active Rep protein (Ni et al. 1994; Urcelay et al. 1995).

The episomal rAAV genome's anomalous mobility may also represent AAV DNA in a circular form, which could arise by annealing the complementary TR sequences (Koczot et al. 1973; Berns and Kelly 1974). A less likely proposal suggests that the concatemerzation of linear genomes occurs through recombination between the TRs of two or more AAV molecules. This possibility seems unlikely, because N. Muzyczka and coworkers have shown that a single AAV molecule is sufficient for transduction

(McLaughlin et al. 1988). As described previously, genotoxic stress can cause a single copy of the TR to promote the *rep*-independent replication of plasmids. This ability would provide a mechanism for amplifying the DNA containing these structures that could lead to the accumulation of multimeric or concatenated circular molecules (Yalkinoglu et al. 1991). The characterization of the detailed episomal structures observed in rAAV-transduced cells will have an impact on the use of these vectors in terms of their ability to stably transduce cells and, potentially, on the understanding of the mechanism of vector integration and persistence.

ADENO-ASSOCIATED VIRUS INTEGRATION

The relevant question for the application of AAV as a gene-therapy vector revolves around the frequency with which the episomes integrate into chromosomal DNA. AAV's potential as an integrating vector and an indication of its relative efficiency were first illustrated by Berns et al. (1975). Detroit 6 cells were infected with wtAAV at doses ranging from 0.0175 to 175 moi. The pools of cells were serially passaged and periodically assayed for the expression of AAV proteins following infection with Ad. Cells dosed at moi 175 were able to rescue AAV for at least 47 passages. Those infected at doses of moi 1.75 and 0.0175 lost their capability to rescue AAV after 34 and 8 passages, respectively. These results suggest either that cells harboring integrated AAV were diluted out of the culture or that episomal copies of AAV were lost by degradation or dilution. In either situation, it is apparent that, when infected at an moi of 1.75, only a small number of cells actually contained provirus.

The authors then further characterized cells that were infected at moi 175 at the clonal level. The cells were plated at limiting dilution to obtain single-cell clones at passage 39 postinfection. When each randomly selected clonal cell line was subsequently challenged with Ad, 18 of 63 (29%) were found to contain rescuable AAV.

Similar experiments were performed by Laughlin et al. (1986), who infected KB cells at moi 200. They were then able to demonstrate latent AAV in approximately 10% of the cloned cells from the infected culture. They went on to show that the AAV infection had no effect on cell viability upon replating. Earlier assays performed by Handa et al. (1997) found that approximately 2% of the KB cells carried latent AAV after infection at moi 500. These experiments revealed that wtAAV can establish latency in a significant proportion of a cell population, but the frequency was low and measurable only at high multiplicity.

RECOMBINANT AAV VECTOR INTEGRATION

Early studies of rAAV utilized vectors that retained the AAV *rep* gene. Unfortunately, these vectors consistently yielded low rates of integration. To illustrate this low yield, Hermonat and Muzyczka (1984) infected D6 cells with 0.1 infectious units per cell of rAAV expressing the *neo* gene. This vector yielded stable colonies that are resistant to G418, representing 0.04% of the cell culture, suggesting that 0.4% of the infecting genomes had integrated. This number of rAAV genomes was remarkably similar to that previously observed with wtAAV at a high multiplicity of infection. In contrast, however, the number of transduced cells rose to only 3% of the culture at moi 1000 when the rAAV vector was titrated to a higher multiplicity (10% if the G418 selection was delayed). Using a *rep*⁺ *neo* vector on 293 and HeLa cells, similar results were obtained by Tratschin et al. (1985). These studies suggested either that high-multiplicity infections with the *rep*⁺ vector or contaminating wtAAV interfered with transduction. Alternatively, only a subpopulation of the cells was susceptible to transduction. To elaborate on this idea, Tratschin et al. duplicated appropriate experiments using a vector preparation comprising 80% wtAAV and found similar transduction rates, arguing against interference from the wild-type virus. The apparent plateau in rAAV *neo* transduction has yet to be sufficiently explained and is unlikely to be related to limitation in the number of cells that can be infected with the virus, since 100% transduction levels have been observed using rAAV–*lacZ* vectors.

McLaughlin et al. (1988) were the first to constructed and characterized a *rep*⁻ rAAV vector containing only the TRs and a polyadenylation signal from the parent virus. The transduction frequencies were found to be far greater in the absence of the *rep* gene when this *neo*-expressing vector was compared to the previously characterized *rep*⁺ vector at low multiplicities of infection. Using the *rep*⁻ vector, approximately 80% of the infectious genomes yielded G418ᴿ colonies at an moi of 0.03. The *rep*⁺ vector again yielded a 0.4% transduction rate in parallel experiments at the same multiplicity of infection. High transduction rates involving greater than 50% of infectious genomes being integrated were also observed by Samulski et al. (1989) using *neo*-expressing rep-rAAV vectors. Further support for these ideas came from Russell and coworkers (1994), who demonstrated *neo* transduction rates in primary human fibroblast cultures of approximately 30% (deduced from the approximate equivalence of 600 physical particles to three infectious units).

It has been asserted that Rep-mediated repression of the heterologous promoters that transcribe the vector genes is responsible for the differences in transduction rates between *rep*⁺ and *rep*⁻. After generating rAAV vector stocks that were free of the contaminating wild-type virus, Samulski et al. (1989) then titrated wild-typeAAV against a constant dose of the recombinant. As the ratios of wild-type to vector approached 1:1, a slight increase in transduction rate resulted, and a decreased rate appeared at a ratio of 10:1. This observation revealed the complex role of *rep* in AAV integration, and, assimilating more recent observations on its effect on specific integration into human chromosome 19, may demonstrate the possibility of alternate integration pathways for *rep*⁺ and *rep* vectors.

In contrast to the preceding reports of efficient *neo* transduction, recent experiments in our laboratory have suggested significantly lower integration rates from rAAV vectors carrying the bacterial *lacZ* gene. Plates of HeLa cells were infected in parallel and evaluated for transient versus stable transduction, which, while transient transduction was high (20–30%), approximately 1% of the total transduced cell population contain integrated vectors (F.K. Ferrari and R.J. Samulski, in prep.). While the sensitivity of transgene detection may account for these differences, many other explanations are possible. In *neo* selection, utilizing G418 may mimic a stress response, facilitating viral integration as mentioned before. The issue of the rate of transient transduction with rAAV vectors lacking the *rep* gene versus the rate of integration deserves a great deal more attention, and should be a primary focus of research in the use of these vectors.

MECHANISM OF AAV INTEGRATION

Although a great deal of investigation and progress has been made, the fundamental mechanism of integration is not well understood for either the wtAAV or the recombinant virus. rAAV integration appears to rely solely on cellular factors, whereas the AAV Rep protein has been implicated as an active factor in wtAAV integration. The chromosome 19 site preference in the *rep*⁺ vector is the most obvious difference in the *rep*⁺ versus *rep*⁻ integration scenarios (Walsh et al. 1992; Kearns et al. 1996). Data from in vitro reconstituted systems demonstrate that the Rep protein can affect the tethering of AAV DNA to the chromosome 19 integration site (Weitzman et al. 1994). Additionally, cell-free in vitro replication systems suggest that Rep's capacity for enzymatic cleavage of a specific site in the chromosomal sequence, analogous to the cleavage site within the AAV terminal repeat, can initiate DNA replication (Urcelay et al. 1995). Analyses of large numbers of AAV integration products (Kotin and Berns 1989; Kotin et al. 1992, Giraud 1994), when taken in context with these observations, have lead to the recent proposals of models for AAV integration (Samulski 1993; Kotin 1996; Linden et al. 1996). Berns and colleagues postulate that a protein–DNA complex involving Rep-AAV catalyses a single-strand nick at the integration site. This nick then serves as primer for a DNA polymerase. Elongation ensues, and the polymerase complex is transferred from the chromosomal template to the viral template, creating the first covalent link between the two sequences. Downstream, a subsequent transfer would complete the integration. While the model outlined above is consistent with observations of viral integration products, it does not address the integration of rep^- minus rAAV vectors. Additionally, it fails to deal with an unusual feature of most proviruses, the appearance of a unique TR flanked (Xiao et al. 1997) at the junction of two tandemly repeated proviral genomes (Samulski 1993). Data suggesting that Rep is covalently associated with the AAV virion support the idea that the Rep–TR complex serves as a catalyst for integration (Prasad and Trempe 1995). Furthermore, several groups have demonstrated that transfection with an AAV plasmid in the presence of a Rep expression plasmid will promote integration of the rAAV molecule into the chromosome 19 target.

Although there are clear differences between wtAAV and rAAV vector integration, most notably chromosome 19 specificity, there are recognizable similarities (Yang et al. 1997). Similar to wtAAV junctions, most rAAV junctions contain deletions of variable length at the ends of the TR sequences. Additionally, microhomologies of two to five base pairs between chromosomal sequences and the viral TR are usually evident at the break points. These observations suggest that the same process mediates the two kinds of integration events. The role of Rep may be to regulate site specific integration and perhaps to inhibit nonspecific integration. This would suggest that rAAV vectors have all of the *cis* elements required for integration and that host factors can carry out this step, but that the efficiency of integration and targeting may be modulated by the Rep protein (Yang et al. 1997). It also should be noted that, although the Rep protein is covalently associated with a fraction of the packaged virions, extensively purified rAAV and wild-type virions have no detectable Rep protein (Yang et al. 1994).

TRANSDUCTION IN VIVO

Efficient Transduction of Nondividing Cells

The lesson from ex vivo transduction experiments with established or primary cells has been that dividing cells were more efficiently transduced than nondividing cells. In contrast, in vivo transduction with rAAV has shown that nondividing cells are the preferred target. There is now overwhelming evidence that rAAV can efficiently transduce nondividing, terminally differentiated cells with extremely high efficiency in vivo. AAV vectors have been used to transduce adult lung, eye, CNS, skeletal and smooth muscle, liver, and endothelial tissue. Furthermore, experiments designed to correlate rAAV transduction with any specific phase of the cell cycle have indicated that passage through the S phase is neither necessary for transduction nor sufficient to ensure that vector gene expression will take place in an rAAV-infected cell (F.K. Ferrari et al., in prep.). All nondividing cells are not efficiently transduced, and it appears that several steps are required to obtain successful gene transduction in nondividing cells in vivo. These include (1) the primary step of viral attachment through the AAV receptor (Summerford and Samulski 1998), (2) efficient conversion to a double-stranded template (Ferrari et al. 1996; Fisher et al. 1996b), and (3) sufficient expression from the AAV vector, whether episomal or integrated (Afione et al. 1996).

Transduction of the Central Nervous System

Kaplitt et al. (1994) performed one of the earlier experiments demonstrating the utility of rAAV vectors in vivo. Aimed at the transduction of brain tissue in rats, the investigators injected rAAV–CMV–*lacZ* vectors into various regions of adult rat brains and used histochemical staining to observe subsequent expression. From 1 to 3 days, the efficiency of gene transfer was approximately 10%, and gene expression was detected for up to 3 months postinfusion. The authors continued this study, injecting into the striatum rAAV vector coding for tyrosine hydroxylase (TH), a potentially therapeutic gene for Parkinson's disease. Using an immunohistochemical assay for vector expression, TH expression was shown to be primarily in neurons, but also in a few glial cells. However, the number of positive cells diminished over the 2 months following injection. Despite this decrease, the vector gene expression resulted in behavioral improvements in a rat model for Parkinson's disease.

These earlier experiments were done with relatively crude vector stocks that may have contained contaminating adenovirus as well as cellular components. Thus, it was not clear how well pure rAAV reagents would perform. The work described in the following was done with highly purified rAAV that was essentially free of wtAAV and the helper adenovirus. McCown et al. (1996) conducted a more systematic survey of rAAV transduction in the brain. Vectors expressing *lacZ* under the control of the CMV promoter were injected into either adult or neonatal rats, and the results illustrated differential transduction and persistence between different brain areas. For example, the hippocampus and cerebral cortex exhibited high initial levels of transgene expression that declined over time, while the inferior colliculus exhibited stable high levels of transgene expression for up to 1 year (Smith and McCown 1997). Other brain areas exhibited lower levels of transgene expression, where the expression declined, as in the caudate, or remained stable, as in the olfactory tubercle. Using the long-term stable gene expression possible in the inferior colliculus, R. Haberman et al. (in press) has recently showed that one can control gene expression from an rAAV vector. Expression from a tetracycline-regulated green fluorescent protein (GFP) gene lasted for 6 weeks in vivo (the duration of the experiment) with on/off regulation of the GFP by the addition of doxycycline in the drinking water (Fig. 2A). In all of the brain areas, neurons proved to be the predominant transduced cell type, while astrocytes were rarely transduced. It also appeared that only a subset of neurons exhibited transgene expression. Recent studies in the inferior colliculus and hippocampus indicate that viral binding and internalization can explain this differential cellular pattern of transduction (Bartlett et al. 1998). Also, it was proposed that inactivation of the CMV promoter was responsible for the decline in gene expression. As important, however, was that McCown et al. found no indication that an immune response was eliminating the transduced cells and no evidence of neurotoxicity associated with vector administration under any condition.

Peel et al. (1997) and Klein et al. (1998) directly addressed the issue of gene product decline over time. Reasoning that expression might be shut off due to inactivation of the CMV promoter used in earlier studies, Peel et al. (1997) investigated the use of the neural specific enolase (NSE) and platelet-derived growth factor β (PDGFβ) promoters driving a humanized GFP gene (Fig. 2). Both promoters had been shown to be specific for neuronal cells, and vectors containing these promoters efficiently transduced spinal cord neurons in adult rats. Maximal expression required up to 2 weeks but was subsequently stable up to 15 weeks. Other experiments have demonstrated that expression from the PDGF promoter is stable for up to 9 months (A.L. Peel, in prep.); similar results were obtained by Klein et al. (1998). This group compared CMV-driven GFP to vectors using the NSE promoter, by injecting a vector into adult rat hippocampus or substantia nigra. Both regions of the brain were efficiently transduced by the NSE vector. Injections of approximately 6×10^5 infectious units produced 3000–15,000 GFP-positive cells in septum and substantia nigra, respec-

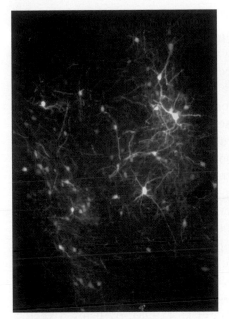

Figure 2. (*Left*) An AAV dual cassette vector carrying GFP and the tet *trans*-activator genes under the tet responsive promoter were assayed by GFP fluorescence for vector transduction in the inferior colliculus 6 weeks postinjection of 6 x 10^5 infectious units; magnification 100x. (*Right*) Transduction of neurons in rat hippocampus 3 weeks after injecting rAAV. Sections were processed with GFP antibody. One ml of NSE promoter-driven GFP rAAV (pTR-NSE; 6 x 10^5 iu) produced transgene expression in CA1 and dentate gyrus. Bar, 200 mm.

tively, with no apparent decrease in expression after 3 months. In contrast, transduction with the CMV vector was less efficient, and the number of transduced cells decreased significantly with time. Although the NSE promoter is believed to be active in most adult neurons, Klein et al. (1998) saw pronounced selectivity in the hippocampal cells that were transduced. Hippocampal expression occurred predominantly in multipolar neurons, while neurons with glutamatergic morphology were poorly transduced. This suggested that some other barrier to transduction, perhaps viral entry or second strand synthesis, existed in some neuronal cell types.

In an effort to quantitate the in vivo transduction efficiency, Pcel et al. (1997) defined a new measure, the transduction ratio. Because the in vivo application of a vector usually involves injection of the virus into solid tissue at a unique position, the cells immediately adjacent to the site of injection receive significantly higher doses than the distal cells. Thus, it is virtually impossible to calculate how many cells were exposed to virus and to calculate a conventional transduction efficiency. With respect to this observation, Peel et al. simply divided the number of input infectious units (iu) by the number of infected cells to determine the transduction ratio. Approximately 15 iu were needed to produce one transduced spinal-cord neuron with the NSE promoter. Klein et al. (1998) achieved similar transduction ratios, 5–20 in septum or substania nigra. Given the fact that a substantial amount of virus is essentially wasted on the cells immediately surrounding the injection point, this suggests that rAAV vectors are extremely efficient for gene transfer in the CNS. Furthermore, the total number of cells transduced, up to 15,000, should be sufficient to achieve physiological effects with certain proteins that may be therapeutic. To date, systematic dose-response measurements have not been reported, and so it is quite possible that significantly higher numbers of cells can be transduced from a single injection. Additionally, rAAV may be a useful basic research tool for generating virus-targeted transgenics. Two studies have demonstrated this potential: Xiao et al. (1997), by using an antisense strategy, and Klein et al. (1998), by using a vector-derived protein function in a specific subset of neurons. This new technology should advance the study of current research problems related to the brain and development.

Several groups are currently exploring the possibility of rAAV mediated therapy. Xiao et al. (1997) demonstrated that delivery of GABA receptor antisense constructs to the inferior colliculus increased seizure duration by 137%. Klein et al. (1998) demonstrated the delivery of BDNF, and more recently, NGF and the Trk receptor (R. Klein, in prep.) to the hippocampus. All of these genes are potentially useful for Alzheimer's therapy. Mandel et al. (1997) delivered the GDNF gene to rat substantia nigra and demonstrated the protection of nigral neurons in the rat 6-hydroxydopamine model of Parkinson's. More recently, Mandel et al. (1998) transduced denervated rat striatum with vectors carrying the tyrosine hydroxylase gene and the GTP-cycohydrolase I gene, two key enzymes that are necessary for dopamine synthesis. This group was able to demonstrate the synthesis of substantial amounts of L-DOPA following instillation of the vectors and showed expression of the transgenes in striatum for up to one year. A double-blinded study in nonhuman primates has supported these results from the rat model. Histochemical, enzymatic, and behavioral changes were confirmed in monkeys that received rAAV carrying the tyrosine hydroxylase gene when compared to rAAV–*lacZ* controls (During et al. 1998). Finally, Phillips and his colleagues (1997) demonstrated the use of rAAV carrying antisense to the angiotensin receptor for reducing hypertension in the naturally hypertensive rat model.

Transduction of the Eye and Other Sensory Organs.

Gene transfer to the retina provides a dramatic illustration of how a judicious choice of promoter can mean the difference between robust expression and no transduction. Most attempts at retinal gene therapy, thus far, have involved subretinal injections, in which the virus is injected between the photoreceptor and the pigmented epithelial (PE) cell layers. The small bolus of virus (1–3 µl) causes a retinal detachment at the site of injection, which reattaches within 2 hours. Early attempts to use AAV for retinal gene therapy used the CMV promoter to drive expression of either the GFP or β-galactosidase genes. Transduction was readily observed in PE cells, but rarely in photoreceptor cells (Zolotukhin et al. 1996). Hauswirth and his colleagues solved the problem by switching to a truncated mouse opsin promoter, a highly regulated promoter expressed exclusively in photoreceptor cells (Flannery et al. 1997). Approximately 20% of the photoreceptor cells in the retina were transduced in a single injection of the rat or guinea pig retina, and virtually 100% of the photoreceptor cells were transduced at the site of injection. In the original report, expression was stable for 2 months; this has now been extended to 15 months (W.W. Hauswirth, pers. comm.). The transduction ratio was approximately 3–4 rAAV iu per transduced photoreceptor cell. The virus used in these studies was essentially free of wtAAV and adenovirus. Thus, regardless of the observations in cell culture with primary cells, no adenovirus coinfection is needed for the efficient transduction of retinal tissue or neuronal tissue. It is worth noting, however, that like the CNS injections discussed above, the onset of expression in the retina was slow, taking at least 3 weeks to achieve maximum transduction. The slow onset of expression, therefore, may reflect an equally slow second-strand synthesis of input genomes. Definitive experiments to characterize the rAAV genome within retinal or neuronal cells have not yet been reported. Specifically, no information is available yet about the length of time that the rAAV genome remains single-stranded, when it is converted to duplex linears, and if it integrates in host chromosomes. Finally, even though the pigmented epithelial cell layer was also exposed to the virus, no expression was seen in the PE cells. Thus, the opsin promoter was capable of targeting expression to a single cell-type, photoreceptors.

The ability to deliver genes to the photoreceptor or pigmented epithelial cell layers of the retina opens a number of therapeutic opportunities. Retinal degeneration, either due to an inherited trait or associated with another disease (e.g., diabetic retinopathy), is a major problem, and the level of retinal transduction seen with the AAV is theoretically sufficient to arrest or slow retinal degeneration. Two groups have reported initial attempts to deliver therapeutic transgenes. Jomary et al. (1997) used rAAV to transfer the cGMP phosphodiesterase (PDE) gene to the *rd/rd* mouse, an experimental model of recessive retinopathy due to the absence of the PDE gene. Compared to age-matched controls, treated animals showed increased numbers of photoreceptors and a twofold increase in sensitivity to light as measured by in vitro electroretinography. Lewin and his colleagues (Drenser et al. 1998; A. Lewin et al., in prep.) focused on the more difficult problem of treating autosomal dominant retinitis pigmentosa (ADRP) caused by the P23H mutant in rhodopsin. Their strategy was to design a ribozyme that preferentially cleaved the mutant P23H containing mRNA, thereby increasing the intracellular concentration of the wild-type rhodopsin gene product (Drenser et al. 1998). Delivery of this ribozyme to the transgenic rat model of ADRP markedly slowed the rate of photoreceptor degeneration for at least 3 months, as judged from photoreceptor cell survival and electroretinography (A. Lewin et al., in prep.). This may be the first example in which a dominant negative defect has been treated successfully with a ribozyme. Finally, Lalwani et al. (1996, 1998) successfully used rAAV to transduce cochlear cells of the guinea pig ear.

Transduction of Skeletal Muscle

Stable transduction of murine muscle tissue with an rAAV vector was reported initially by two groups (Kessler et al. 1996; Xiao et al. 1996). Xiao et al. (1996) reported delivery of the vector by local injection to hind leg muscles in both 3-week-old and adult immunocompetent mice. The mice were killed at various times postinjection, and tissue was sectioned and stained for expression of the AAV–LacZ marker. The average number of positive staining muscle fibers was substantially greater than in mice treated with an Ad–LacZ vector and did not change in mice examined from 4 days postinjection to 19 months. Nearly identical results were obtained in adult mice, suggesting that the majority of transduction events (detected on day 4) went on to form stable episomes or integrants. This was the first example of AAV-persistent long-term expression without significant immuno response to the vector-transduced cells. In this report, a neutralizing antibody to the varion shell was observed.

Some analysis of the fate of the rAAV vector DNA was done by Southern blotting of total cellular DNA from the transduced muscle tissue at various time points. Uncut DNA yielded a smear of LacZ hybridization signal of high molecular weight (greater than 12 kb). Digestion with a single-cut restriction enzyme generated several products, including a monomeric vector length fragment, a diffuse background hybridization from approximately 3 to 12 kb, and a fragment consistent with the product of a tail-to-tail concatemer of the vector. The monomeric digestion product suggested that the proviral DNA consisted primarily of a tandem array of head-to-tail vector sequences. Despite previous observations in lung epithelial cells by Flotte et al. (1994), no evidence of dimer-sized molecules was found. The vector-specific sequences were linked to high-molecular-weight DNA, which may have been large episomes consisting of tandem arrays or concatemeric circles or, simply, chromosomal. The persistence of a background of hybridization after digestion with the single-cut enzyme is consistent with some or all of the vector-specific sequences being linked to chromosomal DNA.

Kessler et al. (1996) also transduced skeletal muscle with a CMV-driven LacZ vector. Once again, vector expression was present over many months, with minimal or no inflammatory responses, as seen by the study of Xiao et al. (1996). This study also examined the expression of a secreted protein, human erythropoietin (h-Epo), from an AAV vector in skeletal muscle. The results of these studies indicate that levels of h-Epo in the serum were sufficient to result in a dramatic increase in hematocrit that was dose-dependent and stable for over 10 months. Full expression of the transgene, however, took at least 4 weeks.

Fisher et al. (1997) provided supporting evidence that the rAAV genome adopts a head-to-tail concatemeric structure following injection into muscle. Additionally, they saw no cellular or humoral immune response to the β-galactosidase transgene used in their study, which supports the conclusions described by Xiao et al. (1996). Snyder et al. (1998) confirmed by Southern analysis that rAAV genomes injected into muscle become associated with high-molecular-weight DNA. They also addressed the question of whether dividing or quiescent muscle cells are more efficiently transduced by the rAAV. By using $BaCl_2$ to induce muscle degeneration and regrowth, they demonstrated that dividing myoblasts are transiently transduced, whereas resting myoblasts and myotubes are permanently transduced. Manning et al. (1998) investigated the potential for administering rAAV to muscle multiple times. Like other vectors, readministration of the rAAV was limited because of the host's immune response to viral components of the vector. These observations were expected from previous studies (Afione et al. 1996; Xiao et al. 1996). Transient immunosuppression of mice by treatment with antibody to CD4 at the time of the primary infection allowed transgene expression after readministration, while treatment with the CD40L antibody had only a modest effect. The ability to readminister rAAV was inversely correlated with AAV neutralizing antibodies.

The ease of intramuscular injections has prompted a number of studies designed to test the possibility of using skeletal muscle as a reservoir for genes whose protein products can be secreted. Herzog et al. (1997) demonstrated that Factor IX could be secreted to achieve therapeutic plasma levels following a single injection. T. Daley et al. (in prep.) have shown secretion of β-glucuronidase following intramuscular injection. Perhaps the most impressive demonstration of the power of this approach was provided by Murphy et al. (1997). This group transduced muscle in the *ob/ob* mouse, which is genetically deficient in leptin and exhibits a phenotype that includes obesity and noninsulin-dependent diabetes mellitus. A single injection of vector prevented obesity and diabetes. Over the 6-month course of the study, the treated animals showed normalization of hyperglycemia, insulin resistance, impaired glucose tolerance, and lethargy. Studies with rAAV in the heart have succeeded with long-term gene expression in a large animal model (porcine) (Kaplitt et al. 1996), suggesting that this vector will transduce the muscles of many different species. In addition, Monahan et al. (1998) were able to demonstrate in a large animal model (canine hemophilia B) direct muscle injection of rAAV vectors with persistent expression, supporting earlier studies using this vector in rodents.

Lung Applications

The first in vivo application of an AAV vector was carried out by Flotte and Carter and their colleagues. Over the past 8 years, this group has been involved in the development of AAV vectors for gene therapy of lung diseases, focusing initially on cystic fibrosis (CF). In their earlier work, they demonstrated that an AAV–CFTR vector in which the ITR alone was present without a promoter produced significant levels of CFTR chloride channel expression (as judged from patch-clamping and isotope tracer efflux assays) (Egan et al. 1992; Flotte et al. 1993). Subsequent CAT reporter gene analysis showed that, in certain cell types, the ITR was able to express at levels approximating those seen with the p5 promoter, while in others it was totally inactive. A careful review of the "d" sequence of the ITR revealed the presence of a consensus initiator (*in*r) site, which, along with a series of SP1 sites present in the ITR, would be predicted to have some promoter activity (Flotte et al. 1993).

These AAV–CFTR vectors were used to evaluate the potential for in vivo gene transfer with AAV (Flotte et al. 1993). Rabbits were treated with 10^{10} total particles (about 10^8 iu) of AAV–CFTR delivered by fiberoptic bronchoscopy to the right lower lobe and examined at time points ranging from 3 to 180 days after vector instillation for vector DNA transfer and RNA and protein expression. Vector expression peaked at 10 days after instillation, but was still present at the 180-day time point. Importantly, there was no evidence of inflammation or any other vector-related toxicity.

It was initially hypothesized that vector integration was responsible for this persistence. However, subsequent studies have been performed in rhesus macaques that tend to indicate that persistence in the bronchial epithelium may have been due to episomal persistence (Afione et al. 1996). A combination of fractionated Southern blots and fluorescence in situ hybridization (FISH) was used to determine the state of the vector DNA, and these studies indicated that the vector DNA was present as a stable DNA episome for up to 6 months after vector delivery. While vector DNA was rescuable from the bronchial epithelial cells in culture, mobilization of the vector DNA in vivo occurred only when a very large inoculum of wtAAV and Ad were administered to the same site as the vector. It is important to note that vector RNA expression was also detectable by RT-PCR and by RNase protection assays (RPA) for 6 months after instillation of the vector, although the levels of RNA expression were low (in the range of 1 copy per cell) (Conrad et al. 1996). Once again, there was no indication of vector-related toxicity as judged from histopathology, chest X-rays, pulmonary function testing, or analysis of bronchoalveolar lavage fluid for inflammatory cells and proinflammatory cytokines.

On the basis of these data, two phase I clinical trials of AAV–CFTR administration were initiated in CF patients. The first is a phase I trial of AAV–CFTR administration to the nasal and bronchial epithelium that was begun at Johns Hopkins University and has since been extended to the University of Florida (Flotte et al. 1996). The second is a trial of maxillary sinus instillation, which has recently been completed at Stanford University. Preliminary results from both trials suggest that stable gene transfer has been observed without any evidence of vector-related toxicity (T. Flotte, pers. comm.).

Transduction of Bone Marrow, Liver, and Vascular Endothelium

It is still not clear whether rAAV vectors will be useful for the long-term transduction of hematopoietic stem cells. Transduction of stem cells differs from the applications described above in two respects: They involve ex vivo gene therapy of the primary cells in culture, and the target cells are a dividing cell population. Several groups have shown that the transduction of established cell lines of hematopoietic origin is possible (Goodman et al. 1994; Miller et al. 1994; Einerhand et al. 1995; Zhou et al. 1996). These groups have also shown that erythroid-specific promoter elements, such as the globin locus control regions, will function correctly in the context of AAV. Another pattern that has emerged is that, in contrast to nondividing tissues, hematopoietic progenitors transduced with rAAV tend to express the transgene within the first 24 to 48 hours after infection. However, the number of transduced cells decreases with time in culture (Bertran et al. 1996; Malik et al. 1997; Qing et al. 1997). This result suggests that second-strand synthesis in this cell type is much more rapid than in targets such as muscle and brain, leading to rapid expression of the transgene, and that episomal rAAV copies are then progressively lost during cell division. In fact, the loss of transgene expression has generally been correlated with the loss of vector DNA. Thus, whether rAAV can achieve long-term transduction in hematopoietic precursors appears to be a function of whether the recombinant genome integrates before it is lost. Indeed, it has been suggested that the AAV might be more useful in combination with ecotropic retrovirus vectors. The idea would be to use the AAV to deliver an ecotropic retrovirus receptor transiently, so as to increase retrovirus transduction efficiency, then use the retrovirus to achieve long-term transduction with a therapeutic gene.

Nevertheless, several groups have reported reasonably high transduction efficiencies with rAAV as judged from in vitro clonogenic assays (Zhou et al. 1993, 1994; Fisher-Adams et al. 1996). For example, Zhou et al. (1993) observed 5–40% CFU–GM colony transduction and as high as 80% BFU-E colony transduction when murine bone marrow cells were targeted. Fisher-Adams et al. (1996) made similar observations using CD34[+]-enriched human hematopoietic cells. This group also demonstrated by FISH analysis that at least some of the rAAV genomes had integrated into the host chromosomes, but the frequency of integration was not clear. Finally, two groups have tried to reconstitute murine or primate hematopoietic systems with AAV ex vivo modified stem cells. Ponnazhagan et al. (1997a) reported experiments in which ex vivo modified stem cells were transplanted into lethally irradiated congenic mice. The transgene was detectable for 6 months in primary recipients and for 3 months in some of the secondary recipients. Schimmenti et al. (in prep.) reported the transfer of vector-modified stem cells to primates and showed that 15 months after transplantation, approximately 1 in 10^5 hematopoietic cells carried the transgene. The vector was detected in cells in the bone marrow, granulocytes, and purified populations of B and T cells, suggesting multilineage repopulating by vector-transduced cells. Thus, it would appear that rAAV has the potential for ex vivo bone marrow transduction, but the key question is whether the efficiency will be much better than that seen with previously studied retroviral vectors.

As yet, relatively little information is available about liver cell targets. Ponnazhagan et al. (1997b) reported a detectable transduction of liver hepatocytes following the intravenous injection of a vector. Koeberl et al. (1997) saw significant expression of Factor IX in hepatocytes following intravenous infection, but argued that both wild-type AAV, a contaminant in their stocks, and gamma irradiation were needed to obtain significant transduction. In contrast, Snyder et al. (1997) reported that therapeutic levels of Factor IX were produced in mice following liver transduction, and this occurred in the absence of wild-type AAV, the adenovirus, or any other kind of adjuvant.

Endothelial cells also appear to be capable of rAAV transduction. Bahou and his colleagues (Gnatenko et al. 1997) have done in vitro and in vivo experiments that suggest that vascular endothelial cells are efficiently transduced.

ENHANCEMENT OF rAAV TRANSDUCTION IN VIVO

As previously discussed, the ability of the infecting rAAV genome to generate its complementary strand appears to be an important factor in the efficiency of transduction in vitro. Ostensibly, a similar mechanism with similar rate limitations should exist in vivo. Investigators have used the principles learned through enhancement of rAAV transduction in vitro and applied them to animal models. Using γ-radiation treatment to overcome the transduction barrier, Alexander et al. (1996) injected an alkaline phosphatase (AP) expressing vector into the brains of rats that had received localized cranial irradiation; they simultaneously injected mock-treated animals. The hippocampus, the ventricular system, and the scalp intradermis were the three sites chosen for injections.

Following injection into the lateral ventricle, the γ-irradiation substantially augmented cellular transduction. Irradiated animals displayed numerous transduced cells in the pia-arachnoid and choroid epithelium, while nonirradiated animals had undetectable levels of AP expression. Interestingly, the pattern of transduced cells appeared to follow the path of cerebral spinal fluid flow through the side of the brain that was injected. These results were different than the conclusions of McCown et al. (1996), who observed little transduction of the ventricular epithelium beyond the local injection site. However, this discrepancy may be explained by an increase in sensitivity gained by the radiation-mediated enhancement, or simply to differences in injection procedures. To discern the duration of expression, the vector stability was evaluated at 3 months postinjection. In most animals, it was found that the increase in the initial transduction did not necessarily lead to increases in the duration of transgene expression, based on barely detectable levels of AP activity. This result corroborates evidence from other laboratories indicating that UV-mediated enhancement of rAAV transduction does not lead to linear increases in stable transduction in HeLa cells (F.K. Ferrari and R.J. Samulski, in prep.). Similarly, the γ-irradiation treatment also substantially increased the number of transduced cells following intradermal scalp injection. However, transduction was confined to the striated muscle cells at this site, and the duration of the transgene expression was not determined in this tissue.

As had been noted previously, AP expression was readily detectable 12 days postinjection in the hippocampus (Alexander et al. 1996). Surprisingly, and in contrast to the other brain tissues studied, the γ-irradiation treatment did not enhance the rate of transduction in this tissue. By 6 weeks postinjection, the level of expression had dropped tenfold in some animals and had become undetectable in others. No differences were observed in the duration of expression between the irradiated and nonirradiated animals. Again, it was not ascertained whether the loss of transgene expression was due to the instability of the vector or a reduction in promoter activity. It was also demonstrated that an rAAV vector expressing a secreted form of AP could transduce the epithelial cells of the ventricular system and that this gene product could be detected in the cerebral spinal fluid. While promising, it should be noted that the doses of radiation used in these experiments would preclude use in clinical applications (Alexander et al. 1996). Furthermore, other results implying the efficient transduction of neuronal tissue question the need for radiation and perhaps raise the issue of whether contaminants in the rAAV stock (such as wtAAV) decrease transduction in these experiments (Koeberl et al. 1997).

The previously discussed report from Fisher et al. (1996b), characterizing the effect of Ad E4 on rAAV transduction, also included two experiments in vivo. In the first experiment following infection with adenovirus, liver tissue in mice were transduced with 10^{11} particles of rAAV–LacZ. The treatment with Ad before rAAV transduction, and at a multiplicity sufficient to infect approximately 25% of the liver hepatocytes, allowed rAAV-mediated transduction in 10–25% of the liver. With mutants lacking efficient E4 expression or without Ad coinfection, the transduction rate was less than 0.01%. This finding suggested that the hepatocytes had initially taken up the rAAV vector and needed only a secondary event to trigger expression from the virion DNA. This idea was substantiated further by showing that a rAd vector expressing alkaline phosphatase could be used to induce a LacZ expression from the rAAV. Additionally, the two markers colocalized in most transduced hepatocytes. Instillation of a mixture of rAAV and Ad in mouse lungs generated similar results. However, it is worth noting again that these reports have been contradicted by other groups that have seen the efficient transduction of both liver and lung using rAAV free of wtAAV and Ad (Snyder et al. 1997; T. Fiotte and B. Carter [presented at the Cystic Fibrosis Foundation Williamsburg Gene Therapy Conference, June 1997]). These observations raise the possibility that the interpretations of earlier results are confounded by contaminants in the vector preparation. Nevertheless, treatments such as UV and γ-irradiation may become useful adjuvants for AAV transduction in some cell types. The use of Ad-free rAAV preps should help resolve these issues.

CONCLUSIONS

We are still in the early stages of understanding and fully exploiting the potential of rAAV gene delivery vectors. Conflicting information still exists as to the transduction and integration efficiencies of these vectors in various cell types. This uncertainty spans a range between less than 1% and greater than 80%, depending, in part, on the presence of the AAV *rep* gene and the method of detection. This frustrating situation largely reflects our limited understanding of the AAV integration mechanism, with or without the presence of the Rep protein. Therefore, two of the most pressing problems in the use of rAAV vectors will be to better define the mechanism of integration and to understand the nature and limitations of persistent expression in the absence of integration.

Increasing the efficiency of transduction using various enhancing agents, many of which may have genotoxic activity, is a challenge related to these questions. Although inherent risks in administering these treatments exist, the potential gain is great. In fact, the use of some of these agents has been warranted and amply justified in therapies for cancer. A major undertaking involves adapting these effective in vitro treatments to acceptable treatments that can be applied to clinical situations.

Nevertheless, many of the reports discussed in this chapter begin to reveal the value of rAAV vectors. It is now clear that efficient and long-term transduction can be achieved with rAAV in several tissue targets including muscle, CNS, eye, and lung. We probably have not yet achieved the greatest potential that rAAV has to offer toward transduction efficiency, yet these results hold great promise for long-term gene delivery using rAAV in vivo. AAV, in fact, is the first vector that has been shown to be capable of fulfilling the goal of long-term gene therapy. The obstacle to be overcome now is learning to use it to its maximum potential.

ACKNOWLEDGMENTS

This work was supported by grants from the National Institutes of Health (NHLB1 51818) and the Muscular Dystrophy Association.

REFERENCES

Afione S.A., Conrad C.K., Kearns W.G., Chunduru S., Adams R., Reynolds T.C., Guggino W.B., Cutting G.R., Carter B.J., and Flotte T.R. 1996. In vivo model of adeno-associated virus vector persitence and rescue. *J. Virol.* **70:** 3235–3241.

Alexander I.E., Russell D.W., Spence A M., and Miller A.D. 1996. Effects of gamma irradiation on the transduction of dividing and nondividing cells in brain and muscle of rats by adeno-associated virus vectors. *Hum. Gene Ther.* **7:** 841–850.

Bachmann P.A., Hoggan M.D., Kurstak E., Melnick J.L., Pereira H.G., Tattersall P., and Vago C. 1979. Parvoviridae: Second report. *Intervirology* **11:** 248–254.

Bartlett J.S., Samulski R.J., and McCown T.J. 1998. Selective and rapid uptake of adeno-associated virus Type 2 (AAV-2) in brain. *Hum. Gene Ther.* **9:** 1181–1186.

Beaton A., Palumbo P., and Berns K.I. 1989. Expression from the adeno-associated virus p5 and pl9 promoters is negatively regulated in *trans* by the Rep protein. *J. Virol.* **63:** 4450–4454.

Berns K.I. and Kelly T.J.J. 1974. Visualization of the inverted terminal repetition in adeno-associated virus DNA. *J. Mol. Biol.* **82:** 267–271.

Berns K.I., Pinkerton T.C., Thomas G.F., and Hoggan M.D. 1975. Detection of adeno-associated virus (AAV)-specific nucleotide sequences in DNA isolated from latently infected Detroit 6 cells. *Virology* **68:** 556–560.

Bertran J., Miller J.L., Yang Y., Fenimore-Justman A., Rudea F., Vanin E.F., and Nienhuis A.W. 1996. Recombinant adeno-associated virus-mediated high-efficiency, transient expression of the murine cationic amino acid transporter (ecotropic retroviral receptor) permits stable transduction of human HeLa cells by ecotropic retroviral vectors. *J.*

Virol. **70:** 6759–6766.

Bridge E., Medghalchi S., Ubol S., Leesong M., and Ketner G. 1993. Adenovirus early region 4 and viral DNA synthesis. *Virology* **193:** 794–801.

Charache S., Terrin M.L., Moore R.D., Dover G.J., Barton F.B., Eckert S.V., McMahon R.P., and Bonds D.R. 1995. Effect of hydroxyurea on the frequency of painful sickle cell anemia. *New Engl. J. Med.* **332:** 1317–1322.

Clark K.R., Voulgaropoulou F., Fraley D.M., and Johnson P. R. 1995. Cell lines for the production of recombinant adeno-associated virus. *Hum. Gene Ther.* **6:** 1329–1341.

Conrad C.K., Allen S.S., Afiore S.A., Reynolds T.C., Beck S.E., Fee Maki M., Barrazza Ortiz X., Adams R., Askin F. B., Carter B.J., Guggino W.B., and Flotte T.R. 1996. Safety of single-dose administration of an adeno-associated virus (AAV)-CFTR vector in the primate lung. *Gene Ther.* **3:** 658–668.

de la Maza L.M. and Carter B.J. 1980. Heavy and light particles of adeno-associated virus. *J. Virol.* **33:** 1129–1137.

Dobner T., Horikoshi N., Rubenwolf S., and Shenk T. 1996. Blockage by adenovirus E4orf6 of transcriptional activation by the p53 tumor supressor. *Science* **272:** 1470–1473.

Drenser K., Hauswirth W., and Lewin A. 1998. Ribozyme-targeted destruction of RNAs associated with a autosoma I-dominant retinitis pigmentosa. *Investig. Opthamol. Vis. Science* **39:** (in press).

During M., Samulski R.J., Elsworth J.D., Kaplitt M.G., Leone P., Xiao X., Li J., Freese A., Taylor J.R., Roth R.H., Sladek J.R., O'Malley K.L., and Redmond D.E., Jr. 1998. In vivo expression of therapuetic genes for dopamine production in caudates of MPTP-treated monkeys using an AAV vector. *Gene Ther.* **5:** 820–827).

Egan M., Flotte T., Afione S., Solow R., Zeitlin P.L., Carter B.J., and Guggino W.B. 1992. Defective regulation of outwardly rectifying Cl-channels by protein kinase A corrected by insertion of CFTR (see comments). *Nature* **358:** 581–584.

Einerhand M.P.W., Antoniou M., Zolotukhin S., Muzyczka N., Berns K.I., Grosveld F., and Valerio D. 1995. Regulated high-level human β-globin gene expression in erythroid cells following recombinant adeno-associated virus-mediated gene transfer. *Gene Ther.* **2:** 336–343.

Ferrari F.K., Samulski T., Shenk T., and Samulski R.J. 1996. Second strand synthesis is a rate limiting step for efficient transduction by rAAV vectors. *J. Virol.* **70:** 3227–3234.

Ferrari F.K., Xiao X., McCarty D., and Samulski R.J. 1997. New developments in the generation of Ad-free high titer rAAV gene therapy vectors. *Nat. Med.* **3:** 1295–1297.

Fisher K., Hester C., Burda J., Chen S.-J., and Wilson J. 1996a. Recombinant adenovirus deleted of all viral genes for gene therapy of cystic fibrosis. *Virology* **217:** 11–22.

Fisher K.J., Gao G.P., Weitzman M.D., DeMatteo R., Burda J.F., and Wilson J.M. 1996b. Transduction with recombinant adeno-associated virus for gene therapy is limited by leading-strand synthesis. *J. Virol.* **70:** 520–532.

Fisher K.J., Jooss K., Alston J., Yang Y., Haecker S.E., High K., Pathak R., Raper S.E., and Wilson J.M. 1997. Recombinant adeno-associated virus for muscle directed gene therapy. *Nat. Med.* **3:** 306–312.

Fisher-Adams G., Wong K.K., Jr., Podaskoff, G., Forman S.J., and Chatterjee S. 1996. Integration of adeno-associated virus vectors in CD34+ human hematopoietic progenitor cells after transduction. *Blood* **88:** 492–504.

Flannery J.G., Zolotukhin M., Vaquero M.I., Lavail M.M., Muzyczka N., and Hauswirth W.W. 1997. Efficient photoreceptor-targeted gene expression in vivo by recombinant adeno-associated virus. *Proc. Natl. Acad. Sci.* **94:** 6916–6921.

Flotte T.R., Afione S.A., and Zeitlin P.L. 1994. Adeno-associated virus vector gene expression occurs in nondividing cells in the absence of vector DNA integration. *Am. J. Respir. Cell Mol. Biol.* **11:** 517–521.

Flotte T.R., Barraza-Ortiz X., Solow R., Afione S.A., Carter B.J., and Guggino W.B. 1995. An improved system for packaging recombinant adeno-associated virus vectors capable of in vivo transduction. *Gene Ther.* **2:** 29–37.

Flotte T.R., Afione S.A., Solow R., Drumm M.L., Markakis D., Guggino W.B., Zeitlin P.L., and Carter B.J. 1993. Expression of the cystic fibrosis transmembrane conductance regulator from a novel adeno-associated virus promoter. *J. Bio. Chem.* **268:** 3781–3790.

Flotte T., Carter B., Conrad C., Guggino W., Reynolds T., Rosenstein B., Taylor G., Walden S., and Wetzel R. 1996. A phase I study of an adeno-associated virus-CFTR gene vector in adult CF patients with mild lung disease. *Hum. Gene Ther.* **7:** 1145–1159.

Giraud C., Winoccour E., and Berns K.I. 1994. Site-specific integration by adeno-associated virus is directed by a cellular DNA sequence. *Proc. Natl. Acad. Sci.* **91:** 10039–10043.

Gnatenko D., Arnold T.E., Zolotukhin S., Nuovo G.J., Muzyczka N., and Bahou W.F. 1997. Characterization of recombinant adeno-associated virus-2 as a vehicle for gene delivery and expression into vascular cells. *J. Investig. Med.* **45:** 87–98.

Goodman S., Xiao X., Donahue R.E., Moulton A., Miller J., Walsh C., Young N.S., Samulski R.J., and Nienhuis A.W. 1994. Recombinant adeno-associated virus-mediated gene transfer into hematopoietic progenitor cells. *Blood* **84:** 1492–1500.

Haberman R.P., McCown, T.J., and Samulski R.J. 1998. Inducible long-term gene expression in brain with adeno-associated virus gene transfer. *Gene Ther.* (in press).

Halbert C.L., Alexander I.E., Wolgamot G.M., and Miller A.D. 1995. Adeno-associated virus vectors transduce primary cells much less efficiently than immortalized cells. *J. Virol.* **69:** 1473–1479.

Handa H., Shiroki K., and Shimojo H. 1977. Establishment and characterization of KB cell lines latently infected with adeno-associated virus type 1. *Virology* **82:** 84–92.

Hauswirth W.W. and Berns K.I. 1977. Origin and termination of adeno-associated virus DNA replication. *Virology* **78:** 488–499.

———.1979. Adeno-associated virus DNA replication. *Virology* **93:** 57–68.

Hermonat P.L. and Muzyczka N. 1984. Use of adeno-associated virus as a mammalian DNA cloning vector: Transduction of neomycin resistance into mammalian tissue culture cells. *Proc. Natl. Acad. Sci.* **81:** 6466–6470.

Herzog R.W., Hagstrom J.N., Kung S.H., Tai S.J., Wilson J.M., Fisher K.J., and High K.A. 1997. Stable gene transfer and expression of human blood coagulation factor IX after intramuscular injection of recombinant adeno-associated virus. *Proc. Natl. Acad. Sci.* **94:** 5804–5809.

Holscher C., Horer M., Kleinschmidt J.A., Zentgraf H., Burkle A., and Heilbronn R. 1994. Cell lines inducibly expressing the adeno-associated virus (AAV) rep gene: Requirements for productive replication of rep-negative AAV mutants. *J. Virol.* **68:** 7169–7177.

Jomary C., Vincent K.A., Grist J., Neal M.J., and Jones S.E. 1997. Rescue of photoreceptor function by AAV-mediated gene transfer in a mouse model of inherited retinal degeneration. *Gene Ther.* **4:** 683–690.

Kaplitt M.G., Leone P., Samulski R.J., Xiao X., Pfaff D W., O'Malley K.L., and During M.J. 1994. Long-term gene expression and phenotypic correction using adeno-associated virus vectors in the mammalian brain. *Nat. Genet.* **8:** 148–154.

Kaplitt M.G., Xiao X., Samulski R.J., Li J., Ojamaa K., Klein I., Makimura H., Kaplitt M.J., Strumpf R.K., and Diethrich E.B. 1996. Long-term gene transfer in porcine myocardium after coronary infusion of adeno-associated virus vector. *Ann. Thorac. Surg.* **62:** 1669–1676.

Kearns W.G., Afione S.A., Fulmer S.B., Pang M.C., Erikson D., Egan M., Landrum M.J., Flotte T.R., and Cutting G.R. 1996. Recombinant adeno-associated virus (AAV-CFTR) vectors do not integrate in a site-specific fashion in an immortalized epithelial cell line. *Gene Ther.* **3:** 748–55.

Kessler P.D., Podsakoff G.M., Chen X., McQuiston S.A., Colosi P.C., Matelis L.A., Kurtzmann G.J., and Byrne B.J. 1996. Gene delivery to skeletal muscle results in sustained expression and systemic delivery of a therapeutic protein. *Proc. Natl. Acad. Sci.* **93:** 14082–14087.

Klein R., Meyer E., Peel A., Zolotukhin S., Muzyczka N., Meyers C., and King M. 1998. Neuron-specific transduction in the rat-septohippocampal or nigrostriatal pathway by recombinant adeno-associated virus vectors. *Exp. Neurol.* **150:** 183–194.

Koczot F.J., Carter B.J., Garon C.F., and Rose J.A. 1973. Self-complementarity of terminal sequences within plus or minus strands of adenovirus-associated virus DNA. *Proc. Natl. Acad. Sci.* **70:** 215–219.

Koeberl D.D., Alexander I E., Halbert C.L., Russell D.W., and Miller A.D. 1997. Persistent expression of human clotting factor IX from mouse liver after intravenous injection of adeno-associated virus vectors. *Proc. Natl. Acad. Sci.* **94:** 1426–1431.

Kotin R.M. 1994. Prospects for the use of adeno-associated virus for human gene therapy. *Hum. Gene Ther.* **5:** 793–801.

———.1996. The role of Rep protein in targeted integration of adeno-associated virus DNA. *Curr. Top. Microbiol. Immunol.* **218:** 25–33.

Kotin R.M., and Berns K.I. 1989. Organization of adeno-associated virus DNA in latently infected Detroit 6 cells. *Virology* **170:** 460–467.

Kotin R.M., Linden R.M., and Berns K.I. 1992. Characterization of a preferred site on human chromosome 19q for integration of adeno-associated virus DNA by non-homologous recombination. *EMBO J.* **11:** 5071–5078.

Labow M.A., Hermonat P.L., and Berns K.I. 1986. Positive and negative autoregulation of the adeno-associated virus type 2 genome. *J. Virol.* **60:** 251–258.

Lalwani A.K., Walsh B.J., Reilly P.G., Muzyczka N., and Mhatre A.N. 1996. Development of in vivo gene therapy for hearing disorders: Introduction of adeno-associated virus into the cochlea of the guinea pig. *Gene Ther.* **3:** 588–592.

Lalwani A., Han J., Walsh B., Zolotukhin S., Muzyczka N., and Mhatre A. 1998. Green fluorescent protein as a reporter for gene transfer studies in the cochlea. *Gene Ther.* **5:** 272–276.

Laughlin C.A., Cardellichio C.B., and Coon H.C. 1986. Latent infection of KB cells with adeno-associated virus type 2. *J. Virol.* **60:** 515–524.

Laughlin C.A., Tratschin J.-D., Coon H., and Carter B.J. 1983. Cloning of infectious adeno-associated virus genomes in bacterial plasmids. *Gene* **23:** 65–73.

Li J., Samulski R.J., and Xiao X. 1997. Role of highly regulated *rep* gene expression in adeno-associated virus vector production. *J. Virol.* **71:** 5236–5243.

Linden R.M., Winocour E., and Berns K.I. 1996. The recombination signals for AAV site specific integration. *Proc. Natl. Acad. Sci.* **93:** 7966–7972.

Lusby E., Fife K.H., and Berns K.I. 1980. Nucleotide sequence of the inverted terminal repetition in adeno-associated virus DNA. *J. Virol.* **34:** 402–409.

Malik P., McQuiston S.A., Yu X.J., Pepper K.A., Krall W.J., Podsakoff G.M., Kurtzman G.J., and Kohn D.B. 1997. Recombinant adeno-associated virus mediates a high level of gene transfer but less efficient integration in the K562 human hematopoietic cell line. *J. Virol.* **71:** 1776–1783.

Mamounas M., Leavitt M., Yu, M., and Wong-Staal F. 1995. Increased titer of recombinant AAV vectors by gene transfer with adenovirus coupled to DNA-polylysine complexes. *Gene Ther.* **2:** 429–432.

Mandel R.J., Spratt S.K., Snyder R.O., and Leff S.E. 1997. Midbrain injection of recombinant adeno-associated virus encoding rat glial cell line-derived neurotrophic factor protects nigral neurons in a progressive 6-hydroxydopamine-induced degeneration model of Parkinson's disease in rats. *Proc. Natl. Acad. Sci.* **94:** 14083–14088.

Mandel R.J., Rendahl K., Spratt S., Snyder R., Cohen L., and Leff S. 1998. Characterization of intrastriatal recombinant adeno-associated virus mediated gene transfer of human tyrosine hydroxylase and human GTP-cyclohydrogylase I in a rat model of Parkinson's disease. *J. Neurosci.* **18:** 4271–4284.

Manning W., Zhou S., Bland M., Escobedo J., and Dwarki V. 1998. Transient immunosuppression allows transgene expression following readministration of adeno-associated viral vectors. *Hum. Gene Ther.* **9:** 477–485.

McCarty D.M., Christensen M., and Muzyczka N. 1991. Sequences required for the coordinate induction of the AAV p19 and p40 promoters by the Rep protein. *J. Virol.* **65:** 2936–2945.

McCarty D.M., Ryan J.H., Zolotukhin S., Zhou X., and Muzyczka N. 1994. Interaction of the adeno-associated virus Rep protein with a sequence within the A palindrome of the viral terminal repeat. *J. Virol.* **68:** 4998–5006.

McCown T., Xiao X., Li J., Breese G., and Samulski R. 1996. Differential and persistent expression patterns of CNS gene transfer by an adeno-associated virus (AAV) vector. *Brain Res.* **713:** 99–107.

McLaughlin S.K., Collis P., Hermonat P.L., and Muzyczka N. 1988. Adeno-associated virus general transduction vectors: Analysis of proviral structures. *J. Virol.* **62:** 1963–1973.

Miller J.L., Walsh C.E., Ney P.A., Samulski R.J., and Nienhuis A.W. 1994. Single copy transduction and expression of human γ-globin in K562 erythroleukemia cells using recombinant adeno-associated virus vectors: The effect of mutations in NF-E2 and GATA-1 binding motifs within the HS2 enhancer. *Blood* **82:** 1900–1906.

Mizukami H., Young N., and Brown K. 1996. Adeno-associated virus type 2 binds to a 150-kilodalton cell membrane glycoprotein. *Virology* **217:** 124–130.

Monahan P.E., Tazelaar J., Xiao X., Nichols T.C., Bellinger D.A., Read M.S., Walsh C.E., and Samulski R.J. 1998. Direct intramuscular injection with recombinant AAV vectors results in sustained expression in a dog model of hemophilia. *Gene Ther.* **5:** 40–49.

Murphy J., Zhou S., Giese K., Williams L., Escobedo J., and Dwarki V. 1997. Long-term correction of obesity and diabetes in genetically obese mice by a single intramuscular injection of recombinant adeno-associated virus encoding mouse leptin. *Proc. Natl. Acad. Sci.* **94:** 13921–13926.

Muzyczka N. 1992. Use of adeno-associated virus as a general transduction vector for mammalian cells. *Curr. Top. Microbiol. Immunol.* **158:** 97–129.

Ni T.H., Zhou X., McCarty D.M., Zolotukhin I., and Muzyczka N. 1994. In vitro replication of adeno-associated virus DNA. *J. Virol.* **68:** 1128–1138.

Ni T., McDonald W.F., Zolotukhin I., Melendy T., Waga S., Stillman B., and Muzyczka N. 1998. Cellular proteins required for AAV DNA replication in the absence of adenovirus coinfection. *J. Virol.* **72:** 2777–2787.

Nordqvist K., Ohman K., and Akusjarvi G. 1994. Human adenovirus encodes two proteins which have opposite effects on accumulation of alternatively spliced mRNAs. *Mol. Cell. Biol.* **14:** 437–445.

Ohman K., Nordqvist K., and Akusjarvi G. 1993. Two adenovirus proteins with redundant activities in virus growth facilitates tripartite leader mRNA accumulation. *Virology* **194:** 50–58.

Peel A.L., Zolotukhin S., Schrimsher G.W., Muzyczka N., and Reier P.J. 1997. Efficient transduction of green flourescent protein in spinal cord neurons using adeno-associated virus vectors containing cell type-specific promotors. *Gene Ther.* **4:** 16–24.

Phillips M.I., Mohuczy Dominiak D., Coffey M., Galli S.M., Kimura B., Wu, P., and Zelles T. 1997. Prolonged reduction of high blood pressure with an in vivo, nonpathogenic, adeno-associated viral vector delivery of AT1-R mRNA antisense. *Hypertension* **29:** 374–380.

Ponnazhagan S., Yoder M.C., and Srivastava A. 1997b. Adeno-associated virus type 2-mediated transduction of murine hematopoietic cells with long-term repopulating ability and sustained expression of human globin gene in vivo. *J. Virol.* **71:** 3098–3104.

Ponnazhagan S., Mukherjee P., Yoder M.C., Wang X.S., Zhou S.Z., Kaplan J., Wadsworth S., and Srivastava A. 1997a. Adeno-associated virus 2-mediated gene transfer in vivo: Organ-tropism and expression of transduced sequences in mice. Gene **190:** 203–210.

Ponnazhagan S., Wang X., Woody M.J., Luo, F., Kang L.Y., Madahavi L.N., Munshi C.M., Shang Z.Z., and Srivasta A. 1996. Differential expression in human cells from the p6 promoter of human parvovirus B19 following plasmid transfection and recombinant adeno-associated virus 2 (AAV) infection: Human megacaryocytic leukaemia cells are non-permissive for AAV infection. *J. Gen. Virol.* **77:** 1111–1122.

Prasad K.M. and Trempe J.P. 1995. The adeno-associated virus Rep78 protein is covalently linked to viral DNA in a preformed virion. *Virology* **214:** 360–370.

Qing K., Wang X., Kube D., Ponnazhagan S., Bajpai A., and Srivastava A. 1997. Role of tyrosine phosphorylation of a cellular protein in adeno-associated virus 2-mediated transgene expression. *Proc. Natl. Acad. Sci.* **94:** 10879–10884.

Qing K., Khuntirait B., Mah C., Kube D.M., Wang X., Ponnazhagan S., Zhou S., Dwarki V., Yoder M.C., and Srivasava, A. 1998. Adeno-associated type 2-mediated gene transfer: Correlation of tyrosine phophorylation of the cellular single-stranded D sequence-binding protein with transgene expression in human cells in vitro and murine tissues in vivo. *J. Virol.* **72:** 1593–1599.

Rose J.A. and Koczot F.J. 1972. Adenovirus-associated virus multiplication: VII. Helper requirement for viral deoxyribonucleic acid and ribonucleic acid synthesis. *J. Virol.* **10:** 1–8.

Russell D.W., Alexander I.A., and Miller A.D. 1995. DNA synthesis and topoisomerase inhibitors increase transduction by adeno-associated virus vectors. *Proc. Natl. Acad. Sci.* **92:** 5719–5723.

Russell D.W., Miller A.D., and Alexander I.E. 1994. Adeno-associated virus vectors preferentially transduce cells in S phase. *Proc. Natl. Acad. Sci.* **91:** 8915–8919.

Samulski R.J. 1993. Adeno-associated virus: Integration at a specific chromosomal locus (review). *Curr. Opin. Genet. Dev.* **3:** 74–80.

Samulski R.J. and Shenk T. 1988. Adenovirus ElB 55-M_r, polypeptide facilitates timely cytoplasmic accumulation of adeno-associated virus mRNAs. *J. Virol.* **62:** 206–210.

Samulski R.J., Chang L.-S., and Shenk T. 1989. Helper-free stocks of recombinant adeno-associated viruses: Normal integration does not require viral gene expression. *J. Virol.* **63:** 3822–3828.

Samulski R.J., Berns K.I., Tan M., and Muzyczka N. 1982. Cloning of adeno-associated virus into pBR322: Rescue of intact virus from the recombinant plasmid in human cells. *Proc. Natl. Acad. Sci.* **79:** 2077–2081.

Sarnow P., Hearing P., Anderson C.W., Halbert D.N., Shenk T., and Levine A.J. 1984. Adenovirus early region 1B 58,000-dalton tumor antigen is physically associated with an early region 4 25,000-dalton protein in productively infected cells. *J. Virol.* **49:** 692–700.

Schimmenti S., Boesen J., Classen E., Valerio D., and Einerhand M. 1998. Long-term genetic modification of hematopoietic cells in rhesus monkeys following transplantation of adeno-associated virus vector transduced CD34[+] cells. (in press).

Smith F.E. and McCown T.J. 1997. AAV vectors: General characteristics and potential use in the central nervous system,. In *Gene transfer and therapy for neurological disorders* (ed. E.A. Chiocca, and X.O. Breakfield), pp. 79–88. Humana Press, Totowa, New Jersey.

Snyder R.O., Xiao X., and Samulski R.J. 1996. Production of recombinant adeno-associated viral vectors. In *Current protocols in human genetics* (ed. N. Dracopoli, et al.), pp. 1211–1224 John Wiley and Sons, New York.

Snyder R.O., Im D.S., Ni T., Xiao X., Samulski R.J., and Muzyczka N. 1993. Features of the adeno-associated virus origin involved in substrate recognition by the viral Rep protein. *J. Virol.* **67:** 6096–6104.

Snyder R., Spratt S., Lagarde C., Bohl D., Kaspar B., Sloan B., Cohen L., and Danos O. 1998. Efficient and stable AAV-mediated transduction in the skeletal muscle of adult immunocompetent mice. *Hum. Gene Ther.* (in press).

Snyder R., Miao C., Patijn G., Spratt S., Danos, O., Nagy D., Gown, A., Winter B., Meuse L., Cohen L., Thompson A., and Kay M. 1997. Persistent and therapeutic concentrations of human factor IX in mice after hepatic gene transfer of recombinant AAV vectors. *Nat. Genet.* **16:** 270–276.

Straus S.E., Sebring E.D., and Rose J.A. 1976. Concatemers of alternating plus and minus strands are intermediates in adenovirus-associated virus DNA synthesis. *Proc. Natl. Acad. Sci.* **73:** 742–746.

Summerford C. and Samulski R.J. 1998. Membrane-associated heparan sulfate proteoglycan is a receptor for adeno-associated virus type 2 virions. *J. Virol.* **72:** 1438–1445.

Tamoyose K., Hirai Y., and Shimada T. 1996. A new strategy for large-scale preparation of high-titer recombinant adeno-associated virus vectors by using packaging cell lines and sulfonated cellulose column chromatography. *Hum. Gene Ther.* **7:** 507–513.

Tratschin J.-D., Tal J., and Carter B.J. 1986. Negative and positive regulation in *trans* of gene expression from adeno-associated virus vectors in mammalian cells by a viral *rep* gene product. *Mol. Cell. Biol.* **6:** 2884–2894.

Tratschin J.D., Miller I.L., Smith M.G., and Carter B.J. 1985. Adeno-associated virus vector for high-frequency integration, expression, and rescue of genes in mammalian cells. *Mol. Cell. Biol.* **5:** 3251–3260.

Tratschin J.-D., West M.H.P., Sandbank T., and Carter B.J. 1984. A human parvovirus, adeno-associated virus, as a eukaryotic vector: Transient expression and encapsidation of the prokaryotic gene for chloramphenicol acetyltransferase. *Mol. Cell. Biol.* **4:** 2072–2081.

Urcelay E., Ward P., Wiener S.M., Safer B., and Kotin R.M. 1995. Asymmetric replication in vitro from a human sequence is dependent on adeno-associated virus Rep protein. *J. Virol.* **69:** 2083–2046.

Vincent K.A., Piraino S.T., and Wadsworth S.C. 1997. Analysis of recombinant adeno-associated virus packaging and requirements for *rep* and *cap* gene products. *J. Virol.* **71:** 1897 1905.

Walsh C.E., Liu J.M., Xiao X., Young N. S., Nienhuis A.W., and Samulski R.J. 1992. Regulated high level expression of a human γ-globin gene introduced into erythroid cells by an adeno-associated virus vector. *Proc. Natl. Acad. Sci.* **89:** 7257–7261.

Weiden M.D. and Ginsberg H.S. 1994. Deletion of the E4 region of the genome produces adenovirus DNA concatemers. *Proc. Natl. Acad. Sci.* **91:** 153–157.

Weitzman M.D., Kyostio S.R., Kotin R.M., and Owens R.A. 1994. Adeno-associated virus (AAV) Rep proteins mediate complex formation between AAV DNA and its integration site in human DNA. *Proc. Natl. Acad. Sci.* **91:** 5808–5812.

Xiao X., Li J., and Samulski R.J. 1996. Efficient long term gene transfer into muscle tissue of immunocompetent mice by adeno-associated virus vector. *J. Virol.* **70:** 8098–8108.

———. 1998. Production of high titer recombinant adeno-associated virus vectors in the absence of helper adenovirus. *J. Virol* **72:** 2224–2232.

Xiao X., McCown T.J., Li J., Breese G. R., Morrow A. L., and Samulski R. J. 1997. Adeno-associated virus (AAV) vector antisense gene transfer in vivo decreases GABA(A) alpha1 containing receptors and increases inferior collicular seizure sensitivity. *Brain Res.* **756:** 76–83.

Yakobson B., Koch T., and Winocou E. 1987. Replication of adeno-associated virus in synchronized cells without the addition of a helper virus. *J. Virol.* **61:** 972–981.

Yakobson B., Hrynko T.A., Peak M.J., and Winocour E. 1989. Replication of adeno-associated virus in cells irradiated with UV light at 254 nm. *J. Virol* **63:** 1023–1030.

Yalkinoglu A.O., Zentgraf H., and Hubscher U. 1991. Origin of adeno-associated virus DNA replication is a target of carcinogen-inducible DNA amplification. *J. Virol.* **65:** 3175–3184.

Yalkinoglu A.O., Heilbronn R., Burkle A., Schlehofer J.R., and zur-Hausen H. 1988. DNA amplification of adeno-associated virus as a response to cellular genotoxic stress. *Cancer Res.* **48:** 3123–3129.

Yang C.C., Xiao X., Zhu X., Ansardi D., Epstein N.D., Frey M.R., Matera A.G., and Samulski R.J. 1997. Cellular recombination pathways and viral terminal repeat hairpin structures are sufficient for AAV integration in vivo and in vitro. *J. Virol.* **71:** 9231–9247.

Yang Q., Chen F., and Trempe J. 1994. Characterization of cell lines that inducibly express the adeno-associated virus Rep proteins. *J. Virol.* **68:** 4847–4856.

Yang Y. and Wilson J.M. 1995. Clearance of adenovirus-infected hepatocytes by MHC class I-restricted CD4[+] CTLs in vivo. *J. Immunol.* **155:** 2564–2570.

Yang Y., Li Q., Ertl H.C., and Wilson J.M. 1995. Cellular and humoral immune responses to viral antigens create barriers to lung-directed gene therapy with recombinant adenoviruses. *J. Virol.* **69:** 2004–2015.

Yang Y., Nunes F.A., Berencsi K., Furth E.E., Gonczol E., and Wilson J.M. 1994. Cellular immunity to viral antigens limits E1-deleted adenoviruses for gene therapy. *Proc. Natl. Acad. Sci.* **91:** 4407–4411.

Zhou S.Z., Li Q., Stamatoyannopoulos G., and Srivastava A. 1996. Adeno-associated virus 2-mediated transduction and erythroid cell-specific expression of a human β-globin gene. *Gene Ther.* **3:** 223–229.

Zhou S.Z., Broxmeyer H.E., Cooper S., Harrington M.A., and Srivastava A. 1993. Adeno-associated virus 2-mediated gene transfer in murine hematopoietic progenitor cells. *Exp. Hematol.* **21:** 928–933.

Zhou S.Z., Cooper S., Kang L.Y., Ruggieri L., Heimfeld S., Srivastava A., and Broxmeyer H. E. 1994. Adeno-associated virus 2-mediated high efficiency gene transfer into immature and mature subsets of hematopoietic progenitor cells in human umbilical cord blood. *J. Exp. Med.* **179:** 1867–1875.

Zolotukhin S., Potter M., Hauswirth W.W., Guy J., and Muzyczka N. 1996. A "humanized" green fluorescent protein cDNA adapted for high-level expression in mammalian cells. *J. Virol.* **70:** 4646–4654.

8

Gene-transfer Tool: Herpes Simplex Virus Vectors

Sylvie Laquerre, William F. Goins, Shusuke Moriuchi, Thomas J. Oligino, David M. Krisky, Peggy Marconi, M. Karina Soares, Justus B. Cohen, and Joseph C. Glorioso
Department of Molecular Genetics and Biochemistry
University of Pittsburgh School of Medicine
Pittsburgh, Pennsylvania 15261

David J. Fink
Departments of Neurology and Molecular Genetics and Biochemistry
University of Pittsburgh School of Medicine
Veterans Administration Medical Center
Pittsburgh, Pennsylvania 15240

The clinical application of gene therapy to treatment of human disease depends on the development of effective gene-transfer vectors. Although considerable technical progress has been made in recent years, a number of obstacles remain. These can be considered in the following categories: (1) toxicity of the vector, including direct cytopathic effects and the potential inflammatory or immune responses; (2) transgene expression, including the efficiency of cellular transduction, the level and duration of expression, and the potential to regulate expression; (3) transgene capacity of the vector; and (4) targeting of the transgene expression to specific cell populations. Some vector systems may overcome many of these obstacles for particular tissues or applications, and attention to the unique biological properties of individual vectors has begun to provide opportunities to meet the challenges of effective gene delivery.

The human herpesviruses represent excellent candidate vectors for several types of applications. As a class, they are large DNA viruses with the potential to accommodate large or multiple transgene cassettes, and they have evolved to persist in a lifelong nonintegrated latent state without causing disease in the immune-competent host. Among the herpesviruses, herpes simplex virus type 1 (HSV-1) is an attractive vehicle especially for gene transfer to the nervous system because its natural history of infection in humans includes lifelong persistence in neurons. Latency is characterized by the absence of viral protein expression, and there is no evidence of neurodegeneration caused by latent genomes. The virus contains a unique, neuron-specific promoter system that remains active during latency and can be used to express therapeutic proteins without compromising the latent state. The establishment of latency does not require the expression of viral lytic functions in replication-defective viruses. Thus, removal of genes required for expression of the viral cascade of gene products expressed during lytic replication allows for the creation of a completely replication-defective vector that cannot reactivate from latency. HSV-1 has a broad host range and does not require cell division for infection and gene expression. Accordingly, HSV may be generally useful for gene transfer to a variety of normal and disease tissues, particularly where only short-term transgene expression is required to achieve a therapeutic effect.

In this chapter, we provide a brief overview of the biology of HSV relevant to vector design and summarize recent progress in reducing virus cytotoxicity. Procedures for the construction of multigene vectors and advances in understanding viral entry relevant to targeting vectors to specific cell populations are described. Although HSV amplicon vectors and replicating viruses have been developed for gene therapy, this review will focus on engineered viral genome vectors that are replication-defective.

OVERVIEW OF HSV BIOLOGY RELEVANT TO VECTOR DESIGN

Lytic Phase

Primary HSV-1 infection occurs by direct contact. Following virus attachment, the virus penetrates epithelial cells of the skin or mucosa. Upon entry of the capsid into the cytoplasm, it becomes attached to microtubule filaments along which it is conducted to the nuclear membrane, where the viral DNA is injected through a nuclear pore. In the nucleus, the viral DNA is transported to nuclear domain-10 (ND-10) structures (Maul and Everett 1994; Mullen et al. 1995) where the initial immediate early (IE) genes are expressed. Transcription of the five IE genes (ICP0, ICP4, ICP22, ICP27, and ICP47) does not require de novo viral protein synthesis. Expression of IE genes is controlled by promoters that each contain an enhancer element responsive to the viral tegument protein VP16 (also known as Vmw 65, αTIF), a *trans*-activator that is transported into the nucleus along with the viral DNA (Mackem and

Roizman 1982; Campbell et al, 1984; Preston et al 1988) The IE genes ICP4 and ICP27 encode products that are required for expression of the early (E) and late (L) genes (Dixon and Schaffer 1980; DeLuca and Schaffer 1985; Sacks et al. 1985; McCarthy et al. 1989), the former gene class primarily specifying enzyme functions required for viral DNA synthesis and the latter primarily comprising of structural components. The IE gene products ICP0 and ICP22 contribute to viral gene transcription but are not essential to virus replication (Stow and Stow 1986; Sacks and Schaffer 1987; Rice et al. 1994). ICP0 is a promiscuous *trans*-activator that exerts its effect prior to the transcription initiation event and it is not a DNA-binding protein (Everett 1984). ICP22 has been found to regulate the level of ICP0 expression (Carter and Roizman 1996). ICP47 does not affect transcription but rather has been reported to interfere with a transporter associated with antigen presentation (TAP) that is responsible for loading major histocompatibility complex (MHC) class I molecules with antigenic peptides (York et al. 1994; Hill and Ploegh 1995). Expression of late genes is dependent on both viral DNA synthesis and IE gene functions (Holland et al. 1980; DeLuca and Schaffer 1985; Mavromara-Nazos and Roizman 1987; McCarthy et al. 1989). Following translation of the L gene products, which become structural proteins of the capsid, tegument, and envelope, the viral DNA is packaged into the capsid. Tegument proteins surround the capsid and the immature particle buds through the inner nuclear membrane where the viral glycoproteins are localized. Although it is unclear whether a naked or a double-membrane-enveloped capsid is released from the membrane of the endoplasmic reticulum (ER), a double-membrane-enveloped virus containing membrane glycoproteins modified by the Golgi apparatus enzymes appears in the cytoplasm and fuses with the cell membrane, forming a virus particle with a single-membrane bilayer (Roizman and Batterson 1985). This infectious particle is released from the infected cell. The lytic cycle is rapid and results in cell death within 24 hours after infection.

Latent Phase

During lytic infection of skin or mucous membrane, newly produced HSV-1 comes into contact with the axon terminals of sensory neurons of the peripheral nervous system (PNS) and virus is transported to the nerve-cell body, where the viral DNA enters the nucleus. Although the virus can express lytic functions in sensory neurons, the cells survive this infection and, through a set of largely undefined molecular events, lytic gene expression is curtailed and the virus enters a latent state. During latency, expression of lytic genes is silenced and only an interrelated group of latency-associated transcripts (LATs) is expressed from the repeat regions flanking the long unique segment (U_L) of the genome (Croen et al. 1987; Deatly et al. 1987; Stevens et al. 1987). The LATs are antisense to and overlap the 3´ end of ICP0 mRNA (Fig. 1A,B). Two colinear, non-polyadenylated LAT RNA species of 2.0 kb and 1.5 kb (Deatly et al. 1988; Wagner et al. 1988), which accumulate in the nuclei of latently infected neurons, appear to be stable introns derived from a large, unstable 8.3 kb polyadenylated primary transcript (Farrell et al. 1991; Krummenacher et al. 1997; Rodahl and Haarr 1997; Zablotony et al. 1997). No viral proteins can be detected in latently infected cells (Doerig et al. 1991), although the level of expression of the LAT transcripts varies considerably among latently infected neurons (Ramakrishnan et al. 1994; Sawtell 1997). The biological role of the LAT locus is unclear, and at least in animal models, this locus is not required for the establishment or maintenance of latency (Javier et al. 1988; Ho and Mocarski 1989; Leib et al. 1989; Sedarati et al. 1989; Steiner et al. 1989). However, some LAT mutations appeared to display a decreased efficiency in establishing latency when their effects were examined at the single-cell level (Thompson and Sawtell 1997), suggesting that LAT may function to increase the number of neurons that harbor latent viral genomes in wild-type virus infections. The virus can remain latent for the life of the individual, but some latent wild-type genomes can also be reactivated through the action of a variety of stimuli, including dexamethasone, ultraviolet light, fever, and stress. Upon reactivation, HSV-1 progeny virions are produced in neurons and, following anterograde axonal transport to peripheral tissues, recurrent infection is established at or near the site

Figure 1 HSV-1 genome organization and latency gene expression. (*A*) A schematic of the prototypic HSV-1 genome structure displaying the unique long (U_L) and short (U_S) segments, each flanked by inverted repeats. The viral genes that are essential or nonessential (accessory) for replication in permissive cells in culture are depicted. Specific gene loci that have been targeted for manipulation are denoted with an asterisk. (*B*) The location of the 5´ and 3´ ends of the family of latency-associated transcripts (LATs) are diagramed relative to several viral lytic genes present within the repeat region of the HSV-1 genome. (*C*) The viral latency active promoters LAP1 and LAP2 are depicted in detail, showing the location of *cis*-acting regulatory sequences.

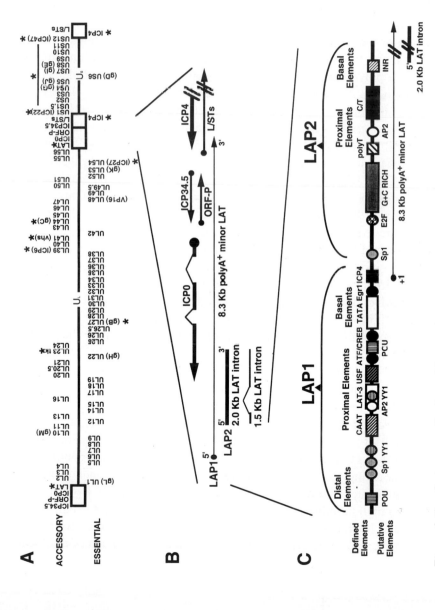

Figure 1 (See facing page for legend.)

of the primary infection. Suppression of effective antiviral immunity can enhance viral spread following reactivation (Mercadal et al. 1993; Mitchell and Stevens 1996). In rare cases, the wild-type virus can infect the central nervous system (CNS), causing encephalitis.

The LAT promoter/regulatory region (Fig. 1C) has been shown to be composed of two latency-active promoters, LAP1 (Fig. 1C) (Dobson et al. 1989; Zwaagstra et al. 1989: Batchelor and O'Hare 1990; Chen et al. 1995) and LAP2 (Nicosia et al. 1993; Goins et al. 1994; Chen et al. 1995). LAP1, which is predominantly responsible for LAT expression during latency (Dobson et al. 1989; Chen et al. 1995), is located immediately upstream of the region for the unstable 8.3 kb-LAT and contains binding sites for many RNA polymerase class II transcription factors functioning with RNA polymerase II, such as a consensus TATA box (Rader et al. 1993; Soares et al. 1996), a CAAT box (Batchelor and O'Hare 1992), an upstream stimulatory factor-1 (USF-1) site (Zwaagstra et al. 1991; Soares et al. 1996; Kenny et al. 1997), and two activating transcription factor/ cycle AMP-responsive (ATF/CREB) elements (Kenny et al. 1994; Soares et al. 1996). Many of the basal promoter elements, including the TATA box and ATF/CREB and USF-1 sites, have been shown to be required for LAT expression during latency (Leib et al. 1989; Rader et al. 1993; Kenny et al. 1994; Soares et al. 1996). Deletion of upstream distal elements believed to be involved in neuron-specific regulation of LAT (Batchelor and O'Hare 1992) dramatically decreases LAT levels during latency in vivo (Wang et al. 1997a; M.K. Soares, unpubl.). LAP2, located 5′ proximal to the LAT introns, is the predominant promoter responsible for LAT expression during lytic infection (Nicosia et al. 1993; Chen et al. 1995); however, it is also capable of expressing LAT at low levels in the absence of LAP1 in latently infected animals, indicating that LAP2 has independent promoter activity (Chen et al. 1995; R. Ramakrishnan, unpubl.). Although LAP2 lacks a TATA box, it contains both a C/T-rich sequence and a poly(T) element (Goins et al. 1994; French et al. 1996 and in prep.; S.W. French, unpubl.) frequently observed in promoters for cellular housekeeping genes (Koller et al. 1991; Kasal et al. 1992; McDermott et al. 1992). We have shown that these elements contribute significantly to LAP2 activity in cell culture (S.W. French, unpubl.) and bind *trans*-acting factors (French et al. 1996 and in prep.) such as nuclease sensitive element protien-1 (NSEP-1) (Postel et al. 1989; Kolluri et al. 1992) and high mobility group factor I(Y) (HMG I[Y]) (Saito et al. 1992; Thanos and Maniatis 1992), respectively. We have mapped an S1-nuclease-sensitive site to this region of LAP2, which perhaps alters the local chromatin conformation to allow long term expression from LAT in the absence of lytic gene expression.

REDUCING HSV VECTOR CYTOTOXICITY

The HSV genes are expressed in a sequential, interdependent order during lytic infection. Consequently, removal of one essential IE gene prevents expression of later genes in the gene-expression cascade, resulting in a defective vector that is incapable of producing virus particles or expressing E or L gene products. However, the remaining IE genes continue to be expressed and, with the exception of ICP47 (DeLuca and Schaffer 1985), have been shown to be individually toxic to cells (Johnson et al. 1992). Accordingly, only the combined elimination of multiple IE genes has been found to reduce the cytotoxicity of HSV-based vectors for tissue culture cells (Johnson et al. 1994; Samaniego et al. 1995; Marconi et al. 1996; Wu et al. 1996; D.M. Krisky, unpubl.). For example, a mutant deleted for ICP4, ICP22, and ICP27 was reported to kill Vero cells in culture at a reduced rate and to arrest DNA synthesis and cell division (Wu et al. 1996). The cytotoxicity of this mutant was dependent on the multiplicity of infection and the cell type (Wu et al. 1996). Mutants deleted for ICP0, ICP4, and ICP27 are even less cytotoxic, and it was found in this background that the ICP0 promoter expressed the *lacZ* gene at low levels for several days in culture (Samaniego et al. 1997). Presumably, removal of all IE genes will eliminate HSV toxicity for cells even at very high multiplicity because UV-irradiated particles that do not express viral functions are not cytotoxic (Leiden et al. 1980; Johnson et al. 1994; Huard et al. 1997). We have recently developed an HSV-1 vector deleted for nine viral functions, including the IE genes ICP4, ICP22, ICP27, and ICP47, to create a mutant background for gene-therapy applications that require short-term high-level expression of multiple products. We did not delete the ICP0 gene because it has been reported that the ICP0 gene product improves transgene expression (Samaniego et al. 1997), but we did remove the ICP47 gene to avoid interference with antigen presentation in applications intended for induction of specific immunity. To increase transgene expression, we also eliminated the virion host shutoff function (vhs) encoded by UL41, which should increase transgene expression because this virion protein interferes with translation of mRNA in infected cells (Read and Frenkel 1983; Kwong et al. 1988; Oroskar and Read 1989). At multiplicities of infection below 10, this vector background can also be used to express transgenes for prolonged times in cultured primary neurons without induction of neuronal cell death (D.M. Krisky, unpubl.).

HSV-VECTOR TRANSGENE CAPACITY

Although gene therapy was initially envisioned for the treatment of monogenetic diseases, the same methods can be utilized for the delivery of multiple genes for treatment of multifactorial or nongenetic pathologic processes to ameliorate disease symptoms or progression, such as cancer or autoimmune or neurodegenerative disease. The delivery of multiple genes requires the development of gene-transfer vehicles that can accommodate extensive foreign genetic sequences.

HSV-1 is well suited for transfer of large amounts of exogenous DNA because of the size of its genome, 152 kb, and the fact that approximately half of its coding sequences are nonessential for virus replication in cell culture and may be deleted to increase transgene capacity without blocking viral replication (Fig. 1A). However, the removal of nonessential genes is not without consequence, because the products of many of these genes contribute to efficient viral growth even in cell culture and their deletion can reduce vector yields. We have simultaneously removed seven nonessential E and L gene functions without compromising viral growth in cell culture. Removal of the nonessential IE genes ICP0 or ICP22 reduces the efficiency of virus replication, thereby reducing the maximum titer that can be conveniently achieved, while other gene products affect titers by reducing infectivity. For example, the UL44 locus, encoding the viral envelope glycoprotein C (gC), is an accessory gene that may serve as a convenient site for transgene insertion. Glycoprotein C has a role in virus attachment by binding to the initial viral receptor heparan sulfate (HS) by a charge interaction (Fuller and Lee 1992; Spear et al. 1992; Herold et al. 1994). Removal of gC reduces the efficiency of virus infection by threefold to tenfold, depending on the cell type, and reduces the effective titer proportionately (discussed later). Vectors designed for in vivo applications may need to retain this function for maximum infectivity. It is not yet clear how important the other nonessential glycoproteins are for virus infection, alone or in combination, but some are required for efficient virus spread in vivo (Balan et al. 1994; Dingwell et al. 1994, 1995). The UL41 gene encodes the vhs protein, a structural protein that is incorporated into the virus tegument (Read et al. 1993). We have found that deletion of UL41 does not lead to an appreciable reduction in viral toxicity but does increase the level of transgene expression (D.M. Krisky, unpubl.). Deletion of the ICP47 gene eliminates virus-mediated suppression of MHC class I expression and should be advantageous for the development of tumor-specific immunization. The viral thymidine kinase (*tk*) gene can be removed without compromising viral replication in vitro (Palella et al. 1989; Mester et al. 1995; Rasty et al. 1995), and other genes that contribute to viral DNA synthesis (e.g., the ribonucleotide reductase gene) are also convenient target loci for transgene incorporation (Desai et al. 1993; Ramakrishnan et al. 1994). The latency region of the virus genome represents approximately 8 kb of sequence that can be removed without reducing gene transfer to cells. The joint region of the virus is composed of 15 kb of redundant sequence that can be eliminated without compromising virus replication (Jenkins et al. 1985). We have removed the complete U_S region of the genome comprising 11 kb of sequence. The only essential gene in this region, glycoprotein D, was introduced into a cell line for vector propagation (S. Laquerre, unpubl.). Taken together, these observations suggest that approximately 44 kb of HSV sequence can be removed if the mutant is grown in cells engineered to complement just three viral functions, but such a combined mutant has not yet been constructed.

Using a rapid gene-insertion procedure, we have developed novel HSV-1 gene vectors with a background suitable for expression of multiple transgenes (Krisky et al. 1997; D.M. Krisky, unpubl.). The model genes chosen for these experiments are potentially synergistic in both the local destruction of tumor cells and the induction of antitumor immunity. The mutant background was designed to take advantage of the reduced cytotoxicity resulting from the deletion of ICP4, ICP22, and ICP27 genes (Samaniego et al. 1995; Wu et al. 1995; D.M. Krisky, unpubl.). This mutant retained ICP0, the reported overexpression of which, in the absence of the other IE genes, would interfere with tumor cell division while enhancing transgene expression. Together, nine viral genes were deleted, resulting in the removal of 11.6 kb of viral DNA that was replaced with multiple transgenes under the control of different promoters. The model multigene vectors were constructed with either four or five independent transgenes at distinct loci (Fig. 2). All of the transgenes were simultaneously expressed for up to 7 days, with maximum expression occurring at 2–3 days postinfection. We demonstrated that new transgenes could be sequentially added to the multigene vector background with high efficiency and that transgene substitutions could be carried out having the potential to create different transgene combinations for studies of their biological interactions either within cells or by stimulating biological changes in the immediate environment. The vectors HX86Z and HX86G (Fig. 2) demonstrate the potential for using HSV-1 vectors

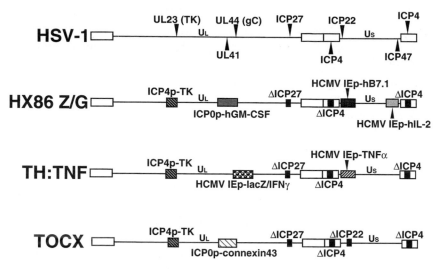

Figure 2 HSV-1 multigene replication-defective vectors. Various multigene replication defective vectors have been engineered in which combinations of HSV-1 IE and other cytotoxic gene products have been deleted from the vector backbone to reduce toxicity and provide space for the insertion of multiple foreign gene cassettes. Three vectors that were engineered for cancer applications include HX86Z/G, TH/TNF, and TOCX, which contain the insertion of multiple therapeutic genes. The location of the various deleted genes compared to a wild-type virus is also depicted.

for the expression of highly complex sets of transgenes. In the future, it may be advantageous to insert transgenes into these loci if the generation of replication-competent virus-carrying foreign transgenes through recombination with wild-type virus in latently infected individuals is to be avoided. That is, for example, if transgene expression cassettes were inserted into the loci of essential gene products such as ICP4 and ICP27, it is likely that recombinational events that could restore these genes would also result in deletion of the transgene, thus alleviating the possibility that a replication-competent virus capable of transgene expression would be generated.

VECTOR TRANSGENE EXPRESSION

Lytic Gene Promoters

We have explored the behavior of foreign promoters in the background of replication-defective mutants (Mester et al. 1995; Rasty et al. 1995; Fink et al. 1996; Oligino et al. 1996; D.M. Krisky, unpubl.). For short-term applications, the use of the IE promoters is convenient and effective. For example, in mutants that lack the IE genes except ICP0, HSV and human cytomegalovirus (HCMV) IE gene promoters produce vigorous transgene expression for up to 1 week after infection (R. Ramakrishnan, unpubl.). Other promoters, such as SV40 and various retroviral LTRs, are also highly active (Mester et al. 1995; Rasty et al. 1996) but transient in these vectors. Although cellular promoters are less explored, particular elements such as the muscle specific enhancer MCK have been shown to work well (G.R. Akkaraju, unpubl.). However, for a number of potential therapeutic applications it would be either necessary or desirable to produce prolonged transgene expression from latent genomes, and for this purpose the native latency gene promoters have been studied.

Latency Gene Promoters

Even though the functional role of LAT remains unknown, the continuous expression of the LAT region during latency suggests that it should be possible to exploit the LAT promoter to express therapeutic genes from persistant genomes in neurons. Both LAP1 and LAP2 have been employed to achieve long-term transgene expression from the HSV-vector genome during latency. Although a LAP1–β-globin recombinant was able to drive transgene expression in murine peripheral neurons during latency (Dobson et al. 1989), the level of product decreased over time (Margolis et al. 1993). Other recombinants with LAP1 driving expression of β-glucuronidase (Wolfe et al. 1992), nerve growth factor (NGF) (Margolis et al. 1993), β-galactosidase (Margolis et al. 1993; Lokensgard et al. 1994;), or murine α-interferon (Mester et al. 1995) either displayed a similar expression pattern or were not active in latently infected animals (Margolis et al. 1993; Lokensgard et al. 1994; Mester et al. 1995), suggesting that LAP1 may lack the *cis* elements required for long-term expression. However, when LAP1 was juxtaposed to the Moloney murine leukemia virus (Mo-MLV) promoter, long-term transgene expression was achieved (Lokensgard et al. 1994; Carpenter and Stevens 1996), unlike mutants constructed to employ either LAP1 or the long terminal repeat (LTR) alone, suggesting that the elements responsible for extended expression lie elsewhere within the LAT promoter/regulatory region and could be complemented by elements within the Mo-MLV promoter. Addition of 1.1 kb of sequence downstream from LAP1, comprising LAP2, restored long-term expression to the LAP1-reporter cassette in the ectopic site within the genome (Lokensgard et al. 1997). The LAP was capable of achieving long-term transgene expression when a *lacZ*

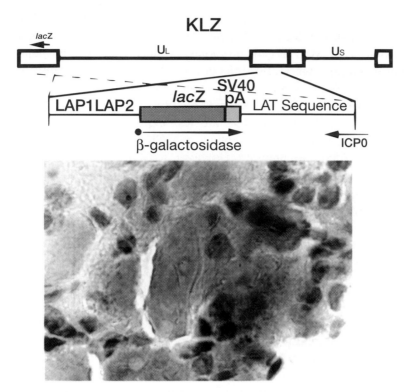

Figure 3 HSV-1-vector-mediated transgene expression in latently infected murine peripeheral neurons. The HSV-1 vector KLZ was capable of long-term (42 days) expression of the β-galactosidase (*lac*Z) gene product in mouse trigeminal ganglia neurons following corneal scarification. Similar vectors have displayed expression of the transgene up to 1-year postinfection.

reporter gene cassette was introduced into the LAT intron in the native LAT locus (Fig. 3) (Ho and Mocarski 1989; Chen et al., in prep.) or when a LAP2-*lac*Z expression cassette was present in an ectopic locus within the viral genome (Goins et al. 1994); we have also shown that a LAP2-NGF cassette present either in the *tk* or Us3 loci of the vector expressed this gene product in latently infected rodent neurons both in culture and in vivo (W.F. Goins, unpubl.). Expression of β-galactosidase from the LAP-*lac*Z vectors could be detected in neurons of the mouse PNS for up to 300 days (Goins et al. 1994), but prolonged expression in the CNS was at very low levels detectable only by reverse transcriptase–polymerase chain reaction (RT-PCR) techniques (X. Chen, unpubl.). In an effort to increase expression from LAP2 in the brain, we have constructed recombinants in which thegreen fluorescent protein (GFP) reporter gene was fused to the internal ribosome entry site (IRES) of the encephalomyocarditis virus and placed within the LAT intron downstream from LAP2 in the native LAT locus; a similar recombinant previously yielded high levels of transgene expression in the PNS (Lachmann and Efstathiou 1997). Our recombinant produced long-term expression in trigeminal ganglion neurons in culture (D.P. Wolfe, unpubl.). These and other modifications to the LAPs should increase and stabilize the level of therapeutic gene expression required for CNS applications.

VECTOR TARGETING

Introduction

Virus targeting, the cell-specific expression of the therapeutic gene, can be accomplished at the level of infection or transduction. Transductional targeting exploits either a cell-specific promoter or a particular characteristic of the targeted cell, such as active cell division in the case of retroviruses (Culver et al. 1992; Ram et al. 1993) or the expression of a selectable marker (Jia et al. 1994); this approach will not be considered here. Cell-specific infection is a more general approach that relies on the capacity of the virus to recognize cell-specific surface receptors for infection and delivery of therapeutic genes. To accomplish targeted delivery following systemic administration, vectors must be developed that seek out specific cell types based on recognition of cell-specific surface receptors by engineered viral ligands. Successful development of this technology depends on (1) the existence and identification of cell-specific surface receptor(s) and (2) the ability to modify viral surface proteins for recognition of novel receptors without compromising infectivity. Because the mechanism of viral entry also impacts on reengineering virus infection, a sufficient understanding of the natural process of virus entry into the cell is also essential.

Various attempts to alter the host range of different viruses have been reported. One approach is to construct pseudotype or hybrid vectors by combining elements from different viruses. Examples of this approach include the production of a retroviral (Mo-MLV) or a lentiviral (HIV-1) nucleocapsid having a vesicular stomatitis virus (VSV) envelope (Emi et al. 1991; Naldini et al. 1996). The HIV-based vector with VSV tropism demonstrated efficient transfer, integration, and sustained expression of a transgene in adult rat brains. However, the pseudotype approach is limited by the availability of cell-type-specific viral surface proteins.

Modification of purified virions by chemical attachment of a ligand capable of binding to a specific cell type has also been explored. Ligands were either chemically attached to viral surface proteins (Neda et al. 1991) or antibodies against viral surface protein were biochemically linked to a second antibody that recognized a cell-surface receptor (Roux et al. 1989). Recently, Rogers et al. (1997) cross-linked a Fab fragment from a neutralizing monoclonal antibody specific for adenoviral knob protein to basic fibroblast growth factor (FGF) to allow infection, albeit inefficiently, of cell lines expressing the FGF receptor. A Sindbis virus displaying a protein A:*env* chimeric protein has also been engineered in a manner that generalizes the applicability of monoclonal specificity for targeted infection. The recombinant virus had a strong affinity for the Fc region of various mammalian IgGs, and cell-specific targeting was accomplished according to the specificity of the antibody (Ohno et al. 1997). Although this procedure was highly efficient in vitro, the stability of the protein A–antibody interaction in the bloodstream has not been evaluated.

Recombinant viruses have been engineered in an attempt to improve effective vector targeting and reduce effort in vector production. A chimeric Mo-MLV with both a wild-type *env* gene and an *env*-erythropoietin (EPO) fusion gene displayed infectivity for human cells bearing the EPO receptor (Kasahara et al. 1994). Likewise, it has been shown that a single-chain antibody Fsv fragment fused to the Mo-MLV *env* gene product recognized its target epitope and that a corresponding viral vector could transduce cells that were resistant to infection by the parental Mo-MLV (Russell et al. 1993; Somia et al. 1995; Marin et al. 1996). Although these recombinant vectors demonstrated new tropism, they showed low levels of infectivity. Recombinant adenovirus (Ad)-encoding chimeric proteins were also engineered (Wickham et al. 1997). One of the two vectors redirected virus binding to α1 integrin (Wickham et al. 1995) and the other redirected binding to heparan sulfate (Wickham et al. 1996) by fusion of the Ad fiber with RGD or polylysine sequences, respectively. Although binding to the fiber receptor still occur, these viruses demonstrated increased transduction in multiple cell types lacking high levels of Ad fiber receptor (Wickham et al. 1997).

The engineering of redirected HSV vectors for infection of particular cell types will no doubt prove to be a formidable task. Although the envelope structure and mechanism of HSV attachment and entry have been studied in some detail, the complexity of the virus envelope and the observation that multiple glycoproteins are required for virus attachment and entry has made it difficult to assign specific and cooperative functions to individual glycoproteins in these processes. Nevertheless, modifications affecting virus attachment, penetration, and spread might still be used to direct viral infection and transgene expression. Described below following are the salient features of molecular biology of HSV infection and early experiments to alter the virus tropism.

HSV Attachment and Penetration

The envelope of HSV-1 contains at least ten glycoproteins (gB-gE, gG-gJ, gL, and gM) and four nonglycosylated integral membrane proteins (products of the UL20, UL34, UL45, and UL49.5 genes). Of the ten glycoproteins, gB, gD, gH, and gL are essential for viral infection (Cai et al. 1988; Desai et al. 1988; Ligas and Johnson 1988; Hutchinson et al. 1992a), whereas gC, gE, gG, gI, gJ, and gM are dispensable for infection in vitro (Spear 1993a,b; Steven and Spear 1997). Of the four nonglyco-sylated integral membrane proteins, the UL20 and UL45 gene products are nonessential envelope proteins (Baines and Roizman 1991; MacLean et al. 1991; Visalli and Brandt 1993; Haanes et al. 1994; Pyles and Thompson 1994), whereas the UL34 and UL49.5 products are essential structural proteins that may have roles in the assembly of virus-specific glycoproteins into mature virions (Purves et al. 1991, 1992; Barker and Roizman 1992; Visalli and Brandt 1993; Ward et al. 1994; Liang et al. 1996). Most HSV glycoproteins appear to be functionally independent, but they closely associate with one another in the particle (e.g., gB and gD) (Zhu and Courtney 1988, 1994; Handler et al. 1996), which may be important for infection and entry into the host cell. Some of them form functional homo-oligomers (e.g., gB) (Sarmiento and Spear 1979; Claesson-Welsh and Spear 1986; Highlander et al. 1991; Laquerre et al. 1996), whereas others form either functional or structural hetero-oligomers (e.g., gI/gE and gH/gL, respectively) (Hutchinson et al. 1992b; Roop et al. 1993; Westra et al. 1997). The envelope glycoproteins mediate infection of the host cell, which takes place in two identifiable stages: (1) attachment to the cell surface and (2) fusion of the viral envelope with the cell surface membrane, resulting in virus entry. In cell culture and in animals, the virion is also capable of infecting neighboring cells by moving transcellulary across cell membranes. This process, referred to as cell-to-cell or lateral spread, can be distinguished from infection by exocytosed particles by virtue of its resistance to virus-neutralizing antibodies. In addition, there is evidence that deletion of certain accessory envelope glycoproteins (e.g., gI/gE) prevents lateral spread but not the initial infection (Dingwell et al. 1994, 1995).

Virus attachment is mediated by several glycoproteins (Spear 1993a,b; Mettenleiter 1994). Binding of exposed domains of glycoproteins B and C to cell-surface glucosaminoglycans (GAGs), mainly heparan sulfate (HS)

(Wudunn and Spear 1989; Fuller and Lee 1992; Shih et al. 1992; Spear et al. 1992; Gruenheid et al. 1993; Herold et al. 1994), but also dermatan sulfate (Banfield et al. 1995; Williams and Straus 1997), has been described (Szilagyi and Cunningham 1991; Herold et al. 1995; Li et al. 1995). Together, this binding represents roughly 85% of the binding activity to Vero cells, the most commonly used cell type for HSV propagation, most of it through gC (Herold et al. 1994). Deletion of gC, a nonessential protein, and the HS-binding domain of gB, which is an essential protein, impairs binding to normal cells to an extent similar to the reduction in binding of wild-type virus to HS-deficient cells (Gruenheid et al. 1993), but the double deletion does not prevent virus adsorption, indicating that other receptors must be involved (S. Laquerre, unpubl.). The initial binding of a virus to cell-surface HS is followed by gD-mediated binding to secondary receptors, one of which, herpesvirus entry mediator (HVEM), was recently shown to be a member of the tumor necrosis factor-α/nerve growth factor (TNF-α/NGF) receptor family (Montgomory et al. 1996). The HVEM-binding domain of gD has been localized to the external domain within residues 1–275 (Whitbeck et al. 1997), and it was shown that expression of gD within the host cell prior to infection can block infection by wild-type virus. However, gD mutants with single amino acid substitutions at residues 25 or 27 in the external domain fail to block infection (Dean et al. 1994), and HSV-1 strains altered at these residues, such as ANG, KOS-rid1, and KOS-rid2, do not attach to Chinese hamster ovary (CHO) cells transduced with the HVEM receptor for wild-type KOS virus (Nicola et al. 1997; Whitbeck et al. 1997). These strains still infect Vero and Hela cells, suggesting the existence of additional receptors.

The sequential attachment steps in infection result in fusion of the viral envelope with the cell-surface membrane and viral entry into the cell. Although the events in penetration are not well understood, it is clear that multiple viral glycoproteins are involved. A role for gD is supported by evidence that an attached virus can be neutralized by anti-gD antibodies and by the observation that virus mutants deleted for gD attach to cells, but do not penetrate (Fuller and Spear 1985; Highlander et al. 1987). It is possible, however, that the tight binding afforded by gD triggers penetration, but that gD does not participate in execution of the fusion process. Mutants deleted for gH/gL or gB are also blocked for virus penetration, but are not defective in attachment. Both gB and gD are capable of inducing syncytia formation on low-pH treatment when expressed on the cell surface, which supports a possible role in fusion for both molecules (Butcher et al. 1990). Fuller and Lee (1992) have proposed that entry involves a cascade of events in which fusion is initiated by gD, a fusion bridge is formed most likely by the action of gB, and extension of the bridge followed by viral release inside the cell is achieved with the involvement of gH/gL.

Mutants deleted for individual accessory glycoprotein genes often display similar phenotypes in studies in which attachment, entry, and lateral spread are not analyzed separately. However, it has been demonstrated that removal of gI or gE does not affect virus entry of polarized epithelial cells (Balan et al. 1994), but greatly inhibits lateral spread in nonpolarized cells in the presence of a virus-neutralizing antibody; the latter was also observed for gM (Sears et al. 1991; Dingwell et al. 1994, 1995). gI and gE form heterodimers through interactions between their external domains (Johnson and Feenstra 1987; Basu et al. 1995; Whitbeck et al. 1996). Heterodimerized gI/gE binds the Fc portion of human as well as rabbit IgG (Johnson and Feenstra 1987; Dubin et al. 1994; Basu et al. 1995, 1997) and facilitates cell-to-cell virus spread in polarized cells.

Targeting of HSV-1 Vectors

To transform HSV-1 into a vector for cell type-specific gene delivery, its host range should be restricted and redirected. This involves two steps: (1) elimination of the natural tropism of the virus, and (2) expression of new viral ligands capable of binding to cell-surface receptors present mainly or exclusively on the targeted cell (Fig. 4). Our laboratory has pursued the first step by deleting the HS-binding capacity of HSV-1. Previous studies demonstrated that gB/gC double-deletion mutants are highly impaired for the initial adsorption phase of infection, suggesting that gB and gC are the principal mediators of HS binding (Herold et al. 1994). To eliminate HS binding, a mutant virus deleted for gC, the major HS-binding protein, was constructed. Complete deletion of gB, the other HS-binding protein, was not attractive because gB is essential for virus entry, but we identified the HS-binding domain of gB and removed the corresponding genetic information from the HSV-1 genome (S. Laquerre, unpubl.). The double-mutant virus, KgBpK$^-$gC$^-$, demonstrated an 80% reduction in binding to Vero cells compared to wild-type virus. Despite this reduced binding, however, the double-mutant virus was only slightly impaired in its ability to penetrate the host cell, suggesting that HS binding is not required for virus penetration.

In an effort to alter the tropism of the mutant virus, the HS-binding domain of gC (Trybala et al. 1994) was genetically replaced by EPO. Biochemical analysis demonstrated that gC:EPO fusion protein was incorporated in the budding virion and that recombinant virus was specif-

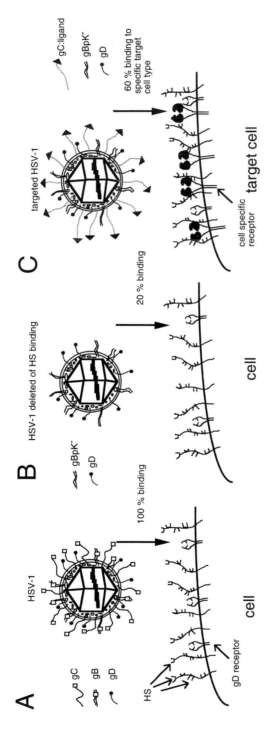

Figure 4 Schematic for targeting HSV-1 to a specific target cell type. (*A*) HSV-1 wild-type adsorption to the cell membrane through binding of gB and gC to heparan sulfate moieties and gD to its receptor. (*B*) Deletion of gC and the heparan-sulfate-binding domain of gB reduced the adsorption of mutant virus to the cell membrane relying on binding of gD to its receptor. (*C*) The binding capacity of HSV-1 mutant virus deleted for heparan sulfate-binding activities was restored (60% binding) to a specific target cell-type through the binding of a chimeric gC:ligand fusion molecule to a cell-specific receptor.

lcally retained on a soluble EPO-receptor column. The EPO virus demonstrated a twofold increase in binding to K562 cells, which expressed the EPO receptor, and a corresponding increase in infectivity (S. Laquerre, unpubl.). These data represent the first evidence that HSV-1 tropism can be redirected.

In an attempt to augment the binding of the recombinant virus to the EPO receptor, the HS-binding domain of gB was also replaced by EPO. The rationale for this strategy was that the amount of EPO expressed on the surface of the viral particle should be proportional to the efficiency of binding to the target cell until all EPO binding sites on the cell are saturated. However, experimental results demonstrated that the gB:EPO fusion protein was not processed through the endoplasmic reticulum and consequently was not incorporated into mature virions. In contrast, experiments to redirect gB binding to the acetylcholine receptor were successful. Snake venom α-bungarotoxin (α-BTX) binds the acetylcholine receptor with very high affinity (10^{-5} M) (Mebs et al. 1971; Liljestrom et al. 1991), and we used a 13-amino-acid α-BTX peptide that contained the acetylcholine-binding domain of full-length α-BTX to replace the HS-binding domain of gB. gB:BTX recombinant virus deleted for gC demonstrated a fivefold increase in binding to cells expressing the acetylcholine receptor compared to a gB/gC-deleted virus (S. Laquerre, unpubl.).

Together, our observations demonstrate that substitution of the HS-binding domains of gB and gC by novel ligands can enhance the binding of virus to cells displaying the corresponding receptors, although virus binding to novel receptors must be greatly improved to be of practical value. Because our targeted vectors retain wild-type gD that continues to recognize the natural virus receptors, specific targeting may be further improved by replacing gD with a gD:ligand fusion molecule.

HSV-VECTOR APPLICATIONS INVOLVING THE NERVOUS SYSTEM

Treatment of Brain Cancer

Strategies to treat cancer by gene therapy can be considered in two categories: (1) tumor-cell destruction by expression of transgenes, the products of which induce cell death, or sensitize the cells to chemotherapy (Moolten 1986) or radiation therapy (Hanna et al. 1997), and (2) tumor vaccination, through expression of transgenes, the products of which recruit, activate, or costimulate immunity or provide tumor antigens. Because these strategies are complementary, it has also been suggested that they can be used in combination. Examples include the use of prodrug-activating genes (Freeman et al. 1996), cytokines (Finke et al. 1997), MHC products such as costimulatory molecules (Chen et al. 1992; Katsanis et al. 1995; Galea-Lauri et al. 1996), allotypic class I or class II molecules (Chen et al. 1992; Katsanis et al. 1995; Galea-Lauri et al. 1996; Schmidt et al. 1996), and tumor antigens (Henderson and Finn 1996; Pecher and Finn 1996), which together may assist in the recruitment and activation of nonspecific inflammatory responses (Herrlinger et al. 1997) or the induction of tumor-specific immunity. We have constructed HSV-1 multigene vectors for experiments involving in vivo tumor therapy, which calls for direct injections of virus preparations. Vectors have also been designed to induce specific immunity by immunization against a tumor antigen displayed by the vector or by direct injection into the tumor for tumor-cell killing and immunization.

The HSV thymidine kinase (tk) gene is a suicide gene commonly employed both in experimental and human clinical protocols (Moolten 1986; Ram et al. 1986; Ezzeddine et al. 1991; Culver et al. 1992; Barba et al. 1993; Caruso et al. 1993). Transfer of tk into tumor cells results in tumor-cell death when the cells are exposed to the antiviral drug gancyclovir (GCV) as tk processes the prodrug into a toxic nucleotide analog, which upon incorporation into nascent DNA results in strand termination during DNA replication. A powerful characteristic of the tk/GVC approach is that the transduction of a small fraction of the tumor cells with the suicide gene can result in significant antitumor activity (bystander effect) (Ram et al. 1986; Short et al. 1990; Ezzeddine et al. 1991; Caruso et al. 1993; Burger and Scheithauer 1994). It has been demonstrated that cell-to-cell transfer of phosphorylated GCV via gap junctions between tk-transduced tumor cells and neighboring unmodified cells is a major mechanism of the bystander effect (Bi et al. 1993; Freeman et al. 1993; Kato et al. 1994; Wu et al. 1994). We have previously tested the ability of tk-overexpressing HSV vectors to act as a treatment for established tumors in rodent glioma models and observed significant increases in survival time. However, we believe that suicide gene therapy has a realistic chance as a cure only if it can be augmented (Freeman et al. 1993; Barba et al. 1994; Boviatsis et al. 1994; Chen et al. 1994; Perez-Cruet et al. 1994).

In an attempt to enhance the cell killing seen in suicide gene therapy, we have taken two approaches. In the first, we created a replication-defective HSV-1-based vector that expresses the human TNF-α gene product in conjunction with the HSV-tk gene (TH:TNF) (Fig. 2). (S. Moriuchi, unpubl.) TNF-α has been demonstrated to possess an array of antitumor activities, including potent cytotoxicity exerted directly on tumor cells (Han et al. 1994), enhancement of the expression of human leukocyte antigens (HLA) (Cao et al. 1997), and immune cell adhesion molecule-1 (ICAM-1) (Watanabe et al. 1989) on tumor cell surfaces, enhancement of interleukin-2 receptors on lymphocytes (Pfizenmaier et al. 1987), and promotion of the activation of such effector cells as natural killer (NK) cells, lymphokine-activated killer (LAK) cells, and cytotoxic T lymphocytes (CTLs) (Ostensen et al. 1987; Plaetinck et al. 1987; Rothlein et al. 1988). However, despite this promising antitumor profile, the clinical use of TNF-α has been constrained by the toxicity of systemic TNF-α delivery (Ranges et al. 1987; Owen-Schaub et al. 1988). The possibility existed though that local production of TNF-α at the site of tumor growth may allow for effective use of this cytokine as an antitumor agent. Indeed, the results of (Fig. 2) both cell culture and in vivo experiments with our TH:TNF vector showed that the effectiveness of suicide gene therapy with HSV-*tk* was enhanced when TNF-α was simultaneously expressed (S. Moriuchi, unpubl.). In the second approach, we introduced the connexin 43 gene into the HSV-*tk* expression vector (Fig. 2). The expression of connexin was found to increase the GCV-mediated bystander effect in tumor cells in which connexin was poorly expressed both in vitro and in vivo (P. Marconi, unpubl.). Experiments are in progress to combine these three genes to further augment GCV-activated killing of tumor cells focusing on animal models of brain cancers.

Treatment of Neurodegenerative Disease

Another class of diseases that might be treated by HSV-1-mediated gene transfer is that of neurodegenerative disease. Because the HSV latent state established in neurons is lifelong, HSV vectors have the potential to treat diseases in which the continued expression of a therapeutic factor might be used to prevent the loss of neurons destined to die by still poorly understood pathologic processes. The achievement of this goal will first require solving the problem of long-term gene expression, which may be achieved in the brain with, for instance, modifications of the LAT promoter region, and second by the identification of specific gene products whose expression would prevent neurodegeneration. As described previously, substantial progress has been made in the identification of latency-active promoter elements, and in laboratories worldwide a number of gene products with therapeutic potential in the treatment of neurodegeneration have been identified.

In short-term experimental animal models, delivery of the trophic factor NGF has been shown to block the loss of the cholinergic phenotype of septal cholinergic neurons following axotomy, and glial-derived neurotrophic factor (GDNF) was shown to prevent the death of nigral dopaminergic neurons induced by 6-hydroxydopamine (6-OHDA) or 1-methyl-4-phenyl-1,2,3,6-tetrahydropyridine (MPTP). In the peripheral nervous system (PNS), administration of the ciliary neurotrophic factor (CNTF) or GDNF has been shown to prevent the axotomy-induced death of motor neurons and the chronic neurodegenerative characteristic of the wobbler mouse (Sagot et al. 1995; Lindsay 1996; Ishiyama et al. 1997). However, in the central nervous system, delivery of these factors to specific regions of interest in the brain is limited by the blood–brain barrier created by the vascular endothelium, and brain–cerebral spinal fluid barriers provided by the ependymal lining limit the diffusion of systemically administered drugs given intraventricularly into brain parenchyma. For the nervous system, a cytokinelike side effect limited the dose of CNTF, which could be administered in therapeutic trials for the treatment of amyotrophic lateral sclerosis, and no beneficial effect was shown in humans despite impressive results in animal models (Cedarbaum and Stambler 1997; McGuire et al. 1997). Theoretically, vector-mediated delivery of the trophic factor gene to a specific brain region, or to a particular group of CNS or PNS neurons, would overcome these limitations, including systemic side effects.

For other applications, it may be necessary to express the transgene directly in an individual cell to achieve a therapeutic effect. For example, in conditions in which cell death results from the expression of a molecule with a toxic gain of function, antisense or ribozyme strategies to eliminate the toxic product would require cell-specific delivery. Similary, cell death occurring through apoptotic pathways could be blocked through the expression of molecules with antiapoptotic activty in these cells, but would require delivery of the transgene to every cell that is to be rescued.

To study the neuroprotective effects of neurotrophic factors in vitro and in vivo, we engineered a series of vectors that express β-NGF from either the human HCMV immediate-early promoter or our novel HSV latency promoter. The vectors employing the latency promoter enabled long-term β-NGF expression in mouse trigeminal ganglia (TG) and were able to ameliorate the toxic effects of hydrogen peroxide neurotoxicity in primary dorsal root ganglions (DRGs) in culture (W.F. Goins, unpubl.). We have employed similar vectors expressing Bcl-2 to protect cortical neurons in culture from apoptotic cell death (T.J. Oligino, unpubl.). We are currently examining the ability of these vectors to ameliorate the PNS neurodegeneration seen in cisplatinum-induced and diabetic neuropathies where LAP-driven therapeutic gene expression should be effective. Other applications for which LAP-directed transgene expression during latency could be effective include the production of factors that regulate the function of local nonneuronal tissues and the transmission of signals to the brain, such as in pain (S. Wilson, unpubl.).

SUMMARY

This overview of HSV biology and gene transfer has focused on the use of highly defective HSV genomic vectors that are blocked very early in the virus lytic cycle. These vectors express few viral functions and are highly reduced in vector toxicity even for primary neurons in culture that are readily killed by less defective HSV vectors. Moreover, these vector backgrounds are suitable for expression of multiple transgenes or single large genes (e.g., dystrophin) (G.R. Akkaraju, unpubl.) in applications where short-term gene expression and multiple-gene products are required to achieve a therapeutic outcome (e.g., tumor-cell killing and vaccination). Expression of these transgenes can be coordinated or even sequential using the same strategies employed by the virus in its cascading of gene expression (Oligino et al. 1996). Expression can also be controlled by drug-sensitive *trans*-activators (T.J. Oligino, unpubl.), which may prove to be important for regulating the timing and duration of transgene expression. HSV vectors may be most suited for expression of genes in the nervous system in which the virus has evolved to remain lifelong in a latent state. The highly defective viruses deleted for multiple IE genes are able to establish latency efficiently in neurons and serve as a platform for long-term gene expression using the latency promoter system. These mutants cannot reactivate from latency and cannot spread to other nerves or tissues following infection of cells. Delivery of these vectors requires direct inoculation of tissue to achieve direct contact with neurons. Ideally, HSV vectors would be most effective if infection could be targeted to specific cell types using enveloped particles defective for their normal receptor recognition ligands, but modified to contain novel attachment and entry functions. This area of research is still very early in development and it remains to be determined as to what extent this will be feasible. Finally, it should be emphasized that viral delivery systems may each become reduced to highly defective viruses retaining only those elements and expressed genes required for vector DNA maintenance and transgene expression. Fortunately, the natural biology of many persistent viruses indicates that long-term vector maintenance will be possible, and we continue to learn from the highly evolved biology of persistent and latent viruses to mimic their strategies for gene transfer and therapy.

REFERENCES

Baines J.D. and Roizman B. 1991. The open reading frames UL3, UL4, UL10 and UL16 are dispensable for the replication of herpes simplex virus 1 in cell culture. *J. Virol.* **65:** 938–944.

Balan P., Davis-Poynter N., Bell S., Atkinson H., Brown H., and Minson T. 1994. An analysis of the in vitro and in vivo phenotypes of mutants of herpes simplex virus type 1 lacking glycoproteins gG, gE, gI or the putative gJ. *J. Gen. Virol.* **75:** 1245–1258.

Banfield B.W., Leduc Y., Esford L., Schubert K., and Tufaro F. 1995. Sequential isolation of proteoglycan synthesis mutants by using herpes simplex virus as a selective agent: Evidence for a proteoglycan-independent virus entry pathway. *J. Virol.* **69:** 3290–3298.

Barba D., Hardin J., Ray J., and Gage F.H. 1993. Thymidine kinase-mediated killing of rat brain tumors. *J. Neurosurg.* **79:** 729–735.

Barba D., Hardin J., Sadelain M., and Gage F.H. 1994. Development of antitumor immunity for following thymidine kinase-mediated killing of experimental brain tumors. *Proc. Natl. Acad. Sci.* **91:** 4348–4352.

Barker D.E. and Roizman B. 1992. The unique sequence of the herpes simplex virus 1 L component contains an additional translated open reading frame designated UL49.5. *J. Virol.* **66:** 562–566.

Basu S., Dubin G., Basu M., Nguyen V., and Friedman H.M. 1995. Characterization of regions of herpes simplex virus type 1 glycoprotein E involved in binding the Fc domain of monomeric IgC and in forming a complex with glycoprotein I. *J. Immunol.* **154:** 260–267.

Basu S., Dubin G., Nagashunmugam T., Basu M., Goldstein L.T., Wang L., Weeks B., and Friedman H.M. 1997. Mapping regions of herpes simplex virus type 1 glycoprotein I required for formation of the viral Fc receptor for monomeric IgG. *J. Immunol.* **158:** 209–215.

Batchelor A.H. and O'Hare P.O. 1990. Regulation and cell-type-specific activity of a promoter located upstream of the latency-associated transcript of herpes simplex virus type 1. *J. Virol.* **64:** 3269–3279.

———. 1992. Localization of *cis*-acting sequence requirements in the promoter of the latency-associated transcript of herpes simplex virus type 1 required for cell-type-specific activity. *J. Virol.* **66:** 3573–3582.

Bi W.L., Parysek L.M., Warnick R., and Stambrook P.J. 1993. In vitro evidence that metabolic cooperation is responsible for the bystander effect observed with HCV Tk retroviral gene therapy. *Hum. Gene Ther.* **4:** 725–731.

Boviatsis E.J., Park J.S., Sena-Esteves M., Kramm C.M., Chase M., Efird J.T., Wei M.X., Breakefield X.D., and Chiocca E.A. 1994. Long-term survival of rats harboring brain neoplasms treated with ganciclovir and a herpes simplex virus vector that retains an intact thymidine kinase gene. *Cancer Res.* **15:** 5745–5751.

Burger P. and Scheithauer B. 1994. Tumors of the central nervous system. In *Atlas of tumor pathology*, Third Series, Fascicle 10, p. 59. Armed Forces Institute of Pathology, Washington, DC.

Butcher M., Raviprakash K., and Ghosh H.P. 1990. Acid pH-induced fusion of cells by herpes simplex virus glycoproteins gB and gD. *J. Biol. Chem.* **265:** 5862–5868.

Cai W., Gu B., and Person S. 1988. Role of glycoprotein B of herpes simplex virus type 1 in viral entry and cell fusion. *J. Virol.* **62:** 2596–2604.

Campbell M.E.M., Palfeyman J.W., and Preston C.M. 1984. Identification of herpes simplex virus DNA sequences which encode a *trans* acting polypeptide responsible for stimulation of immediate early transcription. *J. Mol. Biol.* **180:** 1–19.

Cao G., Kuriyama S., Du P., Sakamoto T., Kong X., Matsui K., and Qi Z. 1997. Complete regression of established murine hepatocellular carcinoma by in vivo tumor necrosis factor α gene transfer. *Gastroenterology* **11:** 270–278.

Carpenter D.E. and Stevens J.G. 1996. Long-term expression of a foreign gene from a unique position in the latent herpes simplex virus genome. *Hum. Gene Ther.* **7:** 1447–1454.

Carter K. and Roizman B. 1996. The promoter and transcriptional unit of a novel herpes simplex virus 1 alpha gene are contained in, and encode a protein in frame with, the open reading frame of the alpha22 gene. *J. Virol.* **70:** 172–178.

Caruso M., Panis Y., Gagandeep S., Houssin D., Salzmann J.L., and Klatzmann D. 1993. Regression of established macroscopic liver metastases after in situ transduction of a suicide gene. *Proc. Natl. Acad. Sci.* **90:** 7024–7028.

Cedarbaum J.M. and Stambler N. 1997. Performance of the Amyotrophic Lateral Sclerosis Functional Rating Scale (ALSFRS) in multicenter clinical trials. *J. Neurol. Sci.* (suppl. 1) **152:** S1–S9.

Chen L., Ashe S., Brady W., Hellstrom I., Hellstrom K.E., Ledbetter J.A., McGowan P., and Linsley P.S. 1992. Costimulation of antitumor immunity by the B7 counterreceptor for the T lymphocyte molecules CD28 and CTLA-4. *Cell* **71:** 1093–1102.

Chen S.H., Shine H.D., Goodman J.C., Grossman R.G., and Woo S.L. 1994. Gene therapy for brain tumors: Regression of experimental gliomas by adenovirus-mediated gene transfer in vivo. *Proc. Natl. Acad. Sci.* **91:** 3054–3057.

Chen X., Schmidt M.C., Goins W.F., and Glorioso J.C. 1995. Two herpes simplex virus type-1 latency active promoters differ in their contribution to latency-associated transcript expression during lytic and latent infection. *J. Virol.* **69:** 7899–7908.

Claesson-Welsh L. and Spear P. G. 1986. Oligomerization of herpes simplex virus glycoprotein B. *J. Virol.* **60:** 803–806.

Croen K.D., Ostrove J.M., Dragovic L.J., Smialek J.E., and Straus S.E. 1987. Latent herpes simplex virus in human trigeminal ganlia. Detection of an immediate early gene "anti-sense" transcript by in situ hybridization. *N. Engl. J. Med.* **317:** 1427–1432.

Culver K., Ram Z., Walbridge S., Ishii H., Oldfield E., and Blaese R. 1992. In vivo gene transfer with retroviral vector-producer cells for treatment of experimental brain tumors. *Science* **256:** 1550–1552.

Dean H.J., Terhune S.S., Shieh M.T., Susmarski N., and Spear P.G. 1994. Single amino acid substitutions in gD of herpes simplex virus 1 confer resistance to gD-mediated interference and cause cell-type-dependent alterations in infectivity. *Virology* **199:** 67–80.

Deatly A.M., Spivack J.G., Lavi E., and Fraser N.W. 1987. RNA from an immediate early region of the HSV-1 genome is present in the trigeminal ganglia of latently infected mice. *Proc. Natl. Acad. Sci.* **84:** 3204–3208.

Deatly A.M., Spivack J.G., Lavi E., O'Boyle D., and Fraser N.W. 1988. Latent herpes simplex virus type 1 transcripts in peripheral and central nervous systems tissues of mice map to similar regions of the viral genome. *J. Virol.* **62:** 749–756.

DeLuca N.A. and Schaffer P.A. 1985. Activation of immediate-early, early, and late promoters by temperature-sensitive and wild-type forms of herpes simplex virus type 1 protein ICP4. *N. Engl. J. Med.* **5:** 1997–2008.

Desai P., Schaffer P., and Minson A. 1988. Excretion of non-infectious virus particles lacking glycoprotein H by a temperature-sensitive mutant of herpes-simplex virus type 1:

Evidence that gH is essential for virion infectivity. *J. Gen. Virol.* **69:** 1147– 1156.

Desai P., Ramakrishnan R., Lin Z.W., Ozak B., Glorioso J.C., and Levine M. 1993. The RR1 gene of herpes simplex virus type 1 is uniquely transactivated by ICP0 during infection. *J. Virol.* **67:** 6125–6135.

Dingwell K., Brunetti C., Hendricks R., Tang Q., Tang M., Rainbow A.J., and Johnson D.C. 1994. Herpes simplex virus glycoproteins E and I facilitate cell-to-cell spread in vivo and across junctions of cultured cells. *J. Virol.* **68:** 834–845.

Dingwell K.S., Doering L.C., and Johnson D.C. 1995. Glycoproteins E and I facilitate neuron-to-neuron spread of herpes simplex virus. *J. Virol.* **69:** 7087–7098.

Dixon R.A.F. and Schaffer P.A. 1980. Fine-structure mapping and functional analysis of temperature-sensitive mutants in the gene encoding the herpes simplex virus type 1 immediate early protein VP175. *J. Virol.* **36:** 189–203.

Dobson A.T., Sederati F., Devi-Rao G., Flanagan W.M., Farrell M.J., Stevens J.G., Wagner E.K., and Feldman L.T. 1989. Identification of the latency-associated transcript promoter by expression of rabbit β-globin mRNA in mouse sensory nerve ganglia latently infected with a recombinant herpes simplex virus. *J. Virol.* **63:** 3844–3851.

Doerig C., Pizer L., and Wilcox C. 1991. Detection of the latency-associated transcript in neuronal cultures during the latent infection with herpes simplex virus type 1. *Virology* **183:** 423–426.

Dubin G., Basu S., Mallory D.L., Basu M., Tal-Singer R., and Friedman H.M. 1994. Characterization of domains of herpes simplex virus type 1 glycoprotein E involved in Fc binding activity for immunoglobulin G aggregates. *J. Virol.* **68:** 2478–2485.

Emi N., Friedmann T., and Yee J. 1991. Pseudotype formation of murine leukemia virus with the G protein of vesicular stomatitis virus. *J. Virol.* **65:** 1202–1207.

Everett R.D. 1984. Transactivation of transcription by herpes simplex virus products: Requirements for two herpes simplex virus type 1 immediate early polypeptides for maximum activity. *EMBO J.* **3:** 3135–3141.

Ezzeddine Z.D., Martuza R.L., Platika D., Short M.P., Malick A., Choi B., and Breakefield X.D. 1991. Selective killing of glioma cells in culture and in vivo by retrovirus transfer of the herpes simplex virus thymidine kinase gene. *New Biol.* **3:** 608– 614.

Farrell M.J., Dobson A.T., and Feldman L.T. 1991. Herpes simplex virus latency-associated transcript is a stable intron. *Proc. Natl. Acad. Sci.* **88:** 790–794.

Fink D.J., Ramakrishnan R., Marconi P., Goins W.F., Holland T.C., and Glorioso J.C. 1996b. Advances in the development of herpes simplex virus-based gene transfer vectors for the nervous system. *Clin. Neurosci.* **3:** 1–8.

Finke S., Trojaneck B., Moller P., Schadendorf D., Neubauer A., Huhn D., and Schmidt-Wolf I.G. 1997. Increase of cytotoxic sensitivity of primary human melanoma cells transfected with the interleukin-7 gene to autologous and allogeneic immunologic effector cells. *Cancer Gene Ther.* **4:** 260–268.

Freeman S.M., Whartenby K.A., Freeman J.L., Abboud C.N., and Marrogi A.J. 1996. In situ use of suicide genes for cancer therapy. *Semin. Oncol.* **23:** 31–45.

Freeman S.M., Abboud C.N., Whartenby K.A., Packman C.H., Koeplin D.S., Moolten F.L., and Abraham G.N. 1993. The "Bystander Effect": Tumor regression when a fraction of the tumor mass is genetically modified. *Cancer Res.* **53:** 5274–5283.

French S.W., Schmidt M.C., and Glorioso J.C. 1996. Involvement of an HMG protein in the transcriptional activity of the herpes simplex virus latency active promoter 2. *Mol. Cell. Biol.* **16:** 5393–5399.

Fuller A.O. and Lee W.C. 1992. Herpes simplex virus type 1 entry through a cascade of virus-cell interactions requires different roles of gD and gH in penetration. *J. Virol.* **66:** 5002–5012.

Fuller A.O. and Spear P.G. 1985. Specificities of monoclonal and polyclonal antibodies that inhibit adsorption of herpes simplex virus to cells and lack of inhibition by potent neutralizing antibodies. *J. Virol.* **55:** 475–482.

Galea-Lauri J., Farzaneh F., and Gaken J. 1996. Novel costimulators in the immune gene therapy of cancer. *Cancer Gene Ther.* **3:** 202–214.

Goins W.F., Sternberg L.R., Croen K.D., Krause P.R., Hendricks R.L., Fink D.J., Straus S.E., Levine M., and Glorioso J.C. 1994. A novel latency-active promoter is contained within the herpes simplex virus type 1 U_L flanking repeats. *J. Virol.* **68:** 2239–2252.

Gruenheid S., Gatzke L., Meadows H., and Tufaro F. 1993. Herpes simplex virus infection and propagation in a mouse L cell mutant lacking heparan sulfate proteoglycans. *J. Virol.* **67:** 93–100.

Haanes E.J., Nelson C.M., Soule C.L., and Goodman J.L. 1994. The UL45 gene product is required for herpes simplex virus type 1 glycoprotein B-induced cell fusion. *J. Virol.* **68:** 5825–5834.

Han S.K., Brody S.L., and Crystal R.G. 1994. Suppression of in vivo tumorigenicity of human lung cancer cells by retrovirus-mediated transfer of the human tumor necrosis factor-α cDNA. *Am. J. Respir. Cell Mol. Biol.* **11:** 270–278.

Handler C.G., Eisenberg R.J., and Cohen G.H. 1996. Oligomeric structure of glycoproteins in herpes simplex virus type 1. *J. Virol.* **70:** 6067–6070.

Hanna N.N., Mauceri H.J., Wayne J.D., Hallahan D.E., Kufe D.W., and Weichselbaum R.R. 1997. Virally directed cytosine deaminase/5-fluorocytosine gene therapy enhances radiation response in human cancer xenografts. *Cancer Res.* **57:** 4205–4209.

Henderson R.A. and Finn O.J. 1996. Human tumor antigens are ready to fly. *Adv. Immunol.* **62:** 217–256.

Herold B., Visalli R., Susmarski N., Brandt C., and Spear P. 1994. Glycoprotein C-independent binding of herpes simplex virus to cells requires cell surface heparan sulfate and glycoprotein B. *J. Gen. Virol.* **75:** 1211–1222.

Herold B.C., Gerber S.I., Polonsky T., Belval B.J., Shaklee P.N., and Holme K. 1995. Identification of structural features of heparin required for inhibition of herpes simplex virus type 1 binding. *Virology* **206:** 1108–1116.

Herrlinger U., Kramm C.M., Johnston K.M., Louis D.N., Finkelstein D., Reznikoff G., Dranoff G., Breakefield X.O., and Yu J. 1997. Vaccination for experimental gliomas using GM-CSF-transduced glioma cells. *Cancer Gene Ther.* **4:** 345–352.

Highlander S.L., Goins W.F., Person S., Holland T.C., Levine M., and Glorioso J.C. 1991. Oligomer formation of the gB glycoprotein of herpes simplex virus type 1. *J. Virol.* **65:** 4275–4283.

Highlander S.L., Sutherland S.L., Gage P.J., Johnson D.C., Levine M., and Glorioso J.C. 1987. Neutralizing monoclonal antibodies specific for herpes simplex virus glycoprotein D inhibit virus penetration. *J. Virol.* **61:** 3356–3364.

Hill A. and Ploegh H. 1995. Getting the inside out: The transporter associated with antigen processing (TAP) and the presentation of viral antigen. *Proc. Natl. Acad. Sci.* **92:** 341–343.

Ho D.Y. and Mocarski E.S. 1989. Herpes simplex virus latent RNA (LAT) is not required for latent infection in the mouse. *Proc. Natl. Acad. Sci.* **86:** 7596–7600.

Holland L.E., Anderson K.P., Shipman C., and Wagner E.K. 1980. Viral DNA synthesis is required for efficient expression of specific herpes simplex virus type 1 mRNA.

Virology **101:** 10–24.

Huard J., Akkaraju G., Watkins S.C., Pike-Cavalcoli M., and Glorioso J.C. 1997. LacZ gene transfer to skeletal mucsle using a replication-defective herpes simplex virus type 1 mutant vector. *Hum. Gene Ther.* **8:** 439–452.

Hutchinson L., Goldsmith K., Snoddy D., Ghosh H., Graham F.L., and Johnson D.C. 1992a. Identification and characterization of a novel herpes simplex virus glycoprotein, gK, involved in cell fusion. *J. Virol.* **66:** 5603–5609.

Hutchinson L., Browne H., Wargent V., Davis-Poynter N., Primorac S., Goldsmith K., Minson A.C., and Johnson D.C. 1992b. A novel herpes simplex virus glycoprotein, gL, forms a complex with glycoprotein H (gH) and affects normal folding and surface expression of gH. *J. Virol.* **66:** 2240–2250.

Ishiyama T., Mitsumoto H., Pioro E.P., and Klinkosz B. 1997. Genetic transfer of the wobbler gene to a C57BL/6J x NZB hybrid stock: Natural history of the motor neuron disease and response to CNTF and BDNF cotreatment. *Exp. Neurol.* **148:** 247–255.

Javier R.T., Stevens J.G., Dissette V.B., and Wagner E.K. 1988. A herpes simplex virus transcript abundant in latently infected neurons is dispensible for establishment of the latent state. *Virology* **166:** 254–257.

Jenkins F.J., Casadaban M.J., and Roizman B. 1985. Application of the mini-Mu-phage for target-sequence-specific insertional mutagenesis of the herpes simplex virus genome. *Proc. Natl. Acad. Sci.* **82:** 4773–4777.

Jia W.W.-G., McDermott M., Goldie J., Cynader M., Tan J., and Tufaro F. 1994. Selective destruction of gliomas in immunocompetent rats by thymidine kinase-defective herpes simplex virus type 1. *J. Natl. Cancer Inst.* **86:** 1209–1215.

Johnson D.C. and Feenstra V. 1987. Identification of a novel herpes simplex virus type 1-induced glycoprotein which complexes with gE and binds immunoglobulin. *J. Virol.* **61:** 2208–2216.

Johnson P., Wang M., and Friedmann T. 1994. Improved cell survival by the reduction of immediate-early gene expression in replication-defective mutants of herpes simplex virus type 1 but not by mutation of the viron host shutoff function. *J. Virol.* **68:** 6347–6362.

Johnson P., Miyanohara A., Levine F., Cahill T., and Friedmann T. 1992. Cytotoxicity of a replication-defective mutant herpes simplex virus type 1. *J. Virol.* **66:** 2952–2965.

Kasahara N., Dozy M., and Kan Y.W. 1994. Tissue-specific targeting of retroviral vectors through ligand-receptor interactions. *Science* **255:** 1373–1376.

Kasai Y., Chen H., and Flint S.J. 1992. Anatomy of an unusual RNA polymerase II promoter containing a downstream TATA element. *Mol. Cell. Biol.* **12:** 2884–2897.

Kato K., Yoshida J., Mizuno M., Sugita K., and Emi N. 1994. Retroviral transfer of herpes simplex thymidine kinase into glioma cells targeting of gancyclovir cytotoxic effect. *Neurol. Med. Chir.* **34:** 339–344.

Katsanis E., Xu Z., Bausero M.A., Dancisak B.B., Gorden K.B., Davis G., Gray G.S., Orchard P.J., and Blazar B.R. 1995. B7-1 expression decreases tumorigenicity and induces partial systemic immunity to murine neuroblastoma deficient in major histocompatibility complex and costimulatory molecules. *Cancer Gene Ther.* **2:** 39–46.

Kenny J., Millhouse S., Wotring M., and Wigdahl B. 1997. Upstream stimulatory factor family binds to the herpes simplex virus type 1 latency-associated transcript promoter. *Virology* **230:** 381–391.

Kenny J.I., Krebs F.C., Hartle H.T., Gartner A.E., Chatton B., Leiden J.M., Hoeffler J.P., Weber P.C., and Wigdahl B.G. 1994. Identification of a second ATF/CREB-like element in the herpes simplex virus type 1 (HSV-1) latency-associated transcript (LAT) promoter. *Virology* **200:** 220–235.

Koller E., Hayman A.R., and Trueb B. 1991. The promoter of the chicken α2(VI) collagen gene has features characteristic of housekeeping genes and of proto-oncogenes. *Nucleic Acids Res.* **19:** 485–491.

Kolluri R., Torrey T.A., and Kinniburgh A.J. 1992. A CT promoter element binding protein: Definition of a double-strand and a novel single-strand DNA binding motif. *Nucleic Acids Res.* **20:** 111–116.

Krisky D., Marconi P., Oligino T., Rouse R., Fink D., and Glorioso J. 1997. Rapid method for construction of recombinant HSV gene transfer vectors. *Gene Ther.* **4:** 1120–1125.

Krummenacher C., Zabolotny J., and Fraser N. 1997. Selection of a nonconsensus branch point is influenced by an RNA stem-loop structure and is important to confer stability to the herpes simplex virus 2-kilobase latency-associated transcript. *J. Virol.* **71:** 5849–5860.

Kwong A.D., Kruper J.A., and Frenkel N. 1988. Herpes simplex virus virion host shutoff function. *J. Virol.* **62:** 912–921.

Lachmann R.H. and Efstathiou S. 1997. Utilization of the herpes simplex virus type 1 latency-associated regulatory region to drive stable reporter gene expression in the nervous system. *J. Virol.* **71:** 3197–3207.

Laquerre S., Person S., and Glorioso J.C. 1996. Glycoprotein B of herpes simplex virus type 1 oligomerizes through the intermolecular interaction of a 28 amino acid domain. *J. Virol.* **70:** 1640–1650.

Leib D.A., Bogard C.L., Kosz-Vnenchak M., Hicks K.A., Coen D.M., Knipe D.M., and Schaffer P.A. 1989. A deletion mutant of the latency-associated transcript of herpes simplex virus type 1 reactivates from the latent infection. *J. Virol.* **63:** 2893–2900.

Leiden J., Frenkel N., and Rapp F. 1980. Identification of the herpes simplex virus DNA sequences present in six herpes simplex virus thymidine kinase-transformed mouse cell lines. *J. Virol.* **33:** 272–85.

Li Y., van Drunen Littel-van den Hurk S., Babiuk L.A., and Liang X. 1995. Characterization of cell-binding properties of bovine herpesvirus 1 glycoproteins B, C, and D: Identification of a dual cell-binding function of gB. *J. Virol.* **69:** 4758–4768.

Liang X., Chow B., Raggo C., and Babiuk L.A. 1996. Bovine herpesvirus 1 UL49.5 homolog gene encodes a novel viral envelope protein that forms a disulfide-linked complex with a second virion structural protein. *J. Virol.* **70:** 1448–1454.

Ligas M. and Johnson D. 1988. A herpes simplex virus mutant in which glycoprotein D sequences are replaced by β-galactosidase sequences binds to but is unable to penetrate into cells. *J. Virol.* **62:** 1486–1494.

Liljestrom P., Lusa S., Huylebroeck D., and Garoff H. 1991. In vitro mutagenesis of a full-length cDNA clone of semliki forest virus: The small 6,000-molecular-weight membrane protein modulates virus release. *J. Virol.* **65:** 4107–4113.

Lindsay R.M. 1996. Therapeutic potential of the neurotrophins and neurotrophin-CNF combinations in peripheral neuropathies and motor neuron diseases. *Ciba Found. Symp.* **196:** 39–48.

Lokensgard J.R., Berthomme H., and Feldman L.T. 1997. The latency-associated promoter of herpes simplex virus type 1 requires a region downstream of the transcription start site for long-term expression during latency. *J. Virol.* **71:** 6714–6719.

Lokensgard J.R., Bloom D.C., Dobson A.T., and Feldman L.T. 1994. Long-term promoter activity during herpes simplex virus latency. *J. Virol.* **68:** 7148–7158.

Mackem S. and Roizman B. 1982. Differentiation between alpha promoter and regulatory regions of herpes simplex virus type 1: The functional domains and sequence of a movable alpha regulator. *Proc. Natl. Acad. Sci.* **79:** 4917–4921.

MacLean C., Esfstathiou S., Elliott M., Jamieson F., and McGeoch D. 1991. Investigation of herpes simplex virus type 1 genes encoding multiply inserted membrane proteins. *J. Gen. Virol.* **72:** 897–906.

Marconi P., Krisky D., Oligino T., Poliani P.L., Ramakrishnan R., Goins W.F., Fink D.J., and Gloriaso J.C. 1996. Replication-defective HSV vectors for gene transfer in vivo. *Proc. Natl. Acad. Sci.* **93:** 11319–11320.

Margolis T.P., Bloom D.C., Dobson A.T., Feldman L.T., and Stevens J.G. 1993. Decreased reporter gene expression during latent infection with HSV LAT promoter constructs. *Virology* **197:** 585–592.

Marin M., Noel D., Valsesia-Wittman S., Brockly F., Etienne-Julan M., Russell S., Cosset F.L., and Piechaczyk N. 1996. Targeted infection of human cells via major histocompatibility complex class I molecules by Moloney murine leukemia virus-derived viruses displaying single-chain antibody fragment-envelope fusion proteins. *J. Virol.* **70:** 2957–2962.

Maul G. and Everett R. 1994. The nuclear location of PML, a cellular member of the C3HC4 zinc-binding domain protein family, is rearranged during herpes simplex virus infection by the C3HC4 viral protein ICP0. *J. Gen. Virol.:* **75:** 1223–1233.

Mavromara-Nazos P. and Roizman B. 1987. Activation of herpes simplex virus 1 γ2 genes by viral DNA replication. *Virology* **161:** 593–598.

McCarthy A.M., McMahan L., and Schaffer P.A. 1989. Herpes simplex virus type 1 ICP27 deletion mutants exhibit altered patterns of transcription and are DNA deficient. *J. Virol.* **63:** 18–27.

McDermott J.B., Peterson C.A., and Piatigorsky J. 1992. Structure and lens expression of the gene encoding chicken βA3/A1-crystallin. *Gene* **117:** 193–200.

McGuire D., Ross M.A., Petajan J.H., Parry G.J., and Miller R. 1997. A brief quality-of-life measure for ALS clinical trials based on a subset of items from the sickness impact profile. The Syntex-Synergen ALS/CNTF Study Group. *J. Neurol. Sci.* **152:** S18–S22.

Mebs D., Narita K., Iwanaga S., Samejima Y., and Lee C. 1971. Amino acid sequence of α-bungarotoxin from the venom of Bungarus Multicinctus. *Biochem. Biophys. Res. Commun.* **44:** 711–716.

Mercadal C.M., Bouley D.M., DeStephano D., and Rouse B.T. 1993. Herpetic stromal keratitis in the reconstituted scid mouse model. *J. Virol.* **67:** 3404–3408.

Mester J.C., Pitha P., and Glorioso J.C. 1995. Anti-viral activity of herpes simplex virus vectors expressing α-interferon. *Gene Ther.* **3:** 187–196.

Mettenleiter T.C. 1994. Initiation and spread of α-herpesvirus infections. *Trends Microbiol.* **2:** 2–3.

Mitchell B.M. and Stevens J.G. 1996. Neuroinvasive properties of herpes simplex virus type 1 glycoprotein variants are controlled by the immune response. *J. Immunol.* **156:** 246–255.

Montgomory R.I., Warner M.S., Lum B.J., and Spear P.G. 1996. Herpes simplex virus 1 entry into cells mediated by a novel member of the TNF/NGF receptor family. *Cell* **87:** 427–436.

Moolten F.L. 1986. Tumor chemosensitivity conferred by inserted herpes thymidine kinase genes: Paradigm for a prospective cancer control strategy. *Cancer Res.* **46:** 5276–5281.

Mullen M.-A., Gerstberger S., Ciufo D.M., Mosca J.D., and Hayward, G.S. 1995.

Evaluation of colocalization interactions between the IE110, IE175, and IE63 transactivator proteins of herpes simplex virus within subcellular punctate structures. *J. Virol.* **69:** 476–491.

Naldini L., Blomer U., Gallay P., Mulligan R., Gage F.H., Verma I.M., and Trono D. 1996. In vivo gene delivery and stable transduction of nondividing cells by a lentiviral vector. *Science* **272:** 263–267.

Neda H., Wu C.H., and Wu G.Y. 1991. Chemical modification of an ecotropic murine leukemia virus results in redirection of its target cell specificity. *J. Biol. Chem.* **266:** 14143–14146.

Nicola A.V., Peng C., Lou H., Cohen G.H., and Eisenberg R.J. 1997. Antigenic structure of soluble herpes simplex virus (HSV) glycoprotein D correlates with inhibition of HSV infection. *J. Virol.* **71:** 2940–2946.

Nicosia M., Deshmane S.L., Zabolotny J.M., Valyi-Nagy T., and Fraser N.W. 1993. Herpes simplex virus type 1 Latency-Associated Transcript (LAT) promoter deletion mutants can express a 2-kilobase transcript mapping to the LAT region. *J. Virol.* **67:** 7276–7283.

Ohno K., Sawai K., Iijima Y., Levin B., and Meruelo D. 1997. Cell-specific targeting of Sindbis virus vectors displaying IgG-binding domains of protein A. *Nat. Biotechnol.* **15:** 763–767.

Oligino T., Poliani P.L., Marconi P., Bender M.A., Schmidt M.C., Fink D.J., and Glorioso J.C. 1996. In vivo transgene activation from an HSV-based gene vector by GAL4: VP16. *Gene Ther.* **3:** 892–899.

Oroskar A. and Read G. 1989. Control of mRNA stability by the virion host shutoff function of herpes simplex virus. *J. Virol.* **63:** 1897–1906.

Ostensen M.E., Thiele D.L., and Lipsky P.E. 1987. Enhancement of human natural killer cell function by the combined effects of tumor necrosis factor α or interleukin-1 and interferon-α or interleukin-2. *J. Biol. Response Modif.* **8:** 53–61.

Owen-Schaub L.B., Gutterman J.U., and Grimm E.A. 1988. Synergy of tumor necrosis factor and interleukin 2 in the activation of human cytotoxic lymphocytes: Effect of tumor necrosis factor α and interleukin 2 in the generation of human lymphokine-activated killer cell cytotoxicity. *Cancer Res.* **48:** 788–792.

Palella T., Hidaka Y., Silverman L., Levine M., Glorioso J., and Kelley W. 1989. Expression of human HPRT mRNA in brains of mice infected with a recombinant herpes simplex virus type 1 vector. *Gene* **80:** 137–144.

Pecher G. and Finn O.J. 1996. Induction of cellular immunity in chimpanzees to human tumor-associated antigen mucin by vaccination with MUC-1 cDNA-transfected Epstein-Barr virus-immortalized autologous B cells. *Proc. Natl. Acad. Sci.* **93:** 1699–1704.

Perez-Cruet M., Trask T.W., Chen S.H., Goodman J.C., Woo S.L, Grossman R.G., and Shine H.D. 1994. Adenovirus-mediated gene therapy of experimental gliomas. *J. Neurosci. Res.* **39:** 506–511.

Pfizenmaier K., Pfizenmaier K., Scheurich P., Schluter C., and Kronke M. 1987. Tumor necrosis factor enhances HLA-A, B, C and HLA-DR gene expression in human tumor cells. *J. Immunol.* **138:** 975–980.

Plaetinck G., Declercq W., Tavernier J., Jabholz M., and Fiers W. 1987. Recombinant tumor necrosis factor can induce interleukin 2 receptor expression and cytolytic activity in a rat x mouse T cell hybrid. *Eur. J. Immunol.* **17:** 1835–1838.

Postel E.H., Mango S.E., and Flint S.J. 1989. A nuclease-hypersensitive element of the human c-myc promoter interacts with a transcription initiation factor. *Mol. Cell. Biol.* **9:** 5123–5133.

Preston C., Frame M., and Campbell M. 1988. A complex formed between cell components and an HSV structural polypeptide binds to a viral immediate early gene regulatory DNA sequence. *Cell* **52:** 425–434.

Purves F.C., Spector D., and Roizman B. 1991. The herpes simplex virus 1 protein kinase encoded by the US3 gene mediates posttranslational modification of the phosphoprotein encoded by the UL34 gene. *J. Virol.* **65:** 5757–5764.

———— 1992. UL34, the target of the herpes simplex virus US3 protein kinase, is a membrane protein which in its unphosphorylated state associates with novel phosphoproteins. *J. Virol.* **66:** 4295–4303.

Pyles R.B. and Thompson R.L. 1994. Mutations in accessory DNA replicating functions alter the relative mutation frequency of herpes simplex virus type 1 strains in cultured murine cells. *J. Virol.* **68:** 4514–4524.

Rader K.A., Ackland-Berglund C.E., Miller J.K., Pepose J.S., and Leib D.A. 1993. In vivo characterization of site-directed mutations in the promoter of the herpes simplex virus type 1 latency-associated transcripts. *J. Gen. Virol.* **74:** 1859–1869.

Ram Z., Culver K.W., Walbridge S., Blaese R.M., and Oldfield E.H. 1986. In situ retroviral-mediated gene transfer for the treatment of brain tumors in rats. *Cancer Res.* **46:** 5276–5281.

————. 1993. In situ retroviral-mediated gene transfer for the treatment of brain tumors. *Cancer Res.* **53:** 83–88.

Ramakrishnan R., Fink D.J., Guihua J., Desai P., Glorioso J.C., and Levine M. 1994. Competitive quantitative polymerase chain reaction (PCR) analysis of herpes simplex virus type 1 DNA and LAT RNA in latently infected cells of the rat brain. *J. Virol.* **68:** 1864–1870.

Ranges G.E., Figari I.S., Espevik T., and Palladino Jr. M.A. 1987. Inhibition of cytotoxic T cell development by transforming growth factor β and reversal by recombinant tumor necrosis factor α. *J. Exp. Med.* **166:** 991–998.

Rasty S., Goins W.F., and Glorioso J.C. 1995. Site-specific integration of multigenic shuttle plasmids into the herpes simplex virus type 1 (HSV-1) genome using a cell-free Cre-lox recombination system. *Methods Mol. Genet.* **7:** 114–130.

Rasty S., Thatikunta P., Gordon J., Khalili K., Amini S., and Glorioso J.C. 1996. Human immunodeficiency virus *tat* gene transfer to the murine central nervous system using a replication-defective herpes simplex virus vector stimulates transforming growth factor β1 gene expression. *Proc. Natl. Acad. Sci.* **93:** 6073–6078.

Read G.S. and Frenkel N. 1983. Herpes simplex virus mutants defective in the virion-associated shutoff of host polypeptide synthesis and exhibiting abnormal synthesis of α (immediate early) viral polypeptides. *J. Virol.* **46:** 498–512.

Read G.S., Karr B.M., and Knight K. 1993. Isolation of a herpes simplex virus type 1 mutant with a deletion in the virion host shutoff gene and identification of multiple forms of the vhs (UL41) polypeptide. *J. Virol.* **67:** 7149–7160.

Rice S., Long M., Lam V., and Spencer C. 1994. RNA polymerase II is aberrantly phosphorylated and localized to viral replication compartments following herpes simplex virus infection. *J. Virol.* **68:** 988–1001.

Rodahl E. and Haarr L. 1997. Analysis of the 2-kilobase latency-associated transcript expressed in PC12 cells productively infected with herpes simplex virus type 1: Evidence for a stable, nonlinear structure. *J. Virol.* **71:** 1703–1707.

Rogers B.E., Douglas J.T., Ahlem C., Buchsbaum D.J., Frincke J., and Curiel D.T. 1997. Use of a novel cross-linking method to modify adenovirus tropism. *Gene Ther.* **4:** 1387–1392.

Roizman G. and Batterson W. 1985. Herpesviruses and their replication. In *Virology* (ed. B.N. Fields), pp. 497–526. Raven Press, New York.

Roop C., Hutchinson L., and Johnson D. 1993. A mutant herpes simplex virus type 1 unable to express glycoprotein L cannot enter cells, and its particles lack glycoprotein H. *J. Virol.* **67:** 2285–2297.

Rothlein R., Czajkowski M., O'Neill M.M., Marlin S.D., and Merluzzi V.J. 1988. Induction of intercellular adhesion molecule 1 on primary and continuous cell lines by pro-inflammatory cytokines. Regulation by pharmacologic agents and neutralizing antibodies. *J. Immunol.* **141:** 1665–1669.

Roux P., Jeanteur P., and Piechaczyk M. 1989. A versatile and potentially general approach to the targeting of specific cell types by retroviruses: Application to the infection of human cells by means of major histocompatibility complex class I and class II antigens by mouse ecotropic murine leukemia virus-derived viruses. *Proc. Natl. Acad. Sci.* **86:** 9079–9083.

Russell S.J., Hawkins R.E., and Winter G. 1993. Retroviral vectors displaying functional antibody fragments. *Nucleic Acids Res.* **21:** 1081–1085.

Sacks W.R. and Schaffer P.A. 1987. Deletion mutants in the gene encoding the herpes simplex virus type 1 immediate-early protein ICP0 exhibit impaired growth in cell culture. *J. Virol.* **61:** 829–839.

Sacks W.R., Greene C.C., Aschman D.P., and Schaffer P.A. 1985. Herpes simplex virus type 1 ICP27 is an essential regulatory protein. *J. Virol.* **55:** 796–805.

Sagot Y., Tan S.A., Baetge E., Schmalbruch H., Kato A.C., and Aebischer P. 1995. Polymer encapsulated cell lines genetically engineered to release ciliary neurotrophic factor can slow down progressive motor neuronopathy in the mouse. *Eur. J. Neurosci.* **7:** 1313–1322.

Saito H., Kouhara H., Kasayama S., Kishimoto T., and Sato B. 1992. Characterization of the promoter region of the murine fibroblast growth factor receptor 1 gene. *Biochem. Biophys. Res. Commun.* **183:** 688–693.

Samaniego L., Naxin W., and DeLuca N. 1997. The herpes simplex virus immediate-early protein ICP0 affects transcription from the viral genome and infected-cell survival in the absence of ICP4 and ICP27. *J. Virol.* **71:** 4614–4625.

Samaniego L., Webb A., and DeLuca N. 1995. Functional interaction between herpes simplex virus immediate-early proteins during infection: Gene expression as a consequence of ICP27 and different domains of ICP4. *J. Virol.* **69:** 5705–5715.

Sarmiento M. and Spear P.G. 1979. Membrane proteins specified by herpes simplex viruses. IV. Conformation of the virion glycoprotein designated VP7(B). *J. Virol.* **29:** 1159–1167.

Sawtell N.M. 1997. Comprehensive quantification of herpes simplex virus latency at the single-cell level. *J. Virol.* **71:** 5423–5431.

Schmidt W., Steinlein P., Buschle M., Schweighoffer T., Herbst E., Mechtler K., Kirlappos H., and Birnstiel M. 1996. Transloading of tumor cells with foreign major histocompatibility complex class I peptide ligand: A novel general strategy for the generation of potent cancer vaccines. *Proc. Natl. Acad. Sci.* **93:** 9759–9763.

Sears A.E., McGwire B.S., and Roizman B. 1991. Infection of polarized MDCK cells with herpes simplex virus 1: Two asymmetrically distributed cell receptors interact with different viral proteins. *Proc. Natl. Acad. Sci.* **88:** 5087–5091.

Sedarati F., Izumi K.M., Wagner E.K., and Stevens J.G. 1989. Herpes simplex virus type 1 latency-associated transcript plays no role in establishment or maintenance of a latent infection in murine sensory neurons. *J. Virol.* **63:** 4455–4458.

Shih M., Wudunn D., Montgomery R., Esko J., and Spear P. 1992. Cell surface receptors for herpes simplex virus are heparan sulfate proteoglycans. *J. Cell Biol.* **116:** 1273–1281.

Short M.P., Choi B.C., Lee J.K., Malick A., Breakefield X.O., and Martuza R.L. 1990. Gene delivery to glioma cells in rat brain by grafting of a retrovirus packaging cell line. *J. Neurosci. Res.* **27:** 427–439.

Soares M.K., Hwang D.-Y., Schmidt M.C., Fink D.J., and Glorioso J.C. 1996. *Cis*-acting elements involved in transcriptional regulation of the herpes simplex virus type-1 latency-associated promoter 1 (LAP1) in vitro and in vivo. *J. Virol.* **70:** 5384–5394.

Somia N.V., Zoppe M., and Verma I.M. 1995. Generation of targeted retroviral vectors by using single-chain variable fragment: An approach to in vivo gene delivery. *Proc. Natl. Acad. Sci.* **92:** 7570–7574.

Spear P. 1993a. Membrane fusion induced by herpes simplex virus. In *Viral fusion mechanisms* (ed. J. Bentz), pp. 201–232. CRC Press, Boca Raton, Florida.

———. 1993b. Entry of alphaherpesviruses into cells. *Semin. Virol.* **4:** 167–180.

Spear P.G., Shieh M.T., Herold B.C., WuDunn D., and Koshy T.I. 1992. Heparan sulfate glycosaminoglycans as primary cell surface receptors for herpes simplex virus. *Adv. Exp. Med. Biol.* **313:** 341–353.

Steiner I., Spivack J.G., Lirette R.P., Brown S.M., MacLean A.R., Subak-Sharpe J.H., and Fraser N.W. 1989. Herpes simplex virus type 1 latency-associated transcripts are evidently not essential for latent infection. *EMBO J.* **8:** 505–511.

Steven A.C. and Spear P.G. 1997. Herpesvirus capsid assembly and envelopment. In *Structural biology of viruses* (ed. W. Chiu et al.), p. 512. Oxford University Press, New York.

Stevens J.G., Wagner E.K., Devi-Rao G B., Cook M.L., and Feldman L.T. 1987. RNA complementary to a herpesviruses α gene mRNA is prominent in latently infected neurons. *Science* **255:** 1056–1059.

Stow N. and Stow E. 1986. Isolation and characterization of a herpes simplex virus type 1 mutant containing a deletion within the gene encoding the immediate early polypeptide Vmw 110. *J. Gen. Virol.* **67:** 2571–2585.

Szilagyi J.F. and Cunningham C. 1991. Identification and characterization of a novel noninfectious herpes simplex virus-related particle. *J. Gen. Virol.* **72:** 661–668.

Thanos D. and Maniatis T. 1992. The high mobility group protein HMG I(Y) is required for NF-κB-dependent virus induction of the human IFN-β gene. *Cell* **71:** 777–789.

Thompson R. and Sawtell N. 1997. The herpes simplex virus type 1 latency-associated transcript gene regulates the establishment of latency. *J. Virol.* **71:** 5432–5440.

Trybala E., Berghtrom T., Svennerholm B., Jeansson S., Glorioso J.C., and Olufsson S. 1994. Localization of the functional site in herpes simplex virus type 1 glycoproteins C involved in binding to cell surface heparan sulfate. *J. Gen. Virol.* **75:** 743–752.

Visalli R.J. and Brandt C.R. 1993. The HSV-1 UL45 18 kDa gene product is a true late protein and a component of the virion. *Virus Res.* **29:** 167–178.

Wagner E.K., Devi-Rao G., Feldman L.T., Dobson A.T., Zhang Y., Flanagan W.M., and Stevens J.G. 1988. Physical characterization of the herpes simplex virus latency-associated transcript in neurons. *J. Virol.* **63:** 1194–1202.

Wang Y., Finan J., Middeldorp J., and Hayward S. 1997. P32/TAP, a cellular protein that interacts with EBNA-1 of Epstein-Barr virus. *Virology* **236:** 18–29.

Ward P.L., Campadelli-Fiume G., Avitabile E., and Roizman B. 1994. Localization and putative function of the UL20 membrane protein in cells infected with herpes simplex virus 1. *J. Virol.* **68:** 7406–7417.

Watanabe Y., Kuribayashi K., Miyatake S., Nishihara K., Nakayama E., Taniyama T., and Sakata T. 1989. Exogenous expression of mouse interferon γ cDNA in mouse neuroblastoma C1300 cells results in reduced tumorigenicity by augmented anti-tumor immunity. *Proc. Natl. Acad. Sci.* **86:** 9456–9460.

Westra D.F., Glazenburg K.L., Harmsen M.C., Tiran A., Jan Scheffer A., Welling G.W., Hauw The T., and Welling-Wester S. 1997. Glycoprotein H of herpes simplex virus type 1 requires glycoprotein L for transport to the surfaces of insect cells. *J. Virol.* **71:** 2285–2291.

Whitbeck J.C., Knapp A.C., Enquist L.W., Lawrence W.C., and Bello L.J. 1996. Synthesis, processing, and oligomerization of bovine herpesvirus 1 gE and gI membrane proteins. *J. Virol.* **70:** 7878–7884.

Whitbeck J.C., Peng C., Lou H., Xu R., Willis S.H., Ponce de Leon M., Peng T., Nicola A.V., Montgomery R.I., Warner M.S., Soulika A.M., Spruce L.A., Moore W.T., Lambris J.D., Spear P.G., Cohen G.H., and Eisenberg R.J. 1997. Glycoprotein D of herpes simplex virus (HSV) binds directly to HVEM, a member of the tumor necrosis factor receptor superfamily and a mediator of HSV entry. *J. Virol.* **71:** 6083–6093.

Wickham T.J., Carrion M.E., and Koveski I. 1995. Targeting of adenovirus penton base to new receptors through replacement of its RGD motif with other receptor-specific peptide motifs. *Gene Ther.* **2:** 750–756.

Wickham T.J., Roelvink P.W., Brough D.E., and Kovesdi I. 1996. Adenovirus targeted to heparan-containing receptors increases its gene delivery efficiency to multiple cell types. *Nat. Biotechnol.* **14:** 1570–1573.

Wickham T.J., Tzeng E., Shears II L.L., Roelvink P.W., Li Y., Lee G.M., Brough D.E. Lizonova A., and Kovesdi I. 1997. Increased in vitro and in vivo gene transfer by adenovirus vectors containing chimeric fiber proteins. *J. Virol.* **71:** 8221–8229.

Williams R.K. and Straus S.E. 1997. Specificity and affinity of binding of herpes simplex virus type 2 glycoprotein B to glycosaminoglycans. *J. Virol.* **71:** 1375–1380.

Wolfe J.H., Deshmane S.L., and Fraser N.W. 1992. Herpesvirus vector gene transfer and expression of β-glucuronidase in the central nervous system of MPS VII mice. *Nat. Genet.* **1:** 379–384.

Wu C.L., Zukerberg L.R., Ngwu C., Harlow E., and Lees J.A. 1995. In vivo association of E2F and DP family proteins. *Mol. Cell. Biol.* **15:** 2536–2546.

Wu J.K., Cano W.G., Meylaerts S.A., Qi P., Vrionis F., and Cherington V. 1994. Bystander tumoricidal effect in the treatment of experimental brain tumors. *Neurosurgery* **35:** 1094–1102.

Wu N., Watkins S.C., Schaffer P.A., and DeLuca N.A. 1996. Prolonged gene expression and cell survival after infection by a herpes simplex virus mutant defective in the immediate-early genes encoding ICP4, ICP27, and ICP22. *J. Virol.* **70:** 6358–6368.

Wudunn D. and Spear P. 1989. Initial interaction of herpes simplex virus with cells is binding to heparan sulfate. *J. Virol.* **63:** 52–58.

York I., Roop C., Andrews D., Riddell S., Graham F., and Johnson D. 1994. A cytosolic herpes simplex virus protein inhibits antigen presentation to CD8+ T lymphocytes. *Cell* **77:** 525–535.

Zablotony J., Krummenacher C., and Fraser, N.W. 1997. The herpes simplex virus type 1 2.0-kilobase latency-associated transcript is a stable intron which branches at a guanosine. *J. Virol.* **71:** 4199–4208.

Zhu Q. and Courtney R.J. 1988. Chemical crosslinking of glycoproteins on the envelope of herpes simplex virus. *Virology* **167:** 377–384.

—— 1994. Chemical cross-linking of virion envelope and tegument proteins of herpes simplex virus type 1. *Virology* **204:** 590–599.

Zwaagstra J., Ghiasi H., Nesburn A.B., and Wechsler S.L. 1989. In vitro promoter activity associated with the latency-associated transcript gene of herpes simplex virus type 1. *J. Gen. Virol.* **70:** 2163–2169.

—— 1991. Identification of a major regulatory sequence in the latency-associated transcript (LAT) promoter of herpes simplex virus type 1 (HSV-1). *Virology* **182:** 287–297.

9

Emerging Viral Vectors

Douglas J. Jolly
Chiron Technologies
Center for Gene Therapy
San Diego, California 92121-1204

Gene therapy is already in clinical use with a number of viral and nonviral vector systems (Marcel and Grausz 1997). Nevertheless, there are a number of perceived and proven issues with some of these systems (e.g., the immunogenicity and toxicity of some versions of adenoviral vectors [Dodge 1995; Knowles et al. 1995], and the insertional activation/inactivation of retrovirus vectors [Temin 1990]). Although some of these issues may be the normal obstacles encountered in trying to develop new human therapies, one response to them has been the active investigation in a number of laboratories and clinical situations of alternative delivery systems. Vector systems are now being developed from additional viruses, based in part on their desirable properties or avoidance of particular safety issues.

The reasons for examining one vector system in particular vary, but they usually fall into the following classes:

1. The system has a particularly attractive property (e.g., very high levels of transgene expression over a limited period of time for alphavirus vectors).
2. The system has a format for which there is extensive clinical experience (e.g., vaccinia vectors).
3. The system seems simple or easy to develop in order to make clinical material in large quantities (e.g., baculovirus or other vectors, such as Avipox, that replicate in nonhuman cells but abortively infect human cells).
4. There is extensive understanding of the viral life cycle and molecular virology (e.g., simian virus 40, herpes simplex virus, and alphavirus).

Many systems combine some of these properties, and in fact a number of laboratories have attempted to create advantageous systems by creating hybrid vectors that may incorporate attractive properties of more than one system (e.g., adenoviral hybrids with retroviral vectors or transposons, with the idea, among others, of combining the ease of manufacture of adenoviruses with the ability to give long-term expression of retroviral vectors).

It is worth noting at this point that a wide variety of viruses have been used in human therapy, so that virally derived gene-delivery vectors have a lengthy, if checkered, lineage. This includes all of the attenuated viral vaccines starting with vaccinia for small pox (Fenner et al. 1988) to the recent chicken pox (varicella) (Gershon et al. 1992) vaccine. However, there is also a history of using replicating viruses as antitumor agents (e.g., parvoviruses, Newcastle disease virus, and rabies [Sinkovics 1989]) and bacteriophages as antibacterials (Barrow and Soothill 1997), although both of these strategies have so far been markedly less successful compared to vaccines.

Finally, as they are being developed, these emerging systems are subject to the same vagaries for evaluation of utility and safety as the more commonly used systems. Chief among these challenges is probably the difficulty of evaluating these systems without a considerable effort to develop efficient methods for manufacturing reproducible and reasonably pure preparations that give reproducible preclinical animal data and that could be used in making material for clinical trials. In general, development of all the systems entails roughly four stages: (1) research material is made in the research laboratory as best as possible and used in tissue culture and small rodent models; (2) some form of scaleable process is designed and implemented with some key characterization tests in place, so that sufficient large animal and preclinical experiments can be performed with single lots or reproducible material; (3) the process is further characterized and material produced under good manufacturing practices (GMP) conditions for toxicology and clinical use; and (4) final scale up for phase III trials and commercial manufacturing. The final stage has not yet been reached by any application for gene therapy. The net result is that any new system requires investment to at least stage 2 before any objective evaluation is possible. It is possible that such consideration may bias the development of the emerging system toward those that apparently pose fewer problems or risks in these stages of design, but this remains to be seen.

This review divides the systems, some of which are not further surveyed, into four classes, as shown in Table 1. These classes are: those vectors for which a method of making helper-independent vector is available; those vectors that are replication-competent in the natural host, but

Table 1 Types of emerging vectors

Replication-incompetent vector	Viral vectors that replicate in their natural host, but abortively infect human cells	Replication-competent viral vectors	Hybrid systems
Alphavirus	SV40	vaccinia	adeno-retroviral
	Avipox	vesicular stomatitis virus	adeno-AAV
	baculovirus	influenza virus	α-retroviral
	bacteriophage	poliovirus	α-VSV
		hepatitis B virus	DNA-retroviral
		Epstein-Barr virus	retroviral-VSVg
		autonomous parvovirus	

are replication-incompetent in human cells; those vectors that are replication competent in human cells; and finally a set of hybrid systems. Although it is not clear whether replication incompetence is completely overriding in importance, for many applications, this has been assumed to be true and hence this type of classification describes systems in roughly an order of depth of knowledge and experience. For each of the systems, the textbook *Fields Virology* provides an in-depth introduction to the parent virus (Fields et al. 1996).

In general, the survey outlines the perceived strengths of the system (including any human experience), a description of the viral life cycle and molecular characteristics of the genome; a description of the vector system, the utility/activity demonstrated in tissue culture, the utility/activity in animals and humans; and issues going forward. A summary of these factors is presented in Table 2 for most of these systems and maps of the smaller (<20 kb) viral genomes corresponding to the more characterized systems are provided in Figure 1.

VECTOR SYSTEMS

Replication-incompetent Vectors

Alphaviruses

Vectors derived from these viruses (Huang 1996; Schlesinger and Schlesinger 1996) are expected to deliver RNA genomes to the cytoplasm of target cells and lead to large amounts of protein production, due to the high levels of vector-generated message encoding the gene of interest. They may have applications as gene-therapy agents to deliver proteins

Table 2 Vector system survey

Vector	Genome	Replication	Strengths	Weaknesses
Vectors with capability to make nonreplicative versions				
Alphaviruses	positive RNA, linear, 12 kb	cytoplasmic, mostly lytic in human cells	very wide host range, high level of expression, some potential for targeting	nonreplicative form—still difficult to scale up, replicative forms seem unstable, probably limited capacity (4–5 kb)
Vectors that are replicative in normal host, nonreplicative in humans				
SV40	DNA, circular double-stranded, 5.2 kb	nuclear, lytic in monkey cells, nonlytic in human cells	very well understood at molecular level, simple genome, some human experience	oncogenic virus, small insert capacity 4–5 kb, no scale-up system
Avipox/Canarypox	DNA, circular double-stranded, 260–300 kb	cytoplasmic, abortively infects human cells	some human use, ability to manufacture, nonreplicative in human cells, large capacity for inserts	limited human use, delivered as vaccine only, possible expression issues in human cells
Baculovirus	DNA, double-stranded, circular, 131 kb (*Autographa californica* nucleopolyhedrovirus, Ac NPV)	nuclear, abortively infects mammalian cells	simple to make, presumed lack of pathogenicity	delivery of new viral genes to patients, efficiency of gene delivery unclear
Bacteriophage	DNA, double/single-stranded, various sizes up to about 30 kb	nuclear in mammalian cells	simple to make, human/animal experience, small particle size, might be targetable	questionable efficiency

Vectors that are replicative or require help from replication-competent helper virus

Virus	Genome	Location	Advantages	Disadvantages
Vaccinia	DNA, linear double-stranded with contiguous strands, 185 kb	cytoplasmic	widely used in humans, standard production process exists, vector system also used in clinical trials, large insert capacity	not totally benign, complicated genome, experience limited to scarification delivery, pre-existing immunity in older populations
VSV	negative-stranded, nonsegmented RNA, 11 kb, 5 ORFs	cytoplasmic	genome is quite plastic and the viral function seems promiscuous, likely useful to kill targeted cells	efficiency unknown, novel replicating agents
Polio	positive RNA, 7.5 kb linear	cytoplasmic	vaccine experience, good molecular characterization	human pathogen, limited insert capacity
Influenza	negative-stranded, segmented RNA, 14 kb total	nuclear	associated with benign disease, good clinical and molecular understanding	difficult to manipulate, limited insert capacity at present
Epstein-Barr virus	DNA, double-stranded, linear, 172 kb	nuclear	proven episomal capability, chronic human infection usually not pathogenic, large potential capacity	human pathogen, preexisting immunity
Autonomous parvovirus (B19, etc.)	DNA 5 kb, linear, partially double-stranded	nuclear	may have preference for tumor cells, has been used in cancer patients	no recognized way to make material
HBV	DNA, 3.5 kb, circular, partially double-stranded	nuclear	well understood molecular virology, hepatotrophic	human pathogen, limited capacity, crowded genome

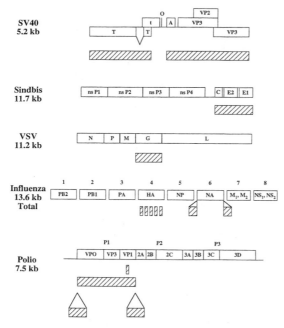

Figure 1 Genome maps of some of the smaller viral genomes that form the basis for vector constructs. The various genomes and coding sequences are as shown, roughly to scale, and are further described in the text. The total size of the genomes is as indicated. The striped boxes represent the regions of the genomes where heterologous sequences have been substituted or inserted. SV40: *O* indicates the origin of replication, *A* the small "agnoprotein" of unknown function. The other symbols are as described in the text. The genome is circular, but is represented here in linear form. Influenza: the segments are not shown to scale but have sizes of 2.3, 2.3, 2.2, 1.8, 1.6, 1.4, 1.0, and 0.9 kb for segments 1–8, respectively.

over a limited period of time and some versions may be very suitable for vaccines or antigen-specific immunotherapy. The most likely members of this genus to be used are Sindbis virus (SIN), Semliki Forest virus (SFV), and Venezuelan equine encephalitis virus (VEE). Appropriate vectors and methods for producing infectious propagation-incompetent recombinant vector particles have been developed for SIN, SFV, and VEE. SIN and SFV are Old World alphaviruses. Attenuated laboratory strains are not infectious for humans. Infection with wild-type SIN is often subclinical, but when symptomatic, it is characterized by fever or mild transient polyarthritis. Symptomatic infection with field strains of SFV is characterized by severe headache. VEE is a New World alphavirus and causes a spectrum of disease ranging from inapparent infection to acute encephalitis.

During epidemic outbreaks, there is about a 4% incidence of encephalitis that develops in people infected with VEE. Neither SIN, SFV, or VEE are endemic in the United States and most of Europe, and therefore patient populations are expected to be naive. An attenuated strain of VEE (TC83) has been used as a vaccine (against infection by the virulent form of the virus) by the United States Army (Pittman et al. 1996).

Alphaviruses are arboviruses and are maintained in nature via an insect-vector cycle. The insect, usually a mosquito, is capable of being chronically infected and likely transmits the virus to humans through bites.

Alphaviruses contain capped and polyadenylated positive-strand RNA genomes that, upon release in the cytoplasm of infected cells, are translated into a polyprotein that is then processed into the nonstructural proteins (nsP1–nsP4). The nonstructural proteins form a viral replicase that copies the positive-strand RNA into a negative strand. The negative strand is then a template from an internal RNA promoter for a tenfold molar excess, compared to genomic length RNA, of copies of the structural protein message. These messages are translated into a polyprotein that is processed into the viral structural proteins, including capsid (C) and envelope glycoproteins E2 and E1. The negative strand is also a template for the production of numerous positive-strand viral genomes. The nonstructural proteins act as the polymerases in these various reactions in a well-controlled manner in which the relative arrangement and processing of the polyprotein programs the appropriate polymerase activity at a given time of the viral replicative cycle. This process of changing viral polymerase specificity based on the temporal availability of the subunits and alterations in their interactions is only partially understood.

Diagrams of this process (Fig. 2) and its adaptation for use as a vector (Fig. 3) are shown. The essence of the process is to substitute the gene of interest for that of the structural proteins. These vectors are known as replicons, due to their property of self-amplification and expression of very high levels of foreign protein. As the virus has rather simple structure, this substitution is conceptually straightforward, but as often is the case, turning this into a useful operational therapeutic system has been, and remains, a challenge.

The first alphaviruses to be cloned were SIN and SFV. The first description of alphavirus vectors appeared in 1989 (Xiong et al. 1989), and the format used was to build constructs with the vector structure shown in Figure 3, part 2. Recombinant vector particles were generated by infection with wild-type virus. As these preparations were contaminated with replication-competent virus, complementary helpers encoding

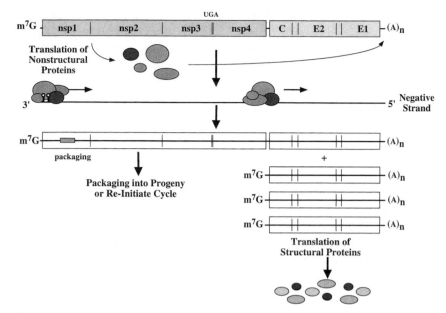

Figure 2 Genome and replication strategy of alphaviruses. The viral positive RNA genome that is directly translated in the target-cell cytoplasm after the initial infection event (*top lines*). It is capped (m^7G) and polyadenylated [$(A)_n$]. The translation products (nsP1–4) form the viral polymerase that synthesizes the negative strand (*second line*) and follows that by synthesis of both positive-strand genomic RNA and multiple copies of the structural protein subgenomic message. The small stipled box in the nsP1 region represents the packaging signal for the viral genome.

the structural proteins under control of the naturally occurring 26S mRNA promoter (junction region [JR], or subgenomic promoter) were developed (Berglund et al. 1993; Bredenbeek et al. 1993). Both cDNA constructs were juxtaposed with bacteriophage promoters and transcribed into RNA in vitro. These RNA preparations were then cotransfected into tissue-culture cells and infectious vector particles harvested from the cell-culture fluids. This method enabled use of the system as a research tool, and in fact, these formats have been available for a number of years from reagent companies. However, it is rather obvious that these methods are extremely cumbersome and likely nonscaleable. This process has allowed experiments to go forward with material that was minimally contaminated with replication-competent virus.

Another class of alphavirus vectors, so-called double subgenomic vectors (Fig. 3, part 1), places a heterologous gene under the control of

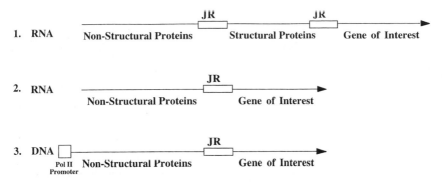

Figure 3 Alphavirus vector configurations. (*1*) A "double-copy" vector, launched by transfection of the in-vitro-transcribed RNA molecule. This is a replication-competent vector, carrying the entire complement of viral proteins. (*2*) A conventional vector configuration, with the gene of interest substituted for the structural gene-coding region. Usually this is made by in vitro transcription and then cotransfected with a complementary defective genome that provides the structural protein-coding region to give vector article production. This vector is replication-incompetent in the absence of a replication competent helper virus. (*3*) DNA-based version of the conventional vector in *2*. The configuration is the same except for the fact that the correctly positioned RNA polymerase II (Pol II) promoter allows generation of the RNA vector after transfection of the DNA molecule. The provision of the complementary defective genome allows production of vector particles after DNA cotransfection.

the 26S mRNA promoter, in the context of infectious virus (Hahn et al. 1992). These vectors tend to be unstable after multiple passage.

One reason that in vitro RNA transcription was used was that a transfectable DNA molecule that would launch the viral life cycle was not available. Previous work with picornaviruses, another positive-stranded RNA virus, suggested that DNA-based expression of RNA viral genomes was inefficient. This turned out to be due to the need for the DNA-encoded RNA to have rather precise ends. This problem was solved in the last few years by at least two groups (Herweijer et al. 1995; Dubensky et al. 1996). This led to the development of two kinds of alphavirus-based vectors: (1) recombinant viral particles derived from packaging cells stably transformed with inducible helper RNA expression cassettes; (2) DNA molecules that in principle, on transduction by whatever means into target cells, lead to much higher levels of protein expression than conventional RNA polymerase II transcription units because of the cytoplasmic amplification of the gene of interest (Fig. 3, part 3). Both of these systems show promise (discussed later).

Packaging cells have been designed that produce titers of 10^9 infectious units (IU)/ml by stably introducing inducible helper RNA expression cassettes into baby hamster kidney (BHK) cells that synthesize the viral structural proteins (sPs) in response to the vector-supplied nsPs. The toxicity of both the viral nsPs and sPs requires the design of inducible packaging cell lines (PCLs). Expression of the nsPs after introduction of replicon vector into PCLs results in a high level of expression of the viral sPs and subsequent production of recombinant vector particles. A remaining problem is that recombination between the vector and structural coding sequence cassette leads to detectable levels of replication-competent alphavirus (RCA). This problem can be avoided by further splitting up the structural polyprotein-encoding cassette into two coding regions (capsid and pE2-E1) each under control of the 26S mRNA promoter (Frolov et al. 1997). Such procedures allow preparation of high (10^9 IU/ml) titer vector particle preparations free of detectable replication-competent virus and are normally achieved by transient transfection techniques. First-generation stable packaging cell lines made on this principle yield lower titers (10^6 to 10^7 IU/ml) at present.

Alphavirus-based vectors have been used in at least three configurations. The replication-competent double copy vectors have been quite widely used in the past as they are relatively simple to produce. A number of foreign antigens have been expressed in these vectors and the major perceived application here is as a vaccine. These vectors can be very efficient and protection has been seen in a mouse influenza challenge model of protection using VEE vectors with as little a 10^4 pfu of double-copy VEE vector-encoding influenza HA gene (Davis et al. 1996). These vectors have also been used to express antigens such as simian immunodeficiency virus (SIV) env in a monkey SIV challenge model (Mossman et al. 1996) in which some evidence of protection from disease was seen.

The DNA versions of the vector have shown some promise in use as an immunotherapeutic or vaccine. In mice, these DNA molecules have shown the ability to induce cellular immune responses with DNA doses that are 10–100-fold decreased with respect to conventional DNA expression vectors (Hariharan et al. 1998). If this property extrapolates to primates, it could reduce some of the variability around induction of these responses in primates, at a feasible dose. The replication-incompetent vector particles may also be used for vaccine applications or for expressing RNA molecules such as anti-human immunodeficiency virus (HIV) ribozymes (Smith et al. 1997).

The alphavirus-based systems are still in need of considerable development. For example, although the simplified description given here leads to the expectation of activity in almost any cell as it takes only one RNA molecule in the cytoplasm to launch activity, in fact this apparently does not usually happen. In addition, the viral nonstructural proteins that are obligatorily expressed in the target cell are usually toxic in the long term. Finally, to be useful, higher titers of material genuinely free of RCA will be needed. It is not completely clear how this will be done. Nevertheless, it is likely that a version of this system will reach the clinic in the next few years.

Viruses That Replicate in Their Natural Host but Abortively Infect Human Cells

There are many viruses that have this property, and only a few are considered here, either because they have been used extensively or they represent a class of agents.

SV40

SV40 (simian virus 40) is a member of the Papovaviridae subfamily Polyomavirinae (Cole 1996) and was actually discovered as a contaminant derived from primary/secondary monkey (rhesus) kidney cultures used to grow poliovirus vaccines. As an unintended consequence, many individuals vaccinated for polio in the 1950s and 1960s were exposed to SV40 (Petricciani 1991). In general, this happened without obvious adverse events occurring, and this circumstance is the basis for believing that this oncogenic virus may be safe in patients. In this context, it is worth noting that 60% of human mesotheliomas have recently been shown to carry SV40 sequences, but this appears to be unlinked to the polio vaccine exposure (Strickler et al. 1998). In the 1960s, 1970s, and 1980s, SV40 was one of the hottest research subjects as a small convenient episomal model for the mammalian chromosome. As a result, the genome is exceptionally well-characterized, there is a good understanding of the lytic life cycle of the virus in tissue culture, and many reagents are available. The virus grows lytically in many monkey cells but abortively infects human cells and leads to transformation and immortalization in about 1 in 10^6 cells in culture. This activity is a function of the large T antigen and transfection of the gene for large T is sufficient for immortalization and transformation of fibroblasts in culture.

The virus is nonenveloped with a capsid of about 50 nm that carries a 5.2-kb double-stranded circular DNA molecule, packaged in cellular histones, but lacking histone H1. A map of the genome is shown in Figure 1. The virus binds to an unknown receptor on the cell surface and the genome is translocated to the nucleus as a complex where it is further uncoated and the viral early genes, small t and large T antigens, transcribed. There appear to be multiple methods of transport to the nucleus, but the major mechanism seems to be transport in pynocytotic cytoplasmic vesicles followed by fusion to the nuclear membrane by the vesicles. The viral genome has the early and late promoters in the same region as the origin of replication, and the genome is divisible into early and late halves, the transcription of which occurs in opposite directions from the common promoter region. After transcription of the T-antigen message, the protein is synthesized in the cytoplasm and then translocates to the nucleus and binds to the SV40 origin of replication. The function of small t antigen is unclear, although it appears to facilitate anchorage-independent growth; it is dispensable for the lytic life cycle in culture. Binding of large T to the origin precipitates both late viral message transcription, leading to synthesis of the capsid proteins VP1, VP2, and VP3, and replication of the viral genome. There are two late messages, 16S, the more abundant, leading to VP1 synthesis, and 19S leading to synthesis of VP2 and VP3. The viral capsid proteins return to the nucleus and encapsidate the newly replicated viral genomes—about 10^5 per nucleus. The virus does not cause cell lysis, but can either be released as the cells disintegrate from viral overload or be actively released from the cell membrane by means of cytoplasmic vesicles. This virus was a major tool in elucidating many of the events in cellular transcription, RNA transport, translation, protein transport, and DNA replication, so that many of these processes have been described in great detail and even represent the archetype of the process.

SV40 was the first virus to be used as a vector for foreign genes following manipulation of the genome in vitro, and in the 1970s and early 1980s, a number of recombinants were constructed (Hamer and Leder 1979; Mulligan et al. 1979; Gruss et al. 1981). These experiments were limited to tissue culture demonstrations of foreign-gene transcription and, occasionally, expression. Usually the late region was partially replaced in the vector and helper viral functions provided either transfection of plasmids providing the complementary functions (VP1-3) or complete SV40 virions with temperature-sensitive large T. In this way, expression of introduced genes, such as insulin and globin, was demonstrated. In 1981, Gluzman (1981) constructed the COS series of cells; these are monkey CV1 cells carrying complete SV40 genomes but with small deletions in the origin of replication, which eliminated viral replication but allowed constitutive early gene expression. These cell lines, in particular COS7, have continued to be used as molecular biology tools. However, these cells can also be used to produce recombinant SV40 vector particles that are free of detectable replication-competent SV40 by introducing constructs with an SV40 origin, and plasmids with other complementing functions such as the late proteins (Oppenheim et al. 1992; Strayer 1996). SV40 has a strict size limit of about one genome plus 10% (about 5.7 kb) for encapsidation and it has been possible to exclude helper plasmids simply by making them larger than this. This system has been used successfully to make viral vector preparations with titers in the range of 10^8 to 10^9 infectious particles/ml.

Recently, an in vitro self-assembling system for making SV40 virions has been described (Sandalon et al. 1997). The capsid proteins were synthesized in an insect expression system and nuclear extracts from the insect cells combined with SV40 or vector DNA. In this system, there was no dependence on the presence of SV40 DNA for encapsidation and the size limit for encapsidation was greater than 5.3 kb. Low titers of infectious virus (10^2 to 10^3/ml) were formed and the non-SV40 DNA was also transmitted to target cells. Given the detailed knowledge of this virus, this is an approach that seems possible, but it remains to be determined whether it can be made into an efficient characterized system and, therefore, clinically useful.

Numerous genes have been expressed in tissue culture with SV40 vector particles over the years including β-galactosidase, globin, insulin, interferon-β (IFN-β), and growth hormone (Hamer and Leder 1979; Mulligan et al. 1979; Gruss et al. 1981; Asano et al. 1985; Villarreal and Soo 1985; Liebhaber et al. 1986).

It has been shown that SV40 vectors can transmit genes to resting human peripheral blood lymphocytes (PBLs) and to mouse bone-marrow cells subsequently reintroduced into mice (Strayer and Milano 1996; Strayer et al. 1997). The genome appears both to exist as an episome and to integrate into the chromosome.

This viral system has some attractive properties (transduction of resting cells, complement resistance, detailed knowledge of virus, potential for good titer production), but its use has been discouraged by its reputation as a transforming virus, despite the polio vaccine experience. However, investigators have been using on-hand laboratory reagents. To go further, there needs to be a concerted effort to build a dedicated production system and good assays to reliably detect replication-competent virus. The in vitro assembly system is interesting, but is at an early stage also.

Avipox

This class of viral vectors includes both fowlpox and canarypox. These are avian pox (Moss 1996) viruses that have homology with vaccinia and are similarly organized. The viruses grow lytically in avian cells (e.g., chicken embryo fibroblasts, CEF) but do not productively infect cells of nonavian species. However, foreign genes incorporated under control of a viral promoter do express and in sufficient quantities that this configuration shows promise for vaccine development. The virus itself does not spread. Thus, the major advantage seen for these vectors is the experience with them as vectors for avian vaccines (Edbauer et al. 1990). This experience can be transferred to making agents suitable for human use. In addition, this experience already exists for experimental nonhuman, nonavian vaccines (e.g., against canine distemper, rabies, and feline leukemia virus) (Cadoz et al 1992; Taylor et al. 1992; Tartaglia et al. 1993). As production methods already exist, this may simplify making clinically acceptable material (although the standards for this and veterinary products are not equivalent). In addition, the similarity of these viruses and vaccinia may allow extrapolation of technology from one system to the other. Finally, as these are nonreplicative in humans, it is possible to consider their use in immunosuppressed individuals, unlike vaccinia.

These are large viruses, about 250 nm by 370 nm and visible at the limit of light microscopy. Much of the information on these viruses is derived by inference compared to vaccinia, and it is often difficult to know whether a particular description is derived from study of the Avipox or by analogy to vaccinia. The genome is linear, double-stranded DNA of about 260 kb to 300 kb. It is assumed to have DNA strands connected to one another at the termini and to possess terminal repeat sequences. The particles carry a virus-specific DNA-dependent RNA polymerase that presumably is responsible for the transcription of viral early genes and any inserted foreign genes. The virus has about 200 open reading frames and completely virus-specific early, intermediate, and late promoter systems. The virus attaches in an undefined way and enters the cell as an intact core particle that then transcribes the early RNA molecules. These are capped and polyadenylated and resemble normal cellular messages. The genome is then uncoated and intermediate gene transcription occurs through the viral polymerase in a complex with cellular proteins. The third stage, late gene transcription, likely occurs from 2–3 hours to 48 hours after infection. Viral DNA replication normally starts about 12–15 hours after infection. The late transcription has a different promoter configuration and a poorly defined complex involving the viral polymerase but also, likely, host RNA polymerase II and other host nuclear factors. The messages have poly(A) capped 5´ ends. After DNA replication, the new and old DNAs are concatemerized, and this is then resolved and the genomes are packaged into infectious particles that are wrapped in Golgi membrane, transported to the cell surface, and released by fusion to the cell membrane. The exported particle retains an outer membrane layer.

Vectors are constructed (Perkus et al. 1993) by cotransfecting plasmids carrying the gene of interest embedded in a chosen (usually the C3, C5, or C6 loci) poxvirus sequence into an infected cell and selecting for viral genomes that have undergone homologous recombination with the plasmid by screening plaques or by genetic means. Variations in which the gene is recombined into a color marker such as bacterial β-galactosidase previously placed in the vector are also used. The promoter can be from Avipox or from related viruses such as vaccinia (e.g., the H6 promoter).

These vectors have been used almost exclusively as vaccines or antigen-specific immunotherapy agents. Mostly, this has been directed at viral diseases, although some anticancer experiments have been performed in rodents (Wang et al. 1995). The widest human use is canarypox expressing HIVenv in HIV vaccine trials (Ferrari et al. 1997). The same constructs have induced protective immunity in macaques against HIV-2 in a laboratory setting (Abimiku et al. 1995). The mechanism of this protection is unclear. The responses in humans have been judged less than desirable, and combination trials involving the vectors with a protein subunit boost are now under way (Pialoux et al. 1995). However, as noted above, in other nonavian species, protection against canine distemper, rabies, and feline leukemia has been seen.

These vectors have also been used with other *genetic immunization* agents in rodent cancer models as both vaccine and therapy with encouraging results (Hodge et al. 1997; Irvine et al. 1997). Finally, this type of vector has been proposed as a vaccine contraceptive using gamete-specific antigens and a boosting cytokine gene.

The most likely role for these agents is as part of an antigen-specific immunotherapy or vaccine, probably in conjunction with other agents. However, they have shown little or no toxicity in clinical trials and so the way is open to try them in other applications. There is little information as to the doses required in humans and how this relates to production capability, but this seems unlikely to be a limiting problem.

Baculoviruses

Baculoviruses are insect viruses (Miller 1996), recombinant versions of which have been used during the past 15 years to express foreign proteins in moderate amounts that can be subsequently purified from the cells or culture medium. This technology is normally used as an intermediate step in recombinant protein production that can give enough protein for extensive preclinical experiments, but it has so far not been used for scale-up for commercial clinical use. As the methods to construct and make recombinant baculovirus are highly developed, several groups have investigated using these as vectors to deliver genes to mammalian cells. The attraction is the preexisting technology, and the prospect of simple production of the agent.

Baculoviruses are insect viruses that have rod-shaped capsids of 250–300 nm in length and 30–60 nm in diameter. The viruses occur in two phenotypes: budded virus (BV), formed when the nucleocapsid buds from the cell membrane early in the course of infection, and polyhedron-derived virus (PDV), which buds from the nucleus and is subsequently occluded in large proteinaceous occlusion bodies termed *polyhedra*. In nature, this structure serves to protect the virus after the death of an infected larva. The genomes are double-stranded circular DNA molecules of 80–150 kb. The most commonly used virus is the *Autographa californica* multicapsid nuclear polyhedrosis virus (AcMNPV) grown in *Spodoptera frugiperda* Sf9 cells derived from the eponymous nocturnal moth. This has a genome of 131 kb. The recombinant viruses form only BV forms as the polyhedrin gene is almost always interrupted by the recombinant gene. The BV form enters cells by receptor-mediated endocytosis, the nucleocapsid is released into the cytoplasm, and the residual viral genome complex travels to the nucleus and enters through the nuclear pores. The early genes are transcribed by host RNA polymerase II, although this requires the viral proteins IE-1 and IE-0 as cofactors in some cases. These proteins are carried in the viral particle. The late phase starts about 6 hours after infection and lasts to about 24 hours, and late gene transcription, viral replication, and viral budding occur during this period.

The vectors are made by cotransfection of *Spodoptera* cells with a transfer vector carrying the desired gene embedded in the polyhedrin gene sequence and purified viral DNA, allowing recombination to take place. Resultant plaques are further purified and screened by testing directly for the desired sequence or by looking for a change in a phenotype such as disappearance of a preengineered bacterial β-galactosidase activity if the introduced gene is targeted to recombine into this gene. Crude lysates made in this way normally have a titer of just over 10^8/ml.

It was originally shown that baculovirus could infect and cause expression of an encoded chloramphenical acetyltransferase (CAT) gene with an Rous sarcoma virus (RSV) long terminal repeat (LTR) promoter in mouse L 929 cells, although this gave levels of expression that were low (<0.05% of the level in *Spodoptera* cells) (Carbonell et al. 1985;

Merrington et al. 1997). Subsequently, it was shown that use of such baculovirus vectors led to quite efficient expression in human HepG2 hepatoma cells and in primary hepatocytes (Boyce and Bucher 1996). This same group then showed that the inclusion of the vesicular stomatitis virus (VSV) G protein in the baculovirus membrane made a large range of cell types in culture take up and express an encoded RSV LTR-driven β-galactosidase gene. As the VSV-G protein presumably facilitates viral entry, this implies that this is the block (Barsoum et al. 1997). However, another group used a very efficient promoter (CAG) to drive the β-galactosidase gene. This promoter is a hybrid cytomegalovirus (CMV) immediate early enhancer combined with the chicken actin promoter and the rabbit β-globin polyadenylation site sequence. These investigators showed that now baculovirus was comparable with adenovirus in terms of expression levels and number of positive cells, in several cell lines including nonhepatocytes, without the use of VSVg (Shoji et al. 1997). These results, although not completely concordant, suggest that the efficiency of expression is such that some form of active transport to the nucleus is taking place in some mammalian cells.

At present, there are no published reports of the use of this system in animals models, but clearly this is the next logical step.

Bacteriophage

There are many kinds and sizes of bacteriophages (Campbell 1996). Usually they have DNA genomes, and commonly used phages have both double-stranded (λ) and single-stranded (M13) genomes. The attraction of these agents is the facility with which they can be manipulated in vitro and the ability to make large numbers of genomes. In general, these are small entities (dimensions in the tens of nanometers), and this is a conceivable advantage for in vivo delivery. In addition, there is a long, somewhat checkered, history of treating patients having bacterial infections with these agents (Lederberg 1996). A good recent review of these experiences exists (Barrow and Soothill 1997). Essentially, interest in this modality faded with the advent of antibiotics and revived with the advent of widespread antibiotic resistance. However, this experience provides an excellent safety record for these agents, and doses such as 10^{10} pfu bacteriophage λ have been given to mice without apparent toxicity. Extrapolating by body weight to humans gives an equivalent dose of about 10^{14}. Over time, it has been shown that virulent phage treatment can save calves and mice from bacterial challenge (Smith and Huggins 1982; Smith et al. 1987; Soothill 1992). In addition, it has recently been possible to select for phage with extended circulatory half-lives for which the mutations mapped to the phage head proteins. In these cases, the dose was 10^{11} pfu. When these mutants were injected into mice, the phage concentration in the bloodstream was more than 10^4-fold higher 16 hours after inoculation compared to results from the same experiment using the potential strain (Merril et al. 1996).

These agents have many different life cycles, but M13 phage, for example, can easily be grown to titers of 10^{10} to 10^{12}/ml of bacteriophage lysate. Lysates are generally made by infection of a growing culture of the host bacterium, and growth continues until lysis occurs. Most of these agents consist of nucleic acid wrapped in a rugged capsid and are remarkably stable on storage.

A number of papers have been published showing in tissue culture that bacteriophage can be used as convenient DNA packages to get DNA into tissue culture cells (Horst et al. 1975; Ishiura et al. 1982; Srivatsan et al. 1984; Okayama and Berg 1985).

A small number of publications describe the design of engineered phage with molecular alterations designed to facilitate entry into mammalian target cells in vivo, but any actual in vivo data appear to be lacking (Hart et al. 1994; Chada and Dubensky 1996).

It remains to be seen where this nascent field will go, but the extensive antibacterial use of these agents suggests that they are not very toxic and their ease of manipulation is attractive.

Replication-competent Viral Vectors

In this brief survey, the replication competent versions of common vectors such as adenovirus- or herpes-derived vectors are not considered.

Vaccinia

This viral system (Fenner 1996; Moss 1996) is attractive because of its extensive use as a smallpox vaccine and the increasingly detailed understanding of its molecular virology. Both the vectors and the virus are stable and easy to transport. The use of this system has been almost exclusively confined to antigen-specific stimulation.

Vaccinia virus is of unknown origin and is now purely a laboratory virus. The viral life cycle is, as far as is known, essentially identical to that described for Avipox (discussed previously). The genome is about 180 kb of linear double-stranded DNA with the strands connected at the ends.

Vectors are usually made by in vivo recombination with a transfer plasmid as described for Avipox. In fact, almost all of the poxvirus-vector technology was developed in vaccinia and then transferred to the other poxviruses. It has also been possible to make molecular chimerae in vitro and transfer these into cells to produce viral vector particles (Merchlinsky and Moss 1992). The leaders in these developments, at least those in the United States, have been the laboratories of Moss (1996) and Paoletti (1996).

A substantial number of antigens have been inserted into these vectors, and these have been examined for their ability to induce immune responses in animals and humans, including human hepatitis B virus (HBV) surface antigen, herpes simplex type 1 gD protein, influenza A hemagglutinin gene (HA), HIV-1 genes, and the nucleocapsid protein of vesicular stomatitis (Smith et al. 1983; MacKett et al. 1985; Flexner et al. 1988). In addition, a recombinant vaccinia rabies vaccine for veterinary use is approved in Europe (Brochier et al. 1995). A considerable effort has been spent trying to use vaccinia recombinants as part of an anti-HIV vaccine without apparent success so far (Hammond et al. 1992; Cooney et al. 1993), but this is being pursued in several ways with prime boost protocols and inclusion of multiple HIV immunogens.

There have also been attempts in rodent cancer models to use vaccinia recombinants expressing tumor-associated antigens or cytokines as protective or immunotherapeutic agents (Qin and Chatterjee 1996; Carroll et al. 1997). There have also been some clinical trials (McAneny et al. 1996). Although some encouraging results have been reported, it is unclear if this will be a useful strategy in humans.

It is possible to summarize a great deal of work by saying that such recombinant vaccinia agents are unlikely to be widely useful as stand alone agents, but that they could well be a component in a multicomponent vaccination regime.

An important impediment to the use of vaccinia is that it is not an innocuous agent, and deaths from disseminated infection in immunosuppressed individuals have been reported (Redfield et al. 1987). The laboratory of Paoletti has therefore undertaken extensive mapping of genes associated with various types of virulence and introduced complete deletions of 18 open reading frames that were implicated in viral virulence. This yields a highly attenuated strain termed NYVAC (Tartaglia et al. 1992), which is still able to induce strong immune responses to extrinsic antigens and protect against challenge against rabies, for example.

It is likely that the future for pox recombinants lies with the attenuated strains. It is possible that these strains may be usable for purposes other than straight immunization, but that remains to be seen.

Vesicular Stomatitis Virus (Rhabdoviruses)

This virus (Wagner and Rose 1996) is of interest for several reasons. It is known to be infectious for human cells in culture but rarely actually infects humans. It infects insects and mammals and causes a nonfatal disease in cattle and swine, which nevertheless has economic importance. The chief attraction is probably the plasticity of this virus and its components, and it is likely that the major useful configurations will be as hybrids with other virus or viral functions.

The virus is a rhabdovirus and has a bullet-shaped virion of about 180 nm length and 75 nm width. It has a 11-kb unsegmented negative-strand RNA genome. This encodes five proteins included in the virion, plus at least two more small proteins that are not. These viral proteins are encoded and have separate, but tightly spaced, open reading frames. The viral proteins (see Fig. 1) are the envelope (G) protein, the matrix (M) protein, the nucleocapsid (N) protein that wraps up the viral genome, and two polymerase proteins (L and P). The nucleocapsid/RNA complex includes the polymerase proteins and is surrounded by the M protein and then a lipid membrane with the G protein embedded. The membrane is plasma-membrane-derived but of different overall composition.

The virus binds to a receptor that is likely phosphatidylserine, and enters through endocytosis followed by release catalyzed by G-protein-mediated fusion in the context of a pH decrease. The negative genome is then transcribed in the cytoplasm into at least the five structural protein messages by the viral polymerase carried in the particle, and these are translated into protein. The L and P proteins are then involved in copying the negative strand into a positive strand and then back to negative to yield progeny viral genomes. The newly translated protein and new viral genomes then complex in a poorly understood fashion that seems to be facilitated by the M protein, and bud off from the membrane with the G protein embedded.

A considerable amount of vector research has also been carried out using the archetypical rhabdovirus, rabies virus. VSV and rabies are discussed interchangeably in the following using VSV terminology. The construction of a recombinant DNA system that could be transfected to launch infection and replication of this or related negative-strand RNA viruses was not straightforward. This was eventually achieved as follows (Schnell et al. 1994; Lawson et al. 1995). The entire VSV genome was placed in a plasmid under the control of the bacteriophage T7 promoter so that transcription yielded the antigenomic (+) strand RNA. Cells were infected with a vaccinia vector encoding the T7 polymerase and transfected with the plasmid encoding the VSV genome plus separate expression plasmids encoding the P, L, and N proteins. The VSV antigenomic RNA is transcribed in the cells by T7 polymerase, and this is then back-transcribed into negative-strand genomic RNA by the P, L, and N proteins, thus generating every entity necessary for virus production. Virus is separated from the vaccinia by filtration through 0.22-μm filters. This process has subsequently been somewhat simplified and the various functions supplied by cotransfecting five separate plasmids.

This process also allowed the engineering of the virus to carry the CAT gene by inserting it between a 3´ truncated G protein message and the start of the L protein-coding sequence (Schnell et al. 1996a). Subsequently, it was shown that some cellular proteins such as CD4 and HIVenv could be incorporated into the VSV envelope (Mebatsion and Conzelmann 1996; Schnell et al. 1996b). This then stimulated two groups to design VSV-derived viruses that did not carry the G protein gene but did carry the genes for the HIV receptors CD4 and CXCR4 (Mebatsion et al. 1997; Schnell et al. 1997). Such viral particles were shown to infect, propagate on, and kill HIV-infected T cells, but not other uninfected cells. Titers of HIV were reduced by up to three orders of magnitude. Although it remains to be seen if this approach is useful in patients, this does raise the exciting prospect of tailoring a cytopathic replicating virus to destroy cells identified by a particular surface protein.

Influenza Virus

The influenza viruses are Orthomyxoviridae (Lamb and Krug 1996), of which there are two genera: influenza A and B viruses, and influenza C virus. All of the viral vectors used so far have been made from type-A viruses. The attractions of this virus as a gene delivery vehicle are the extensive understanding of its virology and molecular biology. In addition, it is known to be very immunogenic (although it escapes antiviral immune responses by mutations in the viral envelope and by reassorting genome segments with other strains). The virus is a negative-strand RNA virus with eight segments and a total nucleic acid length of about 13.6 kb. Each segment has defined 3´ and 5´ sequences of 12 or 13 nucleotides that act as a packaging signal for that segment. The segments encode as follows (see Fig. 1): Segments 1, 2, and 3 encode PB2, PB1, and PA respectively—these are all components of the RNA transcription and replication complex; segment 4 encodes HA, the major surface glycoprotein; segment 5 encodes NP, which binds to RNA to form coiled ribonuclear protein; segment 6 encodes NA–surface glycoprotein with neu-roaminidase activity; segment 7 encodes M1 and M2, matrix proteins; and segment 8 encodes NS1 and NS2 high-abundance nonstructural proteins. The virus is enveloped and the genome is wrapped in NP. The virus enters the cell through endocytosis followed by fusion with the endosomal membrane. The NP-wrapped genomic RNA segments move to the nucleus where they are both transcribed to message and replicated by the viral polymerase.

Recombinant viral vectors are made by transcribing a recombinant segment in vitro, complexing this with NP protein and transfecting this into an influenza-infected cell (Luytjes et al. 1989). The recombinant segment is then incorporated into some particles that can be selected with antibodies or host range mutants. Recently, an alternative system was designed and used that avoids the need to make NP protein (Pleschka et al. 1996). Recombinant segments have been made based on the PB2, HA, NA, NP, and NS encoding segments (Muster et al. 1995).

The majority of inserts, however, have been in the HA (15–20 amino acid peptide segments inserted in at least five different sites) or NA segments. In the NA segment, it is possible to add an extra 200 amino acids worth of sequence. This can be accomplished by means of an internal ribosome entry site (IRES) between the two genes, or a self-cleaving polyprotein can be made. With this technology, the CAT gene (in NA) and epitopes from HIV envelope gene (in HA) have been inserted into vector particles. CAT expression was easily detected (Percy et al. 1994). The hybrid glycoprotein HIVenv-HA epitope segment was shown to incorporate into a purified viral strain and induced anti-gp41 mucosal immunity quite efficiently after nasal administration to mice (Palese et al. 1997).

The future of this system will depend on the points at which it is different from all the other systems. The relative complexity of the current system means that, going forward, an important advance would be to simplify the generation of viral vectors.

Poliovirus

The attraction of using poliovirus vectors is the existence of the Sabin vaccine strains that have been administered to millions of children and hence are presumed to be extremely safe. It is worth noting that these strains have very small but finite reversion rates to the virulent form.

The virus is a picornavirus (Rueckert 1996) that generally infects through the alimentary track, where it replicates in the lymphoid tissue before passing to the reticuloendothelial system and thence, relatively infrequently, to the spinal cord and brain. The genome is a positive-strand RNA molecule of about 7.4 kb, and replication takes place in the cytoplasm. After uncoating, the genome is translated into a polyprotein that has three major initial cleavage products (P1, P2, and P3, reading amino-terminal to carboxy-terminal). P1 is further cleaved by viral proteases to the coat proteins (VP0, VP3, and VP1). P2 and P3 are further cleaved to protease and polymerase molecules. The major polymerase function maps to the P3 cleavage product 3D. The P3 protein also gives rise to 3B or VPg, a small protein that is attached to the 5´ end of the mature viral genome. The virus attaches through a known receptor intercellular adhesion molecule 1 (ICAM 1), and the viral genome is released into the cytoplasm after a change in the conformation of the capsid. The genome is translated and then replicated to the negative strand and back to positive strands that are recycled as translation templates, but can also eventually become viral genomes. Packaging of the virus occurs in the cytoplasm.

The virus has been used as a vector in at least three ways, but in every situation, the motivation has been to make vector for vaccine use. The first and simplest vectors substituted a small amino acid loop in the VP1 capsid protein of the Sabin type-1 virus to give replication-competent vectors with small linear foreign epitopes of 15–20 amino acids. Manipulations were made with DNA; this was in-vitro-transcribed to RNA, and the RNA transfected into target cells (e.g., HEp 2c cells). Examples include an epitope from HIV gp41 (Evans et al. 1989), in which neutralizing antibodies were obtained. Protection in guinea pigs was seen with foot-and-mouth disease (FMD) hybrid (Mattion et al. 1996), but this is an intertypic substitution as FMD is also a picornavirus.

An alternative system is to replace the capsid proteins with the gene of interest, transfect the resultant in-vitro-transcribed RNA, and supply the capsid protein in *trans* with a vaccinia recombinant (Porter et al. 1997). This produces infectious particles that can launch the replicon into a single cell, without further propagation occurring. The vaccinia can be removed by filtration or even freon treatment. The vector can carry up to 1.5 kb of coding sequence, and a hybrid encoding SIV gag-pol sequences has been used to induce immunity in monkeys (Anderson et al. 1997). The course after challenge with a virulent SIV strain has not yet been reported.

The third common type of recombinant is replication-competent with the foreign sequences included either at the very terminus of the polyprotein or between P1 and P2. Viral protease recognition signals are included so that the foreign segment is excised and exists in the cytoplasm, in the hope of generating both cellular and humoral immunity. Up to 400 amino acids could be inserted, although shorter sequences (100 amino acids) seemed more stable. Replication was again launched by transfection of in vitro RNA transcripts. Immunity has been generated in animal models to HIV, influenza, cholera, and rotavirus antigens (Andino et al. 1994; Mattion et al. 1994).

It seems likely that the use of this system will be limited to vaccines, but in fact there are no overwhelming advantages of this system that seem to justify increased attention.

Other Vectors

There are several other vectors for which incidental reports exist; these include vectors based on hepatitis B virus (Chaisomchit et al. 1997), Epstein-Barr virus (Franken et al. 1996; Robertson et al. 1996), and autonomous parvoviruses (Schlehofer et al. 1992; Shaughnessy et al. 1996). The autonomous parvoviruses do have a long, if not very successful, history as wild-type agents in antitumor therapy. The reader is referred to the individual references for more information.

Hybrid Systems

Recently, there have been numerous attempts to design and use hybrid systems that incorporate functions from different viruses in the same vector. The usual motivation for this is to combine the attractive attributes from two different systems while hoping not to create new problems. Although we are unlikely to be as clever at this as Nature and there is danger in thinking that functions can be transplanted in a modular fashion, there are some fascinating prospects. An example of this is the attempt to combine adenovirus with retroviruses or transposons in order to create an easy to make, high-titer integrating hybrid (Bilbao et al. 1997). An alternative motivation is to use one viral system to make another, often because the components for the second system are difficult to make in quantities or are toxic when continuously expressed. An example of this is the launching of adeno-associated virus (AAV) production from an adenovirus system (Fisher et al. 1996).

The list also includes alphavirus-retrovirus hybrids (Wahlfors et al. 1997), alphavirus-VSV hybrids (Rolls et al. 1994), DNA-retrovirus hybrids (Noguiez-Hellin et al. 1996), and all the retroviral VSVg pseudotypes (Emi et al. 1991).

Although it is beyond the scope of this review to examine these in any detail, they represent an interesting picture of the possible future of viral vectors.

CONCLUSION

The viral vector systems described previously are all at a relatively primitive stage of development, and it is likely that introducing new systems beyond the strict research stage will become more and more difficult without obvious strong advantages being evident. This is because of the cost and effort needed to take a new viral system to the clinic and through to commercial production. Nevertheless, it is evident that current individual vector systems have serious limitations so that the effort involved in viral vector research is worthwhile. We can expect to see some of the above systems tried in the clinic and will have to wait to see what the future brings.

ACKNOWLEDGMENTS

The author thanks Dr. Tom Dubensky and Dr. Jiing-Kuan Yee for a critical reading of this manuscript and Ms. Beverly Thacker for help in preparing the manuscript.

REFERENCES

Abimiku A.G., Franchini G., Tartaglia G., Aldrich K., Myagkikh M., Markham P.D., Chong P., Klein M., Kieny M.-P., Paoletti E., Gallo R.C., and Robert-Guroff M. 1995. HIV-1 recombinant poxvirus vaccine induces cross-protection against HIV-2 challenge in rhesus macaques. *Nat. Med.* **1:** 321–329.

Anderson M.J., Porter D.C., Moldoveanu Z., Fletcher III, T., M. McPherson S., and Morrow C.D. 1997. Characterization of the expression and immunogenicity of poliovirus replicons that encode simian immunodeficiency virus SIVmac239 Gag or envelope SU proteins. *AIDS Res. Hum. Retroviruses* **13:**53–62.

Andino R., Silvera D., Suggett S.D., Achacoso P.L., Miller C.J., Baltimore D., and Feinberg M.B. 1994. Engineering poliovirus as a vaccine vector for the expression of diverse antigens. *Science* **265:** 1448–1451.

Asano M., Iwakura Y., and Kawade Y. 1985. SV40 vector with early gene replacement efficient in transducing exogenous DNA into mammalian cells. *Nucleic Acids Res.* **13:** 8573–8586.

Barrow P.A. and Soothill J.S. 1997. Bacteriophage therapy and prophylaxis: Rediscovery and renewed assessment of potential. *Trends Microbiol.* **5:** 268–271.

Barsoum J., Brown R., McKee M., and Boyce F.M. 1997. Efficient transduction of mammalian cells by a recombinant baculovirus having the vesicular stomatitis virus G glycoprotein. *Hum. Gene Ther.* **8:** 2011–2018.

Berglund P., Sjöberg M., Garoff H., Atkins G.J., Sheahan B.J., and Liljeström P. 1993. Semliki forest virus expression system: Production of conditionally infectious recombinant particles. *Bio/Technology* **11:** 916–920.

Bilbao G., Feng M., Rancourt C., Jackson, Jr., W.H. and Curiel D.T. 1997. Adenoviral/retroviral vector chimeras: A novel strategy to achieve high-efficiency stable transduction in vivo. *FASEB J.* **11:** 624–634.

Boyce F.M. and Bucher N.L.R. 1996. Baculovirus-mediated gene transfer into mammalian cells. *Proc. Natl. Acad. Sci.* **93:** 2348–2352.

Bredenbeek P.J., Frolov I. Rice C.M., and Schlesinger S. 1993. Sindbis virus expression vectors: Packaging of RNA replicons by using defective helper RNAs. *J. Virol.* **67:** 6439–6446.

Brochier B., Costy F., and Pastoret P.P. 1995. Elimination of fox rabies from Belgium using a recombinant vaccinia-rabies vaccine: An update. *Vet. Microbiol.* **46:** 269–279.

Cadoz M., Strady A., Meignier B., Taylor J., Tartaglia J., Paoletti E., and Plotkin S. 1992. Immunisation with canarypox virus expressing rabies glycoprotein. *Lancet* **339:** 1429–1432.

Campbell A.M. 1996. Bacteriophages. In *Fields virology*, 3rd edition (ed. B.N. Fields et al.), vol. 2, pp. 587–605. Lippincott-Raven, Philadelphia.

Carbonell L.F., Klowden M.J., and Miller L.K. 1985. Baculovirus-mediated expression of bacterial genes in dipteran and mammalian cells. *J. Virol.* **56:** 153–160.

Carroll M.W., Overwijk W.W., Chamberlain R.S., Rosenberg S.A., Moss B., and Restifo N.P. 1997. Highly attenuated modified vaccinia virus Ankara (MVA) as an effective recombinant vector: A murine tumor model. *Vaccine* **15:** 387–394.

Chada S. and Dubensky Jr., T.W. 1996. Bacteriophage-mediated gene transfer systems capable of transfecting eukaryotic cells. PCT patent application WO 96/21007.

Chaisomchit S., Tyrrell D.L.J., and Chang L.-J. 1997. Development of replicative and nonreplicative hepatitis B virus vectors. *Gene Ther.* **4:** 1330–1340.

Cole, C.N. 1996. Polyomavirinae: The viruses and their replication. In *Fields virology*, 3rd edition (ed. B.N. Fields et al.), vol. 2, pp. 1997–2026. Lipincott-Raven, Philadelphia.

Cooney E.L., McElrath M.J., Corey L., Hu S.L., Collier A.C., Arditti D., Hoffman M., Coombs R.W., Smith G.E., and Greenberg P.L. 1993. Enhanced immunity to human immunodeficiency virus (HIV) envelope elicited by a combined vaccine regimen consisting of priming with a vaccinia recombinant expressing HIV envelope and boosting with gp160 protein. *Proc. Natl. Acad. Sci.* **90:** 1882–1886.

Davis N.L., Brown K.W., and Johnston R.E. 1996. A viral vaccine vector that expresses foreign genes in lymph nodes and protects against mucosal challenge. *J. Virol.* **70:** 3781–3787.

Dodge J.A. 1995. Gene therapy for cystic fibrosis. *Nat. Med.* **1:** 182.

Dubensky T.W. Jr., Driver D., Polo J.M., Belli B.A., Latham E.M., Ibanez C.E., Chada S., Brumm D., Banks T.A., Mento S.J., Jolly D.J., and Chang S.M.W. 1996. Sindbis virus DNA-based expression vectors: Utility for in vitro and in vivo gene transfer. *J. Virol.* **70:** 508–519.

Edbauer C., Weinberg R., Taylor J., Rey-Senelonge A., Bouquet J.-F., Desmettre P., and Paoletti E. 1990. Protection of chickens with a recombinant fowlpox virus expressing the Newcastle disease virus hemagglutinin-neuraminidase gene. *Virology* **179:** 901–908.

Emi N., Friedmann T., and Yee J.-K. 1991. Pseudotype formation of murine leukemia virus with the G protein of vesicular stomatitis virus. *J. Virol.* **65:** 1202–1207.

Evans D.J., McKeating J., Meredith J.M., Burke K.L., Katrak K., John A., Ferguson M., Minor P.D., Weiss R.A., and Almond J.W. 1989. An engineered poliovirus chimaera elicits broadly reactive HIV-1 neutralizing antibodies. *Nature* **339:** 385–388.

Fenner F. 1996. Poxviruses. In *Fields virology*, 3rd edition (ed. B.N. Fields et al.), vol. 2, pp. 2673–2699. Lippincott-Raven, Philadelphia.

Fenner F., Henderson D.A., Arita I., Jezek Z., and Ladnyi I.D. 1988. Smallpox and its eradication. In *Geneva: World Health Organization* booklet.

Ferrari G., Berend C., Ottinger J., Dodge R., Bartlett J., Toso J., Moody D., Tartaglas J., Cox W.I., Paoletti E., and Weinhold K.J. 1997. Replication-defective canarypox (ALVAC) vectors effectively activate anti-human immunodeficiency virus-1 cytotoxic T lymphocytes present in infected patients: Implications for antigen-specific immunotherapy. *Blood* **90:** 2406–2416.

Fields B.N., Knipe D.M., and Howley P.M. 1996. In *Fields virology*, 3rd edition (ed. B.N. Fields et al.), vol. 2. Lippincott-Raven, Philadelphia.

Fisher K.J., Kelley W.M., Burda J.F., and Wilson J.M. 1996. A novel adenovirus-adeno-associated virus hybrid vector that displays efficient rescue and delivery of the AAV genome. *Hum. Gene Ther.* **7:** 2079–2087.

Flexner C., Murphy B.R., Rooney J.F., Wohlenberg C., Yuferov V., Notkins A.L., and Moss B. 1988. Successful vaccination with a polyvalent live vector despite existing immunity to an expressed antigen. *Nature* **335:** 259–262.

Franken M., Estabrooks A., Cavacini L., Sherburne B., Wang F., and Scadden D.T. 1996. Epstein-Barr virus-driven gene therapy for EBV-related lymphomas. *Nat. Med.* **2:** 1379–1382.

Frolov I., Frolova E, and Schlesinger S. 1997. Sindbis virus replicons and Sindbis virus: Assembly of chimeras and of particles deficient in virus RNA. *J. Virol.* **71:** 2819–2829.

Gershon A.A., LaRussa P., Hardy I., Steinberg S., Silverstein S. 1992. Varicella vaccine: The American experience. *J. Infect. Dis.* **166:** S63–S68.

Gluzman Y. 1981. SV40-transformed simian cells support the replication of early SV40 mutants. *Cell* **23:** 175–182.

Gruss P., Efstratiadis A., Karathanasis S., König M., and Khoury G. 1981. Synthesis of stable unspliced mRNA from an intronless simian virus 40 rat preproinsulin gene recombinant. *Proc. Natl. Acad. Sci.* **78:** 6091–6095.

Hahn C.S., Hahn Y.S., Braciale T.J., and Rice C.M. 1992. Infectious Sindbis virus transient expression vectors for studying antigen processing and presentation. *Proc. Natl. Acad. Sci.* **89:** 2679–2683.

Hamer D.H. and Leder P. 1979. Expression of the chromosomal mouse β^{maj}-globin gene cloned in SV40. *Nature* **281:** 35–40.

Hammond S.A., Bollinger B.C., Stanhope P.E., Quinn T.C., Schwartz D., Clements M.L., and Siliciano R.F. 1992. Comparative clonal analysis of human immunodeficiency virus type 1 (HIV-1) specific CD4+ and CD8+ cytolytic T lymphocytes isolated from seronegative humans immunized with candidate HIV-1 vaccines. *J. Exp. Med.* **176:** 1531–1542.

Hariharan M.J., Driver D.A., Townsend K., Brumm D., Polo J.M., Belli B.A., Catton D.J., Hsu D., Mittelstaedt D., McCormack J.E., Karavodin L., Dubensky T.W., Chang S.M.W., and Banks T.A. 1998. DNA immunization against herpes simplex virus: enhanced efficacy using a Sindbis virus-based vector. *J. Virol.* **7:** 950–958.

Hart S.L., Knight A.M., Harbottle R.P., Mistry A., Hunger H.D., Cutler D.F., Williamson R., and Coutelle C. 1994. Cell binding and internalization by filamentous phage displaying a cyclic Arg-Gly-Asp-containing peptide. *J. Biol. Chem.* **269:** 12468–12474.

Herweijer H., Latendresse J.S., Williams P., Zhang G., Danko I., Schlesigner S., and Wolff J.A. 1995. A plasmid-based self-amplifying Sindbis virus vectors. *Hum. Gene Ther.* **6:**

1161–1167.

Hodge J.W., McLaughlin J.P., Kantor J.A., and Schlom J. 1997. Diversified prime and boost protocols using recombinant vaccinia virus and recombinant non-replicating avain pox virus to enhance T-cell immunity and antitumor reponses. *Vaccine* **15:** 759–768.

Horst J., Kluge F., Beyreuther K., and Gerok W. 1975. Gene transfer to human cells: Transducing phage λ *plac* gene expression in GM_1-gangliosidosis fibroblasts. *Proc. Natl. Acad. Sci.* **72:** 3531–3535.

Huang H.V. 1996. Sindbis virus vectors for expression in animal cells. *Curr. Opin. Biotechnol.* **7:** 531–535.

Irvine K.R., Chamberlain R.S., Shulman E.P., Surman D.R., Rosenberg S.A., and Restifo N.P. 1997. Enhancing efficacy of recombinant anticancer vaccines with prime/boost regiments that use two different vectors. *J. Natl. Cancer Inst.* **89:** 1595–1601.

Ishiura M., Hirose S., Uchida T., Hamada Y., Suzuki Y., and Okada Y. 1982. phage particle-mediated gene transfer to cultured mammalian cells. *Mol. Cell. Biol.* **2:** 607–616.

Knowles M.R., Hohneker K.W., Zhou Z., Olsen J.C., Noah T.L., Hu P-C., Leigh M.W., Engelhardt J.F., Edwards L.J., Jones K.R., Grossman M., Wilson J.M., Johnson L.G., and Boucher R.C. 1995. A controlled study of adenoviral-vector-mediated gene transfer in the nasal epithelium of patients with cystic fibrosis. *N. Engl. J. Med.* **333:** 823–831.

Lamb R.A. and Krug R.M. 1996. Orthomyxoviridae: The viruses and their replication. In *Fields virology*, 3rd edition (ed. B.N. Fields et al.), vol. 2, pp. 1353–1395. Lippincott-Raven, Philadelphia.

Lawson N.D., Stillman E.A., Whitt M.A., and Rose J.K. 1995. Recombinant vesicular stomatitis viruses from DNA. *Proc. Natl. Acad. Sci.* **92:** 4477–4481.

Lederberg J. 1996. Smaller fleas ... *ad infinitum*: Therapeutic bacteriophage redux. *Proc. Natl. Acad. Sci.* **93:** 3167–3168.

Liebhaber S.A., Ray J., and Cooke N.E. 1986. Synthesis of growth hormone-prolactin chimeric proteins and processing mutants by the exchange and deletion of genomic exons. *J. Biol. Chem.* **261:** 14301–14306.

Lopez S., Yao J-S., Kuhn R.J., Strauss E.G., and Strauss J.H. 1994. Nucleocapsid-glycoprotein interactions required for assembly of alphaviruses. *J. Virol.* **68:** 1316–1323.

Luytjes W., Krystal M., Enami M., Parvin J.D., and Palese P. 1989. Amplification, expression, and packaging of a foreign gene by influenza virus. *Cell* **59:** 1107–1113.

Mackett M., Yilma T., Rose J.K., and Moss B. 1985. Vaccinia virus recombinants: Expression of VSV genes and protective immunization of mice and cattle. *Science* **227:** 433–435.

Marcel T. and Grausz J.D. 1997. The TMC Worldwide Gene Therapy Enrollment Report, end 1996. *Hum. Gene Ther.* **8:** 775–800.

Mattion N.M., Harnish E.C., Crowley J.C., and Reilly P.A. 1996. Foot-and-mouth disease virus 2a protease mediates cleavage in attenuated Sabin 3 poliovirus vectors engineered for delivery of foreign antigens. *J. Virol.* **70:** 8124–8127.

Mattion N.M., Reilly P.A., DiMichele S.J., Crowley J.C., and Weeks-Levy C. 1994. Attenuated poliovirus strain as a live vector: Expression of regions of rotavirus outer capsid protein VP7 by using recombinant Sabin 3 viruses. *J. Virol.* **68:** 3925–3933.

McAneny D., Ryan C.A., Beazley R.M., and Kaufman H.L. 1996. Results of a phase I trial of a recombinant vaccinia virus that expresses carcinoembryonic antigen in patients with advanced colorectal cancer. *Ann. Surg. Oncol.* **3:** 495–500.

Mebatsion T. and Conzelmann K-K. 1996. Specific infection of CD4$^+$ target cells by recombinant rabies virus pseudotypes carrying the HIV-1 envelope spike protein. *Proc. Natl. Acad. Sci.* **93:** 11366–11370.

Mebatsion T, Finke S., Weiland F., and Conzelmann K.-K. 1997. A CXCR4/CD4 pseudo-type rhabdovirus that selectively infects HIV-1 envelope protein-expressing cells. *Cell* **90:** 841–847.

Merchlinsky M. and Moss B. 1992. Introduction of foreign DNA into the vaccinia virus genome by in vitro ligation: Recombination-independent selectable cloning vectors. *Virology* **190:** 522–526.

Merril C.R., Biswas B., Carlton R., Jensen N.C., Creed G.J., Zullo S., and Adhya S. 1996. Long-circulating bacteriophage as antibacterial agents. *Proc. Natl. Acad. Sci.* **93:** 3188–3192.

Merrington C.L. Bailey M.J., and Possee R.D. 1997. Manipulation of baculovirus vectors. *Mol. Biotechnol.* **8:** 283–297.

Miller L.K. 1996. Insect viruses. In *Fields virology*, 3rd edition (ed. B.N. Fields et al.), vol. 2, pp. 533–556. Lippincott-Raven, Philadelphia.

Moss B. 1996. Poxviridae: The viruses and their replication. In *Fields virology*, 3rd edition (ed. B.N. Fields et al.), vol. 2, pp. 2637–2671. Lippincott-Raven, Philadelphia.

Mossman S.P., Bex F., Berglund P., Arthos J., O'Neil S.P., Riley D., Maul D.H., Bruck C., Momin P., Burny A., Fultz P.N., Mullins J.I., Liljeström P., and Hoover E.A. 1996. Protection against lethal simian immunodeficiency virus SIVsmmPBj14 disease by a recombinant Semliki Forest virus gp160 vaccine and by a gp120 subunit vaccine. *J. Virol.* **70:** 1953–1960.

Mulligan R.C., Howard B.H., and Berg P. 1979. Synthesis of rabbit β-globin in cultured monkey kidney cells following infection with a SV40 β-globin recombinant genome. *Nature* **277:** 108–114.

Muster R., Ferko B., Klima A., Purtscher M., Trkola A., Schulz P., Grassauer A., Engelhardt O.G., Garcia-Sastre A., Palese P., and Katinger H. 1995. Mucosal model of immunization against human immunodeficiency virus type 1 with a chimeric influen-za virus. *J. Virol.* **69:** 6678–6686.

Noguiez-Hellin P., Robert-LeMeur M., Salzmann J.-L., and Klatzmann D. 1996. Plasmoviruses: Nonviral/viral vectors for gene therapy. *Proc. Natl. Acad. Sci.* **93:** 4175–4180.

Okayama H. and Berg P. 1985. Bacteriophage lambda vector for transducing a cDNA clone library into mammalian cells. *Mol. Cell. Biol.* **5:** 1136–1142.

Oppenheim A., Sandalon Z., Peleg A., Shaul O., Nicolis S., and Ottolenghi S. 1992. A cis-acting DNA signal for encapsidation of simian virus 40. *J. Virol.* **66:** 5320–5328.

Palese P., Zavala F., Muster T., Nussenzweig R.S., and Garcia-Sastre A. 1997. Development of novel influenza virus vaccines and vectors. *J. Infect. Dis.* **176:** S45–S49.

Paoletti E. 1996. Applications of pox virus vectors to vacination: An update. *Proc. Natl. Acad. Sci.* **93:** 11349–11353.

Percy N., Barclay W.S., Garcia-Sastre A., and Palese P. 1994. Expression of a foreign protein by influenza A virus. *J. Virol.* **68:** 4486–4492.

Perkus M.E., Kauffman E.B., Taylor J., Mercer S., Smith D., van der Hoeven J., and Paoletti E. 1993. Methodology of using vaccinia virus to express foreign genes in tis-sue culture. *J. Tiss. Cult. Methods.* **15:** 72–81.

Petricciani J.C. 1991. Regulatory philosophy and acceptability of cells for the production of biologicals. *Dev. Biol. Standard.* **75:** 9–15.

Pialoux G., Excler J.L., Riviere Y., Gonzalez-Canali G., Feuillie V., Coulaud P., Gluckman J.C., Matthews T.J., Meigneir B., Kieny M.P., Gonnet P., Diaz I., Méric C., Paoletti E., Tartaglia J., Salomon H., Plotkin S., The AGIS Group, and l'Agence Nationale de Recherche sur le SIDA. 1995. A prime-boost approach to HIV preventive vaccine using a recombinant canarypox virus expressing glycoprotein 160 (MN) followed by a recombinant glycoprotein 160 (MN/LAI). *AIDS Res. Hum. Retroviruses* **11:** 373–381.

Pittman P.R., Makuch R.S., Mangiafico J.A., Cannon T.L., Gibbs P.H., and Peters C.J. 1996. Long-term duration of detectable neutralizing antibodies after administration of live-attenuated VEE vaccine and following booster vaccination with inactivated VEE vaccine. *Vaccine* **14:** 337–343.

Pleschka S., Jaskunas S.R., Engelhardt O.G., Zurcher T., Palese P., and Garcia-Sastre A. 1996. A plasmid-based reverse genetics system for influenza a virus. *J. Virol.* **70:** 4188–4192.

Porter D.C., Wang J., Moldoveanu Z., McPherson S., and Morrow C.D. 1997. Immunization of mice with poliovirus replicons expressing the C-fragment of tetanus toxin protects against lethal challenge with tetanus toxin. *Vaccine* **15:** 257–264.

Qin H. and Chatterjee S.K. 1996. Cancer gene therapy using tumor cells infected with recombinant vaccinia virus expressing GM-CSF. *Hum. Gene Ther.* **7:** 1853–1860.

Redfield R.R., Wright D.C., James W.D., Jones T.S., Brown C., and Burke D.S. 1987. Disseminated vaccinia in a military recruit with human immunodeficiency virus (HIV) disease. *N. Engl. J. Med.* **316:** 673–676.

Robertson E.S., Ooka T., and Kieff E.D. 1996. Epstein-Barr virus vectors for gene delivery to B lymphocytes. *Proc. Natl. Acad. Sci.* **93:** 11334–11340.

Rolls M.M., Webster P., Balba N.H., and Rose J.K. 1994. Novel infectious particles generated by expression of the vesicular stomatitis virus glycoprotein from a self-replicating RNA. *Cell* **79:** 497–506.

Rueckert R.R. 1996. Picornaviridae: the viruses and their replication. In *Fields virology*, 3rd edition (ed. B.N. Fields et al.), vol. 2, pp. 609–645. Lippincott-Raven, Philadelphia.

Sandalon Z., Dalyot-Herman N., Oppenheim A.B., and Oppenheim A. 1997. In vitro assembly of SV40 virions and pseudovirions: Vector development for gene therapy. *Hum. Gene Ther.* **8:** 843–849.

Schlehofer J.R., Rentrop M., and Mönnel D.N. 1992. Parvoviruses are inefficient in inducing interferon-β, tumor necrosis factor-α, or interleukin-6 in mammalian cells. *Med. Microbiol. Immunol.* **181:** 153–164.

Schlesinger S. and Schlesinger M.J. 1996. Togaviridae: The viruses and their replication. In *Fields virology*, 3rd edition (ed. B.N. Fields et al.), vol. 2, pp. 825–841. Lippincott-Raven, Philadelphia.

Schnell M.J., Mebatsion T., and Conzelmann K.-K. 1994. Infectious rabies viruses from cloned cDNA. *EMBO J.* **13:** 4195–4203.

Schnell M.J., Buonocore L., Whitt M.A., and Rose J.K. 1996a. The minimal conserved transcription stop-start signal promotes stable expression of a foreign gene in vesicular stomatitis virus. *J. Virol.* **70:** 2318–2323.

Schnell M.J., Johnson E., Buonocore L., and Rose J.K. 1997. Construction of a novel virus that targets HIV-1 infected cells and controls HIV-1 infection. *Cell* **90:** 849–857.

Schnell M.J., Buonocore L., Kretzschmar E., Johnson E, and Rose J.K. 1996. Foreign glycoproteins expressed from recombinant vesicular stomatitis viruses are incorporated officiently into virus particles. *Proc. Natl. Acad. Sci.* **93:** 11359–11365.

Shaughnessy E., Lu D., Chatterjee S., and Wong K.K. 1996. Parvoviral vectors for the gene therapy of cancer. *Semin. Oncol.* **23:** 159–171.

Shoji I., Aizaki H., Tani H., Ishii K., Chiba T., Saito I., Miyamura T., and Matsuura Y. 1997. Efficient gene transfer into various mammalian cells, including non-hepatic cells, by baculovirus vectors. *J. Gen. Virol.* **78:** 2657–2664.

Sinkovics J.G. (1989). Oncogenes-antioncogenes and virus therapy of cancer. *Anticancer Res.* **9:** 1281–1290.

Smith G.L., Mackett M., and Moss B. 1983. Infectious vaccinia virus recombinants that express hepatitis B virus surface antigen. *Nature* **302:** 490–495.

Smith H.W. and Huggins M.B. 1982. Successful treatment of experimental *Escherichia coli* infections in mice using phage: Its general superiority over antibiotics. *J. Gen. Microbiol.* **128:** 307–318.

Smith H.W., Huggins M.B., and Shaw K.M. 1987. The control of experimental *Escherichia coli* diarrhoea in calves by means of bacteriophages. *J. Gen. Microbiol.* **133:** 1111–1126

Smith S.M., Maldarelli F., and Jeang K.-T. 1997. Efficient expression by an alphavirus replicon of a functional ribozyme targeted to human immunodeficiency virus type 1. *J. Virol.* **71:** 9713–9721.

Soothill J.S. 1992. Treatment of experimental infections of mice with bacteriophages. *J. Med. Microbiol.* **37:** 258–261.

Srivatsan E.S., Stanbridge E.J., Saxon P.J., Stambrook P.J., Trill J.J., and Tischfield J.A. 1984. Plasmid, phage, and genomic DNA-mediated transfer and expression of prokaryotic and eukaryotic genes in cultured human cells. *Cytogenet. Cell Genet.* **38:** 227–234.

Strayer D.S. 1996. SV40 as an effective gene transfer vector in vivo. *J. Biol. Chem.* **271:** 24741–24746.

Strayer D.S. and Milano J. 1996. SV40 mediates stable gene transfer in vivo. *Gene Ther.* **3:** 581 587.

Strayer D.S., Kondo R., Milano J., and Duan L.-X. 1997. Use of SV40-based vectors to transduce foreign genes to normal human peripheral blood mononuclear cells. *Gene Ther.* **4:** 219–225.

Strickler H.D., Rosenberg P.S., Devesa S.S., Hertel J., Fraumeni, Jr., J.F., and Goedert J.J. 1998. Contamination of poliovirus vaccines with simian virus 40 (1955–1963) and subsequent cancer rates. *J. Am. Med. Assoc.* **279:** 292–295.

Tartaglia J., Jarret O., Neil J.C., Desmettre P., and Paoletti E. 1993. Protection of cats against feline leukemia virus by vaccination with a canarypox virus recombinant. *J. Virol.* **67:** 2370–2375.

Tartaglia J., Perkus M.E., Taylor J., Norton E.K., Audonnet J.-C., Cox W.I., Davis S.W., Van Der Hoeven J., Meignier B., Riviere M., Languet B. and Paoletti E. 1992. NYVAC: A highly attenuated strain of vaccinia virus. *Virology* **188:** 217–232.

Taylor J., Weinberg R., Tartaglia J., Richardson C., Alkhatib G., Briedis D., Appel M., Norton E., and Paoletti E. 1992. Nonreplicating viral vectors as potential vaccines: Recombinant canarypox virus expressing measles virus fusion (F) and hemagglutinin (HA) glycoproteins. *Virology* **187:** 321–328.

Temin H. 1990. Safety considerations in somatic gene therapy of human disease with retrovirus vectors. *Hum. Gene Ther.* **1:** 111–123.

Villarreal L.P. and Soo N.J. 1985. Comparison of the transient late region expression of

SV40 DNA and SV40-based shuttle vectors: Development of a new shuttle vector that is efficiently expressed. *J. Mol. Appl. Genetics.* **3:** 62–71.

Wagner R.R. and Rose J.K. 1996. Rhabdoviridae: The viruses and their replication. In*Fields virology*, 3rd edition (ed. B.N. Fields et al.), vol. 2, pp. 1121–1132. Lippincott-Raven, Philadelphia.

Wahlfors J.J., Xanthopoulos K.G., and Morgan R.A. 1997. Semliki Forest virus-mediated production of retroviral vector RNA in retroviral packaging cells. *Hum. Gene Ther.* **8:** 2031–2041.

Wang M., Bronte V., Chen P.W., Gritz L., Panicali D., Rosenberg S.A., and Restifo N.P. 1995. Active immunotherapy of cancer with a nonreplicating recombinant fowlpox virus encoding a model tumor-associated antigen. *J. Immunol.* **154:** 4685–4692.

Xiong C., Levis R., Shen P., Schlesinger S., Rice C.M., and Huang H.V. 1989. Sindbis virus: An efficient, broad host range vector for gene expression in animal cells. *Science* **2:** 1188–1190.

10

Advances in Synthetic Gene-delivery System Technology

Philip L. Felgner, Olivier Zelphati, and Xiaowu Liang

Gene Therapy Systems, Inc.
San Diego, California 92121

Twenty-six years ago in a *Science* article, Friedmann and Roblin (1972) outlined prospects for human gene therapy. This forward-looking review anticipated the development of two alternative gene-delivery systems: viral-gene therapy vectors and synthetic gene-delivery systems using purified gene sequences. As molecular biology techniques matured, the tools to package genes into nonreplicating, recombinant viral vectors became available (Mann et al. 1983), allowing the efficient introduction of recombinant genes into living cells in vitro (cultured cells) and in vivo (animals and humans). During the last several years, we have witnessed an exponential growth in preclinical research and clinical development of recombinant viral vectors for gene-therapy applications. However, the introduction of synthetic, nonviral gene-delivery systems into the clinical gene-therapy repertoire took somewhat longer to develop.

Scientists at Vical, Inc. and the University of Wisconsin made a key discovery that led to increased interest in direct nonviral gene-transfer technology (Wolff et al. 1990). These investigators were the first to show that under certain conditions, muscle tissues could absorb plasmids, leading to expression of the encoded protein persisting for periods of weeks to several months. Improvements in this basic finding led to what has been referred to as "naked DNA" reagents for gene transfer. In addition, numerous laboratories and biotechnology companies are developing other technologies that allow the delivery of DNA directly into nonmuscle tissues, including the use of cationic lipid molecules that facilitate direct absorption of DNA into cells. The plasmids used for these products are chemically well defined and highly purified and can be produced in large quantities by conventional bacterial fermentation. The broad applicability, ease of manufacturing, and potential cost effectiveness of this

- **Advantages**
  - ➤ Straightforward Plasmid Construction
  - ➤ Convenient Plasmid Construction
  - ➤ No Risk of Infection
  - ➤ Low Immunogenicity
- **Disadvantages**
  - ➤ Low Expression

Figure 1 Synthetic gene-delivery approaches. Advantages and disadvantages of synthetic gene-delivery systems.

gene-based drug-therapy approach offers competitive advantages for research and commercialization, compared to the use of viral vectors (Fig. 1) (Felgner and Rhodes 1991).

Although synthetic delivery systems entered the gene-therapy repertoire somewhat later than viral vectors, today the clinical application of nonviral gene-therapy products is rapidly increasing. The direct injection of naked DNA encoding antigens from infectious organisms (Ulmer et al. 1993) is being aggressively pursued clinically as a method of inducing protective immunity in humans against influenza, human immunodeficiency virus (HIV), malaria, hepatitis B, and a long and growing list of additional pathogens. Cationic lipid-based delivery systems are being evaluated in phase I and phase II clinical trials for the treatment of a variety of different types of human cancer and for the treatment of cystic fibrosis. The cationic polymer-based systems have been most widely associated with the generation of receptor-mediated gene-delivery systems, and advanced clinical trials using such systems are under way in Europe. A growing interest in preclinical and basic research directed at improving and controlling the efficacy of synthetic nucleic acid delivery systems is also evident from the increasing number of publications on the topic (Fig. 2), as well as from the participation at scientific conferences devoted to this area.

As the number of investigators and scientific papers on the topic of synthetic gene-delivery systems has increased, the terminology that describes these systems has also expanded. In many cases, different investigators used different terms to describe the same type of system. Recognizing this problem, a committee of investigators active in this area met to define common terminology (Fig. 3) (Felgner et al. 1997). Cationic lipid-mediated transfection was termed *lipofection* and the cationic lipid–DNA complexes that form when cationic liposomes are mixed with DNA were termed *lipoplexes*. Similarly, transfection mediat-

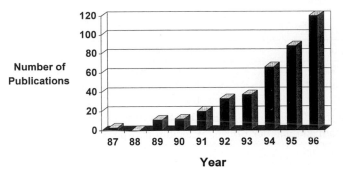

Figure 2 Synthetic gene-delivery systems: Annual publication rate. Synthetic gene-delivery system research is expanding.

ed by hydrophilic polycations such as polylysine or dendrimers was termed *polyfection*, and the complexes that form when these polycations are mixed with DNA are termed *polyplexes*. Most of the synthetic gene-delivery systems under active research and development today are lipoplex, polyplex, lipopolyplex, or naked DNA.

HISTORICAL OVERVIEW OF SYNTHETIC GENE-DELIVERY SYSTEM RESEARCH

In Vitro Transfection

The earliest efforts to identify methods for enhancing delivery of functional purified polynucleotides into living cells were stimulated in the mid-1950s by the results of Alexander and Holland (Alexander et al. 1958; Holland et al. 1959), showing that purified poliovirus RNA was infectious in HeLa cells (Table 1). Since the titer of this purified genom-

- ❑ DEAE Dextran
- ❑ Calcium Phosphate

- ❑ Lipoplex – Cationic Lipid Based
- ❑ Polyplex – Polycationic polymer based
- ❑ Lipopolyplex – Mixtures of lipids and polymers

- ❑ Naked DNA
- ❑ Gene Gun

Figure 3 Synthetic gene-delivery systems. Different types of synthetic gene-delivery systems—Nomenclature.

Table 1 Transfection methods developed to deliver purified infectious viral nucleic acid in vitro

Year	Author	Milestone
1958	Alexander et al.	Purified infectious poliovirus RNA
1959	Holland et al.	Purified infectious poliovirus RNA
1962	Smull and Ludwig	Basic protein mediated
1965	Vaheri and Pagano	DEAE dextran mediated
1973	Graham and van der Eb	Calcium phosphate method

ic RNA stock was extremely low (up to 1 million times less active than the original intact virus), it was questioned whether a vanishingly small quantity of intact virus had survived the purification procedure. They showed that a hypertonic saline solution (1 M NaCl) could enhance infectivity of the purified RNA by about 100-fold. This result, which demonstrated that living virus particles were not required to produce living (i.e., infectious and replicating) viruses, was subsequently confirmed and extended, using improved methods for polynucleotide delivery including calcium phosphate coprecipitation (Graham and van der Eb 1973) and DEAE dextran (Table 1) (Vaheri and Pagano 1965). Broader utility for the calcium phosphate procedure was later demonstrated by experiments showing that transfection of a noninfectious fragment of the herpes simplex virus genome containing the thymidine kinase gene could transform thymidine kinase negative cells (Table 2) (Minson et al. 1978). This result demonstrated that an infectious and proliferating virus was not necessary to induce cellular transformation, leading to a new cellular phenotype. Transfection technology and the emerging recombinant DNA technology converged in the late 1970s when Berg and colleagues applied calcium phosphate, DEAE dextran, and liposome-mediated transfection methodologies to the delivery and expression of recombinant plasmids in cultured mammalian cells (Mulligan et al. 1979; Fraley et al. 1980; Southern

Table 2 Synthetic gene-delivery system applications

Year	Author	Milestone
1978	Minson et al.	Thymidine kinase transformation
1979	Mulligen et al.	Rabbit β-globin—plasmid
1982	Southern and Berg	Neomycin (G418) selection
1983	Mann et al.	Helper-free retrovirus vector
1984	Hwang and Gilboa	Quote: "... retroviral infection is more efficient than ... DNA transfection."

Table 3 Second-generation synthetic gene-delivery systems developed to deliver plasmid in vitro and in vivo

Year	Author	Milestone
1980	Fraley et al.	Liposomes
1987	Felgner et al.	Cationic lipids
1987	Wu and Wu	Polylysine/receptor mediated
1988	Johnston et al.	Gene gun

and Berg 1982). These findings have led to widespread applications for transfection methodology in molecular and cellular biology and in the pharmaceutical industry (Table 2).

In the mid-1980s, a transfection technology emerged from our laboratory that was based on the use of cationic liposomes (Table 3) (Felgner et al. 1987). At that time, there were no chemically stable, liposome-forming cationic lipids available, so we synthesized a series of molecules that could form stable cationic liposomes. The prototype for this approach was a novel positively charged lipid, DOTMA (N[1-(2,3-dioleyloxy) propyl]-N,N,N-trimethylammonium) (Fig. 4). DOTMA forms liposomes that interact spontaneously with DNA or RNA, resulting in a liposome–polynucleotide complex called a *lipoplex*. Lipoplexes are capable of delivering functional nucleic acid molecules into tissue culture cells. Efficacious cationic liposome formulations have three properties that are particularly important to polynucleotide delivery applications (Felgner and Ringold 1989). First, these vesicles spontaneously condense with DNA to form a complex in which up to 100% of the DNA is entrapped. This high entrapment efficiency is not limited by the size of the DNA. Second, virtually all biological surfaces, including cultured cell surfaces, carry a net negative charge. Consequently, positively charged lipid vesicles, complexed with DNA or RNA, interact spontaneously with the negatively charged cell surfaces, delivering the associated polynucleotide to the cell surface. Finally, cationic lipids fuse with cell membranes in a manner that allows the entrapped DNA to enter the cytoplasm and escape from the degradative lysosomal pathway. Although receptor-mediated endocytosis usually enables even conventional liposomes to enter cells in reasonable amounts, these particles do not easily escape the lysosome, within which the liposomes and their encapsulated DNA can be completely degraded. Cationic lipid-mediated fusion depends on the structure of the cationic lipid molecules as well as on the presence of neutral lipids in the final formulation. Today, many different cationic liposome formulations have been found to be effective for enhancing gene delivery (Fig. 4) and more than 30 products are commer-

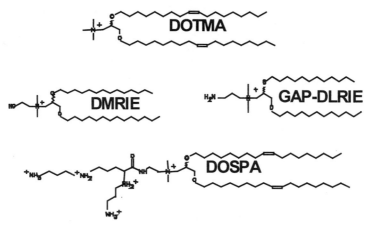

Figure 4 Cationic lipid structures. Representative cationic lipid molecules that are effective for in vitro and in vivo transfection experiments.

cially available for this purpose, including Lipofectin, TransfectAM, LipofectACE, LipofectAMINE, and even LipoTaxi.

In Vivo Gene Delivery: Infectious and Oncogenic Systems

As discussed previously, the earliest successful in vitro transfection experiments were done with purified infectious viral genomes, and the endpoint was the demonstration of infectious virus particles or plaques (Table 1) (Alexander et al. 1958; Holland et al. 1959). Similarly, among the earliest reported in vivo transfections were experiments showing that virus replication could occur following injection of purified or cloned viral genomes in vivo, and the highest levels of infection observed were with particulate formulations of calcium phosphate-precipitated DNA (Table 4) (for review, see Felgner and Rhodes 1991). Israel et al. (1979) showed that polyomavirus DNA was infectious following intraperitoneal injection into mice and hamsters, but at a level of four to five logarithms below that of intact polyoma virions. Dubensky and Bouchard showed that polyomavirus DNA replication efficiency could be improved by using calcium phosphate-precipitated DNA and by treating the tissue with hyaluronidase and collagenase (Dubensky et al. 1984; Bouchard et al. 1984). Both groups also showed that the infectious polyomavirus plasmid produced tumors in a significant percentage of the treated animals. Fung showed that a noninfectious plasmid encoding the *src* oncogene could produce tumors in susceptible chickens (Fung et al. 1983). Gould-Fogerite used liposomes to deliver a noninfectious, oncogenic papylomavirus plasmid DNA frag-

Table 4 In vivo gene transfer with synthetic systems using infectious and oncogenic plasmids

Year	Author	Milestone
1979	Israel et al.	Infectious polyomavirus plasmid
1983	Fung et al.	Oncogenic SRC plasmid
1984	Bouchard et al.	Oncogenic polyoma plasmid
1984	Dubensky et al.	Infectious polyomavirus plasmid
1984	Seeger et al.	Infectious hepatitis virus plasmid

Studies showing that infectious plasmids replicate in vivo and oncogenic plasmids form
tumors in vivo.

ment to induce mouse tumors following direct injection (Gould-Fogerite
et al. 1989). Finally, Varmus and coworkers gave a single 20-μg intrahepatic injection of a cloned infectious hepatitis viral DNA (ground squirrel-
specific virus) into ground squirrels and produced seropositive animals at
11–18 weeks postinjection (Seeger et al. 1984). In all of these cases, the
observed effects were primarily attributed to the ability of the systems
under study to be amplified by replication of an extremely rare, low-fre-
quency infectious or oncogenic event. Discussion of how the apparent low
level of expression obtained could be used to therapeutic benefit was limited. However, the results did show that infectious and oncogenic DNA
sequences could be taken up and expressed by cells following direct in
vivo injection.

In Vivo Gene Expression with Nonreplicating Systems

During the 1980s, several reports were published showing in vivo expression from directly injected noninfectious, nononcogenic plasmid DNA
sequences (Table 5). In vivo gene expression was reported using liposomes, calcium phosphate, a polylysine/protein conjugate, cationic lipids,
and naked DNA. Benvenisty detected chloramphenicol acetyltransferase
(CAT) activity in newborn rat tissues, following intraperitoneal injection
of calcium phosphate-precipitated plasmid DNA (Benvenisty and Reshef
1986). A panel of CAT plasmid constructs containing different eukaryotic viral promoters, a proinsulin gene, and a human growth hormone gene
were all reported to express messenger RNA and/or gene product.

Wu prepared an asialo-orosomucoid/polylysine (AsOR/PL) conjugate that interacted spontaneously with pSV2 CAT plasmid DNA (Table
5) (Wu et al. 1989). Twenty-four hours following intraveneous injection,
CAT activity was detected in rat liver. Activity was only detected when
the pSV2 CAT DNA was complexed with the AsOR/PL conjugate; free

Table 5 Early gene-transfer studies with noninfectious plasmids

Year	Author	Milestone
1983	Nicolau et al.	Insulin gene lowers glucose levels; liposomes
1986	Benvenisty and Reshef	CAT plasmids in peritoneum; calcium phosphate
1987	Wang and Huang	CAT plasmid in peritoneal tumor; liposomes
1989	Wu et al.	CAT plasmids in liver; polylysine/ orosomucoid
1989	Kaneda et al.	Reporter plasmids in liver; liposomes
1989	Brigham et al.	Lung plasmids in lung; cationic lipids
1989	Nabel et al.	β-gal in blood vessel; cationic lipids
1990	Holt et al.	Luciferase in brain; cationic lipids
1990	Yang et al.	Reporter genes—liver skin muscle; gene gun
1990	Wolff et al.	Luciferase and CAT in muscle; naked DNA

plasmid DNA was not active in vivo. The activity declined to baseline levels by 96 hours postdosing. Animals given a partial hepatectomy prior to dosing maintained high levels of CAT activity for up to 11 weeks postdosing. Analysis of DNA extracts from the tissues by Southern blot suggested that the CAT plasmid DNA was integrated into the genomic liver DNA. These results led to the hypothesis that cells undergoing division in regenerating liver permit stable integration of episomal DNA into the host-cell genome and that this newly integrated DNA is active and stable. A third generation of polyplex systems is emerging today with promising results (Table 6). Wagner et al. (1990) have developed greatly enhanced polylysine-based systems derived from the original systems developed by Wu. Polyethylenimine (Boussif et al. 1995) and dendrimers (Tang et al. 1996) have

Table 6 Third-generation synthetic gene-delivery systems

Year	Author	Milestone
1990	Wagner et al.	Polylysine/receptor mediated
1995	Boussif et al.	Polyethylenimine
1996	Tang et al.	Dendrimer

Improved systems for in vitro and in vivo gene delivery.

been shown to been used effectively for in vitro and in vivo gene delivery.

The ability of liposomes to deliver genes in vivo was described in several reports (Table 3). Nicholau reported a transient hypoglycemic effect and elevated insulin levels in the blood, spleen, and liver of rats, following intravenous injection of about 2 µg of a plasmid containing the rat preproinsulin gene (Table 5) (Nicholau et al. 1983). Wang demonstrated CAT activity following intraperitoneal injection of pH-sensitive immunoliposomes containing a CAT plasmid into nude mice bearing an ascites tumor (Wang and Huang 1987). Kaneda described experiments showing expression of SV40 large T antigen in the liver of rats following injection of a pBR SV40 plasmid into the portal vein (Kaneda et al. 1989). The delivery vehicle was a complex mixture of components mixed in a particular order so as to produce complexes of phospholipid vesicles containing encapsulated DNA, ganglioside, Sendai virus fusion proteins, and red-blood-cell membrane proteins.

Cationic lipids were also studied for their ability to deliver genes in vivo. CAT gene expression was detected in the blood and lungs of mice following injection of 15 µg or 30 µg pSV2CAT plasmid complexed with cationic lipid vesicles (Lipofectin™ Reagent, BRL) (Table 5) (Brigham et al. 1989). Injections were intraveneous, intraperitoneal, and intratracheal. Expression was observed for 6 days postinjection. Holt showed that direct injection of CAT and luciferase DNA and RNA into *Xenopus laevis* embryos resulted in expression of luciferase gene product (Holt et al. 1990). Functional expression was shown to be dependent on the presence of the cationic lipid DOTMA, and expression could be enhanced by the coadministration of proteolytic enzymes. Luciferase activity in transfected embryos rose to peak values during the first 48 hours posttransfection and was still detectable 28 days later. And finally, Nabel showed that cationic lipids could enhance expression from plasmids administered into catheterized blood vessels (Nabel et al. 1989).

Interest in the nonviral gene-delivery approach was boosted by the report from Wolff showing in vivo reporter gene expression from plasmids encoding luciferase, CAT, and β-galactosidase genes following direct injection into mouse muscle (Table 5) (Wolff et al. 1990). Up to 1 ng of gene product could be isolated from tissues injected with 10–100 µg of plasmid DNA and expression was shown to persist for more than 6 months. The DNA did not integrate into the host genome, and plasmid essentially identical to the starting material could be recovered from the muscle months after injection. In vitro transcribed messenger RNA was also taken up and expressed in mouse muscle, but the duration of expression was much shorter due to enzymatic breakdown of the message.

Interestingly, no special delivery system was required for these effects. The practical potential of this nonviral approach to gene delivery was further substantiated by the demonstration that animals could be immunized following intramuscular injection of plasmids encoding heterologous antigens (Ulmer et al. 1993). This led to a rapidly expanding interest in the new field of DNA vaccines, which is being actively investigated by immunologists, clinicians, and pharmaceutical companies today. Human clinical trials evaluating DNA vaccines for HIV, influenza, malaria, and hepatitis are currently under way, and at least ten other vaccine candidates are in advanced preclinical development.

CLINICAL PROGRESS IN SYNTHETIC GENE-DELIVERY SYSTEMS

One of the most active areas of gene-therapy clinical research is in cancer (Nabel and Felgner 1993; Felgner et al. 1995). It is possible to consider many different classes of genes, which if administered by a gene-therapy approach either systemically, directly into tumor tissue, or specifically into tumor cells, would exhibit a therapeutic or prophylactic anticancer effect. Three anticancer gene-therapy products are in clinical development at Vical, which utilize (1) IL-2 lymphokine therapy (Leuvectin), (2) HLA-B7 heterologous antigen therapy (Allovectin-7), and (3) an anti-idiotype therapeutic DNA vaccine for B-cell lymphoma (Vaxid).

Allovectin-7 is a gene-based lipoplex product intended for direct injection into tumor lesions of cancer patients. The product contains a gene that encodes a foreign tissue antigen (HLA-B7), which, when injected into tumors, is intended to cause the malignant cells to bear this antigen on their surface. When this foreign antigen is expressed, the patient's immune system, which previously failed to recognize the tumor cells as abnormal, may attack and destroy the cancer cells as if they were foreign tissue.

After a small pilot trial conducted by Nabel, Vical initiated phase I and II clinical trials with approximately 15 patients for each of three advanced cancer indications: renal-cell carcinoma, melanoma, and colorectal carcinoma (Nabel et al. 1995). The trials were designed primarily to test the safety of Allovectin-7 at varying dosage levels and to assess HLA-B7 gene transfer and expression. These studies showed that gene transfer was successful in the majority of patients, the treatment appeared to be safe and well tolerated, and measurable tumor shrinkage was observed in 7 of 14 patients with advanced melanoma. A multicenter phase II clinical trial of Allovectin-7 is currently under way in six tumor types: melanoma, renal-cell carcinoma, colorectal carcinoma, breast carcinoma, non-Hodgkin's lymphoma, and head and neck cancer. More than 100 patients have been treated so far. In addition, Allovectin-7 is being evaluated, either alone or in combination with approved cancer therapeutic agents, in several other phase I or II clinical trials.

Leuvectin contains a gene that encodes the potent immunostimulator, interleukin-2 (IL-2), and is also formulated as a lipoplex to facilitate gene uptake. When injected into tumors, Leuvectin causes the malignant cells to produce and secrete IL-2 in the vicinity of the tumor lesion. The local expression of IL-2 may stimulate the patient's immune system to attack and destroy the tumor cells. Recombinant IL-2 protein is a Food and Drug Administration (FDA)-approved anticancer agent for the treatment of advanced renal cell carcinoma. It has been investigated widely as a cancer immunotherapeutic agent but is frequently associated with serious side effects. Because Vical's gene-based product candidate is designed to deliver IL-2 only at the site of tumor lesions, Leuvectin may provide similar efficacy with fewer side effects than systemic protein-based therapy. This is because the DNA, once introduced into the body, is intended to stimulate the production of an IL-2 at high local concentrations over a prolonged period of time. A major shortcoming of the existing recombinant IL-2 therapy is its short duration of action and the side effects associated with high levels of circulating protein after intravenous administration. By direct injection of the gene into the tumor, a sustained release of the IL-2 within the tumor tissue may be achieved with fewer and less severe side effects.

The initial phase I and II clinical trials were designed primarily to test the safety of Leuvectin at varying dosage levels and to assess IL-2 gene transfer and expression. The initial clinical results showed that gene transfer was effective in the majority of patients, the treatment appeared to be safe and well tolerated, and measurable tumor shrinkage was observed in 5 of 23 patients with various types of advanced malignancies. An additional multicenter phase I and II clinical testing of higher doses of Leuvectin is underway in approximately 45 patients with advanced melanoma, renal cell carcinoma, and soft-tissue sarcoma.

Levy and collaborators at Vical developed a naked DNA anti-idiotype vaccine (Hakim et al. 1996), Vaxid, against low-grade non-Hodgkin's B-cell lymphoma. This type of lymphoma is characterized by a slow growth rate and excellent initial response to chemotherapy or radiotherapy; however, a regular pattern of relapse to a diffuse aggressive lymphoma occurs for which no curative therapy has been identified. Clinical studies involving administration of either monoclonal anti-idiotype antibodies or patient-specific B-cell lymphoma idiotype protein have resulted in prolonged remissions; however, these therapies are limited by the time and effort required to produce the drug product. Vaxid is a DNA plasmid that encodes the patient-specific idiotype of the B-cell tumor immunoglobulin. In preclinical studies, Levy showed that the injection into mice of a murine B-cell lymphoma idiotype plasmid results in strong anti-idiotype immune responses and significant protection against tumor challenge. On the basis of these preclinical studies and additional studies conducted at Vical, the company believes that immunization of postchemotherapy patients with Vaxid could result in the elimination of residual disease and the prevention of the relapse of disease. Vaxid is currently being tested in a human clinical trial.

SYNTHETIC GENE-DELIVERY SYSTEM EFFICIENCY AND EFFICACY

Although it has moved rapidly into clinical testing, the science of synthetic delivery system technology is at the beginning of a discovery and development cycle, which will ultimately lead to much more effective gene-therapy treatments. For comparison, it is instructive to examine the time lines associated with the growth of the recombinant protein industry as it has matured over the past 25 years. The first scientific discoveries launching the biotechnology industry were published in the early 1970s. Genentech, the first recombinant protein biotechnology company, was founded in 1976. Its first successful insulin product was licensed to Eli Lilly and introduced 6 years later. Genentech launched its own growth hormone and tissue plasminogen activator (TPA) products about 10 years after its founding. Then Amgen introduced erythropoietin (EPO), the first real blockbuster recombinant protein product, in 1989, 13 years after the founding of Genentech. It is now more than 20 years since the founding of Genentech, and the recombinant DNA industry is continuing to grow with new recombinant protein products anticipated in the next decade. The scientific activity fueling this expanding technology development effort continues today at an increasing pace.

Nonviral gene-delivery system technology development is much less mature than the recombinant protein industry. A key scientific discovery showing that viruses are not required to deliver functional genes in vivo occurred at Vical and the University of Wisconsin and was published just 8 years ago (Wolff et al. 1990). On the basis of this discovery, Vical's gene-therapy business was started, and today more than ten products based on this technology are in various stages of human clinical evaluation. Comparing this short history with that of the recombinant protein industry, it is reasonable to conclude that this form of gene therapy technology is early in its development cycle and that the technology will continue to mature during the next 15–25 years.

Although nonviral vectors are fundamentally attractive from a pharmaceutical development perspective, the efficacies of the relatively primitive delivery systems in use today are still relatively low. For example, viruses can infect cultured cells with nearly perfect efficiency, wherein 100 infectious viral particles containing 100 viral genomes can successfully infect almost 100 cells. To obtain similar levels of expression with nonviral vectors, it typically takes 100 million copies of plasmid to transfect 100 cells.

A further illustration that nonviral gene-therapy technology is currently in its infancy is shown by the graph in Figure 5. This figure compares the level of gene-product expression obtainable today after in vivo administration of a plasmid encoding a reporter gene, with the level that can be obtained in cultured cells. Here it can be seen that the amount of gene product recovered following intramuscular or intratumor plasmid injection is three to four orders of magnitude lower than that which can be obtained in cultured cells. Thus, there is room to improve in vivo nonviral gene-delivery system technology before it reaches an efficiency level comparable to in vitro transfection. Scientists who succeed in pushing the in vivo expression levels up by one, two, or three orders of magnitude will create new opportunities to demonstrate pharmacological and physiological activities utilizing synthetic gene-delivery systems. The following examples illustrate how increases in in vivo gene expression may expand the clinical gene-therapy opportunities.

EPO is an extremely potent recombinant protein that is used to increase the endogenous production of red blood cells in patients who are experiencing anemia. In humans, it is usually given once per week at 50–500 µg/dose. The corresponding dose of recombinant EPO in a mouse that leads to increased hematocrit is only 25–250 ng. The currently available naked DNA technology allows more than 250 ng of transgene prod-

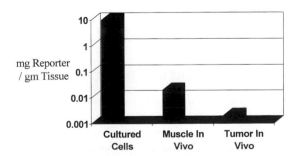

Figure 5 Plasmid expression efficiency. Comparison of in vitro and in vivo transfection efficiencies. There is still room to improve synthetic gene delivery.

uct to be expressed and secreted from the site of a single intramusclular injection of plasmid DNA in a mouse. Therefore, it would be predicted that intramuscular injection of a plasmid secreting EPO from the muscle into the bloodstream would lead to a sufficient amount of EPO to be expressed in order to increase the mouse hematocrit. Indeed, Tripathy showed convincingly that this approach works in the mouse, thus demonstrating the potential for using the muscle as a platform for secretion of a therapeutic protein into the blood (Tripathy et al. 1996). However, in order for this approach to work in humans, who have an approximately 2000-fold larger blood volume than mice, the expression level per administration must be increased by a corresponding 2000-fold. It remains to be determined whether this increased level of expression can be obtained in humans by simply increasing the DNA dose or whether more sophisticated improvements in the delivery system will be needed.

The results of in vivo experiments using Leptin have led to similar conclusions. Intramuscular injection of a plasmid encoding Leptin resulted in a sufficient amount of secreted Leptin to stabilize weight gain and reduce blood glucose levels in genetically obese mice. However, to obtain these results, a tenfold higher DNA dose was required than for EPO. Even at this higher dose, the physiological effect was marginal since the treatment was not sufficient to restore fertility or produce a significant weight reduction. Therefore, to transfer this type of treatment to humans, an increase in expression per administration of greater than 2000-fold would be required. There are several other therapeutic protein delivery applications in which the level of gene-product expression obtained following direct in vivo administration of plasmid is either insufficient or just barely high enough to achieve a physiological response in mice. These applications include insulin to treat type I diabetes and factor IX to treat hemophilia. In these cases, it is even clearer that methods for increasing the level of gene-product expression with synthetic gene-delivery systems are required before practical dosage forms can be successfully evaluated in humans.

OVERCOMING THE BARRIERS TO MORE EFFICIENT GENE DELIVERY

During the last 8 years, we have witnessed rapid progress in the application of synthetic gene-delivery systems; however, the applications remain limited today due to low efficiency of expression. The obstacles to efficient in vivo gene-delivery and expression can be described in terms of "barriers" (Fig. 6). There are extracellular barriers and intracellular barriers. The extracellular barriers refer to those obstacles that the injected gene encounters before it reaches its target cell. The four main extracellular barriers are opsinins, phagocytes, extracellular matrices, and degradative enzymes. Opsinins are proteins that attach themselves to the gene or the delivery system, inactivating the gene and its carrier. Phagocytes are cells that can seek out, engulf, and actively digest the delivery system. Extracellular matrices are zones of polymerized protein and carbohydrate that are present between cells protecting the plasma membrane of the target cell, and it can be difficult for a relatively large DNA carrier system to pass through these barriers. Finally, DNases are present in the extracellular fluid that can rapidly digest unprotected DNA.

The principal intracellular barriers are the plasma membrane, the endosome, and the nuclear membrane. Once the gene-delivery system reaches its target cell, it encounters the plasma membrane, which must be traversed before the gene can be expressed. Under the right circumstances, after the delivery system attaches to the plasma membrane, it can be taken into the cell by a process called endocytosis, but then the delivery system must have a mechanism for escaping from the endosome to avoid being degraded in the lysosomal compartment. Finally, if the gene is able to cross these barriers and enter the cell cytoplasm, it must still have a means of getting across the nuclear membrane.

Delivery system technology development directed at overcoming these barriers involves aspects of molecular biology, DNA condensation technology, and ligand conjugation chemistries. With respect to molecular biology, investigators are looking for ways to increase the level of gene product expressed from plasmids by optimizing promoters, enhancers, introns, terminator sequences, and codon usage. Some investigators are also exploring ways to allow replication of the plasmid inside the nucleus of transfected cells. This approach would theoretically

❑ Extracellular Barriers
  ➢ Opsinins
  ➢ Phagocytic Cells
  ➢ Extracellular Matrix
  ➢ Digestive Enzymes

❑ Intracellular Barriers
  ➢ Plasma Membrane
  ➢ Endosome / Lysosome
  ➢ Nuclear Membrane

Figure 6 Barriers to efficient in vivo gene expression.

increase the number of plasmid copies per cell and also allow expression to persist at high levels within populations of rapidly dividing cells.

DNA condensation is another active area of scientific research. Today, the principal way of accomplishing condensation and packaging of plasmid DNA is with either hydrophobic cations or hydrophilic polycations. Hydrophobic cations form liposomes or micelles that can interact with DNA and reorganize into a cationic lipid–DNA complex called *lipoplex*. The hydrophilic polycations also form complexes when mixed with DNA, and these complexes are called *polyplex*. The hydrophilic polymers are of two general types: the linear polymers, such as polylysine and spermine, and the branched chain, spherical, or globular polycations such as polyethyleneimine or dendrimers. An active area of scientific research involves understanding and controlling the DNA condensation and packaging processes with these agents and determining the structure of the complexes.

Following packaging of the DNA by these methods, investigators are interested in improving gene-delivery efficiency by incorporating ligands into the complexes. These ligands are intended to introduce biological functions into the complexes in order to make them more effective at delivering genes into the target cells. Such ligands may include (1) peptides, which have a specific cell surface receptor so that the complexes will be targeted to specific cells bearing this receptor; (2) nuclear localization signals so that DNA can enter the nucleus more efficiently; (3) pH-sensitive ligands to encourage more efficient endosomal escape; and (4) steric stabilizing agents to avoid interaction with biological factors that would destabilize the complexes after introduction into the biological milieu.

Better methods for studying the biodistribution of plasmid in vitro and in vivo will facilitate synthetic gene-delivery system research. A probe that would allow the simultaneous localization of plasmid DNA and gene-product expression at the cellular and subcellular levels would offer a means to better understand the cellular and molecular barriers to DNA delivery and should provide new insights, leading to more effective plasmid-delivery systems. To date, no technology has been reported that can enable the biodistribution of functionally and conformationally intact plasmid to be followed in real time in living cells while simultaneously monitoring gene expression from the same plasmid. For this purpose, we recently described a nonperturbing plasmid labeling procedure to generate a highly fluorescent plasmid DNA preparation.

To create this system, we used the property of peptide nucleic acids (PNA) to hybridize to nucleic acids in a high affinity and sequence-specific manner. A fluorescent PNA conjugate was hybridized to its complementary sequence on a plasmid. The PNA-binding site was cloned into a region of the plasmid that was not essential for transcription so that PNA binding would not interfere with expression. Fluorescent plasmid prepared in this way was neither functionally nor conformationally altered. PNA binding was sequence-specific, saturable, and extremely stable and did not influence the nucleic acid intracellular distribution. This method was utilized to study conformationally intact plasmid DNA biodistribution in living cells after cationic lipid-mediated transfection. A fluorescent plasmid expressing green fluorescent protein (GFP) enabled simultaneous localization of both plasmid and expressed protein in living cells and in real time. GFP was expressed only in cells containing detectable nuclear fluorescent plasmid.

This detection method offers a way to simultaneously monitor the intracellular localization and expression of plasmid DNA in living cells. This tool should aid in the elucidation of the mechanism of plasmid delivery and its nuclear import with synthetic gene-delivery systems, and to develop more efficient synthetic gene-delivery systems.

Using this system, we were also able to quantitate the uptake of plasmid transfected into cultured cells (Fig. 7). Southern blot analysis showed that in a typical transfection, cells took up 1–10 of the input plasmid. Between 80% and 90% of the plasmid that was isolated in the cell pellet could be removed from the cell surface with a dextran sulfate wash. Five percent of the total input plasmid was taken up into the cells and was stable for at least 48 hours in the cell, and 1–10% of the intracellular plasmid was intranuclear after 48 hours.

In addition to providing a means to obtain biologically active fluorescent DNA, this approach utilizing PNA provides a way to introduce new physical and biological elements onto DNA without perturbing its transcriptional activity. This "gene-chemistry" approach will be used in the future to couple ligands (e.g., nuclear localization signals or other peptides) onto the DNA in order to improve its in vivo bioavailability and expression.

1,000,000 Plasmids / Cell Transfected

300,000 Plasmids / Cell in Pellet

50,000 Plasmids / Cell Intracellular

1,000 Plasmids / Cell Intranuclear

Figure 7 Intracellular transport and plasmid biodistribution after cationic lipid-mediated transfection.

SUMMARY

It has been 40 years since the first studies showing functional delivery of purified nucleic acid into cells. These first studies showed that an intact viral particle was not necessary to produce live, infectious virus, but that only the viral nucleic acid was necessary. The synthetic gene-delivery methods developed in those early years were important to the success of the recombinant-protein science and industry, which still use these methods as routine laboratory tools. The retrovirus vectors that were the first delivery vehicles for gene therapy were generated by using these nonviral gene delivery techniques to introduce the desired genes into the retroviral packaging cell. Finally, a new generation of synthetic gene-delivery systems is being used today to directly administer genes into patients for a host of gene-therapy applications. There has been impressive progress in the development and application of synthetic gene-delivery systems, but there is a long way to go before we will get control over the many variables that influence the activity and practical applicability of these systems.

REFERENCES

Alexander H.E., Koch G., Moran-Mountain I., Sprunt K., and Van Damme O. 1958. Infectivity of ribonucleic acid of poliovirus on HeLa cell monolayers. *J. Exp. Med.* **108:** 493–506.

Benvenisty N. and Reshef L. 1986. Direct introduction of genes into rats and expression of the genes. *Proc. Natl. Acad. Sci.* **83:** 9551–9555.

Bouchard L., Gelinas C., Asselin C., and Bastin M. 1984. Tumorigenic activity of polyoma virus and SV40 DNAs in newborn rodents. *Virology* **135:** 53–64.

Boussif O., Lezoualc'h F., Zanta M.A., Mergny M.D., Scherman D., Demeneix B., and Behr J.P. 1995. A versatile vector for gene and oligonucleotide transfer into cells in culture and in vivo: Polyethylenimine. *Proc. Natl. Acad. Sci.* **92:** 7297–7301.

Brigham K.L., Meyrick B., Christman B., Magnuson M., King G., and Berry L.C., Jr. 1989. In vivo transfection of murine lungs with a functioning prokaryotic gene using a liposome vehicle. *Am. J. Med. Sci.* **298:** 278–281.

Dubensky T.W., Cambell B.A., and Villarreal L.P. 1984. Direct transfection of viral plasmid DNA into the liver or spleen of mice. *Proc. Natl. Acad. Sci.* **81:** 7529–7533.

Felgner P.L. and Rhodes G. 1991. Gene therapeutics. *Nature* **349:** 351–352.

Felgner P.L. and Ringold G.M. 1989. Cationic liposome-mediated transfection. *Nature* **337:** 387–388.

Felgner P.L., Zaugg R.H., and Norman J.A. 1995. Synthetic recombinant DNA delivery for cancer therapeutics. *Cancer Gene Ther.* **2:** 61–65.

Felgner P.L., Gadek T.R., Holm M., Roman R., Chan H.W., Wenz M., Northrop J.P., Ringold G.M., and Danielsen M. 1987. Lipofection: A highly efficient, lipid mediated DNA-transfection procedure. *Proc. Natl. Acad. Sci.* **84:** 7413–7417.

Felgner P.L., Barenholz Y., Behr J.P., Cheng S.H., Cullis P., Huang L., Jessee J.A., Seymour L., Szoka F., Thierry A.R., Wagner E., and Wu G. 1997. Nomenclature for synthetic gene delivery systems. *Hum. Gene Ther.* **8:** 511–512.

Fraley R., Subramani S., Berg P., and Papahadjopoulos D. 1980. Introduction of liposome-encapsulated SV40 DNA into cells. *J. Biol. Chem.* **255:** 10431–10435.

Friedmann T. and Roblin R. 1972. Gene therapy for human genetic disease? *Science* **175:** 949–955.

Fung Y.K., Crittenden L.B., Fadly A.M., and Kung H.J. 1983. Tumor induction by direct injection of cloned v-*src* DNA into chickens. *Proc. Natl. Acad. Sci.* **80:** 353–357.

Gould-Fogerite S., Mazurkiewicz J.E., Raska K., Jr., Voelkerding K., Lehman J.M., and Mannino R.J. 1989. Chimerasome-mediated gene transfer in vitro and in vivo. *Gene* **84:** 429–438.

Graham F.L. and van der Eb A.J. 1973. A new technique for the assay of infectivity of human adenovirus 5 DNA. *Virology* **52:** 456–467.

Hakim I., Levy S., and Levy R. 1996. A nine-amino acid peptide from IL-1β augments antitumor immune responses induced by protein and DNA vaccines. *J. Immunol.* **157:** 5503–5511.

Holland J.J., McLaren L.C., and Syverton J.T. 1959. The mammalian cell-virus relationship. III. Poliovirus production by non-primate cells exposed to poliovirus ribonucleic acid. *Proc. Soc. Exp. Biol. Med.* **100:** 843–845.

Holt C.E., Garlick N., and Cornel E. 1990. Lipofection of cDNAs in the embryonic vertebrate central nervous system. *Neuron* **4:** 203–214.

Hwang L.S. and Gilboa E. 1984. Expression of genes introduced into cells by retroviral infection is more efficient than that of genes introduced into cells by DNA transfection. *J. Virol.* **50:** 417–424.

Israel M.A., Chan H.W., Martin M.A., and Rowe W.P. 1979. Molecular cloning of polyoma virus DNA in *Escherichia coli*: Oncogenicity testing in hamsters. *Science* **205:** 1140–1142.

Johnston S.A., Anziano P.Q., Shark K., Sanford J.S., and Butow R.A. 1988. Mitochondrial transformation in yeast by bombardment with microprojectiles. *Science* **240:** 1538–1541.

Kaneda Y., Iwai K., and Uchida T. 1989. Increased expression of DNA cointroduced with nuclear protein in adult rat liver. *Science* **243:** 375–378.

Mann R., Mulligan R.C., and Baltimore D. 1983. Construction of a retrovirus packaging mutant and its use to produce helper-free defective retrovirus. *Cell* **33:** 153–159.

Minson A.C., Wildy P., Buchan A., and Darby G. 1978. Introduction of the *herpes simplex* virus thymidine kinase gene into mouse cells using virus DNA or transformed cell DNA. *Cell* **13:** 581–587.

Mulligan R.C., Howard B.H., and Berg P. 1979. Synthesis of rabbit β-globin in cultured monkey kidney cells following infection with a SV40 β-globin recombinant genome. *Nature* **277:** 108–114.

Nabel E.G., Plautz G., Boyce F.M., Stanley J.C., and Nabel G.J. 1989. Recombinant gene expression in vivo within endothelial cells of the arterial wall. *Science* **244:** 1342–1344.

Nabel G.J. and Felgner P.L. 1993. Direct gene transfer for immunotherapy and immunization. *Trends Biotechnol.* **11:** 211–215.

Nabel G.J., Yang Z.Y., Nabel E.G., Bishop K., Marquet M., Felgner P.L., Gordon D., and Chang A.E. 1995. Direct gene transfer for treatment of human cancer. *Ann. N.Y. Acad. Sci.* **772:** 227–231.

Nicolau C., Le Pape A., Soriano P., Fargette F., and Juhel M.F. 1983. In vivo expression of rat insulin after intravenous administration of the liposome-entrapped gene for rat insulin I. *Proc. Natl. Acad. Sci.* **80:** 1068–1072.

Seeger C., Ganem D., and Varmus H.E. 1984. The cloned genome of ground squirrel hepatitis virus is infectious in the animal. *Proc. Natl. Acad. Sci.* **81:** 5849–5852.

Smull C.E. and Ludwig E.H. 1962. Enhancement of the plaque-forming capacity of poliovirus ribonucleic acid with basic proteins. *J. Bacteriol.* **84:** 1035–1040.

Southern P.J. and Berg P. 1982. Transformation of mammalian cells to antibiotic resistance with a bacterial gene under control of the SV40 early region promoter. *J. Mol. Appl. Genet.* **1:** 327–341.

Tang M.X., Redemann C.T., and Szoka F.C., Jr. 1996. In vitro gene delivery by degraded polyamidoamine dendrimers. *Bioconjug. Chem.* **7:** 703–714.

Tripathy S.K., Svensson E.C., Black H.B., Goldwasser E., Margalith M., Hobart P.M., and Leiden J.M. 1996. Long-term expression of erythropoietin in the systemic circulation of mice after intramuscular injection of a plasmid DNA vector. *Proc. Natl. Acad. Sci.* **93:** 10876–10880.

Ulmer J.B., Donnelly J.J., Parker S.E., Rhodes G.H., Felgner P.L., Dwarki V.J., Gromkowski S.H., Deck R.R., DeWitt C.M., Friedman A., et al. 1993. Heterologous protection against influenza by injection of DNA encoding a viral protein. *Science* **259:** 1745–1749.

Vaheri A. and Pagano J.S. 1965. Infectious poliovirus RNA : A sensitive method of assay. *Virology* **27:** 434–436.

Wagner E., Zenke M., Cotten M., Beug H., and Birnstiel M.L. 1990. Transferrin-polycation conjugates as carriers for DNA uptake into cells. *Proc. Natl. Acad. Sci.* **87:** 3410–3414.

Wang C.Y. and Huang L. 1987. pH-sensitive immunoliposomes mediate target-cell-specific delivery and controlled expression of a foreign gene in mouse. *Proc. Natl. Acad. Sci.* **84:** 7851–7855.

Wolff J.A., Malone R.W., Williams P., Chong W., Acsadi G., Jani G., and Felgner P.L. 1990. Direct gene transfer into mouse muscle in vivo. *Science* **247:** 1465–1468.

Wu C.H., Wilson J.M., and Wu G.Y. 1989. Targeting genes: Delivery and persistent expression of a foreign gene driven by mammalian regulatory elements in vivo. *J. Biol. Chem.* **264:** 16985–16987.

Wu G.Y. and Wu C.H. 1987. Receptor mediated in vitro gene transformation by a soluble DNA carrier system. *J. Biol. Chem.* **262:** 4429–4432.

Yang N.S., Burkholder J., Roberts B., Martinell B., and McCabe D. 1990. In vivo and in vitro gene transfer to mammalian somatic cells by particle bombardment. *Proc. Natl. Acad. Sci.* **87:** 9568–9572.

11

Receptor-mediated Gene Delivery Strategies

Matt Cotten

Institute for Molecular Pathology
1030 Vienna, Austria

Ernst Wagner

Boehringer Ingelheim R&D
1121 Vienna, Austria

WHAT NEEDS TO BE ACCOMPLISHED FOR SUCCESSFUL GENE DELIVERY

The perceived barriers to gene delivery are summarized in Figure 1. When using these systems in vivo, the complex must avoid inactivation or aggregation by serum components such as complement (Plank et al. 1996). The material must be of the appropriate size to reach the target cells. For example, the fenestrations through which the material must reach the liver parenchyma have a limit of approximately 100 nm. The complex must avoid being recognized by neutralizing antibodies if they are present in the host serum. Furthermore, the charge properties of the complex should be adjusted to avoid nonspecific accumulation on inappropriate surfaces.

Should the complex reach the surfaces of target cells, it is important to include ligands to ensure binding to elements on the surface of the cell. Initially, ligands for receptors known to be rapidly internalized were used, such as asialoglycoproteins or transferrin, but subsequent efforts suggest that the simple binding, aggregation, or accumulation on the surface is sufficient for entry into the vesicle system of the cell.

Once internalized in a vesicle, passage through or across the vesicle membrane into the cytoplasm is required. A large amount of research in membrane disruption has occurred since this barrier was recognized in the early 1990s (see the following). Passage through the cytoplasm is required (or, alternately, disruption of the cytoplasmic organization to promote deposition at the nuclear membrane), passage into the nucleus, and the docking of transfection material in the appropriate nuclear site

Gene delivery barriers

Barrier   Solution

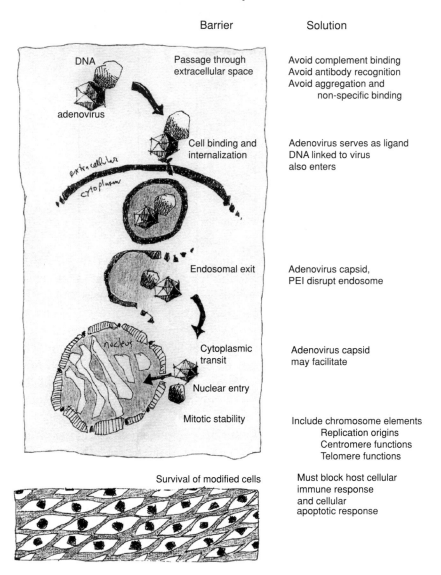

	Barrier	Solution
DNA / adenovirus	Passage through extracellular space	Avoid complement binding Avoid antibody recognition Avoid aggregation and non-specific binding
	Cell binding and internalization	Adenovirus serves as ligand DNA linked to virus also enters
	Endosomal exit	Adenovirus capsid, PEI disrupt endosome
	Cytoplasmic transit	Adenovirus capsid may facilitate
	Nuclear entry	
	Mitotic stability	Include chromosome elements Replication origins Centromere functions Telomere functions
	Survival of modified cells	Must block host cellular immune response and cellular apoptotic response

Figure 1 A summary of the perceived barriers to gene delivery at both the organism and the cellular levels. See text for details.

may be critical. Under some conditions, the passage of the cell through mitosis may be essential for accumulation in the nucleus, although the experimental evidence for this is difficult to obtain. It is also not yet clear where and how the transfected DNA disentangles itself from the carrier polycation. There is evidence from the cationic lipid systems that the dissociation occurs during or after cytoplasmic entry (Zelphati and Szoka 1996); other evidence is available showing the transfection complex in the nucleus (Labat-Moleur et al. 1996).

Once inside the nucleus, transient expression is easy to obtain, but stable maintenance of promoter activity can be difficult to obtain. Maintenance of the DNA in transfected cells after multiple rounds of cell division is also problematic unless the DNA is integrated. Adopting elements of the chromosome that ensure segregation to daughter cells may be fruitful.

Apart from the survival of promoter activity or the DNA itself, all is lost if the transfected cell dies or is eliminated due to the modification. An apoptotic response to transfection is present in many cell types, and this phenomenon can represent a major barrier (Chiocca et al. 1997; Ebert et al. 1997). Immune responses to the modified cell must be considered, and substantial research effort has been spent attempting to elucidate and control these responses, either to inhibit them in gene correction applications or to foster them in a productive manner in vaccine applications. After a first, successful application of transfection complexes, immune responses can arise that limit further application of the complexes.

It is clear then that gene delivery in vivo has turned out to be a more complicated process than perhaps was originally anticipated. However, the great flexibility of the synthetic or conjugates being explored for gene delivery combined with a fair amount of optimism have led to some advances in receptor-mediated gene delivery.

EARLY HISTORY OF RECEPTOR-MEDIATED STRATEGIES

George and Catherine Wu used the DNA condensing activity of polylysine and chemically linked to it a ligand that would target condensed DNA to a receptor present on the target cell. Asialoorosomucoid was used as the ligand, and the hepatocyte-specific asialoglycoprotein receptor was the target. Under certain conditions, these simple conjugates could deliver DNA to the hepatocytes (Wu and Wu 1987, 1988; Wu et al. 1989). The idea captured the attention of a number of groups and variations appeared. Transferrin was used extensively to take advantage of the ubiquitous transferrin receptor pathway (Wagner et al. 1990). As summarized in Table 1, a plethora of ligands has been incorporated into various transfection complexes, including simple synthetic peptides and carbohydrates, peptide growth factors, lectins, and antibodies or antibody frag-

Table 1 Ligands employed in gene delivery systems

Ligand	Comments	References
Asialoglycoprotein	Asialoglycoprotein receptor ASGPR) for hepatocyte delivery	Wu and Wu 1987; (1988); Wu et al. (1989); McKee et al. (1994); Findeis et al. (1994)
Transferrin	Broadly distributed receptor	Cotten et al. (1990); Wagner et al. (1990); Zenke et al. (1990);
Tetra-antenary galactose	ASGPR	Plank et al. (1992); Remy et al. (1995)
Lung surfactants	Lung epithelial cells	Baatz et al. (1994)
Lectin: wheat germ agglutinin	Myoblasts delivery	Cotten et al. (1993)
Other lectins	Tumor delivery	Batra et al. (1994)
Antibodies: anti-CD3, anti-CD4, anti-CD7	T-cell delivery	Buschle et al. (1995)
Antibodies: anti-secretory component	Transcytosis into epithelial cells	Ferkol et al. (1993a, 1995)
Antibody against truncated carbohydrate	Tumor delivery	Thurnher et al. (1994)
Antibody against thrombomodulin (lung epitope)	Lung, tumor delivery	Trubetskoy et al. (1992a,b)
Insulin	Broadly distributed receptor	Huckett et al. (1990); Rosenkranz et al. (1992)
Tri-galactosylated bisacridine	Asgpr	Haensler and Szoka (1993a)
Lactosylated polylysine	Asgpr	Midoux et al. (1993)
Lactosylated albumin	Asgpr	Ferkol et al. (1993b); Perales et al. (1994a,b)
Glycopeptide	Asgpr	Merwin et al. (1994)
RGD peptide	Targets integrins	Hart et al. (1995)
Invasin internalization domain	Targets integrins	Paul et al. (1997)
Transferrin, anti-CD3,	As PEI complexes	Kircheis et al. (1997)
Galactose	As PEI complexes	Zanta et al. (1997)
Peptide derived from alpha 1 anti-trypsin	Serpin enzyme complex receptor, hepatocytes	Ziady et al. (1997)
Anti-Her2 antibody	HER2 (EGF receptor expressed abundantly on certain epithelial tumor types)	Foster and Kern (1997)
Anti-idiotype antibodies	For delivery to B cell lymphomas	Schachtschabel et al. (1996)
Egf	Incorporated into lipid/DNA complex for targeting via EGF receptor	Kikuchi et al. (1996)
Mannosylated polylysine	Monocyte transfection	Erbacher et al. (1996) Ferkol et al. (1996)
Steel factor	Hematopoietic stem cells	Schwarzenberger et al. (1996)

ments. Unfortunately, in most cases, these simple ligand/polycation conjugates work poorly in primary cell types or in vivo unless additional
agents are included. It appeared in many cases that the ligand-polycation
condensed DNA was efficiently internalized but remained trapped in the
vesicular system under most conditions. Topologically, a membrane still
separated the DNA from the cell interior, and methods were sought to sur-
mount this barrier.

ADDITION OF ENDOSOMOLYTIC AGENTS SUCH AS ADENOVIRUS
IMPROVED DELIVERY

Adenovirus, like many other animal viruses, generates a transient perme-
abilization of the host-cell membrane during infection. Thus, the cellular
entry of both fluid phase markers and receptor-bound molecules is
markedly increased in the presence of the adenovirus (Fernández-Puentes
and Carasco 1980; Fitzgerald et al. 1983; for review, see Carrasco 1994,
1995). It was a useful idea to include adenovirus particles in an attempt
to improve the cytoplasmic delivery of receptor-delivered DNA molecules (Curiel et al. 1991) (see Fig. 2). This strategy of using the adenovirus as a transfection enhancer was subsequently pursued by many different groups, supplying the virus in a variety of forms, coupling the virus
to the DNA, and attempting to replace the virus with viral subcomponents
(see the following).

Subcomponents of the virus are reported to function to enhance delivery, but if one examines the results carefully, it becomes clear that only a
fraction of the performance of intact virions is obtained with any of the
subcomponents. These include efforts to demonstrate that the penton base
is membrane-active (Seth et al. 1984; Blumenthal et al. 1986; Seth
1994a,b) as well as efforts to use a penton base/fiber complex to deliver
bound DNA (Fender et al. 1997). The protein capsid of the virus was
thought to be responsible for the enhancement of DNA delivery; a virus
inactivation method using psoralen, which blocked RNA production from
the viral genome while leaving the protein capsid intact, demonstrated
that indeed no viral gene expression was required for the enhancement
(Cotten et al. 1992).

OTHER ENDOSOMOLYTIC AGENTS

Although the adenovirus proved to be a very effective agent for enhancing uptake, chemists sought to simplify the systems by using defined synthetic components. One successful approach employed short synthetic
peptides based on the sequence thought to be important for membrane

fusion by influenza hemagglutinin (Wagner et al. 1992a; Plank et al. 1994). Because of the position of the acidic amino acid side chains, these peptides are largely unstructured at neutral pH, but upon entry in the acid endosomal environment, they adopt an ordered, amphipathic structure that is disruptive to lipid bilayers. Inclusion of these peptides into condensed DNA complexes led to useful improvements in delivery.

Many improvements in gene transfer were obtained once the cellular membrane was recognized as the initial barrier to transfection. This idea was supported by results obtained with a variety of agents that had, as a common property, the ability to disrupt membranes or disrupt the endosome. These include chloroquine, which could accumulate in an acid vesicle and lead to an endosomal swelling (Cotten et al. 1990), and, most recently, the synthetic polycation polyethyleneimine (PEI), which is thought to undergo a low-pH-induced swelling that is endosomolytic (Boussif et al. 1995). For a summary of these agents, see Table 2.

DEVELOPMENT OF VIRUS/DNA COMPLEXES

Although the simple addition of viral particles was effective in many cell culture lines, a very high virus/cell ratio was required for maximum gene delivery enhancement. This was thought to be due to the poor co-internalization of the packaged DNA and the viral particles, with a high concentration of both components required to ensure that a single endosome contained both entities. Methods of physically linking the packaged DNA with the virus were developed and found to be an effective improvement. The most versatile method (and still in extensive use) employed a streptavidin/polylysine conjugate to package the DNA and link it to biotinylated adenovirus particles (Wagner et al. 1992b). Alternate methods are listed in Table 3. Fixing the polycation to the virus by enzymatic or chemical methods was a simple alternative to using streptavidin (Wagner et al. 1992b) or antibody conjugates (Curiel et al. 1992); however, generating a covalent polycation/virus link creates problems of storage and titration. Thus, the most versatile methods employ reagents that allow a simple titration of the virus-to-polycation ratio. PEI has recently emerged as an effective reagent for condensing DNA and linking it to the adenovirus surface (Fig. 2) (Baker et al. 1997). In this system, the PEI-condensed DNA maintains the ability to interact via positive charge groups from the PEI to negative elements on the adenovirus capsid. Subgroup C adenoviruses have a cluster of glutamic acid residues that are conveniently exposed for such a binding (Baker et al. 1997).

Table 2 Agents that enhance cellular entry of DNA complexes

Agent	References	Comments
Adenovirus	Curiel et al. (1991); Wagner et al. (1992b)	
Chicken adenovirus CELO	Cotten et al. (1993)	Naturally defective in human cells
Psoralen-inactivated adenovirus	Cotten et al. (1992, 1994)	No viral gene expression required for DNA delivery
Rhinovirus	Zauner et al. (1995)	RNA virus
Membrane active peptide GALA	Parente et al. (1990)	Membrane-disruptive
Influenza HA peptide and variants	Wagner et al. (1992a); Plank et al. (1994)	Membrane-disruptive
Dendrimers	Haensler and Szoka (1993b); Bielinska et al (1996, 1997); Kukowska-latallo et al. (1996); Tang et al. (1996)	Proton sponge?
Polyethyleneimine (PEI)	Boussif et al. (1995, 1996); Abdallah et al. (1996); Baker et al. (1997); Boletta et al. (1997); Kircheis et al. (1997); Dunlap et al. (1997)	Proton sponge, probably generates an osmotic lysis; also functions to condense DNA
Gramicidin S	Legendre and Szoka (1993)	Amphipathic, cyclic peptide
Chloroquine	Cotten et al. (1990)	Functions to elevate pH, osmotically lyse endosomes, utility limited to a few cell lines
Glycerol	Zauner et al. (1996, 1997)	Alters membrane fluidity
KALA synthetic peptide	Wyman et al. (1997)	DNA binding and membrane active synthetic peptide
Diptheria toxin fragment	Fisher and Wilson (1997)	Membrane-active

REPLACEMENT OF THE ADENOVIRUS WITH SIMPLE COMPONENTS

The initial success with the adenovirus as an enhancing reagent spurred a number of groups to examine this phenomenon in greater detail. It was

Adenovirus/PEI/DNA complex

DNA condensing function

Linkage to adenovirus
    positive charge PEI/DNA to
    negative adenovirus hexon

Endosomolytic agent
    adenovirus particle
    PEI

Ligand
    adenovirus fiber
    charge

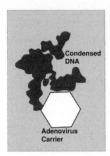 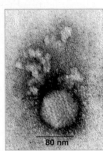

Figure 2 Adenovirus/PEI/DNA complex. An effective receptor-mediated gene delivery system comprising an inactivated adenovirus particle linked, via ionic interactions, with polyethyleneimine-condensed DNA. For additional details, see the text and Baker et al. (1997).

naively hoped that a single protein from the virus capsid would be responsible for the transfection enhancement. Unfortunately, subcomponents of the virus can be shown to contribute only modestly to transfection efficiency, but they never reproduce the full enhancement of the intact virion to any useful level. There are two important conclusions from these efforts: (1) An intact virion with the appropriate architecture is required. Perhaps this is essential for placing the hydrophobic capsid domains in the appropriate spatial organization for lipid disruption. (2) The adenovirus

Table 3 Methods of linking virus to package DNA

Method	Comments	Reference
Streptavidin polylysine/biotin virus	Requires special conjugate	Wagner et al. (1992)
Antibody/polylysine	Requires special bridge conjugate	Curiel et al. (1992)
Transglutaminase linkage	Difficult to store virus/polylysine covalent conjugate, difficult to titrate	Cotten et al. (1996)
Chemical linkage	Difficult to store virus/polylysine covalent conjugate, difficult to titrate	Cristiano et al. (1993); Fisher and Wilson (1994); Wu et al. (1994)
Ionic linkage with PEI condensed DNA	Simple, no special reagents required	Baker and Cotten (1997); Baker et al. (1997)

virion, upon entering the host cell, may be affecting multiple aspects of the cell that will result in a transient increase in transfected gene expression. Focusing on only the membrane passage is not sufficient to understand or to mimic the adenovirion's full effect. These additional effects include an activation of certain promoters by the capsid components. Two hundred and forty copies of protein IX enter the cell as part of each infecting virion, and this protein has recently been found to possess transcriptional activation activity (Lutz et al. 1997). The virion also carries with it approximately 20 copies of a virus-encoded protease which, when expressed in the absence of virus infection, can activate certain classes of promoters (B. Panzenböck and M. Cotten, unpubl.). The adenovirus makes its way promptly to the nucleus of the infected/transfected cell (Greber et al. 1993, 1996, 1997), and changes in the cell that are promoted by the virion to accomplish this movement may also lead to the movement of transfected DNA to the nucleus. Because the nuclear entry of transfected DNA appears to be a significant barrier to transfection success (Zabner et al. 1995), mechanisms that promote this trafficking could lead to potent improvements. Finally, it is becoming clear that the nucleus is not simply a bag of DNA in the center of the cell; it also possesses a variety of organizations. It is likely that certain other components of the adenovirus are important for placing cotransfected DNA in the right compartment of the nucleus for full transcriptional activity. Indeed, there is evidence that including the adenovirus terminal protein on transfected DNA can lead to substantial improvements in the function of transfected or infected DNA, perhaps by promoting the appropriate nuclear presence (Sharp et al. 1976; Jones and Shenk 1978; Schaack et al. 1990; Lieber et al. 1997).

REPLACEMENT OF POLYLYSINE WITH ALTERNATE POLYCATIONS

Naked DNA is an extended molecule that is susceptible to mechanical and enzymatic shear and exists nowhere in nature in an uncondensed form. All natural DNA is associated with some form of polycation, and gene deliverists have also found that condensing DNA with some form of cation is useful, even essential, for useful transfection results. Early forms of receptor-mediated gene delivery employed polylysine extensively, because it is inexpensive and readily available, and because the lysine residues facilitated coupling strategies. Alternate natural polycations have been used with some success, including histone proteins, protamine, spermine or spermidine (Wagner et al. 1991), histone H1 (Fritz et al. 1996), and high mobility group (HMG) proteins (Böttger et al. 1988; Kaneda et al. 1989; Tomita et al. 1992).

Several compounds that have been developed for gene delivery use have both DNA condensation activity and a certain degree of membrane disruption activity. These include the starburst dendrimers synthesized with terminal amino groups (Haensler and Szoka 1993b; Bielinska et al. 1996; Kukowska-Latallo et al. 1996; Tang et al. 1996; Tang and Szoka 1997) and the polyethyleneimines (Boussif et al. 1995; Baker et al. 1997b; Dunlap et al. 1997; Kircheis et al. 1997).

DELIVERY OF VERY LARGE DNA—ARTIFICIAL CHROMOSOME EFFORTS

Mitosis in the target cell is thought to improve transient DNA transfection by increasing the basal levels of the transcription and translation components and by increasing the nuclear entry of cytoplasmic DNA as the daughter nuclei assemble. However, once transfected DNA has reached the nucleus, survival of this DNA in mitotically active cell populations may be difficult to obtain unless the DNA integrates in one of the host chromosomes or possesses signals that ensure replication and segregation to daughter nuclei. Efforts to improve integration are further plagued by the difficulties of maintaining high promoter activity upon integration into random chromosomal sites. Efforts to improve mitotic stability without integration are at an early stage. Again, viruses have provided a model, with DNA molecules that encode Epstein-Barr virus origins plus the EBNA-1 gene, showing substantial mitotic stability (Yates et al. 1985). Subsequent applications of these elements have been made in efforts to establish extrachromosomal replicating vectors (Wohlgemuth et al. 1996; for review, see Calos 1996) and to stabilize large YAC DNA molecules (Simpson et al. 1996).

Among other functions, the EBNA-1 protein binds the oriP element and tethers the complex to host chromosomes, thus ensuring some distribution to both daughter cells. Unfortunately, EBNA-1 has a transforming potential (Wilson et al. 1996), and efforts to identify more benign molecules are under way. One approach involves constructing DNA molecules that contain the appropriate sequence to ensure attachment to a mitotic spindle. Furthermore, it is thought to be important that the molecules be linear, and thus telomeric sequences are included to stabilize the ends and solve the difficulties associated with replicating a linear molecule. One demonstration of this technique is given by Harrington et al. (1997). Centromeric elements are normally in the >100-kb size range, so efforts to develop these ideas have been hampered by an inability to efficiently transfect these molecules with standard transfection reagents. However, recent efforts using an adenovirus carrier showed that large molecules can be readily introduced into mammalian cells (Baker and Cotten 1997). Current efforts to define a mammalian centromere and to construct an artificial chromosome have been reviewed recently (Huxley 1997; Rosenfeld 1997).

SUMMARY

Figure 3 contains a cartoon summarizing an ideal gene transfer reagent. A certain amount of this reagent is hypothetical and not yet realized with our current understanding of DNA transfer. However, the list of features reveals one aspect of the progress that has been made in the last 10 years, and this has been a substantial improvement in our understanding of why gene transfer has not led to rapid clinical success. A clear identification of the phenomena that limit gene transfer in vivo has developed. Future efforts can be made in a more focused manner to solve the relevant problems.

Ideal gene delivery complex

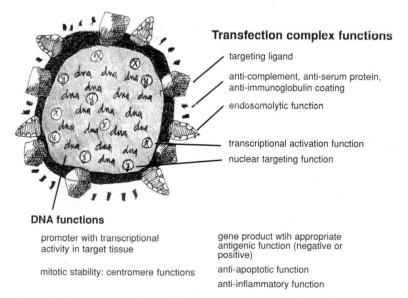

Transfection complex functions

targeting ligand

anti-complement, anti-serum protein, anti-immunoglobulin coating

endosomolytic function

transcriptional activation function

nuclear targeting function

DNA functions

promoter with transcriptional activity in target tissue

mitotic stability: centromere functions

gene product wtih appropriate antigenic function (negative or positive)

anti-apoptotic function

anti-inflammatory function

Figure 3 A summary of an ideal gene-transfer complex. The complex contains hypothetical components that address many of the barriers to gene delivery described in Fig. 1.

ACKNOWLEDGMENTS

We thank Max Birnstiel for his support and for his critical reading of the manuscript and Iris Killisch for her electron microscopy assistance.

REFERENCES

Abdallah B., Hassan A., Benoist C., Goula D., Behr J.P., and Demeneix B.A. 1996. A powerful nonviral vector for in vivo gene transfer into the adult mammalian brain: Polyethylenimine. *Hum. Gene Ther.* **7:** 1947–1954.

Baatz J.E., Bruno M.D., Ciraolo P.J., Glasser S.W., Stripp B.R., Smyth K.L., and Korfhagen T.R. 1994. Utilization of modified surfactant-associated protein B for delivery of DNA to airway cells in culture. *Proc. Natl. Acad. Sci.* **91:** 2547–2551.

Baker A. and Cotten M. 1997. Useful delivery of bacterial artificial chromosomes into mammalian cells using psoralen-inactivated adenovirus carrier. *Nucleic Acids Res.* **25:** 1950–1956.

Baker A., Saltik M., Lehrmann H., Killisch I., Mautner V., Lamm G., Christofori G., and Cotten M. 1997. Polyethylenimine is a simple, inexpensive and effective reagent for condensing and linking plasmid DNA to adenovirus for gene delivery. *Gene Ther.* **4:** 773–782.

Batra R.K., Wang-Johanning F., Wagner E., Garver R.I., and Curiel D.T. 1994. Receptor-mediated gene delivery employing lectin-binding specificity. *Gene Ther.* **1:** 255–260.

Bielinska A.U., Kukowska-Latallo J.F., and Baker J.R., Jr. 1997. The interaction of plasmid DNA with polyamidoamine dendrimers: Mechanism of complex formation and analysis of alterations induced in nuclease sensitivity and transcriptional activity of the complexed DNA. *Biochim. Biophys. Acta* **1353:** 180–190.

Bielinska A., Kukowska-Latallo J.F., Johnson J., Tomalia D.A., and Baker J.R., Jr. 1996. Regulation of in vitro gene expression using antisense oligonucleotides or antisense expression plasmids transfected using starburst PAMAM dendrimers. *Nucleic Acids Res.* **24:** 2176–2182

Blumenthal R., Seth P., Willingham M.C., and Pastan I. 1986. pH-dependent lysis of liposomes by adenovirus. *Biochemistry* **25:** 2231–2237.

Boletta A., Benigni A., Lutz J., Remuzzi G., Soria M.R., and Monaco L. 1997. Nonviral gene delivery to the rat kidney with polyethylenimine. *Hum. Gene Ther.* **8:** 1243–1251.

Böttger M., Vogel F., Platzer M., Kiessling U., Grade K., and Strauss M. 1988. Condensation of vector DNA by the chromosomal protein HMG1 results in efficient transfection. *Biochim. Biophys. Acta* **950:** 221–228.

Boussif O., Zanta M.A., and Behr J.P. 1996. Optimized galenics improve in vitro gene transfer with cationic molecules up to 1000-fold. *Gene Ther.* **3:** 1074–1080.

Boussif O., Lezoualc'h F., Zanta M.A., Mergny M.D., Scherman D., Demeneix B., and Behr J.-P. 1995. A versatile vector for gene and oligonucleotide transfer into cells in culture and in vivo: Polyethylenimine. *Proc. Natl. Acad. Sci.* **92:** 7297–7301.

Buschle M., Cotten M., Kirlappos H., Mechtler K., Schaffner G., Zauner W., Birnstiel M.L., and Wagner E. 1995. Receptor-mediated gene transfer into human T lymphocytes via binding of DNA/CD3 antibody particles to the CD3 T cell receptor complex. *Hum. Gene Ther.* **6:** 753–761.

Calos M.P. 1996. The potential of extrachromosomal replicating vectors for gene therapy. *Trends Genet.* **12:** 463–466.

Carrasco L. 1994, Entry of animal viruses and macromolecules into cells. *FEBS Lett.* **350:** 151–154.

————1995. Modification of membrane permeability by animal viruses. *Adv. Virus Res.* **45:** 61–112.

Chiocca S., Baker A., and Cotten M. 1997. Identification of a novel anti-apoptotic protein, GAM-1, encoded by the CELO adenovirus. *J. Virol.* **71:** 3168–3177.

Cotten M., Wagner E., Zatloukal K., and Birnstiel M.L. 1993. Chicken adenovirus (CELO virus) particles augment receptor-mediated DNA delivery to mammalian cells and yield exceptional levels of stable transformants. *J. Virol.* **67:** 3777–3785.

Cotten M., Baker A., Birnstiel M.L., Zatloukal K., and Wagner E. 1996. Adenovirus polylysine DNA conjugates. In *Current protocols in human genetics* (ed. N.C. Dracopoli et al.), pp. 12.3.1–12.3.33. Wiley, New York.

Cotten M., Saltik M., Kursa M., Wagner E., Maass G., and Birnstiel M. 1994. Psoralen treatment of adenovirus particles eliminates virus replication and transcription while maintaining the endosomolytic activity of the virus capsid. *Virology* **205:** 254–261.

Cotten M., Wagner E., Zatloukal K., Phillips S., Curiel D., and Birnstiel M.L. 1992. High-efficiency receptor-mediated delivery of small and large (48 kb) gene constructs using the endosome-disruption activity of defective or chemically-inactivated adenovirus particles. *Proc. Natl. Acad. Sci.* **89:** 6094–6098.

Cotten M., Längle-Rouault F., Kirlappos H., Wagner E., Mechtler K., Zenke M., Beug H., and Birnstiel M.L. 1990. Transferrin-polycation-mediated introduction of DNA into human leukemic cells: Stimulation by agents that affect the survival of transfected DNA or modulate transferrin receptor levels. *Proc. Natl. Acad. Sci.* **87:** 4033–4037.

Cristiano R.J., Smith L.C., Kay M.A., Brinkley B.R., and Woo S.L. 1993. Hepatic gene therapy: Efficient gene delivery and expression in primary hepatocytes utilizing a conjugated adenovirus-DNA complex. *Proc. Natl. Acad. Sci.* **90:** 11548–11552.

Curiel D.T, Agarwal S., Wagner E., and Cotten M. 1991. Adenovirus enhancement of transferrin-polylysine mediated gene delivery. *Proc. Natl. Acad. Sci.* **88:** 8850–8854.

Curiel D., Wagner E., Cotten M., Birnstiel M.L., Li C., Loechel S., Agarwal S., and Hu P. 1992. High efficiency gene transfer mediated by adenovirus coupled to DNA polylysine complexes via an antibody bridge. *Hum. Gene Ther.* **3:** 147–154.

Dunlap D.D., Maggi A., Soria M.R., and Monaco L. 1997. Nanoscopic structure of DNA condensed for gene delivery. *Nucleic Acids Res.* **25:** 3095–3101.

Ebert O., Finke S., Salahi A., Herrmann M., Trojaneck B., Lefterova P., Wagner E., Kircheis R., Huhn D., Schriever F., and Schmidt-Wolf I.G. 1997. Lymphocyte apoptosis: Induction by gene transfer techniques. *Gene Ther.* **4:** 296–302

Erbacher P., Bousser M.T., Raimond J., Monsigny M., Midoux P., and Roche A.C. 1996. Gene transfer by DNA/glycosylated polylysine complexes into human blood monocyte-derived macrophages. *Hum. Gene Ther.* **7:** 721–729

Fender P., Ruigrok R.W., Gout E., Buffet S., and Chroboczek J. 1997. Adenovirus dodecahedron, a new vector for human gene transfer. *Nat. Biotechnol.* **15:** 52–56.

Ferkol T., Kaetzel C.S., and Davis P.B. 1993a. Gene transfer into respiratory epithelial cells by targeting the polymeric immunoglobulin receptor. *J. Clin. Invest.* **92:** 2394–2400.

Ferkol T., Perales J.C., Mularo F., and Hanson R.W. 1996. Receptor-mediated gene transfer into macrophages. *Proc. Natl. Acad. Sci.* **93:** 101–105.

Ferkol T., Perales J.C., Eckman E., Kaetzel C.S., Hanson R.W., and Davis P.B. 1995. Gene transfer into the airway epithelium of animals by targeting the polymeric immunoglobulin receptor. *J. Clin. Invest..* **95:** 493–502.

Ferkol T., Lindberg G.L., Chen J., Perales J.C., Crawford D.R., Ratnoff O.D., and Hanson R.W. 1993b. Regulation of the phosphoenolpyruvate carboxykinase/human factor IX gene introduced into the livers of adult rats by receptor-mediated gene transfer. *FASEB J.* **7:** 1081–1091.

Fernández-Puentes C. and Carrasco L. 1980. Viral infection permeabilizes mammalian cells to protein toxins. *Cell* **20:**769–775.

Findeis M.A., Wu C.H., and Wu G.Y. 1994. Ligand-based carrier systems for delivery of DNA to hepatocytes. *Methods Enzymol.* **247:** 341–51.

Fisher K.J. and Wilson J.M. 1994. Biochemical and functional analysis of an adenovirus-based ligand complex for gene transfer. *Biochem. J.* **299:** 49–58.

———1997. The transmembrane domain of diphtheria toxin improves molecular conjugate gene transfer. *Biochem. J.* **321:** 49–58

Fitzgerald D.J., Padmanabhan R., Pastan I., and Willingham M.C. 1983. Adenovirus-induced release of epidermal growth factor and *Pseudomonas* toxin into the cytosol of KB cells during receptor-mediated endocytosis. *Cell* **32:**607–617.

Foster B.J. and Kern J.A. 1997. HER2-targeted gene transfer. *Hum. Gene Ther.* **8:**719–727.

Fritz J.D, Herweijer H., Zhang G., and Wolff J.A. 1996. Gene transfer into mammalian cells using histone-condensed plasmid DNA. *Hum. Gene Ther.* **7:** 1395–1404.

Greber U.F., Webster P., Weber J., and Helenius A. 1996. The role of the adenovirus protease on virus entry into cells. *EMBO J.* **15:** 1766–1777.

Greber U.F., Willetts M., Webster P., and Helenius A. 1993. Stepwise dismantling of adenovirus 2 during entry into cells. *Cell* **75:** 477–486.

Greber U.F., Suomalainen M., Stidwill R.P., Boucke K., Ebersold M.W., and Helenius A. 1997. The role of the nuclear pore complex in adenovirus DNA entry. *EMBO J.* **16:** 5998–6007.

Haensler J. and Szoka F.C. 1993a. Synthesis and characterization of a trigalactosylated bisacridine compound to target DNA to hepatocytes. *Bioconjugate Chem* . **4:** 85–93.

———1993b. Polyamidoamine cascade polymers mediate efficient transfection of cells in culture. *Bioconjugate Chem.* **4:** 372–379.

Harrington J.J., Van Bokkelen G., Mays R.W., Gustashaw K., and Willard H.F. 1997. Formation of de novo centromeres and construction of first-generation human artificial microchromosomes. *Nat. Genet.* **15:**345–355.

Hart S.L., Harbottle R.P., Cooper R., Miller A., Williamson R., and Coutelle C. 1995. Gene delivery and expression mediated by an integrin-binding peptide. *Gene Ther.* **2:** 552–554.

Huckett B., Ariatti M., and Hawtrey A.O. 1990. Evidence for targeted gene transfer by receptor-mediated endocytosis: Stable expression following insulin-directed entry of NEO into HepG2 cells. *Biochem. Pharmacol.* **40:** 253–263.

Huxley C. 1997. Mammalian artificial chromosomes and chromosome transgenics. *Trends Genet.* **13:** 345–347.

Jones N. and Shenk T. 1978. Isolation of deletion and substitution mutants of adenovirus type 5. *Cell* **13:** 181–188.

Kaneda Y., Iwai K., and Uchida T. 1989. Increased expression of DNA cointroduced with nuclear protein in adult rat liver. *Science* **243:** 375–378.

Kikuchi A., Sugaya S., Ueda H., Tanaka K., Aramaki Y., Hara T., Arima H., Tsuchiya S., and Fuwa T. 1996. Efficient gene transfer to EGF receptor overexpressing cancer cells by means of EGF-labeled cationic liposomes, *Biochem. Biophys. Res. Commun.* **227:** 666–671.

Kircheis R., Kichler A., Wallner G., Kursa M., Ogris M., Felzmann T., Buchberger M., and Wagner, E. 1997. Coupling of cell-binding ligands to polyethylenimine for targeted gene delivery. *Gene Ther.* **4:** 409–418.

Kukowska-Latallo J.F., Bielinska A.U., Johnson J., Spindler R., Tomalia DA., and Baker J.R., Jr. 1996. Efficient transfer of genetic material into mammalian cells using Starburst polyamidoamine dendrimers. *Proc. Natl. Acad. Sci.* **93:** 4897–4902

Labat-Moleur F., Steffan A.M., Brisson C., Perron H., Feugeas O., Furstenberger P., Oberling F., Brambilla E., and Behr J.P. 1996. An electron microscopy study into the mechanism of gene transfer with lipopolyamines. *Gene Ther.* **3:** 1010–1017.

Legendre J.Y. and Szoka F.C. 1993. Cyclic amphipathic peptide-DNA complexes mediate high-efficiency transfection of adherent cells. *Proc. Natl. Acad. Sci.* **90:** 893–897.

Lieber A., He C.-Y., and Kay M.A. 1997. Adenoviral preterminal protein stabilizes mini-adenoviral genomes in vitro and in vivo. *Nat. Biotechnol.* **15:** 1383–1387.

Lutz P., Rosa-Calatrava M., and Kedinger C. 1997. The product of the adenovirus intermediate gene IX is a transcriptional activator. *J. Virol.* **71:** 5102–5109.

McKee T.D., DeRome M.E., Wu G.Y., and Findeis M.A. 1994. Preparation of asialoorosomucoid-polylysine conjugates. *Bioconjugate Chem.* **5:** 306–311.

Merwin J.R., Noell G.S., Thomas W.L., Chiou H.C., DeRome M.E., McKee T.D., Spitalny G.L., and Findeis M.A. 1994. Targeted delivery of DNA using YEE(GalNAcAH)3, a synthetic glycopeptide ligand for the asialoglycoprotein receptor. *Bioconjugate Chem.* **5:** 612–620.

Midoux P., Mendes C., Legrand A., Raimond J., Mayer R., Monsigny M., and Roche A.C. 1993. Specific gene transfer mediated by lactosylated poly-L-lysine into hepatoma cells. *Nucleic Acids Res.* **21:** 871–878.

Parente R.A., Nir S., and Szoka F.C. 1990. Mechanism of leakage of phospholipid vesicle contents induced by the peptide GALA. *Biochemistry* **29:** 8720–8728.

Paul R.W., Weisser K.E., Loomis A., Sloane D.L., LaFoe D., Atkinson E.M., and Overell R.W. 1997. Gene transfer using a novel fusion protein, GAL4/invasin. *Hum. Gene Ther.* **8:** 1253–1262.

Perales J.C., Ferkol T., Molas M., and Hanson R.W. 1994a. An evaluation of receptor-mediated gene transfer using synthetic DNA-ligand complexes. *Eur. J. Biochem.* **226:** 255–66.

Perales J.C., Ferkol T., Beegen H., Ratnoff O.D., and Hanson R.W. 1994b. Gene transfer in vivo: Sustained expression and regulation of genes introduced into the liver by receptor-targeted uptake. *Proc. Natl. Acad. Sci.* **91:** 4086–4090.

Plank C., Mechtler K., Szoka F.C., Jr. and Wagner E. 1996. Activation of the complement system by synthetic DNA complexes: A potential barrier for intravenous gene delivery. *Hum. Gene Ther.* **7:** 1437–1446.

Plank C., Oberhauser B., Mechtler K., Koch C., and Wagner E. 1994. The influence of endosome-disruptive peptides on gene transfer using synthetic virus-like gene transfer systems. *J. Biol. Chem.* **269:** 12918–12924.

Plank C., Zatloukal K., Cotten M., Mechtler K., and Wagner E. 1992. Gene transfer into hepatocytes using asialoglycoprotein receptor mediated endocytosis of DNA complexed with an artificial tetra-antennary galactose ligand. *Bioconjugate Chem.* **3:** 533–539.

Remy J.S., Kichler A., Mordvinov V., Schuber F., and Behr J.P. 1995. Targeted gene transfer into hepatoma cells with lipopolyamine-condensed DNA particles presenting galactose ligands: A stage toward artificial viruses. *Proc. Natl. Acad. Sci.* **92:** 1744–1748.

Rosenfeld M.A. 1997. Human artificial chromosomes get real. *Nat. Genet.* **15:** 333–335.

Rosenkranz A.A., Yachmenev S.V., Jans D.A., Serebryakova N.V., Murav'ev V.I., Peters R., and Sobolev A.S. 1992. Receptor-mediated endocytosis and nuclear transport of a transfecting DNA construct. *Exp.Cell Res.* **199:** 323–329.

Schaack J., Ho W.Y., Freimuth P., and Shenk T. 1990. Adenovirus terminal protein mediates both nuclear matrix association and efficient transcription of adenovirus DNA. *Genes Dev.* **4:**1197–1208.

Schachtschabel U., Pavlinkova G., Lou D., and Kohler H. 1996. Antibody-mediated gene delivery for B-cell lymphoma in vitro. *Cancer Gene Ther.* **3:** 365–372.

Schwarzenberger P., Spence S.E., Gooya J.M., Michiel D., Curiel D.T., Ruscetti F.W., and Keller J.R. 1996. Targeted gene transfer to human hematopoietic progenitor cell lines through the c-kit receptor. *Blood* **87:** 472–478

Seth P. 1994a. Adenovirus-dependent release of choline from plasma membrane vesicles at an acidic pH is mediated by the penton base protein. *J. Virol.* **68:** 1204–1206.

———1994b. Mechanism of adenovirus-mediated endosome lysis: Role of the intact adenovirus capsid structure. *Biochem. Biophys. Res. Commun.* **205:** 1318–1324.

Seth P., Fitzgerald D., Ginsberg H., Willingham M., and Pastan I. 1984. Evidence that the penton base of adenovirus is involved in potentiation of toxicity of pseudomonas exotoxin conjugated to epidermal growth factor. *Mol. Cell. Biol.* **4:** 1528–1533.

Sharp P.A., Moore C., and Haverty J.L. 1976. The infectivity of adenovirus 5 DNA-protein complex. *Virology* **75:** 442–456.

Simpson K., McGuigan A., and Huxley C. 1996. Stable episomal maintenance of yeast artificial chromosomes in human cells. *Mol. Cell. Biol.* **16:** 5117–5126.

Tang M.X. and Szoka F.C. 1997. The influence of polymer structure on the interactions of cationic polymers with DNA and morphology of the resulting complexes. *Gene Ther.* **4:** 823–832.

Tang M.X., Redemann C.T., and Szoka F.C., Jr. 1996. In vitro gene delivery by degraded polyamidoamine dendrimers. *Bioconjugate Chem.* **7:** 703–714.

Thurnher M., Wagner E., Clausen H., Mechtler K., Rusconi S., Dinter A., Birnstiel M.L., Berger E.G., and Cotten M. 1994. Carbohydrate receptor-mediated gene transfer to human T leukaemic cells. *Glycobiology* **4:** 429–435.

Tomita N., Higaki J., Morishita R., Kato K., Mikami H., Kaneda Y., and Ogihara T. 1992. Direct in vivo gene introduction into rat kidney. *Biochem. Biophys. Res. Commum.* **186:** 129–134.

Trubetskoy V.S., Torchilin V.P., Kennel S., and Huang L. 1992a. Use of N-terminal modified poly(L-lysine)-antibody conjugates as a carrier for targeted gene delivery in mouse lung endothelial cells, *Bioconjugate Chem.* **3:** 323–327.

———1992b. Cationic liposomes enhance targeted delivery and expression of exogenous DNA mediated by N-terminal modified poly(L-lysine)-antibody conjugate in mouse lung endothelial cells. *Biochim. Biophys. Acta.* **1131:** 311–313.

Wagner E., Cotten M., Foisner R., and Birnstiel M.L. 1991. Transferrin-polycation-DNA complexes: The effect of polycations on the structure of the complex and DNA delivery to cells. *Proc. Natl. Acad. Sci.* **88:** 4255–4259.

Wagner E., Plank C., Zatloukal K., Cotten M., and Birnstiel M.L. 1992a. Influenza virus hemagglutinin HA-2 N-terminal fusogenic peptides augment gene transfer by transferrin-polylysine-DNA complexes: Toward a synthetic virus-like gene-transfer vehicle. *Proc. Natl. Acad. Sci.* . **89:** 7934–7938.

Wagner E., Zenke M., Cotten M., Beug H., and Birnstiel M.L. 1990. Transferrin-polycation conjugates as carriers for DNA uptake into cells. *Proc. Natl. Acad. Sci.* **87:** 3410–3414.

Wagner E., Zatloukal K., Cotten M., Kirlappos H., Mechtler K., Curiel D., and Birnstiel M.L. 1992b. Coupling of adenovirus to transferrin-polylysine/DNA complexes greatly enhances receptor-mediated gene delivery and expression of transfected cells. *Proc. Natl. Acad. Sci.* **89:** 6099–6103

Wilson J.B., Bell J.L., and Levine A.J. 1996. Expression of Epstein-Barr virus nuclear antigen-1 induces B cell neoplasia in transgenic mice. *EMBO J.* **15:** 3117–3126.

Wohlgemuth J.G., Kang S.H., Bulboaca G.H., Nawotka K.A., and Calos M.P. 1996. Long-term gene expression from autonomously replicating vectors in mammalian cells. *Gene Ther.* **3:** 503–512.

Wu G.Y. and Wu C.H. 1987. Receptor-mediated in vitro gene transformation by a soluble DNA carrier system. *J. Biol. Chem.* **262:** 4429–4432.

1988. Receptor-mediated gene delivery and expression in vivo. *J. Biol. Chem.* **263:** 14621–14624.

Wu C., Wilson J., and Wu G. 1989. Targeting genes: Delivery and persistent expression of a foreign gene driven by mammalian regulatory elements in vivo, *J. Biol. Chem.* **264:** 16985–16987.

Wu G., Zhan P., Sze L., Rosenberg A., and Wu, C. 1994. Incorporation of adenovirus into a ligand-based DNA carrier system results in retention of original receptor specificity and enhances targeted gene expression. *J. Biol. Chem.* **269:** 11542–11546.

Wyman T.B, Nicol F. Zelphati O. Scaria P.V., Plank C., and Szoka F.C. Jr. 1997. Design, synthesis, and characterization of a cationic peptide that binds to nucleic acids and permeabilizes bilayers. *Biochemistry* **36:** 3008–3017.

Yates J.L., Warren N., and Sugden B. 1985. Stable replication of plasmids derived from Epstein-Barr virus in various mammalian cells. *Nature* **313:** 812–815.

Zabner J., Fasbender A.J., Moninger T., Poellinger K.A., and Welsh M.J. 1995. Cellular and molecular barriers to gene transfer by a cationic lipid. *J. Biol. Chem.* **270:**18997–19007.

Zanta M.A., Boussif O., Adib A., and Behr J.P. 1997. In vitro gene delivery to hepatocytes with galactosylated polyethylenimine. *Bioconjugate Chem.* **8:** 839–844.

Zauner W., Blaas D., Küchler E., and Wagner E. 1995. Rhinovirus mediated endosomal release of transfection complexes. *J. Virol.* **69:** 1085–1092.

Zauner W., Kichler A., Schmidt W., Sinski A., and Wagner E. 1996. Glycerol enhancement of ligand-polylysine/DNA transfection. *BioTechniques* **20:** 905–913.

Zauner W., Kichler A., Schmidt W., Mechtler K., and Wagner E. 1997. Glycerol and polylysine synergize in their ability to rupture vesicular membranes: A mechanism for increased transferrin-polylysine-mediated gene transfer. *Exp. Cell Res.* **232:** 137–145.

Zelphati O. and Szoka F.C. Jr. 1996. Mechanism of oligonucleotide release from cationic liposomes. *Proc. Natl. Acad. Sci.* **93:** 11493–11498.

Zenke M., Steinlein P., Wagner E., Cotten M., Beug H., and Birnstiel M.L. 1990. Receptor-mediated endocytosis of transferrin polycation conjugates: An efficient way to introduce DNA into hematopoietic cells. *Proc. Natl. Acad. Sci.* **87:** 3655–3659.

Ziady A.G., Perales J.C., Ferkol T., Gerken T., Beegen H., Perlmutter D.H., and Davis P.B. 1997. Gene transfer into hepatoma cell lines via the serpin enzyme complex receptor. *Am. J. Physiol.* **273:** G545–G552.

12

Naked DNA Gene Transfer in Mammalian Cells

Jon A. Wolff

Departments of Pediatrics and Medical Genetics
Waisman Center
University of Wisconsin–Madison
Madison, Wisconsin 53705

A common supposition concerning gene transfer into mammalian cells has been that the nucleic acid must be carried within a viral vector or complexed with some type of material within a nonviral vector. However, the DNA basis for genetic transmission, the foundation of molecular genetics and gene therapy, was first demonstrated using naked (pure) DNA. In 1928, Griffith found that nonvirulent pneumococcus became virulent when exposed to dead virulent pneumococcus. Avery, MacLeod, and McCarty purified nucleic acid from the dead virulent pneumococcus and showed that it could transform bacteria. Pure DNA's ability to transmit virulence suggested that DNA was the hereditary material (Avery et al. 1944).

It is well known how Avery's discovery eventually led to the elucidation of the double helix and the molecular basis for DNA's endowment as the hereditary material. Another and less appreciated ramification of Avery's findings is that purified DNA can be transferred and expressed in foreign environments (Miller 1998). In the late 1980s, I became intrigued by the second implication of Avery's discovery: that DNA not within a viral vector could potentially be delivered directly into the whole organism.

The first demonstration of the genetic basis for viral gene transmission in mammalian cells was also demonstrated using naked nucleic acids. In the late 1950s and early 1960s, it was demonstrated that naked RNA or DNA purified from viral material was able to cause viral infection when delivered to mammalian cells both in culture and in the whole animal (Herriott 1961). The efficiency of transfer of the viral nucleic acids was substantially improved by complexation with diethylaminoethyl (DEAE)-dextran or precipitation with calcium phosphate (Vaheri and Pagano 1965; Graham and van der Eb 1973; Szybalski 1992).With the advent of recombinant DNA technologies and the development of plasmid expression vectors, DEAE-dextran or calcium phosphate came into wider use for the transfer of plasmid expression vectors into cells in culture.

The first attempts at direct, nonviral gene transfer into mammals in vivo were reported during the early 1960s in later-discredited articles describing a phenotypic effect following the injection of animal genomes (Wolff and Lederberg 1994). In the recombinant era, plasmid DNA (pDNA) or viral DNA within calcium phosphate precipitates or liposomes was injected directly into animals, but these methods did not gain wide acceptance (Dubensky et al. 1984; Benvenisty and Reshef 1986; Mannino and Gould-Fogerite 1988). Instead, by the late 1980s, the predominant approach to gene therapy involved the reimplantation of cells genetically modified by retroviral vectors.

Direct, nonviral gene transfer into the whole organism, however, remained a desirable goal for gene therapy because it would avoid laborious and costly cell cultures. The gene could be administered as easily as a drug. Nonviral vectors would also be more plastic to engineer than viral vectors, making it possible to attenuate immune and other toxic effects.

After the demonstration that cationic lipids can mediate the efficient transfer of genes into cells in culture (Felgner et al. 1987), I initiated a collaboration with Phillip Felgner to evaluate the ability of cationic lipids to mediate direct gene transfer into animals. Phillip Williams, a technician in my laboratory (University of Wisconsin–Madison), injected intraparenchymally several tissues with in-vitro-transcribed mRNA (prepared by Robert Malone, who initially suggested its use) either complexed with cationic lipids or naked as a control (Wolff et al. 1990). In muscle, it was the control group injected with naked mRNA that contained foreign gene expression (chloramphenicol acetyltransferase [CAT]). Similar results were subsequently demonstrated using pDNA expression vectors. More recent data using intravascular administrations indicate that high levels of foreign gene expression can be obtained with naked DNA in vivo (Budker et al. 1996a).

INTRAMUSCULAR INJECTION

The intramuscular injection of naked pDNAs induces foreign gene expression mainly limited to myofibers (Wolff et al. 1990). This occurs in all types of striated muscle including both type I and type II skeletal myofibers and cardiac muscle cells (Lin et al. 1990; Acsadi et al. 1991a; Buttrick et al. 1992; Jiao et al. 1992; Gal et al. 1993). Muscles such as the rectus femoris or tibialis anterior which are circumscribed by a well-defined epimysium, may enable the highest levels of foreign gene expres-

sion because they provide the best distribution and retention of the injected pDNA. High levels of expression have been observed in the
diaphragm muscle (Davis and Jasmin 1993). In other muscle groups, the
implantation with forceps of pellets of dried pDNA yields higher expression than fluid injection (Jiao et al. 1992; Wolff et al. 1991).

A variety of factors have been shown to affect the levels of foreign
gene expression from intramuscularly injected pDNA. Our initial studies
dissolved the pDNA in 5% sucrose, but similar expression levels were
obtained in a variety of physiologic solutions, such as normal saline
(Wolff et al. 1991). Tris and EDTA solutions, in which pDNA is commonly used to stored, should be avoided because the EDTA damages the
myofibers even though the initial expression is not depressed (Wolff et al.
1991; Manthorpe et al. 1993).

Expression from RNA vectors peaks within the first day after injection, but expression from pDNA vectors peaks at 14 days or even at longer
times (Wolff et al. 1990, 1992a; Davis et al. 1993a; Manthorpe et al. 1993).
This could be the result of delayed pDNA uptake or transcriptional expression. Increased expression is obtained with increasing amounts of pDNA,
but expression plateaus at different amounts of injected pDNA, depending
on the muscle type, species, and pDNA vector. Southern blot analysis indicates that the majority of the injected pDNA is rapidly degraded within
hours, yet a small percentage of the injected pDNA persists in an open, circular form (Wolff et al. 1991; Manthorpe et al. 1993). Injection of linearized pDNA yields much less expression presumably due to its enhanced
degradation (Buttrick et al. 1992; Wolff et al. 1992a).

Expression can also be aided by the enhanced distribution of the
pDNA, which has been accomplished by preinjection of muscles with large
volumes of hypertonic solutions and polymers or by improved injection
techniques such as positioning the needle along the longitudinal axis of the
muscle (Davis et al. 1993a, Levy et al. 1996; Mumper et al. 1996). Such
adaptations can also reduce the large variability in expression (although it
should be noted that many gene transfer techniques such as transfection of
cultured cells have large expression variability as well). Multiple injections
or muscle damage does not particularly aid expression (Wolff et al. 1991;
Manthorpe et al. 1993). Expression is not augmented by electrical stimulation of muscle contractions or muscle denervation (Wolff et al. 1991).

We and others have noted that different batches of pDNA purified by
two CsCl gradients yielded different levels of luciferase expression following intramuscular injection (Wolff et al. 1991; Manthorpe et al. 1993).
This may be due to the variable amounts of bacterial lipopolysaccharide
that copurifies with pDNA, but it is most likely due to other contaminants

(Manthorpe et al. 1993; Wicks et al. 1995). Our laboratory initially observed that muscle expression after injection of pDNA prepared using commercial columns yielded lower and more variable expressions than pDNA prepared by two CsCl gradients. Lately, pDNAs prepared using commercial columns have given similar results as CsCl-purified pDNA. Another laboratory has documented that there is no difference in luciferase expression or vaccination efficacy between pDNAs prepared by these two methods (Davis et al. 1996).

Muscle regeneration induced by myotoxic agents enabled higher levels of expression (Wells and Goldspink 1992; Davis et al. 1993b; Danko et al. 1994; Vitadello et al. 1994). Muscle regeneration induced by ischemia also enhanced pDNA expression (Takeshita et al. 1996a; Tsurumi et al. 1996). Myotoxic agents (such as cardiotoxin) and amide local anesthetics (such as bupivacaine ([Marcaine]) are advantageous because they selectively destroy myofibers without harming myoblasts or the vascular endothelial cells, thus enabling complete recovery of the muscle. Optimal luciferase expression was obtained when the pDNA was injected 3 to 7 days after bupivacaine or cardiotoxin injection or initiation of ischemia—a time when a substantial number of muscle cells have begun to recover from the effects of the myotoxic agent. The increased foreign gene activity may be due to either enhanced pDNA uptake or expression such as from transcriptional activation.

Other studies have shown the importance of the promoter and construct. The direct injection of pDNA constructs provides a quick and easy approach for dissecting the transcriptional elements in tissues in vivo (Kitsis et al. 1991; Kitsis and Leinwand 1992; Vincent et al. 1993; von Harsdorf et al. 1993). The viral promoters Rous sarcoma virus (RSV) and cytomegalovirus (CMV) generally enable higher levels of expression than mammalian promoters, including muscle-specific ones (Manthorpe et al. 1993; Hartikka et al. 1996). Luciferase expression from a CMV immediate-early promoter with a chimeric intron was approximately threefold greater than from an RSV promoter (Manthorpe et al. 1993). Further studies by the same group showed that replacement of the kanamycin gene for the ampicillin gene, removal of the SV40 origin of replication, modification of the plasmid backbone, and improved transcription terminators brought about very large increases in luciferase expressions (Hartikka et al. 1996).

For some applications, regulated expression may be required. The tetracycline-regulation system involving the use of *cis* and *trans* elements has enabled regulated expressions from intramuscular-injected pDNA (Fishman et al. 1994; Liang et al. 1996). Other applications may require a rapid and large rise in pDNA expression. This has been accomplished from a plasmid-based self-amplifying Sindbis virus vector that produces an RNA species which replicates in the large cytoplasmic volume of myofibers (Herweijer et al. 1995b). However, these approaches may have limited human gene therapy utility given the presumed immune response to the foreign regulatory or replicative proteins.

Several studies have shown that newborn animals can express naked plasmid DNA (Hansen et al. 1991; Wolff et al. 1991; Wells and Goldspink 1992; de Luze et al. 1993). In carp, the plasmid expression of CAT in muscle was greater in young fish than in old (Hansen et al. 1991; Tan and Chan 1997). Plasmid expression was deemed efficient after injection into *Xenopus* tadpole muscle (de Luze et al. 1993). Of particular note was the increased CAT expression from the SV40 promoter in 4–6-week-old mice as compared to older mice (Wells and Goldspink 1992).

Using refined plasmid vectors, exceptionally high levels of foreign gene expression were obtained in 2-week-old mouse and rat muscles (Danko et al. 1997). Approximately 50% of the myofibers were intensely blue following the intramuscular injection of a β-galactosidase expression vector in 2-week-old BALB/c mice (Fig. 1). The effects of age, mouse strain, and construct were multiplicative, resulting in over 1000-fold greater luciferase and approximately 20-fold more β-galactosidase-positive cells over our previous results in adult rodents.

A general trend is that expression decreases as the size of the animal increases. Expression is slightly less in rats than in mice, but it is substantially less in animals larger than rodents, such as rabbits, cats, and monkeys (Wolff et al. 1991; Jiao et al. 1992). This was observed in both younger and larger adult animals, including nonhuman primates. In rats, cats, and nonhuman primates, the expression was improved by the implantation of pellets of dried pDNA (Jiao et al. 1992).

The connective tissue in primate muscle may prevent the distribution of the pDNA or its contact the sarcolemma (Wolff et al. 1992b). Histochemical

Figure 1 Histochemical analysis of β-galactosidase expression in a quadricep muscle of a 2-week-old ICR mouse 1 week after intramuscular injection with 10 μg of pCILacZ. Magnification, 40x. (Reprinted, with permission, from Danko et al. 1997 [copyright Oxford University Press].)

and fluorescent stains indicated that primate muscle had substantially more connective tissue within the perimysium. The thicker perimysium in primate muscle may restrict the distribution of the pDNA or may serve as a conduit and an increased potential space in which the pDNA may be dispersed without coming into contact with the myofibers. The latter possibility may explain the increased expression level with implanted pDNA.

INTRAPARENCHYMAL INJECTIONS INTO OTHER TISSUES

Soon after our initial observations in muscle, we intraparenchymally injected several tissues with naked luciferase pDNAs and observed substantially less luciferase activity than in muscle (Acsadi et al. 1991a). Subsequently, other laboratories have reported expression of naked pDNA in other tissues such as liver, skin, thyroid, joints, artery, and melanomas; these results are summarized as follows.

Using pig skin as a model for human skin, Hengge et al. (1995) injected 20 µg of a CMVLacZ pDNA in 50–100 µl at a single site within the superficial subepidermal dermis and observed transient β-galactosidase expression mainly in the epidermis (keratinocytes). Injections into human skin organ cultures and human skin grafts also resulted in epidermal expression (Hengge et al. 1996). Expression is more difficult in the thin skin of mice but can be improved by a tattooing device that contains eight oscillating needles (Ciernik et al. 1996; Hengge et al. 1996). Although the expression of intradermally injected naked pDNA has been sufficient for immunization purposes (Raz et al. 1994) and for neutrophil recruitment (Hengge et al. 1995), a recent report found that expression was inefficient and not enough to correct the histologic abnormalities in regenerated skin from humans with lamellar ichthyosis (Choate and Khavari 1997).

Naked pDNA expression has also been reported following tracheal instillation. One study in mice found that hypotonic solutions did not aid expression, whereas another study in rats found that water instillation did (Meyer et al. 1995; Sawa et al. 1996). In both studies, only scant β-galactosidase-positive cells were observed, suggesting that the levels of expression were relatively low. Nonetheless, the levels of expression from naked pDNA were similar to those achieved using cationic lipids. In fact, a human study found that a CFTR-expressing pDNA partially corrected the electrophysiologic abnormalities within the nasal epithelium of cystic fibrosis patients to the same extent when administered alone or when complexed with a cationic lipid (Zabner et al. 1997).

After the injection of naked pDNA vectors into the knee joints of rats and rabbits, transient foreign gene expression was detected in the synovi-

um (Yovandich et al. 1995, Nita et al. 1996). Use of the human inter-
leukin-1 receptor antagonist or growth hormone genes led to low levels
of the secreted proteins in the synovial fluid.

After the injection of naked pDNA directly into rabbit thyroids, follicular cells expressed β-galactosidase (Sikes et al. 1994). On the basis of
the level of expression in the thyroid and that after injection into the ster-
nocleidomastoid, it was concluded that the efficiency of naked pDNA
expression was the same in both tissues. However, further studies would
be needed to determine whether this is true in other species and other
muscle groups.

Foreign gene expression was also obtained in B16 melanoma solid
tumors grown in mice following intratumor injections of naked pDNA
(Yang and Huang 1996). Expression was enhanced by the use of a large
injection volume and inclusion of 0.01% Triton X-100, but was inhibited
by cationic lipids.

Naked pDNA has also been delivered to arteries by using balloon
angioplasty (Riessen et al. 1993; Isner et al. 1996). The pDNA was
absorbed into a hydrogel polymer coating on the balloon (polyacrylic
acid polymer cross-linked onto the balloon using isocyanate). β-galactosidase expression was mainly in subintimal cells and limited to less
than 0.1% of the cells, although X-gal histochemical staining may under-
estimate the percentage of transfected cells as compared to immunostaining with a β-galactosidase antibody (Couffinhal et al. 1997).

Multiple intraparenchymal injections into the liver also enabled
pDNA expression in hepatocytes (Hickman et al. 1994, 1995; Malone et
al. 1994). Forty-eight hours after the injection of 1 mg and 5 mg of a
pDNA expressing α-1-antitrypsin from the CMV promoter in rats and
cats, respectively, serum levels of α-1-antitrypsin rose to 4–36 ng/ml. Of
note, dexamethasone increased approximately threefold the expression
from the CMV promoter within pDNAs injected into the liver in vivo.

In summary, a variety of tissues express naked pDNA that is delivered intraparenchymally. Although it is difficult to compare expression
efficiencies from different laboratories using different species and injection procedures, the published results suggest that naked pDNA is generally expressed more efficiently in striated muscle than in other tissues. In
light of our intravascular results presented below, this may be more an
effect of the efficiency of delivery than a reflection of the intrinsic capacity of the cell type to take up naked pDNA. Promoter effects also need to
be taken into account. Finally, these studies highlight the importance of
including naked pDNA administration as a control when evaluating nonviral gene transfer methods (Wolff et al. 1990; Levy et al. 1996).

INTRAVASCULAR INJECTIONS INTO THE LIVER AND OTHER INTERNAL ORGANS

The intravascular delivery of genes is attractive because it avoids the necessity for multiple intraparenchymal injections into the target tissue. The gene is disseminated throughout the tissue because the vascular system accesses every cell. Vascular delivery could be systemic or regional in which injections are made into specific vessels that supply a target tissue. The intravascular delivery of adenoviruses or cationic lipid–DNA complexes in adult animals mostly results in expression in vascular-accessible cells such as endothelial cells or hepatocytes reached via the sinusoid fenestrae (Jaffe et al. 1992; Liu et al. 1997).

After our laboratory had developed novel transfection complexes of pDNA and amphipathic compounds and proteins, we sought to deliver them to hepatocytes in vivo (Budker et al. 1996b, 1997). Our control for these experiments was naked pDNA, and we were once again surprised that this experimental group had the highest expression levels (Budker et al. 1996a; Zhang et al. 1997).

Naked pDNA vectors were injected into the afferent and efferent vessels of the liver in mice and rats. Efficient plasmid expression was obtained following delivery via the portal vein, the hepatic vein, and the bile duct. The use of hyperosmotic injection solutions and occlusion of the blood outflow from the liver substantially increased the expression levels. Combining these surgical approaches with improved plasmid vectors enabled uncommonly high levels of foreign gene expression in which over 15 μg of luciferase protein/liver were produced in mice and over 50 μg in rats. Equally high levels of β-galactosidase expression were obtained in that over 5% of the hepatocytes had intense blue staining (Fig. 2). β-galactosidase expression was rarely seen in other types of cells. Expression of luciferase or β-galactosidase was evenly distributed in hepatocytes throughout the entire liver when either of the three routes was injected. Two days after the intraportal injection of 100 μg of pCMVGH, the mean hGH serum concentration was 65 ng/ml ± 26 ($n = 7$), which is approximately 50-fold above the normal baseline levels. Repetitive plasmid administration through the bile duct led to successive events of foreign gene expression. These levels of foreign gene expression are among the highest levels obtained with nonviral vectors.

The use of hyperosmotic injection solutions and occlusion of the blood outflow from the liver was critical for the very high levels of expression. These conditions presumably enhanced the DNA transfer to hepatocytes by transiently opening the hepatic endothelial barrier (Neuwelt and Rapoport 1984; Neuwelt et al. 1987; Robinson and Rapoport 1987). Under

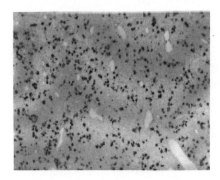

Figure 2 Histochemical analysis of β-galactosidase expression in livers after injection of 100 μg of pCILacZ into mouse portal vein with the hepatic vein clamped. Magnification, 160x. (Reprinted, with permission, from Zhang et al. 1997.)

normal conditions, the approximately 100-nm size of the fenestrae (Nopanitaya et al. 1976; Fraser et al. 1980; Campra and Reynolds 1988) would prevent the exit from the sinusoids of plasmid DNA, which has a gyration radius of approximately 100 nm (Langowski et al. 1992; Vologodskii and Cozzarelli 1994; Fisherman and Patterson 1996). Raising the intraportal osmotic and hydrostatic pressures would transiently enlarge their size and thereby increase the extravasation of the pDNA complexes. In fact, preliminary results using fluorescent-labeled DNA showed that the increased osmotic and hydrostatic pressures were required for movement of the DNA out of the sinusoids and to the hepatocytes.

Luciferase expression has been observed in several other internal organs (such as the spleen, pancreas, kidney, and adrenal glands) following the injection of naked pDNA into blood vessels (J.A. Wolff et al., unpubl.). Studies are in progress to determine the type of cells responsible for the foreign gene expression.

INTRAVASCULAR INJECTIONS INTO MUSCLE

The intravascular delivery of naked pDNA to muscle cells is also attractive particularly because many muscle groups would have to be targeted for intrinsic muscle disorders, such as with Duchenne muscular dystrophy. In addition, an intravascular approach would avoid the limited distribution of pDNA through the interstitial space following intramuscular injection. Muscle has a high density of capillaries (Browning et al. 1996) that are in close contact with the myofibers (Lee and Schmid-Schonbein 1995). Delivery of pDNA to muscle via capillaries puts the pDNA into direct contact with every myofiber and substantially decreases the interstitial space the pDNA must to traverse to access a myofiber. However, the endothelium in muscle capillaries is of the continuous, nonfenestrated type and has low solute permeability, especially to large macromolecules (Taylor and Granger 1984).

The mechanism of macromolecule transendothelial transport is poorly understood. Cell biologists have proposed that it occurs by transcytosis involving plasmalemmal vesicles or by convective transport through transient transendothelial channels formed by the fusion of vesicles (Michel 1996). Physiology experiments suggest that the muscle endothelium has a large number of small pores with radii of about 4 nm and a very low number of large pores with radii of 20–30 nm (Rippe and Haraldsson 1994). Although the radius of gyration of 6-kb pDNA is approximately 100 nm (Fisherman and Patterson 1996), supercoiled DNA in plectonomic form has superhelix dimensions of approximately 10 nm (Rybenkov et al. 1997). This implies that pDNA is capable of crossing microvascular walls by stringing through the large pores. We hypothesized that the rate of pDNA extravasation could be increased by enhancing fluid convection through these large pores by raising the transmural pressure difference in selective regions. Our recent report demonstrates that intravascular pDNA injections under high pressure can in fact lead to high levels of foreign gene expression in muscles throughout a selected hindlimb of an adult rat (Budker et al. 1998).

pCILux (475 µg; a luciferase expression vector utilizing the CMV promoter) in 9.5 ml of normal saline solution (NSS) was injected into the femoral arteries of adult Sprague-Dawley rats while both blood inflow and outflow out of the hindlimb were blocked for 10 minutes. Two days after the pDNA injections, a substantial amount of luciferase activity (over 50 ng of luciferase protein per gram of muscle) was measured in all of the muscles of the hindlimb. There was a critical dependence on the volume and speed of injection, suggesting that either increased hydrostatic pressure, rapid flow, or both are required for efficient expression. Use of fluorescently labeled DNA provided direct evidence that these injection conditions enabled extravasation of the injected DNA.

The type and percentage of the transfected cells in the muscle were determined using the β-galactosidase reporter system (Fig. 3). The vast preponderance of the β-galactosidase-positive cells were myofibers. Very few endothelial cells were stained blue. With the best injection condition, up to 50% of the myofibers expressed β-galactosidase in many areas of the muscles. In the upper anterior leg muscle group, 11–21% of the myofibers were positively stained. Similarly high percentages were observed in the lower posterior leg muscle group. Of the 72,126 myofibers counted in all of the sections in all of the muscle groups of all four animals, 10.1% were β-galactosidase-positive.

These results demonstrate that the intraarterial delivery of pDNA to muscle can be greatly enhanced when injected rapidly, in a large volume, and with all blood vessels leading into and out of the hindlimb occluded. These conditions presumably increase the intravascular hydrostatic pressure and thereby augment the convective outflow that brings the pDNA into contact with the myofibers. The high intravascular pressure may increase the number, size, and permeability of the microvascular pores (Rippe et al. 1985; Wolf et al. 1989). Preliminary studies using collagenase suggest that enzymatic disruption of the vessel's basement membrane or muscle extracellular matrix may increase the delivery of pDNA to the myofibers. Ischemia and papaverine also increase expression moderately.

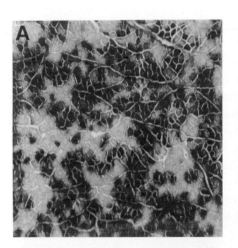
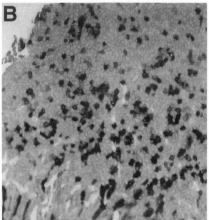

Figure 3 Histochemical analysis of β-galactosidase expression in rat hindlimb muscles 2 days after the intraarterial injection of 475 μg of pCILacZ. Muscle sections (15 μm thick) were stained overnight using X-gal at room temperature and counterstained with eosin. (*A*) A muscle cross section from the posterior lower leg with a high expression level. (*B*) A muscle cross section from the anterior upper leg with a low expression level. (*C*) A longitudinal section from the posterior lower leg. Magnifications were 160x for *A* and *C* and 100x for *B*. (Reprinted, with permission, from Budker et al. 1998.)

Further studies in larger animals will determine the clinical relevance of this study. Preliminary results in rabbits are encouraging. The short time required for occluding blood flow to skeletal muscle should be well tolerated in a human clinical setting because ischemia can be tolerated by muscle for 2 to 3 hours (Gidlof et al. 1988). In fact, a common anesthetic procedure for distal limb surgery (e.g., carpal tunnel repair) involves the placement of a tourniquet to block both venous and arterial blood flow and the intravenous administration of a local anesthetic (e.g., lidocaine) distal to the tourniquet. Surgery in humans can be performed for a couple of hours by using this anesthetic procedure. Similarly, histologic analyses of the rat muscles in our experiments indicated that the ischemia did not cause any myofiber damage.

STABILITY OF EXPRESSION IN MUSCLE

It has been the common assumption that stable expression can only be achieved from the integration of the foreign sequences into chromosomal DNA as occurs with retroviral vectors. After the transfection of pDNA into dividing cells in culture, expression is unstable presumably due to the loss of the pDNA during mitosis (Wolff et al. 1992a). When the nuclear membrane dissolves during mitosis, the pDNA that is initially nuclear is partitioned throughout the cell and is inefficiently retained in the nucleus after the nuclear membrane reforms. Our studies indicated that in post-mitotic myofibers (after intramuscular injection), simple pDNAs stably expressed and persisted in an extrachromosomal state.

After the injection of 100 µg of pRSVL that expresses lux from the RSV promoter into the quadriceps of 6–10-week-old mice, luciferase expression persisted for at least 2 years (Wolff et al. 1992a; Herweijer et al. 1995a). Expression of CAT was also present for at least 4 months after injection of pRSVCAT. Quantitative PCR indicated that the pDNA remained in the muscle. The bacterial methylation pattern of the injected DNA was preserved for at least 19 months, indicating that the pDNA did not replicate in muscle. Southern blot analysis indicated that the majority of the pDNA was of the open, circular form and had not integrated. Chromosomal integration of pDNA was searched for by electroporating the injected muscle DNA into bacteria after restriction enzyme digestion and ligation. No plasmids containing plasmid/chromosome junctions were observed in over 1800 colonies examined. This indicates that integration events must be exceedingly rare.

It is unlikely that specific DNA sequences are required for the persistence of the pDNA. None of the pDNAs studied to date have a known chromosomal origin of replication. Once in the nucleus, the pDNA is most likely treated by the nondividing myofibers as chromosomal DNA because nuclear DNases are not sequence-specific and extrachromosomal DNA exists naturally in several types of mammalian cells (Rush and Misra 1985).

An immune response to the foreign gene product, however, may lead to unstable expression. For example, the percent of β-galactosidase-positive cells decreases several weeks after injection of pRSVLacZ. This could be due to immune-mediated destruction of the β-galactosidase-expressing myofibers, because β-galactosidase is secreted from myofibers (Acsadi et al. 1991a). In fact, a mononuclear cell infiltrate was observed following β-galactosidase expression in bupivacaine treated muscle (Vitadello et al. 1994). Similarly, the expression of particularly immunogenic proteins, such as hepatitis B surface antigen, is unstable (Davis et al. 1997). In this situation, the immune-mediated destruction of myofibers was experimentally associated with the loss of foreign expression.

In cardiac muscle, luciferase expression was unstable (Acsadi et al. 1991a). Immunosuppression of the animals prolonged the expression, suggesting that the luciferase was more immunogenic in cardiac than in skeletal muscle. Also, the body location of the tissue may affect its immunogenicity, because luciferase expression was stable after injection into hearts transplanted into the abdomen of syngeneic Lewis rats (Wang et al. 1992).

In circumstances when the expression levels were particularly high, such as in 2-week-old rodents and in bupivacaine-treated skeletal muscle, the initial high levels did not persist (Danko et al. 1994, 1997). Furthermore, low levels of luciferase also persisted after an initial period of high expression from optimized plasmid constructs in muscle (Manthorpe et al. 1993; Hartikka et al. 1996). Preliminary studies indicate that the high levels of expression achieved from the intravascular delivery of pCILux were also unstable. It appears that the total luciferase gravitates to a low level of expression (1 ng/muscle in our studies) regardless of the initial levels. Perhaps the immune system is responsible for limiting luciferase expression above the threshold of 1 ng/total muscle. Myofibers expressing high amounts of luciferase could be destroyed by the immune system, whereas myofibers expressing less luciferase escape immune detection.

An additional mechanism of unstable expression may be related to the transcriptional suppression of the viral promoters (i.e., CMV and RSV promoters) over time, as has been observed from retroviral vectors in vivo

(Naviaux and Verma 1992). Consistent with this hypothesis is the observation that the amount of pDNA did not change in bupivacaine-treated muscle from 1 to 2 months postinjection over a time when the luciferase expression substantially decreased (Danko et al. 1994). However, such DNA analyses have to be interpreted with some caution because they do not distinguish between extranuclear and nuclear pDNA. Nonetheless, muscles of both 2-week-old mice and mice following bupivacaine treatment contain an increased number of proliferating and fusing myoblasts. This state could enable increased expression at the transcriptional level. Studies are in progress to better define the basis for unstable pDNA expression when the initial levels are high. Preliminary results following intravascular delivery suggest that the unstable expression is due to transcriptional suppression and can be avoided.

STABILITY OF EXPRESSION IN LIVER AND OTHER ORGANS

Unstable expression has also been observed following naked pDNA delivery to the liver, kidney, skin, thyroid, synovium, and artery. In the liver, the hGH reporter gene system was used to explore the stability of expression (Budker et al. 1996a). After intraportal injection of a CMVhGH expression plasmid, mean hGH serum levels decreased from 65 ng/ml at 2 days to approximately 10 ng/ml at 4 days postinjection. In untreated animals, hGH levels were less than 1 ng/ml at 4 weeks after injection. However, hGH levels remained at 6–11 ng/ml for at least 4 weeks in animals that received both dexamethasone and cyclosporine. The increased stability of hGH expression in animals that received both cyclosporine and dexamethasone was most likely due to these agents' combined action on the immune system, but their effect on the transcriptional rate from the human immediate early CMV promoter cannot be excluded.

Loss of foreign gene expression in the liver after the intravascular delivery of naked pDNA appears to be a two-phase process. Plasmid hGH expression falls rapidly after the first day (first phase) and then more slowly after several days (second phase). This pattern was also observed with CMV promoter-driven luciferase expression. On the basis of these preliminary data, our working hypothesis is that the initial decrease in expression during the first week is due mainly to the suppression of the transcription, whereas only a small part of this expression loss is due to hepatocyte death or proliferation. The decrease in expression after the first week is due largely to an immune response. We postulate that, if we prevent transcription suppression and an immune response, then expression would persist much longer. In fact, expression would only be lost as the pDNA is lost from the natural turnover of hepatocytes. Quite possibly, plasmid expression in the liver could persist for the lifetime of a hepatocyte, which has a half-life of up to 1 year in rodents and humans (Leffert et al. 1988; Webber et al. 1994).

MECHANISM OF UPTAKE—CELLULAR ENTRY

The combined intraparenchymal and emerging intravascular data indicate that the uptake and expression of naked DNA are a general property of animal cells within a tissue architecture. They are common to cells of all three lineages: endoderm (e.g., hepatocytes), mesoderm (e.g., muscle), and ectoderm (e.g., skin). This property is lost when the cells are removed and maintained in culture. Tissue disruption and cell isolation may modify the cell so that it can no longer take up naked DNA.

The movement of the plasmid DNA across the plasma membrane has been the focus of a few studies but without any definite conclusions (Wolff et al. 1992b; Hagstrom et al. 1996; Levy et al. 1996). Much of this investigation occurred prior to the intravascular results and needs to be reevaluated in light of the high efficiencies being obtained in several tissues. Nonetheless, possible mechanisms include endocytosis, potocytosis, transient membrane disruptions, or gross membrane disruptions (Wolff et al. 1992b). Several observations suggest that the mechanism of pDNA uptake may involve native cellular uptake processes. The ability for plasmid DNA to be taken up by cells that are distant from the injection site in vivo argues against the role of gross membrane disruptions. It is of interest that high levels of luciferase expression could occasionally be obtained when the DNA was injected into the bile duct in small volumes of isotonic solutions (Zhang et al. 1997). This would suggest that the mechanism of pDNA uptake may in fact involve endogenous cellular pathways.

A receptor-mediated process is suggested by the inhibition of expression (from intramuscularly injected pDNA) by excess salmon sperm DNA or dextran sulfate (Levy et al. 1996). Also, dose-response curves of the amount of pDNA injected versus expression reveals a plateau effect at 25–100 µg of injected pDNA, suggesting receptor saturation (Wolff et al. 1990; Manthorpe et al. 1993).

Our postulated receptor-mediated uptake of pDNA most likely did not evolve specifically for polynucleic acid uptake. Instead, the pDNA has exploited a transport system that evolved for another purpose, such as the uptake of other polyanions (e.g., the scavenger receptor system). On the other hand, polynucleic acid uptake may provide an evolutionary advantage on the basis of its exquisite ability for immune activation. It may also have an important role in clearing the large amount of released polynucleic acid that occurs physiologically and in disease states.

MECHANISM OF UPTAKE—NUCLEAR ENTRY

After traversing the plasma membrane, the pDNA must enter the nucleus because it is highly unlikely that plasmid DNA containing RNA polymerase II promoters could be expressed anywhere else (e.g., mitochondria have a different genetic code). Although it is often assumed that DNA enters the nucleus from the cytoplasm, little is known about the actual nuclear uptake process, despite many advances toward an understanding of protein and RNA nuclear transport. We have been studying the mechanism of DNA nuclear uptake by taking advantage of the ability of primary myotubes to take up and express microinjected DNA (Wolff et al. 1992b). More recently, we have utilized a permeabilized cell system (Hagstrom et al. 1996).

We have found that up to two thirds of the primary rat myotubes that were cytoplasmically microinjected with pSV2nLacZ expressed nuclear β-galactosidase. This provides a system for the study of DNA nuclear transport and irrefutable evidence that DNA can enter the karyoplasm of a postmitotic nucleus with intact membranes.

Light and electron microscopic studies indicated that the movement of cytoplasmic DNA was restricted by its binding to cytoplasmic elements. Gold-labeled DNA was present within or near the nuclear pore complex, suggesting that this was the pathway of DNA entry into the nucleus. Gold-labeled PEG was never found to be closely associated with a nuclear pore complex. The effects of temperature, plasmid concentration, and excess pUC19 DNA on the uptake of labeled DNA and on the expression of LacZ or luciferase plasmids suggested that DNA nuclear uptake was facilitative. β-galactosidase expression and luciferase expression were inhibited by coinjection with the lectin, wheat germ agglutinin (WGA). WGA binds *N*-acetyl glucosamine groups present on nuclear pore proteins and inhibits transport through the pore.

Perhaps the DNA is strung through the nuclear pore by a facilitative process. Nuclear DNA-binding proteins could facilitate this in a process akin to the "diffuse and bind" nuclear transport model for small, karyophilic proteins lacking transport signals (also known as "piggyback"). However, studies using digitonin-permeabilized cells indicate that plasmid DNA transport into the nucleus may be more complicated than this model. Digitonin-permeabilized HeLa cells were exposed for 60 minutes at 37°C to an 853-bp double-stranded DNA molecule (m.w. ≅

563,000) containing covalently linked Texas red molecules (TR-DNA). Confocal fluorescent microscopic analysis revealed that the TR-DNA entered approximately 90% of the nuclei. The predominant staining pattern in the nuclei was multiple punctate signals, but diffuse nuclear staining was also present. The similar nuclear staining pattern obtained with Cy5-labeled DNA or biotin-labeled DNA indicates that the staining pattern was not a consequence of the Texas red molecule. The ability for other labeled DNAs (different sequences) to enter the nuclei with similar efficiencies indicates that this nuclear transport is not sequence-dependent. The nuclei were unbroken, based on studies with other macromolecules such as dextran, and additional studies indicate that uptake was not due to DNase degradation. The ability for DNA molecules to enter intact nuclei was consistent with the ability of microinjected DNA to enter the postmitotic nuclei of myotubes in culture.

A substantial amount of the TR-DNA was sequestered in the cytoplasm. This was also consistent with our previous findings in microinjected myotubes that cytoplasmic elements limit the accessibility of DNA to the nuclei. The inclusion of rabbit reticulocyte lysate in the import buffer greatly increased the nuclear transport of NLS-containing proteins, but not DNA. In fact, inclusion of the extract decreased the nuclear localization of TR-DNA and increased the cytoplasmic staining.

To determine if the fluorescently labeled DNA was accessing the nucleus through the nuclear pore, TR-DNA was incubated with digitonin-permeabilized cells in the presence of WGA-lectin. Addition of this lectin to the import reaction blocked TR-DNA nuclear localization. Nuclear accumulation of TR-DNA was also inhibited when the cells were incubated at 4°C in the presence of 5-mM N-ethylmaleimide (NEM) or without energy. As with WGA, the blockage of nuclear transport by each of these methods resulted in increased nuclear envelope staining. Thus, it is likely that all of these treatments exert their effects by inhibiting the normal transport function of proteins within the nuclear pore complex. This would effectively block DNA translocation to the nucleus without blocking nuclear envelope docking.

Our preliminary model of DNA nuclear uptake is as follows. After cytoplasmic delivery, the small amount of DNA that avoids binding to or sequestration by cytoplasmic elements enters the intact nucleus through the nuclear pore. The relatively rare entry of DNA into the nucleus (in comparison to karyophilic proteins) could be explained by its rapid and substantial cytoplasmic sequestration and its low rate of transport through the nuclear pore. This understanding of DNA nuclear transport provides a basis for future efforts to increase the efficiency of this process and is consistent with efforts to increase the amount of DNA delivered to the cytoplasm. For example, we have recently observed increased nuclear entry in digitonin-permeabilized cells of pDNA containing a covalently attached SV40 T antigen nuclear-localizing signal (Sebestyen et al. 1998).

APPLICATIONS

This review concentrates on the potential applications of pDNA delivery to muscle. The use of DNA injections for immunization purposes, initially proposed by my colleagues and I (Wolff et al. 1990), is reviewed in Ulmer and Liu (this volume). Besides the treatment of intrinsic muscle disorders, muscle could also be used as a platform for the secretion of a therapeutic protein and for clearing a circulating toxic metabolite.

Genetic muscle diseases, such as Duchenne muscular dystrophy (DMD), could be treated by the expression of the respective normal gene in muscle (Acsadi et al. 1991b). To examine the feasibility of transferring the dystrophin gene into dystrophic myofibers by plasmid DNA injection, full-length and Becker-like cDNAs were placed into plasmid vectors and injected into dystrophin-deficient mdx mouse muscle. Dystrophin expression from both cDNAs was observed 1 week after plasmid injection in approximately 1% of the myofibers. Although a greater percentage of myofibers would have to be genetically modified for clinical efficacy, intramuscular injection of plasmid DNA does provide sufficient dystrophin expression to study the correction of the DMD phenotype at the myofiber level. The use of pDNA also enables various modifications in the dystrophin molecule to be expeditiously studied (Fritz et al. 1995).

Full-length and Becker-like dystrophins were localized at the sarcolemmal membrane following plasmid DNA injection. Another important indicator of dystrophin function is that full-length or Becker-like dystrophins are colocalized with dystrophin-associated proteins in individual myofibers after plasmid DNA injection. This indicates that the expression of the human dystrophin enables the expression of the corresponding dystrophin-associated proteins.

Full-length and Becker-like dystrophin gene transfer and expression in the mdx mouse model were shown to prevent myofiber degeneration, one of the pivotal pathologic processes in DMD (Danko et al. 1993). Expression of the full-length and Becker-like dystrophins has been shown to persist for at least 6 months after intramuscular plasmid DNA injection, but luciferase expression is unstable in mdx muscle. However, the coexpression of luciferase with the full-length and Becker-like dystrophins enabled the stable expression of luciferase, suggesting that dystrophin expression prevents myofiber loss.

Studies are in progress to determine whether the high efficiency of the intravascular approach will enable the clinical use of pDNA for intrinsic muscle disorders. The intravascular delivery of naked pDNA to muscle cells is attractive because it avoids the necessity of injecting each muscle directly. Our nonhuman primate studies indicate that there is no immune response against naked pDNA (Jiao et al. 1992). This will enable repeat injections to be performed, as was done for the liver (Zhang et al. 1997). In contrast, repeat injections with viral vectors and some other nonviral vectors may be challenging.

Muscle could also be used as a platform to clear circulating toxic metabolites that accumulate in inborn errors of metabolism. For example, muscle-specific expression of phenylalanine hydroxylase was able to correct the hyperphenylalaninemia in a genetic mouse model for phenylketonuria (Harding et al. 1998).

Another application that has recently attracted much attention is the use of muscle to manufacture and secrete proteins, such as hormones, growth factors, and clotting factors (Barr and Leiden 1991; Dhawan et al. 1991; Dai et al. 1992). Given the low efficiency of expression after intramuscular pDNA injection, applications are limited to the expression of cytokines or hormones for which small levels can have physiologic effects.

After the intramuscular injection of 4000 µg of naked pDNA expressing the human vascular endothelial growth factor (VEGF165) from the CMV promoter in limbs of patients with peripheral artery disease, their ischemic ulcers healed or markedly improved in four of seven limbs (Baumgartner et al. 1998). This clinical improvement was associated with a slight transient rise in serum VEGF levels and new collateral blood vessels as determined by contrast angiography and magnetic resonance angiography. This clinical experience is supported by similar results with the ischemic limbs of animals using intramuscular injection or artery delivery from pDNA on hydrogel-coated catheters (Isner et al. 1996; Takeshita et al. 1996b, 1997; Tsurumi et al. 1996). Such clinical effects may be aided by several features of VEGF biology, such as the up-regulation of VEGF receptors in ischemic tissue and VEGF-induced proliferation of endothelial cells that amplify its further secretion. Also, VEGF expression may have been aided by the muscle ischemia-enhancing pDNA expression (Takeshita et al. 1996a; Tsurumi et al. 1996).

Another cytokine, erythropoietin, has also been used for naked pDNA studies (Tripathy et al. 1996). A single intramuscular injection of an improved pDNA expression vector containing the mouse erythropoietin gene (and the CMV promoter, CMV 5′-untranslated and intron A sequences, and bovine growth hormone polyadenylation signal) raised hematocrits for 90 days, at which time the serum and muscle erythropoietin levels were also elevated. Other cytokine genes such as interferon-α and insulin-like growth factor-I have also been injected intramuscularly (Alila et al. 1997; Lawson et al. 1997).

Hormonal effects from the intramuscular injection of naked pDNA have also been obtained. A single injection of naked pDNA expressing the gene for growth hormone releasing hormone from the avian skeletal α-actin promoter elevated serum growth hormone levels, which led to increased body weight (Draghiaakli et al. 1997). Finally, serum clotting factor VII was detected after the intramuscular injection of naked pDNA in mice (Miller et al. 1995).

One intriguing application is the use of a pDNA intramuscular injection for the generation of animal models of autoimmune disorders (Blechynden et al. 1997). Injection of naked pDNA encoding the autoantigen histidyl-transfer RNA synthetase induced a myositis, although injection of the protein did not.

Studies for applying the use of intravascular delivery of pDNA to the liver have been initiated. Although the initial expression levels should be sufficient to produce phenotypic changes, its instability has hampered our efforts to fully characterize the phenotypic effects in mouse models for phenylketonuria and LDL-receptor deficiency.

HUMAN CLINICAL TRIALS AND SAFETY CONSIDERATIONS

One concern with naked DNA administrations is the generation of anti-DNA antibodies that could cause an autoimmune disorder such as systemic lupus erythematosus (SLE) (Nakamura et al. 1978). High titers of antinative DNA (nDNA) are associated with SLE (Tron and Bach 1977; Whiteside and Dixon 1979). Our studies found that repetitive intramuscular injections of large amounts of pDNA (up to 12,000 μg of pDNA) induced neither antinuclear antibodies (ANA) nor anti-nDNA antibodies (Jiao et al. 1992). This result is consistent with a large body of experimentation on trying to create a mouse model for SLE. These studies indicate that an immune response against DNA can only be elicited in normal mice (not prone to autoimmunity) if it is denatured and complexed with a protein or adjuvant (Gilkeson et al. 1991); however, even then the antibodies are usually against single-stranded DNA, and such antibodies are poorly correlated with autoimmune disorders. In fact, a slight increase in anti-single-stranded DNA but not anti-double-stranded DNA antibodies was detected following repetitive subcutaneous or intramuscular injections of pDNA (Katsumi et al. 1994). Mouse anti-double-stranded DNA monoclonal antibodies have been obtained from autoimmune-prone NZB mice (Eilat et al. 1984) or normal mice that have "natural" autoantibodies (Shefner et al. 1991). The inability of pDNA to elicit ANA and anti-nDNA antibodies in primates suggests that this gene transfer method is unlikely to cause an autoimmune disorder in humans. Also, our experience in performing repetitive injections in the liver indicates that an immune response does not prevent foreign gene expression from repetitive pDNA administrations.

For mass population vaccinations, the issue of pDNA integration into chromosomes has been raised. It is technically challenging to exclude or quantify the frequency of integration events. We developed a sensitive screening method to search for such events and did not detect any (Wolff et al. 1992a). Other workers have also not detected integration (Manthorpe et al. 1993). Even if integration occurs, it is extremely unlikely that it can be detrimental. For vaccination against potentially harmful gene products such as from oncogenes, RNA could be used (Wolff et al. 1990).

A concern for all somatic cell gene therapy approaches is the chance of inadvertent germ line gene modification. The transfer of the injected pDNA to gonadal tissue would be required for such an occurrence. In one study using PCR, intramuscularly injected pDNA was restricted to the muscle, overlying skin, and blood plasma, but not in gonads (Winegar et al. 1996). However, there have been some sketchy notices of the presence of pDNA in gonads (Phillip Noguchi, NIH RAC meeting, December 1997). Even if pDNA is present in gonadal tissue, it would be highly unlikely that this would lead to germline modification.

CONCLUSIONS AND FUTURE PROSPECTS

The uptake and expression of naked pDNA have been described in a wide variety of tissues and species using several different delivery methods. The high expression in several tissues following intravascular delivery suggests that the cellular uptake of naked DNA is a fundamental feature of animal cells. Higher uptake should make the mechanism of uptake easier to elucidate, clarify its possible physiologic and pathophysiologic function, and establish an evolutionary perspective. It could also shed light on the mechanism of uptake of viral nucleic acids, because viruses may use parts of its pathway.

The clinical use of naked DNA has several distinct advantages, given its simplicity and nonimmunogenecity. Expression, especially using an intravascular route and enhanced pDNA vectors, could be sufficient to treat many disorders. The recently reported treatment of ischemic limb disease with an intramuscular injection of naked DNA may represent the first successful clinical gene therapy trial in humans. Further clinical trials involving naked DNA are sure to follow.

ACKNOWLEDGMENTS

The author wishes to thank Kirk Hogan, M.D., for his review comments and Tim Lockie, M.S., for his assistance in preparing the manuscript. This work was supported in part by the Muscular Dystrophy Association (USA), the National Institutes of Health (R01 DK49117), and the Waisman Center for Mental Retardation, University of Wisconsin– Madison.

REFERENCES

Acsadi G., Jiao S., Jani A., Duke D., Williams P., Wang C., and Wolff J. 1991a. Direct gene transfer and expression into rat heart in vivo. *New Biol.* **3:** 71–81.

Acsadi G., Dickson G., Love D.R., Jani A., Walsh F.S., Gurusinghe A., Wolff J.A., and Davies K.E. 1991b. Human dystrophin expression in mdx mice after intramuscular injection of DNA constructs. *Nature* **352:** 815–818.

Alila H., Coleman M., Nitta H., French M., Anwer K., Liu Q.S., Meyer T., Wang J.J., Mumper R., Oubari D., Long S., Nordstrom J., and Rolland A. 1997. Expression of biologically active human insulin-like growth factor-I following intramuscular injection of a formulated plasmid in rats. *Hum. Gene Ther.* **8:** 1785–1795.

Avery O., MacLeod C., and McCarty M. 1944. Studies on the chemical nature of the substance inducing transformation of pneumococcal types. *J. Exp. Med.* **79:** 137–158.

Barr E. and Leiden J.M. 1991. Systemic delivery of recombinant proteins by genetically modified myoblasts. *Science* **254:** 1507–1509.

Baumgartner I., Pieczek A., Manor O., Blair R., Kearney M., Walsh K., and Isner J.M. 1998. Constitutive expression of phVEGF165 following intramuscular gene transfer promotes collateral vessel development in patients with critical limb ischemia. *Circulation* **97:** 1114–1123.

Benvenisty N. and Reshef L. 1986. Direct introduction of genes into rats and expression of the genes. *Proc. Natl. Acad. Sci.* **83:** 9551–9555.

Blechynden L.M., Lawson M.A., Tabarias H., Garlepp M.J., Sherman J., Raben N., and Lawson C.M. 1997. Myositis induced by naked DNA immunization with the gene for histidyl-tRNA synthetase. *Hum. Gene Ther.* **8:** 1469–1480.

Browning J., Hogg N., Gobe G., and Cross R. 1996. Capillary density in skeletal muscle of Wistar rats as a function of muscle weight and body weight. *Microvas. Res.* **52:** 281–287.

Budker V., Zhang G., Knechtle S., and Wolff J.A. 1996a. Naked DNA delivered intraportally expresses efficiently in hepatocytes. *Gene Ther.* **3:** 593–598.

Budker V., Gurevich V., Hagstrom J., Bortzov F., and Wolff J. 1996b. pH-sensitive, cationic liposomes: A new synthetic virus-like vector. *Nat. Biotechnol.* **14:** 760–764.

Budker V., Zhang G., Danko I., Williams P., and Wolff J.A. 1998. The efficient expression of intravascularly delivered DNA in rat muscle. *Gene Ther.* **5**: 272–276.

Budker V., Hagstrom J.E., Lapina O., Eifrig D., Fritz J., and Wolff J.A. 1997. Protein/amphipathic polyamine complexes enable highly efficient transfection with minimal toxicity. *BioTechniques* **23**: 139–147.

Buttrick P.M., Kass A., Kitsis R.N., Kaplan M.L., and Leinwand L.A. 1992. Behavior of genes directly injected into the rat heart in vivo. *Circ. Res.* **70**: 193–198.

Campra J.L. and Reynolds T.B. 1988. The hepatic circulation. In *The liver: Biology and pathobiology* (ed. I.M. Arias et al.), pp 911–930. Raven Press, New York.

Choate K.A. and Khavari P.A. 1997. Direct cutaneous gene delivery in a human genetic skin disease. *Hum. Gene Ther.* **8**: 1659–1665.

Ciernik I.F., Krayenbuhl B.H., and Carbone D.P. 1996. Puncture-mediated gene transfer to the skin. *Hum. Gene Ther.* **7**: 893–899.

Couffinhal T., Kearney M., Sullivan A., Silver M., Tsurumi Y., and Isner J.M. 1997. Histochemical staining following *lacZ* gene transfer underestimates transfection efficiency. *Hum. Gene Ther.* **8**: 929–934.

Dai Y., Roman M., Naviaux R., and Verma I. 1992. Gene therpay via primary myoblasts: Long-term expression of factor IX protein following transplantation *in vivo*. *Proc. Natl. Acad. Sci.* **89**: 10892–10895.

Danko I., Fritz J.D., Jiao S., Hogan K., Latendresse J.S., and Wolff J.A. 1994. Pharmacological enhancement of *in vivo* foreign gene expression in muscle. *Gene Ther.* **1**: 114–121.

Danko I., Fritz J.D., Latendresse J.S., Herwijer J., Schultz E., and Wolff J.A. 1993. Dystrophin expression improves myofiber survival in mdx muscle following intramuscular plasmid DNA injection. *Hum. Mol. Genet.* **2**: 2055–2061.

Danko I., Williams P., Herweijer H., Zhang G., Latendresse J.S., Bock I., and Wolff J.A. 1997. High expression of naked plasmid DNA in muscles of young rodents. *Hum. Mol. Genet.* **6**: 1435–1443.

Davis H.L. and Jasmin B.J. 1993. Direct gene transfer into mouse diaphragm. *FEBS Lett.* **333**: 146–150.

Davis H.L., Millan C.L., and Watkins S.C. 1997. Immune-mediated destruction of transfected muscle fibers after direct gene transfer with antigen-expressing plasmid DNA. *Gene Ther.* **4**: 181–188.

Davis H.L., Whalen R.G., and Demeneix B.A. 1993a. Direct gene transfer into skeletal muscle in vivo: Factors affecting efficiency of transfer and stability of expression. *Hum. Gene Ther.* **4**: 151–159.

Davis H.L., Demeneix B.A., Quantin B., Coulombe J., and Whalen R.G. 1993b. Plasmid DNA is superior to viral vectors for direct gene transfer into adult mouse skeletal muscle. *Hum. Gene Ther.* **4**: 733–740.

Davis H.L., Schleef M., Moritz P., Mancini M., Schorr J., and Whalen R.G. 1996. Comparison of plasmid DNA preparation methods for direct gene transfer and genetic immunization. *BioTechniques* **21**: 92–94.

de Luze A., Sachs L., and Demeneix B. 1993. Thyroid hormone-dependent transcriptional regulation of exogenous genes transferred into *Xenopus* tadpole muscle *in vivo*. *Proc. Natl. Acad. Sci.* **90**: 7322–7326.

Dhawan J., Pan L.C., Pavlath G.K., Travis M.A., Lanctot A.M., and Blau H.M. 1991. Systemic delivery of human growth hormone by injection of genetically engineered myoblasts. *Science* **254**: 1509–1512.

Draghiaakli R., Li X.G., and Schwartz R.J. 1997. Enhanced growth by ectopic expression of growth hormone releasing hormone using an injectable myogenic vector. *Nat. Biotechnol.* **15:** 1285–1289.

Dubensky T., Campbell B., and Villarreal L. 1984. Direct transfection of viral and plasmid DNA into the liver or spleen of mice. *Proc. Natl. Acad. Sci.* **81:** 7529–7533.

Eilat D., Hochberg M., Pumphrey J., and Rudikoff S. 1984. Monoclonal antibodies to DNA and RNA from NZB/NZW F1 mice: Antigenic specificities and NH2 terminal amino acid sequences. *J. Immunol.* **133:** 489–494.

Felgner P.L., Gadek T.R., Holm M., Roman R., Chan H.W., Wenz M., Northrop J.P., Ringold G.M., and Danielsen M. 1987. Lipofection: A highly efficient, lipid-mediated DNA-transfection procedure. *Proc. Natl. Acad. Sci.* **84:** 7413–7417.

Fisherman D.M. and Patterson G.D. 1996. Light scattering studies of supercoiled and nicked DNA. *Biopolymers* **38:** 535–552.

Fishman G.I., Kaplan M.L., and Buttrick P.M. 1994. Tetracycline-regulated cardiac gene expression in vivo. *J. Clin. Invest.* **93:** 1864–1868.

Fraser R., Bowler L., Day W., Dobbs B., Johnson H., and Lee D. 1980. High perfusion pressure damages the sieving ability of sinusoidal endothelium in rat livers. *Br. J. Exp. Pathol.* **61:** 222–228.

Fritz J.D., Danko I., Roberds S.L., Campbell K.P., Latendresse J.S., and Wolff J.A. 1995. Expression of deletion-containing dystrophins in mdx muscle: Implications for gene therapy and dystrophin function. *Pediatr. Res.* **37:** 693–700.

Gal D., Weir L., Leclerc G., Pickering J.G., Hogan J., and Isner J.M. 1993. Direct myocardial transfection in two animal models. Evaluation of parameters affecting gene expression and percutaneous gene delivery. *Lab. Invest.* **68:** 18–25.

Gidlof A., Lewis D.H., and Hammersen F. 1988. The effect of prolonged total ischemia on the ultrastructure of human skeletal muscle capillaries. A morphometric analysis. *Int. J Microcirc. Clin. Exp.* **7:** 67–86.

Gilkeson G.S., Pritchard A.J., and Pisetsky D.S. 1991. Specificity of anti-DNA antibodies induced in normal mice by immunization with bacterial DNA. *Clin. Immunol. Immunopathol.* **59:** 288–300.

Graham F. and van der Eb A. 1973. A new technique for the assay of infectivity of human adenovirus 5 DNA. *Virology* **52:** 456–467.

Griffith F. 1928. The significance of pneumococcal types. *J. Hyg.* **27:** 113–159.

Hagstrom J.E., Rybakova I.N., Staeva T., Wolff J.A., and Ervasti J.M. 1996. Non-nuclear DNA binding proteins in striated muscle. *Biochem. Mol. Med.* **58:** 113–121.

Hansen E., Fernandes K., Goldspink G., Butterworth P., Umeda P.K., and Chang K.C. 1991. Strong expression of foreign genes following direct injection into fish muscle. *FEBS Lett.* **290:** 73–76.

Harding C.O., Wild K., Chang D., Messing A., and Wolff J.A. 1998. Metabolic engineering as therapy for inborn errors of metabolism—Development of mice with phenylalanine hydroxylase expression in muscle. *Gene Ther.* **5:** 677–683.

Hartikka J., Sawdey M., Cornefert-Jensen F., Margalith M., Barnhart K., Nolasco M., Vahlsing H.L., Meek J., Marquet M., Hobart P., Norman J., and Manthorpe M. 1996. An improved plasmid DNA expression vector for direct injection into skeletal muscle. *Hum. Gene Ther.* **7:** 1205–1217.

Hengge U.R., Walker P.S., and Vogel J.C. 1996. Expression of naked DNA in human, pig, and mouse skin. *J. Clin. Invest.* **97:** 2911–2916.

Hengge U.R., Chan E.F., Foster R.A., Walker P.S., and Vogel J.C. 1995. Cytokine gene expression in epidermis with biological effects following injection of naked DNA.

Nat. Genet. **10:** 161–166.

Herriott R.M. 1961. Infectious nucleic acids, a new dimension in virology. *Science* **134:** 256–260.

Herweijer H., Fritz J., Hagstrom J., and Wolff J. 1995a. Direct gene transfer in vivo. In *Somatic gene therapy* (ed. P.L. Chang), pp. 183–202. CRC Press, Ann Arbor, Michigan.

Herweijer H., Latendresse J.S., Williams P., Zhang G., Danko I., Schlesinger S., and Wolff J.A. 1995b. A plasmid-based self-amplifying Sindbis virus vector. *Hum. Gene Ther.* **6:** 1161–1167.

Hickman M.A., Malone R.W., Sih T.R., Akita G.Y., Carlson D.M., and Powell J.S. 1995. Hepatic gene expression after direct DNA injection. *Adv. Drug Delivery Rev.* **17:** 265–271.

Hickman M.A., Malone R.W., Lehmann-Bruinsma K., Sih T.R., Knoell D., Szoka F.C., Walzem R., Carlson D.M., and Powell J.S. 1994. Gene expression following direct injection of DNA into liver. *Hum. Gene Ther.* **5:** 1477–1483.

Isner J.M., Walsh K., Symes J., Pieczek A., Takeshita S., Lowry J., Rosenfield K., Weir L., Brogi E., and Jurayj D. 1996. Arterial gene transfer for therapeutic angiogenesis in patients with peripheral artery disease. *Hum. Gene Ther.* **7:** 959–988.

Jaffe H., Danel D., Longenecker G., Metzger M., Setoguchi Y., Rosenfeld M., Gant T., Thorgeirsson S.S., Stratford-Perricaudet L.D., Perricaudet M., Pavirane A., Lecocq J.-P., and Crystal R.G. 1992. Adenovirus-mediated in vivo gene transfer and expression in normal rat liver. *Nat. Genet.* **1:** 372–378.

Jiao S., Williams P., Berg R.K., Hodgeman B.A., Liu L., Repetto G., and Wolff J.A. 1992. Direct gene transfer into nonhuman primate myofibers *in vivo. Hum. Gene Ther.* **3:** 21–33.

Katsumi A., Emi N., Abe A., Hasegawa Y., Ito M., and Saito H. 1994. Humoral and cellular immunity to an encoded protein induced by direct DNA injection. *Hum. Gene Ther.* **5:** 1335–1339.

Kitsis R.N. and Leinwand L.A. 1992. Discordance between gene regulation in vitro and in vivo. *Gene Expr.* **2:** 313–318.

Kitsis R.N., Buttrick P.M., McNally E.M., Kaplan M.L., and Leinwand L.A. 1991. Hormonal modulation of a gene injected into rat heart in vivo. *Proc. Natl. Acad. Sci.* **88:** 4138–4142.

Langowski J., Kremer W., and Kapp U. 1992. Dynamic light scattering for study of solution conformation and dynamics of superhelical DNA. *Methods Enzymol.* **211:** 430–448.

Lawson C.M., Yeow W.S., Lee C.M., and Beilharz M.W. 1997. In vivo expression of an interferon-alpha gene by intramuscular injection of naked DNA. *J. Interferon Cytokine Res.* **17:** 255–261.

Lee J. and Schmid-Schonbein G.W. 1995. Biomechanics of skeletal muscle capillaries: Hemodynamic resistance, endothelial distensibility, and pseudopod formation. *Ann. Biomed. Eng.* **23:** 226–246.

Leffert H.L., Koch K.S., Lad P.J., Shapiro I.P., Skelly H., and de Hemptinne B. 1988. Hepatocyte regeneration, replication, and differentiation. In *The liver: Biology and pathobiology* (ed. I.M. Arias et al.), pp 833–850. Raven Press, New York.

Levy M.Y., Barron L.G., Meyer K.B., and Szoka F.C., Jr. 1996. Characterization of plasmid DNA transfer into mouse skeletal muscle: Evaluation of uptake mechanism, expression and secretion of gene products into blood. *Gene Ther.* **3:** 201–211.

Liang X., Hartikka J., Sukhu L., Manthorpe M., and Hobart P. 1996. Novel, high expressing and antibiotic-controlled plasmid vectors designed for use in gene therapy. *Gene Ther.* **3:** 350–356.

Lin H., Parmacek M.S., Morle G., Bolling S., and Leiden J.M. 1990. Expression of recombinant genes in myocardium in vivo after direct injection of DNA. *Circulation* **82:** 2217–2221.

Liu Y., Mounkes L.C., Liggitt H.D., Brown C.S., Solodin I., Heath T.D., and Debs R.J. 1997. Factors influencing the efficiency of cationic liposome-mediated intravenous gene delivery. *Nat. Biotechnol.* **15:** 167–173.

Malone R.W., Hickman M.A., Lehmannn-Bruinsma K., Sih T.R., Walzem R., Carlson D.M., and Powell J.S. 1994. Dexamethasone enhancement of gene expression after direct hepatic DNA injection. *J. Biol. Chem.* **269:** 29903–29907.

Mannino R.J. and Gould-Fogerite S. 1988. Liposome mediated gene transfer. *BioTechniques* **6:** 682–690.

Manthorpe M., Cornefert-Jensen F., Hartikka J., Felgner J., Rundell A., Margalith M., and Dwarki V. 1993. Gene therapy by intramuscular injection of plasmid DNA: Studies on firefly luciferase gene expression in mice. *Hum. Gene Ther.* **4:** 419–431.

Meyer K.B., Thompson M.M., Levy M.Y., Barron L.G., and Szoka F.C., Jr. 1995. Intratracheal gene delivery to the mouse airway: Characterization of plasmid DNA expression and pharmacokinetics. *Gene Ther.* **2:** 450–460.

Michel C.C. 1996. Transport of macromolecules through microvascular walls. *Cardiovasc. Res.* **32:** 644–653.

Miller G., Steinbrecher R.A., Murdock P.J., Tuddenham E.G., Lee C.A., Pasi K.J., and Goldspink G. 1995. Expression of factor VII by muscle cells in vitro and in vivo following direct gene transfer: Modelling gene therapy for haemophilia. *Gene Ther.* **2:** 736–742.

Miller R.V. 1998. Bacterial gene swapping in nature. *Sci. Am.* **278:** 66–71.

Mumper R.J., Duguid J.G., Anwer K., Barron M.K., Nitta H., and Rolland A.P. 1996. Polyvinyl derivatives as novel interactive polymers for controlled gene delivery to muscle. *Pharm. Res.* **13:** 701–709.

Nakamura R.M., Greenwald C.A., Peebles C.L., and Tan E.M. 1978. *Autoantibodies to nuclear antigens (ANA): Immunochemical specificities and significance in systemic rheumatic disease* Educational Products Division, American Society of Clinical Pathologists, Chicago, Illinois.

Naviaux R.K. and Verma I.M. 1992. Retroviral vectors for persistent expression in vivo. *Curr. Opin. Biotechnol.* **3:** 540–547.

Neuwelt E.A. and Rapoport S.I. 1984. Modification of the blood-brain barrier in the chemotherapy of malignant brain tumors. *Fed. Proc.* **43:** 214–219.

Neuwelt E.A., Specht H.D., Barnett P.A., Dahlborg S.A., Miley A., Larson S.M., Brown P., Eckerman K.F., Hellström K.E., and Hellström I. 1987. Increased delivery of tumor-specific monoclonal antibodies to brains after osmotic blood-brain barrier modification in patients with melanoma metastatic to the central nervous system. *Neurosurgery* **20:** 885–895.

Nita I., Ghivizzani S.C., Galea-Lauri J., Bandara G., Georgescu H.I., Robbins P.D., and Evans C.H. 1996. Direct gene delivery to synovium. An evaluation of potential vectors in vitro and in vivo. *Arthritis Rheum.* **39:** 820–828.

Nopanitaya W., Lamb J., Grisham J., and Carson J. 1976. Effect of hepatic venous outflow obstruction on pores and fenestration in sinusoidal endothelium. *Br. J. Exp. Pathol.* **57:** 604–609.

Raz E., Carson D.A., Parker S.E., Parr T.B., Abai A.M., Aichinger G., Gromkowski S.H., Singh M., Lew D., Yankauckas M.A., et al. 1994. Intradermal gene immunization: the possible role of DNA uptake in the induction of cellular immunity to viruses. *Proc. Natl. Acad. Sci.* **91:** 9519–9523.

Riessen R., Rahimizadeh H., Blessing E., Takeshita S., Barry J.J., and Isner J.M. 1993. Arterial gene transfer using pure DNA applied directly to a hydrogel-coated angioplasty balloon. *Hum. Gene Ther.* **4:** 749–758.

Rippe B. and Haraldsson B. 1994. Transport of macromolecules across microvascular walls: The two-pore theory. *Physiol. Rev.* **74:** 163–219.

Rippe B., Haraldsson B., and Folkow B. 1985. Evaluation of the "stretched pore phenomenon" in isolated rat hindquarters. *Acta Physiol. Scand.* **125:** 453–459.

Robinson P.J. and Rapoport S.I. 1987. Size selectivity of blood-brain barrier permeability at various times after osmotic opening. *Am. J. Physiol.* **253:** R459–R466.

Rush M.G. and Misra R. 1985. Extrachromosomal DNA in eucaryotes. *Plasmid* **14:** 177–191.

Rybenkov V.V., Vologodskii A.V., and Cozzarelli N.R. 1997. The effect of ionic conditions on the conformations of supercoiled DNA. I. Sedimentation analysis. *J. Mol. Biol.* **267:** 299–311.

Sawa T., Miyazaki H., Pittet J.F., Widdicombe J.H., Gropper M.A., Hashimoto S., Conrad D.J., Folkesson H.G., Debs R., Forsayeth J.R., Fox B., and Wiener-Kronish J.P. 1996. Intraluminal water increases expression of plasmid DNA in rat lung. *Hum. Gene Ther.* **7:** 933–941.

Sebestyen M.G., Ludtke J.J., Bassik M.C., Zhang G., Budker V., Lukhtanov E.A., Hagstrom J.E., and Wolff J.A. 1998. DNA vector chemistry: The covalent attachment of signal peptides to plasmid DNA. *Nat. Biotechnol.* **16:** 80–85.

Shefner R., Kleiner G., Turken A., Papazian L., and Diamond B. 1991. A novel class of anti-DNA antibodies identified in BALB/c mice. *J. Exp. Med.* **173:** 287–296.

Sikes M.L., O'Malley B.W., Jr., Finegold M.J., and Ledley F.D. 1994. In vivo gene transfer into rabbit thyroid follicular cells by direct DNA injection. *Hum. Gene Ther.* **5:** 837–844.

Szybalski W. 1992. Use of the HPRT gene and the HAT selection technique in DNA mediated transformation of mammalian cells: First steps toward developing hybridoma techniques and gene therapy. *BioEssays* **14:** 495–500.

Takeshita S., Isshiki T., and Sato T. 1996a. Increased expression of direct gene transfer into skeletal muscles observed after acute ischemic injury in rats. *Lab. Invest.* **74:** 1061–1065.

Takeshita S., Tsurumi Y., Couffinahl T., Asahara T., Bauters C., Symes J., Ferrara N., and Isner J.M. 1996b. Gene transfer of naked DNA encoding for three isoforms of vascular endothelial growth factor stimulates collateral development in vivo. *Lab. Invest.* **75:** 487–501.

Takeshita S., Isshiki T., Mori H., Tanaka E., Tanaka A., Umetani K. Eto K., Miyazawa Y., Ochiai M., and Sato T. 1997. Microangiographic assessment of collateral vessel formation following direct gene transfer of vascular endothelial growth factor in rats. *Cardiovasc. Res.* **35:** 547–552.

Tan J.H. and Chan W.K. 1997. Efficient gene transfer into zebrafish skeletal muscle by intramuscular injection of plasmid DNA. *Mol. Mar. Biol. Biotechnol.* **6:** 98–109.

Taylor A.E. and Granger D.N. 1984. Exchange of macromolecules across the microcirculation. In *Handbook of physiology: The cardiovascular system microcirulation* (ed. S.R. Geiger et al.), pp 467–520. American Physiological Society, Bethesda, Maryland.

Tripathy S.K., Svensson E.C., Black H.B., Goldwasser E., Margalith M., Hobart P.M., and

Leiden J.M. 1996. Long-term expression of erythropoietin in the systemic circulation of mice after intramuscular injection of a plasmid DNA vector. *Proc. Natl. Acad. Sci.* **93:** 10876–10880.

Tron F. and Bach J.F. 1977. Relationships between antibodies to native DNA and glomerulonephritis in systemic lupus erythematosus. *Clin. Exp. Immunol.* **28:** 426–432.

Tsurumi Y., Takeshita S., Chen D., Kearney M., Rossow S.T., Passeri J., Horowitz J.R., Symes J.F., and Isner J.M. 1996. Direct intramuscular gene transfer of naked DNA encoding vascular endothelial growth factor augments collateral development and tissue perfusion. *Circulation* **94:** 3281–3290.

Vaheri A. and Pagano J.S. 1965. Infectious poliovirus RNA: A sensitive method of assay. *Science* **175:** 434–436.

Vincent C.K., Gualberto A., Patel C.V., and Walsh K. 1993. Different regulatory sequences control creatine kinase-M gene expression in directly injected skeletal and cardiac muscle. *Mol. Cell. Biol.* **13:** 1264–1272.

Vitadello M., Schiaffino M.V., Picard A., Scarpa M., and Schiaffino S. 1994. Gene transfer in regenerating muscle. *Hum. Gene Ther.* **5:** 11–18.

Vologodskii A. and Cozzarelli N. 1994. Conformational and thermodynamic properties of supercoiled DNA. *Annu. Rev. Biophys. Biomol. Struct.* **23:** 609–643.

von Harsdorf R., Schott R.J., Shen Y.T., Vatner S.F., Mahdavi V., and Nadal-Ginard B. 1993. Gene injection into canine myocardium as a useful model for studying gene expression in the heart of large mammals. *Circ. Res.* **72:** 688–695.

Wang J., Jiao S., Wolff J., and Knechtle S. 1992. Gene transfer and expression into rat cardiac transplants. *Transplantation* **53:** 703–705.

Webber E.M., Wu J.C., Wang L., Merlino G., and Fausto N. 1994. Overexpression of transforming growth factor alpha causes liver enlargement and increased hepatocyte proliferation in transgenic mice. *Am. J. Pathol.* **145:** 398–408.

Wells D.J. and Goldspink G. 1992. Age and sex influence expression of plasmid DNA directly injected into mouse skeletal muscle. *FEBS Lett.* **306:** 203–205.

Whiteside T.L. and Dixon J.A. 1979. Clinical usefulness of the *Crithida luciliae* test for antibodies to native DNA. *Am. J. Clin. Pathol.* **72:** 829–835.

Wicks I.P., Howell M.L., Hancock T., Kohsaka H., Olee T., and Carson D.A. 1995. Bacterial lipopolysaccharide copurifies with plasmid DNA: Implications for animal models and human gene therapy. *Hum. Gene Ther.* **6:** 317–323.

Winegar R.A., Monforte J.A., Suing K.D., Oloughlin K.G., Rudd C.J., and Macgregor J.T. 1996. Determination of tissue distribution of an intramuscular plasmid vaccine using PCR and in situ DNA hybridization. *Hum. Gene Ther.* **7:** 2185–2194.

Wolf M.B., Porter L.P., and Watson P.D. 1989. Effects of elevated venous pressure on capillary permeability in cat hindlimbs. *Am. J. Physiol.* **257:** H2025–2032.

Wolff J.A. and Lederberg J. 1994. An early history of gene transfer and therapy. *Hum. Gene Ther.* **5:** 469–480.

Wolff J.A., Ludtke J., Acsadi G., Williams P., and Jani A. 1992a. Long-term persistence of plasmid DNA and foreign gene expression in mouse muscle. *Hum. Mol. Genet.* **1:** 363–369.

Wolff J.A., Williams P., Acsadi G., Jiao S., Jani A., and Chong W. 1991. Conditions affecting direct gene transfer into rodent muscle in vivo. *BioTechniques* **4:** 474–485.

Wolff J.A., Dowty M.E., Jiao S., Repetto G., Berg R.K., Ludtke J.J., and Williams P. 1992b. Expression of naked plasmids by cultured myotubes and entry of plasmids into T tubules and caveolae of mammalian skeletal muscle. *J. Cell Sci.* **103:** 1249–1259.

Wolff J.A., Malone R.W., Williams P., Chong W., Acsadi G., Jani A., and Felgner P.L. 1990. Direct gene transfer into mouse muscle in vivo. *Science* **247:** 1465–1468.

Yang J.P. and Huang L. 1996. Direct gene transfer to mouse melanoma by intratumor injection of free DNA. *Gene Ther.* **3:** 542–548.

Yovandich J., O'Malley B., Jr., Sikes M., and Ledley F.D. 1995. Gene transfer to synovial cells by intra-articular administration of plasmid DNA. *Hum. Gene Ther.* **6:** 603–610.

Zabner J., Cheng S.H., Meeker D., Launspach J., Balfour R., Perricone M.A., Morris J.E., Marshall J., Fasbender A., Smith A.E., and Welsh M.J. 1997. Comparison of DNA-lipid complexes and DNA alone for gene transfer to cystic fibrosis airway epithelia in vivo. *J. Clin. Invest.* **100:** 1529–1537.

Zhang G., Vargo D., Budker V., Armstrong N., Knechtle S., and Wolff J.A. 1997. Expression of naked plasmid DNA injected into the afferent and efferent vessels of rodent and dog livers. *Hum. Gene Ther.* **8:** 1763–1772.

13

Delivery Systems and Adjuvants for DNA Vaccines

Jeffrey B. Ulmer and Margaret A. Liu

Chiron Corporation
Emeryville, California

HISTORICAL BACKGROUND

The expression of antigens by host cells is an effective means of antigen presentation by major histocompatibility complex (MHC) class I molecules leading to the induction of cytotoxic T lymphocyte (CTL) responses. This can be accomplished by the infection of cells with pathogens, such as viruses, or by the transduction of cells with live recombinant vectors or plasmid DNA. The impetus to consider using plasmid DNA for the purpose of vaccination was suggested by two papers that demonstrated protein expression in situ after the administration of plasmid DNA. Benvenisty and Reshef (1986) showed that calcium phosphate precipitated DNA encoding reporter genes and hormones inoculated intraperitoneally in mice resulted in expression. Wolff et al. (1990) administered plasmid DNA encoding reporter genes, such as luciferase and β-galactosidase, by various routes and found reporter protein expression in the tissues injected. However, expression in myocytes after intramuscular (i.m.) injection yielded the highest levels. This expression was subsequently shown to be long-lived (Wolff et al. 1992a), despite the fact that the plasmid remained episomal (Wolff et al. 1992a; Nichols et al. 1995). These observations led several groups of researchers to investigate plasmid DNA inoculation as a potential vaccine strategy. In 1992, the induction of antibody responses was observed by Tang et al. (1992) using a gene gun to deliver DNA-coated gold beads into skin cells. In 1993, the induction of CTL responses by a DNA vaccine was first reported by Ulmer et al. (1993) after the i.m. injection of DNA encoding influenza nucleoprotein (NP). In addition, this study demonstrated initial proof of the concept that a plasmid DNA could be used as a vaccine, in that mice inoculated with NP DNA were protected from a lethal challenge with a heterosubtypic influenza virus strain. Later in 1993, several other groups reported on the use of DNA vaccines in other animal models, such as influenza in chickens (Robinson et al. 1993), human immunodeficiency virus (HIV) in mice (Wang et al. 1993), hepatitis B in mice (Davis et al. 1993a), and bovine herpesvirus in cattle (Cox et al. 1993). Since then, numerous examples of the immunogenicity and efficacy of DNA vaccines in animals have been demonstrated (for review, see Donnelly et al. 1997). These successes with naked DNA (i.e., formulated in saline solution) in preclinical studies have led to early phase I human clinical testing of DNA vaccines.

The purpose of this chapter is to present and discuss recent advancements in DNA vaccine formulation, such as vector development, delivery systems and adjuvants, and mechanisms of antigen presentation leading to the induction of CTLs. An understanding of these processes, and how the modification and manipulation of DNA vaccines affect the quantity and quality of immune responses, will enable the development of DNA vaccines of greater potency and tailored to diseases where specific types of immune responses are desired (e.g., humoral, cellular, or mucosal).

DNA VACCINE VECTORS

There are several ways in which DNA vaccines may be modified. The simplest would be to make changes to the vector itself, which would ultimately not increase the complexity of the vaccine. These changes could include (1) efforts to increase the levels of expression of the encoded antigen; (2) manipulation of the gene to target the gene product to specific intracellular or extracellular locations; (3) modifications to the vector backbone, such as those designed to increase plasmid production in *Escherichia coli* or address potential safety (e.g., integration) or regulatory issues (e.g., antibiotic resistance gene); and (4) immunostimulatory sequences (see the subsection on immunostimulatory nucleotide sequences). Only those issues relating to expression are discussed here.

A reasonable assumption on the relative potency of DNA vaccines is that more antigen produced in situ will lead to higher levels of immune responses. In general, this has been borne out experimentally (J.J. Donnelly, unpubl.). Hence, DNA vaccine vectors have been designed to yield high levels of antigen expression. This has been accomplished by using strong mammalian promoters, such as those from the Rous sarcoma virus (Manthorpe et al. 1993; Ulmer et al. 1993), the immediate early gene 1 of cytomegalovirus (CMV) containing the intron A (Chapman et al. 1991; Cheng et al. 1993; Manthorpe et al. 1993; Ulmer et al. 1993), simian virus 40 (Cheng et al. 1993), and the adenovirus 2 major late gene (Cheng et al. 1993). The combination of the CMV promoter and bovine growth hormone terminator seems to yield the highest levels of expression (Montgomery et al. 1993; Hartikka et al. 1996). Where tissue-specific expression is desired, nonviral promoters may be useful. As examples, the use of the albumin promoter-mediated expression of DNA vaccines in hepatocytes (Cheng et al. 1993) and use of the immunoglobulin promoter and enhancer elements resulted in preferential expression in B cells (Xiong et al. 1997). A second reason one might consider nonviral promoters is the potential of cytokine-mediated down-regulation of viral promoters such as CMV, because certain cytokines can decrease the in vitro expression of proteins driven by CMV promoters (Qin et al. 1997). A comparison of vectors containing CMV and MHC class I promoters showed that interferon-γ decreased the expression of the gene product in the CMV vector, but increased the gene expression in the MHC class I vector (Harms and Splitter 1995). Little work has been reported on inducible promoters for DNA vaccines, such as those responsive to tetracycline (Furth et al. 1994; Dhawan et al. 1995; Liang et al. 1996) or RU486 (Delort and Capecchi 1996), but these may prove to be very important, particularly for the coexpression of immunologically active molecules such as cytokines (see the subsection on cytokines). All of the described promoters require that the expression vector be transported into the nucleus for transcription to take place. This requirement may be obviated by the use of a bacteriophage T7 promoter system, where expression of the T7 RNA polymerase can drive the expression of the antigen controlled by the T7 promoter without the need for host-cell transcription machinery (Chen et al. 1994). This approach may also be useful for expressing proteins that require the involvement of other gene products (e.g., *rev*-dependent expression of HIV env). Such a system has been used in the context of HIV DNA vaccines to induce CTLs in mice (Selby et al. 1997).

Vector modification can also include changes to the gene insert itself, to increase expression or alter the targeting of the gene product. Increased expression levels have been achieved by the inclusion of a eukaryotic signal sequence (Chapman et al. 1991; Montgomery et al. 1997) and could be achieved by changing the codons to reflect usage by the eukaryotic cells (Haas et al. 1996). More commonly, though, modifications are geared toward altering the nature of the antigen expressed. Targeting of the antigen to specific compartments has the potential to induce immune responses that are qualitatively different. For example, secreted and plasma membrane proteins should be more readily available for the induction

encoding hepatitis B surface antigen induced higher levels of humoral and cellular immune responses in mice than did naked DNA. The immunogenicity of HIV *rev* and *env* DNA vaccines delivered by various routes of administration has also been enhanced by formulation in cationic lipids (Ishii et al. 1997; Toda et al. 1997). Certain types of cationic lipid–DNA complexes also seem to have targeting capabilities. McLean et al. (1997) demonstrated that such complexes were internalized preferentially by endothelial cells and hypothesized that uptake may be mediated by a membrane receptor, although it may involve interaction with a protein ligand present in the extracellular milieu. Although the means by which these various cationic lipid complexes enhance immune responses is not understood, it could include the increased transfection of cells in vivo leading to higher levels of antigen production or an adjuvant effect of the complex itself, because cationic lipids have been shown to induce inflammatory responses (Senior et al. 1991).

Polymers

Several types of polymers have been used to enhance gene expression in vitro and in vivo after administration of DNA-polymer complexes. Polyethylenimine (PEI) is a cationic polymer that binds DNA and has a high pH-buffering capacity that may facilitate neutralization of the acidic environment of endosomes/lysosomes and release of the complex into the cytoplasm. PEI has been shown to enhance the transfection of cells in vitro and in vivo (Boussif et al. 1995) and is amenable to the coupling of DNA for the targeting of cell surface receptors (Kircheis et al. 1997) and for the linking of DNA to adenoviral vectors (Baker et al. 1997). Other polymers that may be useful for DNA delivery include polyamidoamine dendrimers, which bind DNA and result in a unique spherical structure, and can efficiently transfect cells in vitro (Bielinska et al. 1997), and erodable microspheres, which can absorb various molecules including DNA (Mathiowitz et al. 1997). Polymers that have been used for in vivo transfection include polyvinyl pyrrolidone, where the expression of a reporter gene was found to be enhanced compared to naked DNA after i.m. injection (Mumper et al. 1996) and a formulation with DNA encoding insulin-like growth factor expressed biologically active protein (Alila et al. 1997), and a nonlipid, glucaramide-based polyamino polymer, where direct intratumoral injection resulted in expression levels that were comparable to a recombininant adenoviral vector (Goldman et al. 1997). Although the utility of these various polymers has not been reported for DNA vaccine delivery, it is reasonable to assume that higher levels of antigen expression in vivo should lead to higher levels of immune responses.

Proteins and Peptides

DNA plasmids have been formulated with various proteins and peptides to facilitate compaction of the DNA, to target DNA to cell surface receptors, to facilitate the escape of DNA from the endosome/lysosome, or to target DNA for uptake by nuclei. Compaction of DNA from a mean diameter size of approximately 200–300 nm to less than 100 nm is thought to aid in the entry of DNA into cells by endocytosis or pinocytosis. Polylysine has commonly been used for this purpose, and it is amenable to derivatization for the addition of ligands that may target cell surface receptors. Wu et al. (1991) used this approach to deliver DNA to hepatocytes via asiaolglycoprotein receptors. Cationic peptides have also been used to compact DNA, and these complexes were shown to more efficiently transfect cells in vitro (Wadhwa et al. 1997). Interestingly, a different cationic peptide with α-helical properties also enhanced transfection, but, conversely, this activity correlated with the ability to form large aggregates of DNA (0.5–5 μm in diameter) (Niidome et al. 1997). Hence, there may be more than one mechanism of DNA entry by cells.

Peptides have also been used to destabilize membranes of the endosome/lysosome, to release DNA from this degradative compartment. Native peptides from influenza hemagglutinin (Metchler and Wagner 1997) and artifical peptides (Kichler et al. 1997) have been shown to be membrane-active and to increase the transfection of cells in vitro. Finally, the uptake of DNA by nuclei may be a limiting step in the transfection of cells. This may be particularly true for cells that do not divide, because the barrier provided by the nuclear membrane remains intact. Dean (1997) examined this issue by studying the uptake of DNA plasmids in nondividing cells and found that certain nucleotide sequences present in the SV40 origin of replication facilitated uptake. A possible explanation for these data is that proteins, such as transcription factors, could transport DNA plasmids into the nucleus by utilizing uptake by nuclear pores. Proteins containing nuclear localization sequences, such as histones and high-mobility group nuclear proteins, have been used to target the uptake of DNA by nuclei in this way (Kaneda et al. 1995; Zaitsev et al. 1997; Sebestyén et al. 1998). However, the potential of inducing antinuclear or anti-DNA antibodies using this approach should be considered.

The use of proteins and peptides to facilitate several aspects of cell transfection appears promising. However, the data so far are limited to in vitro applications, and it is not yet clear whether this would be applicable to transfection in vivo. In addition, the utility of immunogenic entities, such as proteins and peptides, as DNA delivery vehicles may be limited by the induction of antibody responses that could interfere with subsequent inoculations. The immunogenicity of the complex may be circumvented by the use of a multicomponent complex that would include a nonimmunogenic coat (e.g., polymer) to sequester the DNA complex until it is internalized by cells. Such scenarios may involve highly complicated formulation issues that could preclude their use as vaccines.

Mucosal Immunization

Pathogens, in general, enter hosts via mucosal surfaces. Although protective immunity can be conferred by systemic immune responses, even against mucosal pathogens such as the influenza virus, it may be reasonable to assume that a mucosal component to the host immune response is desirable in many cases. In addition, the mucosal administration of vaccines, that is, orally or intranasally, is a less invasive procedure compared to injection with a syringe. For these reasons, the potential of mucosal DNA vaccination has been investigated. Ideally, mucosal vaccination should result in both mucosal and systemic immune responses, including serum antibodies and CTLs. The first suggestion that mucosal immunization with DNA could result in the induction of immunity was reported by Fynan et al. (1993) after intranasal inoculation of mice with influenza hemagglutinin DNA. They found that, despite no detectable antibody responses, the mice were protected from virus challenge. Using a herpes simplex virus model, intranasal administration of gB DNA resulted in the induction of a mucosal IgA response, as well as serum IgG antibodies and cellular immune responses, as indicated by a delayed-type hypersensitivity reaction (Kuklin et al. 1997). These responses were enhanced when the DNA was delivered with the mucosal adjuvant cholera toxin. Similarly, the inoculation of influenza hemagglutinin DNA via the intranasal, oral, and intrajejunal routes resulted in systemic cellular immune responses (i.e., CTLs), and the magnitude of the responses was enhanced by cholera toxin (Etchart et al. 1997). HIV DNA vaccines have also been administered through the mucosal routes intranasally (Asakura et al. 1997) and intravaginally (Wang et al. 1997) with the subsequent induction of mucosal and systemic immune responses. A potential caveat of mucosal vaccination is the potential of inducing tolerance, because it has been shown that the oral administration of protein antigens can result in a state of immunological nonresponsiveness (Chen et al. 1996).

Delivery of DNA via mucosal routes, particularly oral, poses some significant challenges with respect to the uptake of DNA at inductive sites and to the harsh environment through which the DNA must be transported. Complexes that are particulate in nature and maintain a barrier between the DNA vaccine and the digestive enzymes may provide a solution. Poly(DL-lactide-co-glycolide) (PLG) microspheres have been used sucessfully for the delivery of protein-based vaccines, and preliminary evidence suggests that they may be useful for DNA vaccines as well. Jones et al. (1997) administered PLG-encapsulated luciferase DNA to mice via the oral route and found antigen-specific IgG antibodies in the serum and IgA antibodies in the stools. Cationic lipids have been complexed with DNA for the systemic administration of DNA vaccines, as previously discussed, but have also been used for mucosal delivery. Expression of reporter genes in cells of the respiratory mucosa has been observed after aerosol administration of DNA–cationic lipid complexes (Stribling et al. 1992). In addition, both systemic and mucosal immune responses were observed after intranasal inoculation of a DNA–cationic lipid complex (Klavinskis et al. 1997). Therefore, the prospects for induction of broad-based immunity by mucosal administration of DNA vaccines appears promising.

ADJUVANTS FOR DNA VACCINES

One of the attractions of DNA vaccines is their simplicity, that is, they consist of purified plasmid DNA formulated in an aqueous solution. In many animal models, naked DNA has been sufficiently potent to induce protective immunity. Yet, the magnitude of immune responses induced by DNA vaccines can be increased by the inclusion of agents that have effects on the immune system, such as cytokines, immunostimulatory nucleotide sequences, and adjuvants.

Cytokines

The potential use of immunoregulatory molecules (e.g., cytokines, chemokines, and costimulatory molecules) to enhance or modulate the immune response to antigens encoded by DNA vaccines has received much attention. This includes the coexpression of cytokines in situ, through the use of DNA plasmids encoding such molecules, and the coadministration of recombinant cytokines with DNA vaccines, although much less has been reported on the latter.

The granulocyte–macrophage colony-stimulating factor (GM-CSF) is a pluripotent molecule that stimulates the growth and differentiation of various progentior cells and, hence, can have consequences on both humoral and cellular immune responses. The most widespread use of DNA vectors encoding cytokines has been with GM-CSF, with reported enhancements of antibody, helper T-cell, and CTL responses (Xiang and Ertl 1995; Conry et al. 1996; Geissler et al. 1997; Iwasaki et al. 1997a; Kim et al. 1997; Okada et al. 1997; Svanholm et al. 1997). Other cytokine gene constructs that have been demonstrated to increase the potency of immune responses to DNA vaccines include IL-2 (Chow et al. 1997; Geissler et al. 1997), which is a Th1-type cytokine that stimulates the growth and differentiation of T cells and immunoglobulin production from B cells; IL-12 (Kim et al. 1997; Maecker et al. 1997; Okada et al. 1997), which promotes cellular immune respones through the differentiation of Th1 cells; and IL-1β (Hakim et al. 1996; Maeckler et al. 1997), which has a wide variety of effects on immune and inflammatory responses. In addition, a DNA construct encoding the β-chemokine TCA-3, which is a chemotattractant for inflammatory cells, was shown to increase the immunogenicity of an HIV DNA vaccine (Tsuji et al. 1997a). DNA vectors expressing cytokines not only enhance the immunogenicity of DNA vaccines, but can also modulate the type of response. DNA delivery via the gene gun method tends to induce more of a Th2-type of helper T-cell response than does the i.m. injection of DNA (Feltquate et al. 1997). However, this can be altered by the codelivery of DNA constructs encoding IL-2, IL-7, or IL-12, resulting in a stronger Th1-type of response (Prayaga et al. 1997). Recombinant cytokines given together with DNA vaccines have also been tested with some success. rIL-12 has been shown to enhance the antigen-specific immunogenicity of a *Leishmania* DNA vaccine (Gurunathan et al. 1997) and the anti-tumor effects of a DNA vaccine (Irvine et al. 1996).

A different approach has been to use DNA constructs encoding costimulatory molecules (B7-1/B7-2 or CD80/CD86) that are known to be important for providing a signal to T cells during the engagement of T cells with APCs. DNA plasmids encoding B7-1, delivered either as a coexpression plasmid with an antigen or as separate plasmids, augmented the immune responses of the coexpressed antigens (Conry et al. 1996; Iwasaki et al. 1997a; Tsuji et al. 1997b). It is not known how the expression of costimulatory molecules together with DNA vaccines enhances the antigen-specific immune response, but possibilities include (1) by transforming cells that do not normally express costimulatory molecules into APCs, and (2) by increasing the costimulatory potential of the APCs. Along these lines, a DNA construct encoding CD40 ligand (CD154) augmented immune responses against a protein expressed by a coadministered DNA vaccine (Mendoza et al. 1997). CD40 ligand is normally expressed transiently on T cells, and interaction with APCs bearing CD40 promotes antigen presentation to T cells. These results, therefore, implicate CD40-bearing cells in the presentation of antigens expressed by DNA vaccines (see the last section).

The wide variety of cytokines, chemokines, and costimulatory molecules, introduced either as recombinant molecules or as DNA expression vectors, that have shown some effects on DNA vaccines suggests that this is an attractive approach to altering the quantity or quality of the immune response to DNA vaccines. However, a significant issue that needs to be addressed before this can be considered practical for prophylactic vaccination is the consequence of expressing cytokines or chemokines in an unregulated manner. If this poses a problem, it may be necessary to use an inducible promoter system to more tightly regulate the timing and magnitude of cytokine expression.

Immunostimulatory Nucleotide Sequences

The immunostimulatory effects of bacterial DNA have been known for some time. However, during the past 5 years, the nature of these effects has been studied in detail. A specific nucleotide sequence motif, consisting of purine-purine-C-G-pyrimidine-pyrimidine, is a potent inducer of lymphocyte proliferation in vitro that results in the secretion of cytokines such as IL-6, IL-12 and interferon-γ (Messina et al. 1991; Yamamoto et al. 1992; Krieg et al. 1995; Klinman et al. 1996; Yi et al. 1996). Coadministration of such oligonucleotides with protein vaccines can enhance antigen-specific immune responses and modulate the response to include CTLs (Carson and Raz 1997; Lipford et al. 1997; Roman et al. 1997; Wooldridge et al. 1997). These results suggest that the presence of CpG motifs in DNA vaccines may account for the bias toward the induction of Th1-type responses. That plasmid DNA vaccines have immunostimulating capabilities was suggested by the observations that antigen-specific immune responses induced by an influenza HA DNA vaccine could be enhanced by the inclusion of additional noncoding plasmid (Donnelly et al. 1994). In addition, the coinoculation of noncoding plasmid DNA with a protein antigen can modulate the immunoglobulin isotype profile of antibodies against the protein (Donnelly et al. 1994; Leclerc et al. 1997). The potential role of CpG motifs in the effectiveness of DNA vaccines has been suggested by Sato et al. (1996), who demonstrated that the inclusion of the specific motif AACGTT in a DNA vector improved its potency. Furthermore, the immunostimulatory effect can be abrogated if the

cination may be sufficient to produce the CTL responses observed. Second, the transfection of APCs does appear to occur after gene gun inoculation, as suggested by the observation that cells with morphology resembling Langerhans cells expressed green fluorescent protein after gene gun inoculation of DNA (Condon et al. 1996). Interestingly, these cells appeared to migrate from the site of transfection (i.e., skin) to the draining lymph node, which could account for how at least some of the antigen is presented to T cells. Third, CTL responses were observed in mice that had their DNA-injected muscles surgically removed after vaccination, suggesting that cells distal to the injection site were transfected (Torres et al. 1997). Therefore, the transfection of even a small number of APCs, either at or distant from the DNA inoculation site, has the potential to induce CTL responses.

In summary, the data suggest that both the transfer of antigen and the transfection of APCs can lead to the induction of CTLs after DNA vaccination. A reasonable working hypothesis is that these two processes combine to yield the robust CTL responses seen in DNA-vaccinated mice. An understanding of the relative importance of these two mechanisms will enable vaccine developers to design methods of facilitating either or both pathways. For example, if antigen transfer is desired, then better insight must be gained on the nature of this transfer, such as the molecules involved in antigen processing and transfer, and the intracellular and extracellular compartments where this takes place. If, on the other hand, the direct transfection of APCs is necessary, then means of DNA targeting to and expression within such cells will be important.

FUTURE PROSPECTS

The utility of naked DNA vaccines has been amply demonstrated in animal models. The simplicity of these vaccines is one of the features that has the potential to facilitate their development for human application. However, the potency of naked DNA vaccines in humans and the effect of DNA formulation are not yet known. In any case, though, increasing the potency of DNA vaccines may prove to be important, because it could lower the dose of DNA required for protective immunity. This would be beneficial for vaccine manufacturing and may lessen potential safety issues relating to the quantity of DNA in the vaccine. However, the benefit any formulation may provide must be weighed against the increase in complexity to the vaccine, which may pose additional considerations of feasibility, manufacturing, and safety.

REFERENCES

Alila H., Coleman M., Nitta H., French M., Anwer K., Liu Q.S, Meyer T., Wang J.J., Mumper R., Oubari D., Long S., Nordstrom J., and Rolland A. 1997. Expression of biologically active human insulin like growth factor I following intramuscular injection of a formulated plasmid in rats. *Hum. Gene Ther.* **8:** 1785–1795.

An L.-L. and Whitton J.L. 1997. A multivalent minigene vaccine, containing B cell, cytotoxic T cell lymphocyte, and T_h epitopes from several microbes, induces appropriate responses in vivo and confers protection against more than one pathogen. *J. Virol.* **71:** 2292–2302.

Asakura Y., Hinkula J., Leanderson A.-C., Fukushima J., Okuda K., and Wahren B. 1997. Induction of HIV-1 specific mucosal immune responses by DNA vaccination. *Scand. J. Immunol.* **46:** 326–330.

Baker A., Saltik M., Lehrmann H., Killisch I., Mautner V., Lamm G., Christofori G., and Cotten M.R.A. 1997. Polyethylenimine (PEI) is a simple, inexpensive and effective reagent for condensing and linking plasmid DNA to adenovirus for gene delivery. *Gene Ther.* **4:** 773–782.

Benvenisty N. and Reshef L. 1986. Direct introduction of genes into rats and expression of the genes. *Proc. Natl. Acad. Sci.* **83:** 9551–9555.

Bevan M.J. 1976. Minor H antigens introduced on H-2 different stimulating cells cross-react at the cytotoxic T cell level during in vivo priming. *J. Immunol.* **117:** 2233–2238.

Bielinska A.U., Kukowska-Latallo J.F., and Baker J.R.A. 1997. The interaction of plasmid DNA with polyamidoamine dendrimers: Mechanism of complex formation and analysis of alterations induced in nuclease sensitivity and transcriptional activity of the complexed DNA. *Biochim. Biophys. Acta* **1353:** 180–190.

Boczkowski D., Nair S.K., Snyder D., and Gilboa E. 1996. Dendritic cells pulsed with RNA are potent antigen-presenting cells in vitro and in vivo. *J. Exp. Med.* **184:** 465–472.

Boussif O., Lezoualc'h F., Zanta M.A., Mergny M.D., Scherman D., Demeneix B., and Behr J.P. 1995. A versatile vector for gene and oligonucleotide transfer into cells in culture and in vivo: Polyethylenimine. *Proc. Natl. Acad. Sci.* **92:** 7297–7301.

Carson D.A. and Raz E. 1997. Oligonucleotide adjuvants for T helper 1 (Th1)-specific vaccination. *J. Exp. Med.* **186:** 1621–1622.

Casares S., Brumeanu T.-D., Bot A., and Bona C.A. 1997a. Protective immunity elicited by vaccination with DNA encoding for a B cell and a T cell epitope of the A/PR/8/34 influenza virus. *Viral Immunol.* **10:** 129–136.

Casares S., Inaba K., Brumeanu T.-D., Steinman R.M., and Bona C.A. 1997b. Antigen presentation by dendritic cells after immunization with DNA encoding a major histocompatibility complex class II-restriced viral epitope. *J. Exp. Med.* **186:** 1481–1486.

Chapman B.S., Thayer R.M., Vincent K.A., and Haigwood N.L. 1991. Effect of intron A from human cytomegalovirus (Towne) immediate-early gene on heterologous expression in mammalian cells. *Nucleic Acids Res.* **19:** 3979–3986.

Chen X., Yunsheng L., Keyong X., and Wagner T.E. 1994. A self-initiating eukaryotic transient gene expression system based on cotransfection of bacteriophage T7 RNA polymerase and DNA vectors containing a T7 autogene. *Nucleic Acids Res.* **22:** 2114–2120.

Chen Y., Inobe J., Kuchroo V.K., Baron J.L., Janeway C.A., and Weiner H.L. 1996. Oral tolerance in myelin basic protein T-cell receptor transgenic mice: Suppression of autoimmune encephalomyelitis and dose-dependent induction of regulatory cells. *Proc. Natl. Acad. Sci.* **93:** 388–391.

Cheng L., Ziegelhoffer P.R., and Yang N.S. 1993. In vivo promoter activity and transgene expression in mammalian somatic tissues evaluated by using particle bombardment. *Proc. Natl. Acad. Sci.* **90:** 4455–4459.

Chow Y.H., Huang W.L., Chi W.K., Chu Y.D., and Tao M.H. 1997. Improvement of hepatitis B virus DNA vaccines by plasmids coexpressing hepatitis B surface antigen and interleukin-2. *J. Virol.* **71:** 169–178.

Ciernik I.F., Berzofsky J.A., and Carbone D.P. 1996. Induction of cytotoxic T lymphocytes and antitumor immunity with DNA vaccines expressing single T cell epitopes. *J. Immunol.* **156:** 2369–2375.

Condon C., Watkins S.C., Celluzzi C.M., Thompson K., and Falo L.D., Jr. 1996. DNA-based immunization by in vivo transfection of dendritic cells. *Nat. Med.* **2:** 1122–1128.

Conry R.M., Widera G., Lobuglio A.F., Fuller J.T., Moore S.E., Barlow D.L., Turner J., and Curiel D.T. 1996. Selected strategies to augment polynucleotide immunization. *Gene Ther.* **3:** 67–74.

Corr M., Lee D.J., Carson D.A., and Tighe H. 1996. Gene vaccination with naked plasmid DNA: Mechanism of CTL priming. *J. Exp. Med.* **184:** 1555–1560.

Cox G., Zamb T.J., and Babiuk L.A. 1993. Bovine herpesvirus 1: Immune responses in mice and cattle injected with plasmid DNA. *J. Virol.* **67:** 5664–5667.

Davis H.L., Michel M.L., and Whalen R.G. 1993a. DNA-based immunization induces continuous secretion of hepatitis B surface antigen and high levels of circulating antibody. *Hum. Mol. Genet.* **2:** 1847–1851.

Davis H.L., Whalen R.G., and Demeneix B.A. 1993b. Direct gene transfer into skeletal muscle in vivo: Factors affecting efficiency of transfer and stability of expression. *Hum. Gene Ther.* **4:** 151–159.

Dean D.A. 1997. The import of plasmid DNA into the nucleus is sequence specific. *Exp. Cell Res.* **230:** 293–302.

Delort J.P. and Cappechi M.R. 1996. TAXI/UAS: A molecular switch to control expression of genes in vivo. *Hum. Gene Ther.* **7:** 809–820.

Dhawan J., Rando T.A., Elson S.L., Bujard H., and Blau H.M. 1995. Tetracycline-regulated gene-expression following direct gene-transfer into mouse skeletal-muscle. *Somat. Cell Mol. Genet.* **21:** 233–240.

Doe B., Selby M., Barnett S., Baenziger J., and Walker C.M. 1996. Introduction of cytotoxic T lymphocytes by intramuscular immunization with plasmid DNA is facilitated by bone-marrow-derived cells. *Proc. Natl. Acad. Sci.* **93:** 8578–8583.

Donnelly J.J., Ulmer J.B, Shiver J.W., and Liu M.A. 1997. DNA vaccines. *Ann. Rev. Immunol.* **15:** 617–648.

Etchart N., Buckland R., Liu M.A., Wild T.F., and Kaiserlian D.R.A. 1997. Class I restricted CTL induction by mucosal immunization with naked DNA encoding measles virus haemagglutinin. *J. Gen. Virol.* **7:** 1577–1580.

Fasbender A., Marshall J., Moninger T.O., Grunst T., Cheng S., and Welsh M.R.A. 1997. Effect of co lipids in enhancing cationic lipid mediated gene transfer in vitro and in vivo. *Gene Ther.* **4:** 716–725.

Felgner P.L., Tsai Y.J., Sukhu L., Wheeler C.J., Manthorpe M., Marshall J., and Cheng S.H. 1995. Improved cationic lipid formulations for in vivo gene therapy. *Ann. N.Y. Acad. Sci.* **772:** 126–139.

Feltquate D., Heaney S., Webster R.G., and Robinson H.L. 1997. Different T helper cell types and antibody isotypes generated by saline and gene gun DNA immunization. *J. Immunol.* **158:** 2278–2284.

Fu T.-M., Ulmer J.B., Caulfield M.J., Deck R.R., Friedman A., Wang S., Liu X., Donnelly J.J., and Liu M.A. 1997. Priming of cytotoxic T lymphocytes by DNA vaccines: Requirement for professional antigen presenting cells and evidence for antigen transfer from myocytes. *Mol. Med.* **3:** 362–371.

Furth P.A., St. Onge L., Boger H., Gruss P., Gossen M., Kistner A., Bujard H., and Henninghausen L. 1994. Temporal control of gene expression in transgenic mice by a tetracycline-responsive promoter. *Proc. Natl. Acad. Sci.* **91:** 9302–9306.

Fynan E.F., Webster R.G., Fuller D.H., Haynes J.R., Santoro J.C., and Robinson H.L. 1993. DNA vaccines: Protective immunizations by parenteral, mucosal, and gene-gun inoculations. *Proc. Natl. Acad. Sci.* **90:** 11478–11482.

Geissler M., Gesien A., Tokushige K., and Wands J.R. 1997. Enhancement of cellular and humoral immune responses to hepatitis C virus core protein using DNA-based vaccines augmented with cytokine-expressing plasmids. *J. Immunol.* **158:** 1231–1237.

Gerloni M., Billetta R., Xiong S., and Zanetti M. 1997. Somatic transgene immunization with DNA encoding an immunoglobulin heavy chain. *DNA Cell Biol.* **16:** 611–625.

Goldman, C.K., Soroceanu L., Smith N., Gillespie G.Y., Shaw W., Burgess S., Bilbao G., and Curiel D.T., 1997. In vitro and in vivo gene delivery mediated by a synthetic polycationic amino polymer. *Nat. Biotechnol.* **15:** 462–466.

Gregoriadis, G., Safie R., de Souza J.B., 1997. Liposome-mediated DNA vaccination. *FEBS Lett.* **402:** 107–110.

Gurunathan S., Sacks D.L., Brown D.R., Reiner S.L., Charest H., Glaichenhaus N., and Seder R.A. 1997. Vaccination with DNA encoding the immunodominant LACK parasite antigen confers protective immunity to mice infected with Leishmania major. *J. Exp. Med.* **186:** 1137–1147.

Haas J., Park E.C., and Seed B. 1996. Codon usage limitation in the expression of HIV-1 envelope glycoprotein. *Curr. Biol.* **6:** 315–324.

Hakim I., Levy S., and Levy R. 1996. A nine-amino acid peptide from IL-1β augments antitumor immune responses induced by protein and DNA vaccines. *J. Immunol.* **157:** 5503–5511.

Harms J.S. and Splitter G.A. 1995. Interferon-gamma inhibits transgene expression driven by SV40 or CMV promoters but augments expression driven by the mammalian MHC-I promoter. *Hum. Gene Ther.* **6:** 1291–1297.

Hartikka J., Sawdey M., Cornefert-Jensen F., Margalith M., Barnhart K., Nolasco M., Vahlsing H.L., Meek J., Marquet M., Hobart P., Norman J., and Manthorpe M. 1996. An improved plasmid DNA expression vector for direct injection into skeletal muscle. *Hum. Gene Ther.* **7:** 1205–1217.

Huang A.Y.C., Golumbek P., Ahmadzadeh M., Jaffee E., Pardoll D., and Levitsky H. 1994. Role of bone marrow-derived cells in presenting MHC class I-restricted tumor antigens. *Science* **264:** 961–965.

Inaba K., Metlay J.P., Crowley M.T., Witmer-Pack M., and Steinman R.M. 1990. Dendritic cells as antigen presenting cells in vivo. *Int. Rev. Immunol.* **6:** 197–206.

Inchauspé G., Vitvitski L., Major M.E., Jung G., Spengler U., Maisonnas M., and Trepo C. 1997. Plasmid DNA expressing a secreted or a nonsecreted form of hepatitis C virus nucleocapsid: Comparative studies of antibody and T-helper responses following genetic immunization. *DNA Cell Biol.* **16:** 185–195.

Irvine K.R., Rao J.B., Rosenberg S.A., and Restifo N.P. 1996. Cytokine enhancement of DNA immunization leads to effective treatment of established pulmonary metastases. *J. Immunol.* **156:** 238–245.

Ishii N., Fukushima J., Kaneko T., Okada E., Tani K., Tanaka S.-I., Hamajima K., Xin K.-Q., Kawamoto S., Koff W., Nishioka K., Yasuda T., and Okuda K. 1997. Cationic liposomes are a strong adjuvant for a DNA vaccine of human immunodeficiency virus type 1. *Aids Res. Hum. Retroviruses* **13:** 1421–1427.

Iwasaki A., Stiernholm B.J.N., Chan A.K., Berinstein N.L., and Barber B.H. 1997a. Enhanced cytotoxic T-lymphocyte responses mediated by plasmid DNA immunogens encoding co-stimulatory molecules and cytokines. *J. Immunol.* **158:** 4591–4601.

Iwasaki A., Torres C.A., Ohashi P.S., Robinson H.L., and Barber B.H. 1997b. The dominant role of bone marrow-derived cells in CTL induction following DNA immunization at different sites. *J. Immunol.* **159:** 11–14.

Jones D.H., Corris S., MacDonald S., Clegg J.C.S., and Farrar G.H. 1997. Poly(DL-lactide-co-glycolide)-encapsulated plasmid DNA elicits systemic and mucosal antibody responses to encoded protein after oral administration. *Vaccine* **15:** 814–817.

Kaneda Y., Morishita R., and Tomita N. 1995. Increased expression of DNA cointroduced with nuclear protein in adult rat liver. *J. Mol. Med.* **73:** 289–297.

Kichler A., Mechtler K., Behr J.-P., and Wagner E. 1997. Influence of membrane-active peptides on lipospermine/DNA complex mediated gene transfer. *Bioconjugate Chem.* **8:** 213–221.

Kim J.J., Ayyavoo V., Bagarazzi M.L., Chattergoon M.A., Dang K., Wang B., Boyer J.D., and Weiner D.B. 1997. In vivo engineering of a cellular immune response by coadministration of IL-12 expression vector with a DNA immunogen. *J. Immunol.* **158:** 816–826.

Kircheis R., Kichler A., Wallner G., Kursa M., Ogris M., Felzmann T., Buchberger M., and Wagner E. 1997. Coupling of cell-binding ligands to polyethylenimine for targeted gene delivery. *Gene Ther.* **4:** 409–418.

Klavinskis L.S., Gao L., Barnfield C., Lehner T., and Parker S. 1997. Mucosal immunization with DNA-liposome complexes. *Vaccine* **15:** 818–820.

Klinman D.M., Yamshchikov G., and Ishigatsubo Y. 1997. Contribution of CpG motifs to the immunogenicity of DNA vaccines. *J. Immunol.* **158:** 3635–3639.

Klinman D.M., Yi A.K., Beaucage S.L., Conover J., and Krieg A.M. 1996. CpG motifs present in bacteria DNA rapidly induce lymphocytes to secrete interleukin 6, interleukin 12, and interferon gamma. *Proc. Natl. Acad. Sci.* **93:** 2879–2883.

Krieg A.M., Yi A.K., Matson S., Waldschmidt T.J., Bishop G.A., Teasdale R., Koretzky G.A., and Klinman D.M. 1995. CpG motifs in bacterial DNA trigger direct B-cell activation. *Nature* **374:** 546–549.

Kuklin N., Daheshia M., Karem K., Manickan E., and Rouse B. 1997. Induction of mucosal immunity against herpes simplex virus by plasmid DNA immunization. *J. Virol.* **71:** 3138–3145.

Leclerc C., Deriaud E., Rojas M., and Whalen R. 1997. The preferential induction of a Th1 immune response by DNA-based immunization is mediated by the immunostimulatory effect of plasmid DNA. *Cell. Immunol.* **179:** 97–106.

Li S. and Huang L. 1997. In vivo gene transfer via intravenous administration of cationic lipid-protamine-DNA (LPD) complexes. *Gene Thereapy* **4:** 891–900.

Liang X., Hartikka J., Sukhu L., Manthorpe M., and Hobart P. 1996. Novel, high expressing and antibiotic-controlled plasmid vectors designed for use in gene therapy. *Gene Ther.* **3:** 350–356.

Lipford G.B., Bauer M., Blank C., Reiter R., Wagner H., and Heeg K. 1997. CpG-containing synthetic oligonucleotides promote B and cytotoxic T cell responses to protein antigen: A new class of adjuvants. *Eur. J. Immunol.* **27:** 2340–2344.

Maecker H.T., Umetsu D.T., Dekruyff R.H., and Levy S. 1997. DNA vaccination with cytokine fusion constructs biases the immune response to ovalbumin. *Vaccine* **15:** 1687–1696.

Major M.E., Vitvitski L., Mink M.A., Schleef M., Whalen R.G., Trepo C., and Inchauspe G. 1995. DNA-based immunization with chimeric vectors for the induction of immune responses against the hepatitis C virus nucleocapsid. *J. Virol.* **69:** 5798–5805.

Manthorpe M., Cornefert-Jensen F., Hartikka J., Felgner J., Rundell A., Margalith M., and Dwarki V. 1993. Gene therapy by intramuscular injection of plasmid DNA: Studies on firefly luciferase gene expression in mice. *Hum. Gene Ther.* **4:** 419–431.

Mathiowitz E., Jacob J.S., Jong Y.S., Carino G.P., Chickering D.E., Chaturvedi P., Santos C.A., Vijayaraghavan K., Montgomery S., Bassett M., and Morrell C. 1997. Biologically erodable microspheres as potential oral drug delivery systems. *Nature* **386:** 410–412.

McLean J.W., Fox E.A., Baluk P., Bolton P.B., Haskell A., Pearlman R., Thurston G., Umemoto E.Y., and MacDonald D.M. 1997. Organ-specific endothelial cell uptake of cationic liposome-DNA complexes in mice. *Am. J. Physiol.* **273:** H387–H404.

Mendoza R.B., Cantwell M.J., and Kipps T.J. 1997. Cutting edge: Immunostimulatory effects of a plasmid expressing CD40 ligand (CD154) on gene immunization. *J. Immunol.* **159:** 5777–5781.

Messina J.P., Gilkeson G.S., and Pisetsky D.S. 1991. Stimulation of in vitro murine lymphocyte proliferation by bacterial DNA. *J. Immunol.* **147:** 1759–1764.

Metchler K. and Wagner E. 1997. Gene transfer mediated by influenza virus peptides: The role of peptide sequences. *New J. Chem.* **21:** 105–111.

Montgomery D.L., Huygen K., Yawman A.M., Deck R.R., Dewitt C M., Content J., Liu M.A., and Ulmer J.B. 1997. Induction of humoral and cellular immune responses by vaccination with M. tuberculosis antigen 85 DNA. *Cell. Mol. Biol.* **43:** 285–292.

Montgomery D.L., Shiver J.W., Leander K.R., Perry H.C., Friedman A., Martinez D., Ulmer J.B., Donnelly J.J., and Liu M.A. 1993. Heterologous and homologous protection against influenza A by DNA vaccination: Optimization of DNA vectors. *DNA Cell Biol.* **12:** 777–783.

Mumper R.J., Duguid J.G., Anwer K., Barron M.K., Nitta H., and Rolland A.P. 1996. Polyvinyl derivatives as novel interactive polymers for controlled gene delivery to muscle. *Pharm. Res.* **13:** 701–709.

Nichols W.W., Ledwith B.J., Manam S.V., and Troilo P.J. 1995. Potential DNA vaccine integration into host cell genome. *Ann. N.Y. Acad. Sci.* **772:** 30–39.

Niidome T., Ohmori N., Ichinose A., Wada A., Mihara H., Hirayama T., and Aoyagi H. 1997. Binding of cationic α-helical peptides to plasmid DNA and their gene transfer abilities into cells. *J. Biol. Chem.* **272:** 15307–15312.

Okada E., Sasaki S., Ishii N., Aoki I., Yasuda T., Nishioka K., Fukushima J., Miyazaki J., Wahren B., and Okuda K.R.A. 1997. Intranasal immunization of a DNA vaccine with IL-12 and granulocyte macrophage colony stimulating factor (GM-CSF) expressing plasmids in liposomes induces strong mucosal and cell mediated immune responses against HIV 1 antigens. *J. Immunol.* **159:** 3638–3647.

Prayaga S.K., Ford M.J., and Haynes J.R.A. 1997. Manipulation of HIV-1 gp120-specific immune responses elicited via gene gun-based DNA immunization. *Vaccine* **15:** 1349–1352.

Qin L., Ding Y., Pahud D.R., Chang E., Imperiale M.J., and Bromberg J.S. 1997. Promoter attenuation in gene therapy: Interferon-γ and tumor necrosis factor-α inhibit transgene expression. *Hum. Gene Ther.* **8:** 2019–2029.

Robinson H.L., Hunt L.A., and Webster R.G. 1993. Protection against a lethal influenza virus challenge by immunization with a haemagglutinin-expressing plasmid DNA. *Vaccine* **11:** 957–960.

Rodriguez F., Zhang J., and Whitton J.R.A. 1997. DNA immunization: Ubiquitination of a viral protein enhances cytotoxic T lymphocyte induction and antiviral protection but abrogates antibody induction *J. Virol.* **71:** 8497–8503.

Roman M., Martin-Orozco E., Goodman J.S., Nguyen M.-H., Sao Y., Ronaghy A., Kornbluth R.S., Richman D.D., Carson D.A., and Raz E. 1997. Immunostimulatory DNA sequences function as T helper-1-promoting adjuvants. *Nat. Med.* **3:** 849–854.

Sasaki S., Takashi T., Hamajima K., Fukushima J., Ishii N., Kaneko T., Xin K.-Q., Mohri H., Aoki I., Okubo T., Nishka K., and Okuda K. 1997. Monophosphoryl lipid A enhances both humoral and cell-mediated immune responses to DNA vaccination against human immunodeficiency virus type 1. *Infect. Immun.* **65:** 3520–3528.

Sato Y., Roman M., Tighe H., Lee D., Corr M., Nguyen M.D., Silverman G.J., Lotz M., Carson D.A., and Raz E. 1996. Immunostimulatory DNA sequences necessary for effective intradermal gene immunization. *Science* **273:** 352–354.

Schwartz D.A., Quinn T.J. Thorne P.S., Sayeed S., Yi A.K., and Krieg A.M. 1997. CpG motifs in bacterial DNA cause inflammation in the lower respiratory tract. *Clin. Invest.* **100:** 68–73.

Sebestyén M.G., Ludtke J.J., Bassik M.C., Zhang G., Budker V., Lukhtanov E.A., Hagstrom J.E., and Wolff J.A. 1998. DNA vector chemistry: The covalent attachment of signal peptides to plasmid DNA. *Nat. Biotechnol.* **16:** 80–85.

Selby M.J., Doe B., and Walker C.M. 1997. Virus specific cytotoxic T lymphocyte activity elicited by coimmunization with human immunodeficiency virus type 1 genes regulated by the bacteriophage T7 promoter and T7 RNA polymerase protein. *J. Virol.* **71:** 7827–7831.

Senior J.H., Trimble K.R., and Maskiewicz R. 1991. Interaction of positively charged liposomes with blood: Implications for their application in vivo. *Biochim. Biophys. Acta.* **1070:** 173–179.

Spellerberg M.B., Zhu D., Thompsett A., King C.A., Hamblin T.J., and Stevenson F.K. 1997. DNA vaccines against lymphoma: Promotion of anti-idiotypic antibody responses induced by single chain Fv genes by fusion to tetanus toxin fragment C. *J. Immunol.* **159:** 1885–1892.

Stopeck A.T., Hersh E.M., Akporiaye E.T., Harris D.T., Grogan T., Unger E., Warneke J., Schluter S.F., and Stahl S. 1997. Phase I study of direct gene transfer of an allogenic histocompatibility antigen, HLA-B7, in patients with metastatic melanoma. *J. Clin. Oncol.* **15:** 341–349.

Stribling R., Brunette E., Liggitt D., Gaensler K., and Debs R. 1992. Aerosol gene delivery in vivo. *Proc. Natl. Acad. Sci.* **89:** 11277–11281.

Svanholm C., Lowenadler B., and Wigzell H. 1997. Amplification of T cell and antibody responses in DNA based immunization with HIV-1 Nef by co-injection with a GM-CSF expression vector. *Scand. J. Immunol.* **46:** 298–303.

Tang D.C., De Vit M., and Johnston S.A. 1992. Genetic immunization is a simple method for eliciting an immune response. *Nature* **356:** 152–154.

Templeton N.S., Lasic D.D., Frederik P.M., Strey H.H., Roberts D.D., and Pavlakis G.N. 1997. Improved DNA: Liposome complexes for improved systemic delivery and gene expression. *Nat. Biotechnol.* **15:** 647–655.

Thierry A.R., Rabinovich P., Peng B., Mahan L.C., Bryant J.L., and Gallo R.C. 1997. Characterization of liposome-mediated gene delivery: Expression, stability and pharmacokinetics of plasmid DNA. *Gene Ther.* **4:** 226–237.

Tobery T. and Siliciano R. 1997. Targeting of HIV-1 antigens for rapid intracellular degradation enhances cytotoxic T lymphocyte (CTL) recognition and the induction of de novo CTL responses in vivo after immunization. *J. Exp. Med.* **185:** 909–920.

Toda S., Ishii N., Okada E., Kusakabe K.I., Arai H., Hamajima K., Gorai I., Nishioka K., and Okuda K.R.A. 1997. HIV 1 specific cell mediated immune responses induced by DNA vaccination were enhanced by mannan coated liposomes and inhibited by anti interferon gamma antibody. *Immunology* **92:** 111–117.

Torres C., Iwasaki A., Barber B.H., and Robinson H.L. 1997. Differential dependence on target site tissue for gene gun and intramuscular DNA immunizations. *J. Immunol.* **158:** 4529–4532.

Tsuji T., Fukushima J., Hamajima K., Ishii N., Aoki I., Bukawa H., Ishigatsubo Y., Tani K., Okubo T., Dorf M.E., and Okuda K. 1997a. HIV-1-specific cell-mediated immunity is enhanced by co-inoculation of TCΛ3 expression plasmid with DNA vaccine. *Immunology* **90:** 1–6.

Tsuji T., Hamajima K., Ishii N., Aoki I., Fukushima J., Xin K.Q., Kawamoto S., Sasaki S., Matsunaga K., Ishigatsubo Y., Tani K., Okubo T., and Okuda K. 1997b. Immunomodulatory effects of a plasmid expressing B7-2 on human immunodeficiency virus-1-specific cell-mediated immunity induced by a plasmid encoding the viral antigen. *Eur. J. Immunol.* **27:** 782–787.

Ulmer J., Deck R.R., DeWitt C.M., Donnelly J.J., and Liu M. 1996. Generation of MHC class I-restricted cytotoxic T lymphocytes by expression of a viral protein in muscle cells: Antigen presentation by non-muscle cells. *Immunology* **89:** 59–67.

Ulmer J.B., Donnelly J.J., Parker S.E., Rhodes G.H., Felgner P.L., Dwarki V.J., Gromkowski S.H., Deck R.R., De Witt C.M., Friedman A., Hawe L.A., Leander K.R., Martinez D., Perry H.C., Shiver J.W., Montgomery D.L., and Liu M.A. 1993. Heterologous protection against influenza by injection of DNA encoding a viral protein. *Science* **259:** 1745–1749.

Wadhwa M.S., Collard W.T., Adami R.C., McKenzie D.L., and Rice K.G. 1997. Peptide-mediated gene delivery: Influence of peptide structure on gene expression. *Bioconjugate Chem.* **8:** 81–88.

Wang B., Dang K., Agadjanyan M.G., Srikantan V., Li F., Ugen K.E., Boyer J., Merva M., Williams W.V., and Weiner D.B. 1997. Mucosal immunization with a DNA vaccine induces immune responses against HIV-1 at a mucosal site. *Vaccine* **15:** 821–825.

Wang B., Ugen K.E., Srikantan V., Agadjanyan M.G., Dang K., Refaeli Y., Sato A.I., Boyer J., Williams W.V., and Weiner D.B. 1993. Gene inoculation generates immune responses against human immunodeficiency virus type 1. *Proc. Natl. Acad. Sci.* **90:** 4156–4160.

Wettstein P.J., Haughton G., and Frelinger J.A. 1977. H-2 effects on cell-cell interaction in the response to single non-H-2 alloantigens. I. Donor H-2D region control of H-7.1-immunogenicity and lack of restriction in vivo. *J. Exp. Med.* **146:** 1346–1355.

Wolff J.A., Ludtke J.J., Acsadi G., Williams P., and Jani A. 1992a. Long-term persistence of plasmid DNA and foreign gene expression in mouse muscle. *Hum. Mol. Genet.* **1:** 363–369.

Wolff J.A., Williams P., Acsadi G., Jiao S., Jani A., and Chong W. 1991. Conditions affecting direct gene transfer into rodent muscle in vivo. *BioTechniques* **11:** 474–485.

Wolff J.A., Malone R.W., Williams P., Chong W., Acsadi G., Jani A., and Felgner P.L. 1990. Direct gene transfer into mouse muscle in vivo. *Science* **247:** 1465–1468.

Wolff, J.A., Dowty M.E., Jiao S., Repetto G., Berg R.K., Ludtke J.J., Williams P., and Slautterback D.B. 1992b. Expression of naked plasmids by cultured myotubes and entry of plasmids into T tubules and caveolae of mammalian skeletal muscle. *J. Cell Sci.* **103:** 1249–1259.

Wooldridge J.E., Ballas Z., Krieg A.M., and Weiner G.J. 1997. Immunostimulatory oligodeoxynucleotides containing CpG motifs enhance the efficacy of monoclonal antibody therapy of lymphoma. *Blood* **89:** 2994–2998.

Wu Y. and Kipps T.J. 1997. Deoxyribonucleic acid vaccines encoding antigens with rapid proteasome-dependent degradation are highly efficient inducers of cytolytic T lymphocytes. *J. Immunol.* **159:** 6037–6043.

Wu G.Y., Wilson J.M., Shalaby F., Grossman M., Shafritz D.A., and Wu C.H. 1991. Receptor-mediated gene delivery in vivo. Partial correction of genetic analbuminemia in Nagase rats. *J. Biol. Chem.* **266:** 14338–14342.

Xiang Z. and Ertl H.C. 1995. Manipulation of the immune response to a plasmid-encoded viral antigen by coinoculation with plasmids expressing cytokines. *Immunity* **2:** 129–135.

Xiong S., Gerloni M., and Zanetti M. 1997. In vivo role of B lymphocytes in somatic transgene immunization. *Proc. Natl. Acad. Sci.* **94:** 6352–6357.

Yamamoto S., Yamamoto T., Shimada S., Kuramoto E., Yano O., Kataoka T., and Tokunaga T. 1992. DNA from bacteria, but not from vertebrates, induces interferons, activates natural killer cells, and inhibits tumor growth. *Microbiol. Immunol.* **36:** 983–997.

Yi A.K., Chace J.H., Cowdery J.S., and Krieg A.M. 1996. IFN-γ promotes IL6 and IgM secretion in response to CpG motifs in bacterial DNA and oligodeoxynucleotides. *J. Immunol.* **156:** 558–564.

Zaitsev, S.V., Haberland A., Otto A., Vorob'ev V.I., Haller H., and Böttger M. 1997. H1 and HMG17 extracted from calf thymus nuclei are efficient DNA carriers in gene transfer. *Gene Ther.* **4:** 586–592.

14

Targets for Gene Therapy

Theodore Friedmann

Center for Molecular Genetics
UCSD School of Medicine
La Jolla, California 92093-0634

Arno G. Motulsky

Departments of Medicine and Genetics
University of Washington
Seattle, Washington 98195

THE MEDICAL NEED FOR GENE THERAPY

Human gene therapy can be defined as the directed expression of exogenous genes or other nucleotide sequences for the purpose of correcting a disease-causing mutation and thereby modifying a disease or other phenotype. This revolutionary new approach to treatment has been fueled by the explosive growth of molecular and cellular genetics and to a lesser extent by our improved understanding of disease pathophysiology (Orkin and Williams 1988; Anderson 1992; Friedmann 1992, 1997; Weatherall 1995). The concept of gene therapy is the logical extension of this evolving knowledge, and, after an initial period of uncertainty and contention (Fletcher 1983a,b; Rifkin 1985; Walters 1986), there is now broad acceptance of the underlying scientific concepts. Its implementation toward clinical reality has been much more difficult than the establishment of the concept itself (see http://www.nih.gov/news/panelrep.html).

This novel approach to the therapy of human disease is based on the realization that previous medical treatments have usually been directed at the consequences of causative defects or at disease symptoms rather than at the underlying causes. All doctors realize that our approaches of treating many human diseases often fall short in bringing about definitive and permanent disease correction. Even therapies for the monogenic inborn errors of metabolism that are based on elegant modern biochemical and cellular knowledge often are inadequate when applied to real patients with real diseases (Treacy et al. 1995). For most human disease, there remains a deep conceptual and temporal gulf between an understanding of disease pathogenesis and truly effective therapy.

An excellent example is the distance that continues to separate our profound understanding of the molecular biology and cellular physiology of globin synthesis, structure, and function from our ability to treat the hemoglobinopathies (including the thalassemias) successfully. There are probably no proteins that are as well understood as the globins. Yet most therapy for these disorders remains largely supportive and certainly not curative.

More recently, drug treatment to improve the expression of β-globin or to elevate fetal hemoglobin levels has shown promise (Zeng et al. 1995; Fucharoen et al. 1996; Schechter and Rodgers 1996). Gene therapy of these disorders is a logical step in the progression of attempts at treatment. Anemia in these disorders was first treated by the transfer of hemoglobin-carrying red cells by blood transfusion. More modern approaches include the use of hemoglobin-producing erythropoietic cells in bone marrow transplantation and the induction of erythropoiesis with agents such as erythropoietin. The next step, the use of the hemoglobin gene itself, offers the prospect of more definitive amelioration and even cure for these disorders (Motulsky1974; Bank et al. 1988; Beuzard 1996).

TRADITIONAL TREATMENT OF DISEASE

With the exception of surgical therapies and the cell-killing treatment of infections and malignancies, most human diseases were treated in the past by manipulations of the disordered pathophysiology or by the alleviation of symptoms, rather than by a definitive correction of the underlying causative defects. The principal reason for this approach was the obvious fact that, until the past several decades, the fundamental defects responsible for most human disease had not been identified. There was no definitive pathogenetic target to shoot at. The concept of inborn errors of metabolism developed by Sir Archibald Garrod (Garrod 1908, 1923; Childs 1970) provided a more definitive conceptual framework for the treatment of some disorders. The work of Garrod's successors (Scriver et al. 1995a) identified potential sites along the pathway of aberrant disease-causing metabolic steps that might be suitable for therapy. In simplest terms, the Garrodian scheme of inborn errors tells us that an underlying genetic defect leads to an aberration in the function of the metabolic pathways that, in turn, leads to disturbed production of the required products distal to a metabolic block or accumulation of damaging metabolic products proximal to a block. This concept can also be extended to hormonal and enzyme treatment of various disorders that are not caused by monogenic inborn errors. Treatment using these concepts has been approached through the following:

1. Replacement of missing "metabolites." This approach received its initial and most impressive proof of principle in the studies by Banting and Best on the treatment of pancreatectomized diabetic dogs with pancreatic extract and was then used with great (but not absolute) success in human diabetes. Success in the treatment of diseases resulting from a number of hormone deficiencies, such as those involving the growth hormone, the thyroid hormone, adrenal hormones, and others, was often obtained. Specific treatments with missing nutrients for anemias caused by a deficiency of vitamin B_{12} (Fenton and Rosenberg 1995) or iron are other examples. Successful use of the hormone erythropoietin produced by gene technology is a more recent example of this approach for the therapy of the anemia of endstage renal disease.

2. Removal of toxic metabolites. The principle of this mode of pathogenesis is the accumulation to toxic levels of metabolites proximal to a metabolic block. The most effective application of this therapeutic concept has involved chelation therapy with penicillamine for Wilson's disease (Danks 1995) and cystinuria (Segal and Thier 1995). However, even in these disorders, treatment is far from ideal.

3. Reduction of the substrate for a defective pathway. The most effective use of dietary restriction for disease treatment and prevention has come in the therapy of phenylketonuria (PKU) (Scriver et al. 1995b) and to a lesser extent in galactosemia (Segal and Berry 1995) and in urea cycle defects such as ornithine decarbamylase deficiency (Brusilow and Horwich 1995). Even in the case of PKU, which is probably the most successful application of dietary restriction for disease correction, the dietary approach suffers from its inability to completely restore cognitive function in patients and from the requirement for re-treatment of previously treated pregnant women to avoid mental retardation and other birth defects in their offspring.

4. Drug-induced changes in defective metabolic pathways. This approach drives much of the modern pharmaceutical industry and has had many important therapeutic successes. Anticoagulants in the treatment and prevention of thrombosis and the use of antihypertensives that lower blood pressure are impressive examples of this approach to treatment. The use of drugs to inhibit steps in the endogenous production of cholesterol in patients with familial and

acquired forms of hypercholesterolemia (Goldstein et al. 1995; Motulsky and Brunzell 1992) or to increase fetal hemoglobin levels in patients with thalassemia or sickle cell anemia (Zeng et al. 1995; Fucharoen et al. 1996; Schechter and Rodgers 1996) are additional examples of fairly successful therapy.

5. Restoration of enzyme function (enzyme therapy). This area of therapy has long been envisioned but has been burdened with major technical difficulties involving enzyme purification, in vivo stability, and efficient delivery. Nevertheless, it has now achieved clinical credibility in several inherited disorders, including adenosine deaminase deficiency (Hershfield et al. 1993; Hershfield and Mitchell 1995) and Gaucher's disease (Beutler and Grabowski 1995). Success has also been achieved in acquired disorders such as the use of thrombolytic enzymes (streptokinase and others) for the management of myocardial infarctions and strokes.

6. Selective toxicity. Cancer represents the primary area for the use of cytotoxic agents. The principle underlying this form of treatment is the development of drugs that cause more cellular damage to the tumor cells than to normal cells. Its successes are occasionally impressive, but the difficulties and limitations are severe.

7. Tissue transplantation to restore organ function. Tissue transplantation has been used with great effect, not only to correct organ malformations, but also to treat various genetic and other disorders. Transplantation of organs such as the kidney, liver, heart, and, to a lesser extent, lung and pancreas is becoming the established standard for the therapy of many kinds of diseases. Bone-marrow transplantation can be an effective therapy for some patients with thalassemia, immunodeficiency diseases, and other genetic diseases such as chronic granulomatous disease, Gaucher's disease, metachromatic leucodystrophy and adrenoleukodystrophy, and several of the mucopolysaccharidoses (Sciver et al. 1995a).

Although the various approaches to therapy described here are logical in principle, they usually are neither fully effective nor curative. The fact that only 12% of 65 monogenic inborn errors with well-understood basic lesions were reported in 1995 to respond fully to therapy is sobering (Treacy et al. 1995), particularly if one considers the fact that the identical full cure rate of 12% was identical to that found 10 years earlier (Hayes et al. 1985). These facts suggest that gene therapy directed at the fundamental gene defect in monogenic diseases or at specific abnormalities of gene expression in other conditions could possibly be a more effective way of restoring normal function.

TARGET DISEASES FOR GENE THERAPY

Inborn Errors of Metabolism

The earliest disease models for which gene therapy was envisioned were those simple, monogenic, inborn errors of metabolism in which a homozygous genetic defect resulted in the reduced expression of a required gene product and consequent metabolic aberration. When gene therapy was first conceived, these disorders seemed to be good candidates, since relatively low levels of gene product in heterozygotes or even in homozygotes (with some residual enzymic activity) often protected against some or all of the features of clinical disease. These facts imply that even relatively inefficient genetic correction might be of considerable therapeutic value, even out of proportion to the degree of genetic correction. Furthermore, the involved genes responsible for these disorders generally encode "housekeeping" functions that may not require the stringent regulation of gene expression. With this reasoning, diseases such as adenosine deaminase deficiency (Belmont et al. 1986; Kantoff et al. 1986; McIvor et al. 1987), Lesch Nyhan disease (Miller et al. 1983, 1984; Friedmann 1987; Szybalski 1992), cystic fibrosis (Anderson et al. 1991; Stanley et al. 1991; Krauss et al. 1992; Porteous and Dorin 1993; Alton and Geddes 1994), familial hypercholesterolemia (Miyanohara et al. 1988; Chowdhury et al. 1991; Grossman and Wilson 1992; Grossman et al. 1995), Gaucher's disease (Choudary et al. 1986; Sorge et al. 1986; Fink et al. 1990; Beutler 1992; Xu et al. 1994; Brady and Barton 1996), and Duchenne's muscular dystrophy (Karpati and Acsadi 1994; Hauser and Chamberlain 1996) became targets of early in vitro gene transfer studies. In fact, it was the powerful selection properties for and against cells expressing the HPRT enzyme that made the Lesch Nyhan syndrome one of the most useful inborn errors for the development of the early concepts of gene transfer as a mode of therapy (Friedmann 1987; Szybalski 1992). Because of the very encouraging data developed from clinical in vitro studies and preclinical applications in animal model systems for these diseases, clinical applications in human patients were very quickly undertaken. Results from these human clinical studies, however, have been largely discouraging and were limited in most cases to demonstrations of a degree of gene transfer that was insufficient to bring about sufficient gene expression to produce a compelling therapeutic effect (Blaese et al. 1995; Bordignon et al. 1995; Grossman et al. 1995; Knowles et al. 1995). It seems likely that these and other early unfavorable clinical results have been partly due to inefficient and untargeted gene transfer as well as to inadequately refined methods for grafting genetically modified cells in ex vivo gene therapy studies. Nevertheless, such facts do not detract from the relevance of these diseases as models for gene therapy. When improved viral and nonviral vector systems emerge, these and many other disorders should be far more amenable to effective gene therapy than they were with earlier vectors and grafting techniques.

Technical obstacles usually included difficulties in efficient and specific vector delivery, instability of transgene expression in vivo, and host immune response to the vector and to the transgene product. The rapid and impressive growth of new knowledge of transgene expression in vivo and the availability of powerful new tools suggest that these barriers will soon be overcome.

The bone marrow has been one of the most intensively studied target organs for the correction of metabolic errors such as adenosine deaminase (ADA) deficiency and Gaucher's disease, as well as for a variety of hematopoietic diseases. The main difficulties with gene therapy in the bone marrow are the continuing uncertainty about the nature of marrow hematopoietic stem cells, difficulties in their isolation, and the inefficiency of gene transfer into such cells by virtually all current vectors. Interestingly, bone marrow has also recently been found to contain cells with properties of the mesenchymal stem cells that are able to repopulate nonhematopoietic tissues, such as bone and skeletal muscle (Ferrari et al. 1998). The existence of such cells opens the door to the genetic correction of connective tissue as well as of bone and muscle disease by bone-marrow transplantation with genetically modified cells.

Multifactorial and Somatic Genetic Disease

With the successful introduction of the concepts and methods of molecular biology into the biomedical sciences in general, the initial choice of single gene defects as targets for gene therapy has now been augmented and, to some extent, replaced by a variety of more complex disease targets. The major factors responsible for this shift are the increasing characterization of disease at the cellular and molecular levels with the identification of potential target genes and the involvement of the biotechnology industry (see the following). Furthermore, the application of modern molecular genetic techniques together with powerful new methods in statistical and epidemiologic genetics has led to the identification of genetic components in many multifactorial diseases such as hypertension, coronary artery disease, obesity, diabetes mellitus, and others, eventually making them potential targets for gene therapy approaches. Pathogenesis of the common diseases is characterized by etiologic heterogeneity and by the involvement of several genes, as well as by gene-environmental and gene-gene interaction. Genes alone never appear to explain these conditions, as judged from the lack of 100% concordance of disease or other phenotypes, even in identical twins who are genetically alike. These complexities may, in some cases, make for slower progress in designing genetic approaches to therapy than would be the case with monogenic disorders. However, gene therapy approaches may be as valid for multifactorial diseases as for single gene defects, and in some cases, such as in the prevention of restenosis in coronary arteriosclerosis by the use of anti-arteriosclerotic genes or in cancer therapy (see the following), no more difficult to achieve. The pace of work in human genomics as well as the increasing use of molecular biology in the biomedical sciences promises that the elucidation of the genetics of complex human disease with ensuing applications to gene therapy will grow.

Cancer

Not surprisingly, cancer has become one of the most important disease types in the design of therapeutic approaches using gene therapy. This development occurred not only because of the dire nature of the disease, but also because of the modern view that cancer is a genetic disease at the somatic cell level and the growing understanding of genetic mechanisms of oncogenesis and its control (reviewed in later chapters of this book). The role of the host immune system in the development of tumors, and the identification of methods to enhance a host immune cellular response to resident tumor cells through vaccination with autologous tumor cells transduced with appropriate cytokine genes both ex vivo and in vivo (Rosenberg 1991; Nabel et al. 1992; Gansbacher 1994; Dranoff and Mulligan 1995; see chapter by Pardoll and Nabel, this volume) have all been used to design promising therapeutic approaches. In fact, the cancer models for gene therapy are now among the leading candidates for some clinical efficacy in human gene therapy studies. Other gene therapy approaches involving the reconstitution of the tumor suppressor function (Huang et al. 1988; Stanbridge 1989; Oren 1992; Roth 1996; Ruth et al. 1996), modification of the expression of genes regulating apoptosis and cell-cycle events, and the induction of sensitivity in cancer cells to a variety of drug inhibitors and chemotherapeutic agents are also being taken.

Similarly, because of the increasing understanding of the role of genetic mechanisms in atherosclerosis and the role of vascular growth factors in normal and pathological angiogenesis, genetic modification is assuming an increasingly important role in the design of gene-based treatments for many forms of cardiovascular disease. These include atherosclerosis resulting from low-density lipoprotien (LDL)-receptor deficiency, amelioration of the peripheral arterial, and even coronary arteriosclerosis with the introduction and local expression of foreign angiogenic genes such as VEGF (Finkel and Epstein 1995; Isner et al. 1995, 1996; Nabel 1995; Melillo et al. 1997; Vassalli and Dichek 1997), retinal disease (Ali et al. 1996; Miyoshi et al. 1997), and others.

Less compelling results have been obtained with other multifactorial disorders such as diabetes mellitus, in which exquisite control of gene expression and rapid and authentic sugar-sensing functions would be required. The discovery of leptin and the development of leptin knockout mice have provided the opportunity to study mechanisms of obesity and the potential for leptin reconstitution as an approach to therapy (G.X. Chen et al. 1996; Muzzin et al. 1996; Murphy et al. 1997). Indeed, the expression of a leptin transgene in mice has been reported to produce a prolonged weight reduction. The exquisite regulation of gene expression will be essential in many disorders (such as in insulin or erythropoietin deficiency, and obesity) and poses difficult problems because gene therapy would not be workable unless accompanied by the introduction of appropriate regulatory signals together with the missing gene—a difficult task!

Role of Industry

Another important factor in the shift of emphasis to common diseases has been the increasing role being played by the biotechnology industry in basic and clinical gene therapy research. Most monogenic diseases such as inborn errors are rare, and the market for gene therapy would be small. In contrast, the market for diseases such as cancer, coronary artery disease, diabetes, asthma, and the common major psychiatric diseases is enormous, and there is a high potential not only for widespread benefit to patients, but also for profits. It is therefore understandable that the major interests in the commercial sector will be gene therapy for the common and most burdensome diseases of our society. The main emphasis for industry will not be "single-shot" gene therapy treatments such as might be achieved by hematopoietic stem-cell therapy in hematologic and related diseases, but rather diseases that would require repetitive applications of the therapeutic genes. Nevertheless, studies to work out "proof of principle" to demonstrate that gene therapy is possible can often be carried out better with the rarer monogenic diseases and therefore should, and to some extent are, be supported by industry.

Other Nongenetic Diseases

Even when there are no major causative genetic components, therapy for some human diseases can be approached at the genetic level, as long as there are target genes whose modified expression can affect a disease phenotype. The most prominent of the current nongenetic disease models is potential gene therapy for AIDS, using a variety of approaches. The treatment of this and other infectious diseases might be approached by the manipulation of the relevant viral genes responsible for pathogenesis. Virus infections could theoretically be prevented by a suitable vaccination or by the expression of genetic elements such as intracellular antibodies or ribozymes that interfere with viral replication or with the expression of deleterious viral genes (Marasco et al. 1993; S.-Y. Chen et al. 1994; Yu et al. 1994; Wong-Staal 1995; J.D. Clen et al. 1996; Poeschla et al. 1996). Promising, but as yet preliminary, results are being obtained with these approaches in animal studies.

Gene therapy need not be restricted to diseases in which there are clearly identified microbial or genetic targets. As long as a therapeutic cellular or biochemical function can be supplied to defective cells or tissues from a transgene, target diseases can include disorders in which there are few, if any, relevant genetic defects. Degenerative neurological disorders represent such a candidate disease category. The neurodegenerative diseases such as Parkinson's and Alzheimer's are caused by the degeneration of specific neuron populations and by the destruction of specific pathways in the brain. The precise causes of the cellular degeneration are far from clear. However, growing knowledge of the regional specificities of mechanisms of neurotransmission and of the role of supportive neurotrophic factors are making genetic approaches theoretically feasible even for these largely nongenetic disorders. For instance, the restoration of dopamine expression or an augmented expression of the neurotrophic factors in parts of the brain have shown promise in preventing neuron cell loss after injury and in reversing some of the neurological and behavioral aberrations in animal models (Rosenberg et al. 1988;

Wolff et al. 1989; Ebstein et al. 1996; Freese 1997; Haase et al. 1997; Horellou and Mallet 1997). Similarly, some aspects of the acquisition and retention of memory have been shown to be affected by the transfer of neurotrophic factors into the rat brain (Chen and Gage 1995).

Other chronic diseases, such as rheumatoid arthritis, are also approached through genetic means. The introduction of genetically modified synovial cells expressing anti-inflammatory gene products has shown promising therapeutic effects on the development of experimental arthritis in animal models (Kang et al. 1997).

In addition to diseases, gene-based treatments of traumatic tissue injury such as the use of trophic factors for the prevention of neuron loss and for the stimulation of appropriate axon regeneration after spinal cord injury have been proposed (Tuszynski and Gage 1995; Eming et al. 1997).

Vaccination, Infectious and Neoplastic Disease

The expression of foreign genetic material for the purpose of vaccinating humans and other mammals against infectious and other diseases is under intensive study. An early approach to a vaccination for a disorder such as AIDS has taken the more or less traditional route of exposing subjects to suitably defective or inactivated retroviral vectors expressing a viral antigen (see chapter on AIDS gene therapy by Poeschla and Wong-Staal). An alternative approach involves the production of intracellular antibodies targeted to a specific viral function (see above). More recent and apparently successful vaccinations against a number of infectious diseases have taken advantage of the immunogenicity of gene products expressed after injecting the host with naked DNA in the form of a plasmid expression vector (Tang et al. 1992; Sedegah et al. 1994; Shiver et al. 1996; Schrijver et al. 1997). The simplicity of this approach, the lack of a need for prolonged gene expression for effective immunization, and the absence of potentially deleterious virus vectors and other viral gene products make this approach attractive. It seems likely that, in the future, vaccination schemes will increasingly involve the use of plasmid vectors rather than attenuated or inactivated viruses or even viral gene transfer expression vectors.

In addition, vaccination-based treatment has been extended to studies of some noninfectious diseases, particularly cancer. As described elsewhere in this volume, cancer is envisioned to result from an inability of the unaided host immune system alone to mount an effective response to immunogenic signals that are presented to the host immune system by tumors. However, cancer cells genetically modified to produce enhanced levels of appropriate cytokines or to augment antigen presentation to the immune system have been shown to be targeted and destroyed by cells of an "awakened" host immune system (Rosenberg 1991; Nabel et al. 1992; Gansbacher 1994; Dranoff and Mulligan 1995). Promising studies of immunotherapy for cancers such as malignant melanoma and renal cell carcinoma are under way in several centers.

NONDISEASE TRAITS AS TARGETS FOR GENETIC MANIPULATION

Most discussions of the goals of gene therapy have been limited so far to traits that most reasonable observers would agree is "disease." Few would argue, except from the societally unpopular theological and metaphysical positions, that we ought not undertake to reduce the frequency of disease (such as cancer, heart disease, neurological disease, etc.) and therefore relieve suffering. And yet, just as is true of other forms of medical intervention, the argument for the alleviation of suffering can readily be extended to other human traits that do not clearly represent "disease" but rather constitute various characteristics that one might wish to modify, even in "normal" people—cognition, memory, height, athletic ability, visual acuity, musical ability, skin color, and other physical or personality traits. It has often been argued that genetic manipulation for the purpose of "mere enhancement" ought not to be permitted—that it crosses an impermissible ethical boundary (Anderson 1989). These previous discussions of genetic enhancement have used the word "enhancement" pejoratively and have not taken fully into account the medical and ethical urgencies that may support such uses. Furthermore, it was possible not so long ago to take comfort in the notion that these kinds of manipulations were far off in the future and were limited by severe technical difficulties and lack of knowledge of genetic targets for such traits. It now seems likely that the technology used for the correction of genetic "disease" will be just as readily applicable to these other traits as to simple inborn errors of metabolism. The fact that the first Gene Therapy Policy Conference sponsored by the National Institutes of Health in the fall of 1997 was concerned with genetic enhancement suggests that the time is ripe for a more thorough examination of the potentials and problems associated with genetic modification for nondisease traits.

As a society, we have largely tolerated, accepted, and even at times embraced, the concepts and applications of medical intervention for nondisease traits. We have accepted cosmetic surgery and drug therapy of mild depression and even of moderate obesity. Increasingly, we tend to accept similar approaches to conditions that are at the very imprecise borderline of disease and normality or, more correctly, lie along the spectrum between normality and pathology. Often there is no clear threshold in the bell-shaped distribution curve of a given characteristic beyond which a person might be declared diseased. The exact blood pressure level beyond which someone is considered hypertensive is an example. The recent approval by the FDA of the first drug for male impotence is sure to open the door for extended use of this and similar classes of agents not only for those suffering from the vascular diseases that lead to impotence, but also in the easily imagined applications that could benefit many "normals." The effect is thereby to blur the already vague distinction between pathology and shades of normality. Pharmaceutical and biotechnology firms are investing huge resources in the development of drugs to affect the cognitive, mood, and memory deficits of Alzheimer's sufferers. They do so with the clear understanding that there will be opportunities, some justifiable and some not, for the use of such agents to enhance those mental functions in normal people in what could even be abusive, "recreational settings." We already see full-page advertisements for the general public calling attention to prescription drugs for Alzheimer's disease.

A great deal of work suggests that most human cognitive and personality traits are under significant genetic influences (Brunner et al. 1993; Plomin et al. 1994; Svensson et al. 1994; Benjamin et al. 1996; Ebstein et al. 1996; Lesch et al. 1996; Lijam et al. 1997) involving the interaction of multiple genes with the environment. The "normal" and variant forms of the involved genes and their interactions ultimately will come to be identified. It requires little imagination to predict that these kinds of changes in complex interactive functions will raise novel problems. Many therapeutic dilemmas will arise once these properties are elucidated and become amenable to genetic manipulation.

It is probably unrealistic to think that genetic tools, once available, will not be used outside of purely therapeutic settings. There will be temptations and even justified pressures to use genetic agents, like pharmacological agents, in quasi-therapeutic settings. For instance, genetic methods to enhance hair growth in people would be an obviously therapeutic application in patients undergoing cancer chemotherapy or in patients with severe alopecia. But just as obviously, such techniques would be sought by men and women "suffering" or inconvenienced by male-pattern-baldness, hair loss from aging, and other nondisease settings. The problem is not that therapy melts imperceptibly and inevitably into justifiable quasi-therapy, but rather that such manipulations may continue to move into frivolous applications. Much genetic manipulation at the level of somatic gene therapy is not really very different from most forms of medical intervention, except for those instances where gene insertion into a given cell type would be permanent and not reversible.

SOMATIC VERSUS GERM-LINE THERAPY

From the onset, almost all work and discussion of gene therapy have centered around somatic cells as target cells. It has often been stated that there are few, if any, medical therapeutic indications for germ-cell modification. It has been assumed, tacitly or overtly, that stable and safe gene transfer into germ cells or into embryos for the purpose of modifying the germ cells would be both technically difficult and ethically undesirable. As in the case of genetic enhancement, the concepts and technology have now evolved sufficiently to require a reexamination of the medical, technical, and ethical issues (Walters 1986; Fowler et al. 1989; Lappe 1991; Brinster and Avarbock 1994; Walters and Palmer 1997). A prime force in encouraging this reevaluation was the report in 1997 of the birth of the sheep clone Dolly (Campbell et al. 1996; Kass 1997; Mirsky and Rennie 1997). Even though this apparent solution to what seemed to be formidable technical problems standing in the way of mammalian cloning is still unconfirmed, the technique requires confirmation in sheep but has clearly given new impetus to the issues of germ-cell genetic modification, and a number of national organizations are establishing panels to study the issue.

Although most scientists working on gene therapy are extremely hesitant to embark on studies of human germinal gene therapy, public opinion polls suggest that a majority of the public would accept genetic research to correct abnormal genes in parents so as to prevent fatal diseases in their offspring (Walters and Palmer 1997). Medical indications for germinal gene therapy would be the normalization of mutant gene functions early during embryonic development, before organ differentiation and early organ damage. Somatic gene therapy of differentiated respiratory cells, as is currently practiced, does not affect the normal function of affected tissues in all affected organs. Several diseases of the central nervous system, such as Lesch-Nyhan syndrome, cannot at this time be approached easily by somatic gene therapy. They would require very early interventions because brain damage occurs during the first 3 months of fetal life. Similarly, all offspring of the very rare matings between two homozygotes for an identical recessive disease will be affected, an outcome that could be prevented by germinal gene therapy. For the much more frequent carrier x carrier matings in common autosomal recessive disease, prenatal diagnosis with selective abortion or preimplantation diagnosis with implantation of normal embryos would seem a more feasible alternative, an approach that is, however, not universally acceptable nor universally available to all who might benefit from it. Of course, all therapies, from antibiotics to germinal gene therapy, are unfortunately likely to remain more accessible to the affluent than to the poor.

Germ-line therapy would have the further advantage that the children of those effectively treated would no longer carry the mutant genes. The frequency of genetic disease in the long term would be reduced, as would be the need for repetitive somatic gene therapy in the descendants of patients who had previously been successfully treated with this approach.

Despite reasons to consider germinal gene therapy seriously, one of us (AGM) wrote in 1996 that, "the introduction of germ-line manipulation would be a revolutionary step for the human species because it alters genes directly and permanently, thereby affecting future generations in human evolution. Much more needs to be known about human genetics before such a step can be undertaken. Most importantly, before ever considering germ-line manipulations, human society must determine how and by whom decisions regarding germinal therapy are to be made" (Vogel and Motulsky 1996). These discussions have now started.

ACKNOWLEDGMENT

The authors would like to thank Dr. Mark Kay for discussions regarding this chapter of the book.

REFERENCES

Ali R.R., Reuchel M.B., Thrasher A.J., Levinsky R.J., Kinnon C., Kanuga N., Hunt D.M., and Bhattacharya S.S.1996. Gene transfer into the mouse retina mediated by an adeno-associated viral vector. *Hum. Mol. Gen.* **5:** 591–594.

Alton E. and Geddes D. 1994. A mixed message for cystic fibrosis gene therapy. *Nat. Genet.* **8:** 8–9.

Anderson M.P., Rich D.P., Gregory R.J., Smith A.E., and Welsh M.J. 1991. Generation of cAMP-activated chloride currents by expression of CFTR. *Science* **251:** 679–682.

Anderson W.F. 1989. Human gene therapy: Why draw a line. *J. Med. Philos.* **14:** 681–693.

———1992. Human gene therapy. *Science* **256:** 808–813.

Bank A., Donovan-Peluso M., LaFlamme S., Rund D., and Lerner N. 1988. Approaches to gene therapy for beta-thalassemia. *Birth Defects* **23:** 339–346.

Belmont J.W., Henkel-Tigges J., Wager-Smith K., Chang S.M., and Caskey C.T. 1986. Towards gene therapy for adenosine deaminase deficiency. *Ann. Clin. Res.* **18:** 322–326.

Benjamin J., Li L., Patterson C., Greenberg B.D., Murphy D.L, and Hamer D.H. 1996. Population and familial association between the D4 dopamine receptor gene and measures of novelty seeking. *Nat. Genet.* **12:** 81–84.

Beutler E. 1992. Gaucher disease: New molecular approaches to diagnosis and treatment. *Science* **256:** 794–799.

Beutler E. and Grabowski G.A. 1995. Gaucher disease. *The metabolic and molecular bases of inherited disease* (ed. C.R. Scriver et al.), pp. 2641–2670. McGraw Hill, New York.

Beuzard Y. 1996. Towards gene therapy of hemoglobinopathies. *Semin. Hematol.* **33:** 43–52.

Blaese R.M., Culver K.W., Miller A.D., Carter C.S., Fleischer T., Clerici M., Shearer G., Chang L., Chiang Y., Tolstoshev P., Greenblatt J.J., Rosenberg S.A., Klein H., Berger M., Mullen C.A., Ramsey W.J., Muul L., Morgan R.A., and Anderson W.F. 1995. T lymphopcyte-directed gene therapy for ADA-SCID: Initial trial results after 4 years. *Science* **270:** 475–480.

Bordignon C., Notarangelo L.D., Nobili N., Ferrari G., Casorati G., Panina P., Mazzolari E., Maggioni D., Rossi C., Servida P., Ugazio A.G., and Mavilo F. 1995. Gene therapy in peripheral blood lymphocytes and bone marrow for ADA-immunodeficient patients. *Science* **270:** 470–475.

Bothwell T.H., Charlton R.W., and Motulsky A.G. 1995. Hemochromatosis. *The metabolic and molecular bases of inherited disease* (ed. C.R. Scriver et al.) pp. 2237–2269. McGraw Hill, New York.

Brady R.O. and Barton N.W. 1996. Enzyme replacement and gene therapy for Gaucher's disease. *Lipids* (Suppl.) **31:** S137–S139.

Brinster R.L. and Avarbock M.R. 1994. Germline transmission of donor haplotype following spermatogonial transplantation. *Proc. Natl. Acad. Sci.* **91:** 11303–11307.

Brunner H.G., Nelen M., Breakefield X.O., Ropers H.H., and van Oost B.A. 1993. Abnormal behavior associated with a point mutation in the structural gene form monamine oxidase. *Science* **262:** 578–580.

Brusilow S.W. and Horwich A.L. 1995. Urea cycle enzymes. *The metabolic and molecular bases of inherited disease* pp. 1187–1232 (ed. C.R. Scriver et al.) McGraw Hill, New York.

Campbell K.H., McWhir J., Ritchie W.A., and Wilmut I. 1996. Sheep cloned by nuclear transfer from a cultured cell line. *Nature* **380:** 64–66.

Chen G.X., Koyama K., Yuan X., Lee Y., Zhou Y.T., O'Doherty R., Newgard C.B., and Unger R.H. 1996. Disappearance of body fat in normal rats induced by adenovirus-mediated leptin gene therapy. *Proc. Natl. Acad. Sci.* **93:** 14795–14799.

Chen J.D., Yang Q.C., Yang A.G., Marasco W.A., and Chen S.Y. 1996. Intra- and extracellular immunization against HIV-1 infection with lymphocytes transduced with an AAV vector expressing a human anti-gp120 antibody. *Hum. Gene Ther.* **7:** 1515–1525.

Chen K.S. and Gage F.H. 1995. Somatic gene transfer of NGF to the aged brain: Behavioral and morphological amelioration. *J. Neurosci.* **15:** 2819–2825.

Chen S.-Y., Bagley J., and Marasco W.A. 1994. Intracellular antibodies as a new class of therapeutic molecules for gene therapy. *Hum. Gene Ther.* **5:** 595–601.

Childs B. 1970. Sir Archibald Garrod's conception of chemical individuality: A modern appreciation. *N. Engl. J. Med.* **282:** 71 77.

Choudary P.V., Tsuji S., Martin B.M., Guild B.C., Mulligan R.C., Murray G.J., Barranger J.A., and Ginns E.I. 1986. The molecular biology of Gaucher disease and the potential for gene therapy. *Cold Spring Harbor Symp. Quant. Biol.* **51:** 1047–1052.

Chowdhury J.R., Grossman M., Gupta S., Chowdhury N.R., Baker J.R. Jr., and Wilson J.M. 1991. Long-term improvement of hypercholesterolemia after ex vivo gene therapy in LDLR-deficient rabbits. *Science* **254:** 1802–1805.

Danks D.M. 1995. Disorders of copper transport. *The metabolic and molecular bases of inherited disease* (ed. C.R. Scriver et al.), pp. 2211–2235. McGraw Hill, New York.

Dranoff G. and Mulligan R.C. 1995. Gene transfer as cancer therapy. *Adv. Immunol.* **58:** 417–454.

Ebstein R.P., Novik O., Umansky R., Priel B., Osher Y., Blaine D., Bennett E.R., Nemanov L., Katz M., and Belmaker R.H. 1996. Dopamine D4 receptor (D4DR) exon III polymorphism associated with the human personality trait of novelty seeking. *Nat. Genet.* **12:** 78–80.

Eming S.A., Morgan J.R., and Berger A. 1997. Gene therapy for tissue repair: Approaches and prospects. *Br. J. Plast. Surg.* **50:** 491–500.

Fenton W.A. and Rosenberg L.E. 1995. Inherited disorders of cobalamin transport and metabolism. In *The metabolic and molecular bases of inherited disease* (ed. C.R. Scriver et al.), pp. 3129–3149, McGraw Hill. New York.

Ferrari G., Angelis C.-D., Coletta M., Paolucci E., Stornaiuolo A., Cossu G., and Mavilio F. 1998. Muscle regeneration by bone marrow-derived myogenic progenitors. *Science* **279:** 1528–1530.

Fink J.K., Correll P.H., Perry L.K., Brady R.O., and Karlsson S. 1990. Correction of glucocerebrosidase deficiency after retroviral-mediated gene transfer into hematopoietic progenitor cells from patients with Gaucher disease. *Proc. Natl. Acad. Sci.* **87:** 2334–2338.

Finkel T. and Epstein S.E. 1995. Gene therapy for vascular disease. *FASEB J.* **9:** 843–851.

Fletcher J.C. 1983a. Moral problems and ethical issues in prospective human gene therapy. *Va. Law Rev.* **69:** 515–546.

———1983b. Ethics and trends in applied human genetics. *Birth Defects Orig. Artic. Ser.* **19:** 143–158.

Fowler G., Juengst E.T., and Zimmerman B.K. 1989. Germ-line gene therapy and the clinical ethos of medical genetics. *Theor. Med.* **10:** 151–165.

Freese A. 1997. Special issue—Gene therapy for Parkinson's disease: Rationale, prospects, and limitations. *Exp. Neurol.* **144:** 1.

Friedmann T. 1987. Model studies toward human gene therapy. *New approaches to genetic disease* (ed. T. Sasazuki), pp. 177–186. Academic Press, New York.

———1992. A brief history of gene therapy. *Nat. Genet.* **2:** 93–98.

———1997. Overcoming the obstacles to gene therapy. *Sci. Am.* **276:** 96–101.

Fucharoen S., Siritanaratkul N., Winichagoon P., Chowthaworn J., Siriboon W., Muangsup W., Chaicharoen S., Poolsup N., Chindavijak B., Pootrakul P., Piankijagum A., Schechter A.N., and Rodgers G.P. 1996. Hydroxyurea increases hemoglobin F levels and improves the effectiveness of erythropoiesis in beta-thalassemia/hemoglobin E disease. *Blood* **87:** 887–892.

Gansbacher B. 1994. Clinical application of immunostimulatory gene transfer. *Eur. J. Cancer* **30A:** 1187–1190.

Garrod A.E. 1908. The Croonian lectures on inborn errors of metabolism. *Lancet* **2:** 1–7, 73–79, 142–148, 214–220.

———1923. *Inborn errors of metabolism.* Oxford University Press, Oxford, United Kingdom.

Goldstein J.L., Hobbs H.H., and Brown M.S. 1995. Familial hypercholesterolemia. *The metabolic and molecular bases of inherited disease* (ed. C.R. Scriver et al.), pp. 1981–2030. McGraw Hill, New York.

Grossman M. and Wilson J.M. 1992. Frontiers in gene therapy: LDL receptor replacement for hypercholesterolemia. *J. Lab. Clin. Med.* **119:** 457–460.

Grossman M., Rader D.J., Muller D.W.M., Kolansky D.M., Kozarsky K., Clark B.J., III, Stein E.A., Lupien P.J., Brewer H.B., Jr., Raper S.E, and Wilson J.M. 1995. A pilot study of ex vivo gene therapy for homozygous familial hypercholesterolemia. *Nat. Med.* **1:** 1148–1154.

Haase G., Kennel P., Pettmann B., Vigne E., Akli S., Revah F., Schmalbruch H., and Kahn A. 1997. Gene therapy of murine motor neuron disease using adenoviral vectors for neurotrophic factors. *Nat. Med.* **3:** 429–436.

Hauser M.A. and Chamberlain J.S. 1996. Progress towards gene therapy for Duchenne muscular dystrophy. *J. Endocrinol.* **149:** 373–378.

Hayes A., Costa T., Scriver C.R., and Childs B. 1985. The effect of Mendelian disease on human health II: Response to treatment. *Am. J. Med. Genet.* **21:** 243–255.

Hershfield M.S. and Mitchell B.S. 1995. Immunodeficiency diseases caused by adenosine deaminase deficiency and purine nucleoside phosphorylase deficiency. *The metabolic and molecular bases of inherited disease* (ed. C.R. Scriver et al.), pp. 1725–1768. McGraw Hill, New York.

Hershfield M.S., Chaffee S., and Sorensen R.U. 1993. Enzyme replacement therapy with polyethylene glycol-adenosine deaminase in adenosine deaminase deficiency: Overview and case reports of three patients, including two now receiving gene therapy. *Pediatr. Res.* (Suppl.) **33:** S42–S48.

Horellou P. and Mallet J. 1997. Gene therapy for Parkinson's disease. *Mol. Neurobiol.* **15:** 241–256.

Huang H.-J. S., Yee J.-K., Shew J.-Y., Chen P.-L., Bookstein R., Friedmann T., Lee E.Y.-H.P., and Lee W.-H. 1988. Suppression of the neoplastic phenotype by replacement of the RB gene in human cancer cells. *Science* **242:** 1563–1566.

Isner J.M., Walsh K., Rosenfield K., Schainfeld R., Asahara T., Hogan K., and Pieczek A. 1996. Arterial gene therapy for restenosis. *Hum. Gene Ther.* **7:** 989–1011.

Isner J.M., Walsh K., Symes J., Pieczek A., Takeshita S., Lowry J., Rossow S., Rosenfield K., Weir L., Brogi E., and Schainfeld R. 1995. Arterial gene therapy for therapeutic angiogenesis in patients with peripheral artery disease. *Circulation* **91:** 2687–2692.

Kang R., Ghivizzani S.C., Herndon J.H., Robbins P.D., and Evans C.H. 1997. Gene therapy for arthritis: Principles and clinical practice. *Biochem. Soc. Trans.* **25:** 533–537.

Kantoff P.W., Kohn D.B., Mitsuya H., Armentano D., Sieberg M., Zwiebel J.A., Eglitis M.A., McLachlin J.R., Wiginton D.A., Hutton J.J., and Anderson W.F. 1986. Correction of adenosine deaminase deficiency in cultured human T and B cells by retrovirus-mediated gene transfer. *Proc. Natl. Acad. Sci.* **83:** 6563–6567.

Karpati G. and Acsadi G. 1994. The principles of gene therapy in Duchenne muscular dystrophy. *Clin. Invest. Med.* **17:** 499–509.

Kass L.R. 1997. The wisdom of repugnance. *The New Republic* **June 2:** 17–26.

Knowles M.R., Hohneker K.W., Zhou Z., Olsen J.C., Noah T.L., Hu P.-C., Leigh M.W., Engelhardt J.F., Edwards L.J., Jones K.R., Grossman M., Wilson J.M., Johnson L.G., and Boucher R.C. 1995. A controlled study of adenoviral-mediated gene transfer in the nasal epithelium of patients with cystic fibrosis. *N. Engl. J. Med.* **333:** 823–831.

Krauss R.D., Bubien J.K., Drumm M.L., Zheng T., Peiper S.C., Collins F.S., Kirk K.L., Frizzell R.A., and Rado T.A. 1992. Transfection of wild-type CFTR into cystic fibrosis lymphocytes restores chloride conductance at G1 of the cell cycle. *EMBO J.* **11:** 875–883.

Lappe M. 1991. Ethical issues in manipulating the human germ line. *J. Med. Philos.* **16:** 621–639.

Lesch K.P., Bengel D., Heils A., Sabol S.Z., Greenberg B.D., Petri S., Benjamin J., Muller C.R., Hamer D.H., and Murphy D.L. 1996. Association of anxiety-related traits with a polymorphism in the serotonin transporter gene regulatory region. *Science* **274:** 1527–1531.

Lijam N., Paylor R., McDonald M.P., Crawley J.N., Deng C.-X., Herrup K., Stevens K.E., Maccaferri G., McBain C.J., Sussman D.J., and Wynshaw-Boris A. 1997. Social interaction and sensorimotor gating abnormalities in mice lacking Dvl1. *Cell* **90:** 895–905.

Marasco W.A., Haseltine W.A., and Chen S.-Y. 1993. Design, intracellular expression and activity of a human anti-human immunodeficiency virus type 1 gp120 single chain antibody. *Proc. Natl. Acad. Sci.* **90:** 7889–7893.

McIvor R.S., Johnson J.M., Miller A.D., Pitts S., Williams S.R., Valerio D., Martin D.W., Jr., and Verma I.M. 1987. Human purine nucleoside phosphorylase and adenosine deaminase: Gene transfer into cultured cells and murine hematopoietic stem cells by using recombinant amphotropic retroviruses. *Mol. Cell. Biol.* **7:** 838–846.

Melillo G., Scoccianti M., Kovesdi I., Safi J., Jr., Riccioni T., and Capogrossi M.C. 1997. Gene therapy for collateral vessel development. *Cardiovasc. Res.* **35:** 480–489.

Miller A.D., Jolly D.J., Friedmann T., and Verma I.M. 1983. A transmissible retrovirus expressing human hypoxanthine phosphoribosyltransferase (HPRT): Gene transfer into cells obtained from humans deficient in HPRT. *Proc. Natl. Acad. Sci.* **80:** 4709–4713.

Miller A.D., Eckner R.J., Jolly D.J., Friedmann T., and Verma I.M. 1984. Expression of a retrovirus encoding human HPRT in mice. *Science* **225:** 630–632.

Mirsky S. and Rennie J. 1997. What cloning means for gene therapy. *Sci. Am.* **276:** 122–123.

Miyanohara A., Sharkey M.F., Witztum J.L, Steinberg D., and Friedmann T. 1988. Efficient expression of retroviral vector-transduced human low density lipoprotein (LDL) receptor in LDL receptor-deficient rabbit fibroblasts in vitro. *Proc. Natl. Acad. Sci.* **85:** 6538–6542.

Miyoshi H., Takahashi M., Gage F.H., and Verma I.M. 1997. Stable and efficient gene transfer into the retina using an HIV-based lentiviral vector. *Proc. Natl. Acad. Sci.* **94:** 10319–10323.

Motulsky A.G. 1974. Brave new world. *Science* **185:** 653–663.

Motulsky A.G. and Brunzell J.D. 1992. The genetics of coronary atherosclerosis. In *Genetic basis of common diseases* (ed. R.A .King et al.), chap. 1, pp. 3–18. Oxford University Press, New York.

Murphy J.E., Zhou S.Z., Giese K., Williams L.T., Escobedo J.A., and Dwarki V.J. 1997. Long-term correction of obesity and diabetes in genetically obese mice by a single intramuscular injection of recombinant adeno-associated virus encoding mouse leptin. *Proc. Natl. Acad. Sci.* **94:** 13921–13926.

Muzzin P., Eisensmith R.C., Copeland K.C., and Woo S.L.C. 1996. Correction of obesity and diabetes in genetically obese mice by leptin gene therapy. *Proc. Natl. Acad. Sci.* **93:** 14804–14808.

Nabel E.G. 1995. Gene therapy for vascular diseases. *Atherosclerosis* (Suppl.) **118:** S51–S56.

Nabel E.G., Plautz G., and Nabel G.J. 1992. Transduction of a foreign histocompatibility gene into the arterial wall induces vasculitis. *Proc. Natl. Acad. Sci.* **89:** 5157–5162.

Oren M. 1992. p53—The ultimate tumor suppressor gene? *FASEB J.* **6:** 3169–3176.

Orkin S.H. and Williams D.A. 1988. Gene therapy of somatic cells: Status and prospects. *Prog. Med. Genet.* **7:** 130–142.

Plomin R., Owen M.J., and McGuggin P. 1994. The genetic basis of complex human behaviors. *Science* **264:** 1733–1739.

Poeschla, E., Corbeau P., and Wong-Staal F. 1996. Development of HIV vectors for anti-HIV gene therapy. *Proc. Natl. Acad. Sci.* **93:** 11395–11399.

Porteous D.J. and Dorin J.R. 1993. Gene therapy for cystic fibrosis—Where and when? *Hum. Mol. Genet.* **2:** 211–212.

Rifkin J. 1985. *Declaration of a heretic.* Routledge & Kegan Paul, Boston.

Rosenberg M.B., Friedmann T., Robertson R.C., Tuszynski M., Wolff J.A., Breakefield X.O., and Gage F.H. 1988. Grafting genetically modified cells to the damaged brain: Restorative effects of NGF expression. *Science* **242:** 1575–1578.

Rosenberg S.A. 1991. Immunotherapy and gene therapy of cancer. *Cancer Res.* **51:** 5074s–5079s.

Roth J.A. 1996. Gene replacement strategies for cancer. *Isr. J. Med. Sci.* **32:** 89–94.

Roth J.A., Nguyen D., Lawrence D.D., Kemp B.L., Carrasco C.H., Ferson D.Z., Hong W.K., Komaki R., Lee J.J., Nesbitt J.C., Pisters K.M.W., Putnam J.B., Schea R., Shin D.M., Walsh G.L., Dolormente M.M., Han C.I., Martin F.D., Yen N., Xu K., Stephens L.C., McDonnell T.J., Mukhopadhyay T., and Cai D. 1996. Retrovirus-mediated wild-type p53 gene transfer to tumors of patients with lung cancer. *Nat. Med.* **2:** 985–991.

Schechter A.N. and Rodgers G.P. 1996. Hydroxyurea in sickle cell disease. *N. Engl. J. Med.* **334:** 333.

Schrijver R.S., Langedijk J.P.M., Keil G.M., Middel W.G.J., Maris-Veldhuis M., Van Oirschot J.T., and Rijsewijk F.A.M. 1997. Immunization of cattle with a BHV1 vector vaccine or a DNA vaccine both coding for the G protein of BRSV. *Vaccine* **15:** 1908–1916.

Scriver C.R., Beaudet A.L., Sly W.S., and Valle D., eds. 1995a. *The Metabolic and molecular bases of inherited disease.* McGraw Hill, New York.

Scriver C.R., Kaufman S., Eisensmith R.C., and Woo S.L.C. 1995b. The hyperpenylalaninemias. In *The metabolic and molecular bases of inherited disease* (ed. C.R. Scriver et al.), pp. 1015–1075. McGraw Hill, New York.

Sedegah M., Hedstrom R., Hobart P., and Hoffman S. 1994. Protection against malaria by immunization with plasmid DNA encoding circumsporozite protein. *Proc. Natl. Acad. Sci.* **91:** 9866–9870.

Segal S. and Berry G.T. 1995. Disorders of galactose metabolism. In *The metabolic and molecular bases of inherited disease* (ed. C.R. Scriver et al.), pp. 967–1000. McGraw Hill, New York.

Segal S. and Thier S.O. 1995. Cystinuria. In *The metabolic and molecuilar bases of inherited disease* (ed. C.R. Scriver et al.), pp. 3581–3601. McGraw Hill, New York.

Shiver J.W., Ulmer J.B., Donnelly J.J., and Liu M.A. 1996. Humoral and cellular immunities elicited by DNA vaccines: Application to the human immunodeficiency virus and influenza. *Adv. Drug Deliv. Rev.* **21:** 19–31.

Sorge J., Kuhl W., West C., and Beutler E. 1986. Gaucher disease: Retrovirus-mediated correction of the enzymatic defect in cultured cells. *Cold Spring Harbor Symp. Quant. Biol.* **51:** 1041–1046.

Stanbridge E.J. 1989. The reemergence of tumor suppression. *Cancer Cells* **1:** 31–33.

Stanley C., Rosenberg M.B., and Friedmann T. 1991. Gene transfer into rat airway epithelial cells using retroviral vectors. *Somatic Cell Mol. Genet.* **17:** 185–190.

Svensson K., Carlsson A., Huff R.M., Kling-Petersen T., and Waters N. 1994. Behavioral and neurochemical data suggest functional differences between dopamine D2 and D3 receptors. *Eur. J. Pharmacol.* **263:** 235–243.

Szybalski W. 1992. Use of the HPRT gene and the HAT selection technique in DNA-mediated transformation of mammalian cells: First steps toward developing hybridoma techniques and gene therapy. *BioEssays* **14:** 495–500.

Tang D.-C., DeVit M., and Johnston S.A. 1992. Genetic immunization is a simple method for eliciting an immune response. *Nature* **356:** 152–154.

Treacy E., Childs B., and Scriver C.R. 1995. Response to treatment in hereditary metabolic disease: 1993 survey and 10 year comparison. *Am. J. Hum. Genet.* **56:** 359–367.

Tuszynski M.H. and Gage F.H. 1995. Maintaining the neuronal phenotype after injury in the adult CNS: Neurotrophic factors, axonal growth substrates, and gene therapy. *Mol. Neurobiol.* **10:** 151–166.

Vassalli G. and Dichek D.A. 1997. Gene therapy for arterial thrombosis. *Cardiovasc. Res.* **35:** 459–469.

Vogel F. and Motulsky A.G. 1996. *Human genetics: Problems and approaches,* 3rd edition, p. 17. Springer-Verlag, Berlin.

Walters L. 1986. The ethics of human gene therapy. *Nature* **320:** 225–227.

Walters L. and Palmer J.G. 1997. *The ethics of human gene therapy.* Oxford University Press, Oxford, United Kingdom.

Weatherall D.J. 1995. Scope and limitations of gene therapy. *Br. Med. Bull.* **51:** 1–11.

Wolff J.A., Fisher L.J., Xu L., Jinnah H.A., Langlais P.J., Iuovone P.M., O'Malley K.L., Rosenberg M.B., Shimohama S., Friedmann T., and Gage F.H. 1989. Grafting fibroblasts genetically modified to produce L-dopa in a rat model of Parkinson disease. *Proc. Natl. Acad. Sci.* **86:** 9011–9014.

Wong-Staal F. 1995. Ribozyme gene therapy for HIV infection—Intracellular immunization of lymphocytes and CD34+ cells with an anti-HIV-1 ribozyme gene. *Adv. Drug Delivery Rev.* **17:** 363–368.

Xu L., Stahl S.K., Dave H.P.G., Schiffmann R., Correll P.H., Kessler S., and Karlsson S. 1994. Correction of the enzyme deficiency in hematopoietic cells of Gaucher patients using a clinically acceptable retroviral supernatant transduction protocol. *Exp. Hematol.* **22:** 223–230.

Yu M., Poeschla E., and Wong-Staal F. 1994. Progress towards gene therapy for HIV infection. *Gene Ther.* **1:** 13–26.

Zeng Y.T., Huang S.Z., Ren Z.R., Lu Z.H., Zeng F.Y., Schechter A.N., and Rodgers G.P. 1995. Hydroxyurea therapy in beta-thalassemia intermedia: Improvement in haematological parameters due to enhanced beta-globin synthesis. *Br. J. Haematol.* **90:** 557–563.

WWW RESOURCE

http://www.mih.gov/news/panelrep.html Orkin S.H. and Molulsky A.G. 1995. Report and recommendations of the panel to assess the NIH investment in research on gene therapy. Submitted to the meeting of the Advisory Committee to the NIH Director, December 7, 1995, pp. 1–49.

15

The Hematopoietic System as a Target for Gene Therapy

Brian P. Sorrentino and Arthur W. Nienhuis

St. Jude Children's Research Hospital
Memphis, Tennessee 38101

The ability to transfer a gene into repopulating stem cells and to achieve lineage-specific expression in differentiating hematopoietic cells would create many therapeutic opportunities. The hemoglobin disorders, severe β-thalassemia and sickle cell anemia, because of their frequency world-wide and their severe morbidity and mortality, were early targets for human gene therapy. Molecular cloning of the globin genes in the late 1970s (Efstratiadis et al. 1980; Lawn et al. 1980) fueled interest in these diseases as targets for genetic intervention. Demonstration that DNA fragments could be incorporated into the genome of eukaryotic cells by calcium-phosphate-mediated DNA uptake (Pellicer et al. 1978) provided the methodology for attempted gene transfer into stem cells. Reported success in a murine model (Cline et al. 1980) prompted an attempt to treat patients with severe β-thalassemia by introduction of a globin gene into hematopoietic cells using calcium-phosphate-mediated transfection. In retrospect, this attempt carried little, if any, risk to the patients, but it was widely discredited because the very low efficiency of transduction and gene integration achieved by this methodology rendered success very unlikely. Furthermore, the investigators failed to obtain prior approval from the appropriate regulatory bodies before undertaking clinical trials (Anderson and Fletcher 1980).

The field of human gene therapy has evolved substantially from this uncertain beginning. Studies in the murine model provided a sound experimental basis for subsequent work in larger animal models and ulti-mately for the initiation of human clinical trials. Many additional diseases that are potentially amenable to gene-therapy intervention have been characterized. The first portion of this chapter contains a brief review of this phase in the development of stem-cell-targeted gene therapy, which lasted about a decade, extending from approximately 1983 to 1993. During this decade, substantial progress was made, but a number of barriers to human-stem-cell-targeted gene transfer were encountered. An account is provided here of more recent work that has focused on overcoming these barriers by achieving a better understanding of the biology of hematopoietic stem cells and by developing additional vector systems. Progress in a number of disease-specific applications of stem-cell-targeted gene transfer is then reviewed.

THE FIRST DECADE

Retroviral Vectors and Gene Transfer into Murine Hematopoietic Cells

Needed for effective gene therapy for hematopoietic diseases is the ability to insert genetic information into stem cells. Stem cells are defined by their capacity to repopulate hematopoietic tissues following intravenous injection into recipients who have received a myleoablative insult, either lethal irradiation or chemotherapy. Stem cells are rare, representing about 1 in 10^4 to 10^5 of the bone-marrow population, and generally are quiescent, although capable of self-renewal divisions (Fig. 1). Effective gene therapy requires integration of the transgene into this self-renewing stem-cell population. The stem-cell pool sustains hematopoiesis by differentiation into multilineage and then single-lineage progenitors that are defined by their ability to form colonies of maturing hematopoietic cells in semisolid culture medium. Maturing precursors constitute the majority of cells in the bone marrow. The minimum marrow transit time from stem cells to mature neutrophils in humans during bone-marrow regeneration is roughly 21–28 days, although it may take many weeks for full lymphohematopoietic reconstitution.

Retroviruses have a unique capacity to convert their single-stranded RNA genome into a double-stranded DNA molecule, which then is integrated through the action of viral proteins, particularly integrase, into the host-cell chromosome (Varmus 1982). Productive infection of target cells results in the expression of viral genes and the formation of new infectious viral particles, thereby completing the retrovirus life cycle. Acutely transforming retroviruses were found to have incorporated a cellular proto-oncogene, which, in mutated form, acts as an oncogene. RNA tumor viruses thus represent naturally arising vectors for the transfer of oncogenes, and out of this was born the concept that a defective retroviral genome could serve as a vector for the transfer of useful genetic information (Tabin et al. 1982).

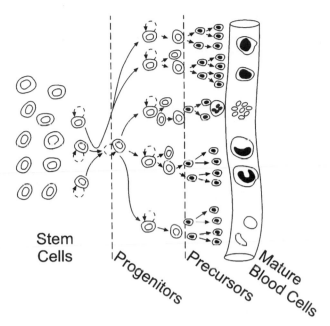

Stem
Cells

Progenitors   *Precursors*   *Mature Blood Cells*

Figure 1 Production of hematopoietic cells. Long-lived, but multipotential stem cells that are usually quiescent but capable of self-renewal are required to sustain blood cell formation. The immediate derivatives of stem cells are multilineage progenitors that retain limited self-renewal potential. Progressive differentiation results in formation of single lineage progenitors that give rise to maturing precursor cells of the individual hematopoietic lineages. The vast majority of cells in the bone marrow and other hematopoietic organs are maturing precursors; progenitors represent about 1% and stem cells about 0.01% of the total hematopoietic cell population.

A growing understanding of the organization of the genome of murine leukemia viruses (MLVs) suggested strategies for deriving vector particles free of replication-competent retrovirus that were potentially useful for gene-therapy applications. Deletion of the sequences between the 5′ long terminal repeat (LTR) and the start of the coding sequences for *gag* and *pol* genes eliminated the packaging signal that facilitates incorporation of the RNA genome into viral particles (Mann et al. 1983; Watanabe and Temin 1983). Packaging cell lines were derived by the integration of such defective genomes into murine fibroblasts by DNA transfection (Fig. 2). These lines synthesize the viral proteins needed to package RNA transcripts from an independently integrated vector genome into infectious particles.

Subsequent routine use of split function packaging cell lines in which the *gag* and *pol* genes were encoded on a separate transcriptional unit from that which encoded the envelope gene reduced the risk of generation of repli-

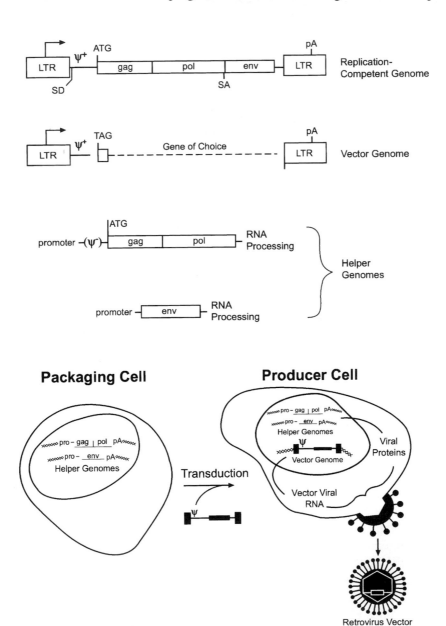

Figure 2 (See facing page for legend.)

cation-competent retroviruses by homologous recombination within vector-producing clones (Watanabe and Temin 1983; Markowitz et al. 1988; Danos and Mulligan 1988). Second- and third-generation vector genomes contained an extended packaging signal that improved vector production (Bender et al. 1987; Miller and Rossman 1989). Despite the extended overlap with the packaging genome, the risk of emergence of replication-competent retrovirus by homologous recombination was reduced by the introduction of mutations into the vector genome (Miller 1990) and by the use of split-function packaging lines.

Retroviral vectors that are based on the Moloney virus and other oncogenic murine viruses require cell division for genome integration. Particle uptake is initiated by the interaction of the viral envelope protein with cell surface molecules that act as a receptor. Different classes of viruses use different proteins for receptors. For example, ecotropic viruses that infect murine and other rodent cells utilize a cationic amino acid transporter, whereas amphotropic viruses, which have a broader host range that includes human cells, utilize a sodium-dependent phosphate transporter (Miller 1996). Movement of the viral genome with associated matrix proteins and enzymatic activities into the nucleus requires dissolution of the nuclear membrane during mitosis (Roe et al. 1993; Lewis and Emerman 1994). Thus, susceptibility of different cell populations to

Figure 2 Production of MLV vector particles. The MLV genome encodes three sets of proteins, matrix proteins (Gag), various enzymes including reverse transcriptase (Pol), and the envelope proteins (Env). The coding sequences for these proteins are deleted from the vector genome, leaving the long terminal repeats (LTRs) that contain the promoter, RNA processing, genome replication and integration functions, and the signals required for genome encapsidation that began at about the boundary of the left-hand LTR and extend into the gag coding sequences. Packaging lines are created by integration of genetic units that encode the viral proteins under the control of retroviral or heterologous transcriptional control elements. The potential for generation of replication-competent retrovirus via recombination with the vector genome in subsequently derived producer cells is reduced by separating the genes encoding viral proteins on two genetic units to create "split packaging" lines. Introduction of the vector genome into packaging cells by retroviral transduction results in derivation of producer clones with an unrearranged vector genome and a greater capacity for vector-particle production. Typically, ecotropic or amphotropic viral particles are generated in a transient expression system and introduced via transduction into amphotropic or ecotropic packaging cells, respectively, which are then screened to identify clones with an intact vector genome and high titer.

retrovirus-mediated gene transfer depends on the level of receptor expression and an active state of cellular proliferation.

A major step in the development of stem-cell-targeted gene transfer was the report in 1984 of the use of retroviral vectors to transduce progenitor cells (colony-forming unit-spleen, CFU-S) capable of forming macroscopic, multilineage hematopoietic colonies in the spleens of irradiated mice (Williams et al. 1984). High-efficiency gene transfer was achieved into this physiologically relevant cell population. CFU-S lack the ability to reconstitute the hematopoietic system, but modifications of the experimental design allowed demonstration that murine retroviral vectors were also capable of gene transfer into the repopulating hematopoietic stem cells of mice (Dick et al. 1985; Eglitis et al. 1985; Keller et al. 1985; Lemischka et al. 1986).

During the next decade, a number of refinements of the basic methodology for stem-cell-targeted gene transfer including strategies to derive high-titer vector producer clones routinely (Miller 1990) and the use of complex cytokine mixtures and extended 4–5-day cultures (Bodine et al. 1989; Luskey et al. 1992) to induce stem-cell division, have resulted in a predictably high efficiency of gene transfer into murine stem cells (Fig. 3) (Persons et al. 1997). Modifications in vector design have overcome many of the problems of sustaining gene expression in the murine model (Hawley et al. 1994; Challita et al. 1995; Riviere et al. 1995). Work in mice is now focused on issues of functional gene expression, often designed to correct specific genetic deficiencies in unique, gene knockout mouse strains (Wolfe et al. 1992; Bjorgvinsdottir et al. 1997; Bunting et al. 1998). Due in part to the availability of these murine models of human disease, the field of murine stem-cell-targeted gene transfer remains an important experimental tool.

Development and Use of Animal Models

The early successes achieved in murine models stimulated an interest in developing large-animal models to test the emerging principles for stem-cell-targeted gene transfer. Nonhuman primates including rhesus (Bodine et al. 1990, 1993; van Beusechem et al. 1992) and cynomolgus macaque (Kantoff et al. 1987; Kohn et al. 1997) monkeys, as well as baboons (Kiem et al. 1997), dogs (Schuening et al. 1991; Carter et al. 1992), and sheep (Ekhterae et al. 1990) have been used for these studies. Survival of these animals during reconstitution with autologous cells following otherwise lethal irradiation has been achieved by the development of relatively sophisticated regimens of supportive care. Early results in these

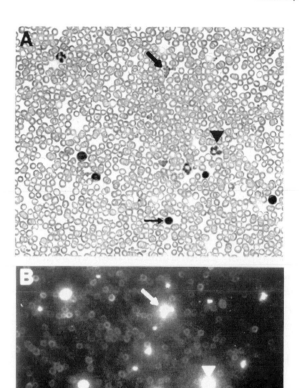

Figure 3 Retrovirus-mediated gene expression in mouse blood cells. Detection of genetically modified cells expressing bacterial green fluorescent protein (GFP) are readily detected in the blood of mice reconstituted with bone marrow that had been transduced with a bicistronic vector that encodes GFP and a trimetrexate resistant variant of dihydrofolate reductase (Persons et al. 1997). The unstained slide was first photographed under fluorescent microscopy (*lower panel, B*) and then stained with hematoxilin-eosin and photographed again using conventional microscopy (*upper panel, A*). The *heavy arrow* (upper part of the field) identifies a clump of brightly fluorescent platelets, the *triangle* (middle part of the field) identifies a GFP-positive neutrophil, and the *thin arrow* (lower part of the field) identifies a lymphocyte that is negative for the GFP marker. The GFP marker in the bicistronic vector system allows ready detection and quantitation of genetically modified cells that express both GFP and a gene of physiological or therapeutic relevance (Persons et al. 1997).

animal models indicated that gene-transfer efficiency, as reflected by the frequency of genetically modified cells in peripheral blood, was far lower than that achieved in the murine models. The percentage of peripheral blood cells that contained the proviral genome was generally less than 1% or 2%. The refractoriness of the stem cells in larger animals to retrovirus-mediated gene transfer was thought to be due, at least in part, to their quiescent status. An exception to the generally poor transduction of stem cells in large animals is the report that 20% of hematopoietic cells contained the vector genome in dogs that had received bone marrow exposed to fresh vector-containing medium weekly for 3 weeks under "long-term culture" conditions that might trigger stem-cell self-renewal (Bienzle et al. 1994). To date, this potentially important observation has not been reproduced in other species (Tisdale et al. 1997).

In general, in vitro assays of gene transfer into clonogenic hematopoietic progenitors have given a misleadingly high estimate of the efficiency of stem-cell gene transfer. Progenitors are proliferating cells that are amenable to retroviral uptake and integration, but progenitor cells lack the capacity for self-renewal and therefore are incapable of sustaining long-term hematopoiesis or contributing to hematopoietic reconstitution (Fig. 4). Assays for human stem cells have been developed in immunodeficient mice (Dick et al. 1991; Nolta et al. 1994; Bock et al. 1995; Larochelle et al. 1996). The gene-transfer efficiency into the cells capable of establishing long-term human hematopoiesis in such animals appears to approximate that seen in repopulating stem cells in large-animal models and in human clinical trials (discussed later). Thus, the immunodeficient mouse models may be useful for optimizing gene-transfer efficiency into primitive cells.

Peripheral blood lymphocytes and lymph node cells have also been explored as potential targets for therapeutic gene transfer. Lymphocytes can be induced into an active state of proliferation with interleukin-2 and mitogens such as phytohemagglutinin. For certain clinical applications, such as the immunodeficiency disorders (discussed later), lymphocyte-targeted gene transfer was thought to be potentially effective and offered an alternative approach that avoided the barriers to stem-cell transduction. In the rhesus model, transduction of stimulated lymphocytes was achieved in vitro and reinfused gene-marked lymphocytes persisted in vivo for up to 2 years (Culver et al. 1990).

Early studies in primate species supported the safety of retroviral vectors for gene transfer. None of the animals that were utilized in the early experiments experienced untoward effects that could be attributed directly to the gene-transfer procedure. Replication-competent retroviruses,

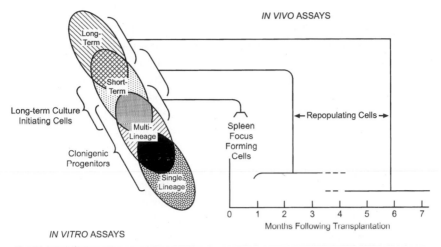

Figure 4 Evaluation of hematopoiesis. Blood-forming tissues contain a hierarchy of stem and progenitor cells that begins with self-renewing, pluripotent stem cells and culminates in single-lineage progenitors that differentiate into precursors that mature into circulating blood cells (Fig. 1). Substantial efforts have been invested in developing assays to quantitate and characterize cells within this hierarchy. Progenitors are readily detected and evaluated in clonogenic cultures that use a semisolid medium, whereas stem cells can only be evaluated in the context of in vivo hematopoietic reconstitution. Long-term-culture initiating cells (LTC-IC) rely on stromal cell support and can be detected and quantitated by virtue of their ability to give rise to clonogenic progenitors after 5 or more weeks in culture. LTC-IC assays have been proposed as a surrogate for quantitation of stem-cell activity; current evidence suggests that the population of cells detected in an LTC-IC assay overlaps, but is not identical to, the population of cells that reconstitute hematopoiesis in vivo. Cells capable of establishing human hematopoiesis in NOD/SCID mice also overlap with the long-term repopulating, stem-cell pool. Of relevance to the field of gene therapy is the fact that progenitors are cycling cells that express the amphotropic receptor and can be transduced with MLV-based retroviral particles, whereas the stem cell pool is quiescent, metabolically inactive, relatively devoid of the amphotropic receptor and refractory to transduction with conventional retroviral vectors. Accordingly, assays that rely on gene transfer into clonogenic progenitors, although useful for evaluating vector-mediated gene expression, are not at all predictive of the ability of a particular vector system to transduce repopulating hematopoietic stem cells.

which may potentially arise within the vector producer clones (Muenchau et al. 1990), are known to cause malignancies in mice by insertional activation of proto-oncogenes or insertional interactivation of tumor-suppressor genes (Tsichlis and Lazo 1991), but injection of large amounts of replication-competent retrovirus into immunosuppressed animals failed to result in any significant disease manifestations or tissue pathology (Cornetta et al. 1990, 1991). However, the development of aggressive lymphomas in rhesus monkeys that had undergone complete ablation of the hematopoietic system by external irradiation and then received stem cells exposed to vector preparations contaminated by replication-competent retroviruses provided the first evidence of the pathogenicity of MLV in primates (Donahue et al. 1992; Vanin et al. 1994). These observations, which were made subsequent to the initiation of human studies, triggered steps to enhance the stringency for screening vector preparations used in clinical trials for replication-competent retroviruses.

In summary, studies in large-animal models indicated that stem-cell-targeted gene transfer was quite inefficient under conditions that worked well in the murine model. Mature lymphocytes were identified as a potential alternative target for therapeutic gene transfer. The procedure of ex vivo gene transfer with vector preparations that were free of replication-competent retroviruses and subsequent infusion of genetically modified cells generally seemed safe in these animal models.

Human Diseases That May Be Amenable to Treatment with Genetically Modified Hematopoietic Cells

Any disease that can be cured by allogeneic bone-marrow transplantation is a potential candidate for treatment by stem-cell-targeted gene therapy. Progress in understanding the immunobiology of transplantation, better control of graft-versus-host disease, and improved supportive care, including the effective treatment of opportunistic infections, have improved the outcome of allogeneic transplantation and encouraged its expanded use for the treatment of genetic diseases. Table 1 lists a variety of hematologic and immune disorders that may be amenable to treatment with genetically modified hematopoietic cells (Kohn 1997). Some of these diseases can be cured with allogeneic transplant, but effective gene therapy would offer a potentially safer and more broadly applicable approach. Other disorders are not suitable for allogeneic transplantation, so that gene therapy could provide a novel curative approach.

Initial Clinical Experience

Initial clinical trials were developed in the context of the known limitations of retrovirus-mediated gene transfer. Marking studies that are relatively independent of the absolute frequency of gene transfer were used to track the behavior of reinfused populations of autologous cells. Therapeutic trials focused on adenosine deaminase (ADA) deficiency, for

Table 1 Hematologic and immune disorders potentially amenable to hematopoietic cell-directed gene therapy[a]

Disorders	Therapeutic genes	Target cells
Inherited immunodeficiency		Stem cells or
Severe combined immuno-deficiency (SCID)	Common γ chain, Janus Kinase 3, adenosine deaminase	T lymphocytes
Combined immunodeficiency	Purine nucleoside	
X-linked agammaglobulinemia	phosphorylase	
Wiskott-Aldrich	Bruton's tyrosine kinase	
Chronic granulomatous	WAS	
disease	Components of phago-cytic oxidase system (Phox 91,47,67,22)	Stem cells
Acquired Immunodeficiency		Stem cells or
AIDS	Transdominant mutants, antisense RNA, or ribozymes directed against HIV gene products	T lymphocytes
Lysosomal storage diseases		Stem cells
Gaucher	Glucocerebrosidase	
GM1 gangliosidosis	β-galactosidase	
Galactosialidos	Protective protein	
Hurlers	α-Idurondase	
MPS-VI	Arylsulfase B	
Fabry	β-galactosidase A	
Hemoglobinopathies		Stem cells
Sickle cell anemia	β-like globins, erythroid transcription factors	
Thalassemia	α or β globins	
Fanconi Anemia	FAA, FAC	Stem cells
Relapsed CML	Thymidine kinase from herpes simplex virus	Allogeneic T lymphocytes
Chemotherapy-induced hematopoietic suppression	Drug-resistance genes, such as MDR1, DHFR, or MGMT	Stem cells

[a]Modified, with permission, from Kohn 1997 © Blackwell Science.

which there seems to be potential for selective amplification of gene-corrected cells. Lymphocytes that can be induced to proliferate and therefore are more amenable to retrovirus-mediated gene transfer were utilized as the therapeutic target in some studies.

Tumor-infiltrating lymphocytes were the targets in the first human trial, in which genetically modified cells were returned to patients after retroviral transduction ex vivo. In this study, the proviral genome and its encoded drug resistance marker allowed the detection of genetically modified cells in peripheral blood and biopsy specimens of tumor lesions. The infused tumor-infiltrating lymphocytes were documented to have survived for several weeks in patients and to have infiltrated tumor deposits (Rosenberg et al. 1990). More recent studies have utilized retroviral marking to track the behavior of ex-vivo-generated cytotoxic T lymphocytes (CTLs) specifically reactive with cells infected with the Epstein-Barr virus (EBV). These CTLs, which are derived from the bone-marrow donor, provide effective prophylaxis or treatment for EBV-induced lymphoproliferative disorders in recipients of allogeneic, T-cell-depleted bone marrow (Rooney et al. 1995). The marker studies demonstrated long-term persistence of EBV virus-specific CTLs and their infiltration into virus-induced tumors (Heslop et al. 1996).

Marking studies were also performed to evaluate the potential of tumor cells in bone-marrow grafts to contribute to relapse in children with malignancies who received autologous bone-marrow transplantation as part of their therapy (Brenner et al. 1993b; Rill et al. 1994). Genetically modified tumor cells were detected after relapse in children with neuroblastoma (Rill et al. 1994) or acute myeloid leukemia (Brenner et al. 1993b), unequivocally establishing that tumor cells in marrow had to be removed if the transplant procedure was to be therapeutically successful. Similar trials have now been performed in adults with myeloid leukemia (Deisseroth et al. 1994; Cornetta et al. 1996), breast cancer, or multiple myeloma (Dunbar et al. 1995). Only in chronic myeloid leukemia have marked tumor cells been documented to contribute to subsequent relapse (Deisseroth et al. 1994). In the future, marking studies are likely to continue to be useful for evaluating the effectiveness of various purging methodologies.

These initial marking studies provided an opportunity to evaluate retrovirus-mediated gene-transfer efficiency into human hematopoietic stem cells (Brenner et al. 1993a; Dunbar et al. 1995). In general, transfer efficiency, as reflected by the proportion of genetically modified cells in peripheral-blood or bone-marrow samples, was quite low ($\leq 1\%$). The highest efficiency was achieved in children whose marrow grafts were harvested after several cycles of chemotherapy.

The first test of the therapeutic efficacy of retrovirus-mediated gene transfer was in the treatment of ADA deficiency, which emerged as the primary candidate for the initial gene-therapy trials in the late 1980s (Kantoff et al. 1988). Although administration of the ADA enzyme conjugated with polyethylene glycol results in clinical improvement, allogeneic bone-marrow transplantation is the only curative therapy (Kantoff et al. 1988). This allogeneic transplant experience not only demonstrates that engraftment with corrected stem cells can be curative, but further suggests that correction of the defect in T cells may also be sufficient for clinical benefit. Many allogeneic transplant recipients are improved despite having T cells as the only donor cell type, suggesting that the transduction of T cells was an alternative approach to stem-cell targeting (Culver et al. 1991). Another attractive feature of ADA deficiency was that low levels of enzyme activity in a minority of ADA-deficient lymphocytes were thought to be sufficient for therapeutic benefit. In contrast, gene correction for the more prevalent hemoglobin disorders is likely to require regulated, high-level gene expression in a majority of cells.

Three trials of gene therapy for ADA deficiency have been performed. In the first, ex-vivo-expanded peripheral blood lymphocytes were the targets for retrovirus-mediated gene transfer (Blaese et al. 1995), whereas in the second trial, both lymphocytes and bone-marrow stem cells were transduced (Bordignon et al. 1995). In the third trial, three infants with ADA deficiency received autologous, cord-blood cells that had been transduced with a retroviral vector encoding the ADA gene (Kohn et al. 1995). Gene-modified cells have persisted in all patients for periods of up to 4 years. Clinical improvement was documented, although the concurrent administration of exogenous, ADA enzyme confounded the interpretation of the therapeutic benefit of gene transfer as discussed in more detail in a later section.

Summary of the First Decade

Recognition that retroviral vectors might be useful for therapeutic applications and their use in human clinical trials occurred over a 10-year period. This significant achievement required refinements in vector and packaging cell design to eliminate the risk of emergence of replication-competent retroviruses. Mouse models proved useful for developing experimental strategies for stem-cell-targeted gene transfer, whereas larger-animal models defined the limitations of these protocols while establishing the overall safety of this approach.

Much useful information has come from the initial clinical studies. Marking protocols have established the importance of purging in the context of autologous bone-marrow transplantation for the treatment of various malignancies. Although no patient was unequivocally cured of ADA deficiency by the first clinical trials, many insights were gained on which to build future work. The patients who participated in those trials have not been harmed and indeed several have experienced significant clinical improvement in the context of a gene therapy trial. Gene transfer into mature lymphocytes, although feasible, is not likely to supplant the interest in achieving long-term corrections of genetic defects that affect hematopoietic cells by gene transfer into stem cells.

STRATEGIES TO OVERCOME THE BARRIERS TO STEM-CELL-TARGETED GENE TRANSFER

A review of the field of human gene therapy in 1995, commissioned by the Director of the National Institutes of Health, cautioned against early initiation of clinical trials and concluded that substantial basic research was necessary for the development of this field. The importance of animal models for testing the principles of gene therapy was emphasized. Needed for successful stem-cell-targeted gene therapy are a better understanding of the biology of hematopoietic stem cells, more knowledge about the cytokines and other microenvironmental factors that influence their behavior in vivo and ex vivo, and a much greater understanding of the strengths and weaknesses of various vector systems for gene insertion into repopulating stem cells. Many new investigators have been attracted to this field, and new knowledge that may help overcome the barriers to stem-cell-targeted gene therapy is accumulating at a rapid rate. This new knowledge is now being used to design more efficient gene-transfer techniques for hematopoietic-cell-directed human gene therapy. Figure 5 summarizes some of the known barriers to stem-cell-targeted gene transfer. Current work focuses around two general strategies: (1) manipulating stem cells to render them more susceptible to transduction with conventional vectors by inducing self-renewal divisions and by overcoming the limitation in receptor expression, or alternatively (2) developing other vector systems that are more capable of gene transfer into quiescent stem cells.

Retrovirus-mediated Gene Transfer and the Biology of Human Stem Cells

A growing understanding of the biochemical and physiological characteristics of stem cells suggests several strategies to enhance the efficiency of stem-cell-targeted gene transfer. Specific approaches include (1) exploiting biological differences in stem cells isolated from various sources and developing culture conditions that favor self-renewal, (2) overcoming the limitation of low receptor expression, (3) developing

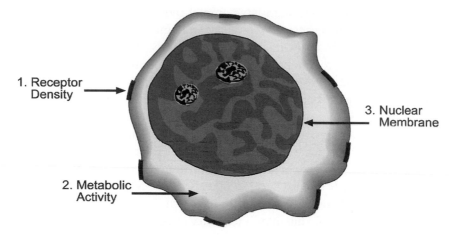

Figure 5 Barriers to gene transfer into human hematopoietic stem cells with MLV retroviral vectors. Cell populations that contain repopulating cells express low levels of the sodium-dependent, phosphate transporter that acts as the receptor for amphotropic viruses (Orlic et al. 1996) and lack the metabolic activity required to complete reverse transcription of the vector genome (Sinclair et al. 1997). Furthermore, stem cells are quiescent, and therefore their nuclear membrane constitutes a barrier to integration of the proviral genome (Lewis and Emerman 1994). Needed to achieve gene transfer into the stem-cell population are strategies to alter these barriers to retrovirus-mediated gene transfer while preserving the fundamental properties of stem cells, namely, self-renewal and multipotency and/or utilization of alternative vector systems that circumvent these barriers.

novel in vitro culture conditions to increase stem-cell transduction, and (4) utilizing drug treatment of transplant recipients for the in vivo expansion of genetically modified stem cells.

The G_0 Block to Transduction with Murine Retroviral Vectors

Conventional retroviral vectors, which are based on oncoretroviruses, require cell division for genome integration. With advances in stem-cell purification, it is becoming increasingly apparent that the majority of hematopoietic stem cells are quiescent. This quiescent state is characteristic of murine and human stem cells, as well as the stem cells of other species used in gene-transfer experiments. One obvious question is why mouse stem cells can be effectively transduced with murine retroviruses, whereas stem cells of larger species, including humans, are generally refractory to transduction.

Experimental protocols in large animal models have attempted, with varying degrees of success, to reproduce strategies that had been found to be successful in transducing murine stem cells. For example, 5-fluo-rouracil (5-FU) is routinely given 48 hours before marrow harvest from syngeneic murine bone-marrow donors because it is known to deplete the hematopoietic tissues of cycling cells effectively and to initiate the activation of otherwise quiescent stem cells (Bodine et al. 1991). Quantitative assays of stem and progenitor cell numbers and the predictable pharmacokinetics of 5-FU in mice are not available in larger species, so that this strategy, which increases murine stem-cell transduction tenfold, has not been developed for effective use in other species. Similarly, coculture of murine bone marrow with vector-producing cells increases transduction efficiencies severalfold but allows recovery of only approximately 25% of the stem cells (Bodine et al. 1991). These nonadherent stem cells may be the mitotically active subset that is most amenable to retrovirus-mediated gene transfer. Substantial loss of stem cells during processing is acceptable in murine models in which the equivalent of multiple syngeneic donors can be used to reconstitute a single recipient. In larger animals, the number of autologous cells that are recovered after transduction is a critically important variable that influences the rate and extent of hematopoietic reconstitution following myeloablation. In addition to these technical issues, there may be important biological variables that influence the relative transduction efficiency of murine versus primate stem cells, as will be discussed in more detail later.

Transduction in murine systems is clearly facilitated by the addition of early acting hematopoietic cytokines. The use of interleukin-3 (IL-3) and interleukin-6 (IL-6) (Bodine et al. 1989) and later stem cell factor (SCF) (Luskey et al. 1992) was shown to enhance mouse stem cell transduction, presumably by supporting stem-cell viability and inducing cell cycle progression. Quantitative estimates obtained using a competitive repopulation assay suggest that murine stem cells double in number during the conventional 2 days of prestimulation followed by 2 days of culture with retroviral vector particles (Bodine et al. 1991). Recent work has documented a threefold expansion of murine stem cells in serum-free medium containing SCF, Flk-3 ligand, and thrombopoietin (Miller and Eaves 1997).

Cytokine administration to donor mice prior to stem-cell collection increases transduction frequency with murine retroviral vectors. The mechanism for this effect may be multifactorial, but it is thought at least in part to be due to the induction of stem-cell cycling. When mice are treated with a 5–day course of SCF and granulocyte–colony-stimulating factor (G-CSF), there is an in vivo expansion of stem cells, first within the peripheral blood (Bodine et al. 1994) and then later within the bone marrow (Bodine et al. 1996). The expanded peripheral blood stem cells were highly transducible and did not require in vitro cytokine prestimulation for efficient transduction. Likewise, the stem cells in the bone marrow of cytokine-treated mice were easily transduced with retroviral vectors (Dunbar et al. 1996). This approach has been tested in the rhesus monkey transplant model. Cytokine treatment of rhesus monkeys prior to stem cell harvest resulted in relatively high transduction frequencies of peripheral blood, and bone-marrow-derived stem cells, resulting in 5% of peripheral blood cells containing the vector for at least a year following transplantation (Dunbar et al. 1996).

Primitive human hematopoietic cells, purified by selection for expression of the CD34 antigen, can be induced to proliferate with various mixtures of cytokines, resulting in more than a 10–100-fold amplification of clonogenic progenitors and differentiating precursor cells (Lyman and Jacobsen 1998 and references therein). Of more relevance to stem-cell-targeted gene transfer is the behavior of cells capable of hematopoietic reconstitution. These cells are within the CD34$^+$ and CD38$^-$ population, a highly quiescent subset amounting to 1–2% of the total CD34$^+$ population. A small CD34$^-$ subset of quiescent cells with repopulating potential can be induced to become CD34$^+$ in culture (Goodell et al. 1997). CD34$^+$ and CD38$^-$ cells, when induced to divide in vitro, usually differentiate into cells with the CD34$^+$ and CD38$^+$ phenotype with loss of long-term proliferative potential. Only recently have conditions been found in which there is a two- to fourfold amplification of the stem cells as assayed by a delayed, high proliferative potential in long-term culture (Shah et al. 1996; Zandstra et al. 1997) or the ability to populate nonobese diabetic NOD–severe combined immunodeficiency (SCID) mice with human hematopoietic cells (Bhatia et al. 1997; Conneally et al. 1997). This amplification occurs over 4–5 days in culture and requires high concentrations of SCF and Flk-3 ligand and lower concentrations of one or more cytokines, which may include thrombopoietin, IL-3, IL-6, G-CSF, or granulocyte macrophage–colony-stimulating factor (GM-CSF) (Bhatia et al. 1997; Borge et al. 1997; Nordon et al. 1997; Zandstra et al. 1997).

Analogous to results in animal models, human stem cells obtained under different conditions or from various sources may vary with respect to their susceptibility to retrovirus-mediated gene transfer. Supporting this interpretation is the relatively high transduction efficiency of repopulating cells in bone marrow of children harvested after three cycles of myelosuppressive chemotherapy (Brenner et al. 1993a). Formal studies in humans using various cytokine combinations for the mobilization of stem cells for gene transfer have yet to be reported. Cord blood contains higher concentrations of clonogenic progenitors that readily proliferate ex vivo than either adult bone marrow or peripheral blood (Hao et al. 1995). Cord blood also contains a higher concentration of NOD/SCID repopulating cells (Wang et al. 1997). Although cord-blood progenitors appeared to be more susceptible to retroviral transduction than those in adult bone marrow (Moritz et al. 1993), cord-blood cells that are capable of establishing long-term hematopoiesis in NOD/SCID mice were poorly transducible (Larochelle et al. 1996). Thus, the issue as to whether a source can be found that yields human stem cells that are readily transduced with conventional retroviral vectors remains to be resolved.

Retroviral Receptor Expression on Primitive Hematopoietic Cells

Important differences exist in the expression patterns of ecotropic and amphotropic receptors on mammalian primitive hematopoietic populations. Both murine and human stem cells express relatively little amphotropic receptor mRNA (Orlic et al. 1996). In contrast, ecotropic receptor expression is relatively high in murine stem cells, correlating with the increased transduction frequency of mouse stem cells seen with ecotropic versus amphotropic vectors (Orlic et al. 1996). Human stem cells express a homolog of the ecotropic receptor, but, because of interspecies polymorphic variation, this molecule is not recognized by ecotropic viruses. These findings may in part explain the differences in transduction efficiency between murine stem cells and those of other species, including human.

There are a variety of approaches being explored to overcome the limiting levels of amphotropic receptor expression on human primitive hematopoietic cells. Murine stem-cell populations that express relatively increased amounts of amphotropic receptor mRNA have been isolated by counterflow elutriation (Orlic et al. 1996). This result suggests it might be possible to purify and target human stem cells that express sufficient amounts of the amphotropic receptor. An alternate possibility is the direct up-regulation of amphotropic receptor expression using in vitro manipulations. Culture of primitive hematopoietic cells in the presence of cytokines increases binding of the amphotropic virus, presumably by upregulating expression of the receptor (Crooks and Kohn 1993). Another potential means of increasing amphotropic receptor expression is culture in phosphate-depleted media (Bunnell et al. 1995; Chien et al. 1997). Because the amphotropic receptor is a sodium-dependent phosphate transporter, phosphate depletion appears to result in a feedback up-regulation of receptor expression.

An alternate strategy to bypass the limitation in expression of the amphotropic receptor on primitive hematopoietic cells is the creation of pseudotyped vector particles in which the amphotropic envelope protein is replaced with that of another virus. This capacity for the envelope proteins of certain viruses to participate in the encapsidation of the genome and core components of other viruses allows derivation of chimeric or pseudotyped virions with an extended host range. Characterized most extensively to date with respect to hematopoietic-cell transduction have been MLV vector particles pseudotyped with the envelope protein of Gibbon ape leukemia virus (GaLV) or the G protein of the vesicular stomatitis virus (VSV-G). Vector particles pseudotyped with several other envelope proteins are also being characterized (Miller 1996; Miller and Chen 1996).

The receptor for GaLV is Glvr1, a sodium-dependent phosphate transporter that is closely related structurally to Ram1, the receptor for amphotropic viruses (D.G. Miller et al. 1994). However, the two proteins have different patterns of expression. A useful packaging cell line (PG13) for vectors bearing the GaLV envelope is available (Miller et al. 1991). Many hematopoietic cells express higher levels of Glvr1 than Ram1, and improved transduction of human progenitor cells with a GaLV-pseudotyped vector preparation has been reported (Bauer et al. 1995). A competitive repopulation study in baboons, each of which received two separate aliquots of autologous cells, one transduced with amphotropic pseudotyped particles and the other with GaLV-pseudotyped particles, suggested that gene transfer into repopulating cells was somewhat more efficient with the GaLV vector preparation (Kiem et al. 1997). Transduction was relatively low with both vector preparations (≤5% genetically modified cells following reconstitution), indicating that utilization of the GaLV-pseudotyped vector, although advantageous, will not fully overcome the barrier to stem-cell-targeted, retrovirus-mediated gene transfer.

The receptor for the VSV-G envelope protein is thought to be a ubiquitously expressed cell-surface phospholipid, so that VSV-G-pseudotyped particles have a broad host range (Yee et al. 1994). The VSV-G protein is toxic to cells, in part because of its ability to induce cell fusion, but useful packaging cell lines that take advantage of the tetracycline-modulated promoter system (Gossen and Bujard 1992) have been derived using both mouse and human cells (Yang et al. 1995; Ory et al. 1996). VSV-G-pseudotyped particles have the additional advantage that they can be concentrated by ultracentrifugation, yielding preparations in defined media that have titers up to 10^9 infectious particles per milliliter. Experience in the use of VSV-G-pseudotyped vector preparations with respect to transduction of human hematopoietic cells has been variable with some investigators reporting a somewhat higher transduction efficiency than that achieved with conventional amphotropic particles (von Kalle et al. 1994). Other work has shown that, despite the ability of VSV-G particles to bind and fuse to CD34$^+$ and CD38$^-$ cells, transduction efficiency may be limited by the inability to initiate reverse transcription in these very primitive, quiescent cells (Sinclair et al. 1997). This barrier may potentially be overcome by the use of VSV-G-pseudotyped lentiviral particles, as will be described in more detail later.

Development of In Vitro Conditions That Facilitate Retrovirus-mediated Gene Transfer

One focus for improving stem-cell-targeted gene transfer has been to define the optimal culture conditions for transduction. Although it has been difficult to extrapolate conditions derived from murine transplant experiments, gains are now being made in large-animal systems, in part due to empirically derived protocols, but also partly due to an increased understanding and exploitation of biologic issues relevant to the stem cells. One example is the use of autologous stromal cells during transduction protocols. Mesenchymal stromal elements in the bone marrow mediate a number of complex interactions with resident stem cells, many via direct cell-to-cell contacts. For example, marrow fibroblasts express a membrane-bound form of SCF that confers a response to stem cells that is distinct from that mediated by soluble SCF (Toksoz et al. 1992). In an attempt to recapitulate these interactions in vitro, investigators have tested if autologous stromal cells, derived from bone-marrow aspirates, can be used to increase retrovirus-mediated stem cell transduction. These studies have shown that coculture in the presence of adherent stromal cells increases the relative transduction frequency of human LTC-ICs (Moore et al. 1992) and of short-term human repopulating cells (Hanania et al. 1996). An important caveat is that stroma cells alone may be insufficient to facilitate gene transfer (Emmons et al. 1997), in that additional soluble cytokines may be required.

Another advantage of stromal support is that stem cells can be maintained ex vivo for relatively prolonged periods of time. On the basis of the hypothesis that prolonged culture in the presence of stromal cells will increase the proportion of cycling stem cells that are exposed to vectors, canine bone-marrow cells have been maintained under long-term culture conditions for 3 weeks with weekly replacement of the culture medium with fresh medium containing vector particles. After reinfusion into unablated dogs, up to 20% of bone-marrow cells were found to be transduced at distant time points after transplant (Carter et al. 1992; Bienzle et al. 1994). This result is especially remarkable given the engraftment disadvantage expected for transduced cells infused into recipients that have received no myeloablation. Unfortunately, efforts to reproduce these observations in nonhuman primates have been unsuccessful to date (Tisdale et al. 1997).

Extracellular matrix molecules arc another component of the bone-marrow microenvironment that can be used to enhance gene transfer. One particular matrix molecule useful in this regard is fibronectin, which was originally identified via its role in adhesion and homing (Williams et al. 1991). Coating of culture plates with fibronectin significantly enhances retrovirus-mediated gene transfer into clonogenic hematopoietic progenitors using clinically relevant transduction protocols (Moritz et al. 1994, 1996). Fibronectin is thought to promote colocalization of primitive hematopoietic cells with recombinant vector particles as the binding sites, for both virus and hematopoietic cells have been mapped to adjacent sites on the fibronectin peptide (Hanenberg et al. 1996). The availability of recombinant fibronectin fragments will greatly facilitate the application of this technology for clinical studies.

In Vivo Selection for Enrichment of Genetically Modified Hematopoietic Cells

A process called *in vivo selection* is a promising strategy for overcoming the limitations imposed by low stem cell transduction efficiencies. The principles of this approach are outlined in Figure 6. A gene that confers a dominant selective advantage on transduced cells is incorporated into the vector, most commonly a gene conferring resistance to a cytotoxic drug. Recipients that are chimeric for transduced and unmodified hematopoietic cells are then treated with the relevant drug. This drug treatment may cause enrichment and expansion of the subset of hematopoietic progenitors and stem cells that express the transferred dominant selectable marker (Fig. 6). Repetitive courses of drug selection would then cause the progressive enrichment of vector-expressing cells. This approach could theoretically be applied to a variety of diseases by inserting a second therapeutic gene into the selectable vector system.

The feasibility of this approach was first demonstrated using retroviral vectors expressing the human multidrug resistance 1 (MDR1) gene.

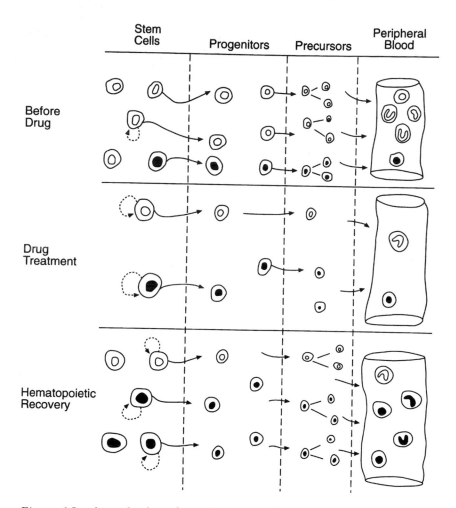

Figure 6 In vivo selection of genetically modified hematopoietic stem cells. In large animals transplanted with retrovirally transduced CD34$^+$ cells, there are very few engrafted stem cells that contain the retroviral vector (*cells with black nucleus, Before Drug*). As a result, there are low numbers of progenitors, precursors, and circulating peripheral blood cells that contain and express the transferred vector. If the vector contains a dominant selectable marker, such as a resistance gene to a myelotoxic drug, then drug treatment will result in killing of unmodified cells, with a relative sparing of cells that express the resistance-conferring vector. This enrichment will result in an amplification of transduced cells during hematopoietic recovery from the acute drug toxicity. If selection has been imposed at the level of the hematopoietic stem cell, then increased numbers of transduced cells will be stably present in differentiating compartments, as well as in the peripheral blood (*black cells*). (Modified, with permission, from Sorrentino 1996.)

The MDR1 gene product, P-glycoprotein (Pgp), which confers resistance to a large number of myelosuppressive drugs (Pastan and Gottesman 1991), was transduced into murine repopulating cells. Following reconstitution with Pgp-expressing cells, taxol treatment resulted in large increases in provirally marked peripheral blood leukocytes in some mice (Podda et al. 1992; Sorrentino et al. 1992). Other drugs that are Pgp substrates have been used for in vivo selection in this mouse model (Aksentijevich et al. 1996), raising the important question about which drugs are optimal for in vivo selection. Recent evidence suggests that the selection seen with taxol and vinca alkyloids may be occurring at a relatively late stage of hematopoietic development (Blau et al. 1997a), predicting that enrichment in genetically modified cells with taxol could be transient in nature. Pretreatment of mice with SCF can sensitize early cells to these drugs (Blau et al. 1997a) and may be a useful strategy for enhancing the selection process.

On the basis of these preclinical data, clinical trials have been initiated to test the feasibility of using the MDR1 system for in vivo selection in cancer patients. Two centers have reported early results from these trials, in which patients have been transplanted with autologous CD34[+] cells that were transduced with amphotropic MDR1 retroviral vectors. Very low levels of MDR1-transduced cells were initially detected after hematopoietic reconstitution, averaging 1 in 1000 cells or less (Hanania et al. 1996; Hesdorffer et al. 1998). Some of these patients have been treated with taxol, but no enrichment of MDR1-modified cells has been seen (A.B. Deisseroth, pers. comm.). Similarly negative results have been obtained in a rhesus monkey model, where very low initial numbers of MDR1-marked cells were not increased by taxol treatment (Tisdale et al. 1997). There are several potential explanations for the lack of selection seen in these studies. It is likely that very few, if any, MDR1-transduced stem cells had engrafted after transplant, as evidenced by the very low numbers of marked cells. There may be a minimum number of transduced cells necessary for selection to occur, and this threshold requirement may be achievable in the murine models but not in large animals. A second possibility is that the MDR1 provirus was rearranged in the clinical studies, a problem that has been demonstrated in the murine model (Sorrentino et al. 1995). If this were the case, the polymerase chain reaction (PCR) assay used for detection of proviral sequences in the human and rhesus monkey studies may have been detecting cells lacking functional Pgp expression. The availability of a more stable MDR1 cDNA could offer an important advantage in this regard (Galipeau et al. 1997).

Another system being explored for in vivo selection utilizes variants of dihydrofolate reductase (DHFR) to confer resistance to antifolates. Mutations in the active site of DHFR inhibit the binding of antifolates such as methotrexate (MTX) and thereby confer drug resistance to cells expressing these variants. The early availability of such variants (Simonsen and Levinson 1983) is the reason why the DHFR vectors were the first to be tested for in vivo selection capabilities. In the first of these studies, mice were transplanted with bone-marrow cells that had been transduced with a retroviral vector expressing a murine DHFR variant (Corey et al. 1990). Subsequent MTX treatment failed to demonstrate convincing selection of DHFR-transduced myeloid cells. Although a later study suggested that MTX could be used for the selection of DHFR-transduced myeloid progenitors (Zhao et al. 1994), this result has been called into question based on the fact that MTX does not exert a cytotoxic effect within the myeloid progenitor compartment (Blau et al. 1996).

A number of recent advances have rekindled interest in DHFR variants for use in selection systems. New active site mutations in the human DHFR gene have been discovered that have kinetic advantages for use as dominant selectable markers (Schweitzer et al. 1989; Lewis et al. 1995). One of these variants, containing a leucine to tyrosine substitution in codon 22 (L22Y-DHFR), confers extremely high levels of cellular resistance to the nonclassical antifolate, trimetrexate (TMTX) (Spencer et al. 1996). In contrast to the relatively modest increase in MTX resistance conferred by the original murine L22R variant, the human L22Y gene results in a 100-fold increase in TMTX resistance. A second advance has been the development of novel pharmacologic approaches for the selection of transduced stem cells. Antifolates alone do not kill myeloid progenitors or stem cells and therefore cannot exert selective pressure on early hematopoietic cells (Blau et al. 1996). This intrinsic antifolate resistance is mediated by nucleotide salvage pathways, which allow stem cells to overcome drug-induced DHFR blockade by importing purine and pyrimidine precursors from the serum (Allay et al. 1997c). By blocking thymidine transport with nucleoside transport inhibitors such as nitrobenzylmercaptopurine riboside phosphate (NBMPR-P), unmodified stem cells and progenitors become very susceptible to killing with TMTX (Allay et al. 1997c). This result predicts that if stem cells are expressing a DHFR resistance vector, they should be efficiently selected by treatment with a combination of TMTX and NBMPR-P (Fig. 7).

Our group has recently tested this hypothesis using a mouse transplant model (Allay et al. 1997b). Bone-marrow cells were transduced

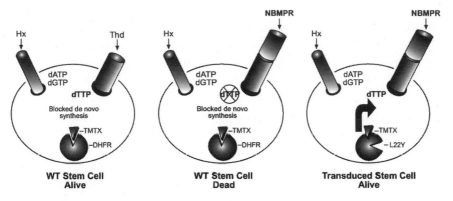

Figure 7 Selection of stem cells expressing DHFR variants using trimetrexate and NBMPR. Unmodified wild-type (WT) stem cells cannot be killed with TMTX alone (*left*). Although DHFR is inhibited, leading to a block in de novo synthesis of dTTP, dATP, and dGTP, stem cells can salvage hypoxanthine (Hx) and thymidine (Thd) from the extracellular environment to reconstitute these nucleotide pools. The inhibition of thymidine transport with nitrobenzylmercaptopurine riboside (NBMPR) sensitizes stem cells to TMTX by effectively depleting intracellular stores of dTTP, thereby leading to killing of stem cells expressing wild-type DHFR (*middle*). In contrast, stem cells transduced with resistance-conferring DHFR constructs (L22Y) can survive exposure to TMTX and NBMPR due to their ability to maintain dTTP pools through continued de novo biosynthesis (*right*). This triad of TMTX, NBMPR, and the L22Y-DHFR gene has been shown to enable efficient selection of genetically modified stem cells in mice.

with bicistronic vectors containing the L22Y-DHFR variant linked to either of two different reporter genes. Mice were transplanted and the proportion of peripheral blood cells that expressed the vector was determined by flow cytometry. Mice were then treated with TMTX and NBMPR-P, allowed to recover, and reassayed for peripheral blood cells that expressed the vector. Increases in vector-expressing cells were detected in all myeloid and lymphoid lineages. Three sequential courses of drug selection resulted in serial increases in marked cells. The proportion of vector-expressing erythrocytes and granulocytes increased from approximately 10% to 80% in some mice. Secondary transplant experiments confirmed that selection had occurred at the level of repopulating stem cells. Assuming that these results can be reproduced in large-animal models, this strategy should prove very useful for increasing the number of transduced hematopoietic cells in vivo.

There is one other system for in vivo selection with demonstrated efficacy in an animal model. Transfer of methylguanine methyltransferase (MGMT) genes can protect hematopoietic cells from the toxicity of nitrosourea drugs such as bischloroethylnitrosourea (BCNU) and cyclohexylchloroethylnitrosourea (CCNU) (Harris et al. 1995; Moritz et al. 1995; Allay et al. 1995; 1997a). One advantage of this system is the significant toxicity of nitrosoureas on stem cells, which should provide a selective advantage for transduced cells in primitive hematopoietic compartments. A direct demonstration of the capacity for in vivo selection using MGMT retroviral vectors has been obtained in a murine gene-transfer model. Mice treated with BCNU-containing regimens displayed an increase in MGMT-transduced cells in the bone marrow (Allay et al. 1997d). A similar result has recently been obtained using MGMT mutants that are resistant to inactivation with O^6 benzylguanine (Davis et al 1997), suggesting that combined treatment with BCNU and O^6 benzylguanine may enhance selection efficiency. Although a relative disadvantage of this approach is the potential for significant nonhematopoietic toxicity and DNA-damaging effects caused by nitrosoureas, this system is clearly worth pursuing as a means to protect the marrow in cancer patients receiving nitrosourea-based chemotherapy.

The potential for toxicity to normal tissues is an intrinsic disadvantage to all selection systems utilizing cytotoxic drugs. A new approach intended to circumvent this limitation is the use of synthetic drugs without toxicity to confer a direct proliferative advantage on genetically modified hematopoietic cells (Spencer et al. 1993). Cell lines expressing a chimeric cytokine receptor, composed of the cytoplasmic domain of the erythropoietic receptor linked to the FK506 drug-binding domain (FKBP12), can be induced to proliferate by adding a dimeric form of FK506 (FK1012) to the media (Blau et al 1997b). These data suggest that a similar effect may be achieved by administering FK1012 to mice transplanted with transduced hematopoietic cells expressing these drug-responsive cytokine receptors. More work will need to be done to determine which cytokine receptor fragments are most active in these constructs, and which of several available pharmacologic dimerization strategies will be most effective. Of relevance is the fact that primitive hematopoietic cells seem to require several cytokines in high concentration to achieve self-renewal, so it remains to be demonstrated whether the signal transmitted by a single cytokine receptor will be adequate to stimulate proliferation of genetically modified stem cells.

Alternative Vector Systems for Stem-cell-targeted Gene Transfer

Defined barriers to stem-cell-targeted gene transfer by MLV-based virus-es are (1) the low levels of expression of the receptor for amphotropic vector particles, (2) the relative metabolic inactivity of primitive hematopoietic cells, which may preclude such early steps as reverse tran-scription, and (3) the nuclear membrane of these quiescent cells, which prevents access of the MLV preintegration complex to the host-cell chro-mosomes (Fig. 5). Vector systems based on lentiviruses or adeno-associ-ated viruses have been investigated for their ability to overcome these barriers and therefore to transfer genes into primitive hematopoietic cells. Nonintegrating viruses such as the adenovirus and herpesvirus have also been studied in the context of attempting to achieve functional, albeit transient, gene expression in early hematopoietic cells.

Lentiviruses

Human immunodeficiency virus (HIV), a lentivirus, has the capacity to infect quiescent cells productively (Naldini et al. 1996; Poeschla et al. 1996; Reiser et al. 1996). In contrast to the nucleoprotein, preinitiation complex of MLV, which requires dissolution of the nuclear membrane during mitosis to access cellular DNA, the preinitiation complex of HIV is transported through the nuclear membrane by virtue of nuclear local-ization signals present in the p19 matrix (MA) and Vpr accessory proteins (Bukrinsky et al. 1993; Heinzinger et al. 1994; Naldini et al. 1996). Either signal will suffice to allow HIV to infect quiescent cells. Carboxy-termi-nal phosphorylation of MA by a cellular kinase during particle assembly facilitates interaction with integrase and is necessary for the karophylic properties of the HIV preinitiation complex (Gallay et al. 1995a,b; Camaur et al. 1997).

Another key property of lentiviruses with respect to their ability to infect nondividing cells is their ability to form a relatively stable tran-scription intermediate (Zack et al. 1990, 1992). In quiescent T lympho-cytes, a partial reverse transcriptase product is formed that can be res-cued by cell activation for at least 7–8 days postinfection. In contrast, the MLV nucleoprotein complex disappears within 24 hours of infection of nondividing cells (Lewis and Emerman 1994; Naldini et al. 1996).

HIV-based vectors have been derived by removing coding sequences for the structural and accessory proteins leaving the LTRs and the viral

RNA packaging signal located between the 5′ LTR and the initiation codon of the Gag polyprotein (Naldini et al. 1996; Poeschla et al. 1996). Inclusion of the Rev response element (RRE) and a second polypurine track may facilitate the generation of vector particles (Parolin et al. 1996). The coding sequences for marker proteins or potential therapeutic products can be put under the control of internal, heterologous promoters or, alternatively, driven by the HIV LTR if the Tat-coding sequences have been retained in the vector. Very high levels of gene expression can be achieved with the Tat-dependent, HIV LTR promoter system (Parolin et al. 1996). A packaging genome encoding the viral Gag and Pol proteins as well as the accessory proteins under the control of heterologous transcriptional control elements is used to generate infectious vector particles (Naldini et al. 1996; Poeschla et al. 1996). Derivations of stable packaging lines and producer clones have proved challenging. To date, vector preparations have been derived by the cotransfection of Cos or 293 T cells with the vector, packaging, and envelope plasmids. Recent work has shown that the accessory proteins Tat, Vif, Vrp, Vpu, or Nef are not required for the formation of vector particles that are capable of transducing nondividing cells (Zufferey et al. 1997; Kim et al. 1998).

HIV vector particles are readily pseudotyped with either the MLV amphotropic or VSV-G proteins. VSV-G-pseudotyped vector particles can be readily concentrated by ultracentrifugation. HIV vectors readily transduce neurons in vivo and primary macrophages in vitro (von Schwedler et al. 1994; Naldini et al. 1996; Blomer et al. 1997). Cell lines that are growth arrested by contact inhibition or cytostatic agents retain the ability to be transduced by HIV vectors, although cellular activation may be required to complete the steps needed for integration and expression of the vector genome. Similarly, functional gene expression in quiescent, peripheral blood lymphocytes after HIV-vector-mediated transduction requires cellular activation, as would be predicted based on the resistance of such cells in the quiescent state to integration of the wild-type genome and the production of viral particles.

Preliminary data suggest that human hematopoietic cells can be efficiently transduced with VSV-G-pseudotyped HIV vectors. In one study, transduction of quiescent, CD34$^+$ cells with vector particles followed by culture in cytokines (IL-3, IL-6, and SCF) for 3 days resulted in expression of the HIV-vector-encoded marker protein in more than 80% of viable cells (Reiser et al. 1996). In a second study, a high proportion of progenitor-derived colonies were found to contain the proviral genome (Akkina et al. 1996). Remaining to be determined is whether HIV vectors will transduce primitive cells capable of long-term hematopoietic reconstitution in an immunodeficient mouse model or a nonhuman primate. On the basis of the known biology of HIV and HIV-based vector systems, cytokine-stimulated activation of such cells will be required to achieve vector genome integration and expression. In addition, the relative stability of the HIV preinitiation complex may allow its persistence, after transduction ex vivo, until stem cell activation occurs following transplantation in the course of marrow repopulation.

Reservations have been expressed regarding the safety and acceptability of HIV-based vector systems for human gene therapy. Elimination of nonessential elements such as the genes for accessory proteins, use of split packaging genomes, and the avoidance of overlap between vector and packaging sequences are steps that have been taken to improve the safety of HIV vectors. Indeed, the vector preparations in current use may readily be shown to be devoid of replication-competent, infectious HIV particles. Hopefully, the decision regarding the acceptability of HIV vector systems will be based on objective estimates of relative benefit versus risk as reflected by the desired functional properties of such vectors versus potential for harm from their use. Analogous vector systems based on other lentiviruses, such as the simian immunodeficiency virus, may also be developed in the future.

Adeno-associated Virus Vectors

Adeno-associated virus (AAV) is a replication defective parvovirus that depends on a helper virus, either adenovirus or herpesvirus, for virus production during lytic infection (Muzyczka 1992; Kotin 1994; Flotte and Carter 1995; Berns and Giraud 1996). Although 80% of humans have detectable serum antibodies, no disease has been associated with this virus. AAV particles are relatively stable, permitting their purification and concentration by physical techniques. AAV has a single-stranded DNA genome of 4680 nucleotides, which includes 145-bp inverted terminal repeats (ITRs) and coding sequences for nonstructural (Rep) and structural (Cap) proteins. The virus has a broad host-cell range. Infection of susceptible cells by wild-type AAV in the absence of a helper virus often results in the integration and persistence of a latent genome. During latent infection, many but not all genome integration events occur within a preferred site (AAVS1) on human chromosome 19 (Kotin et al. 1990). Site-specific integration is mediated by DNA sequence-specific interaction of the higher-molecular-weight Rep protein species, p78 and p68, with sequences in the AAV ITRs and the host-cell AAV site (Weitzman et al. 1994; Chiorini et al. 1995; Urcelay et al. 1995; Linden et al. 1996; Surosky et al. 1997).

The ITRs are the only *cis*-active elements required for recombinant AAV (rAAV) genome replication and encapsidation (Samulski et al. 1987, 1989; Xiao et al. 1997). Because site-specific integration of the AAV genome depends on Rep functions, rAAV vectors without the *rep* gene integrate randomly (Walsh et al. 1992; Kearns et al. 1996). Indeed, the rAAV genome may persist and be expressed for extended periods in an episomal state, although genome integration is also well-documented (Bertran et al. 1996; Malik et al. 1997). The efficiency of both rAAV transduction and episomal-mediated gene expression as well as the frequency of genome integration are dependent on the ratio of viral particles to target cells (multiplicity of infection [moi]). Higher multiplicities of infection are generally required to achieve stable genome integration (Hargrove et al. 1997).

The broad host range of wild-type AAV is mirrored by the diversity of cell types that may be transduced by rAAV. Relative susceptibility to transduction by rAAV is highly variable, although resistance to transduction may sometimes be overcome by increasing the multiplicity of infection. The factors contributing to the variable target susceptibility to rAAV transduction have not been fully defined but may include receptor density (Mizukami et al. 1996) and the availability of host-cell factors to facilitate viral uncoating and conversion of the single-stranded to a double-stranded genome capable of supporting transcription, functions that during lytic infection are provided by the helper virus (Ferrari et al. 1996; Fisher et al. 1996). rAAV has the capacity to transduce quiescent cells, although transduction of cycling cells is more efficient (D.W. Russell et al. 1994). DNA-damaging agents enhance the transduction of nondividing cells (Alexander et al. 1994). This capacity to transduce noncycling cells is observed in vivo, as prolonged rAAV-mediated gene expression has been observed in brain and muscle following local injection and in the liver following intravenous injection of vector particles (Kaplitt et al. 1994; Xiao and Samulski 1996; Fisher et al. 1997; Koeberl et al. 1997; Snyder et al. 1997).

Development of rAAV vectors for therapeutic applications has been limited by relatively inefficient, cumbersome methodology for vector preparation that includes infection of HeLa cells or human kidney cells with adenovirus and cotransfection with a vector plasmid containing one or more transcriptional units flanked by AAV ITRs and a helper genome consisting of the AAV *rep* and *cap* genes without AAV ITRs. This limitation may be overcome by the recent development of a strategy for deriving packaging cell lines (Clarke et al. 1995; Tamayose et al. 1996). Wild-type AAV that can arise by nonhomologous recombination can be eliminated by separating the *rep* and *cap* genes on separate plasmids (Allen et al. 1997). Most rAAV preps are heavily contaminated with adenovirus, which may be removed by column chromatography or cesium chloride–buoyant density centrifugation. Recently, a plasmid encoding the adenoviral functions required for rAAV production, but lacking the adenovirus late genes, has been derived from the adenoviral genome (Ferrari et al. 1997; Xiao et al. 1998). Utilization of this plasmid in a cotransfection protocol with the vector and helper plasmids results in the production of increased numbers of rAAV particles without contamination with adenoviral particles.

Although rAAV appears to hold great promise for certain therapeutic applications, e.g., treatment of hemophilia due to factor IX deficiency, their utility for stem-cell-targeted gene therapy remains uncertain. Early studies of the ability of rAAV to transduce human clonogenic hematopoietic progenitors utilized crude vector preparations that may contain biologically active marker proteins that transfer to target cells, resulting in "pseudotransduction" (Alexander et al. 1997). Detection of the rAAV-encoded mRNA in colonies derived from transduced progenitors by reverse transcriptase–PCR avoids this potential artifact. This strategy has been used to document the transduction of erythroid progenitors and the subsequent expression of an rAAV-encoded globin gene in maturing human erythroid cells (J.L. Miller et al. 1994; Hargrove et al. 1997). rAAV encoding the gene product that is missing in a subset of Fanconi anemia patients complements the defects in the patients' clonogenic cells, allowing the development of colonies in vitro that are resistant to DNA-damaging agents (Walsh et al. 1994b). Thus, there is unequivocal evidence that rAAVs are able to transduce human clonogenic hematopoietic progenitors, although high multiplicities of infection are required, particularly when purified vector particles are used. A recent report suggests that there may be significant variability among progenitors from different donors with respect to their ability to be transduced by rAAV (Ponnazhagan et al. 1997b)

Another issue with respect to transduction of human hematopoietic cells by rAAV is the status of the vector genome. The results obtained in short-term (14 days) clonal hematopoietic cultures could reflect gene expression from an episomal genome. Indeed, transient episomal-mediated gene expression can readily be demonstrated within a cultured hematopoietic-cell line (Hargrove et al. 1997). We have found that a regulatory element located 3′ to the human $^A\gamma$ globin gene, which contains sequences that bind to the nuclear matrix, facilitates rAAV genome integration and enhances the propensity of rAAV to integrate as a head-to-tail tandem array (Hargrove et al. 1997).

Transduction and integration in human progenitors and long-term culture-initiating cells by rAAV have been reported, as has the transduction of murine repopulating stem cells (Fisher-Adams et al. 1996; Lubovy et al. 1996; Ponnazhagan et al. 1997a). Attempts to achieve rAAV-mediated gene transfer into rhesus repopulating cells have been unsuccessful, however (A.W. Nienhuis and R. Donahue, unpubl.). Thus, the utility for this vector system for stem-cell-targeted gene transfer remains uncertain. Perhaps better methods of vector production and a growing understanding of the interaction, uptake, and transduction of cells with rAAV will allow this vector system to be exploited in the future for stem-cell-targeted gene transfer. The ability of the vector genome to persist for several days in an episomal state might allow quiescent stem cells to be transduced ex vivo with subsequent integration of the vector genome during cell division in vivo in the course of hematopoietic reconstitution.

Foamy Virus Vectors

Human foamy viruses, a spumavirus, has been proposed as the basis for vector development (Russell and Miller 1996; Bodem et al. 1997). This virus appears to be nonpathogenic, and it has a broad host range. Its complex genome is coming to be understood, and replication-defective vector particles have been developed (Bieniasz et al. 1997). The human foamy virus LTR depends on a virally encoded, *trans*-activator, bel1, which can be provided in *trans* during virus production so that the LTR is silent in target cells that have been successfully transduced with vector particles. Transduction of human clonogenic hematopoietic progenitors with a foamy virus vector occurred with moderate efficiency, although direct contact between the producer and target populations was required (Hirata et al. 1996). Although the foamy virus may be less dependent on cell division than oncoretroviruses, direct comparison suggests that it is less able to infect G_1/S- or G_2-arrested cells than HIV (Bieniasz et al. 1995). Thus, the future value of foamy virus for human stem-cell-targeted gene therapy application cannot be meaningfully assessed at this time.

Nonintegrating Vectors

Transient expression of a gene product in stem cells might be useful. For example, overexpression of the HOX4B gene has been shown to enhance the self-renewal of primitive hematopoietic cells (Sauvageau et al. 1995). Expression of the murine ecotropic receptor protein renders primitive hematopoietic cells susceptible to ecotropic retrovirus transduction (Qing et al. 1997; D. Persons and E.F. Vanin, unpubl.). Such considerations have prompted exploratory experiments designed to evaluate the ability of adenoviral and herpesvirus vectors to express genes in primitive hematopoietic cells. These vectors give high levels of transient gene expression but their genomes remain episomal.

Adenoviral vectors encoding β-galactosidase or alkaline phosphotase have been shown to transduce primitive hematopoietic progenitors (Neering et al. 1996; Watanabe et al. 1996). Of interest is the fact that quiescent (G_0) cells among the CD34$^+$ and CD38$^-$ populations also expressed the adenoviral-vector-encoded marker protein. Transduction occurred over 24 hours and appeared to plateau at about 30–35% of the CD34$^+$ population of cells. Relatively high multiplicities of infection of 100–500 infectious particles per cell were required; higher multiplicities of infection were toxic to cells, resulting in a decrease in plating efficiency for clonogenic colony formation. Functional expression of the ecotropic receptor in clonogenic progenitor cells has been documented in that subsequent transduction with ecotropic retroviral vectors allowed detection of the retroviral genome in about 10% of colonies derived from dually transduced progenitors (D. Persons and E.F. Vanin, unpubl.). Of interest is the report that a molecular conjugate of inactive adenoviral particles, plasmid DNA, and stem-cell factor (kit ligand) efficiently transduced c-*kit* receptor-positive hematopoietic cells (Schwarzenberger et al. 1996). This strategy might also be useful for creating cytokine conjugates with active vector particles to facilitate the adenoviral vector transduction of hematopoietic cells.

A disabled infectious herpesvirus vector capable of only a single cycle of vector replication in permissive cells results in the release of noninfectious vector particles (Dilloo et al. 1997). Nonpermissive cells may also be transduced, but they do not release virus. The genome of this disabled infectious single-cycle herpes simplex virus (DISC-HSV) has been modified to eliminate a gene (gH) that is essential for viral replication; this gene can be replaced with a marker or therapeutic gene in the DISC-HSV genome. DISC-HSV is produced in a helper cell line, engineered to express the gH protein. Human primary hematopoietic cells and leukemic blasts are readily transduced with a DISC-HSV-vector-encoding β-galactosidase (Dilloo et al. 1997). Peak expression occurs at 2 days and declines over 2 weeks during the expansion of primary hematopoietic cells in culture. A DISC-HSV-vector-encoding murine GM-CSF enhanced the immunogenicity and retarded the growth of murine leukemic cells in a mouse model (Dilloo et al. 1997). Because of their efficiency in transducing human leukemic blasts, DISC-HSV vectors are being developed for use in human vaccine trials.

A potential problem with the use of nonintegrating vectors to achieve therapeutic gene expression is their ability to invoke an immune response in vivo. Expression of adenoviral proteins has been shown to result in the elimination of vector-transduced cells. For certain applications, e.g., tumor vaccines, vector-protein-enhanced immunogenicity may be beneficial, whereas for other applications, e.g., stem-cell-targeted gene transfer, elimination of transduced cells would be detrimental. Ultimately, these issues will require resolution in appropriately designed animal studies or human clinical trials.

Summary of Recent Efforts to Overcome Barriers to Stem-cell-targeted Gene Transfer

The last 5 years have been characterized by a diversity of approaches to achieve stem-cell-targeted gene transfer. Several relevant parameters have been identified. Cytokine-mobilized stem cells are more susceptible to gene transfer. Use of retroviral vectors pseudotyped with alternative envelope proteins may overcome the low expression of amphotropic receptor expression on primitive hematopoietic cells. Modification of culture conditions, e.g., use of fibronectin fragments and high concentrations of kit-ligand and FL3 ligand in complex cytokine mixtures, may increase gene-transfer efficiency. Lentiviral vectors offer the intrinsic advantages of a relatively stable reverse transcription intermediate and the ability of their preinitiation complex to transverse the nuclear membrane of quiescent cells. To date, these independent advances have yet to be integrated into a unified protocol for stem-cell-targeted gene transfer. On the basis of preliminary data with each approach, achievement of stem-cell transduction efficiency of at least 10% with a unified protocol seems to be potentially feasible. Work in the murine model has clearly shown that this proportion of genetically modified cells can be amplified if gene transfer provides an intrinsic selective advantage or affords the opportunity for drug selection; a 10% transduction efficiency of stem cells might be translated into a 50% or more proportion of genetically modified hematopoietic cells in vivo. AAV vectors offer an alternative approach. The relative stability of the rAAV episomal genome and its ultimate integration as a transcriptionally active complex are encouraging, but the relative resistance of primitive hematopoietic cells to rAAV transduction is a barrier that must be overcome. Overall progress in several areas of research encourage optimism that therapeutically useful gene transfer into hematopoietic stem cells will be achieved in the next 5–10 years.

DISEASE-SPECIFIC APPLICATIONS OF STEM-CELL-TARGETED GENE TRANSFER

Hematopoietic-cell-directed gene therapy is an attractive approach for the large number of human diseases that are potentially curable with allogeneic bone-marrow transplantation and known to be caused by single-gene defects (Karlsson 1991). Treatment with gene-corrected, autologous cells would increase the number of patients eligible for definitive treatment and would eliminate the complications due to transplant histoincompatibility. Gene-therapy strategies are also being tested for cancer patients for which the goal is biologic modulation of a neoplastic process and its treatment rather than simple gene replacement. The subset of these strategies for treatment of cancer that specifically target hematopoietic cells typically employs the transfer of genes that alter pharmacologic processes within the modified cells. As will be discussed, some disease-specific applications have already shown promise in clinical trials, although gene therapy for other targeted disorders will clearly require improvements in currently achievable levels of gene transfer and expression.

Immunodeficiency Disorders

As a broad category, immunodeficiency syndromes occupy a major focus for human gene therapy for several reasons. Vectors are available for many of these diseases, as the list of etiologic single gene defects has been steadily growing since the early 1980s. Second, although allogeneic transplant can be curative in some patients, many patients do not have a histocompatible donor, and mismatched transplants often result in only partial immune reconstitution (Fischer et al. 1990; Buckley et al. 1993). Even when matched transplants are available, the risks of allogeneic transplantation are prohibitive for the clinically milder syndromes. Therefore, alternative forms of therapy are clearly needed. Finally, in some of these syndromes, a naturally occurring selective advantage exists for corrected cells (Conley et al 1990; Hirschhorn et al. 1996), so that even small numbers of corrected stem cells may be sufficient for significant phenotypic correction.

Adenosine Deaminase Deficiency

Severe combined immunodeficiency (SCID) resulting from ADA deficiency was the first disease to be targeted for clinical gene therapy. Deficiency of the ADA enzyme results in accumulation of deoxyadenosine triphosphate, a compound that is toxic to developing lymphocytes.

As a result, there is a severe depletion in T- and B-cell numbers, with corresponding defects in cellular and humoral immunity. Despite its rarity, this disorder has a number of unique advantages for a gene-therapy approach. Tight regulation of a transferred ADA gene is probably not necessary, given that reconstitution of enzyme levels to 10% of normal or greater confers significant phenotypic correction. Furthermore, the toxic metabolites that accumulate in cells and within the extracellular space are freely diffusable and thus may be metabolized by a minority of cells that have been genetically corrected. There is also flexibility in the choice of cellular targets. ADA-transduced lymphocytes have been shown to persist for long periods of time in experimental animal models (Culver et al. 1990, 1991; Ferrari et al. 1991). The feasibility of targeting more primitive cells with lympho-myeloid repopulating potential has also been demonstrated in a variety of preclinical models (Kantoff et al. 1987; Cournoyer et al. 1991; van Beusechem et al. 1992; Bodine et al. 1993

On the basis of these considerations, several clinical trials of ADA gene replacement have been undertaken. In a study begun in late 1990 by investigators at the National Institutes of Health, two children with ADA-deficient SCID were treated with multiple infusions of retrovirally transduced autologous T lymphocytes (Blaese et al. 1995). In this study, a human ADA cDNA was under the transcriptional control of the Mo-MLV LTR. In one patient, the proportion of circulating mononuclear cells that contained the vector reached approximately 30% during the second protocol year, whereas the second patient had only between 0.1% and 1% of her cells marked. With infusions of transduced lymphocytes, both patients displayed significantly increased T-cell counts, increased cellular expression of ADA, and improved humoral immune responses. In both of these patients, the doses of PEG-ADA, which is standard therapy for this disorder, were reduced by more than 50% without a decline in immune function.

In an Italian study initiated in 1992, two patients were treated with simultaneous infusions of transduced lymphocytes and bone-marrow cells, each targeted with genetically distinguishable ADA vectors (Bordignon et al. 1995). Both vectors contained two copies of a human ADA cDNA expressed from the normal ADA promoter. Transduced cells were repetitively administered over a 10- or 24-month period. Sixteen months after treatment, the frequency of vector-expressing T lymphocytes was between 2% and 5% and that of myeloid progenitors was between 17% and 25%. ADA activity in these transduced cells reached normal levels. Both patients had clinically significant improvements in humoral and cellular immunity despite a decrease in their dose of PEG-ADA. Interestingly, analysis of T-cell clones over time showed a progressive emergence of clones derived from targeted bone-marrow cells with a reciprocal fall in clones that originated from transduced lymphocytes. These data show that transduction of more primitive lympho-myeloid stem cells can provide long-term reconstitution with ADA-corrected cells.

This latter point has been confirmed in a third trial that exclusively targeted CD34$^+$ umbilical-cord blood cells from three infants with prenatally diagnosed ADA deficiency (Kohn et al. 1995). On the fourth day of life, the patients received a single infusion of transduced cells. Stable levels of vector-marked cells were detected in the peripheral blood and bone marrow for up to 18 months; however, the frequency of marked cells was only 1 in 3000 to 1 in 100,000. Eighteen months after gene transfer, substantial expression of the ADA vector was documented in transduced bone-marrow progenitors. To test the hypothesis that transduced lymphocytes would have a selective advantage on the withdrawal of enzyme replacement therapy, PEG-ADA infusions were recently stopped in one of these patients four years after gene therapy (Kohn et al. 1997). Although a relative enrichment for transduced T lymphocytes was seen, the absolute numbers of T cells declined by 50%, with more precipitous drops in the numbers of B-lymphocytes and natural killer (NK) cells. Clinical immune function deteriorated in this patient, mandating the reinstitution of PEG-ADA therapy. The reasons for failure upon PEG-ADA withdrawal are currently unclear, but are probably related to the low initial proportion of corrected cells and potentially inadequate expression level of the transferred ADA in mature lymphocytes.

These pioneering ADA gene-therapy trials have demonstrated that correction of defective hematopoietic cells can be clinically achieved using a gene-therapy approach. The next challenge will be to determine if treatment with gene-corrected autologous cells can be as effective as enzyme replacement or as allogeneic transplantation. It is reasonable to expect that this goal will be met as improved gene-therapy protocols lead to a higher proportion of transduced cells after transplant and to enhanced vector expression in vivo. It is with this expectation that investigators are pursuing parallel strategies for the treatment of other immunodeficiency syndromes.

Defects in Signal Transduction as Targets for SCID Gene Therapy

Recent work has shown that defects in signal transduction pathways are the most common cause of SCID (Fig. 8). Mutations in the common γ chain (γ_c), a transmembrane peptide that is a shared component of the IL-2, IL-4, IL-7, IL-9, and IL-15 receptors, are the cause of X-linked SCID (XSCID) (Noguchi et al. 1993) and account for approximately

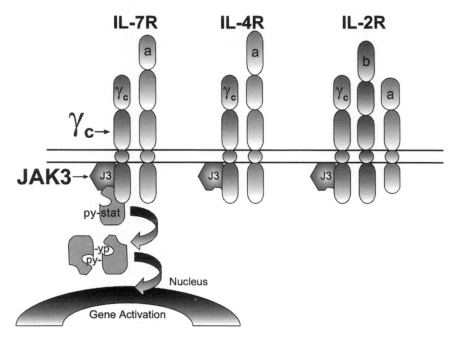

Figure 8 Cytokine signaling pathways as gene-therapy targets for SCID. The cellular receptors for interleukins 7 (IL-7R), 4 (IL-4R), and 2 (IL-2R) are required for the normal development and function of lymphocytes. Each of these receptors is composed of multiple and distinct polypeptide chains; however, they all contain a common γ chain (γ_c) that is required for signal transduction. Mutations affecting the γ_c cause X-linked SCID, the most common cause of the SCID syndrome. For mitogenic signaling to occur, the γ_c must associate with a cytoplasmic tyrosine kinase called Janus Kinase 3 (JAK3). Deficiency of JAK3 causes an autosomal recessive SCID syndrome that is otherwise indistinguishable from a γ_c deficiency. Upon ligand binding to the receptor, JAK3 becomes phosphorylated and leads to the phosphorylation and activation of STAT proteins. This cascade ultimately leads to activation of downstream genes in the nucleus. Both the γ_c and JAK3 are candidates for a gene-replacement approach to SCID.

half of all SCID cases (Buckley et al. 1997). These patients present with a distinct immunophenotype showing severe deficiency of T and NK cells but normal or high numbers of poorly functioning B cells. It is known that γ_c must associate with JAK3, a cytoplasmic tyrosine kinase expressed predominantly in hematopoietic cells, for signal transduction to occur following receptor activation (Miyazaki et al. 1994; S.M. Russell et al. 1994). Ligand binding to the receptor results in JAK3 phosphorylation and activation of downstream effector molecules such as the STAT proteins (Ihle 1995). Because the γ_c–JAK3 interaction is highly specific and nonredundant, it is not surprising that mutations in patients which abrogate JAK3 expression cause an SCID syndrome that is highly similar to that seen in XSCID (Russell et al. 1995; Macchi et al. 1995). In part because JAK3-deficient SCID has an autosomal recessive pattern of inheritance, it is a rare disorder accounting for about 7% of all SCID cases (Buckley et al. 1997).

Preclinical evidence suggests that JAK3- and γ_c-deficient SCID may be amenable to a gene-therapy approach. Retroviral vectors containing these cDNAs have been used to reconstitute signal transduction pathways in EBV-immortalized B cells from patients with XSCID (Candotti et al. 1996a; Taylor et al. 1996a) and JAK3 deficiency (Candotti et al. 1996b). These initial in vitro studies left several important questions unanswered. Early lymphoid development could require tightly regulated expression of JAK3 or the γ_c receptor chain that would not necessarily be recapitulated with expression from an exogenous retroviral vector. Second, potentially disregulated expression of either JAK3 or the γ_c could result in abnormal lymphoproliferation or even leukemia. Other questions are whether a significant number of corrected cells can be achieved in vivo and whether a vector-reconstituted immune system would respond normally to antigenic challenge.

Our group has recently used an animal model of JAK3-deficient SCID to address some of these questions. Mice bearing homozygous disruptions of the JAK3 allele display an SCID-like syndrome, characterized by very low numbers of T and B lymphocytes and severe defects in both cellular and humoral immunity (Nosaka et al. 1995). Reintroduction of the JAK3 gene into stem cells from JAK3-deficient mice and transplantation of these modified cells into JAK3-deficient recipients resulted in significant levels of immune reconstitution (Bunting et al. 1998). Transplanted mice displayed significant increases in the numbers of T and B cells in the peripheral blood (Fig. 9). These JAK3-corrected cells were functionally restored, as evidenced by the reconstitution of serum IgG and IgA levels, and the ability of these mice to mount a specific antibody response to neoantigens. There was no evidence of abnormal hematopoiesis, showing that exogenous JAK3 gene expression appears to be safe and does not cause abnormal cellular proliferation. These experiments also verified a large selective advantage for JAK3-corrected lymphoid cells in vivo, suggesting that a relatively small number of transduced stem cells may be sufficient for reconstituting physiologically significant levels of immunity. On the basis of these animal data, gene therapy seems to be a promising approach for JAK3 deficiency and perhaps for γ_c deficiency as well.

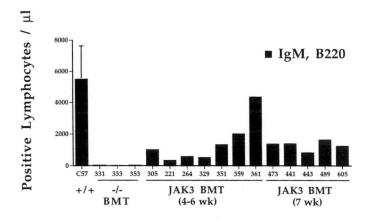

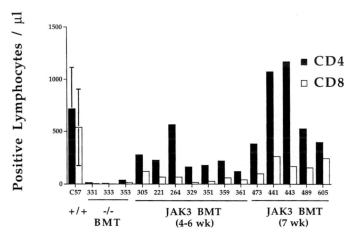

Figure 9 Peripheral blood T- and B-cell counts in JAK3-deficient mice treated with retrovirus-mediated gene transfer. The number of peripheral-blood B cells (*top panel*) and T cells (*bottom panel*) was measured in wild-type mice (+/+), in control JAK3-deficient mice (–/– BMT), and in JAK3-deficient mice transplanted with bone-marrow cells that had been transduced with a JAK3-expressing retroviral vector (JAK3 BMT). Two independent groups of JAK3 BMT mice were analyzed between 4 and 7 weeks after transplant. B cells were defined as cells staining positive for both surface IgM and the B220 antigen. T cells were identified by staining with either anti-CD4 (*dark bars*) or anti-CD8 antibodies (*open bars*). For +/+ mice, the bar represents the average of multiple mice and the lines one standard deviation from the mean. Note that all mice showed at least partial correction of the near absolute lymphopenia seen in JAK3-deficient mice. (Adapted from Bunting et al. 1998.)

Chronic Granulomatous Disease

Another type of immunodeficiency that is the focus of gene-therapy efforts is chronic granulomatous disease (CGD). This disease is characterized by a failure of myeloid-derived phagocytic cells to produce superoxide, resulting in a syndrome of recurrent pyogenic infections and granuloma formation. Defects in any one of four genes can lead to CGD by causing dysfunction in the superoxide-generating enzyme complex NADPH oxidase. Two thirds of the cases are caused by mutations in the X-linked gene called gp91^{phox} (Dinauer et al. 1987). A rarely occurring autosomal recessive form is caused by alterations in the gene encoding gp22^{phox}, and defects in gp47^{phox} and gp67^{phox} account for the remainder of autosomal recessive cases.

In vitro studies have shown the potential of retrovirus-mediated gene transfer for correcting these biochemical defects (Cobbs et al. 1992; Sekhsaria et al. 1993; Kume and Dinauer 1994; F. Li et al. 1994; Weil et al. 1997). The availability of mouse models for these syndromes (Pollock et al. 1995; Jackson et al. 1995) has allowed further testing of this approach in an in vivo setting. In a murine model for X-linked CGD, retrovirus-mediated transfer of the murine gp91^{phox} gene resulted in correction of 50–80% of the circulating neutrophils (Bjorgvinsdottir et al. 1997). These mice were significantly protected from pulmonary infections with *Aspergillus fumigatus*, an important cause of infections in CGD patients. In this study, mice with between 1% and 2% transduced cells were not protected from fungal disease, whereas X-CGD mice containing between 5% and 8% wild-type granulocytes were resistant to infection. Analogous studies have been done in gp47^{phox}-deficient mice (Mardiney et al. 1997). Recipient mice received G-CSF and then subablative (500 cGy) total body irradiation. These animals were then treated with a relatively large dose of gene-corrected progenitors that had been transduced with a gp47^{phox} retroviral vector. Four weeks after transplantation, about 12% of the neutrophils were oxidase-positive, with this number falling to about 3% at 14 weeks. Between 12 and 14 weeks after transplant, these mice were challenged with a relevant bacterial pathogen, *Burkholderia cepacia*. In mice receiving transduced cells, bacteremia was greatly reduced, with an associated significant increase in survival. These results suggest that subablative conditioning regimens may allow reconstitution with a low but clinically useful number of gene-corrected phagocytes.

Both of these animal studies suggest that correction of greater than 5% of circulating neutrophils will have therapeutic benefit, a figure that agrees well with observations in carrier females with skewed X-chromosome inactivation patterns. A human phase I clinical study has been undertaken to determine if this degree of marking can be achieved without the undesirable toxicity of myeloablative conditioning. Autologous CD34$^+$ peripheral blood cells from CGD patients were transduced with an amphotropic retroviral p47^{phox} vector and reinfused without any radiation or cytotoxic drug conditioning (Malech et al. 1996). In five patients treated on this protocol, corrected neutrophils were detected in the peripheral blood for several months, but the frequency was low, ranging between 1 per 2000 and 1 per 50,000. Although these results are encouraging and demonstrate vector function in vivo in transplanted patients, the number of corrected cells is far too low to confer clinical benefit. It is likely that some form of subablative conditioning therapy, perhaps combined with an in vivo selection system, will be required to achieve clinical efficacy.

Gene-therapy vectors are being developed and tested for a number of other genetically defined immunodeficiency syndromes. These include ZAP-70 deficiency (Taylor et al. 1996b), the Wiskott–Aldrich syndrome, and X-linked agammaglobulinemia. In general, these systems are in relatively early stages of exploration and await animal model data to assess their potential for effective gene therapy.

Metabolic Storage Diseases

Metabolic storage diseases are a relatively diverse group of disorders in which deficiencies in enzyme systems lead to toxic substrate accumulations in a variety of tissues. A relatively large number of single gene defects have been identified as the cause of these various disorders. Because many of these diseases are treatable with allogeneic bone-marrow transplantation, hematopoietic cell gene therapy is being studied as an alternative approach.

Gaucher disease is due to a genetic deficiency of glucocerebrosidase (GC), which results in the deleterious accumulation of glucocerebroside in macrophages and macrophage-derived cells within various organs. Pathologic changes are typically noted in the bone marrow, spleen, liver, and brain. In part because this is the most common human lysosomal storage disorder, a number of groups have pursued gene-therapy strategies targeting the hematopoietic stem cell. A number of retroviral vectors containing GC cDNAs have been used for the in vitro correction of cells from Gaucher-disease patients (Fink et al. 1990; Bahnson et al. 1994; Nimgaonkar et al. 1994; Xu et al. 1994). In animal model studies, transfer of GC vectors into repopulating stem cells has been demonstrated, with physiologically significant levels of expression documented in hematopoietic organs (Correll et al. 1990; 1992, 1994; Xu et al. 1995).

For Gaucher disease, successful gene therapy will require the engraftment of genetically corrected cells in various nonhematopoietic organs. In particular, pathology involving the central nervous system (CNS) microglial cells, which are derived from bone-marrow macrophages, leads to severe neurologic dysfunction in type-2 and -3 Gaucher patients. An important question is whether the transduced cells will migrate into the CNS after stem cell transplantation, and thereby confer a beneficial effect on the neurological manifestations. Murine transplant experiments have shed some light on this issue. In mice transplanted with bone-marrow cells transduced with a GC retroviral vector, a rapid appearance of marked macrophages was seen in the bone marrow, spleen, peripheral blood, liver, and lung (Krall et al. 1994). In contrast, engraftment of microglial cells in the brain and spinal cord occurred relatively late, with approximately 21% of the microglial cells in the brain parenchyma expressing the GC vector 6–8 months after transplant. The distribution of modified microglial cells within the CNS is somewhat controversial, with a second study showing that most were located in perivascular and leptomeningeal sites, with very few marked cells in the parenchyma of the brain (Kennedy and Abkowitz 1997). These studies, together with data from allogeneic transplant trials, suggest that most disease manifestations could be effectively treated with the transplant of modified hematopoietic stem cells but that the efficacy for treating neurologic manifestations is uncertain.

Two pilot clinical trials have been initiated to test the safety and feasibility of gene therapy for Gaucher disease. Both trials utilize retrovirally transduced autologous $CD34^+$ cells, obtained from the blood or bone marrow, for transfusion into nonablated patients. Early results from these studies have shown that fewer than 1% of circulating leukocytes contained the GC vector after transplant (Barranger et al. 1997; Dunbar et al. 1997). Surprisingly, high levels of enzyme activity and apparent clinical benefit were reported in one patient, despite the low levels of corrected cells (Barranger et al. 1997). Presumably, higher numbers of transduced cells will be required for consistent benefit. One possibility for achieving this aim is the use of selectable bicistronic vectors, which allow for the enrichment of GC-expressing cells prior to transplant (Migita et al. 1995; Medin et al. 1996) or the in vivo selection of corrected cells after transplant (Aran et al. 1994).

Gene-therapy approaches are being studied for a number of other metabolic storage diseases. Each of these syndromes has unique biologic variables that may influence the potential success of a gene-replacement approach. For example, in mucopolysaccharidosis type II (Hunter syndrome), transduced cells can "cross-correct" defective fibroblasts in vitro (Braun et al. 1993). These results have led to lymphocyte-targeted approaches, deducing that transduced peripheral blood lymphocytes could mediate the correction of other affected cells types in *trans* (Braun et al. 1996). The rapid emergence of animal models for storage disorders (Hahn et al. 1997) will greatly facilitate the elucidation of many of these disease-specific issues.

Bone-marrow Failure Syndromes

Fanconi anemia (FA) is an autosomal recessive disorder characterized by bone-marrow failure, congenital skeletal abnormalities, and an increased incidence of malignancy. Using cell hybridization studies, at least five complementation subtypes have been identified (A–E). The genes for complementation groups A and C have been cloned (Strathdee et al. 1992; Lo et al. 1996), and mutations in these alleles account for 15% and 60% of cases, respectively (Whitney et al. 1993; Joenje et al. 1995). The standard curative therapy for severe FA is allogeneic transplantation. Although relatively good results have been obtained in the minority of patients with a matched sibling donor, patients transplanted with alternative donors do significantly less well (Gluckman et al. 1995). Gene therapy is therefore being pursued as an alternative therapeutic approach.

Several in vitro studies have demonstrated that gene replacement can correct the cellular biochemical defects in FA. Retroviral vectors expressing either the FAC or FAA protein have been used to transduce both lymphoblastoid cell lines and primary myeloid progenitors from FA patients. These studies have shown that transduced cells display normalization of cell cycle kinetics, resistance to drug-induced chromosomal breakage, and an in vitro survival advantage relative to unmodified parental FA cells (Walsh et al. 1994a; Fu et al. 1997). Analogous results have been obtained using an AAV vector that expressed the FAC protein (Walsh et al. 1994b). The observed in vitro survival advantage suggests that corrected stem-cell clones could have a natural selective advantage in vivo.

On the basis of this hypothesis, three FA group C patients have been treated with multiple infusions of retrovirally transduced peripheral blood progenitor cells, given without any preceding myeloablation (Liu et al. 1997). Transient increases in bone-marrow cellularity were seen together with an expansion in myeloid-progenitor numbers. The retroviral genome was transiently detected in circulating peripheral blood cells. The lack of long-term reconstitution with transduced cells suggests that the survival advantage of corrected cells is not sufficient to overcome the very low engraftment efficiency that occurs in the absence of conditioning therapy. Further work will be necessary to determine the most appropriate clinical approach for using genetically corrected cells in this disorder.

Gene-therapy studies are just beginning for another unrelated syndrome that leads to bone-marrow failure, paroxysmal nocturnal hemoglobinuria (PNH). Patients are typically afflicted with intravascular hemolysis, increased thrombosis, and a variable degree of bone-marrow failure. This clonal disorder of hematopoietic stem cells is caused by a partial or complete absence of phosphatidyl inositol glycan (GPI) biosynthesis, leading to deficient expression of GPI-linked proteins on the surface of hematopoietic cells. The molecular defect in PNH involves various lesions in the X-linked PIG-A gene, which encodes a protein required for the biosynthesis of GPI. Retroviral vectors containing the PIG-A cDNA have been used to transduce hematopoietic cells from PNH patients. Preliminary results have shown that these vectors reconstitute cell-surface expression of GPI-linked proteins both in patient-derived cell lines and in primary hematopoietic cells (Nishimura et al. 1997), providing the rationale for further study of this approach.

Hemoglobin Disorders

Severe β-thalassemia and sickle cell anemia, diseases that provided the impetus for initiating the field of gene therapy, remain important targets for therapeutic gene transfer into hematopoietic stem cells (Sadelain 1997). These diseases are among the most common single gene disorders worldwide, and their molecular pathogenesis has been thoroughly studied. Bone-marrow transplantation with genetically normal donor cells is curative, ensuring that transplantation of gene-corrected, autologous stem cells would be beneficial. Simple replacement of the defective or missing β-globin gene in severe β-thalassemia could be curative. High-level expression of a normal γ-globin gene in the erythroid cells of sickle-cell-anemia patients is likely to be beneficial because of the preferential assembly of α and γ tetramers and the relative instability of β^S chains. HbF or mixed tetramers ($\alpha_2\beta^S\gamma$) are not incorporated into sickle hemoglobin polymers. An alternative approach to the transfer of a globin gene, namely, transfer and expression, of a genetic element that enhances γ over β^S globin gene expression, could be of therapeutic benefit, in patients with sickle-cell anemia or severe β-thalassemia.

Effective gene therapy for hemoglobin disorders is likely to require very high levels of globin gene expression. Achieving this high level of expression has represented a unique barrier to the development of gene therapy of hemoglobin disorders in addition to the other barriers to stem-cell-targeted gene transfer. Vectors containing the β-globin gene in reverse orientation to the LTR-driven, retroviral genomic transcript so as to preserve the intronic structure of the globin gene readily transduced murine hematopoietic stem cells (Dzierzak et al. 1988; Karlsson et al. 1988; Bender et al. 1989). However, globin gene expression was either absent or very low, namely, less than 1% of that of a mouse genomic globin gene.

Discovery of the locus control region (LCR) upstream of the human β-globin gene locus (Grosveld et al. 1987) initiated a new effort to make useful globin gene retroviral vectors. The core elements from the LCR, detected as hypersensitive sites (HS) in nuclear chromatin, confer high-level, relatively position-independent expression of globin genes in transgenic mice (Orkin 1995). Unfortunately, the incorporation of HS elements into retroviral vectors renders such vectors unstable during passage in vitro or during attempted transduction of primitive hematopoietic cells (Novak et al. 1990; Chang et al. 1992; Plavec et al. 1993; Leboulch et al. 1994; Ren et al. 1996). Enormous efforts have been invested in attempting to find a vector design that preserves stability and allows transmission of the unrearranged vector genome into hematopoietic, target-cell populations.

Only recently have vectors been described that can be used to transduce mouse erythroleukemia cells or primary murine hematopoietic progenitors. Variability in the level of expression of the β-globin transgene was found, indicating that the truncated core HS elements of the LCR are unable to confer true position-independent expression (Sadelain et al. 1995; Takekoshi et al. 1995). LTR-mediated silencing of a β-globin gene linked to LCR elements has been documented in transgenic mice (McCune and Townes 1994). Thus, the recent, successful stem-cell-targeted transfer and long-term expression of the human β-globin gene in a few mice (Raftopolous et al. 1997), while encouraging, remain to be evaluated in the context of a larger experience.

rAAV vectors have also been evaluated for their ability to transfer globin genes into hematopoietic cells. Transduction of human erythroid progenitors and expression of a mutationally marked $^{A}\gamma*$ gene in maturing erythroblasts have been well documented (J.L. Miller et al. 1994; Hargrove et al. 1997). Using human erythroleukemia cells as a model, inclusion of the regulatory element from 3´ to the human chromosomal $^{A}\gamma$ gene has been shown to facilitate the integration of an rAAV genome in the form of head-to-tail tandem arrays. These tandem arrays are stable during extended cell proliferation, and the integrated globin gene copies are expressed at levels comparable to those of a normal chromosomal globin gene. Although rAAV has been reported to transduce primitive hematopoietic cells (Fisher-Adams et al. 1996; Lubovy et al. 1996; Ponnazhagan et al. 1997a), the high multiplicities of infection required for transduction with the globin gene rAAV have precluded extensive evaluation of their ability to transduce primitive human hematopoietic cells with the potential for bone-marrow reconstitution.

Hematopoietic Cell Gene Therapy for Cancer

The majority of clinical gene-therapy trials that have opened to date are focused on the treatment of malignant disease. The main strategies being employed are insertion of immunomodulatory genes into tumor cells to induce antitumor immunity and insertion of suicide genes to induce drug sensitivity. Because these approaches are usually used to treat solid tumors and are further discussed in another chapter in this volume, they will not be reviewed here. Instead, we will focus on strategies that specifically modify hematopoietic cells to gain an advantage in cancer treatment. One specific approach that has already yielded promising clinical results is the modulation of allogeneic immune effects in patients transplanted for hematologic malignancy. The second area to be reviewed is the insertion of drug-resistant genes into stem cells, to protect against the myelosuppressive effects of cytotoxic anticancer agents. We discuss the results from the first clinical attempts at genetic chemoprotection and the extensive preclinical studies that have led to these trials.

Modulation of Alloreactivity in Bone-marrow Transplantation

Allogeneic bone-marrow transplantation is the treatment of choice for several hematologic malignancies. The curative potential of this treatment relies, in part, on the allogeneic immune reactivity established by the donor hematopoietic system. It is now well established that the donor graft can recognize and destroy residual malignant host cells in what has been termed a *graft-versus-malignancy* effect. This effect has been utilized, through the delayed infusion of donor lymphocytes, to treat patients whose malignancy relapses following allogeneic transplantation. These donor lymphocyte infusions have resulted in complete remissions in relapsed chronic myelogenous leukemia, acute leukemia, lymphoma, and multiple myeloma. Unfortunately, the *allogeneic immune advantage* conferred by donor lymphocytes is often accompanied by an attack on normal host tissues, causing the common but serious syndrome of graft-versus-host disease (GvHD). This complication is a major limitation both to the use of donor lymphocyte infusions and to allogeneic bone-marrow transplantation in general. At present, there is no effective way to segregate the GvHD response from the beneficial antimalignant response.

It has recently been demonstrated that a gene-therapy approach can be used to modulate this allogeneic reactivity in a clinically beneficial way. In an effort to control the unwanted GvHD effect, a suicide vector was used to enable elimination of donor lymphocytes as a specific therapy for GvHD (Bonini et al. 1997b). This retroviral vector contained a herpes simplex virus–thymidine kinase (HSV-TK) gene. HSV-TK confers ganciclovir sensitivity to transduced lymphocytes. The vector also contained a truncated nerve growth factor gene (tNGF) that allowed isolation of transduced lymphocytes based on tNGF cell-surface expression. In eight patients with a variety of posttransplant malignancies, allogeneic lymphocytes were collected from the original transplant donor, transduced with the suicide vector, immunopurified for tNGF expression, and infused at various doses into these patients. The proportion of marked cells in the peripheral circulation ranged from less than 0.001% to 13.4%, and marked cells were detectable in vivo for up to 12 months after the last treatment. Five of the eight patients developed clinical antitumor responses following infusion of the gene-modified donor lymphocytes. Three of these responding patients subsequently developed GvHD and were treated with ganciclovir to eliminate the transduced donor lymphocytes. In two patients, the proportion of transduced lymphocytes fell below the level of PCR detection within 24 hours of ganciclovir administration. This disappearance of transduced cells was accompanied by a complete regression of all signs of acute GvHD. These patients remained in full hematologic remission from their malignancy despite the lysis of allogeneic lymphocytes. One patient with chronic GvHD displayed a partial clinical improvement with ganciclovir treatment, with a reduction but not elimination of transduced donor lymphocytes.

These results show that the allogeneic response can be modulated in a clinically beneficial way for the treatment of relapsed hematologic malignancy. It may be possible to extend this approach to the initial allogeneic transplant procedure by depleting the graft of T cells, modifying the cells with an HSV-TK vector, and reinfusing a stem cell graft supplemented with these modified lymphocytes. Several critical questions regarding this approach will need to be addressed in subsequent clinical trials. Despite the antitumor activity seen in this pilot trial, it will need to be established in a larger group of patients that genetically modified donor lymphocytes are not compromised in their ability to elicit the graft-versus-malignancy response. A second open question is the response rate for treating GvHD with ganciclovir. The lack of complete elimination of transduced lympho-cytes in one patient has been ascribed to cell cycle dependence for ganci-clovir killing. Contaminating untransduced donor lymphocytes could also be the cause of refractory GvHD. For these reasons, the efficacy in GvHD modulation will need to be established in a larger patient group. Finally, several of these patients have developed an immune response to the HSV-TK/neo^R gene product that resulted in the elimination of engineered cells (Bonini ct al. 1997a). Deletion of the neo^R gene from the vector may reduce the incidence of this complication. International multicenter collaborative clinical trials have been initiated to address these issues and to test the effi-cacy of this novel gene-therapy application further.

Hematopoietic Chemoprotection

Failure in cancer chemotherapy is most commonly due to the emergence of drug resistance in tumor cells. In many instances, this resistance can be avoided or overcome by increasing the doses of the cytotoxic drugs being used. Dose intensification is often restricted by toxicity in normal tissues, most commonly myelosuppression. This narrow therapeutic ratio is one of the fundamental limitations of cancer chemotherapy (Fig. 10).

Current clinical strategies for attenuating drug-induced myelosup-pression have clearly been useful, but they are associated with a number of limitations. One strategy that is widely used in clinical oncology is the use of hematopoietic cytokines such as G-CSF. One disadvantage is that these cytokines usually affect only a single hematopoietic lineage, so that severe cytopenias in other lineages may still occur. Second, because cytokine administration acts by hastening hematopoietic regeneration, cytopenic nadirs are not eliminated or diminished in magnitude, but sim-ply shortened in length. In part because of these problems, there are rela-tively few specific instances in which cytokines are of proven benefit (Anonymous 1994). A second strategy for overcoming dose limitations due to hematopoietic toxicity is the use of autologous stem cell rescue. With this approach, patients are given a drug dose that would otherwise cause lethal myelosuppression, but are then rescued by a subsequent infu-sion of stored autologous stem cells. While this allows high doses of myelosuppressive drugs to be given over several days, high-dose therapy is generally administered only once and subsequent chemotherapy is usu-ally poorly tolerated. For these reasons, stem cell transplant does not enable significant dose intensification over time.

The transfer of drug resistant genes to hematopoietic stem cells cir-cumvents many of these limitations (Sorrentino 1996). Transplantation of

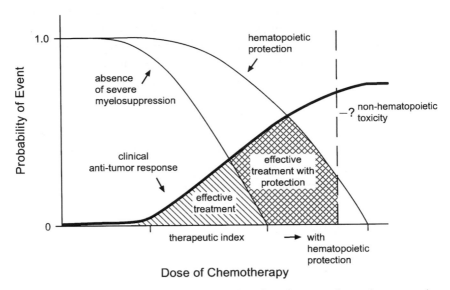

Figure 10 Widening the therapeutic ratio of anticancer drugs by protecting hematopoiesis with drug-resistance vectors. With increases in the dose intensity of chemotherapuetic drugs (*rightward along the x axis*), the probablity of antitumor responses increase (*bold line*). The ability to increase the delivered dose is limited by toxic effects on normal tissues, often dose-limiting myelosuppression. If hematopoietic protection is afforded with drug-resistance vectors, higher doses of drugs may be safely used. The resulting increase in the therapeutic index may potentially lead to an increased probability of effective treatment (cross-hatched area). In animal models, absolute protection against myelosuppression has been achieved with this approach. In this circumstance, dose escalation then becomes limited by nonhematopoietic toxicities in other organ systems. This second barrier to dose escalation occurs at different points with various drugs and remains to be defined for some systems. (Adapted from Sorrentino 1996.)

pluripotent stem cells that have been transduced with a resistance gene may protect all hematopoietic lineages. Because stem-cell progeny are protected at all developmental stages, cytopenic nadirs may be completely eliminated. This hematopoietic protection may be durable with repeated drug-treatment cycles, due to the long-lived nature and self-renewal capacity of stem cells. Finally, because drug-resistant hematopoiesis may be established after engraftment, post-transplant dose intensification may be possible. These advantages have all been validated in murine animal model studies, where the efficiency of stem-cell transduction is not limiting (Spencer et al. 1996).

Dihydrofolate reductase variants for antifolate resistance. Antifolate drugs such as MTX are commonly used in oncologic therapy. Their main toxicity is myelosuppression and gastrointestinal damage. High doses of MTX can be safely given when reduced folate, in the form of leucovorin, is subsequently administered for rescue. One problem with leucovorin rescue is that tumor cells can be rescued along with normal host cells (Sirotnak et al. 1978), so that the specificity and benefit of this approach have been questioned (Ackland and Schilsky 1987). The availability of drug-resistant variants of DHFR (Simonsen and Levinson 1983) suggested the possibility of using gene transfer for hematopoietic protection. In fact, the first successful demonstration of gene-therapy-based chemoprotection was done using a DHFR vector to protect transplanted mice from MTX (Williams et al. 1987). Subsequent studies in murine models have confirmed the ability of DHFR gene transfer to protect hematopoiesis from various antifolates and have shown that this protection confers tolerance to otherwise lethal drug doses (Corey et al. 1990; M.X. Li et al. 1994; Zhao et al. 1994; Spencer et al. 1996). A recent study has shown that this increased tolerance to high doses of MTX leads to improved curability of human breast tumor xenografts (Zhao et al. 1997). DHFR vectors have also been shown to protect human hematopoietic cells from MTX in vitro (Flasshove et al. 1995), suggesting their possible use in human gene-therapy trials.

There are a number of remaining questions regarding the clinical translation of this approach. One important issue is which DHFR genes will be most effective for hematopoietic protection. Since the original description of the murine L22R variant (Simonsen and Levinson 1983), other DHFR mutations have been identified that confer higher levels of drug resistance (Schweitzer et al. 1989; Dicker et al. 1993; Banerjee et al. 1994; Nakano et al. 1994; Lewis et al. 1995). Recent work shows that DHFR cDNAs bearing double active-site substitutions (Patel et al. 1997) or vectors expressing both DHFR together with viral thymidine kinase (Mineishi et al. 1997) may be optimal for antifolate protection. A second issue is whether gastrointestinal toxicity will quickly become dose-limiting when hematopoiesis is protected (Fig. 10). A recent study suggests that this will not be the case. In MTX-treated mice transplanted with bone-marrow cells expressing a DHFR transgene, hematopoietic protection was associated with secondary protection from gastrointestinal toxicity (May et al. 1995).

Efficacy in clinical trials will require achieving an adequate number of transduced cells for hematopoietic protection. We have recently shown that significant protection can be achieved when more than 10% of circulating blood cells in mice express a DHFR vector (Allay et al. 1998). Although this threshold requirement has not yet been achieved in clinical trials, recent experiments in nonhuman primates suggest that it is within reach (Dunbar et al. 1996). Another consideration is that antifolates are usually combined with other myelosuppressive drugs as part of a combination chemotherapy regimen. For most diseases, adequate hematopoietic protection will require resistance to multiple drugs. We have recently shown that this goal can be achieved with vectors containing multiple drug-resistant genes. A retroviral vector expressing both a DHFR variant and the human MDR1 gene protected hematopoietic cells in vitro from a wide variety of drugs, including simultaneous exposure to taxol and trimetrexate (Galipeau et al. 1997).

MDR1 gene transfer. Another widely explored chemoprotection system uses the human MDR1 gene to protect against myelotoxic anticancer drugs. P-glycoprotein (Pgp), the MDR1 gene product, is an energy-dependent transmembrane efflux pump and extrudes a specific set of cytotoxic drugs from the cell's interior. Substrates for Pgp include many commonly used myelosuppressive drugs such as taxol, daunorubicin, vinblastine, and etoposide. The first evidence that the MDR1 gene could be used for hematopoietic protection came from in vitro studies of retrovirally transduced murine myeloid progenitors (McLachlin et al. 1990) and from studies in MDR1 transgenic mice (Mickisch et al. 1991a,b). This protective capacity was later confirmed in mice that were transplanted with retrovirally transduced stem cells. In these experiments, significant protection from drug-induced myelosuppression was seen in mice expressing the MDR1 vector (Sorrentino et al 1992; Hanania et al. 1995a; Aksentijevich et al. 1996). Human CD34[+] hematopoietic cells from a variety of sources have also been transduced with MDR1 retroviral vectors and studied for drug resistance. One question was whether MDR1 vectors could increase the overall level of Pgp expression in primitive myeloid cells, given that CD34[+] cells have relatively high basal levels of endogenous Pgp expression (Chaudhary and Roninson 1991). These studies showed that MDR1 vectors did increase the overall Pgp expression in transduced CFU-Cs and that transduced progenitors were significantly more drug resistant than their untransduced counterparts (Bertolini et al. 1994; Ward et al. 1994, 1996; Hanania et al. 1995b).

Several clinical trials have been initiated to test the feasibility of transferring MDR1 retroviral vectors into reconstituting hematopoietic cells obtained from cancer patients. In a trial at M.D. Anderson Cancer Center, 20 patients with either ovarian or breast cancer were transplanted with MDR1-transduced CD34$^+$ cells obtained from either bone marrow or peripheral blood, respectively (Hanania et al. 1996). Three to four weeks after transplant, MDR1-transduced bone-marrow cells were detected in five of eight patients who received cells transduced on stromal monolayers. The concentration of transduced cells in these patients was about 1 in 1000. No marked cells were seen in any of the ten patients who received cells transduced in suspension without stromal cell support. A second study at Columbia University has yielded similar results (Hesdorffer et al. 1998). CD34^1 hematopoietic cells obtained from patients with advanced cancer were transduced with an amphotropic retroviral MDR1 vector. Following high-dose chemotherapy, these cells were reinfused together with unmanipulated cells in a 1-to-2 ratio. Although between 20% and 50% of CFU-GM contained the vector immediately prior to transplant, marked cells were detected in only two of five evaluable patients following hematopoietic recovery. In these two patients, the proportion of marked cells was less than 1 in 1000 when analyzed between 3 and 10 weeks after transplant.

The small numbers of MDR1-transduced hematopoietic cells seen in these initial clinical trials will not be sufficient for attenuating myelosuppression. It is clear that improved gene-transfer protocols will be necessary to confer significant protection, and a number of areas can be identified for potential improvement. The scaled-up vector preparations used in these studies had relatively low titers, ranging from 2×10^4 to 1×10^5 cfu/ml, so that higher-titer vectors should increase the number of transduced repopulating cells. Second, the low numbers of transduced cells in the Columbia University study may have been due to, at least in part, the fact that unmanipulated cells were concurrently infused. These cells would be predicted to have a competitive advantage over their transduced counterparts (Peters et al. 1996) given the manipulations that are inherent to the transduction protocol. A third area that can be optimized is the stability of the vector genome. The MDR1 cDNA is prone to rearrangements due to the presence of cryptic RNA splicing sites within the coding sequence (Sorrentino et al. 1995). In these clinical studies, it was not clear what proportion of transduced cells were bearing nonfunctional rearranged genomes. The availability of alternate MDR1 cDNAs that yield more stable retroviral vectors should provide an advantage in this regard (Galipeau et al. 1997). Finally, an issue that was not addressed in these studies is the level of Pgp expression obtained in vivo in transduced cells. Newer vectors that direct increased Pgp expression in primitive hematopoietic cells may be advantageous in this regard (Baum et al. 1995; Eckert et al. 1996).

Alkyltransferase gene transfer for nitrosourea protection. A third major system that is being evaluated for hematopoietic protection involves modulation of DNA repair mechanisms. Nitrosoureas, such as BCNU and CCNU, are DNA-damaging agents that alkylate guanine residues at the O^6 position, ultimately leading to cytotoxic interstrand DNA cross-links. These drugs are toxic to hematopoietic stem cells, causing prolonged bone-marrow suppression as the major dose-limiting toxicity. Cellular protection from nitrosoureas can be conferred by expression of the protein O^6-alkylguanine DNA alkyltransferase (AGT). This protein functions as a scavenger for alkyl residues at O^6 guanine and removes DNA adducts before cytotoxic cross-linking events can occur. Overexpression of the human AGT, also referred to as methylguanine DNA methyltransferase (MGMT), is an important cause of nitrosourea resistance in human tumors (Erickson et al. 1980). Therefore, it has been hypothesized that the therapeutic ratio of nitrosoureas can be increased by protecting hematopoietic cells from toxicity, by sensitizing tumor cells to killing, or preferably by the simultaneous attainment of both of these goals.

Hematopoietic progenitors are sensitive to nitrosoureas because they express very low levels of endogenous AGT (Gerson et al. 1985, 1996). This observation has led investigators to explore the possibility of using enforced AGT expression to achieve chemoprotection. Retroviral vectors have been used to transfer the human *MGMT* cDNA into repopulating murine hematopoietic cells. After transplant, mice were engrafted with drug-resistant myeloid progenitors (Allay et al. 1995) and were protected from the myelosuppressive effects of BCNU (Moritz et al. 1995). Stem-cell transfer of *MGMT* also protected lymphopoiesis and thereby ameliorated the severe immunodeficiency caused by BCNU treatment (Maze et al. 1997). These protective effects on hematopoiesis significantly increased survival in heavily BCNU-treated mice (Maze et al. 1996), showing that otherwise lethal dose increases can be achieved using this strategy. In vitro resistance to BCNU has also been demonstrated in human CD34+ bone-marrow cells transduced with an amphotropic *MGMT* retroviral vector (Allay et al. 1996).

This field has most recently focused on the use of mutant *MGMT* genes that encode for proteins resistant to O^6-benzylguanine (BG), an inactivator of wild-type *MGMT*. BG strongly potentiates the tumoricidal effects of nitrosoureas by stoichiometrically depleting cells of *MGMT* activity (Gerson et al. 1993). The attractiveness of using BG to sensitize tumor cells to BCNU has been dampened by the observation that BG also sensitizes hematopoietic cells (Gerson et al. 1996), predicting that the use of this drug combination would be limited by severe myelotoxicity. There is now evidence that this limitation can be circumvented by transfer of *MGMT* genes bearing single amino acid substitutions (Crone et al. 1994) that confer resistance to BG-induced inactivation. One such mutant, bearing a glycine to alanine substitution at amino acid 156, has been tested in the context of a retroviral vector. Human peripheral blood CD34$^+$ cells, transduced with a G156A *MGMT* vector, were significantly more resistant in vitro to BG plus BCNU than cells transduced with a wild-type *MGMT* vector (Reese et al. 1996). This vector has recently been tested in a mouse transplant model, where it was shown to protect from the lethal effects of BG and BCNU administration (Davis et al. 1997). Altogether, these promising results suggest that it may be possible to sensitize tumor cells to nitrosoureas, while simultaneously protecting against hematopoietic toxicity using a gene-transfer approach. Clinical trials testing *MGMT* gene transfer are just beginning at the University of Indiana and at Case Western Reserve University and should determine the feasibility of this approach.

Summary of the Potential of Hematopoietic Cell Gene Transfer as a Clinical Therapy

With the rapid pace of new discoveries regarding the molecular basis of hematopoietic disorders, gene transfer is increasingly being considered as a potential treatment option. For any given disease, there are a number of specific biologic issues that are relevant to the development of a gene-therapy approach. For example, in various SCID syndromes, there exists a naturally occurring selective advantage for corrected cells so that therapeutic benefit may be possible with current gene-transfer methodologies. Even for chronic granulomatous disease, where corrected cells have no survival advantage in vivo, the 3 5% level of corrected cells required for therapeutic benefit may be obtainable by optimizing currently available protocols. Given these considerations, and the limitations associated with allogeneic transplantation, it seems likely that within the near future gene therapy will become a commonly used treatment for these immunodeficiency disorders.

In contrast, effective gene therapy for thalassemia and sickle cell disease will take more time to develop. Effective therapy for these disorders is likely to require the production of high levels of a normal globin polypeptide in a large proportion of maturing erythroblasts. More progress is needed in the development of globin gene vectors. Once available, they will need to be used in conjunction with improved transduction protocols which, together with methodology for the in vivo amplification of transduced stem cells, may allow attainment of an adequate number of corrected erythrocytes. Given the pace of progress in these areas, gene therapy for hemoglobinopathies is a realistic goal over the next decade. Considering the prevalence and worldwide health impact of these disorders, the continued pursuit of this challenging problem seems well worthwhile.

Genetic chemoprotection of hematopoiesis is a promising area for cancer treatment that has been extensively validated in preclinical models. The main barrier to clinical application is achieving an adequate number of transduced stem cells in patients. The minimum proportion of genetically modified hematopoietic cells needed for protection is likely to vary among the different drug resistance systems, so that although some of these strategies may be effectively applied in the near future, others will require improved gene-transfer methodology for success. Finally, modulation of alloreactivity with suicide vectors is a promising approach with established clinical feasibility. Further clinical studies should define the utility of this strategy and determine in which transplant settings it can be applied.

ACKNOWLEDGMENTS

This work was supported in part by National Heart, Lung, and Blood Institute Program Project grant PO1 Hl53749, Cancer Center Support CORE grant P30 Ca21765, the ASSISI Foundation of Memphis grant, and the American Lebanese Syrian Associated Charities (ALSAC). We thank Mrs. Jean Johnson for her invaluable assistance in organizing and preparing this manuscript. This work is dedicated to the memory of Dr. Joseph Sorrentino, whose love and dedication to both science and medicine will always be an inspiration to me (B.P.S.).

REFERENCES

Ackland S.P. and Schilsky R.L.1987. High-dose methotrexate: A critical reappraisal. *J. Clin. Oncol.* **5:** 2017–2031.

Akkina R.K., Walton R.M., Chen M.L., Li Q.-X., Planelles V., and Chen I.S.Y. 1996. High-efficiency gene transfer into CD34+ cells with a human immunodeficiency virus type 1-based retroviral vector pseudotyped with vesicular stomatitis virus envelope glycoprotein G. *J. Virol.* **70:** 2581–2585.

Aksentijevich I., Cardarelli C.O., Pastan I., and Gottesman M.M. 1996. Retroviral transfer of the human MDR1 gene confers resistance to bisantrene-specific hematotoxicity. *J. Clin. Cancer Res.* **2:** 973–980.

Alexander I.E., Russell D.W., and Miller A.D. 1994. DNA-damaging agents greatly increase the transduction of nondividing cells by adeno-associated virus vectors. *J. Virol.* **68:** 8282–8287.

———1997. Transfer of contaminants in adeno-associated virus vector stocks can mimic transduction and lead to artifactual results. *Hum. Gene Ther.* **8:** 1911–1920.

Allay J.A., Davis B.M., and Gerson S.L. 1997a. Human alkyltransferase-transduced murine myeloid progenitors are enriched in vivo by BCNU treatment of transplanted mice. *Exp. Hematol.* **25:** 1069–1076.

Allay J.A., Galipeau J., Blakley R.L., and Sorrentino B.P. 1998. Retroviral vectors containing a variant dihydrofolate reductase gene for drug protection and in vivo selection of hematopoietic cells. *Stem Cells* (suppl. 1) **16:** 223–234.

Allay J.A., Koc O.N., Davis B.M., and Gerson S.L. 1996. Retroviral-mediated gene transduction of human alkyltransferase complementary DNA confers nitrosourea resistance to human hematopoietic progenitors. *Clin. Cancer Res.* **2:** 1353–1359.

Allay J.A., Dumenco L.L., Koc O.N., Liu L., and Gerson S.L. 1995. Retroviral transduction and expression of the human alkyltransferase cDNA provides nitrosourea resistance to hematopoietic cells. *Blood* **85:** 3342–3351.

Allay J.A., Galipeau J., Persons D.A., Ashmun R.A., Blakley R.L., and Sorrentino B.P. 1997b. In vivo selection of DHFR-modified hematopoietic cells following trimetrexate and NBMPR-P treatment. *Blood* **90:** 242a

Allay J.A., Spencer H.T., Wilkinson S.L., Belt J.A., Blakley R.L., and Sorrentino B.P. 1997c. Sensitization of hematopoietic stem and progenitor cells to trimetrexate using nucleoside transport inhibitors. *Blood* **90:** 3546–3554.

Allen J.M., Debelak D.J., Reynolds T.C., and Miller A.D. 1997. Identification and elimination of replication-competent adeno-associated virus (AAV) that can arise by nonhomolgous recombination during AAV vector production. *J. Virol.* **71:** 6816–6822.

Anderson W.F. and Fletcher J.C. 1980. Sounding boards: Gene therapy in human beings: When is it ethical to begin? *N. Engl. J. Med.* **303:** 1293–1297.

Anonymous 1994. American Society of Clinical Oncology. Recommendations for the use of hematopoietic colony-stimulating factors: Evidence-based, clinical practice guidelines. *J. Clin. Oncol.* **12:** 2471–2508.

Aran J.M., Gottesman M.M., and Pastan I. 1994. Drug-selected coexpression of human glucocerebrosidase and P glycoprotein using a bicistronic vector. *Proc. Natl. Acad. Sci.* **91:** 3176–3180.

Bahnson A.B., Nimgaonkar M., Fei Y., Boggs S.S., Robbins P.D., Ohashi T., Dunigan J., Li J., Ball E.D., and Barranger J.A. 1994. Transduction of CD34⁺ enriched cord blood and Gaucher bone marrow cells by a retroviral vector carrying the glucocerebrosidase gene. *Gene Ther.* **1:** 176–184.

Banerjee D., Schweitzer B.I., Volkenandt M., Li M.X., Waltham M., Mineishi S., Zhao S.C., and Bertino J.R. 1994. Transfection with a cDNA encoding a Ser31 or Ser34 mutant human dihydrofolate reductase into Chinese hamster ovary and mouse marrow progenitor cells confers methotrexate resistance. *Gene* **139:** 269–274.

Barranger J.A., Rice E., Sansieri C., Bahnson A., Mohney T., Swaney W., Takiyama N., Dunigan J., Beeler, M., Lucot S., Schierer-Fochler S., and Ball E. 1997. Transfer of the glucocerebrosidase gene to CD34 cells and their autologous transplantation in patients with Gaucher disease. *Blood* **90:** 405a.

Bauer Jr., T.R., Miller A.D., and Hickstein D.D. 1995. Improved transfer of the leukocyte integrin CD18 subunit into hematopoietic cell lines by using retroviral vectors having a gibbon ape leukemia virus envelope. *Blood* **86:** 2379–2387.

Baum C., Hegewisch-Becker S., Eckert H.G., Stocking C., and Ostertag W. 1995. Novel retroviral vectors for efficient expression of the multidrug resistance (mdr-1) gene in early hematopoietic cells. *J. Virol.* **69:** 7541–7547.

Bender M.A., Gelinas R.E., and Miller A.D. 1989. A majority of mice show long-term expression of a human β-globin gene after retrovirus transfer into hematopoietic stem cells. *Mol. Cell. Biol.* **9:** 1426–1434.

Bender M.A., Palmer T.D., Gelinas R.E., and Miller A.D. 1987. Evidence that the packaging signal of Moloney murine leukemia virus extends into the *gag* region. *J. Virol.* **61:** 1639–1646.

Berns K.I. and Giraud C. 1996. Biology of adeno-associated virus. *Curr. Top. Microbiol. Immunol.* **218:** 1–23.

Bertolini F., de Monte L., Corsini C., Lazzari L., Lauri E., Soligo D., Ward M., Bank A., and Malavasi F. 1994. Retrovirus-mediated transfer of the multidrug resistance gene into human haemopoietic progenitor cells. *Br. J. Haematol.* **88:** 318–324.

Bertran J., Miller J.L., Yang Y., Fenimore-Justman A., Rueda F., Vanin E.F., and Nienhuis A.W. 1996. Recombinant adeno-associated virus-mediated high-efficiency, transient expression of the murine cationic amino acid transporter (ecotropic retroviral receptor) permits stable transduction of human HeLa cells by ecotropic retroviral vectors. *J. Virol.* **70:** 6759–6766.

Bhatia M., Bonnet D., Kapp U., Wang J.C.Y., Murdoch B., and Dick J.E. 1997. Quantitative analysis reveals expansion of human hematopoietic repopulating cells after short-term ex vivo culture. *J. Exp. Med.* **186:** 619–624.

Bieniasz P.D., Weiss R.A., and McClure M.O. 1995. Cell cycle dependence of foamy retrovirus infection. *J. Virol.* **69:** 7295–7299.

Bieniasz P.D., Erlwein O., Aguzzi A., Rethwilm A., and McClure M.O. 1997. Gene transfer using replication-defective human foamy virus vectors. *Virology* **235:** 65–72.

Bienzle D., Abrams-Ogg A.C., Kruth S.A., Ackland-Snow J., Carter R.F., Dick J.E., Jacobs R.M., Kamel-Reid S., and Dube I.D. 1994. Gene transfer into hematopoietic stem cells: Long-term maintenance of in vitro activated progenitors without marrow ablation. *Proc. Natl. Acad. Sci.* **91:** 350–354.

Bjorgvinsdottir H., Ding C., Pech N., Gifford M.A., Li L.L., and Dinauer M.C. 1997. Retroviral-mediated gene transfer of gp91phox into bone marrow cells rescues defect in host defense against *Asperfillus fumigatus* in murine X-linked chronic granulomatous disease. *Blood* **89:** 41–48.

Blaese R.M., Culver K.W., Miller A.D., Carter C.S., Fleisher T., Clerici M., Shearer G., Chang L., Chiang Y., Tolstoshev P., Greenblatt J.J., Rosenberg S.A., Klein H., Berger M., Mullen C.A., Ramsey W.J., Muul L., Morgan R.A., and Anderson W.F. 1995. T lymphocyte-directed gene therapy for ADA⁻ SCID: Initial trial results after 4 years. *Science* **270:** 475–480.

Blau C.A., Neff T., and Papayannopoulou T. 1996. The hematological effects of folate analogs: Implications for using the dihydrofolate reductase gene for in vivo selection. *Hum. Gene Ther.* **7:** 2069–2078.

—— 1997a. Cytokine prestimulation as a gene therapy strategy: Implications for using the MDR1 gene as a dominant selectable marker. *Blood* **89:** 146–154.

Blau C.A., Peterson K.R., Drachman J.G., and Spencer D.M. 1997b. A proliferation switch for genetically modified cells. *Proc. Natl. Acad. Sci.* **94:** 3076–3081.

Blomer U., Naldini L., Kafri T., Trono D., Verma I.M., and Gage F.H. 1997. Highly efficient and sustained gene transfer in adult neurons with a lentivirus vector. *J. Virol.* **71:** 6641–6649.

Bock T.A., Orlic D., Dunbar C.E., Broxmeyer H.E., and Bodine D.M. 1995. Improved engraftment of human hematopoietic cells in severe combined immunodeficient (SCID) mice carrying human cytokine transgenes. *J. Exp. Med.* **182:** 2037–2043.

Bodem J., Löchelt M., Yang P., and Flügel R.M. 1997. Regulation of gene expression by human foamy virus and potentials of foamy viral vectors. *Stem Cells* (Suppl. 1) **15:** 141–147.

Bodine D.M., Karlsson S., and Nienhuis A.W. 1989. Combination of interleukins 3 and 6 preserves stem cell function in culture and enhances retrovirus-mediated gene transfer into hematopoietic stem cells. *Proc. Natl. Acad. Sci.* **86:** 8897–8901.

Bodine D.M., Seidel N.E., and Orlic D. 1996. Bone marrow collected 14 days after in vivo administration of granulocyte colony-stimulating factor and stem cell factor to mice has 10-fold more repopulating ability than untreated bone marrow. *Blood* **88:** 89–97.

Bodine D.M., McDonagh K.T., Seidel N.E., and Nienhuis A.W. 1991. Survival and retrovirus infection of murine hematopoietic stem cells in vitro: Effects of 5-FU and method of infection. *Exp. Hematol.* **19:** 206–212.

Bodine D.M., Seidel N.E., Gale M.S., Nienhuis A.W., and Orlic D. 1994. Efficient retrovirus transduction of mouse pluripotent hematopoietic stem cells mobilized into the peripheral blood by treatment with granulocyte colony-stimulating factor and stem cell factor. *Blood* **84:** 1482–1491.

Bodine D.M., McDonagh K.T., Brandt S.J., Ney P.A., Agricola B., Byrne E., and Nienhuis A.W. 1990. Development of a high-titer retrovirus producer cell line capable of gene transfer into rhesus monkey hematopoietic stem cells. *Proc. Natl. Acad. Sci.* **87:** 3738–3742.

Bodine D.M., Moritz T., Donahue R.E., Luskey B.D., Kessler S.W. Martin D.I., Orkin S.H., Nienhuis A.W., and Williams D.A. 1993. Long-term in vivo expression of a murine adenosine deaminase gene in rhesus monkey hematopoietic cells of multiple lineages after retroviral mediated gene transfer into CD34+ bone marrow cells. *Blood* **82:** 1975–1980.

Bonini C., Verzeletti S., Marktel S., Zappone E., Servida P., Rossini S., Traversari C., and Bordignon C. 1997a. Factors affecting efficacy of HSV-tk-transduced donor peripheral blood lymphocytes after allo-BMT. *Blood* **90:** 238a.

Bonini C., Ferrari G., Verzeletti S., Servida P., Zappone E., Ruggieri L., Ponzoni M., Rossini S., Mavilio F., Traversari C., and Bordignon C. 1997b. HSV-TK gene transfer into donor lymphocytes for control of allogeneic graft-versus-leukemia. *Science* **276:** 1719–1724.

Bordignon C., Notarangelo L.D., Nobili N., Ferrari G., Casorati G., Panina P., Mazzolari E., Maggioni D., Rossi C., Servida P., Ugazio A.G., and Mavilio F. 1995. Gene therapy in peripheral blood lymphocytes and bone marrow for ADA⁻ immunodeficient patients. *Science* **270:** 470–475.

Borge O.J., Ramsfjell V., Cui L., and Jacobsen S.E. 1997. Ability of early acting cytokines to directly promote survival and suppress apoptosis of human primitive $CD34^+CD38^-$ bone marrow cells with multilineage potential at the single-cell level: Key role of thrombopoietin. *Blood* **90:** 2282–2292.

Braun S.E., Aronovich E.L., Anderson R.A., Crotty P.L., McIvor R.S., and Whitley C.B. 1993. Metabolic correction and cross-correction of mucopolysaccharidosis type II (Hunter syndrome) by retroviral-mediated gene transfer and expression of human iduronate-2-sulfatase. *Proc. Natl. Acad. Sci.* **90:** 11830–11834.

Braun S.E., Pan D., Aronovich E.L., Jonsson J.J., McIvor R.S., and Whitley C.B. 1996. Preclinical studies of lymphocyte gene therapy for mild Hunter syndrome (mucopolysaccharidosis type II). *Hum. Gene Ther.* **7:** 283–290.

Brenner M.K., Rill D.R., Moen R.C., Krance R.A., Mirro Jr. J., Anderson W.F., and Ihle J.N. 1993a. Gene-marking to trace origin of relapse after autologous bone-marrow transplantation. *Lancet* **341:** 85–86.

Brenner M.K., Rill D.R., Holladay M.S., Heslop H.E., Moen R.C., Buschle M., Krance R.A., Santana V.M., Anderson W.F., and Ihle J.N. 1993b. Gene marking to determine whether autologous marrow infusion restores long-term haemopoiesis in cancer patients. *Lancet* **342:** 1134–1137.

Buckley R.H., Schiff R.I., Schiff S.E., Markert M.L., Williams L.W., Harville T.O., Roberts J.L., and Puck J.M. 1997. Human severe combined immunodeficiency: Genetic, phenotypic, and functional diversity in one hundred eight infants. *J. Pediatr.* **130:** 378–387.

Buckley R.H., Schiff S.E., Schiff R.I., Roberts J.L., Markert M.L., Peters W., Williams L.W., and Ward F.E. 1993. Haploidentical bone marrow stem cell transplantation in human severe combined immunodeficiency. *Semin. Hematol.* **30:** 92–101.

Bukrinsky M.I., Haggerty S., Dempsey M.P., Sharova N., Adzhubel A., Spitz L., Lewis P., Goldfarb D., Emerman M., and Stevenson M. 1993. A nuclear localization signal within HIV-1 matrix protein that governs infection of non-dividing cells. *Nature* **365:** 666–669.

Bunnell B.A., Muul L.M., Donahue R.E., Blaese R.M., and Morgan R.A. 1995. High efficiency retroviral-mediated gene transfer into human and non-human primate peripheral blood lymphocytes. *Proc. Natl. Acad. Sci.* **92:** 7739–7743.

Bunting K.D., Sangster M.Y., Ihle J.N., and Sorrentino B.P. 1998. Restoration of lymphocyte function in Janus kinase 3-deficient mice by retroviral-mediated gene transfer. *Nat. Med.* **4:** 58–64.

Camaur D., Gallay P., Swingler S., and Trono D. 1997. Human immunodeficiency virus matrix tyrosine phosphorylation: Characterization of the kinase and its substrate requirements. *J. Virol.* **71:** 6834–6841.

Candotti F., Johnston J.A., Puck J.M., Sugamura K., O'Shea J.J., and Blaese R.M. 1996a. Retroviral-mediated gene correction for X-linked severe combined immunodeficiency. *Blood* **87:** 3097–3102.

Candotti F., Oakes S.A., Johnston J.A., Notarangelo L.D., O'Shea J.J., and Blaese R.M. 1996b. In vitro correction of JAK3-deficient severe combined immunodeficiency by retroviral-mediated gene transduction. *J. Exp. Med.* **183:** 2687–2692.

Carter R.F., Abrams-Ogg A.C., Dick J.E., Kruth S.E., Valli V.E., Kamel-Reid S., and Dube I.D. 1992. Autologous transplantation of canine long-term marrow culture cells genetically marked by retroviral vectors. *Blood* **79:** 356–364.

Cashman J., Bockhold K., Hogge D.E., Eaves A.C., and Eaves C.J. 1997. Sustained proliferation, multi-lineage differentiation and maintenance of primitive human haemopoietic cells in NOD/SCID mice transplanted with human cord blood. *Br. J. Haematol.* **98:** 1026–1036.

Challita P.M., Skelton D., el-Khoueiry A., Yu X.J., Weiberg K., and Kohn D.B. 1995. Multiple modifications in *cis* elements of the long terminal repeat of retroviral vectors lead to increased expression and decreased DNA methylation in embryonic carcinoma cells. *J. Virol.* **69:** 748–755.

Chang J.C., Liu D., and Kan Y.W. 1992. A 36-base-pair core sequence of locus control region enhances retrovirally transferred human β-globin gene expression. *Proc. Natl. Acad. Sci.* **89:** 3107–3110.

Chaudhary P.M. and Roninson I.B. 1991. Expression and activity of P-glycoprotein, multidrug efflux pump, in human hematopoietic stem cells. *Cell* **66:** 85–94.

Chien M.L., Foster J.L., Douglas J.L., and Garcia J.V. 1997. The amphotropic murine leukemia virus receptor gene encodes a 71-kilodalton protein that is induced by phosphate depletion. *J. Virol.* **71:** 4564–4570.

Chiorini J.A., Yang L., Safer B., and Kotin R.M. 1995. Determination of adeno-associated virus Rep68 and Rep78 binding sites by random sequence oligonucleotide selection. *J. Virol.* **69:** 7334–7338.

Clark K.R., Voulgaropoulou F., Fraley D.M., and Johnson P.R. 1995. Cell lines for the production of recombinant adeno-associated virus. *Hum. Gene Ther.* **6:** 1329–1341.

Cline M.J., Stang H., Mercola K., Morse L., Ruprecht R., Brown J., and Salser W. 1980. Gene transfer in intact animals. *Nature* **284:** 422–425.

Cobbs C.S., Malech H.L., Leto T.L., Freeman S.M., Blaese R.M., Gallin J.I., and Lomax K.J. 1992. Retroviral expression of recombinant p47phox protein by Epstein-Barr virus-transformed B lymphocytes from a patient with autosomal chronic granulomatous disease. *Blood* **79:** 1829–1835.

Conley M.E., Buckley R.H., Hong R., Guerra-Hanson C., Roifman C.M., Brochstein J.A., Pahwa S., and Puck J.M. 1990. X-linked severe combined immunodeficiency. Diagnosis in males with sporadic severe combined immunodeficiency and clarification of clinical findings. *J. Clin. Invest.* **85:** 1548–1554.

Conneally E., Cashman J., Petzer A., and Eaves C. 1997. Expansion in vitro of transplantable human cord blood stem cells demonstrated using a quantative assay of their lymph-myeloid repopulating activity in nonobese diabetic-scid/scid mice. *Proc. Natl. Acad. Sci.* **94:** 9836–9841.

Corey C.A., DeSilva A.D., Holland C.A., and Williams D.A. 1990. Serial transplantation of methotrexate-resistant bone marrow: Protection of murine recipients from drug toxicity by progeny of transduced stem cells. *Blood* **75:** 337–343.

Cornetta K., Morgan R.A., Gillio A., Strum S., Baltrucki L., O'Reilly R., and Anderson W.F. 1991. No retroviremia or pathology in long-term follow-up of monkeys exposed to a murine amphotropic retrovirus. *Hum. Gene Ther.* **2:** 215–219.

Cornetta K., Moen R.C., Culver K., Morgan R.A., McLachlin J.R., Strum S., Selegue J., London W., Blaese R.M., and Anderson W.F. 1990. Amphotropic murine leukemia retrovirus is not an acute pathogen for primates. *Hum. Gene Ther.* **1:** 15–30.

Cornetta K., Srour E.F., Moore A., Davidson A., Broun E.R., Hromas R., Moen R.C., Morgan R.A., Rubin L., Anderson W.F., Hoffman R., and Tricot G. 1996. Retroviral gene transfer in autologous bone marrow transplantation for adult acute leukemia. *Hum. Gene Ther.* **7:** 1323–1329.

Correll P.H., Colilla S., and Karlsson S. 1994. Retroviral vector design for long-term expression in murine hematopoietic cells in vivo. *Blood* **84:** 1812–1822.

Correll P.H., Colilla S., Dave H.P., and Karlsson S. 1992. High levels of human glucocerebrosidase activity in macrophages of long-term reconstituted mice after retroviral infection of hematopoietic stem cells. *Blood* **80:** 331–336.

Correll P.H., Kew Y., Perry L.K., Brady R.O., Fink J.K., and Karlsson S. 1990. Expression of human glucocerebrosidase in long-term reconstituted mice following retroviral-mediated gene transfer into hematopoietic stem cells. *Hum. Gene Ther.* **1:** 277–287.

Cournoyer D., Scarpa M., Mitani K., Moore K.A., Markowitz D., Bank A., Belmont J.W., and Caskey C.T. 1991. Gene transfer of adenosine deaminase into primitive human hematopoietic progenitor cells. *Hum. Gene Ther.* **2:** 203–213.

Crone T.M., Goodtzova K., Edara S., and Pegg A.E. 1994. Mutations in human O6-alkylguanine-DNA alkyltransferase imparting resistance to O6-benzylguanine. *Cancer Res.* **54:** 6221–6227.

Crooks G.M. and Kohn D.B. 1993. Growth factors increase amphotropic retrovirus binding to human CD34$^+$ bone marrow progenitor cells. *Blood* **82:** 3290–3297.

Culver K.W., Morgan R.A., Osborne W.R.A., Lee R.T., Lenschow D., Able C., Cornetta K., Anderson W.F., and Blaese R.M. 1990. *In vivo* expression and survival of gene-modified T lymphocytes in rhesus monkeys. *Hum. Gene Ther.* **1:** 399–410.

Culver K., Cornetta K., Morgan R., Morecki S., Aebersold P., Kasid A., Lotze M., Rosenberg S.A., Anderson W.F., and Blaese R.M. 1991. Lymphocytes as cellular vehicles for gene therapy in mouse and man. *Proc. Natl. Acad. Sci..* **88:** 3155–3159.

Danos O. and Mulligan R.C. 1988. Safe and efficient generation of recombinant retroviruses with amphotropic and ecotropic host ranges. *Proc. Natl. Acad. Sci.* **85:** 6460–6464.

Davis B.M., Reese J.S., Koc O.N., Lee K., Schupp J.E., and Gerson S.L. 1997. Selection for G156A O^6–methylguanine DNA methyltransferase gene-transduced hematopoietic progenitors and protection from lethality in mice treated with O^6–benzylguanine and 1,3-*bis*(2-chloroethyl)-1-nitrosourea. *Cancer Res.* **57:** 5093–5099.

Deisseroth A.B., Zu Z., Claxton D., Hanania E.G., Fu S., Ellerson D., Goldberg L., Thomas M., Janicek K., Anderson W.F., Hester J., Korbling M., Durett A., Moen R., Berenson R., Heimfeld S., Hamer J., Calvert L., Tibbits P., Talpaz M., Kantarjian H., Champlin R., and Reading C. 1994. Genetic marking shows that Ph+ cells present in autologous transplants of chronic myelogenous leukemia (CML) contribute to relapse after autologous bone marrow in CML. *Blood* **83:** 3068–3076.

Dick J.E., Kamel-Reid S., Murdoch B., and Doedens M. 1991. Gene transfer into normal human hematopoietic cells using in vitro and in vivo assays. *Blood* **78:** 624–634.

Dick J.E., Magli M.C., Huszar D., Phillips R.A., and Bernstein A. 1985. Introduction of a selectable gene into primitive stem cells capable of long-term reconstitution of the hemopoietic system of W/W^V mice. *Cell* **42:** 71–79.

Dicker A.P., Waltham M.C., Volkenandt M., Schweitzer B.I., Otter G.M., Schmid F.A., Sirotnak F.M., and Bertino J.R. 1993. Methotrexate resistance in an in vivo mouse tumor due to a non-active-site dihydrofolate reductase mutation. *Proc. Natl. Acad. Sci..* **90:** 11797–11801.

Dilloo D., Rill D., Entwistle C., Boursnell M., Zhong W., Holden W., Holladay M., Inglis S., and Brenner M. 1997. A novel herpes vector for the high-efficiency transduction of normal and malignant human hematopoietic cells. *Blood* **89:** 119–127.

Dinauer M.C., Orkin S.H., Brown R., Jesaitis A.J., and Parkos C.A.1987. The glycoprotein encoded by the X-linked chronic granulomatous disease locus is a component of the neutrophil cytochrome b complex. *Nature* **327:** 717–720.

Donahue R.E., Kessler S.W., Bodine D., McDonagh K., Dunbar C., Goodman S., Agricola B., Byrne E., Raffeld M., Moen R., Bacher J., Zsebo K.M., and Nienhuis A.W. 1992. Helper virus induced T cell lymphoma in nonhuman primates after retroviral mediated gene transfer. *J. Exp. Med.* **176:** 1125–1135.

Dunbar C.E., Seidel N.E., Doren S., Sellers S., Cline A.P., Metzger M.E., Agricola B.A., Donahue R.E., and Bodine D.M. 1996. Improved retroviral gene transfer into murine and rhesus peripheral blood or bone marrow repopulating cells primed in vivo with stem cell factor and granulocyte colony-stimulating factor. *Proc. Natl. Acad. Sci.* **93:** 11871–11876.

Dunbar C.E., Cottler-Fox M., O'Shaughnessy J.A., Doren S., Carter C., Berenson R., Brown S., Moen R.C., Greenblat J., Stewart F.M., Leitman S.F., Wilson W.H., Cowan K., Young N.S., and Nienhuis A.W. 1995. Retrovirally marked CD34-enriched peripheral blood and bone marrow cells contribute to long-term engraftment after autologous transplantation. *Blood* **85:** 3048–3057.

Dunbar C.E., Kohn D.B., Schiffmann R., Barton N.W., Nolta J.A., Wells S., Esplin J.,

Pensiero M., Emmons R.V., Leitman S., Kreps C.B., Carter C., Kimball J., Young N.S., Brady R.O., and Karlsson S. 1997. Retroviral gene transfer of the glucocerebrosidase gene into PB or BM CD34+ cells from patients with Gaucher disease: Results from a clinical trial. *Blood* **90**: 237a.

Dzierzak E.A., Papayannopoulou T., and Mulligan R.C. 1988. Lineage-specific expression of a human β-globin gene in murine bone marrow transplant recipients reconstituted with retrovirus-transduced stem cells. *Nature* **331**: 35–41.

Eckert H.G., Stockschlader M., Just U., Hegewisch-Becker S., Grez M., Uhde A., Zander A., Ostertag W., and Baum C. 1996. High-dose multidrug resistance in primary human hematopoietic progenitor cells transduced with optimized retroviral vectors. *Blood* **88**: 3407–3415.

Efstratiadis A., Posakony J.W., Maniatis T., Lawn R.M., O'Connell C., Spritz R.A., DeRiel J.K., Forget B.G., Weissman S.M., Slightom J.L., Blechl A.E., Smithies O., Baralle F.E., Shoulders C.C., and Proudfoot N.J. 1980. The structure and evolution of the human β-globin gene family. *Cell* **21**: 653–668.

Eglitis M.A., Kantoff P., Gilboa E., and Anderson W.F. 1985. Gene expression in mice after high efficiency retroviral-mediated gene transfer. *Science* **230**: 1395–1398.

Ekhterae D., Crumbleholme T., Karson E., Harrison M.R., Anderson W.F., and Zanjani E.D. 1990. Retroviral vector-mediated transfer of the bacterial neomycin resistance gene into fetal and adult sheep and human hematopoietic progenitors in vitro. *Blood* **75**: 365–369.

Emmons R.V., Doren S., Zujewski J., Cottler-Fox M., Carter C.S., Hines K., O'Shaughnessy J.A., Leitman S.F., Greenblatt J.J., Cowan K., and Dunbar C.E. 1997. Retroviral gene transduction of adult peripheral blood or marrow-derived CD34⁺ cells for six hours without growth factors or on autologous stroma does not improve marking efficiency assessed in vivo. *Blood* **89**: 4040–4046.

Erickson L.C., Laurent G., Sharkey N.A., and Kohn K.W. 1980. DNA cross-linking and monoadduct repair in nitrosourea-treated human tumour cells. *Nature* **288**: 727–729.

Ferrari F.K., Samulski T., Shenk T., and Samulski R.J. 1996. Second-strand synthesis is a rate-limiting step for efficient transduction by recombinant adeno-associated virus vectors. *J. Virol.* **70**: 3227–3234.

Ferrari F.K., Xiao X., McCarthy D., and Samulski R.J. 1997. New development in the generation of Ad-free, high-titer rAAV gene therapy vectors. *Nat. Med.* **3**: 1295–1297.

Ferrari G., Rossini S., Giavazzi R., Maggioni D., Nobili N., Soldati M., Ungers G., Mavilio F., Gilboa E., and Bordignon C. 1991. An in vivo model of somatic cell gene therapy for human severe combined immunodeficiency. *Science* **251**: 1363–1366.

Fink J.K., Correll P.H., Perry L.K., Brady R.O., and Karlsson S. 1990. Correction of glucocerebrosidase deficiency after retroviral-mediated gene transfer into hematopoietic progenitor cells from patients with Gaucher disease. *Proc. Natl. Acad. Sci.* **87**: 2334–2338.

Fischer A., Landais P., Friedrich W., Morgan G., Gerritsen B., Fasth A., Porta F., Griscelli C., Goldman S.F., Levinsky R., and Vossen J. 1990. European experience of bone-marrow transplantation for severe combined immunodeficiency. *Lancet* **336**: 850–854.

Fisher K.J., Gao G.P., Weitzman M.D., DeMatteo R., Burda J.F., and Wilson J.M. 1996. Transduction with recombinant adeno-associated virus for gene therapy is limited by leading-strand synthesis. *J. Virol.* **70**: 520–532.

Fisher K.J., Jooss K., Alston J., Yang Y., Haecker S.E., High K., Pathak R., Raper S.E., and Wilson J.M. 1997. Recombinant adeno-associated virus for muscle directed gene therapy. *Nat. Med.* **3**: 306–312.

Fisher-Adams G., Wong Jr. K.K., Podsakoff G., Forman S.J., and Chatterjee S. 1996. Integration of adeno-associated virus vectors in CD34+ human hematopoietic progenitor cells after transduction. *Blood* **88:** 492–504.

Flasshove M., Banerjee D., Mineishi S., Li M.X., Bertino J.R., and Moore M.A. 1995. Ex vivo expansion and selection of human CD34+ peripheral blood progenitor cells after introduction of a mutated dihydrofolate reductase cDNA via retroviral gene transfer. *Blood* **85:** 566–574.

Flotte T.R. and Carter B.J. 1995. Adeno-associated virus vectors for gene therapy. *Gene Ther.* **2:** 357–362.

Fu K., Foe J.R., Joenje H., Rao K.W., Liu J.M., and Walsh C.E. 1997. Functional correction of Fanconi anemia group A hematopoietic cells by retroviral gene transfer. *Blood* **90:** 3296–3303.

Galipeau J., Benaim E., Spencer H.T., Blakley R.L., and Sorrentino B.P. 1997. A bicistronic retroviral vector for protecting hematopoietic cells against antifolates and P-glycoprotein effluxed drugs. *Hum. Gene Ther.* **8:** 1773–1783.

Gallay P., Swingler S., Aiken C., and Trono D. 1995a. HIV-1 infection of nondividing cells: C-terminal tyrosine phosphorylation of the viral matrix protein is a key regulator. *Cell* **80:** 379–388.

Gallay P., Swingler S., Song J., Bushman F., and Trono D. 1995b. HIV nuclear import is governed by the phosphotyrosine-mediated binding of matrix to the core domain of integrase. *Cell* **83:** 569–576.

Gerson S.L., Miller K., and Berger N.A. 1985. O6 alkylguanine-DNA alkyltransferase activity in human myeloid cells. *J. Clin. Invest.* **76:** 2106–2114.

Gerson S.L., Phillips W., Kastan M., Dumenco L.L., and Donovan C. 1996. Human CD34+ hematopoietic progenitors have low, cytokine-unresponsive O6-alkylguanine-DNA alkyltransferase and are sensitive to O6-benzylguanine plus BCNU. *Blood* **88:** 1649–1655.

Gerson S.L., Zborowska E., Norton K., Gordon N.H., and Willson J.K. 1993. Synergistic efficacy of O6-benzylguanine and 1,3-*bis*(2-chloroethyl)-1-nitrosourea (BCNU) in a human colon cancer xenograft completely resistant to BCNU alone. *Biochem. Pharmacol.* **45:** 483–491.

Gluckman E., Auerbach A.D., Horowitz M.M., Sobocinski K.A., Ash R.C., Bortin M.M., Butturini A., Camitta B.M., Champlin R.E., and Friedrich W. 1995. Bone marrow transplantation for Fanconi anemia. *Blood* **86:** 2856–2862.

Goodell M.A., Rosenzweig M., Kim H., Marks D.F., DeMaria M., Paradis G., Grupp S.A., Sieff C.A., Mulligan R.C., and Johnson R.P. 1997. Dye efflux studies suggest that hematopoietic stem cells expressing low or undetectable levels of CD34 antigen exist in multiple species. *Nat. Med.* **3:** 1337–1345.

Gossen M. and Bujard H. 1992. Tight control of gene expression in mammalian cells by tetracycline-responsive promoters. *Proc. Natl. Acad. Sci.* **89:** 5547–5551.

Grosveld F., van Assendelft G.B., Greaves D.R., and Kollias G. 1987. Position-independent, high-level expression of the human β-globin gene in transgenic mice. Cell **51:** 975–985.

Hahn C.N., del Pilar M., Schroder M., Vanier M.T., Hara Y., Suzuki K., and d'Azzo A. 1997. Generalized CNS disease and massive GM1-ganglioside accumulation in mice defective in lysosomal acid β-galactosidase. *Hum. Mol. Genet.* **6:** 205–211.

Hanania E.G., Fu S., Roninson I., Zu Z., Deisseroth A.B., and Gottesman M.M. 1995a. Resistance to taxol chemotherapy produced in mouse marrow cells by safety-modified retroviruses containing a human MDR-1 transcription unit. *Gene Ther.* **2:** 279–284.

Hanania E.G., Fu S., Zu Z., Hegewisch-Becker S., Korbling M., Hester J., Durett A.,

Andreeff M., Mechetner E., Holzmayer T., Roninson I.B., Giles R.E., Berenson R., Heimfeld S., and Deisseroth A.B. 1995b. Chemotherapy resistance to taxol in clonogenic progenitor cells following transduction of CD34 selected marrow and peripheral blood cells with a retrovirus that contains the MDR-1 chemotherapy resistance gene. *Gene Ther.* **2:** 285–294.

Hanania E.G., Giles R.E., Kavanagh J., Fu S.Q., Ellerson D., Zu Z., Wang T., Su Y., Kudelka A., Rahman Z., Holmes F., Hortobagyi G., Claxton D., Bachier C., Thall P., Cheng S., Hester J., Ostrove J.M., Bird R.E., Chang A., Korbling M., Seong D., Cote R., Holzmayer T., Deisseroth A.B., and Mechetner E. 1996. Results of MDR-1 vector modification trial indicate that granulocyte/macrophage colony-forming unit cells do not contribute to posttransplant hematopoietic recovery following intensive systemic therapy. *Proc. Natl. Acad. Sci.* **93:** 15346–15351.

Hanenberg H., Xiao X.L., Dilloo D., Hashino K., Kato I., and Williams D.A. 1996. Colocalization of retrovirus and target cells on specific fibronectin fragments increases genetic transduction of mammalian cells. *Nat. Med.* **2:** 876–882.

Hao Q.L., Shah A.J., Theimann F.T., Smogorzewska E.M., and Crooks G.M. 1995. A functional comparison of CD34+ CD38− cells in cord blood and bone marrow. *Blood* **86:** 3745–3753.

Hargrove P., Vanin E.F., Kurtzman G., and Nienhuis A.W. 1997. High level globin gene expression mediated by a recombinant adeno-associated virus genome which contains the 3´ λ globin gene regulatory element and integrates as tandem copies in erythroid cells. *Blood* **89:** 2167–2175.

Harris L.C., Marathi U.K., Edwards C.C., Houghton P.J., Srivastava D.K., Vanin E.F., Sorrentino B.P., and Brent T.P. 1995. Retroviral transfer of a bacterial alkyltransferase gene into murine bone marrow protects against chloroethylnitrosourea cytotoxicity. *J. Clin. Cancer Res.* **1:** 359–1368.

Hawley R.G., Lieu F.H., Fong A.Z., and Hawley T.S. 1994. Versatile retroviral vectors for potential use in gene therapy. *Gene Ther.* **1:** 136–138.

Heinzinger N.K., Bukinsky M.I., Haggerty S.A., Ragland A.M., Kewalramani V., Lee M.A., Gendelman H.E., Ratner L., Stevenson M., and Emerman M. 1994. The Vpr protein of human immunodeficiency virus type 1 influences nuclear localization of viral nucleic acids in nondividing host cells. *Proc. Natl. Acad. Sci.* **91:** 7311–7315.

Hesdorffer C., Ayello J., Ward M., Kaubisch A., Vahdat L., Balmaceda C., Garrett T.D., Fetell M., Reiss R., Bank A., and Antman K. 1998. Phase I trial of retroviral-mediated transfer of the human MDR1 gene as marrow chemoprotection in patients undergoing high-dose chemotherapy and autologous stem-cell transplantation. *J. Clin. Oncol.* **16:** 165–172.

Heslop H.E., Ng C.Y., Li C., Smith C.A., Loftin S.K., Krance R.A., Brenner M.K., and Rooney C.M. 1996. Long-term restoration of immunity against Epstein-Barr virus infection by adoptive transfer of gene-modified virus-specific T lymphocytes. *Nat. Med.* **2:** 551–555.

Hirata R.K., Miller A.D., Andrews R.G., and Russell D.W. 1996. Transduction of hematopoietic cells by foamy virus vectors. *Blood* **88:** 3654–3661.

Hirschhorn R., Yang D.R., Puck J.M., Huie M.L., Jiang C.K., and Kurlandsky L.E. 1996. Spontaneous in vivo reversion to normal of an inherited mutation in a patient with adenosine deaminase deficiency. *Nat. Genet.* **13:** 290–295.

Ihle J.N. 1995. Cytokine receptor signalling. *Nature* **377:** 591–594.

Jackson S.H., Gallin J.I., and Holland S.M. 1995. The p47phox mouse knock-out model of chronic granulomatous disease. *J. Exp. Med.* **182:** 751–758.

Joenje H., Lot F.J., Oostra A.B., van Berkel C.G., Rooimans M.A., Schroeder-Kurth T., Wegner R.D., Gille J.J., Buchwald M., and Arwert F. 1995. Classification of Fanconi anemia patients by complementation analysis: Evidence for a fifth genetic subtype. *Blood* **86:** 2156–2160.

Kantoff P.W., Freeman S.M., and Anderson W.F. 1988. Prospects for gene therapy for immunodeficiency diseases. *Annu. Rev. Immunol.* **6:** 581–594.

Kantoff P.W., Gillio A., McLachlin J.R., Bordignon C., Eglitis M.A., Kernan N.A., Moen R.C., Kohn D., Yu S.-F., Karson E., Karlsson S., Zwiebel J.A., Gilboa E., Blaese R.M., Nienhuis A., O'Reilly R.J., and Anderson W.F. 1987. Expression of human adenosine deaminase in non-human primates after retroviral mediated gene transfer. *J. Exp. Med.* **166:** 219–233.

Kaplitt M.G., Leone P., Samulski R.J., Xiao X., Pfaff D.W., O'Malley K.L., and During M.J. 1994. Long-term gene expression and phenotypic correction using adeno-associated virus vectors in the mammalian brain. *Nat. Genet.* **8:** 148–154.

Karlsson S. 1991. Treatment of genetic defects in hematopoietic cell function by gene transfer. *Blood* **78:** 2481–2492.

Karlsson S., Bodine D.M., Perry L., Papayannopoulou T., and Nienhuis A.W. 1988. Expression of the human β-globin gene following retroviral-mediated transfer into multipotential hematopoietic progenitors of mice. *Proc. Natl. Acad. Sci.* **85:** 6062–6066.

Kearns, W.G., Afione, S.A., Fulmer, S.B., Pang, M.C., Erikson, D., Egan, M., Landrum, M.J., Flotte, T.R., and Cutting, G.R. 1996. Recombinant adeno-associated virus (AAV-CFTR) vectors do not integrate in a site-specific fashion in an immortalized epithelial cell line. *Gene Ther.* **3:** 748–755.

Keller G., Paige C., Gilboa E., and Wagner E.F. 1985. Expression of a foreign gene in myeloid and lymphoid cells derived from multipotent haematopoietic precursors. *Nature* **318:** 149–154.

Kennedy D.W. and Abkowitz J.L. 1997. Kinetics of central nervous system microglial and macrophage engraftment: Analysis using a transgenic bone marrow transplantation model. *Blood* **90:** 986–993.

Kiem H.P., Heyward S., Winkler A., Potter J., Allen J.M., Miller A.D., and Andrews R.G. 1997. Gene transfer into marrow repopulating cells: Comparison between amphotropic and gibbon ape leukemia virus pseudotyped retroviral vectors in a competitive repopulation assay in baboons. *Blood* **90:** 4638–4645.

Kim V.N., Mitrophanous K., Kingsman S.M., and Kingsman A.J. 1998. Minimal requirement for a lentivirus vector based on human immunodeficiency virus type 1. *J. Virol.* **72:** 811–816.

Koeberl, D.D., Alexander I.E., Halbert C.L., Russell D.W., and Miller A.D. 1997. Persistent expression of human clotting factor IX from mouse liver after intravenous injection of adeno-associated virus vectors. *Proc. Natl. Acad. Sci.* **94:** 1426–1431.

Kohn D.B. 1997. Gene therapy for haematopoietic and lymphoid disorders. *Clin. Exp. Immunol.* **107:** 54–57.

Kohn D.B., Weinberg K.I., Shigeoka A., Carbonaro D., Brooks J., Smogorzewska E.M., Barsky L.W., Annett G., Nolta J.A., Kapoor N., Crooks G.M., Elder M., Wara D., Bowen T., Muul L., Hershfield M., Blaese R.M., and Parkman R. 1997. PEG-ADA reduction in recipients of ADA gene-transduced autologous umbilical cord blood CD34+ cells. *Blood* **90:** 404a.

Kohn D.B., Weinberg K.I., Nolta J.A., Heiss L.N., Lenarsky C., Crooks G.M., Hanley M.E., Annett G., Brooks J.S., el-Khoureiy A., Lawrence K., Wells S., Moen R.C., Bastian J., Williams-Herman D.E., Elder M., Wara D., Bowen T., Hershfield M.S., Mullen C.A., Blaese R.M., and Parkman R. 1995. Engraftment of gene-modified umbilical cord blood cells in neonates with adenosine deaminase deficiency. *Nat. Med.* **1:** 1017–1023.

Kotin R.M. 1994. Prospects for use of adeno-associated virus as a vector for human gene therapy. *Hum. Gene Ther.* **5:** 793–801.

Kotin R.M., Siniscalco M., Samulski R.J., Zhu X.D., Hunter L., Laughlin C.A., McLaughlin S., Muzyczka N., Rocchi M., and Berns K.I. 1990. Site-specific integration by adeno-associated virus. *Proc. Natl. Acad. Sci.* **87:** 2211–2215.

Krall W.J., Challita P.M., Perlmutter L.S., Skelton D.C., and Kohn D.B. 1994. Cells expressing human glucocerebrosidase from a retroviral vector repopulate macrophages and central nervous system microglia after murine bone marrow transplantation. *Blood* **83:** 2737–2748.

Kume A. and Dinauer M.C. 1994. Retrovirus-mediated reconstitution of respiratory burst activity in X-linked chronic granulomatous disease cells. *Blood* **84:** 3311–3116.

Larochelle A., Vormoor J., Hanenberg H., Wang J.C., Bhatia M., Lapidot T., Moritz T., Murdoch B., Xiao XL, Williams D.A., and Dick J.E. 1996. Identification of primitive human hematopoietic cells capable of repopulating NOD/SCID mouse bone marrow: Implications for gene therapy. *Nat. Med.* **2:** 1329–1337.

Lawn R.M., Efstratiadis A., O'Connell C., and Maniatis T. 1980. The nucleotide sequence of the human β-globin gene. *Cell* **21:** 647–651.

Leboulch P., Huang G.M., Humphries R.K., Oh Y.H., Eaves C.J., Tuan D.Y., and London I.M. 1994. Mutagenesis of retroviral vectors transducing human β-globin gene and β-globin locus control region derivatives results in stable transmission of an active transcriptional structure. *EMBO J.* **13:** 3065–3076.

Lemischka I.R., Raulet D.H., and Mulligan R.C. 1986. Developmental potential and dynamic behavior of hematopoietic stem cells. *Cell* **45:** 917–927.

Lewis P.F. and Emerman M. 1994. Passage through mitosis is required for oncoretroviruses but not for the human immunodeficiency virus. *J. Virol.* **68:** 510–516.

Lewis W.S., Cody V., Galitsky N., Luft J.R., Pangborn W., Chunduru S.K., Spencer H.T., Appleman J.R., and Blakley R.L. 1995. Methotrexate-resistant variants of human dihydrofolate reductase with substitutions of leucine 22. Kinetics, crystallography, and potential as selectable markers. *J. Biol. Chem.* **270:** 5057–5064.

Li F., Linton G.F., Sekhsaria S., Whiting-Theobald N., Katkin J.P., Gallin J.I., and Malech H.L. 1994. CD34+ peripheral blood progenitors as a target for genetic correction of the two flavocytochrome b558 defective forms of chronic granulomatous disease. *Blood* **84:** 53–58.

Li M.X., Banerjee D., Zhao S.C., Schweitzer B.I., Mineishi S., Gilboa E., and Bertino J.R. 1994. Development of a retroviral construct containing a human mutated dihydrofolate reductase cDNA for hematopoietic stem cell transduction. *Blood* **83:** 3403–3408.

Linden R.M., Ward P., Giraud C., Winocour E., and Berns K.I. 1996. Site-specific integration by adeno-associated virus. *Proc. Natl. Acad. Sci.* **93:** 11288–11294.

Liu J.M., Kim S., Read E.J., Dokal I., Carter C.S., Leitman S.F., Pensiero M., Young N.S., and Walsh C.E. 1997. Experimental trial of gene therapy for group C Fanconi anemia patients: 2 year follow-up. *Blood* **90:** 239a.

Lo Jr. T.F., Rooimans M.A., Bosnoyan-Collins L., Alon N., Wijker M., Parker L., Lightfoot J., Carreau M., Callen D.F., Savoia A., Cheng N.C., van Berkel C.G., Strunk

M.H., Gille J.J., Pals G., Kruyt F.A., Pronk J.C., Arwert F., Buchwald M., and Joenje H. 1996. Expression cloning of a cDNA for the major Fanconi anaemia gene, FAA. *Nat. Genet.* **14:** 320–232.

Lubovy M., McCune S., Dong J.Y., Prchal J.F., Townes T.M., and Prchal J.T. 1996. Stable transduction of recombinant adeno-associated virus into hematopoietic stem cells from normal and sickle cell patients. *Biol. Blood Marrow Transplant.* **2:** 24–30.

Luskey B.D., Rosenblatt M., Zsebo K., and Williams D.A. 1992. Stem cell factor, interleukin-3, and interleukin-6 promote retroviral-mediated gene transfer into murine hematopoietic stem cells. *Blood* **80:** 396–402.

Lyman S.D. and Jacobsen E.W. 1998. C-kit ligand and Flt3 ligand: Stem/progenitor cell factors with overlapping yet distinct activities. *Blood* **91:** 1101–1134.

Macchi P., Villa A., Gillani S., Sacco M.G., Frattini A., Porta F., Ugazio A.G., Johnston J.A., Candotti F., and O'Shea J.J. 1995. Mutations of Jak-3 gene in patients with autosomal severe combined immune deficiency (SCID). *Nature* **377:** 65–68.

Malech H.L., Sekhsaria S., Whiting-Theobald N., Linton G.F., Vowells S.J., Li F., Miller J.A., Holland S.M., Leitman S.F., Carter C.S., Read E.J., Butz R., Wannebo C., Fleisher T.A., Deans R.J., Spratt S.K., Maack C.A., Rokovich J.A., Cohen L.K., Maples P.B., and Gallin J.I. 1996. Prolonged detection of oxidase-positive neutrophils in the peripheral blood of five patients following a single cycle of gene therapy for chronic granulomatous disease. *Blood* **88:** 486a.

Malik P., McQuiston S.A., Yu X.-J., Pepper K.A., Krall W.J., Podsakoff G.M., Kurtzman G.M., and Kohn D.B. 1997. Recombinant adeno-associated virus mediates a high level of gene transfer but less efficient integration in the K562 human hematopoietic cell line. *J. Virol.* **71:** 1776–1783.

Mann R., Mulligan R.C., and Baltimore D. 1983. Construction of a retrovirus packaging mutant and its use to produce helper-free defective retrovirus. *Cell* **33:** 153–159.

Mardiney M., Jackson S.H., Spratt S.K., Li F., Holland S.M., and Malech H.L. 1997. Enhanced host defense after gene transfer in the murine p47phox-deficient model of chronic granulomatous disease. *Blood* **89:** 2268–2275.

Markowitz D., Goff S., and Bank A. 1988. A safe packaging line for gene transfer: Separating viral genes on two different plasmids. *J. Virol.* **62:** 1120–1124.

May C., Gunther R., and McIvor R.S. 1995. Protection of mice from lethal doses of methotrexate by transplantation with transgenic marrow expressing drug-resistant dihydrofolate reductase activity. *Blood* **86:** 2439–2448.

Maze R., Carney J.P., Kelley M.R., Glassner B.J., Williams D.A., and Samson L. 1996. Increasing DNA repair methyltransferase levels via bone marrow stem cell transduction rescues mice from the toxic effects of 1,3-*bis* (2-chloroethyl)-1-nitrosourea, a chemotherapeutic alkylating agent. *Proc. Natl.Acad. Sci.* **93:** 206–210.

Maze R., Kapur R., Kelley M.R., Hansen W.K., Oh S.Y., and Williams D.A. 1997. Reversal of 1,3-bis(2-chloroethyl)-1-nitrosourea-induced severe immunodeficiency by transduction of murine long-lived hemopoietic progenitor cells using O6-methylguanine DNA methyltransferase complementary DNA. *J. Immunol.* **158:** 1006–1013.

McCune S.L. and Townes T.M. 1994. Retroviral vector sequences inhibit human β-globin gene expression in transgenic mice. *Nucleic Acids Res.* **22:** 4477–4481.

McLachlin J.R., Eglitis M.A., Ueda K., Kantoff P.W., Pastan I.H., Anderson W.F., and Gottesman M.M. 1990. Expression of a human complementary DNA for the multidrug resistance gene in murine hematopoietic precursor cells with the use of retroviral gene transfer. *J. Natl. Cancer Inst.* **82:** 1260–1263.

Medin J.A., Migita M., Pawliuk R., Jacobson S., Amiri M., Kluepfel-Stahl S., Brady R.O., Humphries R.K., and Karlsson S. 1996. A bicistronic therapeutic retroviral vector enables sorting of transduced CD34+ cells and corrects the enzyme deficiency in cells from Gaucher patients. *Blood* **87:** 1754–1762.

Mickisch G.H., Licht T., Merlino G.T., Gottesman M.M., and Pastan I. 1991a. Chemotherapy and chemosensitization of transgenic mice which express the human multidrug resistance gene in bone marrow: Efficacy, potency, and toxicity. *Cancer Res.* **51:** 5417–5424.

Mickisch G.H., Merlino G.T., Galski H., Gottesman M.M., and Pastan I. 1991b. Transgenic mice that express the human multidrug-resistance gene in bone marrow enable a rapid identification of agents that reverse drug resistance. *Proc. Natl. Acad. Sci.* **88:** 547–551.

Migita M., Medin J.A., Pawliuk R., Jacobson S., Nagle J.W., Anderson S., Amiri M., Humphries R.K., and Karlsson S. 1995. Selection of transduced CD34+ progenitors and enzymatic correction of cells from Gaucher patients, with bicistronic vectors. *Proc. Natl. Acad. Sci.* **92:** 12075–12079.

Miller A.D. 1990. Retrovirus packaging cells. *Hum. Gene Ther.* **1:** 5–14.

——— 1996. Cell-surface receptors for retroviruses and implications for gene transfer. *Proc. Natl. Acad. Sci.* **93:** 11407–11413.

Miller A.D. and Chen F. 1996. Retroviral packaging cells based on 10A1 murine leukemia virus for production of vectors that use multiple receptors for cell entry. *J. Virol.* **70:** 5564–5571.

Miller A.D. and Rosman G.J. 1989. Improved retroviral vectors for gene transfer and expression. *BioTechniques* **7:** 980–982.

Miller A.D., Garcia J.V., von Suhr N., Lynch C.M., Wilson C., and Eiden M.V. 1991. Construction and properties of retrovirus packaging cells based on gibbon ape leukemia virus. *J. Virol.* **65:** 2220–2224.

Miller C.L. and Eaves C.J. 1997. Expansion in vitro of adult murine hematopoietic stem cells with transplantable lympho myeloid reconstituting ability. *Proc. Natl. Acad. Sci.* **94:** 13648–13653.

Miller D.G., Edwards R.H., and Miller A.D. 1994. Cloning of the cellular receptor for amphotropic murine retroviruses reveals homology to that for gibbon ape leukemia virus. *Proc. Natl. Acad. Sci.* **91:** 78–82.

Miller J.L., Donahue R.E., Sellers S.E., Samulski R.J., Young N.S., and Nienhuis A.W. 1994. Recombinant adeno-associated virus (rAAV)-mediated expression of a human γ-globin gene in human progenitor-derived erythroid cells. *Proc. Natl. Acad. Sci.* **91:** 10183–10187.

Mineishi S., Nakahara S., Takebe N., Banerjee D., Zhao S.C., and Bertino J.R. 1997. Co-expression of the herpes simplex virus thymidine kinase gene potentiates methotrexate resistance conferred by transfer of a mutated dihydrofolate reductase gene. *Gene Ther.* **4:** 570–576.

Miyazaki T., Kawahara A., Fujii H., Nakagawa Y., Minami Y., Liu Z.J., Oishi I., Silvennoinen O., Witthuhn B.A., Ihle J.N., and Taniguchi T. 1994. Functional activation of Jak1 and Jak3 by selective association with IL-2 receptor subunits. *Science* **266:** 1045–1047.

Mizukami H., Young N.S., and Brown K.E. 1996. Adeno-associated virus type 2 binds to a 150-kilodalton cell membrane glycoprotein. *Virology* **217:** 124–130.

Moore K.A., Deisseroth A.B., Reading C.L., Williams D.E., and Belmont J.W. 1992. Stromal support enhances cell-free retroviral vector transduction of human bone marrow long-term culture-initiating cells. *Blood* **79:** 1393–1399.

Moritz T., Keller D.C., and Williams D.A. 1993. Human cord blood cells as targets for gene transfer: Potential use in genetic therapies of severe combined immunodeficiency disease. *J. Exp. Med.* **178:** 529–536.

Moritz T., Patel V.P., and Williams D.A. 1994. Bone marrow extracellular matrix molecules improve gene transfer into human hematopoietic cells via retroviral vectors. *J. Clin. Invest.* **93:** 1451–1457.

Moritz T., Mackay W., Glassner B.J., Williams D.A., and Samson L. 1995. Retrovirus-mediated expression of a DNA repair protein in bone marrow protects hematopoietic cells from nitrosourea-induced toxicity in vitro and in vivo. *Cancer Res.* **55:** 2608–2614.

Moritz T., Dutt P., Xiao X., Carstanjen D., Vik T., Hanenberg H., and Williams D.A. 1996. Fibronectin improves transduction of reconstituting hematopoietic stem cells by retroviral vectors: Evidence of direct viral binding to chymotryptic carboxy-terminal fragments. *Blood* **88:** 855–862.

Muenchau D.D., Freeman S.M., Cornetta K., Zwiebel J.A., and Anderson W.F. 1990. Analysis of retroviral packaging lines for generation of replication-competent virus. *Virology* **176:** 262–265.

Muzyczka N. 1992. Use of adeno-associated virus as a general transduction vector for mammalian cells. *Curr. Top. Microbiol. Immunol.* **158:** 97–129.

Nakano T., Spencer H.T., Appleman J.R., and Blakley R.L. 1994. Critical role of phenylalanine 34 of human dihydrofolate reductase in substrate and inhibitor binding and in catalysis. *Biochemistry* **33:** 9945–9952.

Naldini L., Blömer U., Gallay P., Ory D., Mulligan R., Gage F.H., Verma I.M., and Trono D. 1996. In vivo gene delivery and stable transduction of nondividing cells by a lentiviral vector. *Science* **272:** 263–267.

Neering S.J., Hardy S.F., Minamoto D., Spratt S.K., and Jordan C.T. 1996. Transduction of primitive human hematopoietic cells with recombinant adenovirus vectors. *Blood* **88:** 1147–1155.

Nimgaonkar M.T., Bahnson A.B., Boggs S.S., Ball E.D., and Barranger J.A. 1994. Transduction of mobilized peripheral blood CD34$^+$ cells with the glucocerebrosidase cDNA. *Gene Ther.* **1:** 201–209.

Nishimura J., Phillips K.L., Ware R.E., Hall S., Howard T.A., Wilson L., Gentry T.L., Howrey R., Galardi C., Kinoshita T., Gilboa E., Rosse W.F., and Smith C.A. 1997. Efficient retrovirus-mediated PIG-A gene transfer and stable restoration of GPI-anchored protein expression in PNH. *Blood* **90:** 274a.

Noguchi M., Yi H., Rosenblatt H.M., Filipovich A.H., Adelstein S., Modi W.S., McBride O.W., and Leonard W.J. 1993. Interleukin-2 receptor gamma chain mutation results in X-linked severe combined immunodeficiency in humans. *Cell* **73:** 147–157.

Nolta J.A., Hanley M.B., and Kohn D.B. 1994. Sustained human hematopoiesis in immunodeficient mice by cotransplantation of marrow stroma expressing human interleukin-3: Analysis of gene transduction of long-lived progenitors. *Blood* **83:** 3041-3051.

Nordon R.E., Ginsberg S.S., and Eaves C.J. 1997. High-resolution cell division tracking demonstrates the Flt3-ligand-dependence of human marrow CD34$^+$CD38$^-$ cell production in vitro. *Br. J. Haematol.* **98:** 528–539.

Nosaka T., van Deursen J.M., Tripp R.A., Thierfelder W.E., Witthuhn B.A., McMickle A.P., Doherty P.C., Grosveld G.C., and Ihle J.N. 1995. Defective lymphoid development in mice lacking Jak3. *Science* **270:** 800–802.

Novak U., Harris E.A., Forrester W., Groudine M., and Gelinas R. 1990. High-level β-globin expression after retroviral transfer of locus activation region-containing human β-globin gene derivatives into murine erythroleukemia cells. *Proc. Natl. Acad. Sci.* **87:** 3386–3390.

Orkin S.H. 1995. Regulation of globin gene expression in erythroid cells. *Eur. J. Biochem.* **231:** 271–781.

Orlic D., Girard L.J., Jordan C.T., Anderson S.M., Cline A.P., and Bodine D.M. 1996. The level of mRNA encoding the amphotropic retrovirus receptor in mouse and human hematopoietic stem cells is low and correlates with the efficiency of retrovirus transduction. *Proc. Natl. Acad. Sci.* **93:** 11097–11102.

Ory D.S., Neugeboren B.A., and Mulligan R.C. 1996. A stable human-derived cell line for production of high titer retrovirus/vesicular stomatitis virus G pseudotypes. *Proc. Natl. Acad. Sci.* **93:** 11400–11406.

Parolin C., Taddeo B., Palú G., and Sodroski J. 1996. Use of *cis-* and *trans*-acting viral regulatory sequences to improve expression of human immunodeficiency virus vectors in human lymphocytes. *Virology* **222:** 415–422.

Pastan I. and Gottesman M.M. 1991. Multidrug resistance. *Annu. Rev. Med.* **42:** 277–286.

Patel M., Sleep S.E., Lewis W.S., Spencer H.T., Mareya S.M., Sorrentino B.P., and Blakley R.L. 1997. Comparison of the protection of cells from antifolates by transduced human dihydrofolate reductase mutants. *Hum. Gene Ther.* **8:** 2069–2077.

Pellicer A., Wigler M., Axel R., and Silverstein S. 1978. The transfer and stable integration of the HSV thymidine kinase gene into mouse cells. *Cell* **14:** 133–141.

Persons D.A., Allay J.A., Allay E.R., Smeyne R.J., Ashmun R.A., Sorrentino B.P., and Nienhuis A.W. 1997. Retroviral-mediated transfer of the green fluorescent protein gene into murine hematopoietic cells facilitates scoring and selection of transduced progenitors in vitro and identification of genetically modified cells in vivo. *Blood* **90:** 1777–1786.

Peters S.O., Kittler E.L., Ramshaw H.S., and Quesenberry P.J. 1996. Ex vivo expansion of murine marrow cells with interleukin-3 (IL-3), IL-6, IL-11, and stem cell factor leads to impaired engraftment in irradiated hosts. *Blood* **87:** 30–37.

Plavec I., Papayannopoulou T., Maury C., and Meyer F. 1993. A human β-globin gene fused to the human β-globin locus control region is expressed at high levels in erythroid cells of mice engrafted with retrovirus-transduced hematopoietic stem cells. *Blood* **81:** 1384–1392.

Podda S., Ward M., Himelstein A., Richardson C., de la Flor-Weiss E., Smith L., Gottesman M., Pastan I., and Bank A. 1992. Transfer and expression of the human multiple drug resistance gene into live mice. *Proc. Natl. Acad. Sci.* **89:** 9676–9680.

Poeschla E., Corbeau P., and Wong-Staal F. 1996. Development of HIV vectors for anti-HIV gene therapy. *Proc. Natl. Acad. Sci.* **93:** 11395–11399.

Pollock J.D., Williams D.A., Gifford M.A., Li L.L., Du X., Fisherman J., Orkin S.H., Doerschuk C.M., and Dinauer M.C. 1995. Mouse model of X-linked chronic granulomatous disease, an inherited defect in phagocyte superoxide production. *Nat. Genet.* **9:** 202–209.

Ponnazhagan S., Yoder M.C., and Srivastava A. 1997a. Adeno-associated virus type 2-mediated transduction of murine hematopoietic cells with long-term repopulating ability and sustained expression of a human globin gene in vivo. *J. Virol.* **71:** 3089–3104.

Ponnazhagan S., Mukherjee P., Wang X.S., Qing K., Kube D.M., Mah C., Kurpad C., Yoder M.C., Srour E.F., and Srivastava A. 1997b. Adeno-associated virus type 2-mediated transduction in primary human bone marrow-derived CD34+ hematopoietic progenitor cells: Donor variation and correlation of transgene expression with cellular differentiation. *J. Virol.* **71:** 8262–8267.

Qing K., Bachelot T., Mukherjee P., Wang X.-S., Peng L., Yoder M.C., Leboulch P., and Srivastava A. 1997. Adeno-associated virus type 2-mediated transfer of ecotropic retrovirus receptor cDNA allows ecotropic retroviral transduction of established and primary human cells. *J. Virol.* **71:** 5663-5667.

Raftopoulos H., Ward M., Leboulch P., and Bank A. 1997. Long-term transfer and expression of the human β-globin gene in a mouse transplant model. *Blood* **90:** 3414–3422.

Reese J.S., Koc O.N., Lee K.M., Liu L., Allay J.A., Phillips W.P.J., and Gerson S.L. 1996. Retroviral transduction of a mutant methylguanine DNA methyltransferase gene into human CD34 cells confers resistance to O6-benzylguanine plus 1,3-*bis*(2-chloroethyl)-1-nitrosourea. *Proc. Natl. Acad. Sci.* **93:** 14088–14093.

Reiser J., Harmison G., Kluepfel-Stahl S., Brady R.O., Karlsson S., and Schubert M. 1996. Transduction of nondividing cells using pseudotyped defective high-titer HIV type 1 particles. *Proc. Natl. Acad. Sci.* **93:** 15266–15271.

Ren S., Wong B.Y., Li J., Luo X.-N., Wong P.M.C., and Atweh G.F. 1996. Production of genetically stable high-titer retroviral vectors that carry a human γ-globin gene under the control of the α-globin locus control region. *Blood* **87:** 2518–2524.

Rill D.R., Santana V.M., Roberts W.M., Nilson T., Bowman L.C., Krance R.A., Heslop H.E., Moen R.C., Ihle J.N., and Brenner M.K. 1994. Direct demonstration that autologous bone marrow transplantation for solid tumors can return a multiplicity of tumorigenic cells. *Blood* **84:** 380–383.

Riviere I., Brose K., and Mulligan R.C. 1995. Effects of retroviral vector design on expression of human adenosine deaminase in murine bone marrow transplant recipients engrafted with genetically modified cells. *Proc. Natl. Acad. Sci.* **92:** 6733–6737.

Roe T., Reynolds T.C., Yu G., and Brown P.O. 1993. Integration of murine leukemia virus DNA depends on mitosis. *EMBO J.* **12:** 2099–2108.

Rooney C.M., Smith C.A., Ng C.Y., Loftin S., Li C., Krance R.A., Brenner M.K., and Heslop H.E. 1995. Use of gene-modified virus-specific T lymphocytes to control Epstein-Barr-virus-related lymphoproliferation. *Lancet* **345:** 9–13.

Rosenberg S.A., Aebersold P., Cornetta K., Kasid A., Morgan R.A., Moen R., Karson E.M., Lotze M.T., Yang J.C., Topalian S.L., Merino M.J., Culver K., Miller A.D., Blaese R.M., and Anderson W.F. 1990. Gene transfer into humans—Immunotherapy of patients with advanced melanoma, using tumor-infiltrating lymphocytes modified by retroviral gene transduction. *N. Engl. J. Med.* **323:** 570–578.

Russell D.W. and Miller A.D. 1996. Foamy virus vectors. *J. Virol.* **70:** 217–222.

Russell D.W., Miller A.D., and Alexander I.E. 1994. Adeno-associated virus vectors preferentially transduce cells in S phase. *Proc. Natl. Acad. Sci.* **91:** 8915–8919.

Russell S.M., Tayebi N., Nakajima H., Riedy M.C., Roberts J.L., Aman M.J., Migone T.S., Noguchi M., Markert M.L., Buckley R.H., O'Shea J.J., and Leonard W.J. 1995. Mutation of Jak3 in a patient with SCID: Essential role of Jak3 in lymphoid development. *Science* **270:** 797–800.

Russell S.M., Johnston J.A., Noguchi M., Kawamura M., Bacon C.M., Friedmann M., Berg M., McVicar D.W., Witthuhn B.A., Silvennoinen O., Goldman A.S., Schmalstieg F.C., Ihle J.N., O'Shea J.J., and Leonard W.J. 1994. Interaction of IL-2R β and γ$_c$ chains with Jak1 and Jak3: Implications for XSCID and XCID. *Science* **266:** 1042–1045.

Sadelain M. 1997. Genetic treatment of the haemogloinopathies: Recombinations and new combinations. *Brit. J. Haematol.* **98:** 247–253.

Sadelain M., Wang C.H.J., Antoniou M., Grosveld F., and Mulligan R.C. 1995. Generation of a high-titer retroviral vector capable of expressing high levels of the human β-globin gene. *Proc. Natl. Acad. Sci.* **92:** 6728–6732.

Samulski R.J., Chang L.S., and Shenk T. 1987. A recombinant plasmid from which an infectious adeno-associated virus genome can be excised in vitro and its use to study viral replication. *J. Virol.* **61:** 3096–3101.

——— 1989. Helper-free stocks of recombinant adeno-associated viruses: Normal integration does not require viral gene expression. *J. Virol.* **63:** 3822–3828.

Sauvageau G., Thorsteinsdottir U., Eaves C.J., Lawrence H.J., Largman C., Lansdorp P.M., and Humphries R.K. 1995. Overexpression of HOXB4 in hematopoietic cells causes the selective expansion of more primitive populations in vitro and in vivo. *Genes Dev.* **9:** 1753–1765.

Schuening F.G., Kawahara K., Miller A.D., To R., Goehle S., Stewart D., Mullally K., Fisher L., Graham T.C., Appelbaum F.R., Hackman R., Osborne W.R.A., and Storb R. 1991. Retrovirus-mediated gene transduction into long-term repopulating marrow cells of dogs. *Blood* **78:** 2568–2576.

Schwarzenberger P., Spence S.E., Gooya J.M., Michiel D., Curiel D.T., Ruscetti F.W., and Keller J.R. 1996. Targeted gene transfer to human hematopoietic progenitor cell lines through the c-kit receptor. *Blood* **87:** 472–478.

Schweitzer B.I., Srimatkandada S., Gritsman H., Sheridan R., Venkataraghavan R., and Bertino J.R. 1989. Probing the role of two hydrophobic active site residues in the human dihydrofolate reductase by site-directed mutagenesis. *J. Biol. Chem.* **264:** 20786–20795.

Sekhsaria S., Gallin J.I., Linton G.F., Mallory R.M., Mulligan R.C., and Malech H.L. 1993. Peripheral blood progenitors as a target for genetic correction of p47phox-deficient chronic granulomatous disease. *Proc. Natl. Acad. Sci.* **90:** 7446–7450.

Shah A.J., Smogorzewska E.M., Hannum C., and Crooks G.M. 1996. Flt3 ligand induces proliferation of quiescent human bone marrow CD34$^+$ CD38$^-$ cells and maintains progenitor cells in vitro. *Blood* **87:** 3563–3570.

Simonsen C.C. and Levinson A.D. 1983. Isolation and expression of an altered mouse dihydrofolate reductase cDNA. *Proc. Natl. Acad. Sci.* **80:** 2495–2499.

Sinclair A.M., Agarwal Y.P., Elbar E., Agarwal R., Ho A.D., and Levine F. 1997. Interaction of vesicular stomatitis virus-G pseudotyped retrovirus with CD34$^+$ and CD34$^+$ CD34$^-$ hematopoietic progenitor cells. *Gene Ther.* **4:** 918–927.

Sirotnak F.M., Moccio D.M., and Dorick D.M. 1978. Optimization of high-dose methotrexate with leucovorin rescue therapy in the L1210 leukemia and sarcoma 180 murine tumor models. *Cancer Res.* **38:** 345–353.

Snyder R.O., Miao C.H., Patijn G.A., Spratt S.K., Danos O., Nagy D., Gown A.M., Winther B., Meuse L., Cohen L.K., Thompson A.R., and Kay M.A. 1997. Persistent and therapeutic concentrations of human factor IX in mice after hepatic gene transfer of recombinant AAV vectors. *Nat. Genet.* **16:** 270–276.

Sorrentino B.P. 1996. Drug resistance gene therapy. In *Gene therapy in cancer* (ed. M.K. Brenner and R.C. Moen), pp. 189–230. Marcel Dekker, New York.

Sorrentino B.P., McDonagh K.T., Woods D., and Orlic D. 1995. Expression of retroviral vectors containing the human multidrug resistance 1 cDNA in hematopoietic cells of transplanted mice. *Blood* **86:** 491–501.

Sorrentino B.P., Brandt S.J., Bodine D., Gottesman M., Pastan I., Cline A., and Nienhuis

A.W. 1992. Selection of drug-resistant bone marrow cells in vivo after retroviral transfer of human MDR1. *Science* **257:** 99–103.

Spencer D.M., Wandless T.J., Schreiber S.L., and Crabtree G.R. 1993. Controlling signal transduction with synthetic ligands. *Science* 262: 1019–1024.

Spencer H.T., Sleep S.E., Rehg J.E., Blakley R.L., and Sorrentino B.P. 1996. A gene transfer strategy for making bone marrow cells resistant to trimetrexate. *Blood* **87:** 2579–2587.

Strathdee C.A., Gavish H., Shannon W.R., and Buchwald M. 1992. Cloning of cDNAs for Fanconi's anaemia by functional complementation. *Nature* **356:** 763–767.

Surosky R.T., Urabe M., Godwin S.G., McQuiston S.A., Kurtzman G.J., Ozawa K., and Natsoulis G. 1997. The adeno-associated virus Rep proteins target DNA sequences to a unique locus in the human genome. *J. Virol.* **71:** 7951–7959.

Tabin C.J., Hoffmann J.W., Goff S.P., and Weinberg R.A. 1982. Adaptation of a retrovirus as a eucaryotic vector transmitting the herpes simplex virus tymidine kinase gene. *Mol. Cell. Biol.* **2:** 426–436.

Takekoshi K.J., Oh Y.H., Westerman K.W., London I.M., and Leboulch P. 1995. Retroviral transfer of a human β-globin/δ-globin hybrid gene linked to β locus control region hypersensitive site 2 aimed at the gene therapy of sickle cell disease. *Proc. Natl. Acad. Sci.* **92:** 3014–3018.

Tamayose K., Hirai Y., and T. Shimada. 1996. A new strategy for large-scale preparation of high-titer recombinant adeno-associated virus vectors by using packaging cell lines and sulfonated cellulose column chromatography. *Hum. Gene Ther.* **7:** 507–513.

Taylor N., Uribe L., Smith S., Jahn T., Kohn D.B., and Weinberg K. 1996a. Correction of interleukin-2 receptor function in X-SCID lymphoblastoid cells by retrovirally mediated transfer of the γ_c gene. *Blood* **87:** 3103–3107.

Taylor N., Bacon K.B., Smith S., Jahn T., Kadlecek T.A., Uribe L., Kohn D.B., Gelfand E.W., Weiss A., and Weinberg K. 1996b. Reconstitution of T cell receptor signaling in ZAP-70-deficient cells by retroviral transduction of the ZAP-70 gene. *J. Exp. Med.* **184:** 2031–2036.

Tisdale J.F., Moscow J., Huang H., Sellers S.E., Agricola B.A., Donahue R.E. McDonagh K., Cowan K., and Dunbar C.E. 1997. Longterm culture retroviral transduction does not improve gene transfer efficiency into rhesus CD34+ PB cells. *Blood* **90:** 237a.

Toksoz D., Zsebo K.M., Smith K.A., Hu S., Brankow D., Suggs S.V., Martin F.H., and Williams D.A. 1992. Support of human hematopoiesis in long-term bone marrow cultures by murine stromal cells selectively expressing the membrane-bound and secreted forms of the human homolog of the steel gene product, stem cell factor. *Proc. Natl. Acad. Sci.* **89:** 7350–7354.

Tsichlis P.N. and Lazo P.A. 1991. Virus-host interactions and the pathogenesis of murine and human oncogene retroviruses. *Curr. Top. Microbiol. Immunol.* **171:** 95–171.

Urcelay E., Ward P., Wiener S.M., Safer B., and Kotin R.M.. 1995. Asymmetric replication in vitro from a human sequence element is dependent on adeno-associated virus Rep protein. *J. Virol.* **69:** 2038–2046.

van Beusechem V.W., Kukler A., Heidt P.J., and Valerio D. 1992. Long-term expression of human adenosine deaminase in rhesus monkeys transplanted with retrovirus-infected bone-marrow cells. *Proc. Natl. Acad. Sci.* **89:** 7640–7644.

Vanin E.F., Kaloss M., Broscius C., and Nienhuis A.W. 1994. Characterization of replication-competent retroviruses from nonhuman primates with virus-induced T-cell lymphomas and observations regarding the mechanism of oncogenesis. *J. Virol.* **68:** 4241–4250.

Varmus H.E. 1982. Form and function of retroviral proviruses. *Science* **216:** 812–820.

von Kalle C., Kiem H.P., Goehle S., Darovsky B., Heimfeld S., Torok-Storb B, Storb R., and Schuening F.G. 1994. Increased gene transfer into human hematopoietic progenitor cells by extended in vitro exposure to a pseudotyped retroviral vector. *Blood* **84**: 2890–2897.

von Schwedler U., Kornbluth R.S., and Trono D. 1994. The nuclear localization signal of the matrix protein of human immunodeficiency virus type 1 allows the establishment of infection in macrophages and quiescent T lymphocytes. *Proc. Natl. Acad. Sci.* **91**: 6992–6996.

Walsh C.E., Grompe M., Vanin E., Buchwald M., Young N.S., Nienhuis A.W., and Liu J.M. 1994a. A functionally active retrovirus vector for gene therapy in Fanconi anemia group C. *Blood* **84**: 453–459.

Walsh C.E., Nienhuis A.W., Samulski R.J., Brown M.G., Miller J.L., Young N.S., and Liu J.M. 1994b. Phenotypic correction of Fanconi anemia in human hematopoietic cells with a recombinant adeno-associated virus vector. *J. Clin. Invest.* **94**: 1440–1448.

Walsh C.E., Liu J.M., Xiao X., Young N.S., Nienhuis A.W., and Samulski R.J. 1992. Regulated high level expression of a human γ-globin gene introduced into erythroid cells by an adeno-associated virus vector. *Proc. Natl. Acad. Sci.* **89**: 7257–7261.

Wang J.C.Y., Doedens M., and Dick J.E. 1997. Primitive human hematopoietic cells are enriched in cord blood compared with adult bone marrow or mobilized peripheral blood as measured by the quantitative in vivo SCID-repopulating cell assay. *Blood* **89**: 3919–3924.

Ward M., Pioli P., Ayello J., Reiss R., Urzi G., Richardson C., Hesdorffer C., and Bank A. 1996. Retroviral transfer and expression of the human multiple drug resistance (MDR) gene in peripheral blood progenitor cells. *Clin. Cancer Res.* **2**: 873–876.

Ward M., Richardson C., Pioli P., Smith L., Podda S., Goff S., Hesdorffer C., and Bank A. 1994. Transfer and expression of the human multiple drug resistance gene in human CD34+ cells. *Blood* **84**: 1408–1414.

Watanabe S. and Temin H.M. 1983. Construction of a helper cell line for avian reticuloendotheliosis virus cloning vectors. *Mol. Cell. Biol.* **3**: 2241–2249.

Watanabe S., Kuszynski C., Ino K., Heimann D.G., Shephard H.M., Yasui Y., Maneval D.C., and Talmadge J.E. 1996. Gene transfer into human bone marrow hematopoietic cells mediated by adenovirus vectors. *Blood* **87**: 5032–5039.

Weil W.M., Linton G.F., Whiting-Theobald N., Vowells S.J., Rafferty S.P., Li F., and Malech H.L. 1997. Genetic correction of p67phox deficient chronic granulomatous disease using peripheral blood progenitor cells as a target for retrovirus mediated gene transfer. *Blood* **89**: 1754–1761.

Weitzman M.D., Kyostio S.R., Kotin R.M., and Owens R.A. 1994. Adeno-associated virus (AAV) Rep proteins mediate complex formation between AAV DNA and its integration site in human DNA. *Proc. Natl. Acad. Sci.* **91**: 5808–5812.

Whitney M.A., Saito H., Jakobs P.M., Gibson R.A., Moses R.E., and Grompe M. 1993. A common mutation in the FACC gene causes Fanconi anaemia in Ashkenazi Jews. *Nat. Genet.* **4**: 202–205.

Williams D.A., Hsieh K., DeSilva A., and Mulligan R.C. 1987. Protection of bone marrow transplant recipients from lethal doses of methotrexate by the generation of methotrexate-resistant bone marrow. *J. Exp. Med.* **166**: 210–218.

Williams D.A., Lemischka I.R., Nathan D.G., and Mulligan R.C. 1984. Introduction of new genetic material into pluripotent haematopoietic stem cells of the mouse. *Nature* **310**: 476–480.

Williams D.A., Rios M., Stephens C., and Patel V.P. 1991. Fibronectin and VLA-4 in haematopoietic stem cell-microenvironment interactions. *Nature* **352**: 438–441.

Wolfe J.H., Sands M.S., Barker J.E., Gwynn B., Rowe L.B., Vogler C.A., and Birkenmeier

E.H. 1992. Reversal of pathology in murine mucopolysaccharidosis type VII by somatic cell gene transfer. *Nature* **360:** 749–753.

Xiao X., Li J., and Samulski R.J. 1996. Efficient long-term gene transfer into muscle tissue of immunocompetent mice by adeno-associated virus vector. *J. Virol.* **70:** 8098–8108.

———— 1998. Production of high-titer recombinant adeno-associated virus vectors in the absence of helper adenovirus. *J. Virol.* **72:** 2224–2232.

Xiao X., Xiao W., Li J., and Samulski R.J. 1997. A novel 165-base-pair terminal repeat sequence is the sole *cis* requirement for the adeno-associated virus life cycle. *J. Virol.* **71:** 941–948.

Xu L., Stahl S.K., Dave H.P., Schiffmann R., Correll P.H., Kessler S., and Karlsson S. 1994. Correction of the enzyme deficiency in hematopoietic cells of Gaucher patients using a clinically acceptable retroviral supernatant transduction protocol. *Exp. Hematol.* **22:** 223–230.

Xu L.C., Karlsson S., Byrne E.R., Kluepfel-Stahl S., Kessler S.W., Agricola B.A., Sellers S., Kirby M., Dunbar C.E., and Brady R.O. 1995. Long-term in vivo expression of the human glucocerebrosidase gene in nonhuman primates after CD34+ hematopoietic cell transduction with cell-free retroviral vector preparations. *Proc. Natl. Acad. Sci.* **92:** 4372–4376.

Yang Y., Vanin E.F., Whitt M.A., Fornerod M., Zwart R., Schneiderman R.D., Grosveld G., and Nienhuis A.W. 1995. Inducible, high level production of infectious murine leukemia retroviral vector particles pseudotyped with vesicular stomatitis virus G envelope protein. *Hum. Gene Ther.* **6:** 1203–1213.

Yee J.K., Friedman T., and Burns J.C. 1994. Generation of high-titer pseudotyped retroviral vectors with very broad host range. *Methods Cell Biol.* **43:** 99–112.

Zack J.A., Haislip A.M., Krogstad P., and Chen I.S.Y. 1992. Incompletely reverse-transcribed human immunodeficiency virus type 1 genomes in quiescent cells can function as intermediates in the retroviral life cycle. *J. Virol.* **66:** 1717–1725.

Zack J.A., Arrigo S.J., Weitsman S.R., Go A.S., Haislip A., and Chen I.S.Y. 1990. HIV-1 entry into quiescent primary lymphocytes: Molecular analysis reveals a labile, latent viral structure. *Cell* **61:** 213–222.

Zandstra P.W., Conneally E., Petzer A.L., Piret J.M., and Eaves C.J. 1997. Cytokine manipulation of primitive human hematopoietic cell self-renewal. *Proc. Natl. Acad. Sci.* **94:** 4698–4703.

Zhao S.C., Banerjee D., Mineishi S., and Bertino J.R. 1997. Post-transplant methotrexate administration leads to improved curability of mice bearing a mammary tumor transplanted with marrow transduced with a mutant human dihydrofolate reductase cDNA. *Hum. Gene Ther.* **8:** 903–909.

Zhao S.C., Li M.X., Banerjee D., Schweitzer B.I., Mineishi S., Gilboa E., and Bertino J.R. 1994. Long-term protection of recipient mice from lethal doses of methotrexate by marrow infected with a double-copy vector retrovirus containing a mutant dihydrofolate reductase. *Cancer Gene Ther.* **1:** 27–33.

Zufferey R., Nagy D., Mandel R.J., Naldini L., and Trono D. 1997. Multiply attenuated lentiviral vector achieves efficient gene delivery in vivo. *Nat. Biotechnol.* **15:** 871–875.

16

Cancer Immunotherapy

Drew Pardoll

Department of Oncology
Johns Hopkins University
Baltimore, Maryland 21205

Gary J. Nabel

Departments of Internal Medicine and Biological Chemistry
Howard Hughes Medical Institute
University of Michigan Medical Center
Ann Arbor, Michigan 48109-0650

Although gene therapy was originally conceptualized as a treatment for heritable genetic diseases via replacement of defective genes, it has come to encompass the much broader goal of utilization of gene transfer to alter biological systems with therapeutic intent. This expanded vision of gene therapy is most dramatically exemplified by the tremendous number of genetic approaches to manipulate the immune system. Genetic manipulation of antitumor immunity has been the major arena with 58 of the first 100 approved protocols in cancer gene therapy targeting the immune system. Genes encoding immune regulatory molecules and tumor antigens have been introduced into tumor cells, as well as various viruses and bacteria, to create cancer vaccines. More recently, in vivo gene-transfer strategies have demonstrated the capacity to turn on or off specific immune responses. Thus, although activation of antitumor immunity has been the focus of the majority of initial approaches, it is now clear that analogous gene-transfer approaches can be effectively utilized to inhibit or suppress certain immune responses as treatment for autoimmune disease or transplant rejection. Through the combination of improved vectorology and the greater understanding of molecular pathobiology, a large proportion of major diseases may thus become amenable to some form of gene therapy.

GENETICALLY ALTERED WHOLE-CELL TUMOR VACCINES

History of Genetically Altered Whole-cell Tumor Vaccines

Whole-cell tumor vaccines mixed with adjuvants have been tested extensively in multiple cancer types for decades with multiple reports of either anecdotal responses or improved outcome relative to historical controls. These vaccines have consisted of either irradiated autologous or allogeneic tumor cells, tumor cell lysates, or irradiated virus-infected tumor cells (Livingston et al. 1985; Berd et al. 1986, 1990; McCune et al. 1990).

One of the limitations of this type of cancer vaccine was that variability and complexity of the adjuvants, which usually consisted of heat-inactivated bacteria such as *Bacillus Calmette Geurin* and *Corynebacterium parvum*, made dissection of immunologic priming mechanisms difficult. The replacement of complex adjuvants with single defined molecular entities produced locally as a consequence of gene transfer has revolutionized the mechanistic study of cell-based cancer vaccines. Initially, the introduction of genes encoding foreign antigens allowed for direct testing of the concept of presenting foreign antigens at the same site as tumor antigens in order to augment antitumor immunity. The first studies demonstrating enhanced immunogenicity of genetically altered tumors were performed 25 years ago, starting with Lindenmann and Klein (1967), who showed that vaccination with influenza-virus-infected tumor cell lysates generated enhanced systemic immune responses against a challenge with the original wild-type tumor cells. Furthermore, these early studies showed that nonvirally infected tumor cell lysates, or tumor cell lysates mixed with the same virus, are not immunogenic and cannot elicit a systemic immune response against challenge with the parental tumor cells. Because adequate immunization against the tumor required that the tumor cells be infected with the virus, Lindenmann and his colleagues hypothesized that weak antigens derived from the tumor cells might become associated with or incorporated into the virus and subsequently become potent immunogens. However, based on what we have learned since then about immune responses, the enhanced immune response generated by the virally infected tumor cells was probably the result of high viral protein expression and subsequent availability of both major histocompatibility complex (MHC) class I and II antigenic epitopes required for priming the CD4[+] and CD8[+] T cell arms of the immune system. The antitumor response that was generated in addition to the antiviral response occurred as a consequence of the expression of tumor antigens at the same site as the viral antigens. These studies therefore confirmed the importance of employing immune modulators in a paracrine fashion to generate antitumor immune responses.

As newer techniques of gene transfer have been developed, infection with infectious virus has been replaced with specific gene transfer in an attempt to regulate more carefully the nature of the genetic alteration in the tumor. Fearon et al. (1988) used gene transfer to introduce the influenza hemagglutinin antigen (HA) gene into BALB/c-derived CT26 colon tumor cells, thereby resulting in the expression of a viral antigen on the tumor's cell surface. This approach, followed by repeated fluorescence-activated cell sorter (FACS) sorting of high-expressing variants, resulted in transfectants that were rejected by BALB/c mice at doses 100 times greater than the parental tumor. Again, these HA transfectants induced a systemic immune response against challenge with parental CT26 tumors. Although only high HA expressers were able to be rejected by syngeneic animals and induce systemic immune responses against the original tumor, there was significant discrepancy among different subclones of HA transfectants expressing similar HA levels, suggesting that other factors besides simply HA expression contributed to immunogenicity of the tumor. Unfortunately, this approach has not been found to generate adequate antitumor responses consistently in other tumor systems.

Modification of Tumor Immunogenicity by Gene Transfer

With the maturation of gene transfer technology and the cloning of numerous immune-modulating genes, there has been increasing experimental activity in murine tumor models aimed at more directly altering the tumor cell's genetic material to enhance immune responses against them. These maneuvers are often described as enhancing the immunogenicity of the tumor. It is important to point out that although many of these strategies were designed with a specific mechanism in mind, it is becoming clear that genetic manipulation to alter expression of even a single gene product can result in a complex cascade of cellular responses in vivo that ultimately may affect multiple aspects of antigen processing, presentation, and costimulation.

There are two ways to genetically modify the tumor cell to augment T-cell-mediated antitumor immunity. The first way is to genetically modify the tumor cell to express cytokines that function as attractants for dedicated antigen-presenting cells (APCs) such as dendritic cells and macrophages. These "professional" APCs provide the costimulation required for activation and proliferation of a systemic T-cell immune response, thereby acting as intermediates for T-cell activation by the tumor cell (Fig. 1). The second way is to genetically modify the tumor cell itself to become a "professional" APC by expressing the required MHC molecules, costimulatory molecules, and cytokines necessary for T-cell activation and proliferation (Fig. 1). As will be discussed later, examples of both have been studied in murine models and in clinical trials.

Both ex vivo and in vivo methods of gene delivery have been employed in the development of whole-cell cancer vaccines. Ex vivo gene

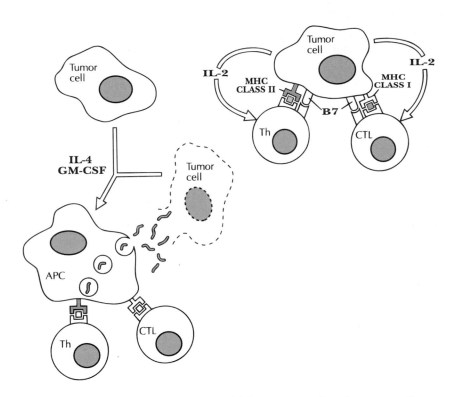

Figure 1 Potential mechanisms by which gene-transduced tumor cells may enhance the activation of tumor-specific T cells. Shown are examples of different classes of gene products that either act as costimulatory signals for T cells or agents that attract and activate APCs for more efficient presentation of tumor antigens to T cells. (Reprinted, with permission, from Pardoll 1993.)

delivery involves the genetic modification of cultured cells. The genetically modified cells are subsequently administered to the host. The most effective way to enhance expression of MHC molecules for improved antigen presentation or to enhance expression of costimulatory molecules such as B7.1 to provide the second signal for T-cell activation is to genetically modify the tumor cell itself. However, when the goal is to deliver cytokines locally in a paracrine fashion, it is not necessary to genetically modify the tumor cells themselves. In fact, in many cases, it is technically difficult to culture primary tumor cells for vaccine production. An alternative is to use a bystander cytokine delivery system such as an autologous fibroblast cell line that has been genetically modified ex vivo and is subsequently mixed with autologous unmodified tumor cells prior to vaccination of the host.

Insertion of MHC Genes into Tumors

Introduction of MHC genes into tumors was one of the first approaches for enhancing the immunogenicity of tumors. Increasing MHC class I expression by gene transfer typically results in decreased tumorigenic capacity in murine tumor models. The decreased tumorigenicity is felt to be due to enhanced presentation of tumor-specific antigens to the CD8[+] cytotoxic T lymphocytes (CTLs) in vivo. However, enhanced cellular expression of MHC class I molecules does not always increase the immunological potency of a tumor, and in certain circumstances has been shown to inhibit natural killer (NK) cells, thereby resulting in a paradoxical increase in tumorigenicity. Tumor immunogenicity has also been enhanced by transfer of allogeneic MHC I genes by ex vivo and in vivo methods (Itaya et al. 1987). In certain cases, the rejection of tumors expressing allogeneic MHC I molecules may result in enhanced systemic immune responses against subsequent challenge with the unmodified parental tumor. This represents an example of the general phenomenon of enhancing tumor-vaccine potency by introducing genes encoding any foreign antigen. This is often referred to as *xenogenization*. The mechanism by which xenogenized tumors enhance systemic immune responses against challenge with parental tumors is still unclear, although the mechanism has been postulated to be the result of nonspecific effects due to increased local cytokine induction.

In some murine tumors, introduction of an MHC II gene into tumor cells decreases their immunogenicity and can result in the generation of enhanced systemic immune responses against the parental MHC-II-negative tumor. It was postulated that the expression of MHC II molecules on tumor cells allowed for the presentation of MHC-II-restricted tumor-specific antigens to CD4[+] T helper cells, which ultimately provided enhanced in vivo help for activation of CTLs via cytokine release. Interestingly, the enhanced immunogenicity in one mouse-tumor model following MHC II gene transfer was dependent on the cytoplasmic tail of the MHC II molecule, implying that MHC-II-dependent signaling in the transfected tumor cell induced secondary genetic events that enhanced immunogenicity by mechanisms other than antigen presentation.

Insertion of the Genes Encoding Costimulatory Molecules into Tumors

B7, originally described as an activation antigen on B cells, is expressed by most APCs and is now known to be the ligand for two receptors expressed on T cells. These are CD28 and the cytotoxic T-lymphocyte-associated antigen (CTLA-4) (Linsley et al. 1991; Azuma et al. 1993). CD28 is now well characterized as a critical receptor for generating costimulatory signals in T-cell activation (Linsley et al. 1991; Thompson et al. 1989). Cross-linking of CD28 has been shown to enhance the level of lymphokine production by $CD4^+$ T cells subsequent to antigen recognition. This enhanced lymphokine production appears to be due to both enhanced transcription and enhanced mRNA stability. Blocking the interaction between B7 and CD28 not only decreases lymphokine production, but can also result in functional anergy to subsequent antigen stimulation. Transfection of B7 into some tumors results in rejection of that tumor, and in some cases, generates systemic immune responses capable of generating protection against challenges of the wild-type tumor at a distant site (Chen et al. 1992; Townsend and Allison 1993). Cointroduction of the B7.1 gene and the gene encoding an MHC II molecule also enhanced the immunogenicity of tumors and generated systemic protection against challenge with the wild-type tumor (Baker et al. 1993). It has been postulated that B7 introduced into tumor cells functions by enhancing the tumor's ability to directly costimulate T cells that recognize tumor antigens presented by MHC molecules. However, evidence also exists that B7 expression by tumors provides a direct target for NK-mediated killing with subsequent release of antigens to bone-marrow-derived APCs (Wu et al. 1995a).

Insertion of Cytokine Genes into Tumor Cells

Genes that encode cytokines are the most common types of genes that have been introduced into tumor cells (Asher et al. 1991; Bannerji et al. 1994; Blankenstein et al. 1991; Bubenik et al. 1990; Colombo et al. 1991; Dranoff et al. 1993; Fearon and Vogelstein 1990; Gansbacher et al. 1990b; Golumbek et al. 1991; Hock et al. 1991; Li et al. 1990; Porgador et al. 1992; Pulaski et al. 1993; Restifo et al. 1991; Rollins and Sunday 1991). Tumor cells transduced with cytokine genes alter the local immunological environment of the tumor cell so as either to enhance presentation of tumor-specific antigens to the immune system or to enhance the activation of tumor-specific lymphocytes. The cytokine is produced at very high concentrations at the vicinity of the tumor, whereas systemic concentrations are relatively low. This paracrine physiology much more closely mimics the natural biology of cytokine action than does the systemic administration of recombinant cytokines. Many cytokine genes have been introduced into tumor cells with varying effects on both tumorigenicity and immunogenicity. Table 1 summarizes the most common cytokines that have been evaluated using this approach and describes the proposed immune mechanisms of tumor rejection generated by these cytokines. Some of these cytokines, when pro-

duced by tumors, induce a local inflammatory response that results in elimination of the injected tumor. This local inflammatory response is most
often predominantly dependent on leukocytes other than classical T cells.
These systems have been used to elucidate the in vivo effects of cytokines
that result in activation of tumoricidal potential by various types of leukocytes. Some of these cytokines, when produced by the tumors, will also
generate a systemic immune response that can reject a challenge of wild-
type parental tumor. In all cases in which systemic immunity against wild-
type tumor challenge has been analyzed, it is mediated by T cells. Varying
results have been reported for certain cytokines, undoubtedly due to large
variations in other critical parameters of immunization. The level of
cytokine expression, location of immunization, and challenge site are crucial parameters affecting vaccine efficacy for any form of genetically engineered tumor vaccine. A large number of cytokines with a range of actions
have been identified. Many of these cytokines have been studied as single
cytokines in one or more tumor vaccine models. However, in order to
develop a human-tumor vaccine, it is critical that these cytokines be compared head-to-head to determine which cytokine or cytokines are most
effective. Also, given that most mouse tumors show significant immunogenicity when simply irradiated, identification of genes that truly enhance
the tumor's immunogenicity significantly above that of irradiated wild-type
tumor cells is important. Only one study has directly compared multiple
cytokines and other genes in murine tumor models (Dranoff et al. 1993).
This study demonstrated, in a number of poorly and moderately immunogenic tumors, that immunization with the tumors transduced with the
cytokine granulocyte-macrophage colony-stimulating factor (GM-CSF)
produced the greatest degree of systemic immunity that was enhanced relative to irradiated nontransduced tumor cells. In vivo depletion of T-cell
subsets demonstrated that this immunity was dependent on both $CD4^+$ and
$CD8^+$ T cells, despite the fact that the tumors did not express MHC class II.
It is now known that GM-CSF is a growth factor for the most potent APCs,
the dendritic cells (Caux et al. 1992; Inaba et al. 1992b).

Development and Testing of Human-gene-transduced Tumor Vaccines

These mouse studies have led the way for clinical trials with genetically
engineered tumor vaccines. However, there are a number of critical
immunologic concerns that require consideration prior to developing
human-gene-transduced tumor vaccines. First, human tumors are hetero-

Table 1 Summary of the preclinical studies evaluating the most commonly tested genetically modified cytokine-secreting tumor vaccines

Cytokine (Ref. no.)	Local rejection	Systemic protection	Systemic cure	Cells mediating systemic immunity
IL-2	yes	yes	yes[a]	CD8+ T cells
IL-3	yes	n.t.	n.t.	n.d.
IL-4	yes	yes	yes	CD4+ and CD8+ T cells
IL-6	yes	yes	no	T cells
IL-7	yes	n.t.[b]	n.t.	n.d.
γ-IFN	yes	some studies yes	no	CD8+ T cells, sometimes NK cells
TNF	yes	no	no	n.d.
GM-CSF	no[c]	yes	yes	CD4+ and CD8+ T cells, NK cells for MHC I neg tumors
IL-10	yes	yes		CD8+ cells, granulocytes, and antibodies
IL-12	yes	yes	yes	CD8+ T cells, CD4+ cells, sometimes NK cells
G-CSF	yes	n.t.	n.t.	n.d.
Flt3-Ligand	yes	yes	n.t..	n.d.

[a]Cures obtained also when combined with systemic IL-12.
[b]IL-7 when coexpressed with the costimulatory molecule B7.1 generates systemic antitumor immunity O.
[c]Vaccine cells must be irradiated to generate systemic immunity.
n.t. = not tested. n.d. = not determined.

geneous and therefore probably express many potential tumor antigens. Therefore, the first important question to consider is, what is the best source of tumor cells that will express the majority of these antigens? The best source is probably the primary tumor rather than a metastasis because metastases likely develop from only one or a few clones of cells that arise from the primary tumor. Therefore, as a result, each patient will require surgical resection of their primary tumor for vaccine development. Second, how can cell selection in culture be avoided during the cell expansion and genetic modification processes required for vaccine development? One way is to use high-efficiency gene-transfer systems to genetically alter these primary tumor lines. Ideally, high-efficiency gene transfer obviates the need for cointroduction of a selection marker, thereby minimizing the amount of in vitro passage and maximally preserving the original antigenic composition of the explanted primary tumor.

Among such vectors, defective retroviral vector systems have been most utilized in the initial stages of human gene therapy. These vectors have the potential for being free of helper virus and therefore are safe. In addition, they have been demonstrated to allow efficient gene transfer into primary human-tumor explants in the absence of cell selection (Mulligan 1993). Other gene-transfer systems such as adenovirus-based systems as well as nonviral gene-transfer systems including liposomes are currently under evaluation. An alternative approach is to use bystander cells such as autologous fibroblasts that have been genetically modified to express cytokines or biodegradable polymers capable of slow release of cytokines that are mixed with the autologous tumor cells and coinjected for immunization. This approach eliminates the need for in vitro tumor-cell culture and therefore prevents in vitro cell selection.

A significant number of gene-therapy studies aimed at augmenting antitumor immune responses against several different cancers have already been completed. The main aim of these studies has been to demonstrate the safety of vaccinating with genetically modified tumor cells or tumor cells mixed with genetically modified bystander cells. The initial studies that have already been completed have demonstrated that genetically modified tumor vaccines are extremely safe and are without any significant toxicities (Simons et al. 1997). In fact, the main side effect that has been described is local inflammatory skin changes consisting of erythema and induration at the site of vaccination that lasts for several days and is self-limiting. In one study, low-grade fevers have also been observed. In addition, the results of a few of these studies have been optimistic in that they have provided evidence of immune priming as measured by delayed-type hypersensitivity (DTH) reactions to autologous tumor cells.

Limitations of Genetically Modified Whole-cell Tumor Vaccines

The majority of first-generation genetically altered tumor vaccines employed autologous tumor cells as the source of tumor antigens for immune priming. However, there are two major limitations to the utility of autologous whole-cell vaccines. First, there are technical limitations to the routine expansion of most human tumors, a necessary first step in the production of any whole-cell vaccine. Second, whole cells are an inefficient antigen source since the majority of antigens expressed by whole tumor cells are not likely to be tumor-rejection antigens. These limitations preclude the use of autologous cellular vaccine therapy for most cancers. However, until specific antigens are defined, allogeneic vaccines provide an alternative whole-cell immunization approach.

Rationale for a Genetically Altered Allogeneic Whole-cell Vaccine Approach

Two recent findings provide the immunologic rationale for an allogeneic vaccine approach. First, the majority of human melanoma antigens identified so far have been demonstrated to be shared among at least 50% of other patients' melanomas (Van der Bruggen et al. 1991; Cox et al. 1992; Traversari et al. 1992; Kawakami et al. 1994; Robbins et al. 1994; Wolfel et al. 1994; Kang et al. 1995; Brichard et al. 1996; Skipper et al. 1996). Second, recent studies have shown that the professional APCs of the host, rather than the vaccinating tumor cells themselves, are responsible for priming CD4[+] and CD8[+] T cells, both of which are required for generating systemic antitumor. Although the antigens recognized by the CD8[+] T cells were not known at the time, these studies demonstrate that tumor-specific CD8[+] T cells are activated via a mechanism referred to as the cross-priming pathway (Huang et al. 1994). These findings suggest that the vaccinating tumor cells do not need to be HLA compatible with the host to generate specific antitumor immunity, thereby providing rationale for a more generalized and feasible whole-cell tumor vaccine approach.

Although the feasibility of administering genetically modified allogeneic tumor vaccines is vastly greater than creating autologous vaccines for each patient, there is a concern that expression of allogeneic MHC molecules at the vaccine site could somehow interfere with or divert immune responses generated against the tumor antigens shared between the vaccine and the patient's tumor. Also, the immune response directed against the allogeneic MHC molecules could eliminate the vaccine cells, thereby perturbing the pharmacokinetics of local cytokine release or costimulatory molecule expression. However, recent reports support the feasibility of an allogeneic vaccine approach (Thomas et al. 1998). Clinical trials have recently been initiated to test cytokine-secreting allogeneic vaccines in patients with various cancers.

RECOMBINANT ANTIGEN-SPECIFIC VACCINE STRATEGIES

The Pathway to Antigen-specific Vaccine Strategies for Cancer

There are three main requirements for the successful design of recombinant vaccine strategies that target specific antigens expressed by tumors. The first is the identification of common T-cell antigenic targets that are expressed by the majority of patient tumors. As mentioned before, for malignant melanoma, it appears that there are antigens recognized by T cells that are expressed by over 50% of patient tumors, regardless of their HLA type (Robbins and Kawakami 1996). Second is the demonstration that a single antigen can serve as an in vivo tumor rejection target. There have been several murine studies that support the concept of immunodominance of tumor-specific antigens expressed by tumor cells (van den Eynde et al. 1989; A.Y.C. Huang et al. 1996). All of these studies demonstrate that although a tumor population expresses multiple antigens, which is not unexpected, the host can successfully eradicate tumor when the predominant T-cell response generated is against one of several antigens expressed by the tumor population. However, both the mouse and human data on immunodominance of tumor antigens are very preliminary. In addition, many of these studies were performed with antigen-specific T-cell lines rather than T-cell clones which might contain T cells specific for more than one tumor antigen. Further studies are required to better demonstrate that common antigens can serve as immunologic rejection targets for the eradication of existing cancer. The third requirement is the development of recombinant vaccine strategies that can successfully generate antigen-specific immunity. A better understanding of (1) the mechanisms of antigen processing and presentation, (2) the mechanisms of T-cell activation, and (3) the antigens expressed by tumor cells that can serve as tumor-rejection targets for T cells will facilitate the design of effective antigen-specific vaccine strategies for the treatment of cancer.

Recombinant Viral Vaccines

The intrinsic immunogenicity of viruses together with the development of standard techniques to engineer recombinant viruses has engendered broad interest in recombinant viral vaccines. On the basis of the early work of Moss and Paoletti (Moss et al. 1984; Cox et al. 1992), recombinant vaccinia and other poxviruses have been the lead agents in cancer (Bernards et al. 1987; Lathe et al. 1987; Wang et al. 1995). More recently, adenoviral and other viruses are being applied to cancer immunotherapy (Inaba et al. 1992b; Juillard et al. 1995; Xiang et al. 1996). The common denominator for all recombinant viral vaccines involves introduction of the gene(s) encoding the antigen into the viral genome using standard recombination and selection approaches adapted to the virus of interest. Two mechanisms underlie the capacity of recombinant viral vaccines to initiate antigen-specific immune responses. First, the cellular damage induced by viral infection produces danger signals that attract and activate bone-marrow-derived APCs that will tend to present antigens in the context of costimulatory signals, leading to T-cell activation. A second mechanism for some recombinant viral vaccines involves direct infection of bone-marrow-derived APCs. Direct infection of bone-marrow-derived APCs allows for efficient processing of endogenously synthesized antigens into the MHC class I pathway.

The concept of direct infection of APCs has led to various modifications of recombinant viral vaccines that enhance the processing and presentation of encoded antigens or incorporate genes encoding costimulatory molecules into the recombinant virus. For example, infection with recombinant vaccinia virus-expressing genes encoding minimal MHC class-I-restricted peptides can lead to enhanced MHC class I presentation of antigen in vivo (Minev et al. 1994; Restifo et al. 1995). Grafting of endosomal lysosomal sorting signals from LAMP-1 onto the gene-encoding antigen have been shown to result in enhanced MHC class II processing and presentation and enhanced CD4+ T-cell activation by recombinant poxviruses (Wu et al. 1995b; Lin et al. 1996). Finally, incorporation of the B7 costimulatory gene and cytokine genes into recombinant poxviruses also have been shown to enhance vaccine potency (Bronte et al. 1995). It has recently been shown that admixture of poxvirus expressing the GM-CSF gene together with poxvirus-expressing antigen leads to enhanced vaccine potency, apparently based on enhanced dendritic cell (DC) differentiation at foci of infection (K. Hung, unpubl.).

Preliminary results of clinical trials with recombinant viral cancer vaccines are just now beginning to be evaluated. A small study of patients with advanced cervical cancer vaccinated with recombinant vaccinia expressing E6 and E7 showed increased human papillomavirus (HPV)-specific antibody responses attributable to vaccination in three of eight vaccines with one of three available patients generating HPV-specific CTL responses postvaccination (Borysiewicz et al. 1996). One of the major barriers to effective vaccination in humans with viruses such as vaccinia and adenovirus is inhibition of vaccine "take" by preexisting neutralizing antibodies. These antibodies, whose titer is extremely variable among different individuals, are the result of previous exposure to cross-reacting viruses (as in the case with adenoviral vaccines) or previous immunization (as in the case with vaccinia). Ultimately, realization of the full clinical value of recombinant viral vaccines will require development of methods to transiently eliminate neutralizing antibodies or development of viral vectors to which most individuals have not yet been exposed.

Recombinant Bacterial Vaccines

One of the most interesting recent recombinant vaccine approaches involves the use of engineered bacteria. A number of bacterial strains including *Salmonella* (Hoiseth and Stocker 1981; Poirier et al. 1988; Sadoff et al. 1988; Curtiss 1990; Strugnell et al. 1990), BCG (Aldovini and Young 1991; Stover et al. 1991), and *Listeria monocytogenes*

(Schafer et al. 1992; Ikonomidis et al. 1994; Goossens et al. 1995; Pan et al. 1995b; Shen et al. 1995) display two characteristics that make them promising vaccine vectors. First, they all possess enteric routes of infection, providing the possibility of oral vaccine delivery. Second, they all have significant cellular tropism for monocytes and macrophages. They therefore have the capacity to target antigens to professional APC types.

Recombinant *L. monocytogenes* (LM) vaccines have been specifically tested in animal models of cancer. LM is a particularly interesting vector because of its two-phase intracellular life cycle (Tilney and Portnoy 1989; Brunt et al. 1990; Falkow et al. 1992). When LM first infects monocytes or macrophages, it occupies the phagolysome for a period of hours. LM then secretes listeriolysin O (LLO), which destabilizes the phagolysomal membrane and allows transit of the bacteria into the cytoplasm. The dual phagolysomal-cytoplasmic life cycle of LM results in efficient processing and presentation of its antigens into the MHC class II pathway during the phagolysomal phase and into the MHC class I pathway during its cytosolic phase. Recombinant LM can be engineered to secrete antigens by linking them to the LLO signal sequence. Vaccination with these recombinant LM engineered to secrete antigens has been shown to result in in vivo activation of antigen specific CD4$^+$ helper cells and CD8$^+$ CTLs. When recombinant LM expressing a model tumor antigen was tested either intraperitoneally or orally as a tumor vaccine, it proved extremely effective in generating antigen-specific antitumor immunity capable of inducing rejection of palpable established tumors (Pan et al. 1995b). Although the mechanism by which recombinant LM induces effective immunity against large established tumors remains elusive, it is likely that potent activation of components of the innate immune response (Newborg and North 1980; Scott et al. 1991; Portnoy 1992) via induction of cytokines such as interleukin-12 (IL-12) contributes to the effective access of antigen-specific T cells into established tumor beds.

Nucleic Acid Vaccines

Vaccines composed of "naked" DNA have recently engendered tremendous interest and activity ever since intramuscular injection of pure recombinant DNA was shown to persist for long periods of time and mediate stable expression of encoded genes (Wolff et al. 1990). Subsequently, DNA injection was shown to mediate both antibody and cellular immune responses against encoded gene products, raising the prospects for naked DNA as a recombinant vaccine (Tang et al. 1992).

Since the original report by Liu and colleagues that naked DNA vaccines encoding influenza nucleoprotein could protect animals from influenza challenge (Montgomery et al. 1993), naked DNA vaccines have been employed in animal models of many infectious diseases and are currently entering clinical testing (Donnelly et al. 1997). Naked DNA vaccines encoding model tumor antigens have been demonstrated to provide some degree of protection against challenge by tumors expressing the immunizing antigen (Irvine et al. 1997).

In general, the potency of naked DNA vaccines is significantly less than other recombinant vaccines such as recombinant vaccinia. This decreased potency is likely due to the fact that naked DNA vaccines do not undergo a replicative amplification as with live recombinant viral and bacterial vaccines, thereby limiting the amount of antigen ultimately presented to the immune system. Also, naked DNA vaccines generate a much smaller inflammatory or "danger" response than a live viral infection, thereby limiting the amount of costimulation available in the priming environment. Nonetheless, it is now clear that injection of nucleic acid does induce local inflammation and some activation of bone-marrow-derived APCs. Chimera experiments have proven definitively that it is indeed bone-marrow-derived APCs that present the antigens encoded by naked DNA vaccines (Fu et al. 1997; Pardoll and Berkerley 1995). It remains to be determined whether the pathway of antigen presentation in bone-marrow-derived cells is the result of direct transduction or transfer of antigen in protein (or RNA) form to bone-marrow-derived APCs similar to the cross-priming observed with whole-cell tumor vaccines. The appreciation that bone-marrow-derived APCs are the critical presenting cell for nucleic acid vaccines allows for the incorporation of similar modifications effecting antigen processing and costimulation that have been employed in recombinant viral vaccines. Indeed, that mixture of plasmids encoding GM-CSF and B7 have been shown to enhance the generation of cellular immune responses induced by DNA vaccines.

An important feature of nucleic acid vaccines comes from the discovery that unmethylated CpG tracts found in bacterial DNA (the source of most recombinant nucleic acid vaccine material) is intrinsically inflammatory and activating to both macrophages and other bone-marrow-derived APCs (Sato et al. 1996; Weiner et al. 1997). It has been demonstrated that unmethylated CpG tracts directly stimulate macrophages to produce proinflammatory cytokines such as IL-12 (Chace et al. 1997). It is likely that the proinflammatory effects of unmethylated CpG tracts in nucleic acid vaccines are absolutely critical to their vaccine capabilities.

Genetically Modified Dendritic Cell Vaccines

More recent attention has focused on DCs as a form of activated APC that is between 100- and 1000-fold more potent than macrophages in simulating antigen-specific T cells. The classic activated DC represents a state of differentiation of bone-marrow-derived mononuclear cells, the differentiation of which is induced by certain cytokines associated with inflammatory responses. As described before, the most important cytokine identified to date responsible for inducing differentiation to the DC lineage is GM-CSF (Caux et al. 1992; Inaba et al. 1992a,b). Other cytokines such as IL-4 and TNF-α (tumor necrosis factor) also synergize with GM-CSF in inducing DC differentiation from $CD34^+$ precursors (Steinman 1991). The unique potency of DCs in activating T cells appears to be related to many factors. Dendritic cells express high levels of surface MHC molecules, particularly MHC class II (50-fold more than activated macrophages). Thus, the density of peptide/MHC ligand on the surface of a DC is much higher than that on other cells, and this is a critical parameter for T-cell receptor engagement. Also, DCs express extremely high levels of intercellular adhesion molecule (ICAM) and B7, important adhesion and costimulatory molecules, respectively, for T-cell activation. More recently, other DC-specific genes, such as one encoding a T-cell-specific chemokine (Adema et al. 1997), add to the list of features that give DCs their unique prowess in initiating T-cell responses.

On the basis of these fundamental studies on DC biology, a number of groups have used DCs grown by culture of bone-marrow progenitors in GM-CSF followed by loading of the DC with antigen in various forms and injection of loaded DCs back into the cancer-bearing animal. The form of antigen by which DCs have been loaded ranges from minimal MHC class-I-restricted peptides (Mayordomo et al. 1997) to protein (Hsu et al. 1966; Paglia et al. 1996) to fusion with whole tumor cells (Gong et al. 1997). More recently, a number of groups have explored ex vivo transduction of DCs using RNA (Boczkowski et al. 1996), replication-defective recombinant retroviral (Specht et al. 1997), or adenoviral vectors (Song et al. 1997) to introduce genes encoding antigen. The use of RNA for gene transfer into DCs carries a number of interesting potential advantages in that cellular RNA encoding a broad spectrum of antigens can be utilized. Also, RNA from small samples of autologous tumor could be potentially amplified prior to transfer into DCs. Loading of DCs with nucleic acid also provides the opportunity to modify the antigen by splicing on targeting signals, which could more efficiently target the antigen to MHC processing pathways.

SUPERANTIGENS

Superantigens derived from bacteria can act as potent activators of T lymphocytes. Upon exposure to these antigens, helper T cells synthesize large amounts of interferon-γ (IFN-γ) and TNF-α. These proteins induce strong T-cell responses by binding to T-cell receptors, producing polyclonal expansion of cells. Because of their potent activity, recent studies have focused on the use of gene transfer to express these proteins in eukaryotic cells. Dow et al. found that the biologic activity of superantigens (staphylococcal enterotoxin A and B [SEA and SEB], and toxic shock syndrome toxin-1 [TSST-1]) expressed in a mammalian system is unchanged by this process, and demonstrated that these proteins can be expressed through in vivo intramuscular injection of the DNA. The strong, local inflammatory reaction can potentially be applied to the treatment of cancers, but the toxicity will need to be addressed before this can be tested clinically (Dow and Potter 1997). Dohlsten et al. fused SEA and the Fab fragment of the C242 monoclonal antibody (Giantonio et al. 1997). In a clinical trial performed by these investigators, a fusion protein incorporating the SEA and the Fab fragment of the C242 monoclonal antibody was tested in 21 patients via a single infusion to determine its effect on cytokine production and its toxicity. Their results indicate that the predominant toxic reactions of fever and hypotension were temporary and were associated with increased production of IL-2 and TNF-α.

CLINICAL STUDIES

Understanding the mechanisms by which metastatic cells evade or interfere with the immune system provides an opportunity to develop therapeutic interventions to prevent and treat cancer. Roth et al. constructed a retroviral vector containing the wild-type p53 gene under the control of a β-actin promoter designed to express this protein in human non-small-cell lung cancers. This vector was administered to nine patients via direct injections. Analyses of posttreatment biopsies showed an increase in the frequency of apoptosis compared to samples observed prior to treatment. In three of these patients, tumors regressed, and growth was stabilized in three others. There was no evidence of a toxic reaction from these treatments (Roth and Cristiano 1997).

Another approach to the treatment of such malignancies involves the enhancement of antitumor immunity. Such immunity may be improved by expression of genes that increase the antigenicity of the tumor through expression of foreign cell surface proteins, introduction of cytokine genes, or by injection of DNA tumor vaccines. When these approaches are used, the requirement for high gene-transfer efficiency may be less critical. Expression of foreign MHC antigens on cell surfaces can stimulate strong, local, immune reactivity. Direct transfection of a DNA-liposome complex generates a response against an allogeneic MHC protein expressed within actively growing tumors, with an accompanying host immune response to weak tumor antigens. In a murine model, foreign MHC genes were directly injected into malignant tumors in vivo, resulting in both a cytotoxic T-cell response and an immune response against tumor-specific antigens on unmodified tumor cells. In these tests, growth of the tumors diminished, and in many cases, there was complete regression. These reports demonstrated that direct gene transfer and expression of recombinant genes using DNA-liposome complexes do not induce generalized autoimmunity, toxicity, or transmission to germ cells (Skinner and Marbrook 1976; Lindahl and Wilson 1977; Hui et al. 1984; Tanaka et al. 1985; Wallich et al. 1985; Parham et al. 1988; Elliott et al. 1989; Gopas et al. 1989; Townsend and Bodmer 1989; Torre-Amione et al. 1990; Cordon-Cardo et al. 1991; Dunn and North 1991; Jardetzky et al. 1991; Pantel et al. 1991; Inge et al. 1992; Wahl et al. 1992; Nabel et al. 1993). These studies provided evidence that direct gene transfer in vivo can induce cell-mediated immunity against specific gene products and were translated into clinical studies that have shown promise for the treatment of cancer.

G.J. Nabel et al. (1993, 1996) have completed two clinical trials using the human MHC class I gene, HLA-B7, complexed to a lipid to establish a safe and effective dose and confirm expression by tumor cells following direct gene transfer in vivo. The immune response to HLA-B7 and tumor antigens and tumor regression were analyzed. In the first trial, five patients with stage IV metastatic melanoma and subcutaneous tumor nodules received intratumoral injections of the human MHC class I gene, HLA-B7, complexed with the DC-cholesterol/DOPE liposome vector (Nabel et al. 1992; G.J. Nabel et al. 1994). Three treatments were given at the lowest dose and three at the next highest dose with cumulative doses of 0.86 μg and 2.58 μg, respectively. One patient with pulmonary metastases received escalating treatments to the right lung delivered by catheter (E.G. Nabel et al. 1994). All five patients tolerated the treatment well, with no pattern of systemic abnormalities detected after treatment (G.J. Nabel et al. 1993). The single patient with pulmonary metastases experienced no toxicities or complications following catheter-based gene delivery (E.G. Nabel et al. 1994). In four patients, DNA was detected within the injected nodule 3–7 days following injection and HLA-B7 was detected by immunohistochemistry in the tumor cells. The patient who received both treatment protocols plus catheter-based gene transfer showed complete regression of the treated nodule and, at the same time, complete regression of metastatic lesions at distant sites. Detection of tumor-specific CTLs in the blood of three patients suggests that expression of a foreign MHC gene after direct gene transfer can alter the reactivity of the immune system.

In a second trial, an HLA-B7 and β_2-microglobulin (Allovectin-7) vector complexed to a different lipid, 1,2-dimyristyloxypropyl-3-dimethyl-hydroxyethyl ammonium bromide (DMRIE) and dioleoyl phosphatidylethanolamine (DOPE), was used to treat patients with metastatic melanoma. Expression of plasmid DNA and RNA was confirmed after direct injection of the complex in nine of ten patients treated, and the immune response was analyzed. A local tumor response was observed, as evidenced by T-cell migration into the treated tumor nodules and increased tumor-infiltrating lymphocyte reactivity. In two patients, there was inhibition of tumor growth in the treated area. There was complete regression of disease in one patient who was subsequently treated with tumor-infiltrating lymphocytes derived from the treated nodule. This study demonstrated that a local antitumor immune response can be elicited from treatment with MHC class I expression vectors (Nabel et al. 1996).

In additional tests of Allovectin-7, Rubin et al. (1997) completed a phase I trial of HLA B7 patients with hepatic metastases from colorectal carcinoma; 93% of the patients expressed DNA and 33% expressed mRNA, and protein was detected in 50% with no significant adverse events. Stopeck et al. (1997) reported similar results when they tested Allovectin-7 in patients with metastatic melanoma; 93% of their post-therapy biopsies had measurable DNA, mRNA, or protein; 50% of their patients showed a decrease of at least 25% in the mass of the injected nodule; and one patient had a complete remission of the treated site.

Dranoff et al. recently completed a phase I study investigating the effects of administering autologous, irradiated melanoma cells containing a gene for human GM-CSF to melanoma patients. Upon treatment, there were measurable T-cell responses to three melanoma-associated antigens, MART-1, MAGE-1, and gp100 as well as evidence of a humoral response. One patient who received increasing amounts of the vaccine exhibited a delayed hypersensitivity reaction and an increase in dendritic cells and CTLs that were highly reactive against the melanoma. There was a short-term, but marked regression and/or stabilization of distant metastases (Dranoff et al. 1997; Ellem et al. 1997). Despite these promising results, the immune response was not sufficiently strong or long-lived to reverse the disease process, and studies are under way to enhance the response.

In another trial, autologous melanoma cells engineered to express IFN-γ were administered to the patient, and the humoral immune response against melanoma-associated antigens was measured. Data collected from this study indicate that both class I and II MHC antigen expression was increased, and IgG reactive with both autologous and allogeneic melanoma was detected in 62% of the patients. This response was correlated with tumor or nodular regression in some patients (Abdel-Wahab et al. 1997).

Immunosuppression

Studies of tumorigenesis indicate that many types of cancers either evade the immune system or inhibit its function. This immune evasion can be mediated by immunosuppressive cytokines and other molecules that may affect the recruitment of immunoreactive cells. These effects on the immune system may contribute to the failure of immune surveillance and genetic immunotherapy for malignancy. Among the factors that may contribute to immune suppression, CD95L or FasL, is a protein normally found in T lymphocytes and macrophages which induces local immune suppression (Brunner et al. 1995; Ju et al. 1995; Chang et al. 1996; Arai et al. 1997a). CD95L has been identified in several types of cancer and postulated to inhibit antitumor immune responses (Hahne et al. 1996; O'Connell et al. 1996; Strand et al. 1996; Tanaka et al. 1996). Fas, a cell-surface receptor that transduces apoptotic signals into cells, has been shown to bind to FasL, causing apoptosis of lymphocytes and immune suppression, inhibiting cytotoxic and antibody responses (Itoh et al. 1991). Analysis of a number of different human carcinomas has shown that the majority are Fas$^+$, and many are susceptible to FasL (Arai et al. 1997b). This concept has been applied to the inhibition of tumor cell growth in vivo. The gene for FasL was introduced into Fast tumor cell lines using an adenoviral construct and was shown to cause rapid apoptosis of tumor cells. Significant regression of tumor, mediated by inflammatory cells, was also, unexpectedly, observed when Fas$^-$ colon carcinoma cells were transfected with FasL. These findings suggest that gene transfer of FasL can be used to induce the regression of malignancies.

Transforming growth factor-β (TGF-β), IL-10, and prostaglandin E_2 (PGE$_2$) have all been identified as modulating the immunogenicity of cancer cells. Among these, TGF-β has the most powerful effect on the immune response, and the likelihood that a tumor will be metastatic directly correlates with the levels produced by the cell (Wojtowiczl-Praga 1997). Morisaki et al. (1996) measured in vivo secretion of TGF-β and IL-10 from human gastric carcinomas and observed that 79% of these tumors expressed TGF-β mRNA compared to 31% of nontumor samples. Sixty-two percent (compared to 17% of normal tissue) expressed IL-10 mRNA (Morisaki et al. 1996). Secretion of both cytokines was more predominant in late-stage and metastatic disease and poorly differentiated adenocarcinomas. Neutralization of TGF-β has been shown to decrease metastases and tumor growth and in some instances has eliminated the tumor entirely (Young et al. 1992).

In other studies, evidence suggests that PGE$_2$ induces IL-10 mRNA transcription in peripheral blood lymphocytes (PBLs). This increase in IL-10 expression has been associated with a decrease in IFN-γ production by PBLs and inhibition of cell-mediated antitumor responses (Wojtowiczl-Praga 1997). This result was also observed when PBLs were maintained in culture with supernatants from human non-small-cell lung tumors (M. Huang et al. 1996). When tumor cells were treated with a prostaglandin inhibitor or anti-IL-10 antibody was added to the cultures, production of IFN-α increased. Similar observations were made in studies of colon carcinoma (Kucharzik et al. 1997). Several immunosuppressive agents have been identified which are associated with specific cancers. The p43 antigen, which is associated with breast cancer, is more active in the early stages of tumor development and has been shown to inhibit lymphocyte proliferation (Rosen et al. 1996). Recently, lipocortin-1 (LC1) was detected in vitro in cells derived from gastric cancer compared to normal cells and was associated with decreased proliferation of peripheral blood mononuclear cells. This effect could be induced by addition of LC1 or ascites from cancer patients to cell cultures, and the addition of anti-LC1 antibody restored normal function (Koseki et al. 1997).

Reduced or aberrant expression of antigens that are involved in immune detection of foreign cells is another factor in tumorigenicity. DCs have been found to be malfunctioning or in inadequate numbers when tumors are present. Garrity et al. (1997) identified an increase in immune-suppressive cells in peripheral blood and tumors of patients with head and neck squamous cell carcinoma. They demonstrated that these cells, when cultured with GM-CSF or TNF-β, were able to differentiate to resemble dendritic cells functionally and phenotypically. Chaux et al. observed that 16 of 25 colorectal tumors failed to express B7-1 and/or B7-2 costimulatory molecules, which are required for induction of an effective immune response (Vieweg and Gilboa 1995; Chaux et al. 1996). MHC molecules, which play a critical role in presentation of tumor-associated antigens, are more often poorly expressed in cancer cells. Decreased CTL-mediated lysis of cells transformed by oncogenes has been reported (Seliger et al. 1996).

SUMMARY

Progress in molecular oncology over the past several years has led to an improved understanding of the genetic defects and immune response in cancer. The fields of cellular and molecular immunology have allowed identification of a variety of cell surface molecules and cytokines whose roles have been increasingly appreciated in oncology. A major feature of malignancy is the genetic instability that occurs as cells progress from the benign to malignant phase. Such instability generates mutant gene products that can serve as targets of the immune system. Although a variety of traditional approaches have proven successful in reducing tumor burden and removing local metastases, a major challenge in clinical oncology remains the elimination of minimal residual disease. This area may provide the greatest potential for immune approaches, which have proven successful in specific malignancies, including melanoma and renal cell carcinoma. As different cytokine proteins have become available, it is clear that a variety of tumor types have been found to be responsive to these treatments. At the same time, it is also evident that different malignancies show various degrees of immunogenicity in terms of their ability to elicit responses that can confer protective immunity in animal models and humans. Gene-transfer approaches offer the ability to introduce a variety of different gene products conveniently and safely at local sites of disease in an effort to stimulate immune responses, which may then be generalized to distant sites. Preclinical and early clinical studies have provided encouraging results, but much remains to be done. Although some more immunologically responsive malignancies respond to molecular gene-transfer approaches, whether the nonimmunogenic cancers will become responsive is uncertain.

A major challenge is to identify the mechanisms by which malignancies, particularly the nonimmunogenic ones, evade immune detection. It is likely that cytokines that suppress lymphocyte proliferation, i.e., TGF-β and CD95L (FasL), are involved in this process, and cytokines that may promote type-2 helper T cell (Th2) responses may allow malignancies to escape immune detection. Finally, posttranslational modifications that mask epitopes and alter the presentation of antigens to the immune system may, in some cases, pose a challenge for immunologic approaches to treatment. At the same time, the tools of gene transfer provide mechanisms to overcome these forms of resistance. In the coming years, the field of genetic immunotherapy will provide opportunities that not only advance the clinical treatment of malignancies, but also allow a better understanding of the mechanisms by which malignancies become resistant to immunologic recognition during molecular oncogenesis.

ACKNOWLEDGMENTS

Dr. Nabel's work is supported, in part, by the Howard Hughes Medical Institute and the National Institutes of Health (P01 CA59327). Dr. Pardoll's work is supported in part by the kindness of The Janney Family, Mrs. Dorothy H. Needle, and Mr. and Mrs. William Topercer. We would like to thank Donna Gschwend for her assistance in the preparation of this paper.

REFERENCES

Abdel-Wahab Z., Weltz C., Hester D., Pickett N., Vervaert C., Barber J.R., Jolly D., and Seigler H.F. 1997. A phase I clinical trial of immunotherapy with interferon-γ gene-modified autologous melanoma cells: Monitoring the humoral immune response. *Cancer* **80:** 401–412.

Adema G.J., Hartgers F., Verstraten R., de Vries E., Marland G., Menon S., Foster J., Xu Y., Nooyen P., McClanahan T., Bacon K.B., and Figdor C.G. 1997. A dendritic-cell-derived C-C chemokine that preferentially attracts naive T cells. *Nature* **387:** 713–717.

Aldovini A. and Young R.A. 1991. Humoral and cell-mediated immune responses to live recombinant BCG-HIV vaccines. *Nature* **351:** 479–482.

Arai H., Chan S.Y., Bishop D.K., and Nabel G.J. 1997a. Inhibition of the alloantibody response by CD95 ligand. *Nat. Med.* **3:** 843–848.

Arai H., Gordon D., Nabel E.G., and Nabel G.J. 1997b. Gene transfer of Fas ligand induces tumor regression *in vivo*. *Proc. Natl. Acad. Sci.* **94:** 13862–13867.

Asher A.L., Mule J.J., Kasid A., Restifo N.P., Salo J.C., Reichert C.M., Jaffe J., Fendly B., Kriegler M., and Rosenberg S.A. 1991. Murine tumor cells transduced with the gene for tumor necrosis factor-alpha. Evidence for paracrine immune effects of tumor necrosis factor against tumors. *J. Immunol.* **146:** 3227–3234.

Azuma M., Ito D., Yagita H., Okumara K., Phillips J.H., Lanier L.L., and Somoza C. 1993. B70 antigen is a second ligand for CLTA-4 and CD28. *Nature* **366:** 76–79.

Baker S., Ostrand-Rosenberg S., Nabavi N., Nadler L.M., Freeman G.J., and Glimcher L.H. 1993. Constitutive expression of B7 restores immunogenicity of tumor cells expressing truncated major histocompatibility complex class II molecules. *Proc. Natl. Acad. Sci.* **90:** 5687–5690.

Bannerji R., Arroyo C.D., Cordon C.C., and Gilboa E. 1994. The role of IL-2 secreted from genetically modified tumor cells in the establishment of antitumor imunity. *J. Immunol.* **152:** 2324–2332.

Berd D., Maguire H.C.J., and Mastrangelo M.J. 1986. Induction of cell-mediated immunity to autologous melanoma cells and regression of metastases after treatment with a melanoma cell vaccine preceded by cyclophosphamide. *Cancer Res.* **46:** 2572–2577.

Berd D., Maguire H.C.J., McCue P., and Mastrangelo M.J. 1990. Treatment of metastatic melanoma with an autologous tumor-cell vaccine: Clinical and immunologic results in 64 patients. *J. Clin. Oncol.* **8:** 1858–1867.

Bernards R., Destree A., McKenzie S., Gordon E., Weinberg R.A., and Panicali D. 1987. Effective tumor immunotherapy directed against an oncogene-encoded product using a vaccinia virus vector. *Proc. Natl. Acad. Sci.* **84:** 6854–6858.

Blankenstein T., Qin Z., and Uberla K. 1991. Tumor suppression after tumor cell-targeted tumor necrosis factor via gene transfer. *J. Exp. Med.* **173:** 1047–1052.

Boczkowski D., Nair S.K., Snyder D., and Gilboa E. 1996. Dendritic cells pulsed with RNA are potent antigen-presenting cells *in vitro* and *in vivo*. *J. Exp. Med.* **184:** 465–472.

Borysiewicz L.K., Fiander A., Nimako M., Man S., Wilkinson G.W., Westermoreland D., Evans A.S., Adams M., Stacey S.N., Boursnell M.E., Rutherford E., Hickling J.K., and Inglis S.C. 1996. A recombinant vaccinia virus encoding human papillomavirus types 16 and 18, E6 and E7 proteins as immunotherapy for cervical cancer. *Lancet* **347:** 1523–1527.

Brichard V.G., Herman J., Van Pel A., Wildmann C., Gaugler B., Wolfel T., Boon T., and Lethe B. 1996. A tyrosinase nonapeptide presented by HLA-B44 is recognized on a human melanoma by autologous cytolytic T lymphocytes. *Eur. J. Immunol.* **26:** 224–230.

Bronte V., Tsung K., Rao J.B., Chen P.W., Wang M., Rosenberg S.A., and Restifo N.P. 1995. IL-2 enhances the function of recombinant poxvirus-based vaccines in the treatment of established pulmonary metastases. *J. Immunol.* **154:** 5282–5292.

Brunner T., Mogil R.J., LaFace D., Yoo N.J., Mahboubi A., Echeverri F., Martin S.J., Force W.R., Lynch D.H., Ware C.F., and Green D.R. 1995. Cell-autonomous Fas (CD95)/Fas-ligand interaction mediates activation-induced apoptosis in T-cell hybridomas. *Nature* **373:** 441–444.

Brunt L.M., Portnoy D.A., and Unanue E.R. 1990. Presentation of *Listeria monocytogenes* to CD8⁺ T-cells requires secretion of hemolysin and intracellular bacterial growth. *J.Immunol.* **145:** 3540–3546.

Bubenik J., Simova J., and Jandlova T. 1990. Immunotherapy of cancer using local administration of lymphoid cells transformed by IL-2 cDNA and constitutively producing IL-2. *Immunol. Lett.* **23:** 287–292.

Caux C., Dezutter-Dambuyant C., Schmitt D., and Banchereau J. 1992. GM-CSF and TNF-α cooperate in the generation of dendritic Langerhans cells. *Nature* **360:** 258–261.

Chace J.H., Hooker N.A., Mildenstein K.L., Krieg A.M., and Cowdery J.S. 1997. Bacterial DNA-induced NK cell IFN-γ production is dependent on macrophage secretion of IL-12. *Clin. Immuno. Immunopatho.* **84:** 185–193.

Chaux P., Moutet M., Faivre J., Martin F., and Martin M. 1996. Inflammatory cells infiltrating human colorectal carcinomas express HLA class I but not B7-1 and B7-2 costimulatory molecules of the T-cell activation. *Lab. Invest.* **74:** 975–983.

Chen L., Ashe S., Brady W.A., Hellstrom I., Hellstrom K.E., Ledbetter J.A., McGowan P., and Linsley P.S. 1992. Costimulation of antitumor immunity by the B7 counterreceptor for the T lymphocyte molecules CD28 and CTLA-4. *Cell* **71:** 1093–1102.

Colombo M.P., Ferrari G., Stoppacciaro A., Parenza M., Rodolfo M., Mavilio F., and Parmiani G. 1991. Granulocyte colony-stimulating factor gene transfer suppresses tumorigenicity of a murine adenocarcinoma *in vivo*. *J. Exp. Med.* **173:** 889–897.

Cordon-Cardo C., Fuks Z., Drobnjak M., Moreno C., Eisenbach L., and Feldman M. 1991. Expression of HLA-A,B,C antigens on primary and metastatic tumor cell populations of human carcinomas. *Cancer Res.* **51:** 6372–6380.

Cox, W.I., J. Tartaglia, and E. Paoletti. 1992. Poxvirus recombinants as live vaccines. In: *Recombinant poxviruses* (ed. M. Binns and G.L. Smith), pp. 123–162. CRC Press, Boca Raton, Florida.

Curtiss, R. 1990. Attenuated *Salmonella* strains as live vectors for the expression of foreign antigens. In: *New generation vaccines* (ed. G.C. Woodrow and M.M. Levine), pp. 161–188. Mercel Dekker, New York.

Donnelly J.J., Ulmer J.B., Shiver J.W., and Liu M.A. 1997. DNA vaccines. *Annu. Rev. Immunol.* **15:** 617–648.

Dow S.W. and Potter T.A. 1997. Expression of bacterial superantigen genes in mice induces localllized mononuclear cell inflammatory responses. *J. Clin. Invest* **99**: 2616–2624.

Dranoff G., Jaffee E., Lazenby A., Golumbek P., Levitsky H., Brose K., Jackson V., Hamada H., Pardoll D., and Mulligan R.C. 1993. Vaccination with irradiated tumor cells engineered to secrete murine granulocyte-macrophage colony-stimulating factor stimulates potent, specific, and long-lasting anti-tumor immunity. *Proc. Natl. Acad. Sci.* **90**: 3539–3543.

Dranoff G., Soiffer R., Lynch T., Mihm M., Jung K., Kolesar K., Liebster L., Lam P., Duda R., Mentzer S., Singer S., Tanabe K., Johnson R., Sober A., Bhan A., Clift S., Cohen L., Parry G., Rokovich J., Richards L., Drayer J., Berns A., and Mulligan R.C. 1997. A phase I study of vaccination with autologous, irradiated melanoma cells engineered to secrete human granulocyte-macrophage colony stimulating factor. *Hum. Gene Ther.* **7**: 111–123.

Dunn P.L. and North R.J. 1991. Effect of advanced aging on the ability of mice to cause tumor regression in response to immunotherapy. *Immunology* **74**: 355–359.

Ellem K.A., O'Rourke M.G., Johnson G.R., Parry G., Misko I.S., Schmidt C.W., Parsons P.G., Burrows S.R., Cross S., Fell A., Li C.L., Bell J.R., Dubois P.J., Moss D.J., Good M.F., Kelso A., Cohen L.K., Dranoff G., and Mullligan R.C. 1997. A case report: Immune responses and clinical course of the first human use of granulocyte/macrophage-colony-stimulating-factor-transduced autologous melanoma cells for immunotherapy. *Cancer Immunol. Immunother.* **44**: 10–20.

Elliott B.E., Carlow D.A., Rodricks A.-M., and Wade A. 1989. Perspectives on the role of MHC antigens in normal and malignant cell development. *Adv. Cancer Res.* **53**: 181–245.

Falkow S., Isberg R.R., and Portnoy D.A. 1992. The interaction of bacteria with mammalian cells. *Annu. Rev. Cell. Biol.* **8**: 333–363.

Fearon E.R. and Vogelstein B. 1990. A genetic model for colorectal tumorigenesis. *Cell* **61**: 759–767.

Fearon E.R., Itaya T., Hunt B., Vogelstein B., and Frost P. 1988. Induction in a murine tumor of immunogenic tumor variants by transfection with a foreign gene. *Cancer Res.* **48**: 2975–2980.

Fu T.M., Ulmer J.B., Caulfield M.J., Deck R.R., Friedman A., Wang S., Liu X., Donnelly J.J., and Liu M.A. 1997. Priming of cytotoxic T lymphocytes by DNA vaccines: Requirement for professional antigen presenting cells and evidence for antigen transfer from myocytes. *Mol. Med.* **3**: 362–371.

Gansbacher B., Bannerji R., Daniels B., Zier K., Cronin K., and Gilboa E. 1990a. Retroviral vector-mediated γ-interferon gene transfer into tumor cells generates potent and long lasting antitumor immunity. *Cancer Res.* **50**: 7820–7825.

Gansbacher B., Zier K., Daniels B., Cronin K., Bannerji R., and Gilboa E. 1990b. Interleukin 2 gene transfer into tumor cells abrogates tumorigenicity and induces protective immunity. *J. Exp. Med.* **172**: 1217–1224.

Garrity T., Pandit R., Wright M.A., Benefield J., Keni S., and Young M.R. 1997. Increased presence of CD34+ cells in the peripheral blood of head and neck cancer patients and their differentiation into dendritic cells. *Int. J. Cancer* **73**: 663–669.

Giantonio B.J., Alpaugh R.K., Schultz J., McAleer C., Newton D.W., Shannon B., Guedez Y., Kotb M., Vitek L., Persson R., Gunnarsson P.O., Kalland T., Dohlsten M., Persson B., and Weiner L.M. 1997. Superantigen-based immunotherapy: A phase I trial of PNU-214565, a monoclonal antibody-staphylococcal enterotoxin A recombinant fusion protein, in advanced pancreatic and colorectal cancer. *J. Clin. Oncol.* **15**: 1994–2007.

Golumbek P.T., Lazenby A.J., Levitsky H.I., Jaffee L.M., Karasuyama H., Baker M., and

Pardoll D.M. 1991. Treatment of established renal cancer by tumor cells engineered to secrete interleukin-4. *Science* **254:** 713–716.

Gong J., Chen D., Kashiwaba M., and Kufe D. 1997. Induction of antitumor activity immunization with fusion of dendritic and carcinoma cells. *Nat. Med.* **3:** 558–561.

Goossens P.L., Milon G., Cossart P., and Saron M.F. 1995. Attenuated *Listeria monocytogenes* as a live vector for induction of CD8+ T cells *in vivo*: A study with the nucleoprotein of the lymphocytic choriomeningitis virus. *Intl. Immunol.* **7:** 797–805.

Gopas J., Rager-Zisman B., Bar-Eli M., Hammerling G.J., and Segal S. 1989. The relationship between MHC antigen expression and metastasis. *Adv. Cancer Res.* **53:** 89–115.

Hahne M., Rimoldi D., Schroter M., Romero P., Schreier M., French L.E., Schneider P., Bornand T., Fontana A., Lienard D., Cerottini J.C., and Tschopp J. 1996. Melanoma cell expression of Fas(Apo-1/CD95) ligand: Implications for tumor immune escape. *Science* **274:** 1363–1366.

Hock H., Dorsch M., Diamanstein T., and Blankenstein T. 1991. Interleukin 7 induces CD4+ T-cell-dependent tumor rejection. *J. Exp. Med.* **174:** 1291–1298.

Hoiseth S.K. and Stocker B.A. 1981. Aromatic-dependent *Salmonella typhimurium* are non-virulent and effective as live vaccines. *Nature* **291:** 238–239.

Hsu F.J., Benike C., Fagnoni F., Liles T.M., Czerwinski D., Taidi B., Engleman E.G., and Levy R. 1966. Vaccination of patients with B-cell lymphoma using autologous antigen-pulsed dendritic cells. *Nat. Med.* **2:** 52–58.

Huang A.Y., Columbek P., Ahmadzadeh M., Jaffee E., Pardoll D., and Levitsky H. 1994. Role of bone marrow-derived cells in presenting MHC class I-restricted tumor antigens. *Science* **264:** 961–965.

Huang A.Y.C., Gulden P.H., Woods A.S., Thomas M.C., Tong C.D., Wang W., Engelhard V.H., Pasternack G., Cotter R., Hunt D., Pardoll D.M., and Jaffee E.M. 1996. The immunodominant major histocompatibility complex class I-restricted antigen of a murine colon tumor derives from an endogenous retroviral gene product. *Proc. Natl. Acad. Sci.* **93:** 9730–9735.

Huang M., Sharma S., Mao J.T., and Dubinett S.M. 1996. Non-small cell lung cancer-derived soluble mediators and prostaglandin E2 enhance peripheral blood lymphocyte IL-10 transcription and protein production. *J. Immunol.* **157:** 5512–5520.

Hui K., Grosveld F., and Festenstein H. 1984. Rejection of transplantable AKR leukemia cells following MHC DNA-mediated cell transformation. *Nature* **311:** 750–752.

Ikonomidis G., Paterson Y., Kos F.J., and Portnoy D.A. 1994. Delivery of a viral antigen to the class I processing and presentation pathway by *Listeria monocytogenes*. *J. Exp. Med.* **180:** 2209–2218.

Inaba K., Inaba M., Romani N., Aya H., Deguchi M., Ikehara S., Muramatsu S., and Steinman R.M. 1992a. Generation of large numbers of dendritic cells from mouse bone marrow cultures supplemented with granulocyte/macrophage colony-stimulating factor. *J. Exp. Med.* **176:** 1693–1702.

Inaba K., Steinmann R.M., Witmer-Pack M., Aya H.A.M., Inaba T., Sudo T., Wolpe S., and Schuler G. 1992b. Identification of proliferating dendritic cell precursors in mouse blood. *J. Exp. Med.* **175:** 1157–1167.

Inge T.H., Hoover S.K., Susskind B.M., Barrett S.K., and Bear H.D. 1992. Inhibition of tumor-specific cytotoxic T-lymphocyte responses by transforming growth factor β_1. *Cancer Res.* **52:** 1386–1392.

Irvine K.R., Chamberlain R.S., Shulman E.P., Surman D.R., Rosenberg S.A., and Restifo N.P. 1997. Enhancing efficacy of recombinant anticancer vaccines with prime/boost regimens that use two different vectors. *J. Natl. Cancer Inst.* **89:** 1595–1601.

Itaya T., Yamagiwa S., Okada F., Oikawa T., Kuzumaki N., Takeichi N., Hosokawa M., and Kobayashi H. 1987. Xenogenization of a mouse lung carcinoma (3LL) by transfection with an allogeneic class I major histocompatibility complex gene (H-2L^d). *Cancer Res.* **47:** 3136–3140.

Itoh N., Yonehara S., Ishii A., Yonehara M., Mizushima S., Sameshima M., Hase A., Seto Y., and Nagata S. 1991. The polypeptide encoded by the cDNA for human cell surface antigen Fas can mediate apoptosis. *Cell* **66:** 233–243.

Jardetzky T.S., Lane W.S., Robinson R.A., Madden D.R., and Wiley D.C. 1991. Identification of self peptides bound to purified HLA-B27. *Nature* **353:** 326–329.

Ju S., Panka D.J., Cui H., Ettinger R., el-Khatib M., Sherr D.H., Stanger B.Z., and Marshak-Rothstein A. 1995. Fas(CD95)/FasL interactions required for programmed cell death after T-cell activation. *Nature* **373:** 444–448.

Juillard V., Villefroy P., Godfrin D., Pavirani A., Venet A., and Guillet J.G. 1995. Long-term humoral and cellular immunity induced by a single immunization with replication-defective adenovirus recombinant vector. *Eur. J. Immunol.* **25:** 3467–3473.

Kang X., Kawakami Y.E.L., Gamil M., Wang R., Sakaguchi K., Yannelli J.R., Appella E., Rosenberg S.A., and Robbins P.F. 1995. Identification of a tyrosinase epitope recognized by HLA-A24-restricted, tumor-infiltrating lymphocytes. *J. Immunol.* **155:** 1343–1348.

Kawakami Y., Eliyahu S., Delgado C.H., Robbins P.F., Sakaguchi K., Appella E., Yannelli J.R., Adema G.J., Miki T., and Rosenberg S.A. 1994. Identification of a human melanoma antigen recognized by tumor-infiltrating lymphocytes associated with *in vivo* tumor rejection. *Proc. Natl. Acad. Sci.* **91:** 6458–6462.

Koseki H., Shiiba K., Suzuki Y., Asanuma T., and Matsuno S. 1997. Enhanced expression of lipocortin-1 as a new immunosuppressive protein in cancer patients and its influence on reduced *in vitro* peripheral blood lymphocyte response to mitogens. *Surg. Today* **27:** 30–39.

Kucharzik T., Lugering N., Winde G., Domschke W., and Stoll R. 1997. Colon carcinoma cell lines stimulate monocytes and lamina propria mononuclear cells to produce IL-10. *Clin. Exp. Immunol.* **110:** 296–302.

Lathe R., Kieny M.P., Gerlinger P., Clertant P., Guizani I., Cuzin F., and Chambon P. 1987. Tumour prevention and rejection with recombinant vaccinia. *Nature* **326:** 878–880.

Li W.Q., Diamantstein T., and Blankenstein T. 1990. Lack of tumorigenicity of interleukin 4 autocrine growing cells seems related to the anti-tumor function of interleukin 4. *Mol. Immunol.* **27:** 1331–1337.

Lin K.Y., Guarnieri F.G., Staveley-O'Carroll K., Levitsky H.I., August J.T., Pardoll D.M., and Wu T.C. 1996. Treatment of established tumors with a novel vaccine that enhances major histocompatibility class II presentation of tumor antigen. *Cancer Res.* **56:** 21–26.

Lindahl K.F. and Wilson D.B. 1977. Histocompatibility antigen-activated cytotoxic T lymphocytes. *J. Exp. Med.* **145:** 508–522.

Lindenmann J. and Klein P.A. 1967. Viral oncolysis: Increased immunogenicity of host cell antigen associated with influenza virus. *J. Exp. Med.* **126:** 93–108.

Linsley P.S., Brady W., Grosmaire L., Arffo A., Damle N.K., and Ledbetter J.A. 1991. Binding of the B cell activation antigen B7 to CD28 costimulates T cell proliferation and interleukin 2 mRNA accumulation. *J. Exp. Med.* **173:** 721–730.

Livingston P.O., Albino A.P., Chung T.J., Real F.X., Houghton A.N., Oettgen H.F., and Old L.J. 1985. Serological response of melanoma patients to vaccines prepared from VSV lysates of autologous and allogeneic cultured melanoma cells. *Cancer* **55:** 713–720.

Mayordomo J.I., Zorina T., Storkus W.J., Zitvogel L., Garcia-Prats M.D., DeLeo A.B., and Lotze M.T. 1997. Bone marrow-derived dendritic cells serve as potent adjuvants for peptide-based antitumor vaccines. *Stem Cells* **15**: 94–103.

McCune C.S., O'Donnell R.W., Marquis D.M., and Sahasrabudhe D.M. 1990. Renal cell carcinoma treated by vaccines for active specific immunotherapy: Correlation of survival with skin testing by autologous tumor cells. *Cancer Immunol. Immunother.* **32**: 62–66.

Minev B.R., McFarland B.J., Spiess P.J., Rosenberg S.A., and Restifo N.P. 1994. Insertion signal sequence fused to minimal peptides elicits specific CD8+ T-cell responses and prolongs survival of thymoma-bearing mice. *Cancer Res.* **54**: 4155–4161.

Montgomery D.L., Shiver J.W., Leander K.R., Perry H.C., Friedman A., Martinez D., Ulmer J.B., Donnelly J.J., and Liu M.A. 1993. Heterologous and homologous protection against influenza A by DNA vaccination: Optimization of DNA vectors. *DNA Cell. Biol.* **12**: 777–783.

Morisaki T., Katano M., Ikubo A., Anan K., Nakamura M., Nakamura K., Sato H., Tanaka M., and Torisu M. 1996. Immunosuppressive cytokines (IL-10, TGF-β) genes expression in human gastric carcinoma tissues. *J. Surg. Oncol.* **63**: 234–239.

Moss B., Smith G.L., Gerin J.L., and Purcell R.H. 1984. Live recombinant vaccinia virus protects chimpanzees against hepatitis B. *Nature* **311**: 67–69.

Mulligan R.C. 1993. The basic science of gene therapy. *Science* **260**: 926–932.

Nabel E.G., Shum L., Pompili V.J., Yang Z.-Y., San H., Shu H.B., Liptay S., Gold L., Gordon D., Derynck R., and Nabel G.J. 1993. Direct gene transfer of transforming growth factor β1 into arteries stimulates fibrocellular hyperplasia. *Proc. Natl. Acad. Sci.* **90**: 10759–10763.

Nabel G.J., Chang A., Nabel E.G., Plautz G., Fox B.A., Huang L., and Shu S. 1992. Clinical protocol: Immunotherapy of malignancy by *in vivo* gene transfer into tumors. *Hum. Gene Ther.* **3**: 399–410.

Nabel G.J., Chang A.E., Nabel E.G., Plautz G.E., Ensminger W., Fox B.A., Felgner P., Shu S., and Cho K. 1994. Clinical protocol: Immunotherapy for cancer by direct gene transfer into tumors. *Hum. Gene Ther.* **5**: 57–77.

Nabel G.J., Gordon D., Bishop D.K., Nickoloff B.J., Yang Z., Aruga A., Cameron M.J., Nabel E.G., and Chang A.E. 1996. Immune response in human melanoma after transfer of an allogeneic class I major histocompatibility complex gene with DNA-liposome complexes. *Proc. Natl. Acad. Sci.* **93**: 15388–15393.

Nabel G.J., Nabel E.G., Yang Z.-Y., Fox B.A., Plautz G.E., Gao X., Huang L., Shu S., Gordon D., and Chang A.E. 1993. Direct gene transfer with DNA-liposome complexes in melanoma: Expression, biologic activity, and lack of toxicity in humans. *Proc. Natl. Acad. Sci.* **90**: 11307–11311.

Newborg M.F. and North R.J. 1980. On the mechanism of T-cell-independent anti-*Listeria* resistance in nude mice. *J. Immunol.* **124**: 571–576.

O'Connell J., O'Sullivan G.C., Collins J.K., and Shanahan F. 1996. The fas counterattack: fas-mediated T cell killing by colon cancer cells expressing fas ligand. *J. Exp. Med.* **184**: 1075–1082.

Paglia P., Chiodoni C., Rodolfo M., and Colombo M.P. 1996. Murine dendritic cells loaded *in vitro* with soluble protein prime cytotoxic T lymphocytes against tumor antigen *in vivo*. *J. Exp. Med.* **183**: 317–322.

Pan Z.K., Ikonomidis G., Pardoll D., and Paterson Y. 1995a. Regression of established tumors in mice mediated by the oral adminstration of a recombinant *Listeria monocytogenes* vaccine. *Cancer Res.* **55**: 4776–4779.

Pan Z.K., Ikonomidis G., Lazenby A., Pardoll D.M., and Paterson Y. 1995b. A recombinant *Listeria monocytogenes* vaccine expressing a model tumour antigen protects mice against lethal tumour cell challenge and causes regression of established tumours. *Nat. Med.* **1:** 471–477.

Pantel K., Schlimok G., Kutter D., Schaller G., Genz T., Wiebecke B., Backmann R., Funke I., and Reithmuller G. 1991. Frequent down-regulation of major histocompatibility class I antigen expression of individual micrometastatic carcinoma cells. *Cancer Res.* **51:** 4712–4715.

Pardoll D.M. 1993. New strategies for enhancing the immunogenicity of tumors. *Curr. Opin. Immunol.* **5:** 719–725.

Pardoll D.M. and Beckerleg A.M. 1995. Exposing the immunology of naked DNA vaccines. *Immunity* **3:** 165–169.

Parham P., Lomen C.E., Lawlor D.A., Ways J.P., Holmes N., Coppin H.L., Salter R.D., Wan A.M., and Ennis P.D. 1988. Nature of polymorphism in HLA-A, -B, -C molecules. *Proc. Natl. Acad. Sci.* **85:** 4005–4009.

Poirier T.P., Kehoe M.A., and Beachey E.H. 1988. Protective immunity evoked by oral administration of attenuated aroA *Salmonella typhimurium* expressing cloned streptococcal M protein. *J. Exp. Med.* **168:** 25–32.

Porgador A., Tzehoval E., Katz A., Vadai E., Revel M., Feldman M., and Eisenbach L. 1992. Interleukin 6 gene transfection into Lewis lung carcinoma tumor cells suppresses the malignant phenotype and confers immunotherapeutic competence against parental metastatic cells. *Cancer Res.* **52:** 3679–3686.

Portnoy D.A. 1992. Innate immunity to a facultative intracellular bacterial pathogen. *Curr. Opin. Immunol.* **4:** 20–24.

Pulaski B.A., McAdam A.J., Hutter E.K., Biggar S., Lord E.M., and Frelinger J.G. 1993. Interleukin 3 enhances development of tumor-reactive cytotoxic cells by a CD4-dependent mechanism. *Cancer Res.* **53:** 2112–2117.

Restifo N.P., Spiess P.J., Karp S.E., Mule J.J., and Rosenberg S.A. 1991. A nonimmunogenic sarcoma transduced with the cDNA for interferon γ elicits CD8[+] T cells against the wild-type tumor: Correlation with antigen presentation capability. *J. Exp. Med.* **175:** 1423–1431.

Restifo N.P., Bacik I., Irvine K.R., Yewdell J.W., McCabe B.J., Anderson R.W., Eisenlohr L.C., Rosenberg S.A., and Bennink J.R. 1995. Antigen processing *in vivo* and the elicitation of primary CTL responses. *J. Immunol.* **154:** 4414–4422.

Robbins P.F. and Kawakami Y. 1996. Human tumor antigens recognized by T cells. *Curr. Opin. Immunol.* **8:** 628–636.

Robbins P.F., el Gamil M., Kawakami Y., Stevens E., Yannelli J.R., and Rosenberg S.A. 1994. Recognition of tyrosinase by tumor-infiltrating lymphocytes from a patient responding to immunotherapy. *Cancer Res.* **54:** 3124–3126.

Rollins B.J. and Sunday M.E. 1991. Suppression of tumor formation *in vivo* by expression of the JE gene in malignant cells. *Mol. Cell. Biol.* **11:** 3125–3131.

Rosen H.R., Ausch C., Reiner G., Reinerova M., Svec J., Tuchler H., Schiessel R., and Moroz C. 1996. Immunosuppression by breast cancer associated p43-effect of immunomodulators. *Breast Cancer Res. Treat.* **41:** 171–176.

Roth J.A. and Cristiano R.J. 1997. Gene therapy for cancer: What have we done and where are we going? *J. Natl. Cancer Inst.* **89:** 21–39.

Rubin J., Galanis E., Pitot H.C., Richardson R.L., Burch P.A., Charboneau J.W., Reading C.C., Lewis B.D., Stahl S., Akporiaye E.T., and Harris D.T. 1997. Phase I study of immunotherapy of hepatic metastases of colorectal carcinoma by direct gene transfer of an allogeneic histocompatibility antigen, HLA-B7. *Gene Ther.* **4:** 419–425.

Sadoff J.C., Ballou W.R., Baron L.S., Majarian W.R., Brey R.N., Hockmeyer W.T., Young J.F., Cryz S.J., Ou J., and Lowell G.H. 1988. Oral *Salmonella typhimurium* vaccine expressing circumsporozoite protein protects against malaria. *Science* **240:** 336–338.

Sato Y., Roman M., Tighe H., Lee D., Corr M., Nguyen M.D., Silverman G.J., Lotz M., Carson D.A., and Raz E. 1996. Immunostimulatory DNA sequences necessary for effective intradermal gene immunization. *Science* **273:** 352–354.

Schafer R., Portnoy D.A., Brassell S.A., and Paterson Y. 1992. Induction of a cellular immune response to a foreign antigen by a recombinant *Listeria monocytogenes* vaccine. *J. Immunol.* **149:** 53–59.

Scott P. and Kaufmann S.H. 1991. The role of T-cell subsets and cytokines in the regulation of infection. *Immunol. Today* **12:** 346–348.

Seliger B., Harders C., Wollscheid U., Staege M.S., Reske-Kunz A.B., and Huber C. 1996. Suppression of MHC class I antigens in oncogenic transformants: Association with decreased recognition by cytotoxic T lymphocytes. *Exp. Hematol.* **24:** 1275–1279.

Shen H., Slifka M.K., Matloubian M., Jensen E.R., Ahmed R., and Miller J.F. 1995. Recombinant *Listeria monocytogenes* as a live vaccine vehicle for the induction of protective anti-viral cell-mediated immunity. *Proc. Natl. Acad. Sci.* **92:** 3987–3991.

Simons J.W., Jaffee E.M., Weber C.E., Levitsky H.I., Nelson W.G., Carducci M.A., Lazenby A.J., Cohen L.K., Finn C.C., Clift S.M., Hauda K.M., Beck L.A., Leiferman K.M., Owens A.H., Jr., Piantadosi S., Dranoff G., Mulligan R.C., Pardoll D.M., and Marshall F.F. 1997. Bioactivity of autologous irradiated renal cell carcinoma vaccines generated by *ex vivo* granulocyte-macrophage colony-stimulating factor gene transfer. *Cancer Res.* **57:** 1537–1546.

Skinner M.A. and Marbrook J. 1976. An estimation of the frequency of precursor cells which generate cytotoxic lymphocytes. *J. Exp. Med.* **143:** 1562–1567.

Skipper J.C., Hendrickson R.C., Gulden P.H., Brichard V., Van Pel A., Chen Y., Shabanowitz J., Wolfel T., Slingluff C.J., Boon T., Hunt D.F., and Engelhard V.H. 1996. An HLA-A2-restricted tyrosinase antigen on melanoma cells results from post-translational modification and suggests a novel pathway for processing of membrane proteins. *J. Exp. Med.* **183:** 527–534.

Song W., Kong H., Carpenter H., Torii H., Granstein R., Rafii S., Moore M.A.S., and Crystal R.G. 1997. Dendritic cells genetically modified with an adenovirus vector encoding the cDNA for a model tumor antigen induce protective and therapeutic anti-tumor. *J. Exp. Med.* **186:** 1247–1256.

Specht J.M., Wang G., Do M.T., Lam J.S., Royal R.E., Reeves M.E., Rosenberg S.A., and Hwu P. 1997. Dendritic cells retrovirally transduced with a model tumor antigen gene are therapeutically effective against established pulmonary metastases. *J. Exp. Med.* **186:** 1213–1221.

Steinman R.M. 1991. The dendritic cell system and its role in immunogenicity. *Annu. Rev. Immunol.* **9:** 271–296.

Stopeck A.T., Hersh E.M., Akporiaye E.T., Harris D.T., Grogan T., Unger E., Warneke J., Schluter S.F., and Stahl S. 1997. Phase I study of direct gene transfer of an allogeneic histocompatibility antigen, HLA-B7, in patients with metastatic melanoma. *J. Clin. Oncol.* **15:** 341–349.

Stover C.K., de la Cruz V.F., Fuerst T.R., Burlein J.E., Benson L.A., Bennett L.T., Bansal G.P., Young J.F., Lee M.H., and Hatfull G.F. 1991. New use of BCG for recombinant vaccines. *Nature* **351:** 456–460.

Strand S., Hofmann W.J., Hug H., Muller M., Otto G., Strand D., Mariani S.M., Stremmel

17

Gene Transfer and Stem-cell Transplantation

Malcolm K. Brenner

Center for Cell and Gene Therapy
Baylor College of Medicine
Houston, Texas 77030

Hematopoietic stem-cell transplantation and gene transfer may be linked in two ways. Since stem-cell transplantation is used to replace and/or repair hematopoietic or lymphopoietic deficiencies in the host, the most obvious linkage is to use a stem cell transplant to deliver a therapeutic gene. This approach, in which gene transfer represents the primary therapy while the stem-cell transplant simply represents the means of delivery, has been described in detail (Nienhuis, this volume). Alternatively, however, stem-cell transplantation itself may be the primary therapy, and the process of gene transfer used to enhance its safety and efficacy. It is to this second usage that this chapter is devoted.

Worldwide, more than 30,000 stem-cell transplants are performed each year, the great majority of which are intended to treat patients with malignant disease (Applebaum 1997). One of the fundamental concepts driving modern oncology is that increased doses of cytotoxic drugs will cure increased numbers of patients. Full implementation of this approach is limited by toxicity to normal tissues, among the most sensitive of which are normal hematopoietic stem cells (HSC). The resulting marrow aplasia may cause death from infection or hemorrhage. However, patients can safely be treated with otherwise lethal doses of chemotherapy/radiation if they are then "rescued" from the consequences by infusion of autologous or allogeneic HSCs derived from marrow or from peripheral or placental blood. The success of this approach has led to a steady increase in the number of patients so treated.

To understand how gene transfer can help improve stem-cell transplantation, it is necessary first to outline the types of stem-cell transplant available.

Improving Reconstitution of the Host Immune System

The above approaches have focused on the recovery of the myeloid compartment of marrow and are therefore most relevant to autologous transplants and placental blood allografts. For other grafts, it is overcoming the delay in immune reconstitution that is the first priority, and a number of options utilizing gene transfer are available.

Gene-modified Cytotoxic T Cells

This approach is most valuable for patients who have received allogeneic stem-cell transplants, particularly if the marrow is from an unrelated or HLA-mismatched donor and/or if it has been depleted of T lymphocytes as GvHD prophylaxis. Some of the most frequent and severe infections in these patients are due to reactivation of herpesviruses, particularly cytomegalovirus (CMV) and Epstein-Barr virus (EBV) (Heslop and Rooney 1997). Disease due to CMV reactivation may be prevented either by pharmacologic agents such as ganciclovir and foscarnet or by infusion of CMV-specific T cells. EBV-induced lymphoma has proved to be a more intractable problem in transplant recipients (Heslop et al. 1994; Papadopoulos et al. 1994; Rooney et al. 1995). EBV infects most individuals and persists in an asymptomatic state by a combination of chronic replication in the mucosa and latency in peripheral blood B cells (Straus et al. 1992). These EBV-infected B cells are highly immunogenic and normally susceptible to killing by specific cytotoxic T lymphocytes (CTLs). However, in immunocompromised patients, the infected B cells may grow unchecked, producing a rapidly progressive lymphoproliferative disease, which usually appears histologically as an immunoblastic lymphoma. This complication occurs in 1–30% of patients receiving immunosuppression after allografts and has a high mortality (Papadopoulos et al. 1994).

If these cells are only able to flourish because of the absence of functional, EBV-reactive CTLs, then administration of normal peripheral blood lymphocytes from EBV-immune donors to patients with lympho-proliferative disease (LPD) should produce resolution. In fact, administration of donor peripheral blood mononuclear cells to recipients after marrow allografting can produce complete clinical and histological responses (Papadopoulos et al. 1994), presumably due to virus-specific T cells within the bulk lymphocyte population. However, since the population also contains many alloreactive T cells, such treatment may also induce severe graft-versus-host disease (Heslop et al. 1994; Papadopoulos et al. 1994).

One way of preventing this adverse effect is to transduce donor T cells with the thymidine kinase (*tk*) gene. If a surface selectable marker gene, such as the truncated nerve growth factor receptor (tNGFR), is included in the vector, then the transduced subpopulation can be enriched by fluorescence flow cytometry prior to infusion (Bonini et al. 1997). The EBV-reactive T cells will remain among the population and be able to expand and extirpate EBV⁺ lymphoblasts. Should GvHD develop, the infused T cells can be destroyed by administration of ganciclovir, since this will be phosphorylated into a toxic nucleoside analog in the *tk*⁺ lymphocytes. Preliminary studies suggest that this may be an effective approach to treatment of EBV disease and that GvHD can be controlled by administration of ganciclovir (Bonini et al. 1997).

An alternative approach is to generate T cells that are specific only for the viral antigens expressed by the EBV⁺ tumor cells. In a study to determine whether EBV-specific CTLs are safe and effective after adoptive transfer to patients following marrow allografts, the cells were marked with the *neo* gene prior to infusion. By tracking the marker in vivo, it was possible to learn about their survival, distribution, and activity (Heslop et al. 1996). Study of more than 50 patients has shown that the infused CTLs produce no adverse effects and are long-lived, the marker gene being detected for more than 3 years. In seven patients with evidence of EBV reactivation, administration of EBV-specific CTLs was rapidly followed by a 1000-fold fall in EBV DNA levels in peripheral blood within 3–4 weeks of infusion. Finally, in patients with biopsy-proven immunoblastic lymphoma, marked T cells accumulated at the sites of disease and induced complete and sustained tumor regression (Rooney et al. 1995; Heslop and Rooney 1997).

It is also feasible to transfer CTLs specific for other pathogens such as adenoviruses that cause severe morbidity after stem-cell transplantation. Here too, gene transfer may contribute. Generation of these CTLs may be simplified by transducing professional antigen-presenting cells with viral genes (Fig. 2), so that the viral proteins are processed and effectively presented to donor T cells. In this way it is possible to make CTLs that will recognize even weakly immunogenic viral epitopes.

The major disadvantage of generating specific CTLs is that they must be prepared individually for each patient, since the phenomenon of MHC restriction means that these cells only function optimally in an individual with closely matched HLA antigens. Efforts are being made to generate a "universal" T cell, in which the normal, MHC-restricted, T-cell receptor is removed or supplemented with a hybrid receptor consisting of an MHC-

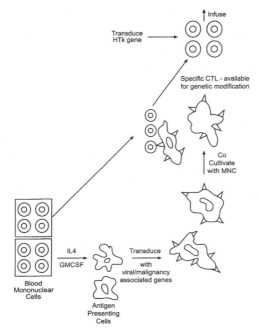

Figure 2 Use of dual gene marking to assess impact of stem-cell treatment on recovery and relapse. Blood mononuclear cells are split into two portions. The first is treated with cytokines such as interleukin-4 (IL-4) and granulocyte macrophage–colony-stimulating factor (GM-CSF) to favor production of antigen–presenting cells (APCs). These APCs are transduced with vectors encoding the genes of interest. Once expression is confirmed, these APCs are cocultured with the remaining mononuclear cells to generate antigen-specific T lymphocytes. These in turn are transduced with a selectable marker gene and a suicide gene such as HTk before they are returned to the patient.

unrestricted component (such as an antibody-binding domain) and the T-cell receptor ζ chain to signal T-cell activation once the hybrid receptor is engaged (Hege et al. 1996). Although T cells of this type appear to be universal in terms of antigen recognition, it has not yet proved possible to make these cells nonimmunogenic: They express their own alloantigens and would likely generate a potent antiallo response following infusion. However, it is likely that it will soon be possible to make these cells non-immunogenic by inducing expression of inhibitory molecules (Wong and Choi 1997) or by suppressing costimulatory proteins (Guinan et al. 1995; Boussiotis et al. 1996; Grewal and Flavell 1996). The availability of universal T cells should simplify our efforts to enhance immune recovery after allografting, and thereby reduce morbidity and mortality.

Relapse

Autologous Transplant

While autologous stem-cell transplants appear to result in an improvement in survival in many malignant diseases, relapse remains the major cause of treatment failure. The possibility that malignant cells may contaminate the stem-cell population and contribute to relapse following reinfusion has led to extensive evaluation of techniques for purging marrow to eliminate residual malignant cells (Champlin 1996). However, it has been unclear whether such maneuvers are necessary or effective. One way of resolving this issue is to mark the marrow at the time of harvest and then find out if the marker gene is present in malignant cells at the time of a subsequent relapse (Brenner 1996).

In the earliest clinical protocols, bone-marrow mononuclear cells were transduced with either the LNL6 or the closely related G1N retroviral vector and then reinfused. Both these vectors encode the neomycin resistance gene, which can subsequently be detected genotypically or phenotypically. In the initial acute myelogenous leukemia (AML) study, 3 of the 12 patients relapsed with marked blast cells (Brenner et al. 1993a) and similar results have been obtained in children with neuroblastoma (Rill et al. 1994) and in adults receiving autologous BMT for chronic myeloid leukemia (Deisseroth et al. 1994). Since retroviruses integrate largely at random, it is possible to analyze the number of individual integration sites in the malignant cells and thereby estimate the number of tumorigenic cells present in the graft that have contributed to relapse. These integrant analyses showed a multiplicity of different integrants, implying that several hundred or thousand malignant cells from marrow contaminate stem cells and contribute to relapse (Rill et al. 1994).

Although the same results have not yet been obtained from peripheral blood stem-cell infusions, it is very likely that these too are contaminated with tumorigenic cells. The implication is that effective purging will be one requirement for improving the outcome of autologous stem-cell transplants for malignant disease. The effectiveness of purging techniques can be assessed by using two distinguishable marker vectors in each patient to compare either marrow purging versus no purging or two different purging techniques (Fig. 1). These investigations are already providing information about the effects of purging on normal hematopoietic recovery (Brenner 1997) and should be able to indicate the frequency with which an individual purging technique fails to remove tumorigenic cells.

To date, all marker studies have been performed using the *neo* gene, which can only be detected by PCR or culture methodology. The recent availability of markers that can be detected by fluorescence-activated cell sorter (FACS) analysis, either on the cell surface (e.g., NGFR or CD24) or within the cytoplasm (e.g., green fluorescent protein), should simplify future marker studies (Pawliuk et al.1994; Bonini et al.1997; Persons et al.1997).

Allogeneic Transplant

Whereas relapse after autologous transplant may result from contamination of the stem-cell product, relapse after allogeneic transplant can only be due to residual disease in the patient. Considerable evidence now exists that the immune system has the potential to eradicate residual hematologic malignancy (Horowitz et al. 1990; Kolb et al. 1995). For example, the presence of graft-versus-host disease may reduce the risk of subsequent relapse of AML, whereas measures that reduce GvHD such as T-cell depletion or the use of an identical twin allograft are associated with an increased risk of disease recurrence. This so-called graft-versus-leukemia (GvL) effect may simply be another manifestation of graft-versus-host disease, in which both normal and malignant host cells share the same host-specific polymorphisms that are a target for alloreactive T lymphocytes (Truitt and Johnson 1995).

It is also possible that donor T cells may recognize discrete target antigens processed and presented by the malignant cell, including mutated oncogenes or fusion proteins generated by chromosomal translocations (Brenner and Heslop 1991; Melief and Kast 1993). If this is true, then a malignancy-specific response could be generated in the absence of any other host reactivity. Although exploration of this possibility and of the effector mechanisms involved is easiest when the target antigens have been identified, it is still possible to use gene transfer to increase the safety of nonspecific donor T cells when these are given for their GvL activity in a manner analogous to that already described for infectious diseases. Thus, patients have received infusions of T lymphocytes transduced with a retroviral vector encoding NGFR and *tk*. After selection of NGFR+ cells, the lymphocytes are infused in an effort to prevent or treat relapse after transplant (Bonini et al. 1997). In the event that GvHD occurs, the patients may be treated with ganciclovir to destroy the transferred gene-modified lymphocytes. A similar protocol using hygromycin selected *tk*-positive T cells will begin soon. In the longer term, efforts are being made to identify tumor-specific antigens that may be presented by malignant cells and to prepare donor CTL that recognize them. Many of the proposed protocols use dendritic cells genetically modified to express these antigens (Fig. 2).

Although none of the previous approaches has been clinically validated, gene transfer offers a means by which the capacity of donor T cells to generate a graft-versus-leukemia effect can be separated from their ability to induce graft-versus-host disease, thereby safely reducing the risk of relapse.

Tumor Vaccines following Autologous and Allogeneic Transplantation

Animal models and some human studies have shown that the transfer of genes encoding immunostimulatory molecules to tumor cells increases their immunogenicity and allows the host to reject preexisting tumor cells (Pardoll 1996). This approach is discussed in detail (Pardoll and Nabel, this volume) and will not be further dissected here. However, it is important to note that the approach may be most effective following stem-cell transplantation. At this time, their disease burden will be low, and any antitumor response will likely have its greatest chance to eliminate the malignant cells.

Graft-versus-Host Disease following Allogeneic Stem-cell Transplantation

Both acute and chronic GvHD remain a major impediment to successful allogeneic stem-cell transplantation. In humans, the disease can cause lethal damage to skin, gut, liver, and the lymphoid organs. The disease and its treatment are associated with profound and prolonged immunosuppression (Ferrara and Deeg 1991; Vogelsang and Hess 1994). Clinical and experimental data have established that the pathogenesis of GvHD requires the transfer of donor-derived T lymphocytes with bone marrow, and a wide array of preventive and active treatment has been directed against these cells (Ferrara and Deeg 1991; Vogelsang and Hess 1994). However, no therapy is absolutely effective, and all current measures for preventing or treating GvHD increase other risks such as rejection, infection, and relapse. One way of preventing graft-versus-host disease while retaining the advantages of host T lymphocytes is to use "conditionally active" T cells. These are functional donor T lymphocytes that can be removed should their benefits be outweighed by the onset of graft-versus-host disease. A clinical protocol has opened to assess the feasibility of this approach (Tiberghien et al. 1997). Donor stem cells are first depleted of their T lymphocytes, which are then activated ex vivo to induce cell division and make them susceptible to transduction with a retroviral vector encoding a selectable marker (in this case *neo*) and the *tk* gene. These cells are then returned with the donor graft to facilitate engraftment, accelerate immune recovery, and reduce the risk of rejection and relapse. Should they also cause GvHD, the patients will be treated with ganciclovir to destroy the effector cells (Tiberghien et al. 1997).

At present, it is not clear how well these transduced T cells will work. Once GvHD is established, T cells may play only a limited part in maintaining the condition, which may instead become a consequence of nonspecific cellular and humoral immune mechanisms. It may be that other approaches independent of gene transfer will be preferable. For example, if donor and recipient mononuclear cells are cocultured, monoclonal antibodies can be used to block the interaction between costimulator molecules and their ligands (e.g., B7.1 and CTLA4) (Guinan et al. 1994; Grewal and Flavell 1996), rendering donor cells unresponsive to host alloantigens while preserving their capacity to respond to third-party stimuli such as viruses and also (it is hoped) malignant cells.

CONCLUSION

Gene transfer is already playing a part in improving autologous and allogeneic stem-cell transplantation. By providing a means of tracking the source of relapse after autologous transplant and by assessing the impact of purging and of growth-factor treatment of stem cells, gene marking should help reduce relapse and accelerate engraftment. After allogeneic transplantation, gene-modified CTLs are helping to reduce the risk of viral infections and may also be useful in preventing relapse and graft-versus-host disease. Future studies should ensure further refinement of these techniques.

ACKNOWLEDGMENTS

This work was supported in part by grants CA-20180, CA-23099, CA-58211, and CA-21765 (CORE) from the National Institutes of Health and by the American Lebanese Syrian Associated Charities (ALSAC).

REFERENCES

Appelbaum F.R. 1997. Allogeneic hematopoietic stem cell transplantation for acute leukemia. *Semin. Oncol.* **24:** 114–123.
Bonini C., Ferrari G., Verzeletti S., Servida P., Zappone E., Ruggieri L., Ponzoni M., Rossini S., Malvilio F., Traversari C., and Bordignon C. 1997. HSV-TK gene transfer into donor lymphocytes for control of allogeneic graft versus leukemia. *Science* **276:** 1719–1724.

Boussiotis V.A., Freeman G.J., Gribben J.G., Nadler L.M. 1996. The role of B7-1/B7-2-CD28/CTLA-4 pathways in the prevention of anergy, induction of productive immunity and down-regulation of the immune response. *Immun Rev* **153:** 5–26.

Brenner M.K. 1996. Gene marking. *Hum. Gene Ther.* **7:** 1927–1936.

———. 1997. Gene marking studies. *Prog. Growth Factors* **3:** 2–6.

Brenner M.K. and Heslop H.E. 1991. Graft-versus-host reactions and bone marrow transplantation. *Curr. Opin. Immunol.* **3:** 752–757.

Brenner M.K., Rill D.R., Moen R.C., Krance R.A., Mirro J., Jr., Anderson W.F., and Ihle J.N. 1993a. Gene-marking to trace origin of relapse after autologous bone marrow transplantation. *Lancet* **341:** 85–86.

Brenner M.K., Rill D.R., Holladay M.S., Heslop H.E., Moen R.C., Buschle M., Krance R.A., Santana V.M., Anderson W.F., and Ihle J.N. 1993b. Gene marking to determine whether autologous marrow infusion restores long-term haemopoiesis in cancer patients. *Lancet* **342:** 1134–1137.

Brugger W., Heimfeld S., Berenson R.J., Mertelsmann R., and Kanz L. 1995. Reconstitution of hematopoiesis after high-dose chemotherapy by autologous progenitor cells generated ex vivo. *N. Engl. J. Med.* **333:** 283–287.

Champlin R.E. 1996. Purging—The separation of normal from malignant cells for autologous transplantation. *Transfusion* **36:** 910–918.

Deisseroth A.B., Zu Z., Claxton D., Hanania E.G., Fu S., Ellerson D., Goldberg L., Thomas M., Janicek J., Anderson W.F., Hester J., Korbling M., Durett A., Moen R., Berenson R., Heimfeld S., Hamer J., Calvert L., Tibbits P., Talpaz M., Kantarjian H., Champlin R., and Reading C. 1994. Genetic marking shows that Ph+ cells present in autologous transplants of chronic myelogenous leukemia (CML) contribute to relapse after autologous bone marrow in CML. *Blood* **83:** 3068–3076.

Dunbar C.E., Cottler-Fox M., O'Shaughnessy J.A., Doren S., Carter C., Berenson R., Brown S., Moen R.C., Greenblatt J., Stewart F.M., Leitman S.F., Wilson W.H., Cowan K., Young N.S., and Nienhuis A.W. 1995. Retrovirally marked CD34-enriched peripheral blood and marrow cells contribute to long-term engraftment after autologous transplantation. *Blood* **85:** 3048–3057.

Emerson S.G., Palsson B.A., Clarke M.F., Silver S.M., Adams P.T., Koller M.R., Van Zant G., Rummel S., Armstrong R.D., and Maluta J. 1995. In vitro expansion of hematopoietic cells for clinical application. *Cancer Treat. Res.* **76:** 215–223.

Ferrara J.L.M. and Deeg H.G. 1991. Graft-versus-host disease. *N. Engl. J. Med.* **324:** 667–674.

Gluckman E., Rocha V., Boyer-Chammard A., Locatelli F., Arcese W., Pasquini R., Ortega J., Souillet G., Ferreira E., Laporte J., Fernandez M., and Chastang C. 1997. Outcome of cord-blood transplantation from related and unrelated donors. Eurocord Transplant Group and the European Blood and Marrow Transplantation Group. *N. Engl. J. Med.* **337:** 373–381.

Grewal I.S. and Flavell R.A. 1996. The role of CD40 ligand in costimulation and T-cell activation. *Immunol. Rev.* **153:** 85–106.

Gribben J.G., Freedman A.S., Neuberg D., Roy D.C., Blake K.W., Woo S.D., Grossbard M.L., Rabinowe S.N., Coral F., Freeman G.J., Ritz J., and Nadler L.M. 1991. Immunologic purging of marrow assessed by PCR before autologous bone marrow transplantation for B-cell lymphoma. *N. Engl. J. Med.* **325:** 1525–1533.

Guinan E.C., Gribben J.G., Boussiotis V.A., Freeman G.J., and Nadler L.M. 1994. Pivotal role of the B7: CD28 pathway in transplantation tolerance and tumor immunity. *Blood*

84: 3261–3282.

Hanania E.G., Giles R.E., Kavanagh J., Ellerson D., Zu Z., Wang T., Su Y., Kudelka A., Rahman Z., Holmes F., Hortobagyi G., Claxton D., Bachier C., Thall P., Cheng S., Hester J., Ostrove JM., Bird RE., Chang A., Korbling M., Seong D., Cote R., Holzmayer T., Mechetner E., Heimfeld S., Berenson R., Burtness B., Edwards C., Bast R., Andreeff M., Champlin R., and Deisseroth A.B. 1996. Results of MDR-1 vector modification trial indicate that granulocyte/macrophage colony-forming unit cells do not contribute to posttransplant hematopoietic recovery following intensive systemic therapy. *Proc. Natl. Acad. Sci.* **93:** 15346–15351.

Hege K.M., Cooke K.S., Finer M.H., Zsebo K.M., and Roberts M.R. 1996. Systemic T cell-independent tumor immunity after transplantation of universal receptor-modified bone marrow into SCID mice. *J. Exp. Med.* **184:** 2261–2269.

Heslop H.E. and Rooney C.M. 1997. Adoptive cellular immunotherapy for EBV lymphoproliferative diseases. *Immunol. Rev.* **157:** 217–222.

Heslop H.E., Brenner M.K., and Rooney C.M. 1994. Donor T cells to treat EBV-associated lymphoma. *N. Engl. J. Med.* **331:** 679–680.

Heslop H.E., Ng C.Y.C., Li C., Smith C.A., Loftin S.K., Krance R.A., Brenner M.K., and Rooney C.M. 1996. Long-term restoration of immunity against Epstein-Barr virus infection by adoptive transfer of gene-modified virus-specific T lymphocytes. *Nat. Med.* **2:** 551–555.

Hongeng S., Krance R.A., Bowman L.C., Srivastava D.K., Cunningham J.M., Horwitz E.M., Brenner M.K., and Heslop H.E. 1997. Outcomes of transplantation with matched-sibling and unrelated-donor bone marrow in children with leukemia. *Lancet* **350:** 767–771.

Horowitz M.M., Gale R.P., Sondel P.M., Goldman J.M., Kersey J., Kolb H., Rimm A.A., Ringden O., Rozman C., Speck B., Truitt RL., Zwaan F.E., and Bortin M.M. 1990. Graft-versus-leukemia reactions after bone marrow transplantation. *Blood* **75:** 555–562.

Kolb H., Schattenberg A., Goldman J.M., Hertenstein B., Jacobsen N., Arcese W., Ljungman P., Ferrant A., Verdonck L., Niederwieser D., van Rhee F., Mittermueller J., de Witte T., Holler E., and Ansari H. 1995. Graft-versus-leukemia effect of donor lymphocyte infusions in marrow grafted patients. European Grup for Blood and Marrow Transplantation working Party Chronic Leukemia. *Blood* **86:** 2041–2050.

Koller M.R., Emerson S.G., and Palsson B.O. 1993. Large-scale expansion of human stem and progenitor cells from bone marrow mononuclear cells in continuous perfusion cultures. *Blood* **82:** 378–384.

Kook H., Goldman F., Padley D., Giller R., Rumelhart S., Holida M., Lee N., Peters C., Comito M., Huling D., Trigg M. 1996. Reconstruction of the immune system after unrelated or partially matched T cell-depleted bone marrow transplantation in children: Immunophenotypic analysis and factors affecting the speed of recovery. *Blood* **88:** 1089–1097.

Kurtzberg J., Laughlin M., Graham M.L., Smith C., Olson J.F., Halperin E.C., Ciocci G., Carrier C., Stevens C.E., and Rubinstein P. 1996. Placental blood as a source of hematopoietic stem cells for transplantation into unrelated recipients. *N. Engl. J. Med.* **335:** 157–166.

Maze R., Hanenberg H., and Williams D.A. 1997. Establishing chemoresistance in hematopoietic progenitor cells. *Mol. Med. Today* **3:** 350–358.

McLachlin J.R., Eglitis M.A., Ueda K., Kantoff P.W., Pastan I.H., Anderson W.F., and

Gottesman M.M. 1990. Expression of a human complementary DNA for the multidrug resistance gene in murine hematopoietic precursor cells with the use of retroviral gene transfer. *J. National Cancer Inst.* **82:** 1260–1990.

Melief C.J. and Kast W.M. 1993. Potential immunogenicity of oncogene and tumor supressor gene products. *Curr. Opin. Immunol.* **5:** 709–713.

Mickisch G.H., Licht T., Merlino G.T., Gottesman M.M., Pastan I. 1991a. Chemotherapy and chemosensitization of transgenic mice which express the human multidrug resistance gene in bone marrow: Efficacy, potency, and toxicity. *Cancer Res.* **51:** 5417–5424.

Mickisch G.H., Merlino G.T., Galski H., Gottesman M.M., and Pastan I. 1991b. Transgenic mice that express the human multidrug-resistance gene in bone marrow enable a rapid identification of agents that reverse drug resistance. *Proc. Natl. Acad. Sci.* **88:** 547–551.

Morrison S.J., Uchida N., and Weissman I.L. 1995. The biology of hematopoietic stem cells. *Immunol. Today* **18:** 156–162.

Murphy D., Crowther D., Renninson J., Prendiville J., Ranson M., Lind M., Patel U., Dougal M., Buckley C.H., and Tindall V.R. 1993. A randomised dose intensity study in ovarian carcinoma comparing chemotherapy given at four week intervals for six cycles with half dose chemotherapy given for twelve cycles. *Ann. Oncol.* **4:** 377–383.

Nienhuis A.W., Walsh C.E., and Liu J. 1993. Viruses as therapeutic gene transfer vectors. In *Viruses and bone marrow* (ed. N.S. Young), pp. 353–414. Marcell Decker Inc., New York.

Papadopoulos E.B., Ladanyi M., Emanuel D., MacKinnon S., Boulad F., Carabasi M.H., Castro-Malaspina H., Childs B.H., Gillio AP., Small T.N., Young J.W., Kernan N.A., and O'Reilly R.J. 1994. Infusions of donor leukocytes to treat Epstein-Barr virus-associated lymphoproliferative disorders after allogeneic bone marrow transplantation. *N. Engl. J. Med.* **330:** 1185–1191.

Pardoll D.M. 1996. Cancer vaccines: A road map for the next decade. *Curr. Opin. Immunol.* **8:** 619–621.

Pastan I. and Gottesman M.M. 1991. Multidrug resistance. *Annu. Rev. Med.* **42:** 277–286.

Pawliuk R., Kay R., Lansdorp P., Humphries R.K. 1994. Selection of retrovirally transduced hematopoietic progenitor cells using CD24 as a marker of gene transfer. *Blood* **84:** 2868–2877.

Persons D.A., Allay J.A., Allay E.R., Smeyne R.J., Ashmun R.A., Sorrentino B.P., and Nienhuis A.W. 1997. Retroviral-mediated transfer of the green fluorescent protein gene into murine hematopoietic cells facilitates scoring and selection of transduced progenitors in vitro and identification of genetically modified cells in vivo. *Blood* **90:** 1777–1786.

Rill D.R., Santana V.M., Roberts W.M., Nilson T., Bowman L.C., Krance R.A., Heslop H.E., Moen R.C., Ihle J.N., and Brenner M.K. 1994. Direct demonstration that autologous bone marrow transplantation for solid tumors can return a multiplicity of tumorigenic cells. *Blood* **84:** 380–383.

Rooney C.M., Smith C.A., Ng C., Loftin S.K., Li C., Krance R.A., Brenner M.K., and Heslop H.E. 1995. Use of gene-modified virus-specific T lymphocytes to control Epstein-Barr virus-related lymphoproliferation. *Lancet* **345:** 9–13.

Shpall E.J., Cagnoni P.J., Bearman S.I., Ross M., and Jones R.B. 1997. Peripheral blood stem cells for autografting. *Annu. Rev. Med.* **48:** 241–251.

Straus S.E., Cohen J.I., Tosato G., and Meier J. 1992. Epstein-Barr virus infections:

Biology, pathogenesis and management. *Ann. Intern. Med.* **118:** 45–58.

Tiberghien P., Cahn J., Brion A., Deconinck E., Racadot E., Herve P., Milpied N., Lioure B., Gluckman E., Bordigoni P., Jacob W., Chiang Y., Marcus S., Reynolds C., and Longo D. 1997. Use of donor T-lymphocytes expressing herpes-simplex thymidine kinase in allogeneic bone marrow transplantation: A phase I-II study. *Hum. Gene Ther.* **8:** 615–624.

To L.B., Haylock D., Simmons P.J., and Juttner C.A. 1997. The biology and clinical uses of blood stem cells. *Blood* **89:** 2233–2258.

Truitt R.L. and Johnson B.D. 1995. Principles of graft-vs.-leukemia reactivity. *Biol. Blood Marrow Transplant.* **1:** 61–68.

Vogelsang G.B. and Hess A.D. 1994. Graft-versus-host disease: New directions for a persistent problem. *Blood* **84:** 206–2067.

Wong B. and Choi Y. 1997. Pathways leading to cell death in T cells. *Curr. Opin. Immunol.* **9:** 358–364.

18

Suicide Genes: Gene Therapy Applications Using Enzyme/Prodrug Strategies

John C. Morris, Renaud Touraine, Oliver Wildner and R. Michael Blaese

Clinical Gene Therapy Branch
National Human Genome Research Institute
National Institutes of Health
Bethesda, Maryland 20892–1851

A major focus of experimental gene-transfer research is an effort to develop efficient technologies to confer selectable traits to cells. Genetic negative-selection systems have found wide applications ranging from the generation of "knockout" animals used to study the role of individual genes on development; and morphogenesis, to use in clinical trials in humans for serious disorders including AIDS, cancer, and graft-versus-host disease. The largest preclinical and therapeutic experience with negative selective gene transfer is the application of "suicide" gene/prodrug therapy for the treatment of cancer and related conditions.

The basic components of a suicide gene/prodrug system include a gene encoding an enzyme ("suicide" gene), usually of viral or prokaryotic origin that is capable of converting an otherwise nontoxic prodrug into an active toxin that causes cell death. In the absence of the prodrug, expression of the suicide gene should be innocuous, with no toxicity or other effects on normal cellular metabolism. The enzyme should have a high catalytic constant (K_{cat}) of the prodrug to allow rapid activation and it should be able to function efficiently at very low substrate concentrations (low K_m) (Connors 1995). The ideal suicide gene is one with no cellular homolog; however, this is not an absolute requirement provided the prodrug is not activated to any significant degree by the native cellular enzyme.

Any potentially useful suicide gene/prodrug combination is constrained by the currently available gene-transfer systems (Zhang and Russell 1996; Wilson 1997). Genetically engineered retroviruses and adenoviruses, the most frequently used gene-transfer vectors, have size limitations on their ability to package the viral genome (Miller and Rosman 1989; Bett et al. 1993; Jolly 1994). If the genome is made too large by the addition of the suicide gene, it will fail to be properly incorporated into the viral capsid or it will undergo spontaneous deletions to allow packaging. This requires that the candidate gene code for a relatively low-molecular-weight protein to facilitate its efficient incorporation into the vector. The encoded enzyme should be a monomer or homomeric protein to avoid problems of incorporating more than one coding sequence into the vector or having the correct relative expression of each subunit to achieve a functional enzyme. In addition, the gene should code for a protein that is readily translated and expressed in mammalian cells, and it should have no species-specific requirements for glycosylation.

The prodrug used in combination with the suicide enzyme should ideally exhibit three characteristics. One, it should be nontoxic or minimally toxic in the absence of expression of the suicide gene. For a suicide gene system to have meaningful in vivo selectivity, it has been estimated that the activated drug should be at least 100-fold more cytotoxic than the inactive prodrug (Connors 1995). Two, it should be an approved drug in clinical use such as an antiviral or antimicrobial agent. The choice of a system that uses an already approved drug significantly reduces the required preclinical testing, development costs, and time to approval before its introduction into clinical trials. Three, as added desirable features, the activated prodrug should have a long half-life ($t_{1/2}$) and be freely diffusable to allow it to affect nearby genetically unmodified cells: A phenomenon known as the "bystander effect." In general, systems that exhibit stronger bystander effects are more desirable for the treatment of cancer (Moolten 1986; Moolten and Wells 1990; Culver et al. 1992). Other characteristics that may be beneficial are activated prodrugs with cell-cycle-specific toxicity. This may be advantageous for the treatment of diseases such as malignant brain tumors. Theoretically, cell-cycle-specific agents will target proliferating neoplastic cells rather than normal brain tissue with its negligible rate of cell division (Maron et al. 1997). The clinical advantage, however, may be slight because the fraction of tumor cells actively dividing at any given time in most neoplasms is relatively small.

The suicide gene/prodrug paradigm represents the prototype; there is, however, an almost endless list of potential activator gene/prodrug systems where a unique enzyme capable of converting a biologically inactive prodrug into an active therapeutic agent may be useful. For example, recombinant tissue plasminogen activator (rTPA) capable of activating the fibrinolytic proenzyme plaminogen could be engineered into vascular tissue grafts to enhance patency and avoid the risks of systemic anticoagulation following vascular bypass grafting. Activator enzyme genes could be over-expressed in privileged sites, such as the central nervous system, to achieve therapeutic concentrations of active drugs that are ordinarily excluded by the blood–brain barrier. The gene for cytochrome P450 could be inserted into cells in organ grafts in order to activate the drug cyclophosphamide. Locally high concentrations of activated cyclophosphamide would kill the alloreactive T cells mediating graft rejection. This would reduce the need for systemic immunosuppression and its serious complications. These examples are but a few of the potential applications of genetic enzyme/prodrug combination therapy to treat human disease. This chapter focuses primarily on the development, current status, and future prospects for suicide gene/prodrug systems for the treatment of neoplastic diseases.

Historically, the development of clinically useful suicide gene therapy was the result of efforts to address the early concern that random integration of retroviral gene-transfer vectors into the host-cell genome could result in insertional mutagenesis and the development of cancer in patients undergoing treatment with these vectors. Retroviruses have long been recognized to be oncogenic in animals and more recently so in humans (Ellerman and Bang 1908; Rous 1911; Moloney 1960; Poiesz et al. 1980). Indeed, an early preclinical trial using a retroviral vector preparation contaminated with recombinant replication-competent retrovirus (RCR) resulted in T-cell lymphomas in nonhuman primates (Donahue et al. 1992). To guard against this possibility, it was envisioned that a suicide gene could be incorporated into the vector design as a "fail safe." If malignant transformation were to occur, a nontoxic prodrug would be administered to the patient that would be activated by the suicide gene and eliminate the gene-modified cells including the malignant clone. Two systems were identified that met these criteria and subsequently underwent extensive testing: the *herpes simplex* virus-1 thymidine kinase enzyme in combination with the antiviral drug ganciclovir, and the *Escherichia coli* cytosine deaminase gene combined with the antifungal 5-fluorocytosine (Moolten 1986; Culver et al. 1992; Mullen et al. 1992). Each has subsequently entered clinical trials, albeit for different indications from those initially envisaged.

SUICIDE GENE SYSTEMS

Herpes Simplex Virus-1 Thymidine Kinase

Early development of the herpes simplex virus-1 thymidine kinase (HSV-tk) suicide gene system grew out of the recognition that the antiviral nucleosides acyclovir (ACV) and ganciclovir (GCV) selectively act against cells infected with herpesvirus (Elion 1982; Field et al. 1983). Not only do these drugs inhibit viral replication, but they are also at least 1000-fold more toxic to infected host cells than to uninfected cells (Schaeffer et al. 1978). This observation lead to the discovery that these agents were selectively toxic because the tk encoded by the HSV is 200-fold more active than mammalian tk in catalyzing the initial phosphorylation of these drugs by ATP (Fyfe et al. 1978; Biron et al. 1985). HSV-2 also encodes a tk enzyme; however, it has kinetic characteristics different from that of HSV-1 and its use as a suicide gene has not been extensively investigated (Cheng 1976). Once ACV or GCV is monophosphorylated (MP) by HSV-tk, endogenous cellular kinases rapidly catalyze the subsequent di- and triphosphorylation of GCV-MP by ATP (Miller and Miller 1980, 1982). The triphosphates of ACV and GCV are highly toxic to proliferating mammalian cells. Acyclovir triphosphate (ACV-TP) lacking a 3´-hydroxyl group acts as an absolute chain terminator once incorporated into DNA, because additional nucleotide bases cannot be added to the nascent DNA strand (Reardon and Spector 1989). Ganciclovir triphosphate (GCV-TP), on the other hand, possess a hydroxyl group that permits continued DNA elongation. The incorporation of this false guanosine nucleotide results in base-pair mismatches, DNA fragmentation, sister chromatid exchange, and lethal genomic instability (Hamzeh and Lietman 1991; Haynes et al. 1996; Thurst et al. 1996). In addition, both ACV-TP and GCV-TP act as inhibitors of the cellular DNA polymerases further blocking DNA synthesis (Ilsley et al. 1995). High-performance liquid chromatography (HPLC) of extracts of mammalian cells expressing HSV-tk exposed to GCV shows the rapid accumulation of high concentrations of intracellular phosphates of GCV (Fig. 1) (St. Clair et al. 1987; Ishii-Morita et al. 1997). In contrast, no detectable levels of phosphorylated GCV are found in the parental cell lines. Incubation of HSV-tk-transduced cells with ^{3}H-labeled GCV shows rapid and stable incorporation of the radiolabel into cellular DNA, while little is detectable in the wild-type cells (Fig. 2) (J.C. Morris, unpubl.).

Moolten et al (1986) proposed the combination of HSV-tk and GCV as a potential anticancer strategy after observing that mouse fibroblasts expressing the HSV-tk gene were sensitized to GCV concentrations 1000 times lower than that of wild-type cells and that GCV treatment induced regression of subcutaneously inoculated K3T3tk[+] sarcomas in BALB/c mice. Moolten subsequently demonstrated that GCV treatment of transgenic mice expressing HSV-tk under the control of the immunoglobulin promoter resulted in complete regression of Abelson murine leukemia virus-induced lymphomas in these animals (Moolten and Wells 1990). This strategy was attractive as a "fail safe" treatment for gene-modified

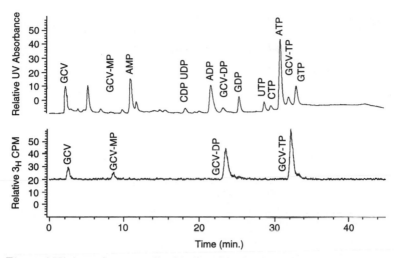

Figure 1 High-performance liquid chromatography (HPLC) analysis of extracts of MC38 murine colon adenocarcinoma cells stably transduced with the STK retroviral vector expressing HSV-tk and selected with G418 after exposure to 10 μM (5 μCi/ml) ^{3}H-labeled ganciclovir (GCV) for 24 hours. Note the high intracellular concentrations of phosphorylated GCV generated in MC38STK cells. No measurable levels of GCV phosphate could be detected in MC38 wild-type cells (data not shown). The ultraviolet HPLC spectrographic analysis (*upper*) indicates the relative position of GCV, GCV monophosphate (GCV-MP) diphosphate (GCV-DP) and triphosphate (GCV-TP) in the cellular nucleoside pools. The scintillation detector (*lower*) confirms the identity of the GCV elution peaks. (AMP) Adenosine monophosphate; (CDP) cytosine diphosphate; (UDP) uridine diphosphate; (GDP) guanosine diphosphate; (UTP) uridine triphosphate; (CTP) cytosine triphosphate; (ATP) adenosine triphosphate; (GTP) guanosine triphosphate. (J.C. Morris and R. Agbaria, unpubl.)

hematopoietic stem cells should leukemia develop in patients after receiving gene therapy due to insertional mutagenesis. The neoplastic cells could be eliminated by GCV administration to rescue the patient from their therapy-induced leukemia even though the nonmalignant gene-modified stem cells would also be lost.

With the development of stable packaging cell lines and the construction of retroviral vectors that express the HSV-tk gene with the neomycin resistance gene (neo^r) as a selectable marker, numerous HSV-tk-expressing cell lines were easily generated (Miller and Buttimore 1986; Moolten and Wells 1990; Ezzeddine et al. 1991). Early studies in BALB/c mice bearing subcutaneous K2STK sarcomas treated with acyclovir confirmed the antitumor effects (S. Freeman and R.M. Blaese,

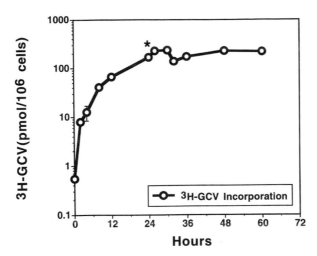

Figure 2 Incorporation of ³H-labeled GCV into the cellular DNA of murine MC38 colon adenocarcinoma cells stably transduced to express HSV-tk with the STK retroviral vector. In this rapidly proliferating tumor cell line, GCV incorporation into the cellular DNA continues to increase while the drug is present up to 24 hours of exposure to the drug and is thereafter unaffected by removing GCV (*) from the culture media. Proliferation of the tumor cells cease by 24 hours, but the cells retain the ability to exclude vital dyes for 48–72 hours. (J.C. Morris, unpubl.)

unpubl.). Further investigation found GCV to be more effective and less toxic than ACV (R.M. Blaese et al., unpubl.). This benefit is likely due to the longer half-life of GCV-TP ($t_{1/2}$ 16.5 hr) compared to that of ACV-TP ($t_{1/2}$ 1–4 hr) and the lower K_m of GCV than that of ACV for HSV-tk (Elionfield et al. 1983; Biron et al. 1985).

An in vitro as well as in vivo GCV dose response was noted in this system (Moolten 1986; Culver et al. 1992; Chen et al. 1994). Higher doses of GCV result in a greater fractional killing of HSV-tk⁺ cells in culture (Fig. 3) as well as higher tumor response rates and survivals of animals bearing HSV-tk⁺ tumors (Fig. 4). Whether the expression of HSV-tk itself exhibits a dose-response effect is controversial. Up to a tenfold difference in GCV sensitivity was found to directly correlate with the level of HSV-tk expression in retrovirus-transduced human colon carcinoma cells (Boucher et al. 1998). Rat C6 glioma cells treated with an adenoviral vector encoding HSV-tk at an increasing multiplicity of infection (moi) showed a direct correlation between moi, assayable levels of HSV-tk activity, and GCV sensitivity (Shewach et al. 1994). This study, however, did not assess the transduction efficiency at each moi. Other groups have reported little or no dose-response effect for expression of HSV-tk.

Chen and colleagues found that subclones of rat 9L cells transduced with an HSV-tk-expressing retroviral vector all had similar GCV sensitivities despite significantly different levels of HSV-tk expression (C.Y. Chen et al. 1995). A comparison of H2122 human lung cancer cells transduced with one of two different retroviral vectors, in which HSV-tk was driven by different promoters, found identical sensitivities to GCV despite a fourfold higher level of expression from the transgene driven by a human cytomegalovirus (CMV) promoter over that of the retroviral long terminal repeat (LTR) (Hoganson et al. 1996). Another study comparing adenoviral vectors in which HSV-tk expression was driven by either the Rous sarcoma virus (RSV) promoter or the human CMV promoter found no difference in antitumor effect, despite 20-fold higher levels of HSV-tk expression from the adenovirus utilizing the CMV promoter (Elshami et al. 1997). The most likely explanation for these conflicting results is that expression of HSV-tk beyond a level that sensitizes cells to GCV by 100–1000-fold does not appear to further enhance cell killing. Measurable levels of GCV phosphates are detectable within 1 hour of exposure of retroviral HSV-tk-transduced cells to GCV and the cells are committed to die within 6–10 hours of exposure (Hamel et al. 1996; Ishii-Morita et al. 1997). Additionally, the sensitivity of cells to HSV-tk/GCV

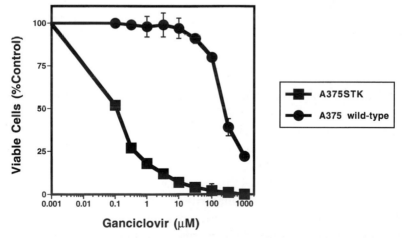

Figure 3 Sensitization of A375 human malignant melanoma cells to increasing doses of ganciclovir following stable gene modification with an HSV-tk expressing retroviral vector (STK). Viable cells remaining in culture 6 days after exposure to GCV are expressed as a percentage of identical cultures grown without exposure to GCV. Expression of the HSV-tk gene renders the tumor cells more than 1000 times more sensitive to GCV than wild-type A375 tumor cells. The IC_{50} for GCV is reduced to 0.1 μM compared with 300 μM for wild-type cells.

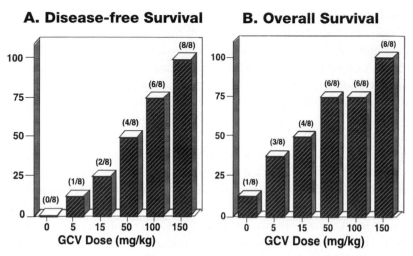

Figure 4 Dose-response effect of ganciclovir (GCV mg/kg) administered twice daily for 7 days on (*A*) disease-free survival and (*B*) overall survival of athymic nude mice bearing at 3 months subcutaneous, HSV-tk expressing human A375 melanomas. These tumors were established using A375 malignant melanoma cells which had been gene modified in vitro using the STK retroviral vector followed by G418 selection. These data demonstrate a dose-response relationship in this system and also illustrate that the antitumor response and cure rate can be substantially improved in vivo with identical degrees of HSV-tk gene-modification by increasing the dose of GCV administered. The dose of GCV employed in most clinical trials to date has been 5 mg/kg administered twice daily for 14 to 21 days.

may not depend solely on the components of the suicide gene system, but may also depend on the activity of other cellular kinases that perform the secondary and tertiary phosphorylations, the rate of cell proliferation, and the cellular DNA polymerases that incorporate GCV-TP into DNA (Ilsley et al. 1995; Boucher et al. 1998).

An early surprising finding of this system was that not every cell need express HSV-tk in order to kill the majority of cells in a culture or achieve complete tumor regression in an experimental animal. This phenomenon is termed the "bystander effect" (BSE) and is a major enhancement of HSV-tk/GCV and other enzyme/prodrug systems. Under optimal circumstances, even the most efficient gene-transfer vectors are only successful in transducing a small minority of cells in a tumor (Moolten 1986; Moolten and Wells 1990; Culver et al. 1992; Moolten et al. 1992; Ram et al. 1993a). Most current evidence suggests that the HSV-tk/GCV BSE is mediated by transfer of phosphorylated GCV between cells through gap junctions. The BSE of HSV-tk/GCV and other suicide gene/prodrug systems are discussed in further detail below and outlined in Table 1. The HSV-tk/GCV system has undergone extensive preclinical testing and was the first suicide gene/prodrug system to receive approval for a phase I clinical trial (Culver et al. 1994; Ram et al. 1997). It is now the subject of multiple ongoing human gene therapy trials.

Varicella Zoster Virus Thymidine Kinase

The causative agent of chicken pox, varicella zoster virus (VZV), another member of the herpesvirus family, expresses a unique thymidine kinase (VZV-tk) that has less than 30% amino acid homology with that of HSV (Sawyer et al. 1988). It also has a substrate activity distinct from HSV-tk (Averett et al. 1991). ACV and GCV are much less effective substrates for VZV-tk than HSV-tk (Biron and Elion 1980; Bevilacqua et al. 1995; Grignet-Debrus and Calberg-Bacq 1997). The prodrugs 6-methoxypurine arabinoside (Ara-M) and (E)-5-(2-bromovinyl)-2′-deoxyuridine (BVdU) are selectively phosphorylated by VZV-tk and subsequently di- and triphosphorylated to adenine arabinonucleoside triphosphate (Ara-ATP) and BVdU-triphosphate, respectively, by cellular enzymes (Huber et al. 1991; Grignet-Debrus and Calberg-Bacq 1997). Wild-type cells are insensitive to Ara-M concentrations up to 1500 μM and BVdU concentrations of 75–250 μM. In cells expressing VZV-tk, concentrations of 1–100 μM Ara-M and 0.06–0.6 μM BVdU are cytotoxic (Degreve et al. 1997). The VZV-tk system has not been extensively studied and its potential as a clinically applicable suicide gene/prodrug therapy remains uncertain.

Cytosine Deaminase

Cytosine deaminase (CD) is an enzyme expressed by *E. coli* and some fungi, but not found in mammalian cells, that mediates the deamination of cytosine to uracil (Andersen et al. 1989; Kilstrup et al. 1989; Danielsen et al. 1992). The antifungal drug 5-fluorocytosine (5FC) is deaminated by CD to generate the anticancer drug 5-fluorouracil (5FU) (Polak and Scholer 1975). Cellular enzymes subsequently metabolize 5FU to 5-fluorouridine-5′-triphosphate and 5-fluoro-2′-deoxyuridine-5′-monophosphate (FdUMP), which inhibit the enzyme thymidylate synthase blocking DNA, RNA, and protein synthesis, thus causing cell death (Diasio and Harris 1989). The CD enzyme of *E. coli* was selected for study because it was found to have better stability than the yeast enzyme (Katsuragi et al. 1987). As a prodrug, 5FC has several advantages. One is that humans

Table 1 Suicide gene/prodrug systems currently under preclinical and clinical evaluation

Suicide gene	Source	Prodrug	Active drug	Bystander effect	Comments	References
Clinical Trials						
Cytosine deaminase (E. coli CD)	E. coli, or Saccharomyces	5-fluorocytosine (5FC)	5-fluorouracil (5FU)	Yes, simple diffusion	Licensed prodrug	1, 2
Thymidine kinase (HSV-tk)	Herpes simplex virus-1	Ganciclovir	Ganciclovir monophosphate	Yes, via gap junctions	Licensed prodrug, phase I clinical trials completed in phase II trial	3–7
Preclinical Investigation						
Carboxylesterase (CE)	Rabbit	CPT-11 (irinotecan)	SN38	Unknown	Licensed prodrug	8, 9
Carboxypeptidase G2 (CPG2)	Pseudomonas	CMDA	CMDA-mustard	Yes, simple diffusion	Cells impermeable to unactivated prodrug, no approved prodrug	10
CYP 2B1 (cytochrome P450)	Rat	Cyclophosphamide, ifosfamide	4HC	Yes, simple diffusion	Licensed prodrugs	11
CYP 4B1 (cytochrome P450)	Rabbit	2-aminoanthracene, 4-ipomeanol	unsaturated or aldehyde furan epoxide?	Yes, enhanced by cell-cell contact	No approved prodrug	12
Deoxycytidine kinase (dCK)	Human	cytosine arabinoside (Ara-C), fludarabine	Ara-CMP, fludarabine-MP	Very weak, via gap junctions?	Licensed prodrug	13, 14

Enzyme (gene)	Source	Prodrug	Active drug	Bystander effect	Approval status	References
Guanosine-xanthine phosphoribosyl transferase (*gpt*)	*E. coli*	6-thioxanthine (6-TX), 6-thioguanine (6-TG)	6-TX, or 6-TG mono-phosphate	"Modified" BSE? probably via gap junctions	BSE detected only by clonogenic assay	15, 16
Nitroreductase (NTR)	*E. coli*	CB1954	Hydroxylamine alkylating agents	Yes, simple diffusion	No approved prodrug	17–19
Purine nucleoside phosphorylase (PNP, *DeoD*)	*E. coli*	6-methylpurine deoxyribonucleoside (6-MeP-dR)	6-methylpurine	Yes, simple diffusion	No approved prodrug	20, 21
Thymidine kinase (VZV-tk)	varicella zoster virus	Ara-M , BVDU	Ara-ATP, or BVDU-triphosphate	Yes, probably via gap junctions	No approved prodrug	22, 23
Thymidine phosphorylase (TP)	*E. coli*	5'-DFUR, tegafur	5'-fluorouracil (5FU)	Yes, simple diffusion	No approved prodrug	24, 25

References: (1) Mullen et al. 1992; (2) Austin and Huber 1993; (3) Moolten 1986; (4) Culver et al. 1992; (5) Chen et al. 1994; (6) Treat et al. 1996; (7) Ram et al. 1997; (8) Satoh et al. 1994; (9) Danks et al. 1998; (10) Marais et al. 1997; (11) S.H. Chen et al. 1995; (12) Rainov et al. 1998; (13) Hoganson et al. 1996; (14) Manome et al. 1996; (15) Tamiya et al. 1996; (16) Ono et al. 1997; (17) Bailey et al. 1996; (18) Bridgewater et al. 1997; (19) Friedlos et al. 1997; (20) Sorcher et al. 1994; (21) Hughes et al. 1995; (22) Huber et al. 1991; (23) Grignet-Debrus et al. 1997; (24) Patterson et al. 1995; (25) Kato et al. 1997.

Abbreviations: SN38, 7-ethyl-10-hydroxycamptothecin; CMDA, 4-([2-chloroethyl][2-mesyloethyl]amino)benzyol-L-glutamic acid; 4HC-4, hydrox-yycyclophosphamide; CB1954, 5-aziridin-2,4-dinitrobenzamidine; 6-MeP-dR, 6-methylpurine deoxyribonucleoside; Ara-M, 6-methoxypurine arabi-noside; BVDU, (E)-5-(2-bromovinyl)-2' deoxyuridine; 5'-DFUR. 5'-deoxy-5-fluorouridine; tegafur, 1-(tetrahydrofuryl)-5-fluorouracil.

do not deaminate 5FC; therefore, it has relatively little toxicity at usual doses (Polak et al. 1976). Indeed, it is believed that most of the toxicity seen with the clinical use of oral 5FC is due to deamination by intestinal bacteria (Diasio et al. 1978). The second advantage is that 5FC is orally well absorbed (>80%) and has good CNS penetration. Additionally, 5FC is an already approved antimycotic agent with a known clinical profile.

In early studies, Nishiyama et al. (1985) observed a reduction in tumor size when 5FC was administered to rats bearing subcutaneous gliomas implanted with membranous tubes containing purified *E. coli* CD. Prolonged measurable levels of 5FU were detectable within the tumor using this technique. Mullen et al (1992) subsequently found that CD-gene-transduced murine fibroblasts, sarcoma and colon carcinoma cell lines were able to convert cytosine to uracil. These cells were killed when exposed to low concentrations of 5FC, whereas the parental cell lines remained resistant. These findings were confirmed by Austin using human WiDr colon carcinoma (Austin and Huber 1993). Human cell lines expressing CD are sensitized to 5FC by 500–2000-fold over wild-type cells (Mullen et al. 1992; Austin and Huber 1993; Ge et al. 1997). Biochemical analysis of mammalian cells expressing CD treated with 5FC shows the generation of measurable intracellular levels of 5FU, inhibition of thymidylate synthase, and extensive formation of RNA adducts (Huber et al. 1994). Cell killing with CD/5FC is slower than that seen with HSV-tk/GCV (Rowley et al. 1996). This is believed to be due to the slower uptake of 5FC by cells and conversion to 5FU than the uptake and phosphorylation of GCV by HSV-tk-expressing cells (West et al. 1982; Plagemenn et al. 1987). Despite slower tumor cell killing, an in vivo study comparing CD/5FC to HSV-tk/GCV using human WiDr colon carcinoma found the CD system to be more active because of a stronger BSE (Trinh et al. 1995). Interestingly, WiDr wild-type or CD$^+$ colon tumors in athymic nude mice do not respond to systemic injections of 5FU, but will regress with CD/5FC or 5FC treatment, respectively (Huber et al. 1993).

Administration of the prodrug 5FC results in a cytotoxic dose-response effect (Mullen et al. 1992; Austin and Huber 1993; Hirschowitz et al. 1995). Higher prodrug concentrations are more effective in killing CD$^+$ cells than lower concentrations. The effect of the level of expression of the CD gene is not as clear and may ultimately depend on the intrinsic sensitivity of the particular cell line to 5FU. In more sensitive cell lines such as colon or lung carcinoma, cytotoxicity directly correlates with the level of CD expression (Hoganson et al. 1996). In sarcoma cell lines that are less sensitive to 5FU, a CD dose response is not as obvious.

Similar to HSV-tk/GCV, the CD/5FC system exhibits a BSE. In contrast to HSV-tk where no phosphorylated GCV can be detected in the media, high concentrations of 5FU can be isolated from the media of CD^+ cell lines exposed to 5FC (Dong et al. 1996). The CD BSE effect is mediated by direct diffusion of the small nonpolar 5FU molecule from cell to cell across the membranes. In the initial studies of the CD/5FC system, Mullen et al. (1992) found only a minimal BSE. This is likely a result of the 5FU-resistant murine sarcoma model used. Subsequent work using colon carcinoma cell lines that are more sensitive to 5FU indicates that the CD/5FC system exhibits a strong in vitro and in vivo BSE. As few as 2% CD^+ cells in a tumor mass may result in complete regression of WiDr tumors in athymic nude mice after treatment with 5FC (Huber et al. 1994).

Other Suicide Gene/Prodrug Systems

A number of other novel suicide gene/prodrug systems are currently under investigation (see Table 1). Most of these systems are not completely characterized or are in early preclinical development. Most lack approved prodrugs slowing their introduction into clinical trials.

Systems utilizing P450 hepatic microsomal enzymes to activate a number of different prodrugs have been described. The P450 enzyme CYP 2B1 isolated from rat liver metabolizes the oxazaphosphorine chemotherapy drugs cyclophosphamide (CPA) and ifosfamide (IFF) to 4-hydroxycyclophosphamide (4HC). 4HC is rapidly converted into the short-lived alkylating agent phosphoramide mustard, which damages cellular DNA. Cells expressing CYP 2B1 are sensitized 15–20-fold to CPA in vivo (Chen and Waxman 1995; L. Chen et al. 1996). Fisher rats subcutaneously inoculated with 9L gliosarcoma cells expressing CYP 2B1 had complete suppression of growth after treatment with CPA, whereas control animals receiving wild-type 9L cells rapidly developed tumors despite CPA treatment. This system exhibits a moderately strong BSE that appears to be mediated by direct 4HC diffusion.

The CYP 2B1/CPA combination offers a potentially attractive strategy for treating cancers that have invaded privileged sites such as the brain or cerebrospinal fluid (carcinomatous meninigitis). CPA freely crosses the blood–brain barrier; however, activation of CPA to 4HC occurs primarily in the liver (Chang et al. 1993). Once activated, 4HC does not readily cross into the CNS (Moore 1991). Vector producer cells or other CYP 2B1 vectors could be used to transduce tumor implants in privileged sites such as the CNS to generate high concentrations of 4HC at the site of the lesions.

In the bone-marrow transplant setting, 4HC is clinically utilized for purging bone marrow of tumor cells. Bone marrow must be rapidly processed and frozen to maintain viability. 4HC is used for purging because only a brief period of exposure of the tumor cells is required for it to be effective. However, 4HC is toxic to normal marrow progenitor cells as well as to tumor cells (Rowley et al. 1987). Adenoviral vectors expressing CYP 2B1 could be used to selectively purge marrows for autologous transplantation for patients with breast or other cancer. Since hematopoietic cells lack the surface receptors for adenoviruses, only the carcinoma cells would be effectively transduced when the bone marrow or stem cells were treated with the vector. The marrow could then be purged of tumor by exposing it to CPA. Only the carcinoma cells that were transduced with CYP 2B1 could generate toxic 4HC and be destroyed while leaving the marrow stem cells essentially unharmed.

Recently, Rianov described the use of the rabbit cytochrome P450 CYP 4B1 in combination with the prodrugs 2-aminoanthracene (2-AA) or 4-ipomeanol (4-IM) (Rainov et al. 1998). CYP 4B1 has selective action converting 2-AA and 4-IM into DNA-alkylating agents. The activated drugs are highly toxic at low concentrations and a significant in vitro BSE was noted with as few as 1% of the cells expressing CYP 4B1.

Human deoxycytidine kinase (dCK) has been suggested as a potential suicide gene based on the observation that clinical responses to the antileukemic nucleoside drugs cytosine arabinoside (Ara-C), 2-chlorodeoxyadenosine (2CdA), and fludarabine correlated with the level of dCK expression in leukemic cells (Colly et al. 1987; Hagenbeek et al. 1987; Kawasaki et al. 1993). The rate-limiting step in the intracellular activation of these prodrugs are their phosphorylation by dCK. Although Ara-C is a potent antitumor agent for hematological malignancies, it has only minimal activity against most solid tumors. One study found that retroviral or adenoviral transfer of dCK into glioma cells sensitized them to Ara-C (Manome et al. 1996). Other investigators, however, could not demonstrate any increased Ara-C sensitivity in two human lung cancer cell lines transduced with a retroviral vector encoding dCK (Hoganson et al. 1996). The possible explanation for these disparate results may be that sensitivity to these agents is determined not only by dCK activity, but also transport mechanisms to bring the drugs into the cell, as well as expression of other necessary enzymes distal in the biochemical pathway to the activity of dCK (Preisler et al. 1985; Wiley et al. 1985; Tanaka and Yoshida 1987).

The *E. coli* guanosine-xanthine phosphoribosyl transferase (*gpt*) gene is unique in that this gene simultaneously encodes sensitivity to thioxanthines and resistance to mycophenolic acid, xanthine, and hypoxanthine, allowing both positive and negative selection of transduced cells. Expression of *gpt* sensitizes cells to 6-thioxanthine (6-TX) and 6-thioguanine (6-TG) which are activated by phosphorylation. Retroviral transduction of rat C6 glioma cells with *gpt* increased their in vitro sensitivity to 6-TX by 50–100-fold (Tamiya et al. 1996). Grafting of *gpt*-retroviral vector producer cells into established intracerebral C6 tumors in athymic nude mice and subsequent treatment with 6-TX resulted in an 80% reduction in mean tumor volume and improved survival over control animals (Ono et al. 1997). Expression of *E. coli gpt* also significantly increases the rate of activation of 6-TG (Tamiya et al. 1996). While not the only enzyme activating this drug, a potential advantage of this system is that 6-TG is in clinical use for treatment of leukemia. Similar to the HSV-tk/GCV system, cell-to-cell contact is required to elicit a BSE because the activated thioxanthines are phosphorylated and are not freely diffusable. The BSE of the gpt system is described as "modified," only detectable by culturing the cells in a clonogenic assay after treatment with 6-TX.

   E. coli nitroreductase (NTR) exhibits a high K_{cat} for the substrate 5-(aziridin-1-yl)-2,4-dinitrobenzamide (CB1954) compared to the homologous human enzyme DT-diaphorase (NAD[P]H dehydrogenase [quinone]) (K_{cat} 360 min^{-1} vs. 0.64 min^{-1}) (Friedlos et al. 1997). CB1954 is rapidly converted to several hydroxylamine alkylating agents by reduction of its nitro groups (Friedlos et al. 1992). Cells expressing NTR are sensitized 500-fold to CB1954 (Bridgewater et al. 1997; Green et al. 1997). The NTR/CB1954 system has a strong BSE that is not dependent on expression of cellular gap junctions, but appears to be due to the diffusion of two major toxic metabolites of CB1954 into the culture media (Bridgewater et al. 1997; Friedlos et al. 1998). NTR also activates the approved antimicrobial agents nitrofurantoin and metronidazol to toxic compounds that can kill the cell. These agents, however, lack significant BSEs limiting their usefulness (Bridgewater et al. 1997).

   The *E. coli* purine nucleoside phosphorylase (PNP) gene (*Deo D*) encodes an enzyme that activates relatively nontoxic 6-methylpurine-deoxyriboside (6-MeP-dR) to highly toxic 6-methylpurine (6-MeP) (Sorcher et al. 1994). Hughes et el. (1995) transduced B16 murine melanoma and 16/C breast cancer cell lines with a retroviral vector expressing the PNP gene driven by the tyrosinase promoter. They were able to show selective killing of melanoma cells exposed to 6-MeP-dR. This system exhibited a strong BSE and was able to kill most of the non-transduced cells when as little as 1% of the population expressed PNP. The activated drug is membrane-permeable and does not require gap junctions for a BSE.

Thymidine phosphorylase (TP) normally catalyzes the phospholytic cleavage of thymidine or deoxyuridine to deoxyribose-1-phosphate. TP can also activate the compounds 5′-deoxy-5-fluorouridine (5′-DFUR) and 1-(tetrahydrofuryl)-5-fluorouracil (tegafur) to the anticancer agent 5-fluorouracil (5FU). This strategy has been shown to sensitize human MCF-7 breast cancer cells 1000-fold to 5′-DFUR (Patterson et al. 1995). The TP system does not require cell-to-cell contact for a BSE because 5FU is freely diffusable (Kato et al. 1997). 5′-DFUR and tegafur are not currently approved and it is not clear if this system offers any advantages over *E. coli* CD, which also produces 5FU as the active prodrug.

Marais et al. (1997) reported a novel cell membrane surface tethered (st) enzyme/prodrug system utilizing the *Pseudomonas* RS16 carboxypeptidase G2 (CPG2) enzyme that cleaves 4-([2-chloroethyl][2-mesyloxyethyl]amino)benzyol-L-glutamic acid (CMDA) to its benzoic acid derivative, which is a potent alkylating agent. CMDA does not easily enter cells and is not well activated by intracellular expression of the CPG2. However, if the enzyme is engineered to be expressed on the external surface of the cell membrane, CMDA is readily activated. The stCPG2/CMDA system exhibits a BSE suggesting that the cells are permeable to the activated drug.

Rabbit hepatic carboxylesterase (CE) has been used to activate the approved chemotherapy drug CPT-11 (irinotecan, 7-ethyl-10-[4-(1-piperidino)-1-piperidino]carbonyloxycamptothecin) as a potential suicide gene/prodrug strategy (Danks et al. 1998). Human rhabdomyosarcoma cells transduced with the rabbit CE gene were greater than eightfold more sensitive to CPT-11 than the parental cell line. The conversion of CPT-11 to its active form SN38 (7-ethyl-10-hydroxycamptothecin) is isoenzyme-specific (Satoh et al. 1994). In humans, hepatic carboxylesterase I is the most active isoenzyme in the metabolism of CPT-11 to SN38 (Satoh and Hosokawa 1995). The monkey, rabbit, pig, and murine hepatic carboxylesterases have significantly higher specific activities for converting CPT-11 to SN-38 than the human enzyme and may represent better choices for a suicide gene.

Two recent studies have suggested that the combination of different suicide gene/prodrug systems may be more effective than treatment with individual suicide genes alone. Enhancement of tumor cell killing was demonstrated in rat 9L gliosarcoma, murine TS/A breast cancer, and human RCC26 renal cell carcinoma models by combining the HSV-tk/GCV and CD/5FC systems (Aghi et al. 1998; Uckert et al. 1998). The production of 5FU from 5FC by CD appears to enhance HSV-tk production of GCV-TP by inhibition of thymidylate synthase, thereby reducing intracellular levels of thymidine that competes with GCV for the active site on the HSV-tk enzyme.

THE BYSTANDER EFFECT AND SUICIDE GENE/PRODRUG SYSTEMS

A surprising observation of the early models of HSV-tk suicide gene therapy was that not every cell in a tumor need express HSV-tk in order to achieve a complete regression with GCV treatment (Moolten 1986; Moolten and Wells 1990; Culver et al. 1992; Ram et al. 1993a). This phenomenon is known as the "bystander effect" (BSE) and can be defined as the ability of genetically modified cells in the presence of the prodrug to result in cytotoxicity and death of cells that lack the suicide gene. In other words, the fraction of cells killed in the total population is in excess of the fraction that actually expresses the suicide gene (Freeman et al. 1993). The BSE is a powerful enhancement of many suicide gene/prodrug systems and it varies by the cell line and by the enzyme/prodrug system being studied (Table 1). It may be nonexistent, or it may be very strong, with as few as 1–5% transduced cells in a tumor resulting in the death of all of the cells, complete regression, and cure of tumor-bearing animals (Moolten and Wells 1990; Culver et al. 1992; Huber et al. 1993; Chen and Waxman 1995). On the other hand, the BSE may be present, but not obvious like that of the *E. coli gpt*/6-TX system, where it is described as "modified" and detected only after replating of the treated cells in a clonogenic assay (Ono et al. 1997). A BSE can be seen between cells derived from different tissues, as well as between cell lines from different mammalian species (Ishii-Morita et al. 1997). Many activated prodrugs such as 5FU are freely diffusable and require no special mechanisms to pass between cells (Mullen et al. 1992; Huber et al. 1994). Others, like GCV-TP, are highly charged molecules that do not readily cross cell membranes and require specialized channels to move between cells to achieve a BSE. The CYP 4B1/2-AA or 4-IM system may use more than one mechanism to transfer activated prodrug (Rainov et al. 1998). Beyond the cellular level, there is strong evidence that the host immune system has a major role in generating the BSE (Freeman et al. 1995; Gagandeep et al. 1996; Ramesh et al. 1996a; Kianmanesh et al. 1997).

Many hypotheses have been advanced as an explanation for the BSE reflecting variables prominent in the different experimental systems studied. In systems such as CD/5FC that generate a soluble diffusable toxin (5FU), the BSE is readily understood. After exposure to 5FC, high levels of

cells through gap junctions (Myhr and DiPaolo 1975). Consistent with this hypothesis, cell-to-cell contact is required for BSE cell killing in the HSV-tk system (Bi et al. 1993). Cell lines and subclones with greater expression of connexins have stronger BSEs (Fick et al. 1995; Elshami et al. 1996; Mesnil et al. 1996, 1997; Shinoura et al. 1996; Vrionis et al. 1997). Cell lines that originally lack a BSE will exhibit both in vitro and in vivo BSEs when transduced with the gene for connexin 43. It is not clear that all members of the family of connexins will mediate the HSV-tk BSE. Current evidence suggests that a BSE can be mediated by gap junctions formed from connexins 43, 37, 32, and 26.

In addition to the "metabolic" BSE, other mechanisms may have a role in contributing to the killing of untransduced cells. Sublethally irradiated and immunodeficient BALB/c mice with tumors derived from mixtures of HSV-tk[+] and wild-type cells have a diminished BSE, suggesting that the immune system may play an important part (Gagandeep et al. 1996; Ramesh et al. 1996a). Animals bearing both HSV-tk-expressing tumors and wild-type tumors implanted at remote sites have experienced regression of not only the transduced lesion, but also the remote nontransduced tumors (Wilson et al. 1996; Bi et al. 1997). This phenomenon has been termed the "bystander effect at a distance" and represents an antitumor immune response (Kianmanesh et al. 1997). It has also been proposed that the BSE is a result of the local release of cytokines by dying HSV-tk[+] cells which recruit cytotoxic T lymphocytes to react against all tumor cells (Freeman et al. 1995; Gagandeep et al. 1996). Several workers have shown that the levels of tumor necrosis factor-α (TNF-α), interleukin-1α (IL-1α), IL-6, interferon-γ (IFN-γ), and granulocyte-macrophage colony-stimulating factor (GM-CSF) increase in HSV-tk[+] cells after exposure to GCV. Additionally, GCV treatment of HSV-tk[+] cells has been shown to increase expression of the cell surface costimulatory molecules ICAM and B7 (Ramesh et al. 1996).

It has also been suggested that phagocytosis of apoptotic vesicles containing GCV-P* from dying HSV-tk[+] cells by wild-type tumor cells is a mechanism for generating the the BSE (Freeman et al. 1993). Although phagocytosis of apoptotic vesicles does occur, it appears to be a late phenomenon and does not account for the early appearance of GCV-P* in wild-type cells (Hamel et al. 1996; Ishii-Morita et al. 1997). Ram et al. (1994) have shown that the vascular endothelium of tumors may also be gene modified, suggesting that direct cytotoxic activity by HSV-tk/GCV on the tumor vasculature may contribute to bystander killing. In situ HSV-tk transduction of intracerebral 9L tumors in rats with vector producer cells resulted in a significant decrease in vascularity after treatment with GCV compared to control tumors or those transduced with the HSV-tk retroviral vector but not receiving GCV.

Efforts are under way to enhance BSE cell killing through pharmacological manipulations to ungate gap junctions or enhance their expression. Retinoids have been reported to increase expression of connexin 43 (Mehta and Lowenstein 1991). Treatment of AB12 malignant mesothelioma cells with retinoic acid was found to increase the expression of connexin 43 and enhance the transfer of small dye molecules between cells (Park et al. 1997). Mice inoculated with AB12 mesothelioma cells consisting of varying proportions of stably transduced HSV-tk$^+$ and wild-type cells had significantly smaller tumors after receiving a concurrent course of retinoic acid and GCV. This effect was noted even at ratios of HSV-tk$^+$ to wild-type cells as low as 1:40. Retinoic acid treatment by itself had no significant effect on tumor size. Although promising, the enhancement of the BSE by retinoids is not universal. Cell lines that do not express connexin 43 fail to respond to retinoids, and the pattern of responses varies in different cell lines of the same histological type, and even in different subclones of the same cell line (Lotan and Nicolson 1977; Elshami et al. 1996; Park et al. 1997). Other agents under investigation may prove effective in enhancing HSV-tk bystander-effect killing. In a mouse intraperitoneal model using MC38 colon carcinoma in which only 10% of the cells were HSV-tk$^+$, survival was significantly improved if the animals were treated with the flavinoid apigenin or the cholesterol synthesis inhibitor lovastatin in addition to GCV (Fig. 6) (R. Touraine, unpubl.). The explanation for this effect is unclear; however, some studies have suggested that lovastatin inhibits activation of cellular tyrosine kinases that function to decrease gap junctional intercellular communication by phosphorylating connexins (Ruch et al. 1993). Apigenin may act to increase the expression of gap junctions. Conversely, it has been found that the cAMP activator forskolin and the calcium channel blocker verapamil decrease the BSE (Samejima and Meruelo 1995; Marini et al. 1996).

TARGETING OF SUICIDE GENE THERAPY

Targeting of suicide gene/prodrug therapy has been approached in three general ways: (1) targeting of the gene delivery vector, (2) targeting of gene expression, and (3) targeting of gene effect (Miller and Vile 1995; Dachs et al. 1997). Transfer of genes by viral vectors is dependent on the trophism or host range of the vector. This in turn is a function of the presence of surface receptors for the vector in the target cell population. A degree of targeting may be achieved by the selection of a vector such as a retrovirus that

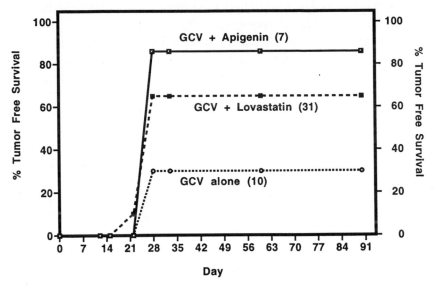

Figure 6 Augmentation of the bystander effect in the HSV-tk/GCV system by the administration of drugs that up-regulate the activity of gap junctions. Apigenin (0.1 mg/kg) or Lovastatin (25 µg/kg) were administered on days 2–3 of a 7-day course of GCV 150 mg/kg twice a day beginning 7 days after inoculation of MC38 tumor cells generated from mixtures consisting of 90% wild-type cells and 10% cells expressing HSV-tk following STK transduction and G418 selection. The BSE found with this tumor normally provides for a 30% response rate. This treatment to augment gap junction activity resulted in substantially improved long-term tumor-free survival demonstrating that the BSE can be manipulated to improve treatment outcome using available pharmacological agents. Numbers of animals in each treatment group are indicated in parentheses.

will transduce only actively proliferating cells, or by use of a HSV-based vector which exhibits a natural trophism for certain tissues such as the brain (Lasner et al. 1996). Another approach has been to modify vectors to alter their host range or tissue specificity. Various manipulations of the commonly used viral vectors have been made to extend or limit their host cell range (Dachs et al. 1997). The Moloney murine leukemia virus (Mo-MLV), from which most retroviral gene-transfer vectors have been derived, normally expresses an ecotropic envelope that restricts host range to murine cells. To broaden the host range, clinically used vectors have an amphotrophic envelope gene substituted into packaging cells from another MLV strain (4070A) (Miller and Rosman 1989; Miller 1990). Mo-MLV-based retroviral vectors have been pseudotyped by substitution the envelope (*env*) gene of Gibbon ape leukemia virus (GaLV) or the g-protein of vesicular stomatitis virus (VSV-g) to alter the range of cells infected (Emi et al. 1991; Miller et al. 1991; Bunnell et al. 1995). Others have taken advantage of the bimodular structure of the retroviral Env protein. The Env protein is composed of two regions: the TM domain that anchors the protein to the surface of the retroviral envelope and the amino-terminal SU domain that mediates binding of the virus to its cell surface receptor. Retroviruses can be targeted to cells expressing a particular surface protein or receptor if this SU domain is replaced with a specific single-chain antibody (scFv) to that protein (Chu et al. 1994).

Attempts to alter the targeting of adenoviral vectors have been made through modification of the fiber protein receptor-binding domain (Michael et al. 1995); adenoviruses are trophic for a broad range of nondividing human cell types. Recently, a putative common Coxsackie B virus/group C adenovirus cellular receptor (CAR) has been cloned (Bergelson et al. 1997). The binding and uptake of adenoviruses are dependent on the presence of the CAR, the $\alpha v \beta 3$ and $\alpha v \beta 5$ vitronectin-binding surface integrins (Goldman and Wilson 1995). Other cell-type-specific integrins can be targeted by altering the repetitive RGD motifs in the penton base of the adenoviral fiber (Wickham et al. 1995). Several groups have exploited various technologies including replacement of one adenovirus group-type fiber with another to alter the receptor specificity, engineering of heparin binding sites into the adenovirus fiber to broaden its trophism, and the addition of short peptide ligands to the knob domain of the fiber while simultaneously blocking the natural ligands with monoclonal antiknob antibody to restrict the vector's range (Michael et al. 1995; Gall et al. 1996; Krasnykh et al. 1996, 1998; Wickham et al. 1996). The redirection of adenoviral vectors to neoplastic cells using a neutralizing antiknob antibody conjugated to folate has been reported (Douglas et al. 1996). It has long been recognized that many malignant cells have a requirement for large amounts of folate and overexpress folic acid receptors on their surface. Folic acid was chemically conjugated to the Fab domain of an adenovirus antiknob monoclonal antibody so that it was presented on the end of the adenovirus fiber as a potential ligand. Tumor cell lines that express high levels of folate receptors were more effectively transduced by these modified adenoviruses. This technique lends itself to substituting a large number of potential ligands to limit or expand the host range of adenoviral gene transfer vectors (Douglas and Curiel 1997).

The most common strategy used to target gene expression is the use of cell-selective or tissue-specific promoters. Parr et al. (1997) reported a novel strategy targeting adenoviral vector gene expression to rapidly proliferating cells. They designed a vector in which the *lacZ* gene was placed under the control of the E2F-1 promoter on the hypothesis that proliferating tumor cells have excess "free" E2F due to disruption of the cellular pRB/E2F complex. When compared to a virus in which the *lac*Z gene was driven by a CMV promoter, *lac*Z expression from the E2F-1 promoter-driven virus was dependent on the number of cells in S phase. Others have designed vectors with suicide genes that target specific organs or tissues by placing the genes under the control of tissue- or tumor-specific promoters. Among the promoters tested are carcinoembryonic antigen (adenocarcinoma), α-fetoprotein promoter and the albumin promoter-enhancer sequence (hepatocellular carcinoma), glial fibrillary acidic protein (GFAP) promoter (brain tumors), the tyrosinase promoter (melanoma), prostatic-specific antigen promoter (prostate cancer), the ErbB2 promoter (breast cancer), the DF3(MUC1) mucin gene promoter (adenocarcinoma) and Epstein-Barr virus responsive promoters (lymphoma) (Manome et al. 1994; Hughes et al. 1995; Davis et al. 1996; Franken et al. 1996; Kanai et al. 1996; Tanaka et al. 1996; Lan et al. 1997; Su et al. 1997; Takakuwa et al. 1997). Each promoter examined has shown some level of selective gene expression in the appropriate tissue or cell type.

The third gene targeting approach is illustrated by the activated drug GCV-TP in the HSV-tk system. GCV-TP is a cell-cycle-specific agent capable of affecting only cells actively synthesizing DNA. In the setting of brain tumors treated with HSV-tk/GCV, normal brain tissue remains unaffected because of its negligible mitotic index, while the tumor cells that have a high proliferation rate are killed by this agent (Ezzeddine et al. 1991). The search for prodrugs that have selective toxicity to some tissues or cell types, but not others, offers a potentially attractive strategy for targeting the effects of suicide-gene therapy.

THE IMMUNE RESPONSE IN SUICIDE GENE/PRODRUG SYSTEMS

Early investigations suggested that induction of antitumor immunity may have a significant role in the in vivo response of tumors to suicide gene therapy (Caruso et al. 1993). Histological evaluation of the tumor site in rats intracerebrally inoculated with retroviral HSV-tk-transduced 9L gliosarcoma cells that underwent regression after GCV treatment revealed inflammatory infiltrates consisting of microglia, macrophages, and T cells (Barba et al. 1994). Long-term disease-free survivors (>90 days) repeatedly rechallenged by subcutaneous or intracerebral injection with 9L cells failed to form tumors, whereas naive control animals quickly developed lesions. In a B16 metastatic melanoma model in immunocompetent mice, in vivo transduction of tumor by tail vein injection of a retroviral vector expressing HSV-tk under the control of the tyrosinase promoter resulted in a significant reduction in the number of lung metastases after GCV treatment (Vile et al. 1994). This reduction was out of proportion to the expected level of tumor transduction. This effect was not seen when the experiment was repeated in athymic nude mice, suggesting an immune component to the response. The immunocompetent mice were resistant to rechallenge by the HSV-tk$^+$ B16 cell line as well as the parental cell line. In other experiments, Ki-*ras*-transformed BALB/c3T3 fibroblasts implanted into athymic nude mice and transduced with HSV-tk by in situ injection of retroviral vector producer cells responded to GCV treatment, but recurred within 2 months (Pavlovic et al. 1996). The same tumors treated in syngenic immunocompetent BALB/c mice did not recur. Others workers found that mice bearing HSV-tk-transduced Renca cell renal carcinoma treated with GCV were resistant to rechallenge with this tumor and had developed a cytotoxic T lymphocyte (CTL) response specific for Renca (Yamamoto et al. 1997). The CTL response against Renca cells was inhibited by incubation of the lymphocytes with CD8$^+$ antisera, but not with CD4$^+$ antibodies. Similar results were obtained in vivo with BALB/c mice bearing HSV-tk$^+$ MC26 colon carcinoma treated with GCV. Posttreatment, pathological examination revealed that the residual tumor was infiltrated with T cells, the majority of which were the CD8$^+$ phenotype (Gagandeep et al. 1996). Control animals bearing tumors transduced with *lacZ* did not respond to GCV treatment and there was no evidence of a cellular immune response to the tumor. Shah (1997) observed that rats bearing subcutaneous or intracerebral 9L gliosarcomas treated with adenovirus-mediated HSV-tk gene transfer and GCV were resistant to rechallenge with 9L cells, and this immunity could be transferred to naive rats by the transfer of splenocytes from successfully treated animals. Untreated control animals, animals receiving the HSV-tk vector but no GCV, those receiving GCV but no vector, animals receiving irradiated 9L cells, or animals undergoing surgical removal of their tumors were not resistant to rechallenge, and their transferred splenocytes provided no protection to naive rats.

This immunization effect is not limited to the HSV-tk/GCV system. It has also been described with the CD/5FC suicide gene system. Mullen et al. (1994) reported that the expression of CD followed by 5FC treatment immunized mice against rechallenge by the same tumor type. CD/5FC reliably induces immunity against the poorly immunogenic TS/A murine breast cancer cell line (Consalvo et al. 1995). The effect of 5FC treatment on CD expressing TS/A breast cancer tumors in immunocompetant BALB/c mice could be reduced by pretreating the animals with anti-CD8⁺ and antigranulocyte antibodies to deplete these cell lineages.

Both HSV-tk and CD, being foreign viral or bacterial proteins, are potentially immunogenic. In a clinical trial, retrovirus-mediated transfer of a hygromycin-HSV-tk fusion gene into anti-HIV CTLs was found to induce immune elimination of these autologous cells on subsequent reinfusion into patients (Riddell et al. 1996). Expression of the costimulatory molecule B7, ICAM, and class I major histocompatability complex antigens (MHC I) is found to be increased in HSV-tk⁺ tumors after treatment with GCV (Ramesh et al. 1996b; Yamamoto et al. 1997). Additionally, viral gene-transfer vectors themselves are immunogenic. Prolonged gene expression is seen after injection of a *lac*Z-expressing adenovirus into the liver of athymic nude mice. However, robust inflammation, liver cell necrosis, and extinction of gene expression is seen within 3–4 weeks when the same experiment is performed in immunocompetent mice (Yang and Wilson 1995; Yang et al. 1996). This immune response appears to be primarily generated against the antigens of the viral vector and not of the transgene. This exuberant immune response is problematic if long-term gene expression is required to achieve a therapeutic effect, such as correction of a chronic inborn error of metabolism. However, this response may be beneficial in cancer gene therapy by acting as adjuvant stimulation to generate antitumor immunity.

Building on these observations, several groups have evaluated the combination of cytokine and suicide gene/prodrug therapy. In early studies, Dranoff et al. (1993) showed that tumor cells engineered to secrete various cytokines, especially GM-CSF, had potent antitumor effects in mice. In a series of experiments by Chen and colleagues, it was found that intrahepatic implants of MC26 colon carcinoma in BALB/c mice treated with adenovirus-mediated mouse IL-2 (AdV.mIL-2) gene transfer failed to generate any tumor responses. Treatment with an adenovirus expressing HSV-tk and systemic GCV treatment with or without the addition of the AdV.mIL-2 vector resulted extensive necrosis and regression of the tumor (S.H. Chen et al. 1995). However, only animals simultaneously treated with both vectors developed a specific CD8⁺ CTL response to MC26 cells and were resistant to rechallenge with tumorigenic doses of parental tumor cells at distant sites. Although the animals developed an immune response, it was found to wane over time and the tumors re-occurred. In a subsequent study, it was found that a combination of AdV.HSV-tk/GCV treatment with injection of AdV.mGM-CSF and AdV.mIL-2 generated long-lasting antitumor immunity and improved the survival (S.H. Chen et al.

1996). Another study found the combination of HSV-tk and IFN-α expression with GCV treatment resulted in regression of distant Friend leukemia cell metastasis (Santodonato et al. 1997).

In contrast to these promising results, other investigators have suggested that no survival improvement could be obtained from the combination of suicide and cytokine gene therapy over that of HSV-tk/GCV alone. In a study by Felzmann et al. (1997), little benefit was found with adenovirus-mediated HSV-tk gene transfer/GCV therapy combined with the adenovirus transfer of the genes for IL-2, IL-6, or B7.1 into mouse tumors over treatment with each vector alone. Others have found benefit when the genes are transferred simultaneously to the same cell using the same vector by cloning them separately into two sites in the vector or having the suicide and cytokine genes cotranscribed by an internal ribosomal re-entry site (IRES) (Castleden et al. 1997).

PRECLINICAL TRIALS

Many animal models of cancer suicide gene therapy have been reported. Included are syngeneic rodent tumors and athymic nude mice bearing human tumor xenografts, including melanomas, gliomas, sarcomas, lung, hepatocellular, breast, and colorectal carcinoma, to name but a few. Because of the large number and varying animal models reported, this discussion will focus only on those that have been translated into human clinical trials in which at least preliminary results have been reported. Early efforts of several groups have focused on development of clinically useful suicide gene/prodrug systems for the treatment of malignant brain tumors. Approximately 17,400 new cases of tumors of the brain and spinal cord are diagnosed each year in the United States, resulting in 13,300 deaths (Landis et al. 1998). Malignant brain tumors are the second leading cause of cancer death in children and 25% of CNS tumors in adults are the high-grade glioblastoma multiforme in which survivals average less than 1 year with current treatments (Chamberlain and Kormanik 1998; Shapiro and Shapiro 1998; Stewart and Cohen 1998).

Ezzeddine et al. (1991) demonstrated the ability to induce tumor regressions in an athymic nude mouse subcutaneous glioma model with retrovirally HSV-tk-transduced rat C6 cells and systemic administration of GCV. There was, however, a high rate of tumor recurrence once the GCV treatment was stopped. This was attributed to slower growing C6 cells in the population that rendered them insensitive to GCV, an inadequate course of GCV, or possibly due to spontaneous loss of HSV-tk expression in some cells. In an attempt to transduce preexisting tumors, this group demonstrated in situ transduction of intracerebral C6 gliomas with *lacZ* by the BAG retroviral vector by engrafting the ψ2-BAG packaging cell line into the tumor bed (Short et al. 1990). Tumor cells were selectively transduced by the vector because of their high proliferation rate compared to that of normal brain. This group subsequently demonstrated enhanced sensitivity of mixed populations of HSV-tk$^+$ and wild-type C6 gliomas by coinfecting HSV-tk$^+$ C6 cells with wild-type Mo-MLV generating the retroviral vector locally by pseudotyping (Takamiya et al. 1992). Working independently, Culver and coworkers reported the injection of viable retroviral vector producer cells (VPCs) directly into 9L brain tumors in rats as a method of generating continuous in situ HSV-tk gene transfer into tumor cells (Culver et al. 1992; Ram et al. 1993a,b). Animals underwent stereotactic intracerebral injection of 9L cells. Five days later, HSV-tk retroviral VPCs were injected at the same coordinates and 5 days after injection of the VPCs, systemic GCV 150 mg/kg twice a day for 5 days was begun. This treatment resulted in HSV-tk transduction of the intracerebral gliosarcomas, complete macroscopic and microscopic response, as well as an improvement in survival of the animals treated with GCV. This work was subsequently translated into a clinical trial evaluating the toxicity and efficacy of this treatment (Culver et al. 1994).

Another approach has been to use an adenovirus-based vector strategy for transduction of brain tumors with HSV-tk. Chen et al. (1994) demonstrated efficient transduction of C6 rat glioma cells in vivo with an adenoviral vector (ADV/RSV-tk). Glioma cells were inoculated intracerebrally by stereotactic injection into athymic nude mice. After 8 days of growth, the tumor was stereotactically injected with 3×10^8 ADV/RSV-tk vector particles. The mice were subsequently treated with intraperitoneal injections of GCV. Animals receiving ADV/RSV-tk had a 500-fold decrease in tumor volume and survivals twice as long as control animals. Although the tumoricidal response was impressive, athymic nude mice lack an intact immune system that appears to have a significant role in this response. In a follow-up study, the activity of ADV/RSV-tk combined with GCV was studied in an intracerebral model in immunocompetent Fisher rats using 9L gliosarcoma cells (Shine and Woo 1996). The 9L cells were stereotactically injected into the caudate nucleus of Fisher rats. Eight days later, ADV/RSV-tk was stereotactically injected into the tumor and GCV treatment was begun the following day. All control animals died within 22 days. In contrast, animals treated with ADV/RSV-tk and GCV survived 150 to 400 days. No animal at necropsy had evidence of residual tumor. No toxic effects were noted in the brain tissue of either mice or rats using adenovirus-mediated HSV-tk gene transfer. To address the concern that the brains of mice and rats may not be truly representative of humans, this therapy was tested on baboons (Goodman et al. 1996). A dose-dependent cytopathic effect was noted at the site of vector injection that appeared was enhanced by GCV treatment on both magnetic resonance imaging (MRI) and necropsy.

On the basis of preclinical models, the University of Pennsylvania recently completed a phase I trial of adenovirus-mediated HSV-tk gene transfer and GCV treatment in patients with malignant mesothelioma (Treat et al. 1996). Malignant mesothelioma, a relatively uncommon cancer of the pleural or peritoneal cavity associated with asbestos exposure, offers an attractive target for suicide gene therapy. Locally advanced at the time of diagnosis, the standard therapeutic modalities of surgery, radiotherapy, and chemotherapy offer patients little chance of response or improved survival (Antman 1989). Even when advanced, mesothelioma infrequently metasta-sizes and tends to remain confined to the pleural space. Additionally, the chest cavity can be easily accessed by thorocentesis for administration of vector and theoretically represents a limited area in which vector can spread. Approaching the problem of achieving efficient in vivo transduction of the maximum amount of tumor possible, Smythe and colleagues reasoned that an adenovirus-vector-based strategy would be superior to that of retrovirus-mediated HSV-tk gene transfer. Preliminary studies found that human mesothelioma cell lines could be easily transduced by an adenoviral HSV-tk vector (Ad.RSVtk) and sensitized to GCV (Smythe et al. 1994). It was also found that most of the mesothelioma cell lines examined exhibited a BSE. In vivo experiments using SCID mice with subcutaneous and intraperitoneal human mesothelioma implants showed that Ad.RSVtk in combination with GCV administration resulted in a marked decrease in tumor volume (Smythe, Hwang et al. 1995). Regression of tumor was seen when as little as 10% of the tumor was composed of HSV-tk[+] cells. At necropsy, control animals exhibited massive intraperitoneal tumors and extensive carcino-matosis with evidence of bowel and hepatic portal obstruction. Ad.RSVtk/GCV-treated animals showed only minimal residual disease ($\leq 2 \times 2$ mm nodules) and benign peritoneal adhesions. Overall, macroscopic dis-ease was eliminated by Ad.RSVtk/GCV treatment in 90% of animals and microscopic disease was eliminated in 80%. Treatment with Ad.RSVtk and GCV significantly improved median survival compared to control animals.

REPLICATING VIRAL VECTORS FOR SUICIDE GENE THERAPY

The major challenge facing the meaningful application of any gene ther-apy strategy that directly targets malignant cells for genetic modification is the limited efficiency of in vivo gene delivery. The presence of meta-

tumors that were also intentionally infected with replication-competent Mo-MLV retrovirus. These replication-competent helper viruses were capable of pseudotyping the HSV-tk vector sequences and spreading the suicide gene by infecting adjacent tumor cells. Miyatake et al. (1997) tested the antitumor effect of defective HSV-tk-expressing herpes simplex vectors mixed with replication-competent, nonneurovirulent HSV mutant helper virus and showed that this combination resulted in significantly decreased tumor sizes in experimental animals compared to controls.

CLINICAL EXPERIENCE

Despite the short history of human gene therapy, there are already several published reports on the use of suicide-gene technology in different clinical settings. Riddell et al. (1996) introduced a suicide gene construct into HIV-gag-specific cytolytic T-cell clones that were being studied as a potential AIDS immunotherapy treatment. It was unknown whether infusions of large numbers of activated anti-HIV CTLs might result in catastrophic damage to HIV-infected organs such as the brain. To provide a mechanism to abort such an attack, the CTL clones were transduced with a retroviral vector delivering a hygromycin-thymidine kinase (HyTk) fusion gene to confer both positive and negative selectable traits. For positive selection, the construct expressed hygromycin phosphotransferase, which ensured that following in vitro hygromycin selection, the gene-modified T cells would also be expressing sufficient levels of the negative selector, HSV-tk, to assure that they could be eliminated by GCV if required. As the study progressed, the suicide gene was found to not be needed. Patients developed immunity to the HyTk-fusion protein expressed by the gene-modified CTLs after the initial treatment and rapidly eliminated the infused T cells during the dose escalation phase of the study.

The success of allogeneic bone-marrow transplantation (BMT) has been facilitated by the development of powerful techniques to deplete mature T cells from donor bone marrows. By greatly reducing the incidence of severe graft-versus-host disease (GvHD) associated with allogeneic BMT, T-cell depletion has even permitted the use of less than fully HLA-matched donors (Sierra and Anasetti 1995). However, the use of T-cell-depleted marrow as a treatment for leukemia is associated with a higher risk of relapse because of the loss of the beneficial role of the graft-versus-leukemia (GvL) effect (Weiden et al. 1979; Mavroudis and Barrett 1996). Furthermore, recipients of T-cell-depleted marrows were found to experience an increased incidence of posttransplant Epstein-Barr virus (EBV)-induced lymphoma (List et al. 1987).

Treatment of leukemic relapse and posttransplant EBV-induced lymphoproliferative disorders with infusions of peripheral blood T cells obtained from the marrow donor have shown success in achieving remission of these malignancies, but at the risk of developing severe or fatal GvHD (Cullis et al. 1992; Szer et al. 1993; Papdopoulus et al. 1994; van Rhee and Kolb 1995). There is currently no satisfactory treatment for severe GvHD (Parkman et al. 1998). Bonini and colleagues approached this problem by modifying the donor T cells with a suicide gene (Bordignon et al. 1995; Bonini et al. 1997). In their study, they used a retroviral vector expressing the HSV-tk gene and the gene for a cell surface low-affinity nerve growth factor receptor (NGFR) that had been truncated so as not to transduce a signal. The initial efficiency of gene transfer with this vector ranged from 20% to 50%. Using a monoclonal antibody to NGFR, these HSV-tk-expressing cells were selectively enriched resulting in a yield of transgene-expressing cells ranging from 95% to 99.6%. These highly selected HSV-tk positive cells were then used to treat eight patients with recurrent malignancy post-BMT in a protocol using escalating doses of the cells. Of the eight transplant patients, three developed significant GvHD. One patient with severe, biopsy-proven GvHD-associated hepatitis 9 weeks after the T-cell infusion rapidly responded to GCV treatment. In fact, the proportion of transgene-marked cells in the blood of this patient dropped from 2.0% pre-GCV to less than 0.0001% within 24 hours of treatment. Another patient with severe cutaneous GvHD responded with a drop in transgene-expressing cells from 13.4% to less than 0.0001% within 24 hours of initiation of GCV treatment and had a complete resolution of their GvHD. A third patient, with chronic GvHD, also responded with a reduction in disease activity. However, with 7 days of GCV infusions, the transgene containing cells in the blood of this patient fell only from 11.9% to 2.8%. Complete elimination of the gene-modified cells could not be achieved despite prolonged GCV treatment. The residual transgene-positive cells from this patient's blood could still be killed by GCV exposure when they were activated in culture, suggesting that they were not actively cycling in vivo and therefore were insensitive to GCV (C. Bordignon, pers. comm.).

The ability to use the power of the GvL effect for prophylaxis of leukemic relapse and EBV-related lymphoproliferative disease, while retaining the capacity to control and eliminate the immunoreactive cells with an incorporated suicide gene, represents a significant addition to the treatment repertoire for these otherwise refractory malignancies.

The largest number of clinical studies employing suicide genes involves attempts to directly kill malignant cells with prodrug administration following gene modification of the tumor in situ. Although multiple clinical trials are under way involving both HSV-tk and CD as suicide genes using retroviral and adenoviral vectors as delivery vehicles, to date the results of only two trials have been reported (Ram et al. 1997; Sterman et al. 1998). In a preclinical study, Culver et al. (1992) demonstrated that implantation of murine cells engineered to produce retroviral vectors (VPC) expressing the HSV-tk gene into tumors induced the regression of experimental brain tumors in rodents after GCV administration. To follow up these observations, in 1992, a clinical trial was initiated testing this procedure for safety and efficacy in 15 patients with progressive recurrent primary or metastatic brain tumors (Ram et al. 1997). Seven days following stereotactic implantation of murine VPCs administered in a series of parallel tracts throughout the enhancing area of the tumor as visualized by MRI, a 2-week course of intravenous GCV was given. Antitumor activity as measured by at least a 50% reduction in the contrast-enhanced area of each tumor was seen in 5 of 19 treated lesions with a 1–3-month duration of response. A notable exception being one patient with two treated lesions who remains in complete remission more than 5 years after treatment. In every case, the antitumor responses were limited to smaller lesions (volume ≥2.4 ml).

Tumor biopsies were obtained from two patients 7 days after stereotactic injection of the VPCs, but before GCV administration. The most striking finding was seen when this biopsy tissue was evaluated for HSV-tk mRNA expression by in situ hybridization. HSV-tk gene expression was seen and transduction of tumor cells by the in situ retroviral vector could be confirmed, but the vast majority of tumor in each biopsy was not transduced, even though the injection tracts were spaced only 3 mm apart. If this result is generalized to each patient, calculations indicate that even in those cases where the antitumor response was complete, less than 5% of the tumor cells were actually HSV-tk gene-modified by this process. These results emphasize the severe limitations to vector spread in vivo, even when the vector is produced locally within the tumor. It is even more remarkable that with this very restricted distribution, significant antitumor responses were seen (Fig. 8). What better illustration could there be of the crucial role of a potent BSE to have any hope for antitumor responses using in situ gene delivery strategies.

Recently, the University of Pennsylvania reported a phase I trial of adenovirus-mediated HSV-tk gene transfer followed by GCV treatment in patients with malignant pleural mesothelioma (Treat et al. 1996; Sterman et al. 1998). Cohorts of patients were treated with increasing doses of vector ranging from 1×10^9 to 3.2×10^{12} pfu injected into the affected pleural cavity. Systemic GCV treatment at 5 mg/kg twice a day was initiated 24 hours later and continued for 2 weeks. Gene transfer was documented in

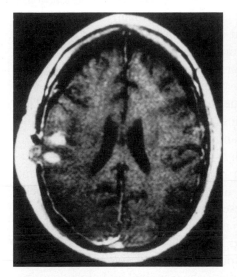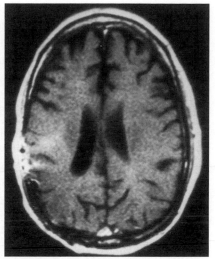

Figure 8 MRI-imaging of a section of the brain of a 47-year-old male with a recurrent glioblastoma multiforme (*left*). Following clinical evaluation and biopsy, the recurrent tumor masses were stereotactically injected with murine PA317 cells producing a HSV-tk expressing retroviral vector (VPC). Seven days following VPC injection to permit time for local production and uptake of the HSV-tk vector within the tumor, the patient was given a 2-week course of ganciclovir 5 mg/kg twice daily. At follow-up examination 12 months later (*right*), no sign of tumor recurrence was detected. It is now 5 years since this patient was treated with HSV-tk suicide gene therapy for his recurrent brain tumor and he remains disease free. (Reprinted, with permission, from Ram et al. 1997.)

11 of 20 evaluable patients, and a strong inflammatory response was generated in the pleural space. Side effects were generally mild and included fever, anemia, localized vesiculobullous skin eruption, granulocytopenia (one patient), minor elevation of serum liver transaminases, and transient hypotension and hypoxemia in the highest-dose cohort. No definitive tumor regression was noted in any patient.

CHALLENGES FOR THE FUTURE OF SUICIDE GENE THERAPY

It is difficult to predict where the use of suicide gene/prodrug therapies will ultimately fit in the overall scope of cancer treatment. Although limited in number, the initial clinical applications of suicide genes for therapy of cancer have illustrated both the promise and the limitations of present technology. In very specific situations where the suicide gene is delivered to cells ex vivo and the transduced cells can be selected by cloning, drug resistance, or physical techniques, clinically valuable treatment approaches are clearly at hand (e.g., controllable GvL). However, for all but the most atypical applications, in situ and preferably systemic suicide gene delivery would be required to effectively control most types of malignancy. If systemic delivery were possible, then mechanisms to restrict expression of the suicide gene to the cancer will be needed. This could be accomplished by restricting gene expression or by targeting delivery of the vector to the malignant cells, thereby excluding injury of neighboring normal tissues. Although work with "tumor-specific" promoters has shown some restriction of gene expression, these systems often fail to address important but limited populations of normal cells which might express genes using these promoters (e.g., tyrosinase expression in the normal retina as well as in malignant melanoma). Furthermore, gene delivery in vivo is so restricted by the physics of diffusion of viral particles in tissue spaces that widespread uniform delivery of vectors is unlikely. Adequate gene delivery remains the principal and most daunting hurdle to be crossed. Even when suicide gene therapy is contemplated only for local treatment, the likelihood of success in delivering the gene to every tumor cell is remote. Thus, there is a critical need for effective amplifying strategies for the bystander effect to help eliminate tumor cells not successfully modified to express the suicide gene.

Replicating viral vectors carrying suicide genes that can spread by secondary rounds of infection through tumors offers one mechanism that may enhance gene delivery. The relative ease by which different viral vectors are able to transduce different tissues and even different tumors of the same histologic type also needs to be factored into the calculation of the most appropriate gene-delivery system to be employed. Also to be considered is whether replication-defective or replication-competent viral vectors would be expected to induce an antiviral immune response that will ultimately limit their ability to transfer genes to tumor cells in vivo. The use of different immunotypes or viruses (adenovirus group types, adeno-associated virus, lentiviruses, poxviruses, retroviruses, herpesviruses, and alphaviruses) may allow repeated treatment if specific antiviral immunity becomes limiting. However, immune response against the transgene has also been seen and therefore a panel of non-cross-reacting suicide genes (e.g., HSV-tk, VZV-tk, *E. coli* CD, and fungal CD) may also be necessary.

Among the various suicide gene strategies, the mechanism of the BSE involved is critically important in the overall rationale of treatment. With the HSV-tk/GCV system, gene therapy is the only way the cytotoxin can be delivered since the activated drug is highly charged and cannot enter cells from the outside. It must be produced inside the cell by interaction of

GCV with the HSV-tk enzyme and passed directly from cell to cell through gap junctions. In contrast, 5FU can readily diffuse into and out of cells and therefore it need not be produced by action of an intracellular gene product on the prodrug. If gene delivery is largely restricted to local areas (i.e., the end of a stainless steel needle) as is the current situation, locally delivered 5FU (via catheter) is probably as effective as genetically produced 5FU—and a much less complex and expensive undertaking.

With HSV-tk and similar suicide gene systems and their unique BSE, improvements in efficacy might be obtained by screening tumors for their suitability for these treatments by documenting the presence of gap junctions just as breast cancers are screened for estrogen receptor status. The use of pharmacological agents to up-regulate or ungate gap junctions, development of new prodrugs that yield derivatives with longer half-lives, optimization of prodrug dosing schedules, combinations of different suicide genes, combinations of suicide genes and conventional cancer treatments, and the use of suicide genes combined with cytokine genes, or other immune enhancing factors all need to be tested in order for enzyme/prodrug gene therapy to fulfill its ultimate promise.

REFERENCES

Aghi M., Kramm C.M., Chou T.C., Breakefield X.O., and Chiocca E.A. 1998. Synergistic anticancer effects of ganciclovir/thymidine kinase and 5-fluorocytosine/cytosine deaminase gene therapies. *J. Natl. Cancer Inst.* **90:** 370–380.

Andersen L., Kilstrup M., and Neuhard J. 1989. Pyrimidine, purine and nitrogen control of cytosine deaminase synthesis in *Escherichia coli* K 12. Involvement of the glnLG and purR genes in the regulation of codA expression. *Arch. Microbiol.* **152:** 115–118.

Antman K.H. 1989. Natural history and staging of malignant mesothelioma. *Chest* **96:** 93S–95S.

Austin E.A. and Huber B.E. 1993. A first step in the development of gene therapy for colorectal carcinoma: Cloning, sequencing, and expression of *Esherichia coli* cytosine deaminase. *Mol. Pharmacol.* **43:** 380–387.

Averett D.R., Koszalka G.W., Fyfe J.A., Roberts G.B., Purifoy D.J., and Krenitsky T.A. 1991. 6-Methoxypurine arabinoside as a selective and potent inhibitor of varicella-zoster virus. *Antimicrob. Agents Chemother.* **35:** 851–857.

Bailey S.M., Knox R.J., Hobbs S.M., Jenkins T.C., Mauger A.B., Melton R.G., Burke P.J., Connors T.A., and Hart I.R. 1996. Investigation of alternative prodrugs for use with *E. coli* nitroreductase in "suicide gene" approaches to cancer therapy. *Gene Ther.* **3:** 1143–1150.

Barba D., Hardin J., Sadelain M., and Gage F.H. 1994. Development of anti-tumor immunity following thymidine kinase mediated killing of experimental brain tumors. *Proc. Natl. Acad. Sci.* **91:** 4348–4352.

Bergelson J.M., Cunningham J.A., Droguett G., Kurt-Jones E.A., Krithivas A., Hong J.S., Horwitz M.S., Crowell R.L., and Finberg R.W. 1997. Isolation of a common receptor for Coxsackie B virus and adenoviruses 2 and 5. *Science* **275:** 1083–1092.

Bett A.J., Prevec L., and Grahman F.L. 1993. Packaging capacity and stability of human adenovirus type 5 vectors. *J. Virol.* **67:** 5911–5921.

Bevilacqua F., Davis-Poynter N., Worrallo J., Gower D., Collins P., and Darby G. 1995. Construction of a herpes simplex virus/varicella-zoster virus (HSV/VZV) thymidine kinase recombinant with the pathogenic potential of HSV and a drug sensitivity profile resembling that of VZV. *J. Gen. Virol.* **76:** 1927–1935.

Bi, W.L., Parysek L.M., Warnick R., and Stambrook P.J. 1993. *In vitro* evidence that metabolic cooperation is responsible for the bystander effect observed with HSV-tk retroviral gene therapy. *Hum. Gene Ther.* **4:** 725–731.

Bi W.L., Kim Y.G., Feliciano E.S., Pavelic L., Wilson K.M., Pavelic Z.P., and Stambrook P.J. 1997. An HSV*tk*-mediated local and distant anti-tumor bystander effect in tumors of head and neck origin in athymic nude mice. *Cancer Gene Ther.* **4:** 246–251.

Biron K.K. and Elion G.B. 1980. In vitro susceptibility of varicella-zoster virus to acyclovir. *Antimicrob. Agents Chemother.* **18:** 443–447.

Biron K.K., Stanat S.C., Sorrell J.B., Fyfe J.A., Keller P.M., Lambe C.U., and Nelson D.J. 1985. Metabolic activation of the nucleoside analog 9-[(2-hydroxy-1-(hydroxymethyl)ethoxy]methyl)guanine in human diploid fibroblasts infected with human cytomegalovirus. *Proc. Natl. Acad. Sci.* **82:** 2473–2477.

Bischoff J.R., Kirn D.H., Williams A., Heise C., Horn S., Muna M., Ng L., Nye J.A., Sampson-Johannes A., Fattaey A., and McCormick F. 1996. An adenovirus mutant that replicates selectively in p53-deficient human tumor cells (see comments). *Science* **274:** 373–376.

Bonini C., Ferrari G., Verzeletti S., Servida P., Zappone E., Ruggieri L., Ponzoni M., Rossini S., Mavilio F., Traversari C., and Bordignon C. 1997. HSV-TK gene transfer into donor lymphocytes for control of allogeneic graft-versus-leukemia (see comments). *Science* **276:** 1719–1724.

Bordignon C., Bonini C., Verzeletti S., Nobili N., Maggioni D., Traversari C., Giavazzi R., Servida P., Zappone E., Benazzi E., and et al. 1995. Transfer of the HSV-tk gene into donor peripheral blood lymphocytes for in vivo modulation of donor anti-tumor immunity after allogeneic bone marrow transplantation. *Hum. Gene Ther.* **6:** 813–819.

Boucher P.D., Ruch R.J., and Shewach D.S. 1998. Differential ganciclovir-mediated cytotoxicity and bystander killing in human colon carcinoma cell lines expressing herpes simplex thymidine kinase. *Hum. Gene Ther.* **9:** 801–814.

Bridgewater J.A., Knox R.J., Pitts J.D., Collins M.K., and Springer C.J. 1997. The bystander effect of the nitroreductase/CB1954 enzyme/prodrug system is due to a cell-permeable metabolite. *Hum. Gene Ther.* **8:** 709–717.

Bunnell B.A., Muul L.M., Donahue R.E., Blaese R.M., and Morgan R.A. 1995. High-efficiency retroviral-mediated gene transfer into human and nonhuman primate peripheral blood lymphocytes. *Proc. Natl. Acad. Sci.* **92:** 7739–7743.

Caruso M., Panis Y., Gagandeep S., Houssin D., Saltzmann J.L., and Klatzmann D. 1993. Regression of established macroscopic liver metastases after *in situ* transduction with a suicide gene. *Proc. Natl. Acad. Sci.* **90:** 7024–7028.

Castleden S.A., Chong H., Garcia-Ribas I., Melcher A.A., Hutchinson G., Roberts B., Hart I.R., and Vile R.G. 1997. A family of bicistronic vectors to enhance both local and systemic antitumor effects of HSVtk or cytokine expression in a murine melanoma model. *Hum. Gene Ther.* **8:** 2087–2102.

Chamberlain M. C. and Kormanik P.A. 1998. Practical guidelines for the treatment of malignant gliomas. *West. J. Med.* **168:** 114–120.

Chang T.K.H., Weber G.F., Crespi C.L., and Waxman D.J. 1993. Differential activation of cyclophosphamide and ifosfamide by cytochromes P-450 2B and 3A in human liver microsomes. *Cancer Res.* **53**: 5629–5637.

Chen C.Y., Chang Y.N., Ryan P., Linscott M., McGarrity G.J., and Chiang Y.L. 1995. Effect of herpes simplex virus thymidine kinase expression levels on ganciclovir-mediated cytotoxicity and the "bystander effect." *Hum. Gene Ther.* **6**: 1467–1476.

Chen L. and Waxman D.J. 1995. Intratumoral activation and enhanced chemotherapeutic effect of oxazaphosphorines following cytochrome P-450 gene transfer: Development of a combined chemotherapy/cancer gene therapy strategy. *Cancer Res.* **55**: 581-589.

Chen L., Waxman D.J., Chen D., and Kufe D.W. 1996. Sensitization of human breast cancer cells to cyclophosphamide and ifosfamide by transfer of a liver cytochrome P450 gene. *Cancer Res.* **56**: 1331–1340.

Chen S.H., Shine H.D., Goodman J.C., Grossman R.G., and Woo S.L.C. 1994. Gene therapy for brain tumors: Regression of experimental gliomas by adenovirus-mediated gene transfer *in vivo*. *Proc. Natl. Acad. Sci.* **91**: 3054–3057.

Chen S.H., Chen X.H., Wang Y., Kosai K., Finegold M.J., Rich S.S., and Woo S.L. 1995. Combination gene therapy for liver metastasis of colon carcinoma in vivo. *Proc. Natl. Acad. Sci.* **92**: 2577–2581.

Chen S.H., Kosai K., Xu B., Pham-Nguyen K., Contant C., Finegold M.J., and Woo S.L. 1996. Combination suicide and cytokine gene therapy for hepatic metastases of colon carcinoma: Sustained antitumor immunity prolongs animal survival. *Cancer Res.* **56**: 3758–3762.

Cheng Y.-C. 1976. Deoxythymidine kinase induced in HeLa TK⁻ cells by herpes simplex virus type I and type II: Substrate specific and kinetic behavior. *Biochem. Biophys. Acta* **452**: 370–381.

Chu T.T., Martinez I., Sheay W., and Dornberg R. 1994. Cell targeting with retroviral particles containing antibody-envelope fusion proteins. *Gene Ther.* **1**: 292–299.

Colly L.P., Peters W.G., Richel D., Arentsen-Honders M.W., Starrenburg C.W., and Willemze R. 1987. Deoxycytidine kinase and deoxycytidine deaminase values correspond closely to clinical response to cytosine arabinoside remission induction therapy in patients with acute myelogenous leukemia. *Semin. Oncol.* **14**: 257–261.

Connors T.A. 1995. The choice of prodrugs for gene directed enzyme prodrug therapy of cancer. *Gene Ther.* **2**: 702–709.

Consalvo M., Mullen C.A., Modesti A., Musiani P., Allione A., Cavallo F., Giovarelli M., and Forni G. 1995. 5-Fluorocytosine-induced eradication of murine adenocarcinomas engineered to express the cytosine deaminase suicide gene requires host immune competence and leaves an efficient memory. *J. Immunol.* **154**: 5302–5312.

Cullis J.O., Jiang Y.Z., Schwarer A.P., Hughes T.P., Barrett A.J., and Goldman J.M. 1992. Donor leukocyte infusions for chronic myeloid leukemia in relapse after allogenic bone marrow transplantation. *Blood* **79**: 1379–1381.

Culver K.W., Ram Z., Wallbridge S., Ishii H., Oldfield E.H., and Blaese R.M. 1992. In vivo gene transfer with retroviral vector-producer cells for treatment of experimental brain tumors (see comments). *Science* **256**: 1550–1552.

Culver K.W., Van Gilder J., Link C.J., Carlstrom T., Buroker T., Yuh W., Koch K., Schabold K., Doornbas S., Wetjen B., and Blaese R.M. 1994. Gene therapy for treatment of malignant tumors with *in vivo* tumor transduction with herpes simplex thymidine kinase/ganciclovir system. *Hum. Gene Ther.* **5**: 343–379.

Dachs GU., Dougherty G.J., Stratford I.J., and Chaplin D.J. 1997. Targeting gene therapy to cancer: A review. *Oncol Res.* **9**: 313–325.

Danielsen S., Kilstrup M., Barilla K., Jochimsen B., and Neuhard J. 1992.

Characterization of the *Escherichia coli* codBA operon encoding cytosine permease and cytosine deaminase. *Mol. Microbiol.* **6:** 1335–1344.

Danks M.K., Morton C.L., Pawlik C.A., and Potter P.M. 1998. Overexpression of a rabbit liver carboxylesterase sensitizes human tumor cells to CPT-11. *Cancer Res.* **58:** 20–22.

Davis B.M., Koc O.N., Lee K., and Gerson S.L. 1996. Current progress in the gene therapy of cancer. *Curr. Opin. Oncol.* **8:** 499–508.

Degreve B., Andrei G., Izquierdo M., Piette J., Morin K., Knaus E.E., Wiebe L.I., Basrah I., Walker R.T., De Clercq E., and Balzarini J. 1997. Varicella-zoster virus thymidine kinase gene and antiherpetic pyrimidine nucleoside analogues in a combined gene/chemotherapy treatment for cancer. *Gene Ther.* **4:** 1107–1114.

Diasio R.B. and Harris B.E. 1989. Clinical pharmacology of 5-fluorouracil. *Clin. Pharmacokinet.* **16:** 215–220.

Diasio R.B., Lakings D.E., and Bennett J.E. 1978. Evidence of conversion of 5-fluorocytosine to 5-fluorouracil in humans: A possible factor in 5-fluorocytosine clinical toxicity. *Antimicrob. Agents Chemother.* **14:** 903–907.

Dion L.D., Goldsmith K.T., and Garver R.I., Jr. 1996a. Quantitative and in vivo activity of adenoviral-producing cells made by cotransduction of a replication-defective adenovirus and a replication-enabling plasmid. *Cancer Gene Ther.* **3:** 230–237.

Dion L.D., Goldsmith K.T., Strong T.V., Bilbao G., Curiel D.T., and Garver R.I., Jr. 1996b. E1A RNA transcripts amplify adenovirus-mediated tumor reduction. *Gene Ther.* **3:** 1021–1025.

Donahue R.E., Kessler S.W., Bodine D., McDonagh K., Dunbar C., Goodman S., Agricola B., Byrne E., Raffeld M., Moen R., Bacher J., Zsebo K.M., and Nienhuis A.W. 1992. Helper virus induced T-cell lymphoma in nonhuman primates after retroviral mediated gene transfer. *J. Exp. Med.* **176:** 1125–1135.

Dong Y., Wen P., Manome Y., Parr M., Hirshowitz A., Chen L., Hirschowitz E.A., Crystal R., Weichselbaum R., Kufe D.W., and Fine H.A. 1996. In vivo replication-deficient adenovirus vector-mediated transduction of the cytosine deaminase gene sensitizes glioma cells to 5-fluorocytosine. *Hum. Gene Ther.* **7:** 713–720.

Douglas J.T. and Curiel D.T. 1997. Strategies to accomplish targeted gene delivery to muscle cells employing tropism-modified adenoviral vectors. *Neuromuscul. Disord.* **7:** 284–298.

Douglas J.T., Rogers B.E., Rosenfeld M.E., Michael S.I., Feng M., and Curiel D.T. 1996. Targeted gene delivery by trophism-modified adenoviral vectors. *Nat. Biotechnol.* **14:** 1574–1578.

Dranoff G., Jaffee E., Lazenby A., Golumbek P., Levitsky H.I., Broze K., Jackson V., Hamada H., Pardoll D., and Mulligan R.C. 1993. Vaccination with irradiated tumor cells engineered to secrete murine granulocyte-macrophage colony-stimulating factor stimulates potent, specific and long lasting anti-tumor immunity. *Proc. Natl. Acad. Sci.* **90:** 3539–3543.

Elion G.B. 1982. Mechanism of action and selectivity of acyclovir. *Am. J. Med.* **73:** 7–13.

———. 1983. The biochemistry and mechanism of action of acyclovir. *J. Antimicrob Chemother.* **12:** 9–17.

Ellerman V. and Bang. O. 1908. Experimentell leukamie bei huhnern. *Zentbl. Bakteriol.* **46:** 595–609.

Elshami A.A., Saavedra A., Zhang H., Kucharczuk J.C., Spray D.C., Fishman G.I., Amin K.M., Kaiser L.R., and Albelda S.M. 1996. Gap junctions play a role in the 'bystander effect' of the herpes simplex virus thymidine kinase/ganciclovir system in vitro. *Gene Ther.* **3:** 85–92.

Elshami A.A., Cook J.W., Amin K.M., Choi H., Park J.Y., Coonrod L., Sun J., Molnar-Kimber K., Wilson J.M., Kaiser L.R., and Albelda S.M. 1997. The effect of promoter strength in adenoviral vectors containing herpes simplex virus thymidine kinase on cancer gene therapy in vitro and in vivo. *Cancer Gene Ther.* **4:** 213–221.

Emi N., Friedmann T., and Yee J.K. 1991. Pseudotype formation of murine leukemia virus with the G protein of vesicular stomatitis virus. *J. Virol.* **65:** 1202–1207.

Ezzeddine Z.D., Martuza R.L, Platika D., Short M.P., Malick A., Choi B., and Breakefield X.O. 1991. Selective killing of glioma cells in culture and in vivo by retrovirus transfer of the herpes simplex virus thymidine kinase genc. *New Biol.* **3:** 608–614.

Felzmann T., W.J. Ramsey, and Blaese R.M. 1997. Characterization of the antitumor immune response generated by treatment of murine tumors with recombinant adenoviruses expressing HSVtk, IL-2, IL-6 or B7-1. *Gene Ther.* **4:** 1322–1329.

Fick J., Barker II, F.G., Dazin P., Westphale E.M., Beyer E.C., and Israel M.A. 1995. The extent of heterocellular communication mediated by gap junctions is predictive of bystander tumor cytotoxicity in vitro. *Proc. Natl. Acad. Sci.* **92:** 11071–11075.

Field A.K., Davies M.E., DeWitt C., Perry H.C., Liou R., Germershausen J., Karkas J.D., Ashton W.T., Johnston D.B., and Tolman R.L. 1983. 9-([2-hydroxy-1-(hydroxymethyl)ethoxy]methyl)guanine: A selective inhibitor of herpes group virus replication. *Proc. Natl. Acad. Sci.* **80:** 4139–4143.

Franken M., Estabrooks A., Cavacini L., Sherburne B., Wang F., and Scadden D.T. 1996. Epstein-Barr virus-driven gene therapy for EBV-related lymphomas. *Nat. Med.* **2:** 1379–1382.

Freeman S.M., Ramesh R., Shastri M., Munshi A., Jensen A.K., and Marrogi A.J. 1995. The role of cytokines in mediating the bystander effect using HSV-TK xenogeneic cells. *Cancer Lett.* **92:** 167–174.

Freeman S.M., Abboud C.N., Whartenby K.A., Packman C.H., Koeplin D.S., Moolten F.L., and Abraham G.N. 1993. The "bystander effect": Tumor regression when a fraction of the tumor mass is genetically modified. *Cancer Res.* **53:** 5274–5283.

Friedlos F., Denny W.A., Palmer B.D., and Springer C.J. 1997. Mustard prodrugs for activation by *Escherichia coli* nitroreductase in gene-directed enzyme prodrug therapy. *J. Med. Chem.* **40:** 1270–1275.

Friedlos F., Quinn J., Knox R.J., and Roberts J.J. 1992. The properties of total adducts and interstrand crosslinks in the DNA of cells treated with CB1954. *Biochem. Pharmacol.* **43:** 1249–1254.

Friedlos F., Court S., Ford M., Denny W.A., and Springer C. 1998. Gene-directed enzyme prodrug therapy: quantitative bystander cytotoxicity and DNA damage induced by CB1954 in cells expressing bacterial nitroreductase. *Gene Ther.* **5:** 105–112.

Fyfe J.A., Keller P.M., Furman P.A., Miller R.L., and Elion G.B. 1978. Thymidine kinase from herpes simplex virus phosphorylates the new antiviral compound, 9-(2-hydroxyethoxymethyl)guanine. *J.Biol. Chem.* **253:** 8721–8727.

Gagandeep S., Brew R., Green B., Christmas S.E., Klatzmann D., Poston G.J., and Kinsella A.R. 1996. Prodrug-activated gene therapy: involvement of an immunological component in the "bystander effect." *Cancer Gene Ther.* **3:** 83–88.

Gall J., Kass-Eisler A., Leinwand L., and Falck-Pedersen E. 1996. Adenovirus type 5 and 7 capsid chimera: Fiber replacement alters receptor tropism without affecting primary immune neutralization epitopes. *J. Virol.* **70:** 2116–2123.

Ge K., Xu L., Zheng Z., Xu D., Sun L., and Liu X. 1997. Transduction of cytosine deaminase gene makes rat glioma cells highly sensitive to 5-fluorocytosine. *Int. J. Cancer.*

71: 675–679.

Gilula N.B., Reeves O.R., and Steinbach A. 1972. Metabolic coupling, ionic coupling, and cell contacts. *Nature* **235:** 262–265.

Goldman M.J. and Wilson J.M. 1995. Expression of alpha v beta 5 integrin is necessary for efficient adenovirus-mediated gene transfer in the human airway. *J. Virol.* **69:** 5951–5958.

Goodman J.C., Trask T.W., Chen S.H., Woo S.L., Grossman R.G., Carey K.D., Hubbard G.B., Carrier D.A., Rajagopalan S., Aguilar-Cordova E., and Shine H.D. 1996. Adenoviral-mediated thymidine kinase gene transfer into the primate brain followed by systemic ganciclovir: Pathologic, radiologic, and molecular studies. *Hum. Gene Ther.* **7:** 1241–1250.

Green N.K., Youngs D.J., Neoptolemos J.P., Friedlos F., Knox R.J., Springer C.J., Anlezark G.M., Michael N.P., Melton R.G., Ford M.J., Young L.S., Kerr D.J., and Searle P.F. 1997. Sensitization of colorectal and pancreatic cancer cell lines to the prodrug 5-(aziridin-1-yl)-2,4-dinitrobenzamide (CB1954) by retroviral transduction and expression of the *E. coli* nitroreductase gene. *Cancer Gene Ther.* **4:** 229–238.

Grignet-Debrus C. and Calberg-Bacq C.M. 1997. Potential of Varicella zoster virus thymidine kinase as a suicide gene in breast cancer cells. *Gene Ther.* **4:** 560–569.

Hagenbeek A., Martens A.C., and Colly L.P. 1987. In vivo development of cytosine arabinoside resistance in the BN acute myelocytic leukemia. *Semin. Oncol.* **14:** 202–206.

Hamel W., Magnelli L., Chiarugi V.P., and Israel M A. 1996. Herpes simplex virus thymidine kinase/ganciclovir-mediated apoptotic death of bystander cells. *Cancer Res.* **56:** 2697–2702.

Hamzeh F.M. and Lietman P.S. 1991. Intranuclear accumulation of subgenomic noninfectious human cytomegalovirus DNA in infected cells in the presence of ganciclovir. *Antimicrob. Agents Chemother.* **35:** 1818–1823.

Haynes P., Lambert T.R., and Mitchell I.D. 1996. Comparative in-vivo genotoxicity of antiviral nucleoside analogues; penciclovir, acyclovir, ganciclovir and xanthine analogue, caffeine, in the mouse bone marrow micronucleus assay. *Mutat. Res.* **369:** 65–74.

Heise C., Sampson-Johannes A., Williams A., McCormick F., Von Hoff D.D., and Kirn D.H. 1997. ONYX-015, an E1B gene-attenuated adenovirus, causes tumor-specific cytolysis and antitumoral efficacy that can be augmented by standard chemotherapeutic agents (see comments). *Nat. Med.* **3:** 639–645.

Hirschowitz E.A., Ohwada A., Pascal W.R., Russi T.J., and Crystal R.G. 1995. In vivo adenovirus-mediated gene transfer of the *Escherichia coli* cytosine deaminase gene to human colon carcinoma-derived tumors induces chemosensivity to 5-fluorocytosine. *Hum. Gene Ther.* **6:** 1055–1063.

Hoganson D.K., Batra R.K., Olsen J.C., and Boucher R.C. 1996. Comparison of the effects of three different toxin genes and their levels of expression on cell growth and bystander effect in lung adenocarcinoma. *Cancer Res.* **56:** 1315–1323.

Hollstein M., Sidransky D., Vogelstein B., and Harris C.C. 1991. p53 mutations in human cancers. *Science* **253:** 49–53.

Huber B.E., Richards C.A., and Krenitsky T.A. 1991. Retroviral-mediated gene therapy for the treatment of hepatocellular carcinoma: An innovative approach for cancer therapy. *Proc. Natl. Acad. Sci.* **88:** 8039–8043.

Huber B.E., Austin E.A., Richards C.A., Davis S.T., and Good S.S. 1994. Metabolism of 5-fluorocytosine to 5-fluorouracil in human colorectal tumor cells transduced with the cytosine deaminase gene: Significant antitumor effects when only a small percentage of tumor cells express cytosine deaminase. *Proc. Natl. Acad. Sci.* **91:** 8302–8306.

Huber B.E., Austin E.A., Good S.S., Knick V.C., Tibbels S., and Richards C.A. 1993. *In vivo* antitumors activity of 5-fluorocytosine on human colorectal carcinoma cells genetically modified to express cytosine deaminase. *Cancer Res.* **53:** 4619–4626.

Hughes B.W., Wells A.H., Bebok Z., Gadi V.K., Garver R.I., and Parker W.B. 1995. Bystander killing of melanoma cells using the human tyrosinase promoter to express the *Escherichia coli* purine nucleoside phosphorylase gene. *Cancer Res.* **55:** 3339–3345.

Ilsley D.D., Lee S.-K., Miller W.H., and Kuchta R.D. 1995. Acyclic guanosine analogs inhibit DNA polymerases α, β, and ε with very different potencies and have unique mechanisms of action. *Biochemistry* **34:** 2504–2510.

Ishii-Morita H., Agbaria R., Mullen C.A., Hirano H., Koeplin D.A., Ram Z., Oldfield E.H., Johns D.G., and Blaese R.M. 1997. Mechanism of 'bystander effect' killing in the herpes simplex thymidine kinase gene therapy model of cancer treatment. *Gene Ther.* **4:** 244–251.

Jolly D. 1994. Viral vector systems for gene therapy. *Cancer Gene Ther.* **1:** 52–64.

Kanai F., Shiratori Y., Yoshida Y., Wakimoto H., Hamada H., Kanegae Y., Saito I., Nakabayashi H., Tamaoki T., Tanaka T., Lan K.H., Kato N., Shiina S., and Omata M. 1996. Gene therapy for alpha-fetoprotein-producing human hepatoma cells by adenovirus-mediated transfer of the herpes simplex virus thymidine kinase gene. *Hepatology* **23:** 1359–1368.

Kato Y., Matsukawa S., Muraoka R., and Tanigawa N. 1997. Enhancement of drug sensitivity and a bystander effect in PC-9 cells transfected with a platelet-derived endothelial cell growth factor thymidine phosphorylase cDNA. *Br. J. Cancer* **75:** 506–511.

Katsuragi T., Sakai T., and Tonomura K. 1987. Implantable enzyme capsules for cancer chemotherapy from bakers' yeast cytosine deaminase immobilized on epoxy-acrylic resin and urethane prepolymer. *Appl. Biochem. Biotechnol.* **16:** 61–69.

Kawasaki H., Carrera C.J., Piro L.D., Saven A., Kipps T.J., and Carson D.A. 1993. Relationship of deoxycytidine kinase and cytoplasmic 5′-nucleotidase to the chemotherapeutic efficacy of 2-chlorodeoxyadenosine. *Blood* **81:** 597–601.

Kianmanesh, A. R., Perrin H., Panis Y., Fabre M., Nagy H.J., Houssin D., and Klatzmann D. 1997. A "distant" bystander effect of suicide gene therapy: regression of nontransduced tumors together with a distant transduced tumor (see comments). *Hum. Gene Ther.* **8:** 1807–1814.

Kilstrup M., Meng L.M., Neuhard J., and Nygaard P. 1989. Genetic evidence for a repressor of synthesis of cytosine deaminase and purine biosynthesis enzymes in *Escherichia coli. J. Bacteriol.* **171:** 2124–2127.

Krasnykh V.N., Mikheeva G.V., Douglas J.T., and Curiel D.T. 1996. Generation of recombinant adenovirus vectors with modified fibers for altering viral tropism. *J. Virol.* **70:** 6839–6846.

Krasnykh V., Dmitriev I., Mikheeva G., Miller C.R., Belousova N., and Curiel D.T.. 1998. Characterization of an adenovirus vector containing a heterologous peptide epitope in the HI loop of the fiber knob. *J. Virol.* **72:** 1844–1852.

Kumar N.M. and Gilula N.B. 1996. The gap junction communication channel. *Cell* **84:** 381–388.

Lan K.H., Kanai F., Shiratori Y., Ohashi M., Tanaka T., Okudaira T., Yoshida Y., Hamada H., and Omata M. 1997. In vivo selective gene expression and therapy mediated by adenoviral vectors for human carcinoembryonic antigen-producing gastric carcinoma.

Cancer Res. **57:** 4279–4284.

Landis S.H., Murray T., Bolden S., and Wingo P.A. 1998. Cancer statistics, 1998. *CA: cancer. J. Clin.* **48:** 6–30.

Lasner, T.M., Kesari S., Brown S.M., Yee V.M.-Y., Fraser N.W., and Trojanowski J.Q. 1996. Therapy of a murine model of pediatric brain tumors using a herpes simplex virus type-1 ICP34.5 mutant and demonstration of viral replication within the CNS. *J. Neuropathol. Exp. Neurol.* **55:** 1259–1269.

List A.F., Greco F.A., and Vogler L.B. 1987. Lymphoproliferative disease in the immuno-compromised host: the role of Epstein-Barr virus. *J. Clin. Oncol.* **5:** 1673–1689.

Lotan R. and Nicolson G.L. 1977. Inhibitory effect of retinoic acid or retinyl acid on the growth of untransformed, transformed and tumor cells *in vitro. J. Natl. Cancer Inst.* **59:** 1717–1722.

Manome Y., Abe M., Hagen M.F., Fine H., and Kufe D. 1994. Enhancer sequences of the DF3 gene regulate expression of the herpes-simplex virus thymidine kinase gene and confer sensitivity of human breast cancer cells to ganciclovir. *Cancer Res.* **54:** 5408–5413.

Manome Y., Wen P.Y., Dong Y., Tanaka T., Mitchell B.S., Kufe D.W., and Fine H.A. 1996. Viral vector transduction of the human deoxycytidine kinase cDNA sensitizes glioma cells to the cytotoxic effects of cytosine arabinoside in vitro and in vivo. *Nat. Med.* **2:** 567–573.

Marais R., Spooner R.A., Stribbling S.M., Light Y., Martin J., and Springer C.J. 1997. A cell surface tethered enzyme improves efficiency in gene-directed enzyme prodrug therapy. *Nat. Biotechnol.* **15:** 1373–1377.

Marini III, F.C., Pan B.F., Nelson J.A., and Lapeyre J.N. 1996. The drug verapamil inhibits bystander killing but not cell suicide in thymidine kinase-ganciclovir prodrug-activated gene therapy. *Cancer Gene Ther.* **3:** 405–412.

Maron A., Havaux N., Le Roux A., Knoops B., Perricaudet M., and Octave J.N. 1997. Differential toxicity of ganciclovir for rat neurons and astrocytes in primary culture following adenovirus-mediated transfer of the HSVtk gene. *Gene Ther.* **4:** 25–31.

Mavroudis D. and Barrett J. 1996. The graft-versus-leukemia effect. *Cur.r Opin. Hematol.* **3:** 423–429.

Mehta P.P. and Lowenstein W.R. 1991. Differential regulation of communication by retinoic acid in homologous and heterologous junctions between normal and trans-formed cells. *J. Cell Biol.* **113:** 371–379.

Mesnil M., Piccoli C., and Yamasaki H. 1997. A tumor suppressor gene, Cx26, also medi-ates the bystander effect in HeLa cells. *Cancer Res.* **57:** 2929–2932.

Mesnil M., Piccoli C., Tiraby G., Willecke K., and Yamasaki H. 1996. Bystander killing of cancer cells by herpes simplex virus thymidine kinase gene is mediated by connex-ins. *Proc. Natl. Acad. Sci.* **93:** 1831–1835.

Michael S.I., Hong J.S., Curiel D.T., and Engler J.A. 1995. Addition of a short peptide lig-and to the adenovirus fiber protein. *Gene Ther.* **2:** 660–668.

Miller A.D. 1990. Retrovirus packaging cells. *Hum. Gene Ther.* **1:** 5–14.

Miller A.D. and Buttimore C. 1986. Redesign of retrovirus packaging cell lines to avoid recombination leading to helper virus production. *Mol. Cell. Biol.* **6:** 2895–2902.

Miller A.D. and Rosman G.J. 1989. Improved retroviral vectors for gene transfer and expression. *BioTechniques* **7:** 980–982.

Miller A.D., Garcia J.V., von Suhr N., Lynch C.M., Wilson C., and Eiden M.V. 1991. Construction and properties of retrovirus packaging cells based on gibbon ape leukemia virus. *J. Virol.* **65:** 2220–2224.

Miller N. and Vile R.. 1995. Targeted vectors for gene therapy. *FASEB J.* **9:** 190–199.

Miller W.H. and Miller R.L. 1980. Phosphorylation of acyclovir (acycloguanosine) monophosphate by GMP kinase. *J. Biol. Chem.* **255:** 7204–7207.

———. 1982. Phosphorylation of acyclovir diphosphate by cellular enzymes. *Biochem. Pharmacol.* **31:** 3879–3884.

Miyatake S., Martuza R.L., and Rabkin S.D. 1997. Defective herpes simplex virus vectors expressing thymidine kinase for the treatment of malignant glioma. *Cancer Gene Ther.* **4:** 222–228.

Moloney J.B. 1960. Biological studies on a lymphoid-leukemia virus extracted from sarcoma 37. I. Origin and introductory investigation. *J. Natl. Cancer Inst.* **24:** 933–951.

Moolten F.L. 1986. Tumor chemosensitivity conferred by inserted herpes thymidine kinase genes: paradigm for a prospective cancer control strategy. *Cancer Res.* **46:** 5276 5281.

Moolten F.L. and Wells J.M. 1990. Curability of tumors bearing herpes thymidine kinase genes transferred by retroviral vectors. *J. Natl. Cancer Inst.* **82:** 297–300.

Moolten F.L., Wells J.M., Heyman R.A., and Evans R.M. 1990. Lymphoma regression induced by ganciclovir in mice bearing a herpes thymidine kinase transgene. *Hum. Gene Ther.* **1:** 125–134.

Moolten F.S., Wells J.M., and Mroz P.J. 1992. Multiple transduction as a means of preserving ganciclovir chemosensitivity in sarcoma cells carrying retrovirally transduced herpes thymidine kinase genes. *Cancer Lett.* **64:** 257–263.

Moore M.J. 1991. Clinical pharmacokinetics of cyclophosphamide. *Clin. Pharmacokinet.* **20:** 194–208.

Mullen C.A., Kilstrup M., and Blaese R.M. 1992. Transfer of the bacterial gene for cytosine deaminase to mammalian cells confers lethal sensitivity to 5-fluorocytosine: A negative selection system. *Proc. Natl. Acad. Sci.* **89:** 33–37.

Mullen C.A., Coale M.M., Lowe R., and Blaese R.M. 1994. Tumors expressing the cytosine deaminase suicide gene can be eliminated *in vivo* with 5-fluorocytosine and induce protective immunity to wild-type tumor. *Cancer Res.* **54:** 1503–1508.

Myhr B.C. and DiPaolo J.A. 1975. Requirement for cell dispersion prior to selection on induced azaguanine-resistent colonies of Chinese hamster cells. *Genetics* **80:** 157–169.

Nishiyama T., Kawamura Y., Kawamoto K., Matsumura H., Yamamoto N., Ito T., Ohyama A., Katsuragi T., and Sakai T. 1985. Antineoplastic effects in rats of 5-fluorocytosine in combination with cytosine deaminase capsules. *Cancer Res.* **45:** 1753–1761.

Norrby E. 1983. Viral vaccines: The use of currently available products and future developments. *Arch. Virol.* **76:** 163–177.

Ono Y., Ikeda K., Wei M.X., Harsh IV, G.R., Tamiya T., and Chiocca E.A. 1997. Regression of experimental brain tumors with 6-thioxanthine and *Escherichia coli gpt* gene therapy. *Hum. Gene Ther.* **8:** 2043–2055.

Papadopoulus E.B., Ladanyi M., Emanuel D., Mackinnon S., Boulad F., Carabasi M.H., Castro-Malaspina H., Childs B.H., Gillio A.P., Small T.N., Young J.W., Kernan N.A., and O'Reilly R.J. 1994. Infusions of donor leukocytes to treat Epstein-Barr virus-associated lymphoproliferative disorders after allogenic bone marrow transplantation. *N. Engl. J. Med.* **330:** 1185–1191.

Park J.Y., Elshami A.A., Amin K., Rizk N., Kaiser L.R., and Albelda S.M. 1997. Retinoids augment the bystander effect in vitro and in vivo in herpes simplex virus thymidine kinase/ganciclovir-mediated gene therapy. *Gene Ther.* **4:** 909–917.

Parkman R. 1998. Chronic graft-versus-host disease. *Curr. Opin. Hematol.* **5:** 22–25.

Parr M.J., Yoshinobu M., Tanaka T., Wen P., Kufe D.W., Kaelin W.G., and Fine H.A. 1997. Tumor-selective transgene expression *in vivo* mediated by an E2F-responsive adenoviral vector. *Nat. Med.* **3:** 1145–1149.

Patterson A.V., Zhang H., Moghaddam A., Bicknell R., Talbot D.C., Stratford I.J., and Harris A.L.1995. Increased sensitivity to the prodrug 5′-deoxy-5-fluorouridine and modulation of 5-fluoro-2′-deoxyuridine sensitivity in MCF-7 cells transfected with thymidine phosphorylase. *Br. J. Cancer* **72:** 669–675.

Pavlovic J., Nawrath M., Tu R., Heinicke T., and Moelling K. 1996. Anti-tumor immunity is involved in the thymidine kinase-mediated killing of tumors induced by activated Ki-ras(G12V). *Gene Ther.* **3:** 635–643.

Plagemenn P.G.W., Woffendin C., Puziss M.B., and Wohlheuter R.M. 1987. Purine and pyrimidine transport and permeation in human erythrocytes. *Biochim. Biophys. Acta* **905:** 17–29.

Poiesz B.J., Ruscetti F.W., Gazdar A.F., Bunn P.A., Minna J.D., and Gallo R.C. 1980. Detection and isolation of type C retrovirus particles from fresh and cultured lymphocytes of a patient with cutaneous T-cell lymphoma. *Proc. Natl. Acad. Sci.* **77:** 7415–7419.

Polak, A. and Scholer, H.J. 1975. Mode of action of 5-fluorocytosine and mechanisms of resistance. *Chemotherapy* **21:** 113–130.

Polak A., Eschenhof E., Fernex M., and Scholer H.J. 1976. Metabolic studies with 5-fluorocytosine-14-C in mouse, rat, rabbit, dog and man. *Pharmacology* **22:** 137-153.

Preisler H.D., Rustum Y., and Priore R.L. 1985. Relationship between leukemic cell retention of cytosine arabinoside triphosphate and the duration of remission un patients with acute nonlymphocytic leukemia. *Eur. J. Cancer Clin. Oncol.* **21:** 23–30.

Rainov N.G., Dobberstein K.-U., Sena-Esteves M., Herrlinger U., Kramm C.M., Philpot R.M., Hilton J., Chiocca E.A., and Breakefield X.O. 1998. New prodrug activation gene therapy for cancer using cytochrome P450 4B1 and 2-aminoanthracene/4-ipomeanol. *Hum. Gene Ther.* **9:** 1261–1273.

Ram Z., Culver K.W., Walbridge S., Blaese R.M., and Oldfield E.H. 1993a. *In situ* retroviral-mediated gene transfer for the treatment of brain tumors in rats. *Cancer Res.* **53:** 83–88.

Ram Z., Culver K.W., Walbridge S., Frank J.A., Blaese R.M., and Oldfield E.H. 1993b. Toxicity studies of retroviral-mediated gene transfer for the treatment of brain tumors. *J. Neurosurg.* **79:** 400–407.

Ram Z., Wallbridge S., Shawker T., Culver K.W., Blaese R.M., and Oldfield E.H. 1994. The effect of thymidine kinase transduction and ganciclovir therapy on tumor vasculature and growth of 9L gliomas in rats. *J. Neurosurg.* **81:** 256–260.

Ram Z., Culver K.W., Oshiro E.M., Viola J.J., DeVroom H.L., Otto E., Long Z., Chiang Y., McGarrity G.J., Muul L.M., Katz D., Blaese R.M., and Oldfield E.H. 1997. Therapy of malignant brain tumors by intratumoral implantation of retroviral vector-producing cells. *Nat. Med.* **3:** 1354–1361.

Ramesh R., Marrogi A.J., Munshi A., Abboud C.N., and Freeman S.M. 1996a. In vivo analysis of the 'bystander effect': a cytokine cascade. *Exp. Hematol.* **24:** 829–838.

Ramesh R., Munshi A., Abboud C.N., Marrogi A.J., and Freeman S.M. 1996b. Expression of costimulatory molecules: B7 and ICAM up-regulation after treatment with a suicide gene. *Cancer Gene Ther.* **3:** 373–384.

Reardon J.E. and Spector T. 1989. Herpes simplex virus type 1 DNA polymerase: Mechanism of inhibition by acyclovir triphosphate. *J. Biol. Chem.* **264:** 7405–7411.

Riddell S.R., Elliott M., Lewinsohn D.A., Gilbert M.J., Wilson L., Manley S.A., Lupton S.D., Overell R.W., Reynolds T.C., Corey L., and Greenberg P.D. 1996. T-cell mediated rejection of gene-modified HIV-specific cytotoxic T lymphocytes in HIV-infected patients (see comments). *Nat. Med.* **2:** 216–223.

Rosenberg S.A., Blaese R.M., Brenner M.K., Deisseroth A.B., Ledley F.D., Lotze M.T.,

Wilson J.M., Nabel G.J., Cornetta K., Economou J.S., Freeman S.M., Riddell S.R., Oldfield E., Gansbacher B., Dunbar C., Walker R.E., Schuening F.G., Roth J.A., Crystal R.G., Welsh M.J., Culver K., Heslop H.E., Simons J., Wilmott R.W., Boucher R.C., and et al. 1997. Human gene marker/therapy clinical protocols. *Hum. Gene Ther.* **8:** 2301–2338.

Rous, P. 1911. A sarcoma of fowl transmissible by an agent separable from tumor cells. *J. Exp. Med.* **13:** 397–411.

Rowley S., Lindauer M., Gebert J.F., Haberkorn U., Oberdorfer F., Moebius U., Herfarth C., and Schackert H.K. 1996. Cytosine deaminase gene as a potential tool for the genetic therapy of colorectal cancer. *J. Surg. Oncol.* **61:** 42–48.

Rowley S., Zuehlsdorf M., Braine H., Colvin O.M., Davis J., Jones R.J., Saral R., Sensenbrenner L.L., Yeager A., and Santon G.W. 1987. CFU-GM content of bone marrow graft correlates with time to hematologic reconstitution following autologous bone marrow transplantation with 4-HC purged bone marrow. *Blood* **70:** 271–275.

Ruch R.J., Madhukar B.V., Trosko J.E., and Klauning J.E. 1993. Reversal of *ras*-induced inhibition of gap-junctional intercellular communication, transformation, and tumorigenesis by lovostatin. *Mol. Carcinog.* **7:** 50–59.

Russell S.J. 1994. Replicating vectors for gene therapy of cancer: Risks, limitations and prospects. *Eur. J. Cancer.* **30A:** 1165–1171.

Samejima Y. and Meruelo D. 1995. 'Bystander killing' induces apoptosis and is inhibited by forskolin. *Gene Ther.* **2:** 50–58.

Santodonato L., D'Agostino G., Santini S.M., Carlei D., Musiani P., Modesti A., Signorelli P., F. Belardelli F., and Ferrantini M. 1997. Local and systemic antitumor response after combined therapy of mouse metastatic tumors with tumor cells expressing IFN-α and HSVtk: Perspectives for the generation of cancer vaccines. *Gene Ther.* **4:** 1246–1255.

Satoh T. and Hosokawa M. 1995. Molecular aspects of carboxylesterase isoforms in comparison to other esterases. *Toxicol. Lett.* **82/83:** 439–435.

Satoh T., Hosokawa M., Atsumi R., Suzuki W., Hakusui H., and Nagai E. 1994. Metabolic activation of CPT-11, 7-ethyl-10-[4-(1-piperidino)-1-piperidino]carbonyloxycamptothecin, a novel antitumor agent, by carboxylesterase. *Biol. Pharm. Bull.* **17:** 662–664.

Sawyer M.H., Inchauspe G., Biron K.K., Waters D.J., Straus S.E., and Ostrove J.M. 1988. Molecular analysis of the pyrimidine deoxyribonucleoside kinase gene of wild-type and acyclovir-resistant strains of varicella-zoster virus. *J. Gen. Virol.* **69:** 2585–2593.

Schaeffer H.J., Beauchamp L., de Miranda P., Elion G.B., Bauer D.J., and Collins P. 1978. 9-(2-hydroxyethoxymethyl) guanine activity against viruses of the herpes group. *Nature* **272:** 583–585.

Shah M.R. 1997. "Tumor immunity following adenovirus mediated herpes simplex thymidine kinase gene transfer to experimental rat gliomas." Ph.D. thesis, Department of Anatomy. Medical College of Virginia, Virginia Commonwealth University, Richmond.

Shapiro W.R. and Shapiro J.R. 1998. Biology and treatment of malignant glioma. *Oncology* **12:** 233–240.

Shewach D.S., Zerbe L.K., Hughes T.L., Roessler B.J.,. Breakefield X.O, and Davidson B.L. 1994. Enhanced cytotoxicity of antiviral drugs mediated by adenovirus directed transfer of the herpes simplex virus thymidine kinase gene in rat glioma cells. *Cancer Gene Ther.* **1:** 107–112.

Shine H.D. and Woo S.L.C. 1996. *Adenovirus-mediated gene therapy of tumors in the central nervous system.* Academic Press, London.

Shinoura N., Chen L., Wani M.A., Kim Y.G., Larson J.J., Warnick R.E., Simon M., Menon A.G., Bi W.L., and Stambrook P.J. 1996. Protein and messenger RNA expression of connexin43 in astrocytomas: Implications in brain tumor gene therapy. *J Neurosurg.* **84:** 839–845.

Short M.P., Choi B.C., Lee J.K., Malick A., Breakefield X.O., and Martuza R.L. 1990. Gene delivery to glioma cells in rat brain by grafting of a retrovirus packaging cell line. *J. Neurosci. Res.* **27:** 427–439.

Sierra J. and Anasetti C. 1995. Marrow transplantation from unrelated donors. *Curr. Opin. Hematol.* **2:** 444–451.

Smith R.R., Huebner R.J., Rowe W.P., Schatten W.E., and Thomas L.B. 1956. Studies on the use of viruses in the treatment of carcinoma of the cervix. *Cancer* **9:** 1211–1218.

Smythe W.R., Hwang H.C., Amin K.M., Eck S.L., Davidson B.L., Wilson J.M., Kaiser L.R., and Albelda S.M. 1994. Use of recombinant adenovirus to transfer the herpes simplex virus thymidine kinase (HSV*tk*) gene to thoracic neoplasms: An effective *in vitro* drug sensitization system. *Cancer Res.* **54:** 2055–2059.

Smythe W.R., Hwang H.C., Elshami A.A., Amin K.M., Eck S.L., Davidson B.L., Wilson J.M., Kaiser L.R., and Albelda S.M. 1995. Treatment of experimental human mesothelioma using adenovirus transfer of the herpes simplex thymidine kinase gene. *Ann. Surg.* **222:** 78–86.

Sorcher E.J., Peng S., Bebok Z., Allan P.W., Bennett L.L.J., and Parker W.B. 1994. Tumor cell killing in colonic carcinoma utilizing the *E. coli DeoD* gene to generate toxic purines. *Gene Ther.* **1:** 233–238.

Southam C. M. 1960. Present status of oncolytic virus studies. *Trans. N.Y. Acad. Sci.* **22:** 657–673.

St. Clair M.H., Lambe C.U., and Furman P.A. 1987. Inhibition by ganciclovir of cell growth and DNA synthesis of cells biochemically transformed with herpesvirus genetic information. *Antimicrob. Agents Chemother.* **31:** 844–849.

Sterman D.H., Treat J., Litzky L.A., Amin K.M., Coonrod L., Molnar-Kimber K., Recio A., Knox L., Wilson J.M., Albelda S.M., and Kaiser L.R. 1998. Adenovirus-mediated Herpes simplex virus thymidine kinase/ganciclovir gene therapy in patients with localized malignancy: Results of a phase I clinical trial in malignant mesothelioma. *Hum. Gene Ther.* **9:** 1083–1089.

Stewart E.S. and Cohen D.G. 1998. Central nervous system tumors in children. *Semin. Oncol. Nurs.* **14:** 34–42.

Su H., Lu R., Chang J.C., and Kan Y.W. 1997. Tissue-specific expression of herpes simplex virus thymidine kinase gene delivered by adeno-associated virus inhibits the growth of human hepatocellular carcinoma in athymic mice. *Proc. Natl. Acad. Sci.* **94:** 13891–13896.

Subak-Sharpe J.H., Burk R.R., and Pitts J.D. 1969. Metabolic cooperation between biochemically marked mammalian cells in culture. *J. Cell Sci.* **4:** 353–367.

Szer J., Grigg A.P., Phillips G.L., and Sheridan W.P. 1993. Donor leukocyte infusions after chemotherapy for patients relapsing with acute leukemia following allogenic BMT. *Bone Marrow Transplant.* **11:** 109–111.

Takakuwa K., Fujita K., Kikuchi A., Sugaya S., Yahata T., Aida H., Kurabayashi T., Hasegawa I., and Tanaka K. 1997. Direct intratumoral gene transfer of the herpes simplex virus thymidine kinase gene with DNA-liposome complexes: growth inhibition of tumors and lack of localization in normal tissues. *Jpn. J. Cancer Res.* **88:** 166–175.

Takamiya Y., Short M.P., Ezzeddine Z.D., Moolten F.L., Breakefield X.O., and Martuza

R.L. 1992. Gene therapy of malignant brain tumors: A rat glioma line bearing the herpes simplex virus type 1-thymidine kinase gene and wild type retrovirus kills other tumor cells. *J. Neurosci. Res.* **33:** 493–503.

Tamiya T., Ono Y., Wei M.X., Mroz P.J., Moolten F.L., and Chiocca E.A. 1996. *Escherichia coli* gpt gene sensitizes rat glioma cells to killing by 6-thioxanthine or 6-thioguanine. *Cancer Gene Ther.* **3:** 155–162.

Tanaka M. and Yoshida S. 1987. Formation of cytosine arabinoside-5´-triphosphate in cultured human leukemic cell lines correlates with nucleoside transport capacity. *Jpn. J. Cancer Res.* **78:** 851–857.

Tanaka T., Kanai F., Okabe S., Yoshida Y., Wakimoto H., Hamada H., Shiratori Y., Lan K., Ishitobi M., and Omata M. 1996. Adenovirus-mediated prodrug gene therapy for carcinoembryonic antigen-producing human gastric carcinoma cells in vitro. *Cancer Res.* **56:** 1341–1345.

Thurst R., Schacke M., and Wutzler P. 1996. Cytogentic genotoxicity of the antiherpes virostatics in Chinese hamster V79-E cells. I. Purine nucleoside analogues. *Antivi. Res.* **31:** 105–113.

Treat J., Kaiser L.R., Sterman D.H., Litzky L., Davis A., Wilson J.M., and Albelda S.M. 1996. Treatment of advanced mesothelioma with the recombinant adenovirus H5.010RSVTK: A phase 1 trial (BB-IND 6274). *Hum. Gene Ther.* **7:** 2047–2057.

Trinh Q.T., Austin E.A., Murray D.M., Knick V.C., and Huber B.E. 1995. Enzyme/prodrug gene therapy: Comparison of cytosine deaminase/5-fluorocytosine versus thymidine kinase/ganciclovir enzyme/prodrug systems in a human colorectal carcinoma cell line. *Cancer Res.* **55:** 4808–4812.

Uckert W., Kammertons T., Haack K., Qin Z., Gebert J., Schendel D.J., and Blankenstein T. 1998. Double suicide gene (cytosine deaminase and herpes simplex virus thymidine kinase) but not single gene transfer allows reliable elimination of tumors cells *in vivo*. *Hum. Gene Ther.* **9:** 855–866.

van Rhee F. and Kolb H.J. 1995. Donor leukocyte transfusions for leukemic relapse. *Curr. Opin. Hematol.* **2:** 423–430.

Vile R.G., Nelson J.A., Casteldon S., Chong H., and Hart I.R. 1994. Systemic gene therapy of murine melanoma using tissue specific expression of the HSV-TK gene involves an immune component. *Cancer Res.* **54:** 6228–6234.

Vrionis F.D., Wu J.K., Qi P., Waltzman M., Cherington V., and Spray D.C. 1997. The bystander effect exerted by tumor cells expressing the herpes simplex virus thymidine kinase (HSVtk) gene is dependent on connexin expression and cell communication via gap junctions. *Gene Ther.* **4:** 577–585.

Weiden P.L., Flournoy N., Thomas E.D., Prentice R., Fefer A., Buckner C.D., and Storb R. 1979. Antileukemic effect of graft-versus-host disease in human recipients of allogeneic-marrow grafts. *N. Engl. J. Med.* **300:** 1068–1073.

West T.P., Shanley M.S., and O'Donovan G.A. 1982. Purification and some properties of cytosine deaminase from *Salmonella typhimurium. Biochim. Biophys. Acta* **719:** 251–258.

Wickham T.J., Carrion M.E., and Kovesdi I. 1995. Targeting of adenovirus penton base to new receptors through replacement of its RGD motif with other receptor-specific peptide motifs. *Gene Ther.* **2:** 750–756.

Wickham T.J., Roelvink P.W., Brough D.E., and Kovesdi I. 1996. Adenovirus targeted to heparin-containing receptors increases its gene delivery efficiency to multiple cell types. *Nat. Biotechnol.* **14:** 1570–1573.

Wiley J.S., Taupin J., Jamieson G.P., Snook M., Sawyer W.H., and Finch L.R. 1985.

Cytosine arabinoside transport and metabolism in acute leukemias and T cell lymphoblastic lymphoma. *J. Clin. Invest.* **75:** 632–642.

Wilson J.M. 1997. Vectors—Shuttle vehicles for gene therapy. *Clin. Exp. Immunol.* **1:** 31–32.

Wilson K.M., Stambrook P.J., Bi W.-L., Pavelic Z.P., Pavelic L., and Gluckman J.L. 1996. HSV-tk gene therapy in head and neck squamous cell carcinoma: enhnacement by the local and distant bystander effect. *Arch. Oto-Rhino-Larygol.* **122:** 746–749.

Yamamoto S., Suzuki S., Hoshino A., Akimoto M., and Shimada T. 1997. Herpes simplex virus thymidine kinase/ganciclovir-mediated killing of tumor cell induces tumor-specific cytotoxic T cells in mice. *Cancer Gene Ther.* **4:** 91–96.

Yang Y. and Wilson J.M. 1995. Clearance of adenovirus-infected hepatocytes by MHC class I-restricted CD4+ CTLs in vivo. *J. Immunol.* **155:** 2564–2570.

Yang Y., Jooss K.U., Su Q., Ertl H.C., and Wilson J.M. 1996. Immune responses to viral antigens versus transgene product in elimination of recombinant adenovirus-infected hepatocytes in vivo. *Gene Ther.* **3:** 137–144.

Zhang J. and Russell S.J. 1996. Vectors for cancer gene therapy. *Cancer Metastasis Rev.* **15:** 385–401.

19

The Logic of Anti-angiogenic Gene Therapy

Judah Folkman, Philip Hahnfeldt,* and Lynn Hlatky*

Department of Surgery, Children's Hospital
*Joint Center for Radiation Therapy and
Dana-Farber Cancer Institute
Harvard Medical School
Boston, Massachusetts 02115

Virtually all current strategies for gene therapy of cancer target the cancer cell. The goal is the destruction of cancer cells in the primary tumor and in its metastases. These approaches include, among others, the introduction of genes that (1) permit tumor cells to express toxic molecules; (2) prevent or correct genetic defects; (3) increase the immunogenicity of tumor cells; or (4) increase the sensitivity of tumor cells to drugs (Friedmann 1996; Blaese 1997; Lowenstein 1997; Roth and Cristiano 1997). Although gene therapy of cancer may be inherently less toxic than conventional chemotherapy, it may still have to overcome at least three other fundamental obstacles that hinder conventional chemotherapy, i.e., limited access to tumor cells, heterogeneity of tumor cells, and emergence of resistant tumor cells. Whereas conventional chemotherapy is directed mainly against tumor cells, anti-angiogenic therapy is directed specifically against microvascular endothelial cells that have been recruited into the tumor bed. In contrast to chemotherapy, specific anti-angiogenic therapy has little or no toxicity, does not require that the therapeutic agent enter any tumor cells nor cross the blood brain barrier, acts independent of tumor cell heterogeneity, and does not induce acquired drug resistance (Boehm et al. 1997). In addition, the effectiveness of anti-angiogenic therapy is independent of tumor type and growth fraction. Therefore, it may be fruitful to consider gene therapy of cancer that is optimized to the endothelial cell population of the tumor bed, e.g., anti-angiogenic gene therapy. We do not view anti-angiogenic gene therapy and anticancer cell gene therapy as mutually exclusive. They may be employed together or separately.

The Development of Human Gene Therapy ©1999 Cold Spring Harbor Laboratory Press 0-87969-528-5/99 **527**

BIOLOGIC BASIS OF ANTI-ANGIOGENIC THERAPY

The Switch to the Angiogenic Phenotype

Most human and animal tumors are not initially able to recruit new cap-illary vessels from the host. Angiogenic capacity appears to emerge rather suddenly during the multistage progression to neoplasia (Folkman et al. 1989; Hanahan and Folkman 1996). Angiogenic activity can occur before or after the onset of frank malignancy. For example, in carcinoma of the cervix, angiogenic activity first appears in the preneoplastic stage of dys-plasia (Smith McCune and Weidner 1994), whereas in many other tumors (e.g., breast carcinoma, melanoma, and prostate carcinoma), the angio-genic phenotype appears after the neoplastic stage.

In either event, before the onset of neovascularization, expansion of tumor mass is absolutely restricted to a microscopic size or to a barely visible in situ lesion. After a tumor has become neovascularized, contin-uous expansion of tumor mass becomes possible and tumor cells can shed into the circulation.

The angiogenic "switch" itself has been characterized as a shift in the net balance of stimulators and inhibitors of angiogenesis (Rastinejad et al. 1989; Bouck 1990; O'Reilly et al. 1994, 1996). The positive regulators of angiogenesis include at least 14 angiogenic proteins (for review, see Folkman 1996). Basic fibroblast growth factor (bFGF) and vascular endothelial growth factor (VEGF) are the most studied and are found in a majority of different types of human tumors. During the angiogenic switch, one or more of these angiogenic factors may be overexpressed (Kandel et al. 1991), and/or mobilized from extracellular matrix, (Folkman et al. 1988), and/or released from macrophages recruited by the tumor (Polverini and Leibovich 1984). It appears, however, that this up-regulation of angiogenic stimulators is accompanied by down-regulation of local tissue inhibitors of angiogenesis. An example of such an inhibitor is thrombospondin, produced by normal human fibroblasts under the con-trol of wild-type p53 (Rastinejad et al. 1989; Dameron et al. 1994). Thrombospondin is down-regulated by about 96% when fibroblasts from cancer-prone Li-Fraumeni patients (which already lack one allele of p53) lose or mutate the second p53 allele during progression to malignancy. Additional evidence suggests that certain tumor suppressor genes may normally code for proteins that inhibit angiogenesis (Van Meir et al. 1994) and that angiogenic factors may be under the control of specific oncogenes (Rak et al. 1995).

Morphologic Interaction between Tumor Cells and Endothelial Cells in a Tumor

To understand how microvascular endothelial cells control tumor mass and how inhibition of capillary blood-vessel growth can bring about tumor regression in experimental animals, it is necessary to appreciate the unique three-dimensional configuration by which tumor cells and endothelial cells co-exist in a tumor. Once angiogenesis has been initiated, capillary sprouts and loops converge upon a microscopic or an in situ tumor. The vascular sprouts lack smooth muscle. Tumor cells form perivascular cuffs of approximately 4–5 cell layers radius, but this can vary among different tumors. This configuration can be observed in confocal microscopic reconstructions of human breast cancer (Folkman 1997) (Figs. 1 and 2). In normal tissue, most parenchymal cells, such as in the liver, form only a single cell layer around each capillary. A tumor of 1 g contains approximately 10^8 to 10^9 tumor cells and approximately 2

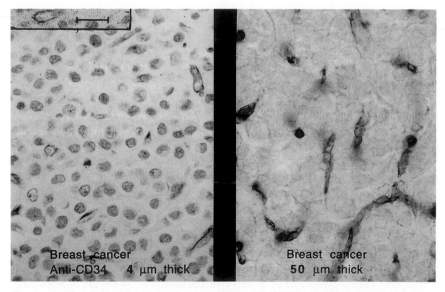

Figure 1 Histology of human breast cancer. (*Left*) Histologic section of human breast cancer, 4 μm thick, stained with antibody to CD34 to highlight blood vessels. (*Right*) Same section, but 50 μm thick–confocal microscopy. Capillary vessels now project in all directions. Although the tumor cells go out of focus in this thick section, it is possible to see that each vessel is surrounded by a microcylinder of three or four layers of tumor cells. (Reprinted, with permission, from Folkman 1997.)

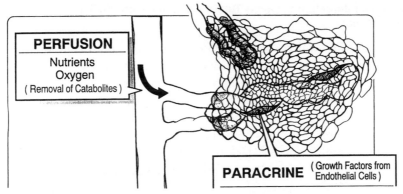

Figure 2 Diagram of perfusion and paracrine effects of tumor neovascularization (Folkman 1997). Vascular endothelial cells produce up to 20 growth factors and survival factors that may drive tumor cell growth (Rak et al. 1996).

$\times 10^7$ endothelial cells (Modzelewski et al. 1994). Therefore, approximately 5–50 or more tumor cells may be supported by the length of capillary vessel occupied by a single endothelial cell.

Amplification of Tumor Cell Killing by Targeting Endothelial Cells

The anatomic configuration in which one endothelial cell appears to support multiple tumor cells reveals a possible basis for amplified arrest or killing of tumor cells when anti-angiogenic therapy is directed specifically against endothelial cells. Therefore, if the growth of one endothelial cell is arrested, the expansion of the cylindrical mass of tumor cells surrounding this vascular segment may also be restricted. If this endothelial cell undergoes apoptosis, its associated microcylinder of tumor cells may die. This endothelial control point of tumor growth is demonstrated in animal tumors that secrete VEGF under the control of a tetracycline promoter (Benjamin and Keshet 1997). After the tumors are large enough to be palpable, VEGF production is shut off. This is followed by a wave of endothelial apoptosis and then by tumor cell apoptosis. In our own experiments, where large tumors (1–2% of body weight) were treated with the specific angiogenesis inhibitor endostatin (Boehm et al. 1997), a wave of endothelial cell apoptosis occurred first and was followed approximately 3 days later by a peak frequency of tumor cell apoptosis (J. Folkman et al., unpubl.). Therefore, gene therapy targeted to the angiogenic process may be significantly amplified, in contrast to gene therapy targeted only to tumor cells. Both therapeutic approaches may benefit from the bystander effect (Xu et al. 1997).

Molecular Interaction between Tumor Cells and Endothelial Cells in a Tumor

The molecular interaction between endothelial cells in microvessels and the tumor cells that surround these vessels is bidirectional. Tumor cells drive endothelial cells by releasing angiogenic proteins. Endothelial cells stimulate tumor growth by providing growth factors and survival factors to tumor cells from the blood (*perfusion* effect) or directly from endothelial cells (*paracrine* effect). This working model of the two major interacting cell populations in a tumor provides a framework for the development of angiogenesis inhibitors as well as for the design of anti-angiogenic gene therapy (Fig. 3).

The Perfusion Effect

Once new capillary loops converge toward a small in situ carcinoma or a microscopic metastasis, the tumor cells are bathed in additional survival factors and growth factors from the circulating blood. The perfusion effect of tumor neovascularization temporarily solves the problem of exchange of nutrients, oxygen, and catabolites in a crowded three-dimensional cell population for which simple diffusion of such molecules across the outer surface of the tiny tumor mass has become inadequate (Folkman et al. 1974; Folkman 1975).

The Paracrine Effect

In addition to their perfusion effect, vascular endothelial cells also produce paracrine survival factors and growth factors for tumor cells. These endothelial-derived paracrine factors include bFGF, platelet-derived growth factor (PDGF-BB), and insulin-like growth factor-1 (IGF-1) among a total of at least 20 such factors (Rak et al. 1996). This paracrine effect can be observed directly by an in vitro experiment in which tumor cells preferentially grow apposed to capillary endothelial tubes, despite the absence of blood flow (Nicosia et al. 1984; Nicosia and Ottinett 1990). If anti-angiogenic therapy arrests the growth of the intratumoral endothelial cell mass or causes this cell population to regress, growth factors and/or survival factors may decrease. In animal tumors (O'Reilly et al. 1994, 1996, 1997; Holmgren et al. 1995) and in human tumors (Lu and Tanigawa 1997), the induction of angiogenesis correlates with at least a sevenfold decrease in apoptosis. In contrast, the proliferation rate of tumor cells remains equally high in the prevascular as well as in the vas-

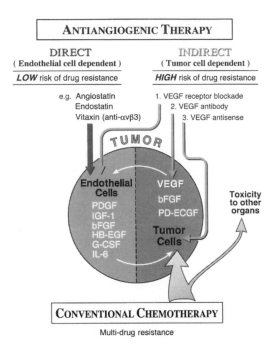

Figure 3 Diagram of the endothelial cell compartment and the tumor-cell compartment in a tumor. Each cell population stimulates growth of the other by releasing mitogens and survival factors. This figure also illustrates the critical difference between "direct" and "indirect" anti-angiogenic therapy in terms of the risk of inducing acquired drug resistance. "Direct" anti-angiogenic therapy, which specifically targets endothelial cells, has a low risk of inducing drug resistance. In contrast, "indirect" anti-angiogenic therapy which depends on blockade of tumor-derived angiogenic activity, has a higher risk of inducing drug resistance because tumor cells may eventually appear that release a different angiogenic factor. For example, a tumor that is producing mainly VEGF may eventually generate a subpopulation of tumor cells that produces bFGF. Conventional chemotherapy, which targets mainly tumor cells, almost always induces drug resistance.

cularized tumor (up to 40% by BrdU). This pattern is observed in a wide range of different tumor types. It indicates that tumor expansion after neovascularization may depend mainly on restriction of tumor cell apoptosis. Analogous systems are found in *myc*-dependent lymphoma cells or in fibroblasts with dysregulated *myc* expression. Both cell types grow in the presence of IGF-1 or IGF-2, respectively, but undergo apoptosis when these factors are withdrawn (Evans 1994). It is not clear whether reduction of tumor growth by blocked angiogenesis may also depend on a mechanism by which endothelial cells are activated to lyse tumor cells directly (Li et al. 1991). Also, it remains to be shown whether angiostatin or endostatin therapy can suppress endothelial production of paracrine growth or survival factors for tumor cells.

LACK OF ACQUIRED DRUG RESISTANCE WITH ANTI-ANGIOGENIC THERAPY

The genetic instability of cancer cells and their relatively high mutation rate often leads to selection of resistant cells under the pressure of a wide array of chemotherapeutic approaches. Endothelial cells, like other normal cells, have a relatively low mutation rate (approximately 10^{-9} per cell division versus 10^{-3} per cell division for tumor cells). Therefore, it is less likely that drug-resistant endothelial cells will be selected from the homogeneous parent endothelial population. In preliminary studies, drug resistance was not inducible in mice bearing Lewis lung carcinomas treated with TNP-470, a selective angiogenesis inhibitor that inhibits tumor growth (Brem et al. 1994). Cancer patients treated with TNP-470 in clinical trials have to date not developed drug resistance (J. Folkman, unpubl.). In 1991, Kerbel proposed a hypothesis that anti-angiogenic therapy could be a strategy to circumvent acquired drug resistance (Kerbel 1991). However, the angiogenesis inhibitors available at that time were not sufficiently potent to test Kerbel's hypothesis in a compelling manner, because these inhibitors (e.g., TNP-470, interferon α-2a, platelet factor 4, and others) could slow tumor growth in animals but could not regress tumors. The discovery of angiostatin (O'Reilly et al. 1994) and endostatin (O'Reilly et al. 1997) provided anti-angiogenic therapy that could not only regress large tumors in animals, but also hold them dormant at a microscopic size (O'Reilly et al. 1996) as long as therapy was continued. With these inhibitors, we have recently shown formal proof that anti-angiogenic therapy does not induce acquired drug resistance in tumor-bearing animals (Boehm et al. 1997). In brief, tumors were allowed to grow to large sizes of 1–2% of body weight (200–400 mm^3) in mice. The mice then received daily subcutaneous injections of purified murine recombinant endostatin (from *Escherichia coli*). Endostatin is a 20-kD carboxy-terminal fragment of collagen XVIII that specifically inhibits endothelial proliferation and has no obvious effect on resting endothelial cells nor on a variety of normal, transformed, or neoplastic cells (O'Reilly et al. 1997). When the tumors had regressed to become barely visible but were still detectable microscopically (i.e., after approximately 2–3 weeks of treatment), therapy was discontinued and the tumors were allowed to regrow. Endostatin therapy was resumed when the tumors had again reached 1–2% of body weight (Fig. 4). Lewis lung carcinoma repeatedly regressed during six such treatment cycles over a period of 185 days before it no longer resumed growth after endostatin treatment was discontinued. This is equivalent to more than 17 human years (on the basis that 1 mouse day is equivalent to 35 human days). A murine fibrosarcoma underwent repeated regression during four cycles of endostatin therapy over a period of 160 days before becoming permanently dormant. B16F10 melanoma (obtained from Isaiah Fidler) regressed during two cycles of therapy over a period of 80 days before becoming dormant off therapy. No drug resistance was demonstrated in

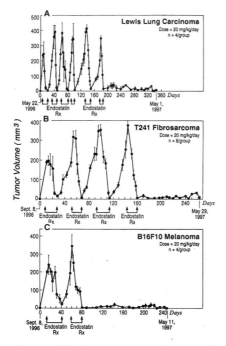

Figure 4 Cycled endostatin therapy of tumors grown in the subcutaneous dorsa of mice. (*A*) Lewis lung carcinoma. When tumor volume reached 250–400 mm^3, endostatin therapy was initiated. Tumors in four untreated animals grew to 10,005 mm^3 ± 131 mm^3 by day 27 (data not graphed). (*B*) T241 fibrosarcoma. Endostatin therapy was initiated when the mean tumor volume was 200 mm^3 and in subsequent cycles when the tumor volume was greater than 300 mm^3. Tumors in four untreated animals grew to 9127 ± 1183 mm^3 by day 28. (*C*) B16F10 melanoma (a subclone of cells supplied by Isaiah Fidler, Houston). Endostatin therapy was initiated when the mean tumor volume was 230 mm^3 for the initial cycle and at 350 mm^3 for the second cycle (endostatin 20 mg per kg per day = 400 µg per 20-g mouse). Tumors in four untreated animals grew to 7431 ± 1122 mm^3 by day 41. (Reprinted, with permission, from Boehm et al. 1997 [copyright Macmillan].)

any of these mice. No mice died and there was no detectable toxicity of any kind. All mice gained weight normally during therapy and had normal hair regrowth and normal white and red blood cell counts.

In contrast, drug resistance developed rapidly when Lewis lung tumors of similar size were treated with cytotoxic chemotherapy (Fig. 5). Cyclophosphamide is the most effective single agent against Lewis lung carcinoma, yet few animals are cured (Hill 1975). We treated mice with Lewis lung carcinomas between 100 mm^3 and 500 mm^3 with cyclophosphamide at the maximum tolerated dose of 150 mg/kg every other day for three doses. Tumors remained stable for approximately 13 days and then resumed growth. It was not possible to repeat the chemotherapy in less than 21 days without high mortality due to bone-marrow suppression (Hill 1975). To allow for bone-marrow recovery, repeated cycles of cyclophosphamide were administered every 21 days. Although survival was prolonged, acquired drug resistance developed after the second cycle and the mice died of rapidly progressing tumors despite continued chemotherapy (Fig. 5). The animals lost weight during therapy, and to reduce gastrointestinal dysfunction from cyclophosphamide, they were pretreated with dexamethasone (Elkins-Sinn, New Jersey) and ondansetron (Glaxo-Wellcome, North Carolina) (Boehm et al. 1997).

These experiments illustrate a fundamental difference between therapy directed specifically against the endothelial population in a tumor verus therapy directed mainly against the tumor cell population. With endostatin therapy, there was no drug resistance and the animals did not die of their tumor or of the treatment. With cyclophosphamide therapy, the tumors became drug resistant after only two cycles of therapy and mice died either of their tumor or of the treatment.

POSSIBLE STRATEGIES FOR ANTI-ANGIOGENIC GENE THERAPY

There are only a few experiments with anti-angiogenic gene therapy of cancer directed against endothelial cells (Harris et al. 1994; Fan et al. 1995; Martiny-Baron and Marme 1995; Warren et al. 1995; Ozaki et al. 1996; Plate 1996; Saleh et al. 1996; Zhang and Russell 1996; Tanaka et al. 1997; Xu et al. 1997; Kong and Crystal 1998), in contrast to hundreds of reports on gene therapy directed against cancer cells. This paucity of work is due in part to our limited knowledge about the growth control of vascular endothelial cells. Only in the past few years have we learned how to turn off endothelial proliferation in the face of potent mitogens (e.g., bFGF or VEGF). The receptors for inhibitors such as angiostatin and endostatin are not yet characterized, and their signal transduction

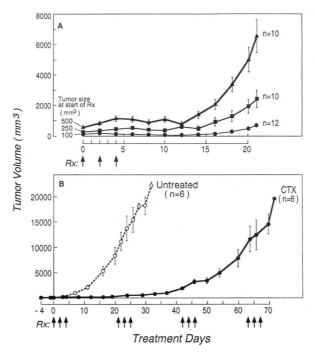

Figure 5 Treatment of Lewis lung carcinoma with cyclophosphamide. (*A*) Cyclophosphamide therapy (150 mg/kg) was initiated when mean tumor volumes reached 100, 250, or 500 mm³ and was repeated every other day for three doses (*arrows*) to a maximum tolerated total dose of 460 mg/kg⁻¹. Tumor volumes are recorded as mean ± S.E.M. (B) Cycled therapy with cyclophosphamide. Mice bearing carcinomas of 100 mm³ were treated with the maximum tolerated dose of cyclophosphamide (CTX). The treatment cycle was repeated every 21 days (*arrows*), the shortest interval of recovery which these mice can tolerate without mortality. Comparable control mice received only saline (*untreated*). Tumor volumes are plotted as mean ± S.E.M (Boeham et al. 1997).

pathways are virtually unknown. Furthermore, the fact that large tumors can be regressed, held dormant at a microscopic size, and even eradicated by angiogenesis inhibitors that are neither toxic nor drug-resistance-inducing has only been demonstrated during the past year (O'Reilly et al. 1996; Boehm et al. 1997). It is not surprising, therefore, that those investigators who specialize in gene therapy may not be fully aware of the potential of anti-angiogenic gene therapy—thus, the reason for this chapter. Because so few preliminary experiments have been carried out with anti-angiogenic gene therapy, it is only possible to speculate about future directions.

Systemic Anti-angiogenic Gene Therapy

In the foreseeable future, anti-angiogenic therapy of cancer may require long-term administration of angiogenesis inhibitors (up to 1 year or more) at frequent intervals (e.g., daily or weekly). If angiostatin or endostatin, for example, could be administered systemically (or locally) by gene therapy at less frequent intervals (e.g., monthly or more), this would be a major advance. It is conceivable that much lower doses of these proteins would be effective if they were continuously released into the bloodstream at low levels, instead of being administered subcutaneously as a large bolus. For example, a Lewis lung carcinoma of 500–1000 mm^3 generates sufficient angiostatin to suppress its metastases as effectively as natural human angiostatin (purified from plasminogen) administered at 12–24 µg per 20-g mouse once per day, by the intraperitoneal route (O'Reilly et al. 1994). The tumor itself probably generates significantly lower quantities than this. Regression of a primary tumor requires recombinant murine angiostatin or endostatin to be administered at up to 400 µg per 20-g mouse per day.

We have carried out preliminary studies in which murine hematopoietic cells have been transduced with retroviral vectors expressing the angiogenesis inhibitors angiostatin and endostatin. Stable retroviral transfers at high viral titers and efficient protein expression/secretion were demonstrated. These cells have been engrafted into the bone marrow (see Pawliuk et al. 1998). This study is designed to determine whether the bone marrow can become a source of circulating anti-angiogenic proteins.

Gene transfer of a cDNA coding for mouse angiostatin into murine T241 fibrosarcoma cells suppressed the growth of primary tumors and their metastases in vivo (Cao et al. 1998). Implantation of stable clones expressing mouse angiostatin in C57BL/6 mice inhibited primary tumor growth by an average of 77%. After removal of primary tumors, the pulmonary micrometastases in about 70% of mice remained in a microscopic dormant state for the duration of the experiments, e.g., 2–5 months. The tumor cells in the dormant micrometastases exhibited a high rate of apoptosis balanced by a high rate of tumor cell proliferation. Transfection of the angiostatin cDNA per se did not inhibit proliferation of tumor cells in vitro.

Anti-angiogenic therapy with angiostatin or endostatin (or both) entails low risk because only growing, migrating vascular endothelial cells in the tumor bed are inhibited. The remaining endothelium in the body, being quiescent, is unaffected.

Local Anti-angiogenic Gene Therapy

Anti-angiogenic gene therapy may also be administered locally or regionally. For example, the cDNA for murine angiostatin has shown efficacy when introduced into mouse brain tumors by a viral vector (Tanaka et al. 1997). We (L. Hlatky et al., unpubl.) have preliminary data in which cDNA for murine endostatin packaged in an adeno-associated virus (AAV) vector has been injected into human glioblastoma xenografts in nude mice, with gradual regression of the tumor after a single intratumoral injection.

Other strategies for anti-angiogenic gene therapy could be based on endothelial targeting, for example, blockade of endothelial receptors for the common angiogenic factors, such as bFGF or VEGF. Alternatively, angiogenic activity of tumor cells could be blocked by preventing or reversing the switch to the angiogenic phenotype or by inducing tumor cells to generate inhibitors such as endostatin. These therapeutic approaches may have reduced toxicity in comparison to conventional chemotherapy. Such strategies must await a more complete understanding of the molecular genetics of the angiogenic phenotype. However, there are subtle but important differences between "direct" anti-angiogenic therapy targeted specifically to endothelial cells (e.g., by angiostatin or endostatin) and "indirect" anti-angiogenic therapy, which is mediated by blockade of production of angiogenic factors by tumors or by interference with the receptor for these angiogenic factors.

DIRECT VERSUS INDIRECT ANTI-ANGIOGENIC THERAPY

"Direct" anti-angiogenic therapy, which specifically targets endothelial cells (e.g., angiostatin, endostatin, and an inhibitor of the $\alpha_v\beta_3$ integrin [vitaxin]), has a low risk of inducing drug resistance (Fig. 3). In contrast, "indirect" anti-angiogenic therapy, which depends on blockade of tumor-derived angiogenic activity may have a higher risk of inducing drug resistance because tumor cells may eventually arise that release a different angiogenic factor. For example, a tumor that is producing mainly VEGF may eventually generate a subpopulation of tumor cells that produces bFGF (R. Jain, pers. comm.). Anti-angiogenic therapy that is based on VEGF receptor blockade or antibody to VEGF, or VEGF antisense technology, may entail the risk of eventual acquired resistance. This risk is theoretical at this writing and cannot be predicted. It should not, however, preclude clinical trials of gene therapy directed at tumor-derived angiogenic factors. These may be valuable for anti-angiogenic therapy of tumors before surgery, induction of regression of primary tumors and their metastases, combination therapy with other modalities, or maintenance therapy. In contrast to direct and indirect anti-angiogenic therapy, conventional chemotherapy, which targets mainly tumor cells, commonly induces drug resistance.

THEORETICAL CONSIDERATIONS OF THE ADVANTAGES OF ANTI-ANGIOGENIC THERAPY, WHETHER ADMINISTERED AS A DRUG OR A GENE

The process of angiogenesis depends mainly on locomotion, proliferation, and tube formation by capillary endothelial cells. Taken together with the endothelium of larger vessels, these cells occupy a monolayer of approximately 1000 m^2, i.e., an area the size of a tennis court. Under physiologic conditions, the vascular endothelium is a quiescent tissue with a very low cell-division rate; turnover times are measured in hundreds of days (Denekamp 1990). In contrast, the average turnover time for bone marrow cells is 5 days, and this is sustained by approximately 6 billion cell divisions per hour. During angiogenesis, endothelial cells emerge from their quiescent state and can proliferate as rapidly as bone-marrow cells, but physiologic angiogenesis is usually focal and of brief duration. For example, a burst of angiogenesis in the ovarian follicle persists for a few days and is then turned off. Wound-healing angiogenesis may last for 1 week or more before it subsides. Tumor angiogenesis, however, persists for an indefinite duration under the stimulation of the growing tumor; it rarely terminates spontaneously and has until recently been difficult to suppress therapeutically. Therefore, the fundamental goal of all anti-angiogenic therapy is to return foci of proliferating microvessels to their normal resting state and to prevent their regrowth.

Tumors, for all of the classic diversity of phenotypes they possess, depend absolutely on delivery of survival factors and removal of catabolites via the comparatively well-organized, low-diversity process called angiogenesis. Therefore, the low-diversity population of proliferating endothelial cells on which tumor cells are dependent can serve as a targetable "weak link" to which the diverse tumor cell population has little or no evasive response. The ability of a tumor to evade a modality is directly related to the diversity of expression by the tumor, because diversity provides alternative avenues of tumor persistence. The greater the number of these avenues, the greater the prospect for continued tumor survival. However, the elimination of nutrient and metabolite exchange, as well as the withdrawal of paracrine growth and survival factors consequent to loss of the low-diversity population of proliferating endothelial cells upon which tumor cells are dependent, is a constraint that cannot be overcome by a tumor no matter how diverse its cell population (Fig. 6).

In summary, accumulated evidence from many laboratories argues that the various phenotypes of cancer, including tumor growth, invasion, metastasis, progression, tumor dormancy, and tumor-cell apoptosis, are not autonomous but are under the tight control of the microvascular endothelial cell. For this reason, anti-angiogenic gene therapy directed against the microvascular endothelial cell may be an important new cancer therapy.

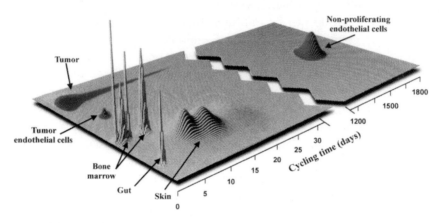

Figure 6 Schematic comparing therapy directed mainly against cancer cells to anti-angiogenic therapy. Shown are mass distributions with respect to turnover time for various tissues. Data-derived three-dimensional histogram of tissue mass with respect to proliferation time of tumor cells (*purple mass*) among the treatment-limiting tissues of gut, bone marrow, and skin (*gray masses*). Conventional therapy targets proliferating cells in the tumor, but its use is limited by coincident injury to normal tissues in accordance with the individual proliferation rates of those tissues. Rapidly cycling tissues succumb first, with more slowly cycling tissues following as the therapy continues. Thus, the effort to eradicate the entire tumor comes at a significant cost to normal tissues due to the wide range of cycling times found in tumors (note tumor tail). Additionally, acquired resistance to treatment stemming from the genetic instability and high mutation rate of tumor cells further reduces the prospect for successful treatment. In contrast, therapy that specifically targets the highly proliferating endothelial cells of the tumor (*red mass, left*) restricts growth of all tumor cells regardless of their cycling state. Drug resistance does not develop because of the genetic stability and low mutation rate of endothelial cells. Moreover, normal endothelial cells (*red mass, right*) are very slowly proliferating and therapeutically distinct. Thus, therapy directed toward the tumor endothelial-cell compartment may potentially circumvent two major therapeutic issues: toxicity to normal tissues and drug resistance.

REFERENCES

Benjamin L.E. and Keshet E. 1997. Conditional switching of vascular endothelial growth factor (VEGF) expression in tumors: Induction of endothelial cell shedding and regression of hemangioblastoma-like vessels by VEGF withdrawal. *Proc. Natl. Acad. Sci.* **94:** 8761–8766.

Blaese R.M. 1997. Gene therapy for cancer. *Sci. Amer.* **276:** 111–115.

Boehm T., Folkman J., Browder T., and O'Reilly M.S. 1997. Anti-angiogenic therapy of experimental cancer does not induce acquired drug resistance. *Nature* **390:** 404–407.

Bouck N. 1990. Tumor angiogenesis: The role of oncogenes and tumor suppressor genes. *Cancer Cells* **2:** 179–185.

Brem H., Goto F., Budson A., Saunders L., and Folkman J. 1994. Minimal drug resistance after prolonged anti-angiogenic therapy with AGM-1470. *Surg. Forum* **45:** 674–677.

Cao Y., O'Reilly M.S., Marshall B., Flynn E., Jie R.-W., and Folkman J. 1998. Expression of angiostatin cDNA in a murine fibrosarcoma suppresses primary tumor growth and produces long-term dormancy of metastases. *J. Clin. Invest.* **101:** 1055–1063.

Dameron K.M., Volpert O.V., Tainsky M.A., and Bouck N. 1994. Control of angiogenesis in fibroblasts by p53 regulation of thrombospondin-1. *Science* **265:** 1582–1584.

Evans G.I. 1994. Old cells never die, they just apoptose. *Trends Cell Biol.* **4:** 191–192.

Denekamp J. 1990. Vascular attack as a therapeutic strategy for cancer. *Cancer Metastasis Rev.* **3:** 267–282.

Fan T.P., Jaggar R., and Bicknell R. 1995. Controlling the vasculature: angiogenesis, anti-angiogenesis and vascular targeting of gene therapy. *Trends Pharmacol. Sci.* **16:** 57–66.

Folkman J. 1975. Tumor angiogenesis. In *Cancer: Comprehensive treatise* (ed. F.F. Becker), pp. 355–388, Plenum Press, New York.

——— 1996. Tumor angiogenesis. In *Cancer medicine*, 4th edition (ed. J.F. Holland et al.), vol. 1, pp. 181–204. Williams and Wilkens, Baltimore, Maryland.

——— 1997. Anti-angiogenic therapy. In *Cancer: Principles & practice of oncology*, 5th edition (ed. V.T. DeVita, Jr., et al.), pp. 3075–3085. Lippincott-Raven, Philadelphia, Pennsylvania.

Folkman J., Hochberg M., and Knighton D. 1974. Self-regulation of growth in three dimensions: The role of surface area limitation. *Cold Spring Harbor Conf. Cell Proliferation* **1:** 833–842.

Folkman J., Watson K., Ingber D., and Hanahan D. 1989. Induction of angiogenesis during the transition from hyperplasia to neoplasia. *Nature* **339:** 58–61.

Folkman J., Klagsbrun M., Sasse J., Wadzinski M., Ingber D., and Vlodavsky I. 1988. A heparin-binding angiogenic protein–basic fibroblast growth factor–is stored within basement membrane. *Am. J. Pathol.* **130:** 393–400.

Friedmann T. 1996. Human gene therapy–an immature genie, but certainly out of the bottle. *Nat. Med.* **2:** 144–147.

Hanahan D. and Folkman J. 1996. Patterns and emerging mechanisms of the angiogenic switch during tumorigenesis. *Cell* **86:** 353–364.

Harris A.L., Fox S., Bicknell R., Leek R., Relf M., LeJeune S., and Kaklamanis L. 1994. Gene therapy through signal transduction pathways and angiogenic growth factors as therapeutic targets in breast cancer. *Cancer* **74:** 1021–1025.

Hill D.L. 1975. *A review of cyclophosphamide.* Charles C. Thomas, Springfield, Illinois.

Holmgren L., O'Reilly M.S., and Folkman, J. 1995. Dormancy of micrometastases: Balanced proliferation and apoptosis in the presence of angiogenesis suppression. *Nat.*

Med. **1:** 149–153.

Kandel J., Bossy-Wetzel E., Radvanyi F., Klagsbrun M., Folkman J., and Hanahan D. 1991. Neovascularization is associated with a switch to the export of bFGF in the multistep development of fibrosarcoma. *Cell* **66:** 1095–1104.

Kerbel R. 1991. Inhibition of tumor angiogenesis as a strategy to circumvent acquired drug resistance to anticancer therapeutic agents. *BioEssays* **13:** 31–36.

Kong H.-L. and Crystal R.G. 1998. Gene therapy strategies for tumor antiangiogenesis. *J. Natl. Cancer Inst.* **90:** 273–286.

Li L.M., Nicolson G.L., and Fidler I.J. 1991. Direct *in vitro* lysis of metastatic tumor cells by cytokine-activated murine vascular endothelial cells. *Cancer Res.* **51:** 245–254.

Lowenstein P.R. 1997. Why are we doing so much cancer gene therapy? Disentangling the scientific basis from the origins of gene therapy. *Gene Ther.* **4:** 755–756.

Lu C. and Tanigawa N. 1997. Spontaneous apoptosis is inversely related to intratumoral microvessel density in gastric carcinoma. *Cancer Res.* **57:** 221–224.

Martiny-Baron G. and Marme D. 1995. VEGF-mediated tumour angiogenesis: A new target for cancer therapy. *Curr. Opin. Biotechnol.* **6:** 675–680.

Modzelewski R.A., Davies P., Watkins S.C., Auerbach R., Chang M-J, and Johnson C.S. 1994. Isolation and identification of fresh tumor-derived endothelial cells from a murine RIF-1 fibrosarcoma. *Cancer Res.* **54:** 336–339.

Nicosia R.F. and Ottinett A. 1990. Growth of microvessels in serum-free matrix culture of rat aorta. A quantitative assay of angiogenesis in vitro. *Lab. Invest.* **63:** 15–122.

Nicosia R., McCormick J., and Bielunas J. 1984. The formation of endothelial webs and channels in plasma clot culture. *Scanning Electron Microsc.* **II:** 793–799.

O'Reilly M.S., Holmgren L., Chen C., and Folkman J. 1996. Angiostatin induces and sustains dormancy of human primary tumors in mice. *Nat. Med.* **2:** 689–692.

O'Reilly M.S., Boehm T., Shing Y., Fukai N., Vasios G., Lane W.S., Flynn E., Birkhead J.R., Olsen B.R., and Folkman J. 1997. Endostatin: An endogenous inhibitor of angiogenesis and tumor growth. *Cell* **88:** 277–285.

O'Reilly M.S., Holmgren L., Shing Y., Chen C., Rosenthal R.A., Moses M., Lane W.S., Cao Y., Sage E.H., and Folkman J. 1994. Angiostatin: A novel angiogenesis inhibitor that mediates the suppression of metastases by a Lewis lung carcinoma. *Cell* **79:** 315–328.

Ozaki K., Yoshida T., Ide H., Saito I., Ikeda Y., Sugimura T., and Terada M. 1996. Use of von Willebrand factor promoter to transduce suicidal gene to human endothelial cells, HUVEC. *Hum. Gene Ther.* **7:** 1483–1490.

Pawliuk R., Bachelot T., Boehm T., Folkman J., and Leboulch P. 1998. Bone marrow engraftment of murine hematopoietic cells transduced with retroviral vectors expressing the angiogenesis inhibitors angiostatin and endostatin. *Proc. Amer. Assoc. Cancer Res.* **39:** 555.

Plate K.H. 1996. Gene therapy of malignant glioma via inhibition of tumor angiogenesis. *Cancer Metastasis Rev.* **15:** 237–240.

Polverini P.J. and Leibovich J.S. 1984. Induction of neovascularization *in vivo* and endothelial proliferation *in vitro* by tumor-associated macrophages. *Lab. Invest.* **51:** 635–642.

Rak J., Filmus J., and Kerbel R.S. 1996. Reciprocal paracrine interactions between tumour cells and enothelial cells: The "angiogenesis progression" hypothesis. *Eur. J. Cancer* **32A:** 2438–2450.

Rak J., Filmus J., Finkenzeller G., Grugel S., Marme D., and Kerbel R.S. l995. Oncogenes as inducers of tumor angiogenesis. *Cancer Metastasis Rev.* **14:** 263–277.

Rastinejad F., Polverini P., and Bouck N.P. 1989. Regulation of the activity of a new inhibitor of angiogenesis by a cancer suppressor gene. *Cell* **56:** 345–355.

Roth J.A. and Cristiano R.J. 1997. Gene therapy for cancer: What have we done and where are we going? *J. Natl. Cancer Inst.* **89:** 21–39.

Saleh M., Stacker S.A., and Wilks A.F. 1996. Inhibition of growth of C6 glioma cells in vivo by expression of antisense vascular endothelial growth factor sequence. *Cancer Res.* **56:** 393–401.

Smith-McCune K.K. and Weidner N. 1994. Demonstration and characterization of the angiogenic properties of cervical dysplasia. *Cancer Res.* **54:** 800–804.

Tanaka T., Manome Y., Wen P., Kufe D.W., and Fine II.A. 1997. Viral vector-mediated transduction of a modified platelet factor 4 cDNA inhibits angiogenesis and tumor growth. *Nat. Med.* **3:** 437–442.

Van Meir E.G., Polverini P.J., Chazin V.R., Huang H.-J.S., de Tribolet N., and Cavanee W.K. 1994. Release of an inhibitor of angiogenesis upon induction of wild type p53 expression in glioblastoma cells. *Nat. Genet.* **8:** 171–176.

Warren R.S., Yuan H., Matli M.R., Gillett N.A., and Ferrara N. 1995. Regulation by vascular endothelial growth factor of human colon cancer tumorigenesis in a mouse model of experimental liver metastasis. *J. Clin. Invest.* **95:** 1789–1797.

Xu M., Kumar D., Srinivas S., Detolla L.J., Yu S.F., Stass S.A., and Mixson A.J. 1997. Parenteral gene therapy with p53 inhibits human breast tumors in vivo through a bystander mechanism without evidence of toxicity. *Hum. Gene Ther.* **8:** 177–185.

Zhang J. and Russell S.J. 1996. Vectors for cancer gene therapy. *Cancer Metastasis Rev.* **15:** 385–401.

20

Apoptosis as a Goal of Cancer Gene Therapy

John C. Reed

The Burnham Institute
La Jolla, California 92037

The recent explosion in information about the somatic genetic alterations that occur in tumors has paved the way for thinking about strategies for novel interventions that seek to correct defects in the function of genes and gene products involved in cell-growth control in cancer cells. Among these strategies is gene therapy. Conceptually, the idea of gene therapy for cancer is extremely attractive, particularly with respect to restoring the function of tumor suppressor genes that become inactivated during the pathogenesis of many tumors. Practically, however, limitations in the efficiency of DNA delivery into cancerous cells in vivo create major impediments to successfully restoring the tumor suppressor gene function for the vast majority of patients who suffer from metastatic cancer. Nevertheless, some situations in clinical oncology present opportunities for loco-regional approaches that increase chances for successful gene therapy, despite the present inefficiency of gene delivery.

Attempts at cancer gene therapy should be guided by two underlying principles. First, the genetic intervention should be cytotoxic and not merely cytostatic for the tumor cell. Therapies that are merely designed to stop cell division are unlikely to make a substantial impact on clinical outcome if viable tumor cells remain in the patient. Second, given the current inefficiency of gene delivery in vivo, the most successful therapies are likely to be those that produce a bystander effect in which not only the genetically modified tumor cells, but also neighboring cancerous cells, are influenced by the gene-transfer event.

In this chapter, I discuss apoptosis regulatory proteins as potential targets and tools for cancer gene therapy. Tumors very commonly develop defects in the genes that normally ensure a limited life span for cells through induction of programmed cell death. Restoration of sensitivity to apoptosis by gene transfer, therefore, represents an attractive approach

The Development of Human Gene Therapy ©1999 Cold Spring Harbor Laboratory Press 0-87969-528-5/99 **545**

for reducing tumor burden in patients with cancer, provided the gene-delivery technology is efficient enough or that the genetic intervention can influence neighboring cells through bystander effects.

DEFECTIVE APOPTOSIS IN CANCERS

In all tissues in which cell division occurs, there exists a need for mechanisms that remove or eliminate cells. Otherwise, cells will accumulate in astounding numbers, given that an average adult produces 50–70 billion cells per day. Indeed, in the course of a typical year, each of us produces and eradicates a mass of cells equivalent to almost our entire body weight. This need for cell turnover is met in vivo by programmed cell death (PCD). PCD represents a mechanism by which cells actively commit suicide, culminating typically (but not always) in a constellation of morphologic changes known as apoptosis, which include chromatin condensation and nuclear fragmentation (pyknosis), cell shrinkage, plasma membrane blebbing, and extrusion of cellular particles (apoptotic bodies). Unlike necrosis in which cell lysis occurs followed by an inflammatory response, apoptotic cells and apoptotic bodies are rapidly cleared through phagocytosis by neighboring viable cells without accompanying inflammation (Kerr et al. 1972; Wyllie et al. 1980).

PCD is controlled by the counteracting influences of a number of genes that either promote or block cell death. Many of the genes involved in the cell-death pathway are genetically conserved from simple multicellular organisms such as the nematode *Caenorhabditis elegans* and the fly *Drosophila melanogaster* to mammals, including humans, thus implying distant evolutionary origins (Hengartner and Horvitz 1994b; Steller 1995). Defects in the structure, expression, or function of apoptosis-regulating genes commonly occur in cancers, contributing to the neoplastic expansion of tumor cells by prolonging their survival. Moreover, defects in apoptosis may create a fertile soil in which other genetic changes that contribute to the multistep process of tumorigenesis are tolerated, given that apoptosis can be induced by (1) certain activated cellular oncogenes involved in driving cell division, such as *C-MYC* and *CYCLIN-D1(BCL-1)*; (2) DNA mutations and chromosomal abnormalities associated with genetic instability; (3) disobeyance of cell cycle checkpoints; and (4) cell detachment from surrounding extracellular matrix proteins during attempts at invasion and metastasis (Bissonnette et al. 1992; Fanidi et al. 1992; Frisch and Francis 1994; Kranenburg et al. 1996; Minn et al. 1996).

In addition to participating in the pathogenesis of tumors, defects in PCD pathways appear to account in large part for failed attempts to treat many types of cancer with chemotherapy and radiation. This is because all conventional anticancer drugs and irradiation ultimately kill tumor cells by inducing apoptosis (Reed 1995a,b). Consequently, tumor cells that possess defects in their apoptosis machinery tend to be relatively more resistant to the cytotoxic actions of chemotherapeutic drugs and radiation.

APOPTOSIS GENES

The identification of genes that control apoptosis has come from many sources, including cancer cytogenetics, investigations of viruses, genetic analysis of developmental programmed cell death in lower organisms, and studies of mutant mouse strains with immune-system abnormalities. A few of the more prominent apoptosis-regulating proteins are described here (see Fig. 1).

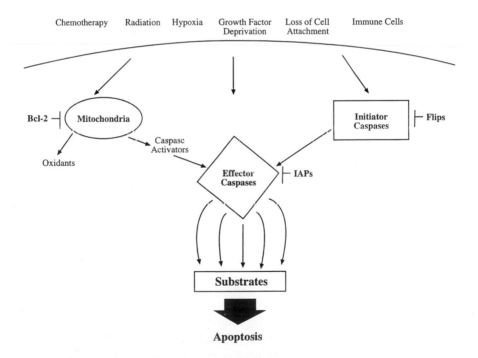

Figure 1 Schematic of cell-death pathway. Some of the major steps involved in the cell-death pathway for apoptosis are depicted. Apoptosis can be suppressed by Flip family proteins, which bind to initiator caspases and prevent their activation; by Bcl-2 family proteins, which protect mitochondria and prevent release of caspase activators; and by IAP-family proteins, which bind to and inhibit effector caspases.

Bcl-2

The anti-apoptosis gene *BCL-2* was first discovered because of its involvement in the t(14;18) chromosomal translocations seen in many non-Hodgkin's B-cell lymphomas (Tsujimoto et al. 1985). The discovery of *BCL-2* in 1985 represents a milestone in oncology research, because *BCL-2* was the first proto-oncogene identified which was found to contribute to neoplastic cell growth not by accelerating cell division, but by halting cell turnover caused by programmed cell death (Vaux et al. 1988). Since then, at least 14 cellular homologs of Bcl-2 have been described, including some that function similar to Bcl-2 as blockers of cell death, and others that do just the opposite and promote or induce apoptosis (Kroemer 1997). Interestingly, many of these Bcl-2 family proteins physically interact with each other, forming a complex network of homo- and heterodimers (Oltvai et al. 1993; Sato et al. 1994). Most Bcl-2 family proteins are anchored in intracellular membranes but oriented toward the cytosol. The predominant intracellular locations of Bcl-2 family proteins include the outer mitochondrial membrane, endoplasmic reticulum, and nuclear envelope (Krajewski et al. 1993; González-Garcia et al. 1994; Yang et al. 1995).

Homologs of Bcl-2 have been discovered in certain DNA viruses, including the E1b 19-kD protein of adenovirus and the BHRF-1 protein of Epstein-Barr virus (Henderson et al. 1993; Chiou et al. 1994). Some lower organisms have also been shown to contain Bcl-2 homologs. The free-living nematode *C. elegans,* for example, contains an antideath gene CED-9 that shares striking amino acid sequence homology with the human Bcl-2 protein (Hengartner and Horvitz 1994a). Loss of CED-9 results in a nonviable animal. Transgenic experiments have shown that human Bcl-2 can rescue CED-9-deficient worms (Vaux et al. 1992), demonstrating the extraordinary conservation of functional similarity across evolution.

The biochemical mechanism of action of Bcl-2 family proteins remains controversial (Reed 1997). Some members of this family of apoptosis-regulating proteins have been reported to share structural homology with the pore-forming domains of certain bacterial toxins and can create ion channels in synthetic membranes (Muchmore et al. 1996; Antonsson et al. 1997; Minn et al. 1997; Schendel et al. 1997; Schlesinger et al. 1997). In addition to pore formation, anti-apoptotic members of the Bcl-2 family protein also bind to a variety of nonhomologous proteins implicated in cell-death control (for review, see Reed 1997), suggesting a role in mediating critical protein-protein interactions involved in apoptosis control.

Many examples of aberrant regulation of *BCL-2* family genes in human cancers have been described. Elevations in the levels of anti-apoptotic Bcl-2 family proteins such as Bcl-2, Bcl-X$_L$, and Mcl-1 have been reported in a wide variety of solid tumors, leukemias, and lymphomas. Conversely, decreases in the expression of pro-apoptotic members of the Bcl-2 family such as Bax and Bak have been identified in many types of malignancies, including adenocarcinomas of the colon, stomach, and breast, as well as leukemias and occasional lymphomas. Mechanisms of *BCL-2* gene activation include chromosomal translocations, gene amplification, and transcriptional dysregulation (Tsujimoto et al. 1985; Miyashita et al. 1994a; Hewitt et al. 1995; Heckman et al. 1997; Monni et al. 1997). *BAX* gene inactivation can occur as a result of frameshift and missense mutations (Meiijerink et al. 1995; Rampino et al. 1997), in addition to other mechanisms.

Caspases

Genetic analysis of developmental PCD in *C. elegans* led to the discovery of a family of cell death proteases, called caspases, that are conserved throughout the animal kingdom (Yuan et al. 1993). These cysteine proteases, which cleave protein substrates following asparatic acid residues, constitute the effectors of apoptosis, clipping a variety of specific target proteins in cells and thus leading to the apoptotic demise of the cell (Martin and Green 1995; Patel et al. 1996; Salvesen and Dixit 1997). At least 11 caspases have been identified in mammalian species. Members of this family of proteases exist as inactive zymogens in virtually all animal cells. Caspases become activated by proteolytic processing at specific asparatic acid residues. Documented or suspected mechanisms responsible for the processing and activation of the proforms of caspases include (1) cleavage by other active caspases; (2) interactions with other proteins that probably entice conformations that allow procaspases to autoprocess themselves; and (3) dimerization or oligomerization of caspases, resulting in *trans*-processing and activation (Salvesen and Dixit 1997)

Like all stimuli that induce apoptosis, successful induction of apoptosis by chemotherapeutic drugs in tumor cells has been documented to involve processing and activation of caspases (Mesner et al. 1997). In some neoplastic cell lines, anticancer drugs can also lead to increases of the expression of particular caspases at the mRNA and protein levels (Kondo et al. 1995). Little effort has been placed thus far on attempts to identify mutations that might inactive specific caspase-encoding genes in cancers, but evidence of loss of caspase expression has been obtained for occasional tumor cell lines.

Bcl-2 family proteins functionally interact with caspases through at least three mechanisms. First, some anti-apoptotic Bcl-2 family proteins (but not pro-apoptotic members) can be coimmunoprecipitated with the inactive proforms of some members of the caspase family (Chinnaiyan et al. 1997a), implying that protein complexes can form that bridge these two types of proteins together. Thus far, three human proteins have been described that can enhance formation of protein complexes that contain both Bcl-2 family proteins and caspases: Mrit, Bap31, and Apaf-1 (Han et al. 1997; Ng et al. 1997; Pan et al. 1998). The last of these, Apaf-1, has a counterpart in *C. elegans* called CED-4 that is essential for developmental PCD in the worm (Yuan and Horvitz 1990). CED-4 directly binds both the Bcl-2 homolog CED-9 and the pro-form of the caspase CED-3 (Chinnaiyan et al. 1997a; Spector et al. 1997; D. Wu et al. 1997). The interaction of CED-9 with CED-4 appears to prevent CED-4 from triggering processing and activation of the CED-3 proprotein (Chinnaiyan et al. 1997b; Seshagiri and Miller 1997). Second, some anti-apoptotic Bcl-2 family proteins are reportedly cleaved by active caspases (Cheng et al. 1997; Xue and Horvitz 1997). The functional significance of this partial proteolysis of Bcl-2 family proteins may vary depending on the circumstances. In human cells, caspase-mediated cleavage of Bcl-2 removes an important amino-terminal domain (called BH4) that is necessary for its anti-apoptotic function, thus converting Bcl-2 from a protector to a killer protein (Cheng et al. 1997). Mutations that ablate this cleavage site have been found in some tumors (Tanaka et al. 1992). In *C. elegans*, however, the presence of caspase cleavage sites in the amino-terminal domain of CED-9 appears to confer added protection from PCD (Xue and Horvitz 1997). Third, Bcl-2 and Bcl-X_L inhibit, whereas Bax induces, the release of caspase-activating proteins from mitochondria into the cytosol (Kluck et al. 1997; Yang et al. 1997; Jurgensmeier et al. 1998; Rosse et al. 1998). It has been speculated that this interesting property of Bcl-2 family proteins may be related to their ability to form pores in membranes, but this hypothesis awaits experimental verification.

Inhibitor of Apoptosis Proteins

A family of anti-apoptotic proteins has been identified that all share a novel zinc-binding fold termed a baculovirus IAP repeat (BIR) domain, reflecting the original discovery of these cell-death-suppressing proteins from insect baculoviruses (Crook et al. 1993). IAP family proteins are conserved across evolution, with homologs present in *C. elegans*, *Drosophila*, and mammals.

Humans contain at least five IAP family genes: cIAP-1, cIAP-2, X-IAP, NAIP, and SURVIVIN (Rothe et al. 1995; Roy et al. 1995; Liston et al. 1996) (Duckett 1996 #1309; Uren 1996 #1490; Ambrosini 1997 #2777). The Survivin protein is widely expressed during fetal development but absent from all but a few normal tissues in adults (Adida et al. 1998). However, most tumors purportedly overexpress Survivin, suggesting that up-regulation of this anti-apoptotic protein is a very common event in the pathogenesis of human cancers (Ambrosini et al. 1997). Little is known about the expression of the other members of the IAP family in cancers. At least one of them, cIAP-2, is up-regulated by NF-κB (Chu et al. 1997), a transcription factor implicated in suppression of apoptosis in several circumstances (Antwerp et al. 1996; Liu et al. 1996).

At least one mechanism by which IAP family proteins can inhibit apoptosis has recently been determined. Three of the known five human IAPs, cIAP-1, cIAP-2, and XIAP, have been reported to bind active caspases directly and potently inhibit them with inhibitory constants (K_is) in the low nanomolar or subnanomolar range (Deveraux et al. 1997; Roy et al. 1997). Interestingly, IAP family proteins inhibit only certain caspases, specifically, caspases 3, 7, and 9 (Deveraux et al. 1997, 1998; Roy et al. 1997). These IAP-inhibitable caspases generally operate in the distal portions of proteolytic cascades involved in PCD, functioning as the final effectors of apoptosis. Other upstream caspases that commonly function as initiators of apoptotic protease cascades, including caspases 1, 8, and 10, are not inhibited by IAPs. Although IAPs can inhibit certain caspases, it has been suggested that these proteins may have other anti-apoptotic functions that remain to be clarified (Harvey et al. 1997; Vucic et al. 1997).

Tumor Necrosis Factor–Family Death Receptors

A subgroup of the tumor necrosis factor (TNF) receptor family proteins has been implicated in the induction of apoptosis through activation of caspases. TNF-R1 (CD120a), Fas (CD95), death receptor-3 (DR3), DR4, and DR5 all share a cytosolic domain that links these receptors to the cellular apoptotic machinery. This so-called death domain binds to other adapter proteins such as Fadd/Mort-1, which contain both a death domain and another protein-protein interaction module known as a death effector domain (DED). Interestingly, DEDs are also found in the amino-terminal prodomains of some caspases (e.g., caspases 8 and 10). Adapter proteins such as Fadd thus serve as bridging proteins that link TNF family death receptors with these caspases, allowing the proforms of the proteases to be recruited to the receptor complexes on binding of ligand (for review, see Wallach et al. 1997). Once recruited to the receptor complex, oligomerization of pro-caspase proteins at the membrane surface results in auto- or trans-processing, removal of the prodomain, and release into the cytosol of the active cell-death proteases. These initiator proteases then begin a cascade of proteolytic events that culminates in processing and activation of downstream effector caspases and ultimately apoptosis.

Changes in the expression of some of these death-inducing TNF family receptors or their ligands have been observed in cancers. Moreover, during apoptosis induced by chemotherapeutic drugs, some tumors appear to up-regulate expression of TNF family receptors (Fas, DR5) or their corresponding ligands (Fas-ligand), thus resulting in an autocrine cell-death mechanism (Owen-Schaub et al. 1995; Friesen et al. 1996; G.S. Wu et al. 1997).

Several mechanisms have been uncovered that can render tumors resistant to the apoptotic actions of TNF family cytokine receptors. For example, a decoy version of the receptors for Trail (a TNF family ligand) has been identified that is differentially expressed on tumor and normal cells. This decoy protein contains an extracellular domain that binds Trail but lacks a significant cytosolic tail (MacFarlane et al. 1997; Pan et al. 1997). Shedding of TNFR1 and Fas from the surface of tumors has also been described, either as a result of extracellular metalloproteases that liberate the ligand-binding domain from the cell surface or through alternative mRNA splicing mechanisms that result in secretion of soluble receptors lacking the usual transmembrane domain (Cheng et al. 1994; Moss et al. 1997). Finally, a family of cellular and viral DED-containing proteins has been discovered (Flips/Flames) that compete with caspases 8 and 10 for binding to Fadd/Mort-1, thus preventing recruitment of these pro-enzymes to TNF-family death receptors and thereby nullifying apoptotic signaling by these receptors (Irmler et al. 1997; Srinivasula et al. 1997; Thome et al. 1997). Elevated levels of the cellular Flip protein have been described in some types of tumors (Irmler et al. 1997).

p53

The tumor suppressor p53 is a multifunctional protein that can induce apoptosis in some cellular contexts. Inactivation of p53 occurs in more than half of all human cancers, making loss of p53 function the most frequent abnormality identified in human malignancies to date (Vogelstein and Kinzler 1992; Levine 1997). How p53 induces apoptosis remains a subject of controversy. Some studies have implicated the *trans*-activation function of p53 in apoptosis, implying that p53 promotes cell death by binding specific consensus decameric DNA sequences in cell-death genes and inducing their transcription (Attardi et al. 1996; X. Chen et al. 1996; Roemer and Mueller-Lantzch 1996; Yonish-Rouach et al. 1996). In this regard, several apoptosis-inducing genes have been identified as direct or indirect transcriptional targets of p53 (see Fig. 2). The first pro-apoptotic target of p53 discovered was *BAX*, a member of the Bcl-2 family (Miyashita et al. 1994b; Miyashita and Reed 1995). The promoter of the *BAX* gene contains four typical p53-binding sites and can be transcriptionally up-regulated by p53 (Miyashita and Reed 1995). Only about half of p53's activity, however, appears to be dependent on Bax, based on experiments in which the effects of p53 were examined in Bax-deficient ("knock-out") mice in an in vivo context where p53 suppresses tumor formation primarily by promoting apoptosis (Yin et al. 1997). Moreover, p53

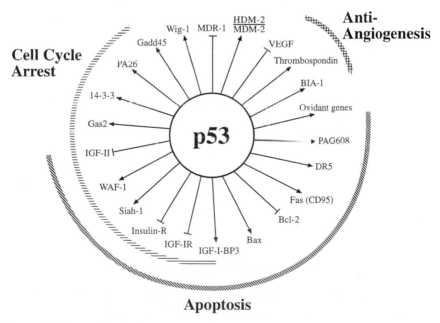

Figure 2 Targets of p53. Multiple genes can be regulated by p53, including genes involved in cell cycle arrest, apoptosis, and inhibition of angiogenesis. Some p53-regulated genes may affect either cell division or cell death, depending on cellular context.

fails to up-regulate *BAX* gene expression in many types of cells (Zhan et al. 1994; Kitada et al. 1996), indicating that tissue-specific or cell-context-dependent factors modulate the ability of p53 to induce this cell-death gene. Other apoptosis-inducing genes that p53 has been reported to up-regulate include (1) the TNF family receptors Fas and DR5; (2) the insulin-like growth factor I binding protein 3 (IGFI-BP3) gene; (3) the zinc-finger proteins PAG608 and Siah-1; and (4) several genes thought to be involved in oxidative stress (Buckbinder et al. 1995; Owen-Schaub et al. 1995; Amson et al. 1996; Israeli et al. 1997; Polyak et al. 1997; G.S. Wu et al. 1997).

As opposed to its role as a *trans*-activator, some investigations of p53 suggest that its function as a *trans*-repressor of gene expression may be more important for apoptosis (Caelles et al. 1994). Among the anti-apoptotic genes whose expression is purportedly suppressed by p53 are Bcl-2, the insulin and IGF-I receptors, and IGF-II (Miyashita et al. 1994a; Webster et al. 1996; Werner et al. 1996; Zhang et al. 1996; Prisco et al. 1997). How p53 suppresses the expression of these and other genes is presently unclear but probably is due to its interactions with other proteins involved in transcriptional regulation rather than through recognition of nucleotide sequences in DNA. Interestingly, Bcl-2 and Elb 19-kD have been reported to relieve p53-mediated transcriptional repression, but the mechanism is unknown (Sabbatini et al. 1995; Shen and Shenk 1994).

Cell death through hypoxia may be indirectly induced by p53 due to its ability to up-regulate expression of the anti-angiogenic factors thrombospondin and BAI-1 and to suppress production of vascular endothelial growth factor (VEGF) (Dameron et al. 1994; Mukopadhyay et al. 1995; Bouck 1996; Nishimori et al. 1997).

The emerging data therefore reveal a complicated picture as to how p53 induces apoptosis, with multiple p53 gene targets potentially contributing and with cellular context determining which specific p53 target genes are quantitatively most important. Probably no one single gene accounts for p53's actions as an inducer of apoptosis.

Gene Therapy Applications of Apoptosis Inducers

Since the ultimate goal of cancer therapy is to kill tumor cells selectively, apoptosis-inducing genes seem likely candidates for this task. Certainly, if one could achieve selective targeting of genes into tumor cells, then introduction of a killer gene would seem a highly logical approach for dealing with tumors, essentially bypassing the need for more indirect approaches predicated on restoring cell-growth control via effects on oncogenes and tumor suppressor genes such as Ras, MYC, and Rb. However, since apoptosis resistance genes are commonly overexpressed in human tumors, knowledge about the molecular characteristics of the tumor target may be necessary to choose the most appropriate pro-apoptotic gene for the job.

Moreover, the current inefficiencies of gene-delivery technology make strategies that produce a bystander effect highly enviable. Many of the current attempts to apply gene-therapy approaches for bolstering immune response to cancer are undertaken with the implicit recognition of this important concept of bystander effects. Gene-transfer interventions that seek to break immune tolerance against tumors or in which tumors are genetically modified to secrete immunostimulatory cytokines and lymphokines, for example, have the potential to extend benefits to nearby and even distant tumor cells despite successful genetic modification typical of only a small percentage of the cancerous cells that comprise the entire tumor burden within the patient.

The strategies discussed later, therefore, are mostly envisioned within the context of loco-regional gene delivery rather than systemic gene therapy. Clinical scenarios where a loco-regional approach is conceivable include (1) inoperable, early-stage lung cancers, where maintenance of airway patency rather than tumor eradication represents the chief goal; (2) ovarian cancers, where intraperitoneal spread represents the major cause of morbidity and mortality; (3) inoperable brain tumors; (4) hepatocellular carcinomas or liver metastases, where administration of gene-delivery vehicles through the portal circulation or hepatic artery is feasible; (5) inoperable squamous cell carcinomas of the head and neck, where local invasion rather than distant metastases is often the most life-threatening problem; and (6) basal cell carcinomas of the skin that arise in cosmetically challenging locations.

A few of the ideas for how apoptosis-inducing genes might be exploited for cancer gene therapy are presented in the following, along with a discussion of some of the pitfalls likely to be encountered.

p53

Among the apoptosis-inducing genes, p53 has received the most attention for cancer gene therapy (for review, see Bookstein et al. 1996; Boulikas 1997). Gene therapy utilizing p53 is attractive for several reasons: First, p53 gene inactivation occurs in more than half of all human cancers; second, p53 induces the expression of several genes implicated in apoptotic cell death, in addition to its effects on cell cycle inhibitors; and third, p53 can induce production of secreted proteins that promote cell death, thus providing opportunities for influencing nearby tumor cells that were unsuccessfully transduced.

These attributes of p53, however, are predicated on the assumption that the introduced wild-type p53 gene can fulfill its mission once expressed in tumor cells. Unfortunately, several obstacles may stand in the way of p53. For example, some tumors overexpress p53 inhibitors, such as the cellular proteins Hdm-2/mdm-2 and p53CP or the viral E6 protein. Hdm-2/mdm-2 and E6 bind to p53, masking its *trans*-activation domain or targeting it for ubiquitination and degradation, respectively (J. Chen et al. 1995, 1996; Maki et al. 1996), whereas p53CP competes with p53 for its DNA-binding sites (Bian and Sun 1997). Moreover, it now appears that p53 requires certain cofactors for its function, including p300/CBP, ATM, HMG-1, p85, and p33 ING-1 (Avantaggiati et al. 1997; Lill et al. 1997; Morgan and Kastan 1997; Garkavtsev et al. 1998; Jayaraman et al. 1998; Yin et al. 1998). Thus, tumors with limiting amounts of these cofactors may be relatively insensitive to p53 gene therapy. Molecular chaperones can also modulate the activity or stability of p53 in complicated ways (Selkirk et al. 1994). Another potential problem concerns the observation that the p53 protein may require activation by phosphorylation, proteolysis, acetylation, or other posttranslational modifications before it can carry out its tumor-suppressive functions (Jamal and Ziff 1995; Lohrum and Scheidtmann 1996; Gu and Roeder 1997; Hu et al. 1997; Shieh et al. 1997; Siliciano et al. 1997; Okorokov et al. 1997; Kobayashi et al. 1998). Alternatively, phosphorylation of the p53 protein at certain sites may partially inactivate the protein (Lohrum and Scheidtmann 1996; Ko et al. 1997).

These potential limitations of p53 protein suggest the possibility of bypassing p53 altogether and instead going downstream to certain p53 target genes that are more immediate mediators of p53's tumor suppressive actions. Some of those p53 targets such as Bax and IGF-IBP3 are discussed later. However, since p53 affects the expression of so many genes, focusing on any one of them may limit the clinical scenarios where this approach will be successful.

A topic that has received little attention thus far is the idea of using genetic engineering to make "super-p53." During the past decade, much has been learned about the structure-function relationships of the p53 protein. This knowledge potentially could be exploited to generate site-specific mutants of p53 or chimeric (fusion) proteins containing regions from p53, thus producing a version of p53 that is more potent than the wild-type protein and that is both constituitively active and impervious to the resistance mechanisms involving binding to other proteins (e.g., Hdm2/Mdm-2, E6), phosphorylation, or other events. More effort should therefore be devoted to improving p53 through genetic engineering.

IGF-I Antagonists

Among the secreted proteins induced by p53 is the IGF-I-binding protein, IGFI-BP3. The IGF-I receptor is a protein tyrosine kinase that has been shown to provide essential signals for cell survival in tumors and a variety of normal types of cells (Baserga et al. 1997). IGF-I receptors are overexpressed in some common types of epithelial cancer and have been correlated with resistance to radiation and chemotherapeutic drugs (Sell et al. 1995; Baserga et al. 1997; Turner et al. 1997). The availability of IGF-I is tightly controlled by a family of IGF-I-binding proteins, which are present in serum and other body fluids and exert inhibitory effects on tumors (Clemmons 1997; McCarty 1997).

Recent studies suggest that the pathway chiefly responsible for this survival effect of IGF-I involves the lipid kinase phosphatidylinositol 3´ kinase (PI3K) and the protein kinase Akt (also known as PKB). Some of the lipid substrates of PI3K become potent second messengers on undergoing phosphorylation by PI3K family proteins. These phosphatidylinositol phosphates bind to a pleckstrin homology (PH) domain in Akt, thus inducing its recruitment to the plasma membrane and subsequent activation (for review, see Hemmings 1997). Once activated, Akt can then directly or indirectly induce phosphorylation of proteins involved in cell-death control, including the pro-apoptotic Bcl-2 family protein BAD. Akt-mediated phosphorylation of the BAD protein prevents it from heterodimerizing with death suppressors such as Bcl-X_L, thus inactivating BAD and promoting cell survival (Datta et al. 1997; del Peso et al. 1997).

The ability of p53 to induce secretion of IGFI-BP3 presumably squelches signaling by IGF-I receptors and would be expected, therefore, to reduce signaling through the PI3K/Akt pathway. It also suggests the possibility of bypassing p53 and simply transducing tumor cells with expression vectors that produce IGF-I-binding proteins or neutralizing anti-IGF-1 antibodies, thus nullifying this cell-survival factor within the local environment of the tumor. Indeed, gene-transfer-mediated local production of IGF-I antagonists might have major advantages over systemic delivery of recombinant proteins or monoclonal antibodies, which conceivably would produce unwanted side effects.

IGF-I, however, is not the only stimulator of cell-survival signal transduction pathways. Indeed, most growth factors trigger the PI3K/Akt pathway and provide signals for cell survival, including EGF-R and C-erbB-2/NEU, which have been implicated in a wide variety of epithelial cancers. Thus, proper selection of tumors for those signs which exhibit absolute dependence on IGF-I might be required to ensure efficacy of this gene-therapy strategy.

Anti-angiogenesis

For either primary or metastatic tumors to exceed a certain size, blood-vessel recruitment is necessary. Similar to normal tissues, abrupt cessation of oxygen-carrying blood flow to tumor deposits of greater than 0.5 cm diameter would be expected to result in ischemia-induced necrosis as well as apoptosis. Indeed, p53-deficient tumors have been reported to be more resistant to death by hypoxia than otherwise identical tumors that retain wild-type p53 (Graeber et al. 1996).

In addition to intrinsic resistance to hypoxia-induced cell death, p53-deficient tumor cells may also benefit from the loss of anti-angiogenesis factors such as thrombospondin and BAI-1 that are normally induced by wild-type p53 (Dameron et al. 1994; Nishimori et al. 1997) and by elevated production of VEGF, whose synthesis is normally suppressed by p53 (Mukopadhyay et al. 1995). Inactivation of p53 therefore presumably removes some of the obstacles to tumor vascularization, allowing for more efficient ingrowth of capillaries. Hence, one of the bystander mechanisms by which p53 gene therapy may suppress tumor growth in vivo may be by stimulating production of anti-angiogenic proteins. This observation suggests the possibility of bypassing p53 by genetically modifying tumors to secrete anti-angiogenic proteins or monoclonal antibodies designed to neutralize VEGF or other angiogenic factors secreted by tumors.

Alternatively, one could imagine transducing tumors with vectors that encode antibodies to the $\alpha_v\beta_3$ integrin complex, which is uniquely expressed on migrating and proliferating endothelial cells at sites of neovascularization. This is because blocking interactions of $\alpha_v\beta_3$ with its extracellular ligands induces apoptosis of proliferative endothelial cells but not of quiescent endothelial cells lining established blood vessels (Stromblad et al. 1996; Varner and Cheresh 1996). The phenomena of cell death occurring upon depriving cells of attachments to extracellular matrix proteins via integrins has been termed *anoikis*, which is derived from the Greek word for homelessness (Frisch and Francis 1994). Anoikis occurs in epithelial, endothelial, neuronal, and muscle cells when integrin-mediated survival signals are absent. Many tumors, however, exhibit resistance to anoikis, accounting for their ability to survive and grow in an anchorage-independent fashion (for review, see Ruoslahti and Reed 1994). Thus, strategies aimed at interfering with integrin signaling may not be particularly useful if targeted against cancer cells but are conceivably worthwhile when aimed against the normal endothelial cells that form the life-supporting vessels of tumors. Theoretically, therefore, recombinant proteins that act as $\alpha_v\beta_3$ antagonists could potentially be employed for cancer gene therapy.

A major limitation of most anti-angiogenesis strategies, however, is that they may be primarily relevant for newly forming tumor vessels and not established ones. Consequently, although gene-transfer-mediated production of anti-angiogenic proteins within cancers may suppress additional tumor growth by preventing recruitment of more blood vessels (provided expression of the transgene is sustained long enough), it may do relatively little to shrink existing tumors that have already formed a vasculature. Therefore, short-term expression of thombospondin, $\alpha_v\beta_3$ antagonists, or other anti-angiogenesis proteins may have limited consequences for established primary tumors and their metastases.

An alternative scheme might envision attempting to attack established vessels in tumors, thus inducing tumor infarction. In this regard, it has recently been shown that even the endothelial cells of established vessels of tumors may express cell-surface proteins that distinguish them from vessels located elsewhere in the body and that therefore permit a targeted attack of the tumor vasculature (Arap et al. 1998). Moreover, small peptide ligands that bind to endothelial cells lining the vessels of tumors have been identified using phage-display technology, raising the possibility of tethering DNA or viral gene-delivery vehicles to them for selective genetic modification of these cells. An attractive idea thus is to deliver apoptosis-inducing genes to tumor endothelium using this approach, although direct delivery of cytotoxic chemicals conjugated to tumor vasculature-homing peptides might be equally or more effective and far less complicated.

Fas Ligand

Although numerous cancer gene therapy strategies have been devised that involve transducing tumors with cytokine-encoding genes (see Pardoll and Nabel, this volume), some of the TNF family ligands provide a novel approach that is designed to kill tumor cells directly rather than encourage the immune system to do the work. For example, gene-transfer-mediated expression of Fas ligand (Fas-L) on the surface of Fas-positive tumors could both induce apoptosis of the genetically modified cancer cells and kill adjacent cells that failed to take up DNA or viral vectors. Of course, this strategy depends on (1) expression of adequate levels of Fas on the tumor cell targets; (2) absence of shedding or secretion of soluble Fas decoy proteins; and (3) a lack for barriers to Fas-induced apoptosis caused by high levels of Flip/Flame or other anti-apoptotic proteins that can interfere with Fas-L-based cytotoxicity. One potentially deleterious effect of Fas-L gene therapy, however, is that if the tumor cells do prove resistant to this TNF family ligand, they might then be endowed with the ability to kill neighboring normal cells or enabled to destroy attacking cytolytic T cells and other immune cells that are typically Fas-sensitive (Hahne et al. 1996; O'Connell et al. 1996).

Bax and Bcl-X$_S$

The *BAX* gene is a direct transcriptional target of p53 (Miyashita and Reed 1995), suggesting that cancer gene therapy based on *BAX* gene transfer could provide another means of bypassing the need for p53 and achieving the apoptotic elimination of cancer cells. In addition, tumors with the microsatellite instability phenotype indicative of defects in DNA mismatch repair enzymes, commonly contain frameshift mutations within the coding region of *BAX* that inactive the gene (Rampino et al. 1997; Yamamoto et al. 1997). Thus, both tumors with p53 gene inactivation and those with mutant *BAX* are obvious candidates for *BAX* gene therapy. One limitation of gene therapy strategies predicated on *BAX*, however, is that they afford no obvious bystander effects and thus are presumably limited in their efficacy to only those few percentage of tumor cells that one generally can expect to transduce successfully in vivo. Moreover, it has recently been suggested that *BAX* resides in a latent state in the cytosol of cells until undefined signals result in its translocation into mitochondrial membranes, thereby triggering apoptosis (Hsu et al. 1997; Wolter et al. 1997). Hence, the mere expression of *BAX* may not guarantee killing of tumors. One obvious way to prompt Bax into action is to combine it with chemotherapy or radiation, thereby providing the apoptotic "signals" needed to active the Bax protein. In this regard, in vitro gene-transfer studies indicate that Bax may be insufficient by itself to induce apoptosis but can sensitize tumor cells to the cytotoxic actions of anticancer drugs and radiation (Sakakura et al. 1996). Indeed, this synergy between gene-transfer-mediated increases in apoptosis-prompting proteins and conventional anticancer therapy is well documented not only for Bax, but also for p53 (Lowe et al. 1993; Fisher 1994), highlighting the significance of combining the gene-therapy approach with stimuli that nudge apoptosis pathways into action. Similar to the arguments just presented for p53, genetic engineering to produce a "super-Bax" protein with constitutive killer activity might improve the efficiency of *BAX* gene therapy for cancer.

In principle, several of the pro-apoptotic Bcl-2 family proteins could be supplanted for Bax in cancer gene-therapy strategies. One of these, Bcl-X_S, has been shown either to induce apoptosis directly or to sensitize tumor cells to the apoptotic effects of chemotherapeutic drugs (Sumantran et al. 1995; Ealovega et al. 1996;). A strategy for purging bone marrow of solid-tumor cells using Bcl-X_S-encoding adenoviruses has been suggested (Clarke et al. 1995) and is predicated on the observation that adenoviruses generally do not infect hematopoietic stem cells, whereas they do efficiently enter epithelial cells. Bcl-X_S is capable of forming heterodimers with many of the cytoprotective members of the Bcl-2 family, including Bcl-2, Bcl-X_L, and Mcl-1 (Sato et al. 1994), presumably acting as a *trans*-dominant inhibitor of these cell-survival proteins.

Caspases

Gene-transfer-mediated overexpression of many, but not all, types of caspases induces apoptosis, suggesting the possibility of directly triggering apoptosis in tumors using these cell-death proteases. Potential obstacles to the successful application of caspase gene therapy, however, include (1) IAP-family proteins; (2) Flip/Flame-family proteins; and (3) inhibitory isoforms of caspases that can arise through alternative mRNA splicing (e.g., Ich-1S [caspase-2S]) and that prevent activation of their "full-length" counterparts (Wang et al. 1994). Although most caspases lack any obvious mechanism for bystander effects, it should be noted that some caspases, including caspases 1, 4, 5, and 11, play the dual role of inducing apoptosis and processing the proforms of proinflammatory cytokines such as interleukin-1β, and the interferon-γ-inducing factor (IGIF; IL-18) (Kuida et al. 1995; Akita et al. 1997; Gu et al. 1997). It is conceivable, therefore, that caspase gene therapy could be used both to induce apoptosis directly in genetically modified tumor cells and to elicit an inflammatory reaction against neighboring cancer cells.

SUMMARY

Cancer gene therapy has traditionally focused on attempts either to restore tumor-suppressor gene function or to entice immune responses to tumors. Restoration of tumor-suppressor gene function can be focused on genes that inhibit cell growth either through effects on cell proliferation or through induction of cell death. Although cell cycle inhibitory genes such as Rb may reduce tumor cell proliferation rates, they generally are not cytotoxic and, in fact, may prevent tumor cell apoptosis under some circumstances (Haupt et al. 1995; Macleod et al. 1996). Since the ultimate goal of cancer therapy is to kill tumor cells selectively, greater attention should be paid to the lessons learned from investigations of the fundamental mechanisms of programmed cell death and the molecular basis for its dysregulation in cancer. Apoptosis-inducing genes can efficiently promote killing of malignant cells but are likely to be most effective when combined with conventional treatments such as chemotherapy and radiation. Proper choice of particular apoptosis-inducing genes for specific types of cancer is an important facet of this strategy and must be done with the recognition that cancer cells often express anti-apoptotic genes that counter the actions of pro-apoptotic proteins. Furthermore, until the efficiency and selectivity of gene-delivery technologies improve, the most successful strategies are likely to be those that induce killing not only of the genetically modified tumor cells, but also of the neighboring cells through the effects of killer proteins expressed on the surface of transduced tumor cells or secreted from them. Also attractive are approaches that seek to exploit the apparently unique characteristics of the endothelial cells that line the vasculature of tumors, promoting death of these normal diploid cells that presumably lack the impediments to apoptosis commonly found in malignant cells.

ACKNOWLEDGMENTS

I thank T. Brown for assistance with manuscript preparation and artwork, and acknowledge the National Cancer Institute, the state of California, the Department of Defense, and CaP-CURE for their generous support.

REFERENCES

Adida C., Crotty P.L., McGrath J., Berrebi D., Diebold J., and Altieri, D.C. 1998. Developmentally regulated expression of the novel cancer anti-apoptosis gene surviving in human and mouse differentiation. *Am. J. Pathol.* **152:** 43–49.

Akita K., Ohtsuki T., Nukada Y., Tanimoto T., Namba M., Okura T., Takakura-Yamamoto R., Torigoe K., Gu Y., Su M.S.S., Fujii M., Satoh-Itoh M., Yamamoto K., Kohno K., Ikeda M., and Kurimoto, M. 1997. Involvement of caspase-1 and caspase-3 in the production and processing of mature human interleukin 18 in monocytic THP.1 cells. *J. Biol. Chem.* **272:** 26595–26603.

Ambrosini G., Adida C., and Altieri D. 1997. A novel anti-apoptosis gene, *survivin*, expressed in cancer and lymphoma. *Nat. Med.* **3:** 917–921.

Amson R.B., Nemani M., Roperch J.-P., Israeli D., Bougueleret L., Le Gall I., Medhioub M., Linares-Cruz G., Lethrosne F., Pasturaud P., Piouffre L., Prieur S., Susini L., Alvaro V., Millasseau P., Guidicelli C., Bui H., Massart C., Cazes L., Dufour F., Bruzzoni-Giovanelli H., Owadi H., Hennion C., Charpak G., Dausset J., Calvo F., Oren M., Cohen D., and Telerman A. 1996. Isolation of 10 differentially expressed cDNAs in p53-induced apoptosis: Activation of the vertebrate homologue of the *Drosphila* seven in absentia gene. *Proc. Natl. Acad. Sci.* **93:** 3953–3957.

Antonsson B., Conti F., Ciavatta A., Montessuit S., Lewis S., Martinou I., Bernasconi L., Bernard A., Mermod J.-J., Mazzei G., Maundrell K., Gambale F., Sadoul R., and Martinou, J.-C. 1997. Inhibition of Bax channel-forming activity by Bcl-2. *Science* **277:** 370–372.

Antwerp D.J., Martin S.J., Kafri T., Green D.R., and Verma I.M. 1996. Suppression of TNF-α-induced apoptosis by NF-κB. *Science* **274:** 787–789.

Arap W., Pasqualini R., and Ruoahti, E. 1998. Cancer treatment by targeted drug delivery to tumor vasculature in a mouse model. Science **279:** 377–380.

Attardi L.D., Lowe S.W., Brugarolas J., and Jacks T. 1996. Transcriptional activation by p53, but not induction of the p21 gene, is essential for oncogene-mediated apoptosis. *EMBO J.* **15:** 3693–3701.

Avantaggiati M.L., Ogryzko V., Gardner K., Giordano A., Levine A.S., and Kelly K. 1997. Recruitment of p300/CBP in p53-dependent signal pathways. *Cell* **89:** 1175–1184.

Baserga R., Hongo A., Rubini M., Prisco M., and Valentinis B. 1997. The IGF-I receptor in cell growth, transformation and apoptosis. *Biochim. Biophys. Acta* **1332:** F105–F126.

Bian J. and Sun Y. 1997. p53CP, a putative p53 competing protein that speciffically binds to the consensus p53 DNA binding sites: A third member of the p53 family? *Proc. Natl. Acad. Sci.* **94:** 14753–14758.

Bissonnette R.P., Exheverri F., Mahboubi A., and Green D.R. 1992. Apoptotic cell death induced by c-*myc* is inhibited by bcl-2. *Nature* **359:** 552–554.

Bookstein R., Demers W., Gregory R., Maneval D., Park J., and Wills K. 1996. p53 gene therapy in vivo of herpatocellular and liver metastatic colorectal cancer. *Semin. Oncol.* **23:** 66–77.

Bouck N. 1996. P53 and angiogenesis, *Biochim Biophys. Acta* **1287:** 63–66.

Boulikas T. 1997. Gene therapy of prostate cancer: p53, suicidal genes, and other targets. *Anticancer Res.* **17:** 1471–1505.

Buckbinder L., Talbott R., Velasco-Miguel S., Takenaka I., Faha B., Seizinger B.R., and Kley N. 1995. Induction of the growth inhibitor IGF-binding protein 3 by p53. *Nature* **377:** 646–649.

Caelles C., Heimberg A., and Karin M. 1994. p53-dependent apoptosis in the absence of transcriptional activation of p53-target genes. *Nature* **370:** 220–224.

Chen J., Lin J., and Levine A.J. 1995. Regulation of transcription functions of the p53 tumor suppressor by the mdm-2 oncogene. *Mol. Med.* **1:** 142–152.

Chen J., Wu X., Lin J., and Levine A.J. 1996. mdm-2 inhibits the G1 arrest and apoptosis functions of the p53 tumor suppressor protein. *Mol. Cell. Biol.* **16:** 2445–2452.

Chen X., Ko L.J., Jayaraman L., and Prives C. 1996. p53 levels, functional domains, and DNA damage determine the extent of the apoptotic response of tumor cells. *Genes Dev.* **10:** 2348–2451.

Cheng E.H.-Y., Kirsch D.G., Clem R.J., Ravi R., Kastan M.B., Bedi A., Ueno K., and Hardwick J.M. 1997. Conversion of Bcl-2 to a Bax-like death effector by caspases.

Science **278:** 1966–1968.

Cheng J., Zhou T., Liu C., Shapiro J.P., Brauer M.J., Kiefer M.C., Barr P.J., and Mountz J.D. 1994. Protection from Fas-mediated apoptosis by a soluble form of the Fas molecule. *Science* **263:** 1759–1762.

Chinnaiyan A.M., O'Rourke K., Lane B.R., and Dixit V.M. 1997a. Interaction of CED-4 with CED-3 and CED-9: A molecular framework for cell death. *Science* **275:** 1122–1126.

Chinnaiyan A., Chaudhary D., O'Rourke K., Koonin E., and Dixit V. 1997b. Role of CED-4 in the activation of CED-3. *Nature* **388:** 728–729.

Chiou S.-K., Tseng C.-C., Rao L., and White E. 1994. Functional complementation of the adenovirus E1B 19-kilodalton protein with Bcl-2 in the inhibition of apoptosis in infected cells. *J. Virol.* **68:** 6553–6566.

Chu Z.L., McKinsey T.A., Liu L., Gentry J.J., Malim M.H., and Ballard D.W. 1997. Suppression of tumor necrosis factor-induced cell death by inhibitor of apoptosis c-IAP2 is under NF-κB control. *Proc. Natl. Acad. Sci.* **94:** 10057–10062.

Clarke M.F., Apel I.J., Benedict M.A., Eipers P.G., Sumantran V., Gonzalez-Garcia M., Doedens M., Fukanaga N., Davidson B., Dick J.E., Minn A.J., Boise L.H., Thompson C.B., Wicha M., and Nunez G. 1995. A recombinant bcl-xs adenovirus selectively induces apoptosis in cancer cells but not in normal bone marrow cells. *Proc. Natl. Acad. Sci.* **92:** 11024–11028.

Clemmons D.R. 1997. Insulin-like growth factor binding proteins and their role in controlling IGF actions. *Cytokine Growth Factor Rev.* **8:** 45–62.

Crook N.E., Clem R.J., and Miller L.K. 1993. An apoptosis-inhibiting baculovirus gene with a zinc finger-like motif. *J. Virol.* **67:** 2168–74.

Dameron K.M., Volpert O.V., Tainsky M.A., and Bouck N. 1994. Control of angiogenesis in fibroblasts by p53 regulation of thrombospondin-1. *Science* **265:** 1582–1584.

Datta S.R., Dudek H., Tao X., Masters S., Fu H., Gotoh Y., and Greenberg M.E. 1997. Akt phosphorylation of BAD couples survival signals to the cell-intrinsic death machinery. *Cell* **91:** 231–241.

del Peso L., González-García M., Page C., Herrera R., and Nunez G. 1997. Interleukin-3-induced phosphorylation of BAD through the protein kinase Akt. *Science* **278:** 687–689.

Deveraux Q., Takahashi R., Salvesen G.S., and Reed J.C. 1997. X-linked IAP is a direct inhibitor of cell death proteases. *Nature* **388:** 300–303.

Deveraux Q.L., Roy N., Stennicke H.R., Van Arsdale T., Zhou Q., Srinivasula M., Alnemri E.S., Salvesen G.S., and Reed J.C. 1998. IAPs block apoptotic events induced by caspase-8 and cytochrome c by direct inhibition of distinct caspases. *EMBO J.* **17:** 2215a–2223.

Duckett, C.S., Nava V.E., Gedrich R.W., Clem R.J., Van Dongen J.L., Gilfillan M.C., Shiels H., Hardwick J.M., and Thompson C.B. 1996. A conserved family of cellular genes related to the baculovirus iap gene and encoding apoptosis inhibitors. *EMBO J.* **15:** 2685–2689.

Ealovega M.W., McGinnis P.K., Sumantran V.N., Clarke M.F., and Wicha M.S. 1996. bcl-Xs gene therapy induces apoptosis of human mammary tumors in nude mice. *Cancer Res.* **56:** 1965–1969.

Fanidi A., Harrington E.A., and Evan G.I. 1992. Cooperative interaction between c-myc and bcl-2 proto-oncogenes. *Nature* **359:** 554–556.

Fisher D.E. 1994. Apoptosis in cancer therapy: crossing the threshold. *Cell* **78:** 539–542.

Friesen C., Herr I., Krammer P.H., and Debatin K.-M. 1996. Involvement of the CD95 (APO-1/Fas) receptor/ligand system in drug induced apoptosis in leukemia cells. *Nat. Med.* **2:** 574–577.

Frisch S.M. and Francis H. 1994. Disruption of epithelial cell-matrix interactions induces apoptosis. *J. Cell Biol.* **124:** 619–626.

Garkavtsev I., Grigorian I.A., Ossovskaya V.S., Chernov M.V., Chumakov P.M., and Gudkov A.V. 1998. The candidate tumour suppressor p33^{ING1} cooperate with p53 in cell growth control. *Nature* **391:** 295–298.

González-Garcia M., Pérez-Ballestero R., Ding L., Duan L., Boise L., Thompson C., and Núñez G. 1994. bcl-XL is the major bcl-x mRNA form expressed during murine development and its product localizes to mitochondria. *Development* **120:** 3033–3042.

Graeber T.G., Osmanian C., Jacks T., Housman D.E., Koch C.J., Lowe S.W., and Giaccia A.J. 1996. Hypoxia-mediated selection of cells with diminished apoptotic potential in solid tumours. *Nature* **379:** 88–91.

Gu W. and Roeder R.G. 1997. Activation of p53 sequence-specific DNA binding by acetylation of the p53 C-terminal domain. *Cell* **90:** 595–606.

Gu Y., Kuida K., Tsutsui H., Ku G., Hsiao K., Fleming M.A., Hayashi N., Higashino K., Okamura H., Nakanishi K., Kurimoto M., Tanimoto T., Flavell R.A., Sato, V., Harding, M.W., Livingston, D.J., and Su M.S. 1997. Activation of interferon-γ inducing factor mediated by interleukin-1β converting enzyme. *Science* **275:** 206–209.

Hahne M., Rimoldi D., Schroter M., Romero P., Schreier M., French L.E., Schneider P., Bornand T., Fontana A., Lienard D., Cerottini J.-C., and Tschopp J. 1996. Melanoma cell expression of Fas(Apo-1/CD95) ligand: Implications for tumor immune escape. *Science* **274:** 1363–1366.

Han D.K.M., Chaudhary P.M., Wright M.E., Friedman C., Trask B.J., Riedel R.T., Baskin D.G., Schwartz S.M., and Hood L. 1997. MRIT, a novel death-effector domain-containing protein, interacts with caspases and BclXL and initiates cell death. *Proc. Natl. Acad. Sci.* **94:** 11333–11338.

Harvey A.J., Bidwai A.P., and Miller L.K. 1997. Doom, a product of the *Drosophila* mod(mdg4) gene, induces apoptosis and binds to baculovirus inhibitor-of-apoptosis proteins. *Mol. Cell. Biol.* **17:** 2835–2843.

Haupt Y., Rowan S., and Oren M. 1995. p53-mediated apoptosis in HeLa cells can be overcome by excess pRB. *Oncogene* **10:** 1563–1571.

Heckman C., Mochon E., Arcinas M., and Boxer L. 1997. The WT1 protein is a negative regulator of the normal bcl-2 allele in t(14;18) lymphomas. *J. Biol. Chem.* **272:** 19609–19614.

Hemmings B.A. 1997. Akt signalling: Linking membrane events to life and death decisions. *Science* **275:** 628–630.

Henderson S., Huen D., Rowe M., Dawson C., Johnson G., and Rickinson A. 1993. Epstein-Barr virus-coded BHRF1 protein, a viral homologue of Bcl-2, protects human B cells from programmed cell death. *Proc. Natl. Acad. Sci.* **90:** 8479–8483.

Hengartner M.O. and Horvitz H.R. 1994a. *C. elegans* cell survival gene ced-9 encodes a functional homolog of the mammlian proto-oncogene bcl-2. *Cell* **76:** 665–676.

———. 1994b. Programmed cell death in *Caenorhabditis elegans. Curr. Opin. Genet. Dev.* **4:** 581–586.

Hewitt S.M., Hamada S., McDonnell T.J., Rauscher F.J., and Saunders G.F. 1995. Regulation of the proto-oncogenes bcl-2 and c-myc by the Wilms' tumor suppressor gene WT1. *Cancer Res.* **55:** 5386–5389.

Hsu Y.-T., Wolter K.G., and Youle R.J. 1997. Cytosol-to-membrane redistibution of members of Bax and Bcl-X_L during apoptosis. *Proc. Natl. Acad. Sci.* **94:** 3668–3672.

Hu M.C.T., Qui W.R., and Wang Y.P. 1997. JNK1, JNK2 and JNK3 are p53 N-terminal serine 34 kinases. *Oncogene* **15:** 2277–2287.

Irmler M., Thome M., Hahne M., Schneider P., Hofmann K., Steiner V., Bodmer J.-L., Schröter M., Burns K., Mattmann C., Rimoldi D., French L.E., and Tschopp J. 1997. Inhibition of death receptor signals by cellular FLIP. *Nature* **388:** 190–195.

Israeli D., Tessler E., Haupt Y., Elkeles A., Wilder S., Amson R., Telerman A., and Oren M. 1997. A novel p53-inducible gene, PAG608, encodes a nuclear zinc finger protein whose overexpression promotes apoptosis. *EMBO J.* **16:** 4384–4392.

Jamal S. and Ziff E.B. 1995. Raf phosphorylates p53 in vitro and potentiates p53-dependent transcriptional transactivation in vivo. *Oncogene* **10:** 2095–2101.

Jayaraman L., Moorthy N.C., Murthy K.G.K., Manley J.L., Bustin M., and Prives C. 1998. High mobility group protein-1 (HMG-1) is a unique activator of p53. *Genes Dev.* **12:** 462–472.

Jurgensmeier J.M., Xie Z., Deveraux Q., Ellerby L., Bredesen D., and Reed J.C. 1998. Bax directly induces release of cytochrome c from isolated mitochondria. *Proc. Natl. Acad. Sci.* **95:** 4997–5002.

Kerr J.F., Wyllie A.H., and Currie A.R. 1972. Apoptosis: a basic biological phenomenon with wide-ranging implications in tissue kinetics. *Br. J. Cancer* **26:** 239–57.

Kitada S., Krajewski S., Miyashita T., Krajewska M., and Reed J.C. 1996. γ-Radiation induces upregulation of Bax protein and apoptosis in radiosensitive cells in vivo. *Oncogene* **12:** 187–92.

Kluck R.M., Bossy-Wetzel E., Green D.R., and Newmeyer D.D. 1997. The release of cytochrome c from mitochondria: A primary site for Bcl-2 regulation of apoptosis. *Science* **275:** 1132–1136.

Ko L.J., Shieh S.Y., Chen X., Jayaraman L., Tamai K., Taya Y., Prives C., and Pan Z.Q. 1997. p53 is phosphorylated by CDK7-cyclin H in a p36^{MAT1}-dependent manner. *Mol. Cell. Biol.* **17:** 7220–7229.

Kobayashi T., Ruan S., Jabbur J.R., Consoli U., Clodi K., Shiku H., Owen-Schaub L.B., Andreeff M., Reed J.R., and Zhang W. 1998. Differential p53 phosphorylation and activation of apoptosis-promoting genes Bax and Fas/APO-1 by irradiation and ara-C treatment. *Cell Death Differ.* (in press).

Kondo S., Barna B.P., Morimura T., Takeuchi J., Yuan J., Akbasak A., and Barnett G.H. 1995. Interleukin-1β-converting enzyme mediates cisplatin-induced apoptosis in malignant glioma cells. *Cancer Res.* **55:** 6166–6171.

Krajewski S., Tanaka S., Takayama S., Schibler M.J., Fenton W., and Reed J.C. 1993. Investigation of the subcellular distribution of the bcl-2 oncoprotein: Residence in the nuclear envelope, endoplasmic reticulum, and outer mitochondrial membranes. *Cancer Res.* **53:** 4701–14.

Kranenburg O., van der Eb A.J., and Zantema A. 1996. Cyclin D1 is an essential mediator of apoptotic neuronal cell death. *EMBO J.*. **15:** 46–54.

Kroemer G. 1997. The proto-oncogene Bcl-2 and its role in regulating apoptosis. *Nat. Med.* **3:** 614–20.

Kuida K., Lippke J.A., Ku G., Harding M.W., Livingston D.J., Su M.S.-S., and Flavell R.A. 1995. Altered cytokine export and apoptosis in mice deficient in interleukin-1β converting enzyme. *Science* **267:** 2000–2003.

Levine A.J. 1997. p53, the cellular gatekeeper for growth and division. *Cell* **88:** 323–331.

Lill N.L., Grossman S.R., Ginsberg D., DeCaprio J., and Livingston D.M. 1997. Binding and modulation of p53 by p300/CBP coactivators. *Nature* **387:** 823–827.

Liston P., Roy N., Tamai K., Lefebvre C., Baird S., Cherton-Horvat G., Farahani R., McLean M., Ikeda J., MacKenzie A., and Korneluk R.G. 1996. Suppression of apoptosis in mammalian cells by NAIP and a related family of IAP genes. *Nature* **379:** 349–353.

Liu Z.-G., Hsu H., Goeddel D.V., and Karin M. 1996. Dissection of TNF receptor 1 effector functions: JNK activation is not linked to apoptosis while NF-κB activation prevents cell death. *Cell* **87:** 565–576.

Lohrum M. and Scheidtmann K.H. 1996. Differential effects of phosphorylation of rat p53 on transactivation of promoters derived from different p53 responsive genes. *Oncogene* **13:** 2527–2539.

Lowe S.W., Ruley H.E., Jacks T., and Houseman D.E. 1993. p53-dependent apoptosis modulates the cytotoxicity of anticancer agents. *Cell* **74:** 957–967.

MacFarlane M., Ahmad M., Srinivasula S., Fernandes-Alnemri T., Cohen G., and Alnemri E. 1997. Identification and molecular cloning of two novel receptors for the cytotoxic ligand TRAIL. *J. Biol. Chem.* **272:** 25417–25420.

Macleod K.F., Hu Y., and Jacks T. 1996. Loss of Rb activates both p53-dependent and independent cell death pathways in the developing mouse nervous system. *EMBO J.* **15:** 6178–6188.

Maki C.G., Huibregtse J.M., and Howley P.M. 1996. In vivo ubiquitination and proteasome-mediated degradation of p53. *Cancer Res.* **56:** 2649–2654.

Martin S.J. and Green D.R. 1995. Protease activation during apoptosis: Death by a thousand cuts? *Cell* **82:** 349–52.

McCarty M.F. 1997. Up-regulation of IGF binding protein-1 as an anticarcinogenic strategy: Relevance to caloric restriction, exercise, and insulin sensitivity. *Med. Hypotheses* **48:** 297–308.

Meijerink J.P.P., Smetsers T.F.C.M., Slöetjes A.W., Linders E.H.P., and Mensink E.J.B.M. 1995. Bax mutations in cell lines derived from hematological malignancies. *Leukemia* **9:** 1828–1832.

Mesner Jr. P.W., Budihardjo I.I., and Kaufmann S.H. 1997. Chemotherapy-induced apoptosis. *Adv. Pharmacol.* **41:** 461–499.

Minn A.J., Boise L.H., and Thompson C.B. 1996. Expression of Bcl-x$_L$ and loss of p53 can cooperate to overcome a cell cycle checkpoint induced by mitotic spindle damage. *Genes Dev.* **10:** 2621–2631.

Minn A.J., Velez P., Schendel S.L., Liang H., Muchmore S.W., Fesik S.W., Fill M., and Thompson C.B. 1997. Bcl-x$_L$ forms an ion channel in synthetic lipid membranes. *Nature* **385:** 353–357.

Miyashita T. and Reed J.C. 1995. Tumor suppressor p53 is a direct transcriptional activator of human BAX gene. *Cell* **80:** 293–299.

Miyashita T., Harigai M., Hanada M., and Reed J.C. 1994a. Identification of a p53-dependent negative response element in the Bcl-2 gene. *Cancer Res.* **54:** 3131–3135.

Miyashita T., Krajewski S., Krajewska M., Wang H.G., Lin H.K., Hoffman B., Lieberman D., and Reed J.C. 1994b. Tumor suppressor p53 is a regulator of BCL-2 and BAX in gene expression in vitro and in vivo. *Oncogene* **9:** 1799–1805.

Monni O., Joensuu H., Franssila K., Klefstrom J., Alitalo K., and Knuutila S. 1997. BCL2 overexpression associated with chromosomal amplification in diffuse large B-cell lymphoma. *Blood* **90:** 1168–1174.

Morgan S.E. and Kastan M.B. 1997. p53 and ATM: Cell cycle, cell death, and cancer. *Adv. Cancer Res.* **71:** 1–25.

Moss M.L., Jin S.L., Becherer J.D., Bickett D.M., Burkhart W., Chen W.J., Hassler D., Leesnitzer M.T., McGeehan G., Milla M., Moyer M., Rocque W., Seaton T., Schoenen F., Warner J., and Willard D. 1997. Structural features and biochemical properties of TNF-α converting enzyme (TACE). *J. Neuroimmunol.* **72:** 127–129.

Muchmore S.W., Sattler M., Liang H., Meadows R.P., Harlan J.E., Yoon H.S., Nettesheim D., Changs B.S., Thompson C.B., Wong S., Ng S., and Fesik S.W. 1996. X-ray and NMR structure of human Bcl-XL, an inhibitor of programmed cell death. *Nature* **381:** 335–341.

Mukopadhyay D., Tsiokas L., and Sukhatme V.P. 1995. Wild-type p53 and v-Src exert opposing influences on human vascular endothelial growth factor gene expression. *Cancer Res.* **55:** 6161–6165.

Ng F.W.H., Nguyen M., Kwan T., Branton P.E., Nicholson D.W., Cromlish J.A., and Shore G.C. 1997. p28 Bap31, a Bcl-2/Bcl-X$_L$-and procaspase-8-associated protein in the endoplasmic reticulum. *J. Cell Biol.* **39:** 327–338.

Nishimori H., Shiratsuchi T., Urano T., Kimura Y., Kiyono K., Tatsumi K., Yoshida S., Ono M., Kuwano M., Nakamura Y., and Tokino T. 1997. A novel brain-specific p53-target gene, BAI1, containing thrombospondin type 1 repeats inhibits experimental angiogensis. *Oncogene* **15:** 2145–2150.

O'Connell J., O'Sullivan G.C., Collins J.K., and Shanahan F. 1996. The Fas counterattack: Fas-mediated T cell killing by colon cancer cells expressing Fas ligand. *J. Exp. Med.* **184:** 1075–1082.

Okorokov A.L., Ponchel F., and Milner J. 1997. Induced N- and C-terminal cleavage of p53: A core fragment of p53, generated by interaction with damaged DNA, promotes cleavage of the N-terminus of full-length p53, whereas ssDNA induces C-terminal cleavage of 53. *EMBO J.* **16:** 6008–6017.

Oltvai Z., Milliman C., and Korsmeyer S.J. 1993. Bcl-2 heterodimerizes in vivo with a conserved homolog, Bax, that accelerates programmed cell death. *Cell* **74:** 609–619.

Owen-Schaub L.B., Zhang W., Cusack J.C., Angelo L.S., Santee S.M., Fujiwara T., Roth J.A., Deissertoth A.B., Zhang W.W., Kruzel E., and Radinsky R. 1995. Wild-type human p53 and a temperature-sensitive mutant induce Fas/APO-1 expression. *Mol. Cell. Biol.* **15:** 3032–3040.

Pan G., O'Rourke K., and Dixit V.M. 1998. Caspase-9, bcl-X$_L$, and apaf-1 form a ternary complex. *J. Biol. Chem.* **273:** 5841–5845.

Pan G., Ni J., Wei Y.F., Yu G., Gentz R., and Dixit V.M. 1997. An antagonist decoy receptor and a death domain-containing receptor for TRAIL. *Science* **277:** 815–818.

Patel T., Gores G.J., and Kaufmann S.H. 1996. The role of proteases during apoptosis. *FASEB J.* **10:** 587–597.

Polyak K., Zweler J.L., Kinler K.W., and Vogelstein B. 1997. A model for p53- induced apoptosis. *Nature* **389:** 300–305.

Prisco M., Hongo A., Rizzo M.G., Sacchi A., and Baserga R. 1997. The insulin-like growth factor I receptor as a physiologically relevant target of p53 in apoptosis caused by interleukin-3 withdrawal. *Mol. Cell. Biol.* **17:** 1084–1092.

Rampino N., Yamamoto H., Ionov Y., Li Y., Sawai H., Reed J.C., and Perucho M. 1997. Somatic frameshift mutations in the BAX gene in colon cancers of the microsatellite mutator phenotype. *Science* **275:** 967–969.

Reed J.C. 1995a. Bcl-2: Prevention of apoptosis as a mechanism of drug resistance.

Hematol. Oncol. Clin. N. Am. **9:** 451–474.

———. 1995b. Regulation of apoptosis by Bcl-2 family proteins and its role in cancer and chemoresistance. *Curr. Opin. Oncol.* **7:** 541–546.

———. 1997. Double identity for proteins of the Bcl-2 family. *Nature* **387:** 773–776.

Roemer K. and Mueller-Lantzch N. 1996. p53 transactivation domain mutant q22, s23 is impaired for repression of promoters and mediation of apoptosis. *Oncogene* **12:** 2069–2079.

Rosse T., Olivier R., Monney L., Rager M., Conus S., Fellay I., Jansen B., and Borner C. 1998. Bcl-2 prolongs cell survival after bax-induced release of cytochrome c. *Nature* **391:** 496–499.

Rothe M., Pan M.-G., Henzel W.J., Ayres T.M., and Goeddel D.V. 1995. The TNFR2-TRAF signaling complex contains two novel proteins related to baculoviral inhibitor of apoptosis proteins. *Cell* **83:** 1243–1252.

Roy N., Deveraux Q.L., Takashashi R., Salvesen G.S., and Reed J.C. 1997. The c-IAP-1 and c IaP-2 proteins are direct inhibitors of specific caspases. *EMBO J.* **16:** 6914–6925.

Roy N., Mahadevan M.S., McLean M., Shutler G., Yaraghi Z., Farahani R., Baird S., Besner-Johnson A., Lefebvre C., Kang X., Salih M., Aubry H., Tamai K., Guan X., Ioannou P., Crawford T.O., de Jong P.J., Surh L., Ikeda J.-E., Korneluk R.G., and MacKenzie A. 1995. The gene for neuronal apoptosis inhibitory protein is partially deleted in individuals with spinal muscular atrophy. *Cell* **80:** 167–178.

Ruoslahti E. and Reed J.C. 1994. Anchorage dependence, integrins, and apoptosis. *Cell* **77:** 477–478.

Sabbatini P., Chiou S.-K., Rao L., and White E. 1995. Modulation of p53-mediated transcriptional repression and apoptosis by the adenovirus E1B 19K protein. *Mol. Cell. Biol.* **15:** 1060–1070.

Sakakura C., Sweeney E.A., Shirahama T., Igarashi Y., Hakomori S., Nakatani H., Tsujimoto H., Imanishi T., Ohgaki M., Ohyama T., Yamazaki J., Hagiwara A., Yamaguchi T., Sawai K., and Takahashi T. 1996. Overexpression of bax sensitizes human breast cancer mcf-7 cells to radiation-induced apoptosis. *Int. J. Cancer* **67:** 101–105.

Salvesen G.S. and Dixit V.M. 1997. Caspases: Intracellular signaling by proteolysis. *Cell* **91:** 443–446.

Sato T., Hanada M., Bodrug S., Irie S., Iwama N., Boise L.H., Thompson C.B., Golemis E., Fong L., Wang H.-G., and Reed J.C. 1994. Interactions among members of the bcl-2 protein family analyzed with a yeast two-hybrid system. *Proc. Natl. Acad. Sci.* **91:** 9238–9242.

Schendel S.L., Xie Z., Montal M.O., Matsuyama S., Montal M., and Reed J.C. 1997. Channel formation by antiapoptotic protein Bcl-2. *Proc. Natl. Acad. Sci.* **94:** 5113–8.

Schlesinger P., Gross A., Yin X.-M., Yamamoto K., Saito M., Waksman G., and Korsmeyer S. 1997. Comparison of the ion channel characteristics of proapoptotic BAX and antiapoptotic BCL-2. *Proc. Natl. Acad. Sci.* **94:** 11357–11362.

Selkirk J.K., Merrick B.A., Stackhouse B.L., and He C. 1994. Multiple p53 protein isoforms and formation of oligomeric complexes with heat shock proteins Hsp70 and Hsp90 in the human mammary tumor, T47D, cell line. *Appl. Theor. Electrophor.* **4:** 11–8.

Sell C., Baserga R., and Rubin R. 1995. Insulin-like growth factor I (IGF-I) and the IGF-I receptor prevent etoposide-induced apoptosis. *Cancer Res.* **55:** 303–306.

Seshagiri S. and Miller L. 1997. *Caenorhabditis elegans* CED-4 stimulates CED-3 processing and CED-3-induced apoptosis. *Curr. Biol.* **7:** 455–460.

Shen Y. and Shenk T. 1994. Relief of p53-mediated transcriptional repression by the adenovirus E1B 19-kDa protein or the cellular bcl-2 protein. *Proc. Natl. Acad. Sci.* **91:** 8940–8944.

Shieh S.Y., Ikeda M., Taya Y., and Prives C. 1997. DNA damage-induced phosphorylation of p53 alleviates inhibition by MDM2. *Cell* **91:** 10.

Siliciano J.D., Canman C.E., Taya Y., Sakaguchi K., Appella E., and Kastan M.B. 1997. DNA damage induces phosphorylation of the amino terminus of p53. *Genes Dev.* **11:** 3471–3481.

Spector M.S., Desnoyers S., Heoppner D.J., and Hengartner M.O. 1997. Interaction between the *C. elegans* cell-death regulators CED-9 and CED-4. *Nature* **385:** 653–656.

Srinivasula S.M., Ahmad M., Ottilie S., Bullrich F., Banks S., Wang Y., Fernandes-Alnemri T., Croce C.M., Litwack G., Tomaselli K.J., Armstrong R.C., and Alnemri E.S. 1997. FLAME-1, a novel FADD-like antiapoptotic molecule that regulates Fas/TNFR1-induced apoptosis. *J. Biol. Chem.* **272:** 18542–18545.

Steller H. 1995. Mechanisms and genes of cellular suicide. *Science* **267:** 1445–1449.

Stromblad S., Becker J.C., Yebra M., Brooks P.C., and Cheresh D.A. 1996. Suppression of p53 activity and p21$^{WAF1/CIP1}$ expression by vascular cell integrin $\alpha v \beta 3$ during angiogenesis. *J. Clin. Invest.* **98:** 426–433.

Sumantran V.N., Ealovega M.W., Nuñez G., Clarke M.F., and Wicha M.S. 1995. Overexpression of Bcl-X$_S$ sentizes MCF-7 cells to chemotherapy-induced apoptosis. *Cancer Res.* **55:** 2507–2510.

Tanaka S., Louie D.C., Kant J.A., and Reed J.C. 1992. Frequent incidence of somatic mutations in translocated BCL2 oncogenes of non-Hodgkin's lymphomas. *Blood* **79:** 229–237.

Thome M., Schneider P., Hofmann K., Fickenscher H., Meinl E., Neipel F., Mattmann C., Burns K., Bodmer J.-L., Schroter M., Scaffidi C., Krammer P.H., Peter M.E., and Tschopp J. 1997. Viral FLICE-inhibitory proteins (FLIPs) prevent apoptosis induced by death receptors. *Nature* **386:** 517–521.

Tsujimoto Y., Cossman J., Jaffe E., and Croce C. 1985. Involvement of the bcl-2 gene in human follicular lymphoma. *Science* **228:** 1440–1443.

Turner B., Haffty B., Narayanan L., Yuan J., Havre P., Gumbs A., Kaplan L., Burgaud J.-L., Carter D., Baserga R., and Glazer P. 1997. Insulin-like growth factor-I receptor overexpression mediates cellular radioresistance and local breast cancer recurrence after lumpectomy and radiation. *Cancer Res.* **57:** 3079–3083.

Uren A.G., Pakusch M., Hawkins C.J., Puls K.L., and Vaux D.L. 1996. Cloning and expression of apoptosis inhibitory protein homologs that function to inhibit apoptosis and/or bind tumor necrosis factor receptor-associated factors. *Proc. Natl. Acad. Sci.* **93:** 4974–4978.

Varner J.A. and Cheresh D.A. 1996. Tumor angiogenesis and the role of vascular cell integrin alphavbeta3. *Important Adv. Oncol.*, pp. 69–87.

Vaux D.L., Cory S., and Adams J.M. 1988. Bcl-2 gene promotes haemopoietic cell survival and cooperates with c-myc to immortalize pre-B cells. *Nature* **335:** 440–442.

Vaux D.L., Weissman I.L., and Kim S.K. 1992. Prevention of programmed cell death in *Caenorhabditis elegans* by human bcl-2. *Science* **258:** 1955–1957.

Vogelstein B. and Kinzler K.W. 1992. p53 function and dysfunction. *Cell* **70:** 523–526.

Vucic D., Kaiser W.J., Harvey A.J., and Miller L.K. 1997. Inhibition of reaper-induced apoptosis by interaction with inhibitor of apoptosis proteins (IAPs). *Proc. Natl. Acad. Sci.* **94:** 10183–10188.

Wallach D., Boldin M., Varfolomeev E., Beyaert R., Vandenabeele P., and Fiers W. 1997. Cell death induction by receptors of the TNF family: Towards a molecular understanding. *FEBS Lett.* **410:** 96–106.

Wang L., Miura M., Bergeron L., Zhu H., and Yuan J. 1994. Ich-1, and ice/ced-3-related gene, encodes both positive and negative regulators of programmed cell death. *Cell* **78:** 739–750.

Webster N.J., Resnik J.L., Reichart D.B., Strauss B., Haas M., and Seely B.L. 1996. Repression of the insulin receptor promoter by the tumor suppressor gene product p53: A possible mechanism for receptor overexpression in breast cancer. *Cancer Res.* **56:** 2781–2788.

Werner H., Karnieli E., Rauscher F.J., and LeRoith D. 1996. Wild–type and mutant p53 differentially regulate transcription of the insulin-like growth factor I receptor gene. *Proc. Natl. Acad. Sci.* **93:** 8318–8323.

Wolter K.G., Hsu Y.T., Smith C.L., Nechushtan A., Xi X.G., and Youle R.J. 1997. Movement of bax from the cytosol to mitochondria during apoptosis. *J. Cell Biol.* **139:** 1281–1292.

Wu D., Wallen H.D., and Nunez G. 1997. Interaction and regulation of subcellular localization of CED-4 by CED-9. *Science* **275:** 1126–1129.

Wu G.S., Burns T.F., McDonald E.R., III, Jiang W., Meng R., Krantz I.D., Kao G., Gan D.D., Zhou J.Y., Muschel R., Hamilton S.R., Spinner N.B., Markowitz S., Wu G., and el-Deiry W.S. 1997. KILLER/DR5 is a DNA damage-inducible p53-regulated death receptor gene. *Nat. Genetics* **17:** 141–143.

Wyllie A.H., Kerr J.F.R., and Currie A.R. 1980. Cell death, the significance of apoptosis. *Int. Rev. Cytol.* **68:** 251–306.

Xue D. and Horvitz H.R. 1997. *Caenorhabditis elegans* CED-9 protein is a bifunctional cell-death inhibitor. *Nature* **390:** 305–308.

Yamamoto H., Sawai H., and Perucho M. 1997. Frameshift somatic mutations in gastrointestinal cancer of the microsatellite mutator phenotype. *Cancer Res.* **57:** 4420–4426.

Yang J., Liu X., Bhalla K., Kim C.N., Ibrado A.M., Cai J., Peng I.-I., Jones D.P., and Wang X. 1997. Prevention of apoptosis by Bcl-2: Release of cytochrome c from mitochondria blocked. *Science* **275:** 1129–1132.

Yang T., Kozopas K.M., and Craig R.W. 1995. The intracellular distribution and pattern of expression of Mcl-1 overlap with, but are not identical to, those of Bcl-2. *J. Cell Biol.* **128:** 1173–1184.

Yin C., Knudson C.M., Korsmeyer S.J., and Van Dyke T. 1997. Bax suppresses tumorigenesis and stimulates apoptosis in vivo. *Nature* **385:** 637–640.

Yin Y., Terauchi Y., Solomon G.G., Aizawa S., Rangarajan P.N., Yazaki Y., Kadowaki T., and Barrett J.C. 1998. Involvement of p85 in p53-dependent apoptotic response to oxidative stress. *Nature* **391:** 707–710.

Yonish-Rouach E., Deguin V., Zaitchouk T., Breugnot C., Mishal Z., Jenkins J.R., and May E. 1996. Transcriptional activation plays a role in the induction of apoptosis by transiently transfected wild-type p53. *Oncogene* **12:** 2197–2205.

Yuan J.Y. and Horvitz H.R. 1990. The *Caenorhabditis elegans* genes ced-3 and ced-4 act cell autonomously to cause programmed cell death. *Dev. Biol.* **138:** 33–41.

Yuan J., Shaham S., Ledoux S., Ellis H.M., and Horvitz H.R. 1993. The *C. elegans* cell death gene ced-3 encodes a protein similar to mammalian interleukin-1 beta-converting enzyme. *Cell* **75:** 641–652.

Zhan Q., Fan S., Bae I., Guillouf C., Liebermann D.A., O'Connor P.M., and Fornace Jr., A.J. 1994. Induction of BAX by genotoxic stress in human cells correlates with normal p53 status and apoptosis. *Oncogene* **9:** 3743–3751.

Zhang L., Kashanchi F., Zhan Q., Zhan S., Brady J.N., Fornace A.J., Seth P., and Helman L.J. 1996. Regulation of insulin–like growth factor II P3 promotor by p53: A potential mechanism for tumorigenesis. *Cancer Res.* **56:** 1367–1373.

21

Advances in Gene Therapy for HIV and Other Viral Infections

Eric M. Poeschla

Department of Medicine
University of California, San Diego
La Jolla, California 92093-0655

Flossie Wong-Staal

Departments of Medicine and Biology
University of California, San Diego
La Jolla, California 92093-0655

Gene therapies and other treatment approaches to human immunodeficiency virus (HIV) disease are rapidly evolving in the context of several recent developments. Although the mechanisms that perturb T-cell homeostasis in HIV disease remain enigmatic, much has been gained from quantitative modelling of the dynamics of HIV replication in vivo (Embretson et al. 1993; Pantaleo et al. 1993; Piatak et al. 1993; Coffin 1995; Ho et al. 1995; Wei et al. 1995; Haase et al. 1996; Perelson et al. 1996, 1997; Cavert et al. 1997). These insights were enabled in part by a second discovery: HIV replication can be potently and durably suppressed in vivo in many patients by combining previously marginally effective reverse transcriptase inhibitors with HIV type 1 (HIV-1) protease inhibitors, which are peptidomimetic transition state analogs that were designed from knowledge of the crystal structure of this enzyme (Gulick et al. 1997; Hammer et al. 1997). In addition, the long-sought coreceptors to CD4 have been identified, a discovery that has yielded a coherent molecular explanation for how HIV infects varied cell types in vivo and imparted conceptual clarity to what was previously a murky phenomenology of viral phenotypes. This burst of discoveries about chemokines and chemokine receptors has also revealed the first convincing evidence for a link between a human genetic mutation and resistance to HIV infection. All of these findings have important implications for the potential of gene therapy for HIV disease.

HIV is a lentiretrovirus with specialized ability to infect nondividing and postmitotic cells permanently. It has therefore always been probable that diverse cellular sanctuaries of latent, infectious virus persist in vivo. Some appear to possess very slow decay rates, and permanent reservoirs may exist (Perelson et al. 1997). If virus in peripheral blood is suppressed below the limits of current RNA detection assays for 1–2 years, withdrawal of drug therapy inevitably leads to resurgence of high levels of replication within days to weeks. Consistent with these observations, when T cells from patients who have been maintained on a year or so of drug suppression with nucleosides and protease inhibitors are activated ex vivo, outgrowth of infectious, drug-sensitive virus is readily observed (Chun et al. 1997a; Finzi et al. 1997; Wong et al. 1997). Replication-competent, integrated HIV persists in comparatively rare, but long-lived, resting T cells in vivo (Chun et al. 1995, 1997). After prolonged, suppressive antiviral therapy, replication-competent provirus was also shown in one study to reside in the resting CD4+ T-cell population and to not decrease with time on therapy (Finzi et al. 1997). These reports are from patients on only 1–2 years of uninterrupted three-drug therapy, so more extended treatment durations could eventually produce different outcomes.

These and other studies indicate that in addition to the problems of long-term compliance with complex medication regimens, drug toxicity, drug resistance, and expense, neither eradication of infection nor restoration of robust immunity may be feasible through prolonged drug therapy alone. It is clear that arresting viral replication is only a first step to the complex task of restoring immune function; suppression of plasma viremia most often results in only slight increases in CD4+ T-cell levels and function (Autran et al. 1997; Connors et al. 1997). Further analyses of T-cell life span and lymphocyte population changes in patients receiving highly active antiretroviral therapy (HAART) suggest that arresting viral replication results first in redistribution to the periphery of memory lymphocytes entrapped within lymphoid tissues, followed by a slower accumulation of newly produced T cells with a naive phenotype (Roederer 1988; Pakker et al. 1998). All of these studies make clear that we only partially understand the mechanisms by which HIV destroys the immune system. For these reasons, the prospect of reducing the genetic susceptibility of cellular targets to HIV by gene therapy remains an important alternative. Synergism between therapeutic strategies is likely to be essential. For example, reducing infected cell burden with combination drug therapy may enhance the potential of gene therapy and/or gene vaccination directed at immunologic preservation (Rosenberg et al. 1997). This chapter reviews selected recent developments in gene therapy for HIV and other infectious diseases and related topics such as adoptive T-cell transfer.

ANTIVIRAL GENE THERAPY: WHAT MUST BE ACCOMPLISHED

Any antiviral strategy must take into account what is known about HIV-1 dynamics in vivo and the associated mathematical imperatives of drug resistance. The situation is no different for gene therapy (Yu et al. 1994; Poeschla and Wong-Staal 1995; Wong-Staal 1996). Like other RNA viruses, HIV displays mutation rates on the order of 10^{-3} to 10^{-5} per nucleotide per replication cycle, approximately a millionfold that of cellular DNA and DNA virus mutation rates (Domingo and Holland 1994). This genetic plasticity is amplified in vivo by prolific, sustained viral replication that is evident from the onset of HIV infection and persists after the development of both humoral and cellular immune responses in most individuals. A synopsis of recent findings are that roughly 10^{10} virions are produced per day 140 viral generations occur per year, productively infected T cells die with a $t_{1/2}$ of approximately 1.6 days, and virions decay with a $t_{1/2}$ of about 6 hours. Even shorter virion half-life has recently been suggested from studies in the simian immunodeficiency virus (SIV) model (P. Johnson, and D. Ho, pers. comm.). Daily turnover of 10^9 infected CD4$^+$ T cells may occur (Embretson et al. 1993; Pantaleo et al. 1993; Piatak et al. 1993; Coffin 1995; Ho et al. 1995; Wei et al. 1995; Haase et al. 1996; Perelson et al. 1996, 1997; Cavert et al. 1997). A reasonable conclusion is that replication drives pathogenesis. Even a single measurement of plasma HIV-1 RNA, which may range from the current detection limit of 50 copies/ml to more than 10^6 copies/ml, correlates well with prognosis for subsequent CD4$^+$ lymphocyte depletion and disease development (Mellors et al. 1996).

Although the mechanisms involved in CD4$^+$ T-cell depletion remain controversial, as is even the question of whether increased destruction or decreased production are dominant (Wolthers et al. 1996; Pakker et al. 1998; Roederer 1988), the foregoing numerical estimates indicate that the pressure for mutation-driven escape of HIV from antiviral agents is high and unremitting. Ongoing clinical data suggest that any strategy that does not result in complete, sustained suppression of viral replication is highly likely to foster emergence of drug resistance and clinical failure (Havlir and Richman 1996). In particular, suppression by combination drug therapy must be uninterrupted since even brief nonadherence can result in permanent loss of efficacy for single drugs or classes of drugs. Such long-term (potentially lifelong) seamless compliance with sometimes toxic triple- or quadruple-drug regimens has no successful precedent in medical therapeutics. Not surprisingly, clinical outcomes outside the artificially supportive environments of clinical trials are showing higher rates of resistance and treatment failure.

Because avoiding resistance is central to clinical success, a potentially crucial advantage of nucleic-acid-based gene-therapy strategies (e.g., antisense RNA and ribozymes) over current antiviral drugs lies in design. Drugs enter the development pipeline as a result of empirical screening or, much less commonly, by design from molecular target modeling. In no case, however, has it been possible to design drug molecules to target regions of the HIV genome specifically that are less prone or insusceptible to viable escape mutations. Nucleic acid modalities can exploit the specificity of Watson-Crick base pairing to target such regions selectively and may therefore be less likely to select for resistance. Although this advantage is compelling in principle, it is not demonstrated by data, since in vivo studies showing lack of HIV resistance development are not yet at hand. In contrast to antiviral drugs, if a limited number of resistance mutations arise within a conserved target, the sequence of an antiviral RNA can be precisely adjusted to bind the mutant target, provided the mutation is not in indispensable residues (see the section on ribozymes).

NEW TARGETS: HIV CORECEPTORS

Targeting the virus itself is the focus of most therapeutic approaches. However, the parallel strategy of targeting or exploiting cellular molecules needed for HIV replication in vivo is also plausible. Until recently, CD4 was the only known lentiviral receptor. The inability of HIV to infect certain nonprimate cells, even if they express human CD4, has long indicated that CD4 alone is insufficient to mediate viral entry. In addition, the classification of HIV strains into laboratory adapted, T-cell-line-tropic (T-tropic) and primary, macrophage-tropic (M-tropic) phenotypes was exhaustively described but poorly understood. The discoveries by Berger and colleagues (Feng et al. 1996) that a seven-transmembrane-domain α-chemokine receptor (CXCR4) mediates fusion of T-tropic HIV strains and of Gallo and colleagues (Cocchi et al. 1995) that β-chemokines inhibit infection by M-tropic HIV led to the subsequent validation by numerous laboratories that chemokine receptors are the long-sought HIV coreceptors. In conjunction with CD4, the β-chemokine receptor CCR5 principally mediates entry of M-tropic HIV (Alkhatib et al. 1996; Choe et al. 1996; Deng et al. 1996; Doranz et al. 1996; Dragic et al. 1996; Trkola et al. 1996; Wu et al. 1996) and CXCR4 principally mediates entry of T-tropic virus (Feng et al. 1996). Additional chemokine receptors appear to be involved in infection in some cell types (Doranz et al. 1996; Deng et al. 1997; He et al. 1997). Remarkably, even feline immunodeficiency virus, one of the five known nonprimate lentiviruses, turns out to also use CXCR4, even the human homolog, as a receptor (Willett et al. 1997; Poeschla and Looney 1998). This is the case even though the nonprimate lentiviruses do not use CD4. Use of CXCR4, and undoubtedly of other chemokine receptors, thus appears to be a fundamental feature of lentivirus biology independent of CD4. This shared cellular link to infection and cytopathicity for distantly related lentiviruses that cause AIDS implicates chemokine receptors as the primordial lentivirus receptors (Poeschla and Looney 1998; for recent reviews, see Berger 1997; Doranz et al. 1997; Moore 1997).

Another remarkable outcome of studies of chemokine receptor biology and HIV epidemiology may be of particular relevance to gene therapy. Some people do not become infected by HIV despite persistent, normally efficient exposure. A subset of these individuals have T cells that are resistant to infection by M-tropic but not T-tropic HIV strains. This resistance has been shown to have a genetic basis: homozygosity for a 32-nucleotide deletion in the gene encoding CCR5 (Dean et al. 1996; Huang et al. 1996; R. Liu et al. 1996; Paxton et al. 1996; Samson et al. 1996; Martinson et al. 1997; Michael et al. 1997; Zimmerman et al. 1997). The $\Delta 32$ CCR5 allele is prevalent in Caucasian populations, where approximately 1% are homozygous and 15–20% of individuals are heterozygous. $\Delta 32$ homozygotes are even protected against direct intravenous challenge, which indicates that the mutation does not simply provide a barrier to establishment of infection in an initial mucosal macrophage or dendritic cell compartment after exposure. Why the CCR5 gene is dispensable in humans and why non-Caucasian populations do not harbor the allele to any extent are unknown, but could reflect ancient disparities in exposure to infectious pathogens.

The HIV coreceptors are intriguing but uncertain therapeutic targets. The finding that $\Delta 32$ CCR5 homozygotes suffer no apparent health problems from the gene defect makes this molecule, as well as other chemokine receptors, a plausible focus for antiviral gene therapy. In addition, the majority of heavily exposed, uninfected individuals are not $\Delta 32$ CCR5 homozygotes, suggesting other, possibly genetic, resistance factors exist (Zimmerman et al. 1997). However, the protection the $\Delta 32$ CCR5 deletion confers to homozygotes is not absolute, since a few HIV-infected homozygotes have been identified (Biti et al. 1997), and these patients are not protected from rapid disease progression once infection is established. As described later, this may reflect use of alternative chemokine receptors, especially CXCR4.

Because they are encoded by cellular genes, coreceptors are much less susceptible to mutation than are HIV genes (Landau 1997). However, a noteworthy precedent exists for the failure of therapeutic strategies based on expression of an HIV receptor, as soluble CD4 was ineffective in inhibiting primary HIV-1 isolates (Daar et al. 1990; Daar and Ho 1991). In addition, even if effective small molecules or gene-therapy strategies that target the major coreceptor for primary HIV-1 isolates (CCR5) are developed, there is the real possibility of simply pushing in vivo evolution toward use of other chemokine receptors (Lu et al. 1997). CXCR4, CCR2, and CCR3 have all been shown to be alternative coreceptors, and CXCR4 use correlates with emergence of syncytium-inducing strains and may herald accelerated disease progression (Choe et al. 1996; Rucker et al. 1996; He et al. 1997; Moore 1997; Scarlatti et al. 1997). Heterozygosity for Δ32-CCR5 may confer resistance to disease progression, but only if CXCR4 use does not emerge (Michael et al. 1997). CCR5 is expressed principally on CD45 RO$^+$ memory T cells, whereas CXCR4 is expressed in T cells with a CD45 RA$^+$ or resting phenotype (Bleul et al. 1997). The transition to CXCR4 use may therefore hugely increase the pool of T cells susceptible to infection. In addition, two more coreceptors have been found for SIV and HIV (Deng et al. 1997), and 50 or more chemokine receptor genes may exist in the human genome (Landau 1997). Nevertheless, since use of chemokine receptors other than CCR5 appears to be selected against in vivo until the advent of immune collapse, CCR5-directed therapy has the most rationale at present (Landau 1997).

Initial attempts to devise gene-therapy strategies that exploit discoveries about HIV coreceptors have been reported. Acting on the premise that preventing CXCR4 use could inhibit disease progression, Chen et al. (1997) incorporated a tetrapeptide endoplasmic reticulum (ER) retention signal (KDEL) in the carboxyl terminus of SDF-1, which is the natural C–X–C chemokine ligand for CXCR4 (Bleul et al. 1996). When expressed from retroviral vectors, this hybrid "intrakine" was retained in the ER and reduced the expression of surface CXCR4, presumably through sequestering the molecule in the ER. Both transduced T-cell lines and primary peripheral blood lymphocytes were relatively resistant to challenge by CXCR4-using HIV strains. The analogous strategy for CCR5, using RANTES and MIP-1 intrakines, has also been reported (A.G. Yang et al. 1997). However, the physiologic consequences of impairing cell surface chemokine receptor expression in vivo remain unclear.

Now that the full complement of molecules needed to permit HIV-1 entry is known for most cell types, it has been possible to show that fusion resulting from binding of HIV and SIV envelope glycoproteins to CD4 and chemokine receptors is directionally independent. Replicating rhabdoviruses (Schnell et al. 1997) encoding CD4 and CXCR4, rhabdoviral particles (Mebatsion et al. 1997) pseudotyped with CD4 and CXCR4, or retroviral vector particles (Endres et al. 1997) pseudotyped with a CD4-chemokine receptor complex relevant to a given HIV/SIV strain can fuse with and infect or transduce infected cells. Such approaches offer a novel way to target selectively only infected cells with vectors expressing antiviral or toxic genes or simply with viruses whose productive phase kills the cell (Schnell et al. 1997). It is conceivable that any enveloped virus could be adapted to reverse the usual polarity of membrane fusion in this manner. Previous work had already shown that CD4 could be incorporated into avian leukosis virus particles (Young et al. 1990) and into vesicular stomatitis virus (VSV) (Schnell et al. 1996; J.E. Johnson et al. 1997). In each case, hybrid receptor/envelope chimerics that permitted selective incorporation into viral particles were instrumental. Schnell et al. (1997) engineered VSV to replicate and express CD4 and CXCR4 instead of the very broadly tropic native VSV glycoprotein G envelope protein. The striking outcome was that this virus (VSVΔG-CC4) continuously propagated through HIV-1-infected cultures, where it selectively destroyed HIV-1-infected cells and reduced levels of infectious HIV and RT activity by $2 \log_{10}$ to $3 \log_{10}$. Both VSVΔG-CC4 and HIV-1 persisted in the culture indefinitely, presumably because HIV-1-infected cells that express levels of gp120 too low to engage the chimeric VSV remain viable. Consistent with this scenario, low-level increases in infectious HIV appeared periodically in the coinfected cultures but were rapidly suppressed.

There are several novel potential advantages to such an "antiviral virus" and several rather formidable drawbacks (Nolan 1997). On the plus side, the chimeric VSV is restricted to targeting only HIV-infected cells, and its replication would parallel HIV infection and be expected to diminish with clearance of HIV-infected cells (Schnell et al. 1997). Unlike drug-resistance mutations, mutations in HIV-1 *env* that confer resistance to infection by the VSVΔG-CC4 virus would entail loss of the HIV's ability to infect cells via these molecules. It is also possible that this virus could bind directly to and fuse with virions themselves. The fact that the virus encodes only "self" molecules on its surface would in theory prevent the development of a neutralizing antibody. Multiple coreceptors might also be employed in one particle to achieve activity against a range of viral phenotypes.

On the minus side, the approach would not prevent rapid development of an ablative cellular immune response as the VSV core proteins will be expected to enter the normal major histocompatibility complex (MHC) class I antigen presentation pathway. Resulting cytotoxic T lymphocyte (CTL) responses would be likely to clear the therapeutic virus while the ability of HIV-1 to persist in the face of vigorous CTL responses is well established. In addition, the chimeric VSV has only been studied in a T-cell line. In the body, most HIV-infected cells may be well on the path to death by the time they express significant amounts of cell surface gp120 (note the cellular half-life estimates discussed at the start of this chapter). Viewed in this light, the strategy is unlikely to advance beyond application to T-cell lines that are resistant to HIV killing in vitro. Other, more complicated, medical issues also await study before these cell culture results can be contemplated for clinical application (Nolan 1997). Potential problems include the uncertain behavior of such a virus in immunocompromised patients, transfer from patients to unintended recipients, and deleterious immunological consequences, e.g., the possibility that continuous antigen presentation of foreign fusion site epitopes would induce anti-self-immune responses to chemokine receptors or to CD4. Nevertheless, because of their specificity, animal model investigations of these live vectors and adaptation of the general approach for in vivo targeting of replication-defective vectors (Mebatsion et al. 1997) or even nonviral liposome devices (Nolan 1997) deserve detailed study.

The discovery of specific human resistance to HIV-1 will probably also make xenotransplantation a less attractive option for HIV-1 infection since it raises the possibility of transplanting bone marrow from Δ32-CCR5 homozygotes or other individuals with as yet to be discovered genotypes. The counterargument is that since Δ32-CCR5 homozygotes are not protected from disease progression once infected (Biti et al. 1997), CXCR4-tropic or other chemokine-receptor-using variants may prevail even if CCR5 is ablated in an infected individual.

Because HIV-1 infects a variety of nondividing cells in vivo, replication-defective viral vectors that can target such cell populations may have relevance to gene therapy of HIV disease and more generally to hematopoietic stem cell gene transfer. Lentiviral vectors based on HIV-1 have been shown to deliver marker genes to postmitotic cells in vitro and in vivo (Naldini et al. 1996). However, HIV-1 vectors raise obvious and complex safety concerns (Emerman 1996). Development of nonprimate lentivirus vectors is one alternative strategy, since the nonprimate lentiviruses cannot infect humans. However, the highly restricted tropisms of these viruses for particular feline and ungulate host species correspondingly raises doubts about feasibility. Recently, a feline immuno-deficiency virus lentiviral vector system capable of transducing nondi-viding human cells has been described (Poeschla et al. 1998).

RIBOZYMES AND ANTISENSE RNAs: MOLECULES THAT EVADE IMMUNE RECOGNITION AND ARE PROSPECTIVELY TARGETABLE TO CONSERVED VIRAL SEQUENCES

Antisense RNAs and ribozymes possess several advantages over antiviral drugs and protein-based gene therapy strategies (Poeschla and Wong-Staal 1994, 1995). First, untranslated RNAs are unlikely to generate humoral or cellular immune responses. In particular, induction of cell-mediated immune responses requires that short peptides be derived from antigens in intracellular proteasomes, loaded into the clefts of class I MHC molecules, and presented at the cell surface to T-cell receptors. The prediction that this highly evolved system for detecting foreign intracel-lular proteins would trigger CTL responses to gene-altered cells express-ing nonhost proteins has been borne out in a human gene-therapy trial. When CD8[+] HIV-specific cytotoxic T cells transduced with retroviral vectors encoding a suicide gene/selectable marker fusion protein were adoptively transferred to HIV-1-infected subjects, five of six developed CTL responses that eliminated the transduced cells (Riddell et al. 1996). The generation of ablative CTL responses by these immunocompromised recipients suggests that successful gene therapy with proteins will require novel approaches to circumventing CTL recognition, all the more so for immunocompetent recipients. Viruses form the main evolutionary selec-tion pressure for the elaborate antigenic peptide presentation system and have necessarily coevolved with it: Since a number of viruses, including HIV-1 and several of the Herpesviridae, have evolved diverse strategies to evade CTL recognition by encoding proteins that interfere with class I MHC presentation (Hill et al. 1995; Schwartz et al. 1996; Fruh et al. 1997), coexpression of such proteins with foreign therapeutic proteins in gene therapy has been proposed (Wiertz et al. 1997). A problem is that these proteins would be expected to render cells generally vulnerable to infection by intracellular parasites.

Ribozymes are RNA enzymes that bind to specific sequences in RNA substrates and catalyze endoribonucleolytic cleavage (Kruger et al. 1982; Symons 1992). Cleavage is normally an intramolecular reaction that occurs naturally in processing of certain introns and in the replication of certain plant viroids and one human pathogen, hepatitis delta virus (Haseloff and Gerlach 1988,1989), but the reaction can be engineered to occur in *trans*. Since the domains, or helices, of ribozymes that base-pair to substrate RNAs are functionally separable from the moieties that effect cleavage, the substrate specificity of ribozymes can be altered within certain constraints to allow catalytic, *trans*-cleavage of many sites in the HIV-1 genome (Dropulic et al. 1993; Poeschla and Wong-Staal 1994; Gibson and Shillitoe 1997). For example, hairpin ribozymes require only a guanosine residue immediately 3′ to the cleavage site (Chowrira et al. 1991), although a GUC sequence is optimal. Twelve residues flanking the cleavage site bind to the substrate RNA to form two base-paired, bimolecular helices; these residues can be modified extensively, allowing versatile targeting (Berzal-Herranz et al. 1993; Joseph and Burke 1993; Anderson et al. 1994; Poeschla and Wong-Staal 1994). The hammerhead ribozyme is even less constrained, requiring only a UN dinucleotide for cleavage (where N is either A, C, or U).

It is this potential to prospectively target short stretches of highly conserved regions of the HIV genome that are less susceptible to mutational escape and the potential to counter precisely specific resistance mutations that may constitute the chief advantages of untranslated RNAs over antiviral proteins (Poeschla and Wong-Staal 1994, 1995; Rossi 1995; Wong-Staal 1996). Both hairpin and hammerhead ribozymes have been applied for the purposes of inhibiting HIV-1 replication (Sarver et al. 1990; Akhtar and Rossi 1996; Ramezani and Joshi 1996; Wong-Staal 1996; Bauer et al. 1997; Weerasinghe et al. 1991; Chatterjee et al. 1992; Rossi et al. 1992; Dropulic et al. 1993; Poeschla and Wong-Staal 1994; Zhou et al. 1994; Rossi 1995; Sun et al. 1995, 1997). Hairpin ribozymes have been shown to inhibit HIV-1 replication in human T-cell lines, in primary peripheral blood lymphocytes, and in the differentiated progeny of $CD34^+$ hematopoietic stem cells (Yu et al. 1993, 1995; Leavitt et al. 1994, 1996; Yamada et al. 1994, 1996a; Heusch et al. 1996; Wong-Staal 1996). Yamada et al. (1994) reported inhibition of both pre- and postintegration steps in the viral life cycle.

Recently, a chimeric HIV-1 minimal Rev response element (RRE)–ribozyme fusion molecule was shown to combine the inhibitory properties of a hairpin ribozyme targeting the HIV-1 U5 element with an RNA decoy effect and to be more efficient than either the ribozyme or the decoy alone (Yamada et al. 1996). In addition, this fusion molecule was inserted in the U3 region of double-copy retrovirus vectors that also internally encoded a separate RRE-ribozyme fusion molecule targeting the *env*/*rev* region (Gervaix et al. 1997). Ribozyme RNA transcripts were estimated by competitive-quantitative reverse transcription (RT) PCR to be 1.2×10^5 1.2×10^4, and 1.5×10^3 copies per cell for the triple-copy, double-copy, and single-copy vectors, respectively. In addition, multiple geographic subtypes of HIV-1 (clades A–E) were susceptible to commensurate levels of inhibition by the three vectors.

These and other studies illustrate that vector design, promoter choice and considerations of promoter interference, intracellular localization, and cell-type-specific expression require individual attention. The particular promoters used and their configuration in retroviral vectors have been shown to have pronounced effects on ribozyme efficacy (Thompson et al. 1995; Bertrand et al. 1997; Good et al. 1997). Subcellular localization of ribozyme transcripts can determine activity (Hormes et al. 1997). Strategies that colocalize ribozymes intracellularly with substrate molecules have been shown to enhance activity markedly (Sullenger 1995; Sullenger and Cech 1993).

In evaluating the potential of gene-therapy strategies such as ribozymes and antisense RNA, the lessons learned from antiviral drug therapies are valuable. Although protease inhibitors lack the advantage of versatile targeting and readily generate resistance, their potency, in terms of $\log_{10}$ suppression of viral replication, exceeds that of RNAs or proteins used in gene therapy studies. Combination antiviral therapy was marginally successful before the achievement of such profound suppression. It is also abundantly clear from many studies that inhibition by ribozymes and antisense RNA molecules is quite sensitive to multiplicity of infection (Akhtar and Rossi 1996) and can be overwhelmed by increased viral input in cell-culture systems studied to date. Further improvements in vector design that allow higher levels of sustained gene expression, as well as the aforementioned colocalization strategies, may help in this regard. In addition, it will be useful if methods can be developed to detect ribozyme cleavage products directly in human cells and to verify catalytic mode of action better. Some progress has been made in regard to the latter issue by use of control (disabled) ribozymes with mutations in the catalytic domain (Yamada et al. 1994, 1996b).

A phase I trial is in progress to evaluate the safety of anti-HIV-1 hairpin ribozymes in vivo using ex vivo T-cell transduction (Leavitt et al. 1996). Three patients have now been infused with a mean of approximately 10^{10} autologous T cells and no significant adverse effects have been apparent (E.M. Poeschla and F. Wong-Staal, unpubl.). Additional patients are currently under study.

Ribozymes that cleave the RNAs of human hepatitis viruses have also been studied, although work on these viruses has proceeded more slowly because of a lack of tissue culture models. Intracellular inhibition of viral production for hepatitis B (Welch et al. 1997) and of infection of cells by hepatitis C has been accomplished using both hammerhead (Lieber et al. 1996; Sakamoto et al. 1996) and hairpin (Welch et al. 1996) ribozymes.

RNA decoys, antisense RNAs, and antisense oligodeoxynucleotides have been studied extensively for anti-HIV activity. The decoy strategy shows promise for a specific mode of action and may avert resistance by targeting conserved *trans*-activation response (TAR) and RRE domains (Sullenger et al. 1990, 1991; Lisziewicz et al. 1993; Chang et al. 1994; Dropulic and Jeang 1994; Gilboa and Smith 1994; Lee et al. 1994, 1995; Lori et al. 1994; Lisziewicz 1996). In many studies, the basis for the antiviral activity of antisense oligonucleotides has been unclear, as a variety of nonspecific effects have been documented (Stein and Cheng 1993; Wagner 1994).

TRANSDOMINANT PROTEINS AND INTRACELLULAR ANTIBODIES

Transdominant mutants competitively interfere with wild-type protein function (Herskowitz 1987). This mechanism was first validated with a mutant of the herpes simplex virus *trans*-activator VP16 (Friedman et al. 1988). In the case of HIV-1, mutants of Rev, a viral regulatory protein indispensable for accumulation of unspliced RNA and hence for viral replication, have received considerable study. One such transdominant, RevM10, has two amino acid substitutions (D78L and L79E) in a conserved carboxy-terminal leucine-rich domain. RevM10 exhibits wild-type RRE binding and Rev/Rev multimerization but has lost the ability to mediate unspliced HIV mRNA transport. Stable T-cell lines and primary human peripheral blood lymphocytes (PBLs) expressing RevM10 are resistant to HIV-1 replication (Malim et al. 1989, 1992; Malim and Cullen 1991; Escaich et al. 1995; Fox et al. 1995). No toxicity has been apparent in cultured cells. RevM10 has now been studied in vivo in a human gene therapy clinical trial (Nabel et al. 1994; Woffendin et al. 1994, 1996). Autologous CD4-enriched T cells from HIV-1-infected subjects were transfected ex vivo with linearized RevM10 plasmid DNA or a deletion mutant control (ΔRevM10) using gold microparticle injection (Woffendin et al. 1996). Three patients were infused with 10^9 to 10^{10} cells, of which approximately 10% had detectable plasmid DNA at infusion. Postinfusion limiting dilution polymerase chain reaction (PCR) analyses indicated that in the two patients who received greater than 10^{10} total cells, a four- to fivefold selective survival advantage was seen for RevM10 over ΔRevM10 and these differences persisted for up to 4 weeks, with RevM10 containing cells still detectable at up to 8 weeks. No toxicity was detected. In vitro, RevM10 also inhibited viral replication in progeny of primitive hematopoietic progenitors (Bonyhadi et al. 1997). Transdominant HIV Tat (Caputo et al. 1996; Orsini and Debouck 1996), protease (Babe et al. 1995; Junker et al. 1996), Gag (Trono et al. 1989) and envelope (Buchschacher et al. 1992; Steffy and Wong-Staal 1993; Chen et al. 1996) mutants have also exhibited antiviral activity in vitro.

The intrakine strategy just discussed is reminiscent of the previously described "intrabody" concept. Single-chain variable region antibody fragments or monovalent Fab molecules can be expressed and confined intracellularly, where they have been shown to bind to viral proteins. Further improvements are likely to accrue from exploitation of powerful technologies that bypass immunization in order to produce high-affinity monoclonal antibodies directly through creative use of antibody libraries and phage display (Burton et al. 1991; Duchosal et al. 1992; Lerner et al. 1992; Chanock et al. 1993; Williamson et al. 1993). Several laboratories have developed intrabodies that target herpes simplex virus or HIV regulatory or structural genes (Marasco et al. 1993; Chen et al. 1994; Duan et al. 1994a,b 1995; Mhashilkar et al. 1995; Richardson and Marasco 1995; Sanna et al. 1995; Levy-Mintz et al. 1996; Pilkington et al. 1996; Shaheen et al. 1996; Levin et al. 1997; Marasco 1997). Sequestration of the HIV-1 envelope by intracellularly retained CD4 is a similar strategy (Buonocore and Rose 1990, 1993), but neutralization of primary isolates is likely to be as problematic as that for soluble CD4 (Daar and Ho 1991).

ADOPTIVE CELL TRANSFER FOR TREATMENT OF VIRAL DISEASES: IMPACT OF GENE TRANSFER TECHNOLOGY AND NEW T-CELL-CULTURE METHODS

Several studies have demonstrated that large numbers of cloned, virus-specific T cells can be transferred to humans with favorable therapeutic effects. This topic has been comprehensively reviewed recently (Riddell et al. 1993, 1994; Greenberg et al. 1994; Riddell and Greenberg 1994, 1995a,b). Here we review selected aspects related to use of gene transfer and advances in cell culture of T cells from HIV-1-infected individuals.

The first human investigations of adoptive T-cell clone transfer were carried out in allogeneic bone-marrow transplant recipients at high risk for cytomegalovirus (CMV) disease (Riddell et al. 1992; Walter et al. 1995). CMV-induced interstitial pneumonia, the most life-threatening manifestation of CMV disease in such patients, occurs at high frequency in individuals who fail to recover CMV-specific CTL responses after transplantation (Li et al. 1994). Approximately $10^9/m^2$ CD8[+] T-cell clones specific for CMV structural proteins were derived from MHC-matched allogeneic donors and adoptively transferred (Riddell et al. 1992). To prevent toxicity, the clones were also selected for lack of cross-reactivity with CMV-uninfected recipient cells. The degree of CMV-specific CTL responses in the recipients was equivalent to or greater than the immunocompetent donors (Riddle et al. 1992; Riddell et al. 1992; Riddell and Greenberg 1995a,b; Walter et al. 1995). Analyses of rearranged T-cell–receptor genes indicated that adoptively transferred clones can persist for at least 12 weeks (Walter et al. 1995). Neither CMV viremia nor CMV disease developed in any recipients and no toxicity was observed. Analogous studies have been carried out for Epstein-Barr virus (EBV)-associated lymphoproliferative syndromes in recipients of T-cell-depleted allogeneic bone-marrow transplant (BMT) (Papadopoulos et al. 1994; Rooney et al. 1995; Heslop et al. 1996).

Adaptation of the adoptive transfer of T-cell clones to the treatment of HIV-1-infected patients has been less successful. The initial attempts have illustrated the potential for ablative immune responses to gene-altered cells expressing foreign proteins, thus highlighting an advantage of untranslated RNA modalities. Use of CTL clones in HIV-positive recipients required consideration of differences in the pathophysiology of HIV compared to CMV infection (Riddell and Greenberg 1995a,b). In particular, since the immunopathogenesis of HIV disease remains uncertain, the safety of infusing large numbers of highly virus-specific CTLs in the presence of ongoing high levels of HIV-1 replication was uncertain. In addition, the greater potential for an RNA virus to mutate antigenic epitopes could lead to immune escape. The decreased levels of T-cell help in $CD4^+$ T-cell-depleted individuals could also lead to reduced persistence of infused $CD8^+$ T-cell clones. For example, in studies of adoptive CTL transfer in patients with CMV disease, recipients deficient in $CD4^+$ T-helper cells specific for CMV experienced a decline in cytotoxic $CD8^+$ T-cell activity, suggesting that helper-T-cell function is required for persistence of transferred $CD8^+$ T cells (Walter et al. 1995).

Because of HIV's high mutation rate and quasispecies diversity, the selection pressure of clonal CTL therapy carries the risk of inducing antigenic escape variants or even accelerating disease progression (Mathieson 1995). These risks were amply demonstrated in a study by Koenig et al. 1995). An HIV-1-infected patient was infused with an expanded, autologous *nef*-specific CTL clone ($\sim 10^{10}$ cells). The unexpected, worrisome result was a fall in circulating $CD4^+$ T cells and an increase in viral load. Subsequent HIV-1 clones recovered from this individual were deleted in the Nef epitope targeted by the CTL clone (Koenig et al. 1995). It appears, therefore, that HIV can escape T-cell clone "monotherapy" as readily as it does single-drug therapy. Moreover, this study suggests that CTLs may contribute to pathogenesis and that HIV-1-specific CTL clone therapy may trigger disease progression. Escape of HIV-1 from immunodominant CTL epitopes is not limited to the adoptive immunotherapy setting, but has also been observed in the course of natural CTL responses (Borrow et al. 1997).

To provide a safeguard against potential immunopathology mediated by cytotoxic T-cell clones in HIV-1-infected recipients, Riddell et al. (1996) transduced the cells with a retroviral vector encoding HyTk, a suicide gene with an inducible phenotype, and selected the cells ex vivo in hygromycin. HyTk is a fusion of the hygromycin phosphotransferase gene (for positive selection ex vivo) and the herpes simplex thymidine kinase gene (for negative selection in vivo; the viral thymidine kinase phosphorylates the prodrug ganciclovir to a toxic metabolite). The intention is to reserve the ability to ablate the cells in vivo if toxicity is observed.

Patients in this study were given four escalating doses (ranging from 0.1×10^9 to 3.3×10^9 cells/m^2 of body surface area per dose) administered at 2-week intervals. No patients required ablation of the cells for reasons of toxicity. However, in five of the six recipients, the adoptively transferred cells persisted only after the first two doses; the larger third and fourth doses were rapidly cleared. Disappearance of the cells correlated with the appearance of HyTk-specific CTL responses, indicating immune-mediated ablation (Riddell et al. 1996). It is noteworthy that such ablative CTL responses have not been seen in previous human gene-therapy trials, but Riddell et al. were the first to use drug selection ex vivo, thereby undoubtedly increasing the proportion of cells expressing the foreign drug resistance marker in vivo. In any case, without specific strategies to counter the generation of CTL responses, it is likely that such immune responses will remain problematic for any gene therapy in which a protein foreign to the host is expressed. The patients studied by Riddell et al. were immunocompromised by HIV infection; immunocompetent recipients may mount even brisker CTL responses. This problem is of significance for the above-reviewed intrakine, intrabody, and transdominant protein strategies.

Some previous studies have also used various degrees of myeloablative conditioning, which may have led to tolerization or to ineffective CTL responses to foreign proteins. Bordignon and colleagues recently used a thymidine kinase–neomycin phosphotransferase suicide fusion gene (*tk-neo*) to control graft versus host disease (GvHD) mediated by donor lymphocytes given to BMT recipients who had relapsed or developed EBV-induced lymphoma after T-cell-depleted BMT (Bonini et al. 1997). Two of three patients who developed GvHD achieved successful ablation of donor cells with ganciclovir, and the GvHD was partially ameliorated in the third. This study is perhaps the most definitive example of successful human gene therapy to date. Persistence of the gene-modified cells in these immunocompromised patients was substantial (>12 months in some cases). An important difference in this regard compared to the study by Riddell et al. (1996) is that selection for retrovirally transduced cells was not performed by drug selection, but by flow cytometry for a truncated version of a human protein (nerve growth factor) that was also encoded by the vector. Nevertheless, one patient still developed a CTL response to Tk-neo after returning to immunocompetence (Bonini et al. 1997).

A different approach to HIV-specific T-cell therapy involves the engineering of CTL clones that express antigen-specific but MHC-unrestricted "universal" receptors. The strategy exploits the observation that the cytoplasmic domain of the invariant ζ chain of the T-cell receptor can mediate signal transduction and T-cell activation when fused to heterologous extracellular receptor domains (Irving and Weiss 1991). Chimeric genes encoding a fusion of the ζ-chain signaling domain to the extracellular and transmembrane domains of either human CD4 (CD4-ζ) or a single-chain antibody to HIV-1 gp41 have been expressed in primary human T cells (Romeo and Seed 1991; Roberts et al. 1994; O.O. Yang et al. 1997) and a human natural killer cell line (Tran et al. 1995). Both kinds of immune effector cells show cytolytic activity against HIV-1-infected target cells or cells transfected with HIV-1 gp120. Cells infected with some primary HIV isolates have also been susceptible to specific lysis (O.O. Yang et al. 1997). A phase I clinical trial to test the approach in vivo is in progress (Walker 1996). Non-T-cell immune system lineages can also mediate HIV cytolytic activity via CD4-ζ. When murine hematopoietic progenitor cells retrovirally transduced with CD4-ζ were transplanted into severe combined immunodeficient (SCID) mice, long-term expression was observed in circulating myeloid and natural killer cells (Hege et al. 1996). These mice resisted challenge with a lethal dose of human leukemic cells expressing gp120.

A major teleologic reason for the evolution of MHC molecules is prevention of nonspecific activation of T cells by soluble antigen (by confining recognition/activation to cell-derived antigens). It is unclear, therefore, what the fate of such MHC-unrestricted T cells will be in vivo in the presence of high-level viremia or shed HIV-1 gp120. In one study, inclusion of up to 50% HIV-1-positive donor serum did not block cytolytic activity (Roberts et al. 1994), but the question of activation by cell-free antigen requires further investigation.

For T-cell gene therapy of HIV infection, unless syngeneic (twin) donors are used (Walker et al. 1993; Morgan and Walker 1996; Walker 1996), current approaches entail culture of large numbers of autologous CD4[+] T cells ex vivo. In addition to achieving adequate transduction efficiency, problems that need to be addressed to make the approach feasible include suppression of HIV-1 replication in the cultures, prevention of CD8$^+$ T-cell overgrowth, the impaired clonogenic potential (Margolick et al. 1985; Maggi et al. 1987; Pantaleo et al. 1990) of T cells from HIV-infected individuals, and blocking apoptosis. Several groups have prevented detectable expansion of HIV-1 in such cultures with combinations of antiviral agents (Wilson et al. 1995; Woffendin et al. 1996). Recently, June and colleagues determined that costimulation with immobilized antibodies to CD3 and CD28 resulted in a much greater log$_{10}$ expansion of CD4$^+$ T cells ex vivo than did conventional stimulation with phyto-hemagglutinin and interleukin-2 (Levine et al. 1996). In addition, the CD28 costimulation resulted in rapid loss of HIV-1 from the cultures and resistance to exogenous HIV-1 infection in the absence of antiviral agents. The expanded cells were also found to maintain a diverse TCR-Vβ repertoire and to secrete large amounts of cytokines indicative of a Th1 phenotype. Subsequent experiments showed that the virus-resistant state is due to essentially complete inhibition of CCR5 expression and that resistance is conferred only to M-tropic HIV (Carroll et al. 1997; Riley et al. 1997). It is also noteworthy that CD28 costimulation inhibits apoptosis in cultures of HIV-infected CD4 cells (Groux et al. 1992; Boise et al. 1995). CD28 costimulation may therefore become a useful technique in large-scale ex vivo expansion and transduction protocols for T-cell gene therapy of HIV-1-infected individuals. However, the consequences of infusing cells with a CXCR4-high/CCR5-low phenotype remain to be seen; the potential problems, such as accelerated replication of CXCR4-tropic variants, have recently been reviewed (Rowland-Jones and Tan 1997). Antiviral drugs will also still be needed in ex vivo cultures if a patient's T cells harbor CXCR4-tropic HIV.

DNA VACCINES

More than 90% of HIV infections occur in developing countries where combination drug therapy is economically insupportable. Neither are any other forms of treatment for established HIV infection, particularly the complicated methods of current gene therapy, that are at present feasible for the vast majority of HIV-1-infected persons. An effective vaccine is likely to be the only realistic long-term solution to the pandemic.

The difficulties inherent in devising and testing an effective vaccine for HIV are severe (Letvin 1993; Haynes et al. 1996; Miller and McGhee 1996; Burton 1997), and initial attempts have not prevented human infection after sexual exposure (McElrath et al. 1996). Live-attenuated vaccines have indisputably been the most successful at inducing protective immunity in the SIV model (Kestler et al. 1991; Daniel et al. 1992; Desrosiers 1994; Putkonen et al. 1995; Wyand et al. 1996, 1997; R.P. Johnson et al. 1997). However, live HIV vaccines present obvious safety concerns that have increased since the finding that multiply attenuated SIV can induce disease in both infant and adult macaques (Baba et al. 1995; Cohen 1997).

The protection afforded by live-virus vaccines may be presumed to derive to a large extent from the natural (intracellular) antigen presentation they permit. An alternate, safer way of producing authentic intracellular antigen presentation is DNA vaccination. Interest in this strategy was stimulated by the findings that injection of plasmid DNA into muscle tissue led to cellular uptake and protein expression (Wolff et al. 1990), and that pure DNA encoding influenza A nucleoprotein could also confer heterologous immunity to widely divergent influenza viruses (Ulmer et al. 1993). Extensive exploration of this approach has been undertaken for HIV and SIV vaccines (Wang et al. 1993, 1994, 1995a,b, 1997; Kumar and Sercarz 1996; M.A. Liu et al. 1996; Yasutomi et al. 1996; Bagarazzi et al. 1997; Boyer et al. 1997; Chattergoon et al. 1997; Kim et al. 1997a,b; Letvin et al. 1997; Lubeck et al. 1997; Ugen et al. 1997).

The mechanisms by which potent CTL responses are induced by DNA injection into muscle have been uncertain, since myocytes are poor antigen-presenting cells and lack important costimulatory molecules. Recent studies have revealed that blood-derived dendritic cells, rather than muscle cells, are probably the principal cell type mediating antigen presentation by DNA vaccines (Raz et al. 1994; Iwasaki et al. 1997). In addition, immunostimulatory prokaryotic DNA sequences contained in plasmid DNA (unmethylated CpG motifs) have been found to be necessary for induction of effective Th1 responses (Krieg et al. 1995; Klinman et al. 1996; Krieg 1996a,b; Roman et al. 1997; Sato et al. 1996). In the situation of gene therapy rather than vaccination, inclusion of such sequences may have the counterproductive effect of inducing proinflammatory cytokines (Sato et al. 1996).

Therapeutic vaccination has been poorly immunogenic and has had little effect on the course of HIV disease (Kelleher et al. 1997). Combination of highly active antiretroviral therapy (HAART) with vaccination might be more efficacious. Rosenberg et al. (1997) recently demonstrated that strong HIV-1-specific, CD4+ helper T-cell responses are associated with low viral load and secretion of interferon-γ and anti-HIV β-chemokines in chronically infected patients. In acutely HIV-1-infected individuals in this study, vigorous CD4+ T-cell proliferative responses were also detected following prompt institution of HAART prior to seroconversion. Since HAART does not appear to result in restoration of strong HIV-1-specific CD4 function in chronic infection (Autran et al. 1997; Connors et al. 1997), DNA vaccination in combination with such therapy should be investigated. It is noteworthy in this regard that vigorous HIV-1-specific CD4[+] T-cell responses can be induced in the SIV macaque model with DNA vaccination (Lekutis and Letvin 1997; Lekutis et al. 1997).

REFERENCES

Akhtar S. and Rossi J.J. 1996. Anti-HIV therapy with antisense oligonucleotides and ribozymes: Realistic approaches or expensive myths? *J. Antimicrob. Chemother.* **38:** 159–165.

Alkhatib G., Combadiere C., Broder C.C., Feng Y., Kennedy P.E., Murphy P.M., and Berger E.A. 1996. CC CKR5: A RANTES, MIP-1α, MIP-1β receptor as a fusion cofactor for macrophage-tropic HIV-1. *Science* **272:** 1955–1958.

Anderson P., Monforte J., Tritz R., Nesbitt S., Hearst J., and Hampel A. 1994. Mutagenesis of the hairpin ribozyme. *Nucleic Acids Res.* **22:** 1096–1100.

Autran B., Carcelain G., Li T.S., Blanc C., Mathez D., Tubiana R., Katlama C., Debre P., and Leibowitch J. 1997. Positive effects of combined antiretroviral therapy on CD4[+] T cell homeostasis and function in advanced HIV disease. *Science* **277:** 112–116.

Baba T.W., Jeong Y.S., Pennick D., Bronson R., Greene M.F., and Ruprecht R.M. 1995. Pathogenicity of live, attenuated SIV after mucosal infection of neonatal macaques. *Science* **267:** 1820–1825.

Babe L.M., Rose J., and Craik C.S. 1995. Trans-dominant inhibitory human immunodeficiency virus type 1 protease monomers prevent protease activation and virion maturation. *Proc. Natl. Acad. Sci.* **92:** 10069–10073.

Bagarazzi M.L., Boyer J.D., Javadian M.A., Chattergoon M., Dang K., Kim G., Shah J., Wang B., and Weiner D.B. 1997. Safety and immunogenicity of intramuscular and intravaginal delivery of HIV-1 DNA constructs to infant chimpanzees. *J. Med. Primatol.* **26:** 27–33.

Bauer G., Valdez P., Kearns K., Bahner I., Wen S.F.H, Zaia J.A., and Kohn D.B. 1997. Inhibition of human immunodeficiency virus-1 (HIV-1) replication after transduction of granulocyte colony-stimulating factor-mobilized CD34[+] cells from HIV-1-infected donors using retroviral vectors containing anti-HIV-1 genes. *Blood* **89:** 2259–2267.

Berger E. 1997. HIV entry and tropism: The chemokine receptor connection. *AIDS* **11:** S3–16.

Bertrand E., Castanotto D., Zhou C., Carbonnelle C., Lee N.S., Good P., Chatterjee S., Grange T., Pictet R., Kohn D., Engelke D., and Rossi J.J. 1997. The expression cassette determines the functional activity of ribozymes in mammalian cells by controlling their intracellular localization. *RNA* **3:** 75–88.

Berzal-Herranz A., Joseph S., Chowrira B.M., Butcher S.E., and Burke J.M. 1993. Essential nucleotide sequences and secondary structure elements of the hairpin ribozyme. *EMBO J.* **12:** 2567–2573.

Biti R., Ffrench R., Young J., Bennetts B., Stewart G., and Liang T. 1997. HIV-1 infection in an individual homozygous for the CCR5 deletion allele. Nat. Med. **3:** 252–253.

Bleul C.C., Wu L., Hoxie J.A., Springer T.A., and Mackay C.R. 1997. The HIV coreceptors CXCR4 and CCR5 are differentially expressed and regulated on human T lymphocytes. *Proc. Natl. Acad. Sci.* **94:** 1925–1930.

Bleul C.C., Farzan M., Choe H., Parolin C., Clark-Lewis I., Sodroski J., and Springer T.A. 1996. The lymphocyte chemoattractant SDF-1 is a ligand for LESTR/fusin and blocks HIV-1 entry. Nature **382:** 829–833.

Boise L.H., Minn A.J., Noel P.J., June C.H., Accavitti M.A., Lindsten T., and Thompson C.B. 1995. CD28 costimulation can promote T cell survival by enhancing the expression of Bcl-XL. *Immunity* **3:** 87–98.

Bonini C., Ferrari G., Verzeletti S., Servida P., Zappone E., Ruggieri L., Ponzoni M., Rossini S., Mavilio F., Traversari C., and Bordignon C. 1997. HSV-TK gene transfer into donor lymphocytes for control of allogeneic graft-versus-leukemia. *Science* **276:** 1719–1724.

Bonyhadi M.L., Moss K., Voytovich A., Auten J., Kalfoglou C., Plavec I., Forestell S., Su L., Bohnlein E., and Kaneshima H. 1997. RevM10-expressing T cells derived in vivo from transduced human hematopoietic stem-progenitor cells inhibit human immunodeficiency virus replication. *J. Virol.* **71:** 4707–4716.

Borrow P., Lewicki H., Wei X., Horwitz M.S., Peffer N., Meyers H., Nelson J.A., Gairin J.E., Hahn B.H., Oldstone M.B., and Shaw G.M. 1997. Antiviral pressure exerted by HIV-1-specific cytotoxic T lymphocytes (CTLs) during primary infection demonstrated by rapid selection of CTL escape virus. *Nat. Med.* **3:** 205–211.

Boyer J.D., Ugen K.E., Wang B., Agadjanyan M., Gilbert L., Bagarazzi M.L., Chattergoon M., Frost P., Javadian A., Williams W.V., Refaeli Y., Ciccarelli R.B., McCallus D., Coney L., and Weiner D.B. 1997. Protection of chimpanzees from high-dose heterologous HIV-1 challenge by DNA vaccination. *Nat. Med.* **3:** 526–532.

Buchschacher G.L., Jr., Freed E.O., and Panganiban A.T. 1992. Cells induced to express a human immunodeficiency virus type 1 envelope gene mutant inhibit the spread of wild-type virus. *Hum. Gene Ther.* **3:** 391–397.

Buonocore L., and Rose J.K. 1990. Prevention of HIV-1 glycoprotein transport by soluble CD4 retained in the endoplasmic reticulum. *Nature* **345:** 625–628.

——— 1993. Blockade of human immunodeficiency virus type 1 production in CD4$^+$ T cells by an intracellular CD4 expressed under control of the viral long terminal repeat. *Proc. Natl. Acad. Sci.* **90:** 2695–2699.

Burton D.R. 1997. A vaccine for HIV type 1: The antibody perspective. *Proc. Natl. Acad. Sci.* **94:** 10018–10023.

Burton D.R., Barbas C.F.,III, Persson M.A., Koenig S., Chanock R.M., and Lerner R.A. 1991. A large array of human monoclonal antibodies to type 1 human immunodeficiency virus from combinatorial libraries of asymptomatic seropositive individuals. *Proc. Natl. Acad. Sci.* **88:** 10134–10137.

Caputo A., Grossi M.P., Bozzini R., Rossi C., Betti M., Marconi P.C., Barbanti-Brodano G., and Balboni P.G. 1996. Inhibition of HIV-1 replication and reactivation from latency by tat transdominant negative mutants in the cysteine rich region. *Gene Ther.* **3:** 235–245.

Carroll R.G., Riley J.L., Levine B.L., Feng Y., Kaushal S., Ritchey D.W., Bernstein W., Weislow O.S., Brown C.R., Berger E.A., June C.H., and St. Louis D.C. 1997. Differential regulation of HIV-1 fusion cofactor expression by CD28 costimulation of CD4$^+$ T cells. *Science* **276:** 273–276.

Cavert W., Notermans D.W., Staskus K., Wietgrefe S.W., Zupancic M., Gebhard K., Henry K., Zhang Z.Q., Mills R., McDade H., Goudsmit J., Danner S.A., and Haase A.T. 1997. Kinetics of response in lymphoid tissues to antiretroviral therapy of HIV-1 infection. *Science* **276:** 960–964.

Chang H.K., Gendelman R., Lisziewicz J., Gallo R.C., and Ensoli B. 1994. Block of HIV-

1 infection by a combination of antisense *tat* RNA and TAR decoys: A strategy for control of HIV-1. *Gene Ther.* **1:** 208–216.

Chanock R.M., Crowe J.E., Jr., Murphy B.R., and Burton D.R. 1993. Human monoclonal antibody Fab fragments cloned from combinatorial libraries: Potential usefulness in prevention and/or treatment of major human viral diseases. *Infect. Agents Dis.* **2:** 118–131.

Chattergoon M., Boyer J., and Weiner D.B. 1997. Genetic immunization: A new era in vaccines and immune therapeutics. *FASEB J.* **11:** 753–763.

Chatterjee S., Johnson P.R., and Wong K.K., Jr. 1992. Dual-target inhibition of HIV-1 in vitro by means of an adeno-associated virus antisense vector. *Science* **258:** 1485–1488.

Chen S.S., Ferrante A.A., and Terwilliger E.F. 1996. Characterization of an envelope mutant of HIV-1 that interferes with viral infectivity. *Virology* **226:** 260–268.

Chen S.Y., Khouri Y., Bagley J., and Marasco W.A. 1994. Combined intra- and extracellular immunization against human immunodeficiency virus type 1 infection with a human anti-gp120 antibody. *Proc. Natl. Acad. Sci.* **91:** 5932–5936.

Chen J.D., Bai X.F., Yang A.G., Cong Y.P., and Chen S.Y. 1997. Inactivation of HIV-1 chemokine co-receptor CXCR-4 by a novel intrakine strategy. *Nat. Med.* **3:** 1110–1116.

Choe H., Farzan M., Sun Y., Sullivan N., Rollins B., Ponath P.D., Wu L., Mackay C.R., LaRosa G., Newman W., Gerard N., Gerard C., and Sodroski J. 1996. The β-chemokine receptors CCR3 and CCR5 facilitate infection by primary HIV-1 isolates. *Cell* **85:** 1135–1148.

Chowrira B.M., Berzal-Herranz A., and Burke J.M. 1991. Novel guanosine requirement for catalysis by the hairpin ribozyme. *Nature* **354:** 320–322.

Chun T.W., Finzi D., Margolick J., Chadwick K., Schwartz D., and Siliciano R.F. 1995. In vivo fate of HIV-1-infected T cells: Quantitative analysis of the transition to stable latency. *Nat. Med.* **1:** 1284–1290.

Chun T.-W., Stuyver L., Mizell S., Ehler L., Mican J., Baseler M., Lloyd L., Nowak M., and Fauci A.S. 1997a. Presence of an inducible HIV-1 latent reservoir during highly active antiretroviral therapy. *Proc. Natl. Acad. Sci.* **94:** 13193–13197.

Chun T.W., Carruth L., Finzi D., Shen X., DiGiuseppe J.A., Taylor H., Hermankova M., Chadwick K., Margolick J., Quinn T.C., Kuo Y.H., Brookmeyer R., Zeiger M.A., Barditch-Crovo P., and Siliciano R.F. 1997b. Quantification of latent tissue reservoirs and total body viral load in HIV-1 infection. *Nature* **387** 183–188.

Cocchi F., DeVico A.L., Garzino-Demo A., Arya S.K., Gallo R.C., and Lusso P. 1995. Identification of RANTES, MIP-1 α, and MIP-1 β as the major HIV-suppressive factors produced by CD8$^+$ T cells. *Science* **270:** 1811–1815.

Coffin J.M. 1995. HIV population dynamics in vivo: implications for genetic variation, pathogenesis, and therapy. *Science* **267:** 483–489.

Cohen J. 1997. Weakened SIV vaccine still kills. *Science* **278:** 24–25.

Connors M., Kovacs J.A., Krevat S., Gea-Banacloche J.C., Sneller M.C., Flanigan M., Metcalf J.A., Walker R.E., Falloon J., Baseler M., Stevens R., Feuerstein I., Masur H., and Lane H.C. 1997. HIV infection induces changes in CD4$^+$ T-cell phenotype and depletions within the CD4$^+$ T-cell repertoire that are not immediately restored by antiviral or immune-based therapies. *Nat. Med.* **3:** 533–540.

Daar E. and Ho D.D. 1991. Relative resistance of primary HIV-1 isolates to neutralization by soluble CD4. *Am. J. Med.* **90:** 22S–26S.

Daar E.S., Li X.L., Moudgil T., and Ho D.D. 1990. High concentrations of recombinant soluble CD4 are required to neutralize primary human immunodeficiency virus type 1 isolates. *Proc. Natl. Acad. Sci.* **87:** 6574–6578.

Daniel M.D., Kirchhoff F., Czajak S.C., Sehgal P.K., and Desrosiers R.C. 1992. Protective effects of a live attenuated SIV vaccine with a deletion in the nef gene. *Science* **258:** 1938–1941.

Dean M., Carrington M., Winkler C., Huttley G.A., Smith M.W., Allikmets R., Goedert J.J., Buchbinder S.P., Vittinghoff E., Gomperts E., Donfield S., Vlahov D., Kaslow R., Saah A., Rinaldo C., Detels R., and O'Brien S.J. 1996. Genetic restriction of HIV-1 infection and progression to AIDS by a deletion allele of the CKR5 structural gene. Hemophilia Growth and Development Study, Multicenter AIDS Cohort Study, Multicenter Hemophilia Cohort Study, San Francisco City Cohort, ALIVE Study. *Science* **273:** 1856–1862.

Deng H.K., Unutmaz D., KewalRamani V.N., and Littman D.R. 1997. Expression cloning of new receptors used by simian and human immunodeficiency viruses. *Nature* **388:** 296–300.

Deng H., Liu R., Ellmeier W., Choe S., Unutmaz D., Burkhart M., Di Marzio P., Marmon S., Sutton R.E., Hill C.M., Davis C.B., Peiper S.C., Schall T.J., Littman D.R., and Landau N.R. 1996. Identification of a major co-receptor for primary isolates of HIV-1. *Nature* **381:** 661–666.

Desrosiers R.C. 1994. Safety issues facing development of a live-attenuated, multiply deleted HIV-1 vaccine [letter]. *AIDS Res. Hum. Retroviruses* **10:** 331–332.

Domingo E. and Holland J. 1994. Mutation rates and rapid evolution of RNA viruses. In *The evolutionary biology of viruses*, ed. S. Morse, pp. 161–184 Raven Press, New York.

Doranz B.J., Berson J.F., Rucker J., and Doms R.W. 1997. Chemokine receptors as fusion cofactors for human immunodeficiency virus type 1 (HIV-1). *Immunol. Res.* **16:** 15–28.

Doranz B.J., Rucker J., Yi Y., Smyth R.J., Samson M., Peiper S.C., Parmentier M., Collman R.G., and Doms R.W. 1996. A dual-tropic primary HIV-1 isolate that uses fusin and the β-chemokine receptors CKR-5, CKR-3, and CKR-2b as fusion cofactors. *Cell* **85:** 1149–1158.

Dragic T., Litwin V., Allaway G.P., Martin S.R., Huang Y., Nagashima K.A., Cayanan C., Maddon P.J., Koup R.A., Moore J.P., and Paxton W.A. 1996. HIV-1 entry into CD4$^+$ cells is mediated by the chemokine receptor CC-CKR-5. *Nature* **381:** 667–673.

Dropulic B. and Jeang K.T. 1994. Gene therapy for human immunodeficiency virus infection: genetic antiviral strategies and targets for intervention. *Hum. Gene Ther.* **5:** 927–939.

Dropulic B., Elkins D.A., Rossi J.J., and Sarver N. 1993. Ribozymes: use as anti-HIV therapeutic molecules. *Antisense Res. Dev.* **3:** 87–94.

Duan L., Zhu M., Bagasra O., and Pomerantz R.J. 1995. Intracellular immunization against HIV-1 infection of human T lymphocytes: utility of anti-rev single-chain variable fragments. *Hum. Gene Ther.* **6:** 1561-1573.

Duan L., Bagasra O., Laughlin M.A., Oakes J.W., and Pomerantz R.J. 1994a. Potent inhibition of human immunodeficiency virus type 1 replication by an intracellular anti-Rev single-chain antibody. *Proc. Natl. Acad. Sci.* **91:** 5075–5079.

Duan L., Zhang H., Oakes J.W., Bagasra O., and Pomerantz R.J. 1994b. Molecular and virological effects of intracellular anti-Rev single-chain variable fragments on the expression of various human immunodeficiency virus-1 strains. *Hum. Gene Ther.* **5:** 1315–134.

Duchosal M.A., Eming S.A., Fischer P., Leturcq D., Barbas C.F.,III, McConahey P.J., Caothien R.H., Thornton G.B., Dixon F.J., and Burton D.R. 1992. Immunization of hu-PBL-SCID mice and the rescue of human monoclonal Fab fragments through combinatorial libraries. *Nature* **355:** 258–562.

Embretson J., Zupancic M., Ribas J.L., Burke A., Racz P., Tenner-Racz K., and Haase A.T. 1993. Massive covert infection of helper T lymphocytes and macrophages by HIV during the incubation period of AIDS. *Nature* **362:** 359–362.

Emerman M. 1996. From curse to cure: HIV for gene therapy? *Nat. Biotechnol.* **14:** 943.

Endres M.J., Jaffer S., Haggarty B., Turner J.D., Doranz B.J., Obrien P.J., Kolson D.L., and Hoxie J.A. 1997. Targeting of HIV- and SIV-infected cells by CD4-chemokine receptor pseudotypes. *Science* **278:** 1462–1464.

Escaich S., Kalfoglou C., Plavec I., Kaushal S., Mosca J.D., and Bohnlein E. 1995. RevM10-mediated inhibition of HIV-1 replication in chronically infected T cells. *Hum. Gene Ther.* **6:** 625–634.

Feng Y., Broder C.C., Kennedy P.E., and Berger E.A. 1996. HIV-1 entry cofactor: Functional cDNA cloning of a seven-transmembrane, G protein-coupled receptor. *Science* **272:** 872–877.

Finzi D., Hermankova M., Pierson T., Carruth L.M., Buck C., Chaisson R.E., Quinn T.C., Chadwick K., Margolick J., Brookmeyer R., Gallant J., Markowitz M., Ho D.D., Richman D.D., and Siliciano R.F. 1997. Identification of a reservoir for HIV-1 in patients on highly active antiretroviral therapy. *Science* **278:** 1295–1300.

Fox B.A., Woffendin C., Yang Z.Y., San H., Ranga U., Gordon D., Osterholzer J., and Nabel G.J. 1995. Genetic modification of human peripheral blood lymphocytes with a transdominant negative form of Rev: Safety and toxicity. *Hum. Gene Ther.* **6:** 997–1004.

Friedman A.D., Triezenberg S.J., and McKnight S.L. 1988. Expression of a truncated viral trans-activator selectively impedes lytic infection by its cognate virus. *Nature* **335:** 452–454.

Fruh K., Ahn K., and Peterson P.A. 1997. Inhibition of MHC class I antigen presentation by viral proteins. *J. Exp. Med.* **75:** 18–27.

Gervaix A., Li X., Kraus G., and Wong-Staal F. 1997. Multigene antiviral vectors inhibit diverse human immunodeficiency virus type 1 clades. *J. Virol.* **71:** 3048–3053.

Gibson S.A. and Shillitoe E.J. 1997. Ribozymes. Their functions and strategies for their use. *Mol. Biotechnol.* **7:** 125–137.

Gilboa E. and Smith C. 1994. Gene therapy for infectious diseases: The AIDS model. *Trends Genet.* **10:** 139–144.

Good P.D., Krikos A.J., Li S.X., Bertrand E., Lee N.S., Giver L., Ellington A., Zaia J.A., Rossi J.J., and Engelke D.R. 1997. Expression of small, therapeutic RNAs in human cell nuclei. *Gene Ther.* **4:** 45–54.

Greenberg P.D., Nelson B., Gilbert M., Sing A., Yee C., Jensen M., and Riddell S.R. 1994. Genetic modification of T cell clones to improve the safety and efficacy of adoptive T cell therapy. *Ciba Found. Symp.* **187:** 212–228.

Groux H., Torpier G., Monte D., Mouton Y., Capron A., and Ameisen J.C. 1992. Activation-induced death by apoptosis in $CD4^+$ T cells from human immunodeficiency virus-infected asymptomatic individuals. *J. Exp. Med.* **175:** 331–340.

Gulick R.M., Mellors J.W., Havlir D., Eron J.J., Gonzalez C., McMahon D., Richman D.D., Valentine F.T., Jonas L., Meibohm A., Emini E.A., and Chodakewitz J.A. 1997. Treatment with indinavir, zidovudine, and lamivudine in adults with human immunodeficiency virus infection and prior antiretroviral therapy. *N. Eng. J. Med.* **337:** 734–739.

Haase A.T., Henry K., Zupancic M., Sedgewick G., Faust R.A., Melroe H., Cavert W., Gebhard K., Staskus K., Zhang Z.Q., Dailey P.J., Balfour H.H., Jr., Erice A., and Perelson A.S. 1996. Quantitative image analysis of HIV-1 infection in lymphoid tissue. *Science* **274:** 985–989.

Hammer S.M., Squires K.E., Hughes M.D., Grimes J.M., Demeter L.M., Currier J.S., Eron J.J., Jr., Feinberg J.E., Balfour H.H., Jr., Deyton L.R., Chodakewitz J.A., and Fischl M.A. 1997. A controlled trial of two nucleoside analogues plus indinavir in persons with human immunodeficiency virus infection and CD4 cell counts of 200 per cubic millimeter or less. AIDS Clinical Trials Group 320 Study Team. *N. Eng. J. Med.* **337:** 725–733.

Haseloff J. and Gerlach W.L. 1988. Simple RNA enzymes with new and highly specific endoribonuclease activities. *Nature* **334:** 585–591.

———— 1989. Sequences required for self-catalysed cleavage of the satellite RNA of tobacco ringspot virus. *Gene* **82:** 43–52.

Havlir D.V. and Richman D.D. 1996. Viral dynamics of HIV: Implications for drug development and therapeutic strategies. *Ann. Intern. Med.* **124:** 984–994.

Haynes B.F., Pantaleo G., and Fauci A.S. 1996. Toward an understanding of the correlates of protective immunity to HIV infection. *Science* **271:** 324–328.

He J., Chen Y., Farzan M., Choe H., Ohagen A., Gartner S., Busciglio J., Yang X., Hofmann W., Newman W., Mackay C.R., Sodroski J., and Gabuzda D. 1997. CCR3 and CCR5 are co-receptors for HIV-1 infection of microglia. *Nature* **385:** 645–649.

Hege K.M., Cooke K.S., Finer M.H., Zsebo K.M., and Roberts M.R. 1996. Systemic T cell-independent tumor immunity after transplantation of universal receptor-modified bone marrow into SCID mice. *J. Exp. Med.* **184:** 2261–2269.

Herskowitz I. 1987. Functional inactivation of genes by dominant negative mutations. *Nature* **329:** 219–222.

Heslop H.E., Ng C.Y., Li C., Smith C.A., Loftin S.K., Krance R.A., Brenner M.K., and Rooney C.M. 1996. Long-term restoration of immunity against Epstein-Barr virus infection by adoptive transfer of gene-modified virus-specific T lymphocytes. *Nat. Med.* **2:** 551–555.

Heusch M., Kraus G., Johnson P., and Wong-Staal F. 1996. Intracellular immunization against SIVmac utilizing a hairpin ribozyme. *Virology* **216:** 241–244.

Hill A., Jugovic P., York I., Russ G., Bennink J., Yewdell J., Ploegh H., and Johnson D. 1995. Herpes simplex virus turns off the TAP to evade host immunity. *Nature* **375:** 411–415.

Ho D.D., Neumann A.U., Perelson A.S., Chen W., Leonard J.M., and Markowitz M. 1995. Rapid turnover of plasma virions and CD4 lymphocytes in HIV-1 infection. *Nature* **373:** 123–126.

Hormes R., Homann M., Oelze I., Marschall P., Tabler M., Eckstein F., and Sczakiel G. 1997. The subcellular localization and length of hammerhead ribozymes determine efficacy in human cells. *Nucleic Acids Res.* **25:** 769–775.

Huang Y., Paxton W.A., Wolinsky S.M., Neumann A.U., Zhang L., He T., Kang S., Ceradini D., Jin Z., Yazdanbakhsh K., Kunstman K., Erickson D., Dragon E., Landau N.R., Phair J., Ho D.D., and Koup R.A. 1996. The role of a mutant CCR5 allele in HIV-1 transmission and disease progression. *Nat. Med.* **2:** 1240–1243.

Irving B.A. and Weiss A. 1991. The cytoplasmic domain of the T cell receptor chain is sufficient to couple to receptor-associated signal transduction pathways. *Cell* **64:** 891–901.

Iwasaki A., Torres C.A., Ohashi P.S., Robinson H.L., and Barber B.H. 1997. The dominant role of bone marrow-derived cells in CTL induction following plasmid DNA immunization at different sites. *J. Immunol.* **159:** 11–14.

Johnson J.E., Schnell M.J., Buonocore L., and Rose J.K. 1997. Specific targeting to CD4[+] cells of recombinant vesicular stomatitis viruses encoding human immunodeficiency virus envelope proteins. *J. Virol.* **71:** 5060–5068.

Johnson R.P., Glickman R.L., Yang J.Q., Kaur A., Dion J.T., Mulligan M.J., and Desrosiers R.C. 1997. Induction of vigorous cytotoxic T-lymphocyte responses by live attenuated simian immunodeficiency virus. *J. Virol.* **71:** 7711–7718.

Joseph S. and Burke J.M. 1993. Optimization of an anti-HIV hairpin ribozyme by in vitro selection. *J. Biol. Chem.* **268:** 24515–24518.

Junker U., Escaich S., Plavec I., Baker J., McPhee F., Rose J.R., Craik C.S., and Bohnlein E. 1996. Intracellular expression of human immunodeficiency virus type 1 (HIV-1) protease variants inhibits replication of wild-type and protease inhibitor-resistant HIV-1 strains in human T-cell lines. *J. Virol.* **70:** 7765–7772.

Kelleher A.D., Emery S., Cunningham P., Duncombe C., Carr A., Golding H., Forde S., Hudson J., Roggensack M., Forrest B.D., and Cooper, D.A. 1997. Safety and immunogenicity of UBI HIV-1MN octameric V3 peptide vaccine administered by subcutaneous injection. *AIDS Res. Hum.Retroviruses* **13:** 29–32.

Kestler H.W.,III, Ringler D.J., Mori K., Panicali D.L., Sehgal P.K., Daniel M.D., and Desrosiers R.C. 1991. Importance of the nef gene for maintenance of high virus loads and for development of AIDS. *Cell* **65:** 651 652.

Kim J.J., Ayyavoo V., Bagarazzi M.L., Chattergoon M., Boyer J.D., Wang B., and Weiner D.B. 1997a. Development of a multicomponent candidate vaccine for HIV-1. *Vaccine* **15:** 879–883.

Kim J.J., Bagarazzi M.L., Trivedi N., Hu Y., Kazahaya K., Wilson D.M., Ciccarelli R., Chattergoon M.A., Dang K., Mahalingam S., Chalian A.A., Agadjanyan M.G., Boyer J.D., Wang B., and Weiner D.B. 1997b. Engineering of in vivo immune responses to DNA immunization via codelivery of costimulatory molecule genes. *Nat. Biotechnol.* **15:** 641–646.

Klinman D.M., Yi A.K., Beaucage S.L., Conover J., and Krieg A.M. 1996. CpG motifs present in bacteria DNA rapidly induce lymphocytes to secrete interleukin 6, interleukin 12, and interferon γ. *Proc. Natl. Acad. Sci.* **93:** 2879–2883.

Koenig S., Conley A.J., Brewah Y.A., Jones G.M., Leath S., Boots L.J., Davey V., Pantaleo G., Demarest J.F., Carter C., et al. 1995. Transfer of HIV-1-specific cytotoxic T lymphocytes to an AIDS patient leads to selection for mutant HIV variants and subsequent disease progression. *Nat. Med.* **1:** 330–336.

Krieg A.M. 1996a. An innate immune defense mechanism based on the recognition of CpG motifs in microbial DNA. *J. Lab. Clin. Med.* **128:** 128–133.

——— 1996b. Lymphocyte activation by CpG dinucleotide motifs in prokaryotic DNA. *Trends Microbiol.* **4:** 73–76.

Krieg A.M., Yi A.K., Matson S., Waldschmidt T.J., Bishop G.A., Teasdale R., Koretzky G.A., and Klinman D.M. 1995. CpG motifs in bacterial DNA trigger direct B-cell activation. *Nature* **374:** 546–549.

Kruger K., Grabowski P.J., Zaug A.J., Sands J., Gottschling D.E., and Cech T.R. 1982. Self-splicing RNA: Autoexcision and autocyclization of the ribosomal RNA intervening sequence of *Tetrahymena. Cell* **31:** 147–157.

Kumar V., and Sercarz, E. 1996. Genetic vaccination: The advantages of going naked (comment). *Nat. Med.* **2:** 857–859.

Landau N.R. 1997. HIV coreceptor identification: Good or bad news for drug discovery? *Curr. Opin. Immunol.* **9:** 628–630.

Leavitt M.C., Yu M., Wong-Staal F., and Looney D.J. 1996. Ex vivo transduction and expansion of CD4+ lymphocytes from HIV + donors: Prelude to a ribozyme gene therapy trial. *Gene Ther.* **3:** 599–606.

Leavitt M.C., Yu M., Yamada O., Kraus G., Looney D., Poeschla E., and Wong-Staal F. 1994. Transfer of an anti-HIV-1 ribozyme gene into primary human lymphocytes. *Hum. Gene Ther.* **5:** 1115–1120.

Lee S.W., Gallardo H.F., Gilboa E., and Smith C. 1994. Inhibition of human immunodeficiency virus type 1 in human T cells by a potent Rev response element decoy consisting of the 13-nucleotide minimal Rev-binding domain. *J. Virol.* **68:** 8254–8264.

Lee S.W., Gallardo H.F., Gaspar O., Smith C., and Gilboa, E. 1995. Inhibition of HIV-1 in CEM cells by a potent TAR decoy. *Gene Ther.* **2:** 377–384.

Lekutis C. and Letvin N.L. 1997. HIV-1 envelope-specific CD4+ T helper cells from simian/human immunodeficiency virus-infected rhesus monkeys recognize epitopes restricted by MHC class II DRB1*0406 and DRB*W201 molecules. *J. Immunol.* **159:** 2049–2057.

Lekutis C., Shiver J.W., Liu M.A., and Letvin N.L. 1997. HIV-1 *env* DNA vaccine administered to rhesus monkeys elicits MHC class II-restricted CD4+ T helper cells that secrete IFN-γ and TNF-α. *J. Immunol.* **158:** 4471–4477.

Lerner R.A., Kang A.S., Bain J.D., Burton D.R., and Barbas C.F., III. 1992. Antibodies without immunization. *Science* **258:** 1313–1314.

Letvin N.L. 1993. Vaccines against human immunodeficiency virus—Progress and prospects. *N. Eng. J. Med.* **329:** 1400–1405.

Letvin N.L., Montefiori D.C., Yasutomi Y., Perry H.C., Davies M.E., Lekutis C., Alroy M., Freed D.C., Lord C.I., Handt L.K., Liu M.A., and Shiver J.W. 1997. Potent, protective anti-HIV immune responses generated by bimodal HIV envelope DNA plus protein vaccination. *Proc. Natl. Acad. Sci.* **94:** 9378–9383.

Levin R., Mhashilkar A.M., Dorfman T., Bukovsky A., Zani C., Bagley J., Hinkula J., Niedrig M., Albert J., Wahren B., Gottlinger H.G., and Marasco W.A. 1997. Inhibition of early and late events of the HIV-1 replication cycle by cytoplasmic Fab intrabodies against the matrix protein, p17. *Mol. Med.* **3:** 96–110.

Levine B.L., Mosca J.D., Riley J.L., Carroll R.G., Vahey M.T., Jagodzinski L.L., Wagner K.F., Mayers D.L., Burke D.S., Weislow O.S., St. Louis D.C., and June, C.H. 1996. Antiviral effect and ex vivo CD4+ T cell proliferation in HIV-positive patients as a result of CD28 costimulation. *Science* **272:** 1939–1943.

Levy-Mintz P., Duan L., Zhang H., Hu B., Dornadula G., Zhu M., Kulkosky J., Bizub-Bender D., Skalka A.M., and Pomerantz R.J. 1996. Intracellular expression of single-chain variable fragments to inhibit early stages of the viral life cycle by targeting human immunodeficiency virus type 1 integrase. *J. Virol.* **70:** 8821–8832.

Li C.R., Greenberg P.D., Gilbert M.J., Goodrich J.M., and Riddell S.R. 1994. Recovery of HLA-restricted cytomegalovirus (CMV)-specific T-cell responses after allogeneic bone marrow transplant: Correlation with CMV disease and effect of ganciclovir prophylaxis. *Blood* **83:** 1971–1979.

Lieber A., He C.Y., Polyak S.J., Gretch D.R., Bar D., and Kay M.A. 1996. Elimination of hepatitis C virus RNA in infected human hepatocytes by adenovirus-mediated expression of ribozymes. *J. Virol.* **70:** 8782–8791.

Lisziewicz J. 1996. TAR decoys and trans-dominant gag mutant for HIV-1 gene therapy. *Antibiot. Chemother.* **48:** 192–197.

Lisziewicz J., Sun D., Smythe J., Lusso P., Lori F., Louie A., Markham P., Rossi J., Reitz M., and Gallo, R.C. 1993. Inhibition of human immunodeficiency virus type 1 replication by regulated expression of a polymeric Tat activation response RNA decoy as a strategy for gene therapy in AIDS. *Proc. Natl. Acad. Sci.* **90:** 8000–80004.

Liu M.A., Yasutomi Y., Davies M.E., Perry H.C., Freed D.C., Letvin N.L., and Shiver J.W. 1996. Vaccination of mice and nonhuman primates using HIV-gene-containing DNA. *Antibiot. Chemother.* **48:** 100–104.

Liu R., Paxton W.A., Choe S., Ceradini D., Martin S.R., Horuk R., MacDonald M.E., Stuhlmann H., Koup R.A., and Landau, N.R. 1996. Homozygous defect in HIV-1 coreceptor accounts for resistance of some multiply-exposed individuals to HIV-1 infection. *Cell* **86:** 367–377.

Lori F., Lisziewicz J., Smyth J., Cara A., Bunnag T.A., Curiel D., and Gallo R.C. 1994. Rapid protection against human immunodeficiency virus type 1 (HIV-1) replication mediated by high efficiency non-retroviral delivery of genes interfering with HIV-1 tat and *gag. Gene Ther.* **1:** 27–31.

Lu Z., Berson J.F., Chen Y., Turner J.D., Zhang T., Sharron M., Jenks M.H., Wang Z., Kim J., Rucker J., Hoxie J.A., Peiper S.C., and Doms, R.W. 1997. Evolution of HIV-1 coreceptor usage through interactions with distinct CCR5 and CXCR4 domains. *Proc. Natl. Acad. Sci.* **94:** 6426–6431.

Lubeck M.D., Natuk R., Myagkikh M., Kalyan N., Aldrich K., Sinangil F., Alipanah S., Murthy S.C., Chanda P.K., Nigida S.M., Jr., Markham P.D., Zolla-Pazner S., Steimer K., Wade M., Reitz M.S., Jr., Arthur L.O., Mizutani S., Davis A., Hung P.P., Gallo R.C., Eichberg J., and Robert-Guroff, M. 1997. Long-term protection of chimpanzees against high-dose HIV-1 challenge induced by immunization. *Nat. Med.* **3:** 651–658.

Maggi E., Macchia D., Parronchi P., Mazzetti M., Ravina A., Milo D., and Romagnani S. 1987. Reduced production of interleukin 2 and interferon-γ and enhanced helper activity for IgG synthesis by cloned CD4+ T cells from patients with AIDS. *Eur. J. Immunol.* **17:** 1685–1690.

Malim M.H. and Cullen B.R. 1991. HIV-1 structural gene expression requires the binding of multiple Rev monomers to the viral RRE: Implications for HIV-1 latency. *Cell* **65:** 241–248.

Malim M.H., Bohnlein S., Hauber J., and Cullen B.R. 1989. Functional dissection of the HIV-1 Rev *trans*-activator—derivation of a *trans*-dominant repressor of Rev function. *Cell* **58:** 205–214.

Malim M.H., Freimuth W.W., Liu J., Boyle T.J., Lyerly H.K., Cullen B.R., and Nabel, G.J. 1992. Stable expression of transdominant Rev protein in human T cells inhibits human immunodeficiency virus replication. *J. Exp. Med.* **176:** 1197–1201.

Marasco W.A. 1997. Intrabodies: Turning the humoral immune system outside in for intracellular immunization. *Gene Ther.* **4:** 11–15.

Marasco W.A., Haseltine W.A., and Chen S.Y. 1993. Design, intracellular expression, and activity of a human anti-human immunodeficiency virus type 1 gp120 single-chain antibody. *Proc. Natl. Acad. Sci.* **90:** 7889–7893.

Margolick J.B., Volkman D.J., Lane H.C., and Fauci A.S. 1985. Clonal analysis of T lymphocytes in the acquired immunodeficiency syndrome. Evidence for an abnormality affecting individual helper and suppressor T cells. *J. Clin. Invest.* **76:** 709–715.

Martinson J.J., Chapman N.H., Rees D.C., Liu Y.T., and Clegg J.B. 1997. Global distribution of the CCR5 gene 32-basepair deletion. *Nat. Genet.* **16:** 100–103.

Mathieson B.J. 1995. CTL to HIV-1: Surrogates or sirens (comment). *Nat. Med.* **1:** 304–305.

McElrath M.J., Corey L., Greenberg P.D., Matthews T.J., Montefiori D.C., Rowen L., Hood L., and Mullins J.I. 1996. Human immunodeficiency virus type 1 infection despite prior immunization with a recombinant envelope vaccine regimen. *Proc. Natl. Acad. Sci.* **93:** 3972–3977.

Mebatsion T., Finke S., Weiland F., and Conzelmann K.K. 1997. A CXCR4/CD4 pseudo-type rhabdovirus that selectively infects HIV-1 envelope protein-expressing cells. *Cell* **90:** 841–847.

Mellors J.W., Rinaldo C.R., Jr., Gupta P., White R.M., Todd J.A., and Kingsley L.A. 1996. Prognosis in HIV-1 infection predicted by the quantity of virus in plasma. *Science* **272:** 1167–1170.

Mhashilkar A.M., Bagley J., Chen S.Y., Szilvay A.M., Helland D.G., and Marasco W.A. 1995. Inhibition of HIV-1 Tat-mediated LTR transactivation and HIV-1 infection by anti-Tat single chain intrabodies. *EMBO J.* **14:** 1542–1551.

Michael N.L., Chang G., Louie L.G., Mascola J.R., Dondero D., Birx D.L., and Sheppard H.W. 1997. The role of viral phenotype and CCR-5 gene defects in HIV-1 transmission and disease progression. *Nat. Med.* **3:** 338–340.

Miller C.J. and McGhee, J.R. 1996. Progress towards a vaccine to prevent sexual transmission of HIV. *Nat. Med.* **2:** 751–752.

Moore J.P. 1997. Coreceptors: Implications for HIV pathogenesis and therapy. *Science* **276:** 51–52.

Morgan R.A. and Walker, R. 1996. Gene therapy for AIDS using retroviral mediated gene transfer to deliver HIV-1 antisense TAR and transdominant Rev protein genes to syngeneic lymphocytes in HIV-1 infected identical twins. *Hum. Gene Ther.* **7:** 1281–1306.

Nabel G.J., Fox B.A., Post L., Thompson C.B., and Woffendin C. 1994. A molecular genetic intervention for AIDS—Effects of a transdominant negative form of *Rev. Hum. Gene Ther.* **5:** 79–92.

Naldini L., Blomer U., Gallay P., Ory D., Mulligan R., Gage F.H., Verma I.M., and Trono D. 1996. In vivo gene delivery and stable transduction of nondividing cells by a lentiviral vector. *Science* **272:** 263–267.

Nolan G.P. 1997. Harnessing viral devices as pharmaceuticals: Fighting HIV-1's fire with fire (comment). *Cell* **90:** 821–824.

Orsini M.J. and Debouck C.M. 1996. Inhibition of human immunodeficiency virus type 1 and type 2 Tat function by transdominant Tat protein localized to both the nucleus and cytoplasm. *J. Virol.* **70:** 8055–8063.

Pakker N.G., Notermans D., De Boer R., Roos M., De Wolf F., Hill A., Leonard J., Danner S., Miedema F., and Schellekens P.T. A. 1998. Biphasic kinetics of peripheral blood T cells after triple combination therapy in HIV-1 infection: A composite of redistribution and proliferation. *Nat. Med.* **4:** 208–214.

Pantaleo G., Koenig S., Baseler M., Lane H.C., and Fauci A.S. 1990. Defective clonogenic potential of CD8+ T lymphocytes in patients with AIDS. Expansion in vivo of a nonclonogenic CD3+CD8+DR+CD25– T cell population. *J. Immunol.* **144:** 1696–1704.

Pantaleo G., Graziosi C., Demarest J.F., Butini L., Montroni M., Fox, C.H., Orenstein J.M., Kotler D.P., and Fauci A.S. 1993. HIV infection is active and progressive in lymphoid tissue during the clinically latent stage of disease. *Nature* **362:** 355–358.

Papadopoulos E.B., Ladanyi M., Emanuel D., Mackinnon S., Boulad F., Carabasi M.H., Castro-Malaspina H., Childs B.H., Gillio A.P., Small T.N., and et al. 1994. Infusions of donor leukocytes to treat Epstein-Barr virus-associated lymphoproliferative disorders after allogeneic bone marrow transplantation. *N. Eng. J. Med.* **330:** 1185–1191.

Paxton W.A., Martin S.R., Tse D., O'Brien T.R., Skurnick J., VanDevanter N.L., Padian N., Braun J.F., Kotler D.P., Wolinsky S.M., and Koup R.A. 1996. Relative resistance to HIV-1 infection of CD4 lymphocytes from persons who remain uninfected despite multiple high-risk sexual exposure. *Nat. Med.* **2:** 412–417.

Perelson A.S., Neumann A.U., Markowitz M., Leonard J.M., and Ho D.D. 1996. HIV-1 dynamics in vivo: Virion clearance rate, infected cell life-span, and viral generation time. *Science* **271:** 1582–1586.

Perelson A.S., Essunger P., Cao Y., Vesanen M., Hurley A., Saksela K., Markowitz M., and Ho D.D. 1997. Decay characteristics of HIV-1-infected compartments during combination therapy. *Nature* **387:** 188–191.

Piatak M., Jr., Saag M.S., Yang L.C., Clark S.J., Kappes J.C., Luk K.C., Hahn B.H., Shaw G.M., and Lifson J.D. 1993. High levels of HIV-1 in plasma during all stages of infection determined by competitive PCR. *Science* **259:** 1749–1754.

Pilkington G.R., Duan L., Zhu M., Keil W., and Pomerantz R.J. 1996. Recombinant human Fab antibody fragments to HIV-1 Rev and Tat regulatory proteins: Direct selection from a combinatorial phage display library. *Mol. Immunol.* **33:** 439–450.

Poeschla E. and Looney D. 1998. CXCR4 requirement for infection and pathogenicity of a non-primate lentivirus: Studies with heterologously expressed FIV. *J. Virol.* **72:** (in press).

Poeschla E. and Wong-Staal F. 1994. Antiviral and anticancer ribozymes. *Curr. Opin. Oncol.* **6:** 601–606.

——— 1995. Gene therapy and HIV disease. *AIDS Clin. Rev.* **2:** 1–45.

Poeschla E., Wong-Staal F., and Looney, D. 1998. Efficient transduction of nondividing cells by feline immunodeficiency virus lentiviral vectors. *Nat. Med.* **4:** 354–357.

Putkonen P., Walther L., Zhang Y.J., Li S.L., Nilsson C., Albert J., Biberfeld P., Thorstensson R., and Biberfeld G. 1995. Long-term protection against SIV-induced disease in macaques vaccinated with a live attenuated HIV-2 vaccine. *Nat. Med.* **1:** 914–918.

Ramezani A. and Joshi S. 1996. Comparative analysis of five highly conserved target sites within the HIV-1 RNA for their susceptibility to hammerhead ribozyme-mediated cleavage in vitro and in vivo. *Antisense Nucleic Acid Drug Dev.* **6:** 229–235.

Raz E., Carson D.A., Parker S.E., Parr T.B., Abai A.M., Aichinger G., Gromkowski S.H., Singh M., Lew D., Yankauckas M.A., et al. 1994. Intradermal gene immunization: The possible role of DNA uptake in the induction of cellular immunity to viruses. *Proc. Natl. Acad. Sci.* **91:** 9519–9523.

Richardson J.H. and Marasco W.A. 1995. Intracellular antibodies: Development and therapeutic potential. *Trends Biotechnol.* **13:** 306–310.

Riddell S.R. and Greenberg P.D. 1994. Therapeutic reconstitution of human viral immunity by adoptive transfer of cytotoxic T lymphocyte clones. *Curr. Top. Microbiol. Immunol.* **189:** 9–34.

——— 1995a. Cellular adoptive immunotherapy after bone marrow transplantation. *Cancer Treat. Res.* **76:** 337–369.

——— 1995b. Principles for adoptive T cell therapy of human viral diseases. *Annu. Rev. Immunol.* **13:** 545–586.

Riddell S.R., Gilbert M.J., and Greenberg P.D. 1993. CD8+ cytotoxic T cell therapy of cytomegalovirus and HIV infection. *Curr. Opin. Immunol.* **5:** 484–491.

Riddell S.R., Walter B.A., Gilbert M.J., and Greenberg P.D. 1994. Selective reconstitution of CD8+ cytotoxic T lymphocyte responses in immunodeficient bone marrow transplant recipients by the adoptive transfer of T cell clones. *Bone Marrow Transplant* (Suppl 4) **14:** S78–S84.

Riddell S.R., Watanabe K.S., Goodrich J.M., Li C.R., Agha M.E., and Greenberg P.D. 1992. Restoration of viral immunity in immunodeficient humans by the adoptive transfer of T cell clones. *Science* **257:** 238–241.

Riddell S.R., Elliott M., Lewinsohn D.A., Gilbert M.J., Wilson L., Manley S.A., Lupton S.D., Overell R.W., Reynolds T.C., Corey L., and Greenberg P.D. 1996. T-cell mediated rejection of gene-modified HIV-specific cytotoxic T lymphocytes in HIV-infected patients. *Nat. Med.* **2:** 216–223.

Riley J.L., Carroll R.G., Levine B.L., Bernstein W., St. Louis D.C., Weislow O.S., and June C.H. 1997. Intrinsic resistance to T cell infection with HIV type 1 induced by CD28 costimulation. *J. Immunol.* **158:** 5545–5553.

Roberts M.R., Qin L., Zhang D., Smith D.H., Tran A.C., Dull T.J., Groopman J.E., Capon D.J., Byrn R.A., and Finer M.H. 1994. Targeting of human immunodeficiency virus-infected cells by CD8+ T lymphocytes armed with universal T-cell receptors. *Blood* **84:** 2878–2889.

Roederer M. 1988. Getting to the HAART of T cell dynamics. *Nat. Med.* **4:** 145–146.

Roman M., Martin-Orozco E., Goodman J.S., Nguyen M.D., Sato Y., Ronaghy A., Kornbluth R.S., Richma D.D., Carson D.A., and Raz, E. 1997. Immunostimulatory DNA sequences function as T helper-1-promoting adjuvants. *Nat. Med.* **3:** 849–854.

Romeo C. and Seed B. 1991. Cellular immunity to HIV activated by CD4 fused to T cell or Fc receptor polypeptides. *Cell* **64:** 1037–1046.

Rooney C.M., Smith C.A., Ng C.Y., Loftin S., Li C., Krance R.A., Brenner M.K., and Heslop H.E. 1995. Use of gene-modified virus-specific T lymphocytes to control Epstein-Barr-virus-related lymphoproliferation. *Lancet* **345:** 9–13.

Rosenberg E.S., Billingsley J.M., Caliendo A.M., Boswell S.L., Sax P.E., Kalams S.A., and Walker B.D. 1997. Vigorous HIV-1-specific CD4(+) T cell responses associated with control of viremia. *Science* **278:** 1447–1450.

Rossi J.J. 1995. Controlled, targeted, intracellular expression of ribozymes: Progress and problems. *Trends Biotechnol.* **13:** 301–306.

Rossi J.J., Elkins D., Zaia J.A., and Sullivan S. 1992. Ribozymes as anti-HIV-1 therapeutic agents: Principles, applications, and problems. *AIDS Res. Hum. Retroviruses* **8:** 183–189.

Rowland-Jones S. and Tan R. 1997. Control of HIV co-receptor expression: Implications for pathogenesis and treatment. *Trends Microbiol.* **5:** 302–303.

Rucker J., Samson M., Doranz B.J., Libert F., Berson J.F., Yi Y., Smyth R.J., Collman R.G., Broder C.C., Vassart G., Doms R.W., and Parmentier M. 1996. Regions in β-chemokine receptors CCR5 and CCR2b that determine HIV-1 cofactor specificity. *Cell* **87:** 437–446.

Sakamoto N., Wu C.H., and Wu G.Y. 1996. Intracellular cleavage of hepatitis C virus RNA and inhibition of viral protein translation by hammerhead ribozymes. *J. Clin. Invest.* **98:** 2720–2728.

Samson M., Libert F., Doranz B.J., Rucker J., Liesnard C., Farber C.M., Saragosti S., Lapoumeroulie C., Cognaux J., Forceille C., Muyldermans G., Verhofstede C., Burtonboy G., Georges M., Imai T., Rana S., Yi Y., Smyth R.J., Collman R.G., Doms R.W., Vassart G., and Parmentier M. 1996. Resistance to HIV-1 infection in caucasian individuals bearing mutant alleles of the CCR-5 chemokine receptor gene. *Nature* **382:** 722–725.

Sanna P.P., Williamson R.A., De Logu A., Bloom F.E., and Burton D.R. 1995. Directed selection of recombinant human monoclonal antibodies to herpes simplex virus glycoproteins from phage display libraries. *Proc. Natl. Acad. Sci.* **92:** 6439–6443.

Sarver N., Cantin E.M., Chang P.S., Zaia J.A., Ladne P.A., Stephens D.A., and Rossi J.J. 1990. Ribozymes as potential anti-HIV-1 therapeutic agents. *Science* **247:** 1222–1225.

Sato Y., Roman M., Tighe H., Lee D., Corr, M., Nguyen, M.D., Silverman, G.J., Lotz, M., Carson D.A., and Raz E. 1996. Immunostimulatory DNA sequences necessary for effective intradermal gene immunization. *Science* **273**: 352–354.

Scarlatti G., Tresoldi E., Bjorndal A., Fredriksson R., Colognesi C., Deng H.K., Malnati M.S., Plebani A., Siccardi A.G., Littman D.R., Fenyo E.M., and Lusso P. 1997. In vivo evolution of HIV-1 co-receptor usage and sensitivity to chemokine-mediated suppression. *Nat. Med.* **3**: 1259–1265.

Schnell M.J., Johnson J.E., Buonocore L., and Rose J.K. 1997. Construction of a novel virus that targets HIV-1-infected cells and controls HIV-1 infection. *Cell* **90**: 849–857.

Schnell M.J., Buonocore L., Kretzschmar E., Johnson E., and Rose J.K. 1996. Foreign glycoproteins expressed from recombinant vesicular stomatitis viruses are incorporated efficiently into virus particles. *Proc. Natl. Acad. Sci.* **93**: 11359–11365.

Schwartz O., Marechal V., Le Gall S., Lemonnier F., and Heard J.M. 1996. Endocytosis of major histocompatibility complex class I molecules is induced by the HIV-1 Nef protein. *Nat. Med.* **2**: 338–342.

Shaheen F., Duan L., Zhu M., Bagasra O., and Pomerantz R.J. 1996. Targeting human immunodeficiency virus type 1 reverse transcriptase by intracellular expression of single-chain variable fragments to inhibit early stages of the viral life cycle. *J. Virol.* **70**: 3392–3400.

Steffy K.R. and Wong-Staal F. 1993. Transdominant inhibition of wild-type human immunodeficiency virus type 2 replication by an envelope deletion mutant. *J. Virol.* **67**: 1854–1859.

Stein C.A. and Cheng Y.C. 1993. Antisense oligonucleotides as therapeutic agents—Is the bullet really magical? *Science* **261**: 1004–1012.

Sullenger B.A. 1995. Colocalizing ribozymes with substrate RNAs to increase their efficacy as gene inhibitors. *Appl.Biochem. Biotechnol.* **54**: 57–61.

Sullenger B.A. and Cech T.R. 1993. Tethering ribozymes to a retroviral packaging signal for destruction of viral RNA. *Science* **262**: 1566–1569.

Sullenger B.A., Gallardo H.F., Ungers G.E., and Gilboa E. 1990. Overexpression of TAR sequences renders cells resistant to human immunodeficiency virus replication. *Cell* **63**: 601–608.

——— 1991. Analysis of *trans*-acting response decoy RNA-mediated inhibition of human immunodeficiency virus type 1 transactivation. *J. Virol.* **65**: 6811–6816.

Sun L.Q., Ely J.A., Gerlach W., and Symonds G. 1997. Anti-HIV ribozymes. *Mol. Biotechnol.* **7**: 241 251.

Sun L.Q., Pyati J., Smythe J., Wang L., Macpherson J., Gerlach W., and Symonds G. 1995. Resistance to human immunodeficiency virus type 1 infection conferred by transduction of human peripheral blood lymphocytes with ribozyme, antisense, or polymeric trans-activation response element constructs. *Proc. Natl. Acad. Sci.* **92**: 7272–7276.

Symons R.H. 1992. Small catalytic RNAs. *Annu. Rev. Biochem.* **61**: 641–671.

Thompson J.D., Ayers D.F., Malmstrom T.A., McKenzie T.L., Ganousis L., Chowrira B.M., Couture L., and Stinchcomb D.T. 1995. Improved accumulation and activity of ribozymes expressed from a tRNA-based RNA polymerase III promoter. *Nucleic Acids Res.* **23**: 2259–2268.

Tran A.C., Zhang D., Byrn R., and Roberts M.R. 1995. Chimeric ζ-receptors direct human natural killer (NK) effector function to permit killing of NK-resistant tumor cells and HIV-infected T lymphocytes. *J. Immunol.* **155**: 1000–1009.

Trkola A., Dragic T., Arthos J., Binley J.M., Olson W.C., Allaway G.P., Cheng-Mayer C., Robinson J., Maddon P.J., and Moore, J.P. 1996. CD4-dependent, antibody-sensitive interactions between HIV-1 and its co-receptor CCR-5. *Nature* **384:** 184–187.

Trono D., Feinberg M.B., and Baltimore D. 1989. HIV-1 Gag mutants can dominantly interfere with the replication of the wild-type virus. *Cell* **59:** 113–120.

Ugen K.E., Boyer J.D., Wang B., Bagarazzi M., Javadian A., Frost P., Merva M.M., Agadjanyan M.G., Nyland S., Williams W.V., Coney L., Ciccarelli R., and Weiner D.B. 1997. Nucleic acid immunization of chimpanzees as a prophylactic/immunotherapeutic vaccination model for HIV-1: Prelude to a clinical trial. *Vaccine* **15:** 927–930.

Ulmer J.B., Donnelly J.J., Parker S.E., Rhodes G.H., Felgner P.L., Dwarki V.J., Gromkowski S.H., Deck R.R., DeWit C.M., Friedman A., et al. 1993. Heterologous protection against influenza by injection of DNA encoding a viral protein. *Science* **259:** 1745–1749.

Wagner R.W. 1994. Gene inhibition using antisense oligodeoxynucleotides. *Nature* **372:** 333–335.

Walker R.E. 1996. A phase I/II pilot study of the safety of the adoptive transfer of syngeneic gene-modified cytotoxic T lymphocytes in HIV-infected identical twins. *Hum. Gene Ther.* **7:** 367–400.

Walker R., Blaese R.M., Carter C.S., Chang L., Klein H., Lane H.C., Leitman S.F., Mullen C.A., and Larson M. 1993. A study of the safety and survival of the adoptive transfer of genetically marked syngeneic lymphocytes in HIV-infected identical twins. *Hum. Gene Ther.* **4:** 659–680.

Walter E.A., Greenberg P.D., Gilbert M.J., Finch R.J., Watanabe K.S., Thomas E.D., and Riddell S.R. 1995. Reconstitution of cellular immunity against cytomegalovirus in recipients of allogeneic bone marrow by transfer of T-cell clones from the donor. *N. Eng. J. Med.* **333:** 1038–1044.

Wang B., Merva M., Dang K., Ugen K.E., Boyer J., Williams W.V., and Weiner D.B. 1994. DNA inoculation induces protective in vivo immune responses against cellular challenge with HIV-1 antigen-expressing cells. *AIDS Res. Hum. Retroviruses* (suppl 2) **10:** S35–S41.

Wang B., Dang K., Agadjanyan M.G., Srikantan V., Li F., Ugen K.E., Boyer J., Merva M., Williams W.V., and Weiner D.B. 1997. Mucosal immunization with a DNA vaccine induces immune responses against HIV-1 at a mucosal site. *Vaccine* **15:** 821–825.

Wang B., Ugen K.E., Srikantan V., Agadjanyan M.G., Dang K., Refaeli Y., Sato A.I., Boyer J., Williams W.V., and Weiner D. B. 1993. Gene inoculation generates immune responses against human immunodeficiency virus type 1. *Proc. Natl. Acad. Sci.* **90:** 4156–4160.

Wang B., Boyer J., Srikantan V., Ugen K., Agadjanian M., Merva M., Gilbert L., Dang K., McCallus D., Moelling K., and et al. 1995a. DNA inoculation induces cross clade anti-HIV-1 responses. *Ann. N.Y. Acad. Sci.* **772:** 186–197.

Wang B., Boyer J., Srikantan V., Ugen K., Gilbert L., Phan C., Dang K., Merva M., Agadjanyan M.G., Newman M., et al. 1995b. Induction of humoral and cellular immune responses to the human immunodeficiency type 1 virus in nonhuman primates by in vivo DNA inoculation. *Virology* **211:** 102–112.

Weerasinghe M., Liem S.E., Asad S., Read S.E., and Joshi S. 1991. Resistance to human immunodeficiency virus type 1 (HIV-1) infection in human CD4$^+$ lymphocyte-derived cell lines conferred by using retroviral vectors expressing an HIV-1 RNA-specific ribozyme. *J. Virol.* **65:** 5531–5534.

Wei X., Ghosh S.K., Taylor M.E., Johnson V.A., Emini E.A., Deutsch P., Lifson J.D., Bonhoeffer S., Nowak M.A., Hahn B.H., et al. 1995. Viral dynamics in human immunodeficiency virus type 1 infection. *Nature* **373:** 117–1122.

Welch P.J., Tritz R., Yei S., Barber J., and Yu M. 1997. Intracellular application of hairpin ribozyme genes against hepatitis B virus. *Gene Ther.* **4:** 736–743.

Welch P.J., Tritz R., Yei S., Leavitt M., Yu M., and Barber J. 1996. A potential therapeutic application of hairpin ribozymes: In vitro and in vivo studies of gene therapy for hepatitis C virus infection. *Gene Ther.* **3:** 994–1001.

Wiertz E.J., Mukherjee S., and Ploegh H.L. 1997. Viruses use stealth technology to escape from the host immune system. *Mol. Med. Today* **3:** 116–123.

Willett B.J., Picard L., Hosie M.J., Turner J.D., Adema K., and Clapham P.R. 1997. Shared usage of the chemokine receptor CXCR4 by the feline and human immunodeficiency viruses. *J. Virol.* **71:** 6407–6415.

Williamson R.A., Burioni R., Sanna P.P., Partridge L.J., Barbas C.F. II, and Burton D.R. 1993. Human monoclonal antibodies against a plethora of viral pathogens from single combinatorial libraries (erratum *Proc. Natl. Acad. Sci.* [1994] **91:** 1193. *Proc. Natl. Acad. Sci.* **90:** 4141–4145.

Wilson C.C., Wong J.T., Girard D.D., Merrill D.P., Dynan, M., An D.D., Kalams S.A., Johnson R.P., Hirsch M.S., D'Aquila R.T., et al. 1995. Ex vivo expansion of CD4 lymphocytes from human immunodeficiency virus type 1-infected persons in the presence of combination antiretroviral agents. *J. Infect. Dis.* **172:** 88–96.

Woffendin C., Ranga U., Yang Z., Xu L., and Nabel G.J. 1996. Expression of a protective gene-prolongs survival of T cells in human immunodeficiency virus-infected patients. *Proc. Natl. Acad. Sci.* **93:** 2889–2894.

Woffendin C., Yang Z.Y., Udaykumar Xu, L., Yang N.S., Sheehy M.J., and Nabel G.J. 1994. Nonviral and viral delivery of a human immunodeficiency virus protective gene into primary human T cells. *Proc. Natl. Acad. Sci.* **91:** 11581–11585.

Wolff J.A., Malone R.W., Williams P., Chong W., Acsadi G., Jani A., and Felgner P.L. 1990. Direct gene transfer into mouse muscle in vivo. *Science* **247:** 1465–1468.

Wolthers K.C., Bea G., Wisman A., Otto S.A., de Roda Husman A.M., Schaft N., de Wolf F., Goudsmit J., Coutinho R.A., van der Zee A.G., Meyaard L., and Miedema F. 1996. T cell telomere length in HIV-1 infection: No evidence for increased CD4+ T cell turnover. *Science* **274:** 1543–1547.

Wong J.K., Hezareh M., Günthard H., Havlir D., Ignacio C., Spina C., and Richman D. 1997. Recovery of replication-competent HIV despite prolonged suppression of plasma viremia. *Science* **278:** 1291–1295.

Wong-Staal F. 1996. Development of ribozyme gene therapy for HIV infection. *Antibiot. Chemother.* **48:** 226–232.

Wu L., Gerard N.P., Wyatt R., Choe H., Parolin C., Ruffing N., Borsetti A., Cardoso A.A., Desjardin E., Newman W., Gerard C., and Sodroski J. 1996. CD4-induced interaction of primary HIV-1 gp120 glycoproteins with the chemokine receptor CCR-5. *Nature* **384:** 179–183.

Wyand M.S., Manson K.H., Lackner A.A., and Desrosiers R.C. 1997. Resistance of neonatal monkeys to live attenuated vaccine strains of simian immunodeficiency virus. *Nat. Med.* **3:** 32–36.

Wyand M.S., Manson K.H., Garcia-Moll M., Montefiori D., and Desrosiers R.C. 1996. Vaccine protection by a triple deletion mutant of simian immunodeficiency virus. *J. Virol.* **70:** 3724–3733.

Yamada O., Kraus G., Leavitt M.C., Yu M., and Wong-Staal F. 1994. Activity and cleavage site specificity of an anti-HIV-1 hairpin ribozyme in human T cells. *Virology* **205:** 121–126.

Yamada O., Kraus G., Luznik L., Yu M., and Wong-Staal F. 1996a. A chimeric human immunodeficiency virus type 1 (HIV-1) minimal Rev response element-ribozyme molecule exhibits dual antiviral function and inhibits cell-cell transmission of HIV-1. *J. Virol.* **70:** 1596–1601.

Yamada O., Kraus G., Sargueil B., Yu Q., Burke J.M., and Wong-Staal F. 1996b. Conservation of a hairpin ribozyme sequence in HIV-1 is required for efficient viral replication. *Virology* **220:** 361–366.

Yamada O., Yu M., Yee J.K., Kraus G., Looney D., and Wong-Staal F. 1994. Intracellular immunization of human T cells with a hairpin ribozyme against human immunodeficiency virus type 1. *Gene Ther.* **1:** 38–45.

Yang A.G., Bai X.F., Huang X.F., Yao C.P., and Chen S.Y. 1997. Phenotypic knockout of HIV type 1 chemokine coreceptor CCR-5 by intrakines as potential therapeutic approach for HIV-1 infection. *Proc. Natl. Acad. Sci.* **94:** 11567–11572.

Yang O.O., Tran A.C., Kalams S.A., Johnson R.P., Roberts M.R., and Walker B.D. 1997. Lysis of HIV-1-infected cells and inhibition of viral replication by universal receptor T cells. *Proc. Natl. Acad. Sci.* **94:** 11478–11483.

Yasutomi Y., Robinson H.L., Lu S., Mustafa F., Lekutis C., Arthos J., Mullins J.I., Voss G., Manson K., Wyand M., and Letvin N.L. 1996. Simian immunodeficiency virus-specific cytotoxic T-lymphocyte induction through DNA vaccination of rhesus monkeys. *J. Virol.* **70:** 678–681.

Young J.A., Bates P., Willert K., and Varmus H.E. 1990. Efficient incorporation of human CD4 protein into avian leukosis virus particles. *Science* **250:** 1421–1423.

Yu M., Poeschla E., and Wong-Staal F. 1994. Progress towards gene therapy for HIV infection. *Gene Ther.* **1:** 13–26.

Yu M., Ojwang J., Yamada O., Hampel A., Rapapport J., Looney D., and Wong-Staal F. 1993. A hairpin ribozyme inhibits expression of diverse strains of human immunodeficiency virus type 1 (erratum *Proc. Natl. Acad. Sci.* [1993] **90:** 8303). *Proc. Natl. Acad. Sci.* **90:** 6340–6344.

Yu M., Poeschla E., Yamada O., Degrandis P., Leavitt M.C., Heusch, M., Yees, J.K., Wong-Staal, F., and Hampel, A. 1995. In vitro and in vivo characterization of a second functional hairpin ribozyme against HIV-1. *Virology* **206:** 381–386.

Zhou C., Bahner I.C., Larson G.P., Zaia J.A., Rossi J.J., and Kohn E.B. 1994. Inhibition of HIV-1 in human T-lymphocytes by retrovirally transduced anti-*tat* and *rev* hammerhead ribozymes. *Gene* **149:** 33–39.

Zimmerman P.A., Buckler-White A., Alkhatib G., Spalding T., Kubofcik J., Combadiere C., Weissman D., Cohen O., Rubbert A., Lam G., Vaccarezza M., Kennedy P.E., Kumaraswami V., Giorgi J.V., Detels R., Hunter J., Chopek M., Berger E.A., Fauci A.S., Nutman T.B., and Murphy P.M. 1997. Inherited resistance to HIV-1 conferred by an inactivating mutation in CC chemokine receptor 5: Studies in populations with contrasting clinical phenotypes, defined racial background, and quantified risk. *Mol. Med.* **3:** 23–36.

22

Progress toward Gene Therapy for Nervous System Diseases

Alberto Martínez-Serrano

Center of Molecular Biology Severo Ochoa
Autonomous University of Madrid–Consejo Superior de Investigaciones Cientificas
Campus Cantoblanco, Madrid 28049, Spain

Anders Björklund

Wallenberg Neuroscience Center
Section of Neurobiology
Sölvegatan 17, Lund 223 62, Sweden

Gene therapy may be defined as the introduction of functional genetic material into somatic cells for therapeutic purposes. During the last decade, we have seen an impressive development of this field, the coining of concepts, the initial application of emerging techniques, and the evolution of new tools based on more refined and complex molecular genetic techniques. In this chapter, we first discuss some of the conceptual and technical aspects of the gene transfer to the nervous system (NS), in order to provide a framework for the discussion of alternative approaches to gene therapy for neurological diseases. Although we take a broad approach, it is not our purpose to review extensively all aspects of gene therapy in the NS, but rather to select a few examples that can serve to illustrate the potential, the problems, and some of the possible strategies that may bring us closer to the goal of effective and safe gene therapy for NS disorders. As a complement, the reader is referred to several recent reviews (and the work cited therein) that have already dealt with the details of some of the important breakthroughs in this field during recent years (Fisher and Ray 1994; Crystal 1995; Karpati et al. 1996; Snyder and Fisher 1996; Doering 1997; Freese 1997; Friedmann 1997; Lanza et al. 1996; Vile 1997; Ho and Sapolski 1997; Kay et al. 1997; Miller and Whelan 1997; Raymon et al. 1997; Verma and Somia 1997).

In this brief introduction, we would also like to make some comments on the terminology used in this chapter. As regards the broadly used term "gene therapy" we only use this term in relation to human subjects when the outcome is, or is expected to be, a therapeutical benefit. Other gene-transfer procedures, whether carried out in human subjects or in experimental disease models, either for gene marking or for preclinical exploratory purposes, will be referred to as "experimental gene therapy." The many examples of gene-transfer techniques, procedures, and experiments carried out in vitro or in vivo in laboratory animals will be collectively termed "gene-transfer." As regards different types of gene-transfer procedures, we distinguish between those carried out in vivo or in a direct way (i.e., the transfer of genetic material by the use of a noncellular vehicle, the "vector," to the living organism), and the ex vivo or indirect gene-transfer procedures (when the genetic material is first introduced into cells that are expanded in culture and, later on, implanted in the organism). Since ex vivo gene transfer is applied to cells in culture, this approach can make use of a broader repertoire of molecular techniques than in vivo gene transfer, thus offering a higher degree of flexibility. In fact, all procedures in current use for in vivo gene transfer may also be applied ex vivo.

POTENTIAL APPLICATIONS OF GENE-THERAPY STRATEGIES IN THE NERVOUS SYSTEM, ACCORDING TO DISEASE CHARACTERISTICS

Gene therapy in the NS may be based on a wide range of biological mechanisms and diverse cellular targets. It is obvious, for example, that stimulation of atrophic neurons does not represent the same problem as halting a neuron death process, stimulating reinnervation of a target area, or replacing lost neurons. Therefore, the design of any gene-transfer approach must take into account the nature of the disease process and the features of the cellular and molecular targets in order to make it possible to tackle as many aspects of the disease as possible. For this reason, an open approach designed to be flexible enough to incorporate more than one genetic "bullet" should always be considered. The following examples, summarized in Figure 1, may serve to illustrate this point:

- The early stages of a slow degeneration process affecting a specific set of neurons may be reversed by the supply of trophic factors (or other proteins needed for neuronal survival and function) at the level of the target region innervated by these neurons, in order to maintain a functional terminal network and/or to promote regrowth of new axonal terminals.

- Acute and rapid degeneration of a specific set of neurons may be blocked by the supply of neurotrophic proteins (or other neuronal survival or anti-apoptotic factors) at the neuronal cell body level.

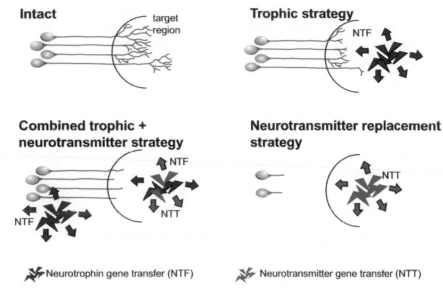

Intact
target region

Trophic strategy
NTF

Combined trophic +
neurotransmitter strategy
NTF
NTF
NTT

Neurotransmitter replacement
strategy
NTT

Neurotrophin gene transfer (NTF)   Neurotransmitter gene transfer (NTT)

Figure 1 Different gene-transfer strategies may be considered, depending on the stage and severity of neurodegenerative process. During early stages of the neurodegenerative process, neurotrophic factors supplied to the terminal region of the damaged neurons may help to promote recovery, either by stimulation of neuronal function or through the induction of axonal sprouting (neurotrophic strategy). At later stages, the supply of neurotrophic factors at both the cell body and the terminal level may help to protect the affected neurons and prevent further loss of functional axonal projections. When neurodegeneration has progressed further, attempts to restore normal neurotransmitter synthesis and release in the target area may help to preserve function (combined trophic and neurotransmitter replacement approach). Finally, when most neurons have degenerated, as is the case, e.g., in advanced Parkinson's disease, gene therapy may attempt to restore normal levels of the neurotransmitter (neurotransmitter replacement strategy).

- Complete loss of a specific set of neurons may be compensated for by the replacement of the lost neurotransmitter within the denervated target region(s) through reintroduction of the biosynthetic machinery for neurotransmitter production. In some cases, the reconstruction of a whole metabolic pathway might be needed, and not just that of a single enzyme.

- In case of widespread cellular malfunction or degeneration, gene therapy is faced with the problem of multifocal gene transfer, as opposed to the previous cases where gene transfer may be confined to a circumscribed target area.

- A special case is represented by postinsult treatment after acute, severe trauma or injury to the NS. In such conditions, experimental designs may involve the prevention of damage caused by the insult by a gene transfer applied prior to the insult as a way to explore possible strategies for therapeutic intervention; however, a more relevant strategy for the clinical situation requires that gene transfer be performed to work after injury has occurred.

- Chronic versus acute damage: The temporal characteristics of a specific disease or insult, whether it represents acute damage or it has a protracted development, is another essential element to consider. This will determine, for example, whether the genetic intervention can be of short duration or if it should allow long-lasting expression of the transduced gene.

- The genetic component of the disease may be important. A gene-therapy strategy designed to correct for a missing or dysfunctioning set of genes would require a targeted somatic cell gene replacement strategy. Moreover, it may be necessary to supply a correct copy of genetic information in a substantial proportion, preferably all, cells of a given type, which appears unattainable with currently available techniques.

GENE TRANSFER TO THE NERVOUS SYSTEM

In this section, we discuss the basic features of the principal gene transfer systems currently in use and illustrate how research and application of these procedures in the NS have evolved. The properties of the major systems and vectors, as well as the cell types used as vehicles of genetic information, are reviewed. The pros and cons of each system are discussed, and we address the reproducibility, safety, and degree of control of each procedure, i.e., aspects of outstanding relevance from the practical clinical point of view. Needless to say, it has not been possible for us to cite all relevant scientific work where original data have been published. For more complete coverage, the reader is referred to the reviews cited in the introduction to this chapter.

General Considerations

One of the important technical issues is how to ensure high transcriptional activity in the living organism (i.e., a high level of synthesis of messenger RNA), necessary for the synthesis of the potentially therapeutic proteins. A multitude of expression constructs that usually work well in vitro, ranging from very simple plasmid expression vectors to genetically engineered complex viral particles, have been used in vivo, with varying degrees of efficiency. Despite considerable efforts, a universal combination of genetic elements that would provide a vector system with a predictable and reproducible, sustained, long-term, and high-level basal expression remains to be identified. The search for a universally applicable expression system that would fulfill all these requirements nevertheless remains an important challenge for future research.

Attenuation of production of transgenic proteins in vivo is not only related to the specific vector design since there also other elements that influence the decrease in efficiency with time. First, in the case of vectors that do not integrate into the genome the genetic material may be lost either by degradation or as a result of division of the transduced cells; second, even in those cases where the genetic material is integrated into the genome the promoter may be attenuated or shut off (for recent discussion, see Paillard 1997a); third, the produced transgenic protein might become inactivated, degraded, or neutralized by the host immune system, or the transduced cells may be removed from the organism. Finally, there are other positive and negative factors influencing the transcriptional activity of an expression cassette, like the activity of elements able to control transcription rates (enhancers, *loci* controlling regions) or the more general and poorly defined effects of the site of integration into the genome, i.e., the chromatin environment surrounding the integrated copy of the vector system.

The design of cell- or tissue-targeted gene-transfer strategies (Harris and Lemoine 1996) has usually been approached in two ways. First, vectors can be designed to penetrate and transduce genetic information only to the cells of interest. This can be accomplished by using viral vectors, which can be engineered to be recognized only by certain receptors at the plasma membrane of specific cell types. Second, vectors may be introduced nonselectively into many types of cells at the time of gene transfer, but later on, selectivity is provided by elements in the vector that respond only to a unique intracellular environment, i.e., through the use of a cell-type-restricted transcriptional promoter, like neuronal or glial-specific promoters. Unfortunately, vectors bearing promoters with these characteristics do not always work as expected when they are in a different chromatin context than the genuine promoter. This complication may also apply to viral promoters when taken out of their natural viral context.

Inducible and Regulated Expression Systems

In recent years, there has been an increasing interest in the use of regulated expression systems since they may offer the investigator the opportunity to externally control the expression of the transgene, and hence the production of protein of interest, in a pharmacological way through the systemic administration of well-known pharmacological agents of common use. Examples of such regulatable expression systems are the tetracycline-based and the hormone-modulated systems (Miller and Whelan 1997; Saez et al. 1997). A new regulatory principle has recently been introduced, illustrating the feasibility of expressing a protein under the control of the normal physiological signals of the host, in this case, the regulation of the expression of transgenic erytrhopoietin by oxygen availability (Rinsch et al. 1997). In another recent example (Varley et al. 1997), the authors have taken advantage of the process of inflammation, common to most neurodegenerative conditions, to express proteins in a pathophysiological context, by using an expression system that is activated in response to inflammatory stimuli.

Regulated expression of a transduced protein would be ideal in many gene-transfer protocols; however, one should keep in mind that these systems are usually designed to work through the concerted action of several, usually two or three, independent expression cassettes, with their corresponding promoters. Recalling the problem to maintain long-term expression from a single promoter, it remains to be seen whether complex and interacting systems of these types can be made to function over sufficiently long periods in vivo.

Ribozymes to Inhibit Gene Function

In many cases of genetically determined diseases, the problem is not to supplement or augment defective genetic information, but rather to eliminate the deleterious effects caused by the expression of the defective genetic information (for instance, a toxic protein produced from the inside of the cell). For these cases, new hopes have recently been raised by the introduction of engineered ribozymes that can destroy the defective mRNA through suicide RNA editing or cleavage. From the work of Sidney Altman and Tom Czecz, we know that there exist RNAs capable of modifying other RNA molecules in a sequence-specific manner. Therefore, molecular engineers have recently started to develop vectors that code for RNA molecules with these RNA modifying properties directed against specific sequences in target cellular RNA molecules. Although this approach is still in its infancy and has at present many limitations, it holds the promise of providing new tools that will allow us to block specific disease processes, once the molecular strategies to target a specific subset of diseased cells have been identified (Thompson et al. 1995; Jones and Sullenger 1997).

The Antisense Approach

Similar to the ribozymes, antisense oligonucleotides or expressed antisense sequences may be used to inhibit or suppress expression of a defective gene by specifically hybridizing to the corresponding mRNA. Antisense oligonucleotides (usually short, ≤50 bp) are chemically synthesized and, after injection into the tissue, internalized by the cells where they may block translation of an mRNA into protein. However, since the injected oligonucleotides are quickly degraded in the cytoplasm, their use is limited to short-term treatment strategies. Expression of antisense sequences from within the cell in a stable manner, on the other hand, may offer the opportunity to obtain a long-lasting blockade of the translation of specific mRNAs, for instance, of genes with inherited mutations producing aberrant proteins. This approach, however, will require a highly efficient targeted in vivo gene transfer of a type that has not yet been achieved.

Protein Processing

Protein processing must always be taken into consideration whenever a gene is introduced into a cell in order to be expressed as a functional protein, and, in the case of secreted proteins, the existence and correct operation of the export machinery for the transduced protein. The biochemical and cellular machinery available in a specific transduced cell type is not guaranteed to work for all proteins. One illustrative example is the processing by proteolysis of inactive preprotein precursors of the neurotrophin family of secreted proteins. Regardless of using in vivo or ex vivo gene transfer techniques, therefore, there is always the need to test for bioactivity of the transgenically produced protein, its enzymatic activity, cellular function, or, when secreted, for their actual presence as a bioactive molecule in the extracellular space. In other words, relying only on indirect evidence (such as mRNA synthesis or immunological detection of the synthesized protein) may lead to a misinterpretation of the efficiency of the gene-transfer system used (Sakaguchi 1997; Seidah and Chrétien 1997).

Immune Response to Foreign Proteins

This immune response to foreign proteins is also important to assess when testing a gene-therapy system designed to work long term. Experiments may initially be carried out under immune suppression of the host. Later on, however, it is advisable to test for stability of the gene-transfer system in the absence of immune suppression, in order to ascertain that the transduced cells do not elicit a deleterious immune response. This may be triggered by several types of immunogens, by the cells used as vehicles for the genetic information (in ex vivo gene transfer), by proteins encoded in the vector (other than the one of interest) but needed for the vector construct (as is the case for instance in adenovirus and herpesvirus vectors), or by the transduced protein itself, when it is derived from a species different from that of the host. One should also be aware of the fact that certain proteins encoded by reporter or selection genes, usually of bacterial origin, may also be immunogenic. For these reasons, an ideal system for human gene therapy may use cloned human genetic information, carried by either cells of human origin or introduced by vectors not transmitting any other polypeptide coding information, or proteins, foreign to the host.

Integrative, Episomal, and Extrachromosomal Replicating Vectors

These vectors are common technical variants used for gene transfer, offering different properties. Different gene-transfer systems differ with respect to the final fate of the DNA introduced into the target cells. Some vector systems will incorporate into the chromatin, a process that sometimes requires cell division. Other vectors may stay as nonintegrated extrachromosomal nuclear entities (episomes). In this situation, the inserted genetic material may replicate and thus be perpetuated with each cell division (as is the case for extrachromosomal replicating plasmids; for further discussion, see Calos 1996; Cooper et al. 1997). In other cases (such as adenovirus or herpes virus vectors), the vectors are stable only in nondividing cells, and will become diluted with cell division. Integrative vectors may appear to be the optimal way of getting genes stably expressed; viral vectors that do not integrate usually show progressive shutoff of their transcriptional activity. On the other hand, integrated DNA is almost always modified by the cell (as a defence mechanism), undergoing methylation and inactivation; moreover, integration of foreign DNA randomly into the cell's genome may carry the risk of altering the expression of other, even distant, cellular genes (Doerfler et al. 1997).

There are also many *molecular tools* that may help to identify transfected cells among their unmodified neighbors in the culture dish. Among these tools, there are reporter proteins whose transcription and expression can be easily identified, in some cases, by their enzymatic activities, for instance, in the case of β-galactosidase, alkaline phosphatase, luciferase, or the jellyfish green fluorescent protein (GFP). All these reporter genes may aid in the identification and selection of transduced cells, both in vitro and in vivo. When cotransfected with the gene of interest, reporter genes may help to select in culture those cellular clones likely to carry the protein of interest, by conferring resistance to a cytotoxic drug. In this situation, one can choose among several selectable genetic markers (each of them with their own advantages and disadvantages), such as adenosine deaminase (ADA), aminoglycoside phosphotransferase (neo, G418, APH), dihydrofolate reductase (DHFR), hygromycin B phosphotransferase (HPH), the forward and backward selection based on thymidine kinase (TK), bleomycin, and puromycin N-acetyl transferase (P1-derived artificial chromosome [PAC]) (Ausubel et al. 1997). These reporter or drug-resistant genes may be combined with the gene(s) of interest in a variety of composite vectors, some of them depicted in Figure 2.

In Vivo Gene-transfer Systems

In this section, we summarize the main features of the in vivo gene-transfer vector systems currently in use, focusing on those aspects that are of particular importance for efficient gene transfer to the mature nervous system.

Retroviral Vectors

The retrovirus carries a RNA genome that can be modified so as to transfer the sequence of a protein of interest (up to 6.5 kb of foreign genetic material) to the chromatin of a eukaryotic cell, once this RNA is copied into DNA and incorporated by the infected cell. Retroviral genomes, usually one per cell, integrate randomly in a process that requires cell division. Therefore, infection and transduction of quiescent cells (as is the case for most cells in the mature nervous system) are precluded. Indeed, retroviruses are quickly inactivated in the presence of complement, so they are degraded quite fast in vivo. Retroviruses, therefore, have found their most common applications either in developmental studies to mark dividing cells and their progeny or in ex vivo gene-transfer, using rapidly dividing cells in culture. The type of retroviral vectors used for gene-transfer purposes are infectious, but replication-deficient, since all viral genes are removed from the vector. They have the advantage that helper-free virus stocks can be easily prepared, since specially manipulated packaging cell lines for the generation of viral particles are available. However, the random integration of the viral genome in its DNA form (provirus) poses the problem of being potentially mutagenic or even tumorigenic. For this reason, in the case of ex vivo retrovirally transduced cells, extensive tests should be carried out in order to reduce the risk that the transduced cells may develop tumors after implantation into the host.

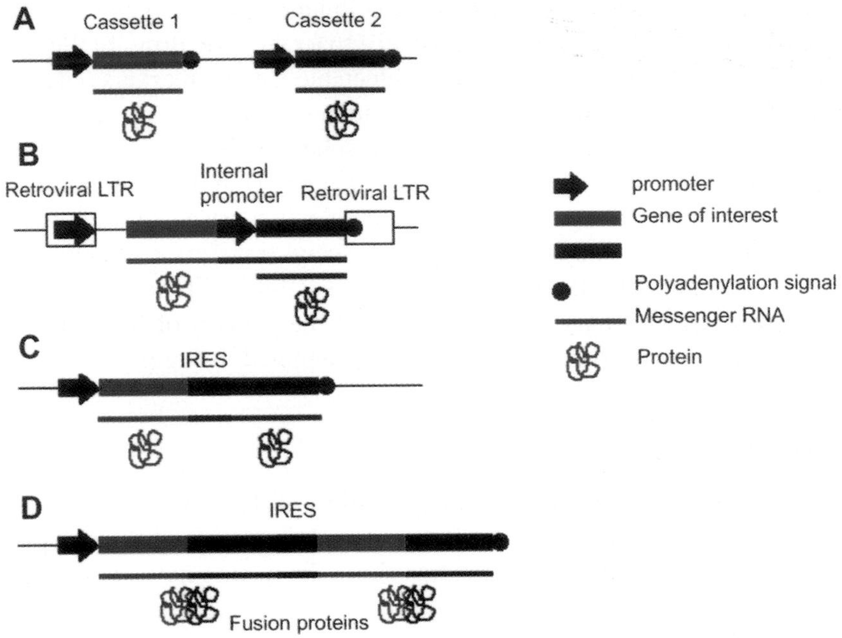

Figure 2 Use of reporter (rep) or drug-resistance (dr) genes in different types of constructs. (*A*) In the initial designs, the cDNAs of the gene of interest and that of the rep or dr genes were transfected into cells using two independent expression cassettes in the same plasmid, or in separate plasmids. Although the probability of finding a cell coexpressing both cDNAs is rather low, around 10% of rep+ or dr+ cells may be positive for the gene of interest. (*B*) Retroviral vectors helped to bring both cDNAs in the same transducing entity; in the example shown, the integrated retroviral genome is transcribed to a full-length RNA (the retroviral RNA genome) that can only express the first cDNA in the form of protein (using the 5´ cap for ribosome assembly), and a shorter mRNA from the internal promoter (which is also capped at 5´) coding for the second protein. (*C*) The use of a genetic element called IRES (internal ribosome entry sites) provides new assembly sites for ribosomes, once they have translated the first cDNA (using the 5´ cap), so that a single transcript can be used for the synthesis of two proteins consecutively. (*D*) In the most complex scenario, constructs may be engineered to use the elements just mentioned, and cDNAs engineered to code for fusion proteins, so that several activities would be simultaneously present in two or more polypeptides, all translated from a single transcript.

Retroviral vectors usually (but not always) undergo a decrease in transcriptional activity by one to two orders of magnitude in vivo, particularly in cells that stop dividing and become quiescent. However, transgene expression may be maintained long term, although at a reduced level.

Adenovirus

Adenoviral vectors are attractive in that they may accommodate larger foreign DNA fragments (up to 8 kb), can be generated free of contaminant replication-competent virus at extremely high titers, and can infect both dividing and nondividing cells. The wild-type adenoviral genome integrates after infection, but since some genes have been removed in the vector constructs, they will remain as a nonintegrated episome in the cell nucleus; therefore, adenovirus vectors are most adequate for nondividing cells. The adenoviral vectors currently in use have the disadvantage that they contain toxic viral proteins in their envelopes and too much of the original viral genetic information, so upon infection there is production by the cell of adenoviral proteins, which may themselves have some cytotoxic adverse effects on the transduced cells and may trigger or contribute to generate an immune reaction against the genetically modified cells (for further discussion, see Tripathy et al. 1996; Wood et al. 1996; Paillard 1997b; Verma and Somia 1997).

Adeno-associated Virus (AAV)

This type of vector may accommodate around 5 kb of foreign material and is also able to infect both dividing and nondividing cells. It has the property of integrating into the chromatin in a nonrandom way, which reduces the risk of mutagenesis and ensures the perpetuation of the vector in the cell's progeny. The AAV vectors have the disadvantage that, for the preparation of viral stocks, there is a need to use replication-competent helper viruses. The final stock, therefore, may be contaminated with the wild-type helper virus, a nonoptimal situation in a therapeutic context (During and Leone 1996).

Herpes Simplex Virus(HSV)

Vectors derived from herpes simplex virus have a large capacity to accommodate foreign genetic information (up to 36 kb). Depending on the specific type used (so-called amplicon or recombinant vectors), the stocks may or may not be contaminated with helper virus. After infection of either dividing or nondividing cells, the HSV vectors do not integrate into the cell's chromatin and will thus remain as an episome. As it happens with the wild-type virus, longevity of gene expression is seriously compromised in most cases, since the vector goes into a latency state where the genetic material remains inactive. Use of latency-associated promoters may help to reduce this problem. In addition, residual viral proteins expressed by the HSV vectors often induce cytotoxicity (Fink et al. 1996; Neve and Geller 1996). As is the case for adenoviruses, human beings have their immune system prepared to react against these kinds of viral proteins, since every person has been infected by adenovirus or herpesvirus previously and mounted an immune response against them (Paillard 1997b; Verma and Somia 1997).

Lentiviral Vectors

These vectors are derived from a family of retroviruses, among which the best known is human immunodeficiency virus 1 (HIV-1). Lentiviral vectors may be highly useful for gene transfer in the nervous system, because they share the properties of retroviral vectors, mentioned previously, and since they can also infect nondividing cells, like neurons, with high efficiency in vivo, and integrate nonrandomly into their genome (Naldini et al. 1996; Verma and Somia 1997). Current work is focused on the safety aspects of these vectors; given the seriousness of HIV-1 infection in humans, the risk of recombination events or contamination with replication-competent wild-type virus is a matter of concern (Fox 1997; Verma and Somia 1997; Zufferey et al. 1997).

Pseudotyped Retroviral Vectors

Because of the limited range of cell types that can be infected with retroviruses (which ultimately depends on the presence or absence of receptors for the retroviral particle in the cell membrane), efforts have been made to modify the retroviral vectors to extend their host range. This can be achieved by packing the retroviral particles in cells that provide other proteins to the viral envelope. A good example is the vector equipped with the vesicular stomatitis virus G envelope protein, which has made retroviral-mediated gene transfer accessible to other cell types otherwise resistant to infection by these types of viruses (Friedmann and Yee 1995).

Hybrid Retroadenoviral Vectors (or Adenoretroviral Vectors)

In this system, two adenoviruses are simultaneously used, one carrying a retroviral vector and the other carrying the genes needed for retrovirus packaging. The cells infected by both adenoviruses will thus become retroviral packaging cells, which can produce retroviruses to infect surrounding dividing cells. Using this strategy a high-level but short-lasting expression system (adenoviral) transforms itself into a long-term, albeit reduced expression level system (retroviral) (Feng et al. 1997). The most immediate application of a system with these characteristics might be that of providing a continuous source of retroviruses in vivo, for instance, designed to kill tumor cells. This system represents a case where the desired features and properties of two different viral vectors have been combined, suggesting the possibility for the near future to design new combined vector systems (Feng et al. 1997; Vile 1997).

Nonviral Systems

Cationic liposomes and naked DNA have been tested in in vivo gene-transfer experiments, resulting in the actual expression of foreign proteins in cells of the nervous system. Although attractive because of their non-viral nature, these methods are still highly inefficient compared to the viral vector systems. Moreover, they share the problems related to DNA stability and long-term gene expression (for further discussions, see Felgner 1997). The nonviral systems are of particular interest since they can be viewed as examples of purely synthetic vectors, which may offer possibilities to circumvent the ethical and safety problems that are inherent in the use of viral vectors.

Ex Vivo Gene Transfer Systems

As noted at the beginning of this chapter, ex vivo systems refer to the modification of cells in culture and their subsequent implantation in the organism, to provide the transgenic protein of interest, or an associated function of it. At first sight, the ex vivo systems may be seen as having the disadvantage of incorporating a new complicating element in the strategy, i.e., the cell to be used as vehicle for the genetic information. These cells need to be carefully chosen or even generated for a specific purpose. However, in contrast to any in vivo strategy, the ex vivo approach offers the opportunity to study the system from many angles in culture, before implantation in the organism, so that the whole strategy can be much better understood and thus better kept under control. Ex vivo strategies broaden the possibilities of gene therapy, because they offer the possibility to combine gene therapy and cell therapy, i.e, the cell chosen to transport the genetic information may have beneficial properties of its own and exert effects that may complement or enhance the result of gene transfer alone (an aspect not covered here in detail, but see Snyder and Macklis 1996; Martínez-Serrano and Björklund 1997). In addition, implantation of ex-vivo-manipulated cells obviously eliminates the concerns associated with genetic modification of the host's own cells, which may be potentially dangerous.

The first interest in using genetically modified cells for transplantation to the central nervous system (CNS) can be traced back to the seminal paper of Gage et al. (1987). After a few initial experiments late in the 1980s, which made use of cells or cell lines with origins outside the NS, it was soon clear that better cells preferably of neural origin were needed. During the last few years, the focus has been on cells of nervous-system origin, such as astrocytes and different types of neural progenitors (see Gage et al. 1995; Martínez-Serrano and Björklund 1996a; Snyder and Macklis 1996).

Based on the experience gained so far, a number of criteria have been established as a minimum set of requirements that cells need to fulfill in order to be useful for ex vivo gene transfer to the NS (Martínez-Serrano and Björklund 1997). One of the requirements is that cells should be generally available, grow under regular tissue culture conditions without problems and at a reasonable speed, should be easy to manipulate using routine molecular techniques, and should beamenable for molecular and cellular characterization before implantation. Another requirement is that, after transplantation, the cells should survive well in the intact or damaged adult nervous system and integrate into the host tissue in a nondisruptive manner without initiating an immune response in the host. The cells should, moreover, stop dividing after implantation, without any risk of tumor formation. Finally, their use should not pose any ethical or biosafety concerns, and for the purpose of gene therapy, they should optimally be of human origin and of clonal nature, to ensure homogeneity among preparations.

The initial experiments, which were carried out mainly in the laboratories of F.H. Gage, T. Friedmann, J. Mallet, and K. Ushida a decade ago, made use of transformed cells, either fibroblasts or cells of endocrine, neuroendocrine, or neural tumor origin. Although the grafted cells did actually form tumors in the brain of the recipient animals, these experiments provided "proof of principle," since they could demonstrate that it was possible to practice gene transfer in the NS, and thus opened the door for future work and developments. From the beginning, interest was concentrated on two model systems, namely, neurotransmitter replacement in animals with toxin-induced parkinsonism (reinstatement of dopamine synthesis by gene transfer of tyrosine hydroxylase, the rate-limiting enzyme of catecholamine biosynthesis), and the supply of neurotrophic factors to block lesion-induced neurodegeneration (in particular, nerve growth factor [NGF] gene transfer, which was shown to be able to rescue cholinergic neurons from axotomy) (see Fig. 1). These two model systems have continued to provide useful in vivo test systems that have helped advancement in the field. Since current work is no longer carried out longer using transformed cells, these are not going to be considered any further here; the reader is referred to other reviews for details on those experiments (Gage et al. 1987; Gage and Fisher 1991; Martínez-Serrano and Björklund 1997).

Among the cellular systems currently in use, we deal with three alternative types: primary cells, immortalized neural stem cells and progenitors, and encapsulated cells.

Primary Cells

Astrocytes, oligodendrocytes, O2-A progenitors, and neural-stem progenitor cells, are primary cells that can be taken and expanded from fetal CNS tissue. Alternatively, nonneuronal cells, such as myoblasts and skin-derived fibroblasts, can be obtained from biopsies from adult subjects. When the cells are derived from the same subject that is going to receive the implant after genetic modification of the cells (e.g., fibroblasts), there is no need to immune-suppress the recipient; otherwise immunosuppression may be necessary as with other types of allogeneic transplants. Primary cells are usually not very easy to grow in culture; in some cases, they divide quite slowly, unless they are propagated under the influence of growth factors, such as epidermal growth factor (EGF) and basic fibroblast growth factor (bFGF) (see Gage et al. 1995, Snyder and Macklis 1996). In some cases, it is known that cells have a limited life span and will undergo senescence in culture, thus limiting the number of possible cell divisions in culture (so far, this process is best understood in the case of fibroblasts). An unlimited supply of primary cells for grafting is by no means granted. In addition, genetic modification and subsequent subcloning of cells in primary culture may be unpractical or even unfeasible for some cell types. As mentioned previously, having an unlimited and homogeneous population of well-characterized and tested cells, seems to be an essential requirement to practice gene therapy; this may be difficult to achieve with primary cells.

The ability of primary cells to integrate into the host neural tissue is not the same for all cell types. Fibroblasts and myoblasts do not integrate at all; they stay isolated from the tissue as a mass of cells, and in the case of fibroblasts, they will be enclosed in a collagen matrix produced by the grafted cells. Astrocytes and oligodendrocytes, of nervous system origin, integrate better. Immature cells, such as neural stem/progenitors, are the ones that exhibit the most consistent and robust integration into the host CNS tissue.

Immortalized Neural Stem Cells and Progenitors

These provide another category of cells that have been successfully used for gene-transfer purposes. These cells are the counterparts of primary neural stem cells and neuroblasts or glioblasts, but behave as regular cell lines in culture without being transformed because of certain genetic manipulations leading to their establishment in culture and acquisition of immortal pheno-type and unlimited growth. Three cells lines, in particular, have been the object of intensive study in recent years (for review, see Martínez-Serrano and Björklund 1997): The brain-derived HiB5 (Renfranz et al. 1991), C17-2 (Snyder et al. 1992), and RN33B cell lines (Whittemore and White 1993), which are derived either from the fetal or early postnatal rodent brain, have been generated through immortalization with *myc*, or through conditional immortalization with a temperature-sensitive variant of the SV40 large T antigen. In all cases, the cells can be expanded under regular culture condi-tions, and they integrate well and differentiate into cells of neuronal and/or glial phenotypes upon transplantation to the rodent brain (Renfranz et al. 1991, Snyder and Macklis 1996, Lundberg et al. 1997). These kinds of cells have not been observed to form tumors in vivo, nor do they require immune suppression of the host for prolonged survival in the nervous system of mice or rats. They are advantageous in that they offer the opportunity for virtual-ly any kind of ex vivo genetic modification, and at the same time show good integrative properties in the recipient nervous system after transplantation. Current efforts (Sah et al. 1997) are aimed at the development of human immortalized neural progenitor cell lines, with properties similar to those of the now available rodent cell lines. Such cells should provide useful models for the study of biological properties of human neural precursors.

Encapsulation Technique

This technique represents an alternative ex vivo gene-transfer method that allows the use of virtually any type of cell, either of human or animal origin, regardless of its integrative properties (Aebischer et al. 1996a,b; Lanza et al. 1996). In this modality, cells are encapsulated within a biopolymer, that protects both the cells from the host (in case of an immune response) and the host from the cells (in case of using trans-formed cells as gene carriers), but allows the free diffusion of nutrients from the host to the encapsulated cells, and of transgenic products pro-duced by the encapsulated cells (although they are delivered in a point-source fashion). An extra advantage from the safety point of view is that the gene-transfer device can be easily removed. Cells tested so far with the encapsulation technique in models of neurodegenerative disease have been transformed cells, due to their easiness of growth (for comments, see Deglon et al. 1996). As discussed later in this chapter, interesting results have been reported in animal models of Huntington's disease and amyotrophic lateral sclerosis (ALS), and some of these procedures have reached the clinic.

GENE TRANSFER IN ANIMAL MODELS OF NERVOUS SYSTEM DISEASE

In this section, we describe some of the attempts to apply in vivo and ex vivo gene-transfer procedures in animal models of nervous system injury or disease, with particular emphasis on cognitive impairments, motor disorders, metabolic disorders, and injuries to the nervous system. Special attention will be given not only to the anatomical and functional aspects of recovery, but also to the clinical relevance of the models used.

Cognitive Impairments

Cognitive impairments are a frequent symptom of several common neurodegenerative conditions, including Alzheimer's, Parkinson's, and Huntington's diseases. For the development of therapeutic procedures aiming at ameliorating deficits in learning and memory, researchers have chosen, among other models, a naturally occurring neurodegenerative condition, the nonpathological process of aging, which is accompanied by a parallel decline in learning and memory abilities (for further discussion, see Martínez-Serrano and Björklund 1996a). In the aged rat, there is a pronounced atrophy of the cholinergic system in the forebrain that correlates with deficits in performance in a variety of cognitive tests (Dunnett and Fibiger 1993; Finch 1993; Gallagher and Colombo 1995), and which can be reversed by the exogenous administration (intracerebroventricular injection) of purified NGF protein (Fischer et al. 1987, 1991; Markowska et al. 1994; Frick et al. 1997). Since NGF, like other trophic proteins, does not readily cross the blood-brain barrier, any treatment based on exogenous NGF supply is limited in its application, since it would require repeated intracranial injections, or infusions of the protein. For this reason, intracerebral delivery of NGF by gene transfer of the NGF gene to the brain has been considered as an effective strategy to improve the behavioral performance of memory-impaired aged animals.

Experimental support for this idea has been obtained in ex vivo gene-transfer experiments, where either immortalized neural progenitor cells (Martínez Serrano et al. 1995a,b) or skin biopsy-derived fibroblasts (Chen and Gage 1995), modified using retroviral vectors coding for NGF, have been transplated to the basal forebrain in cognitively impaired aged rats. Albeit working with different strains of animals and slightly different vectors, both groups reported that NGF, administered via gene transfer, was able to reverse the age-dependent cholinergic atrophy in the aged animals and ameliorate their learning and memory deficits, as tested in the Morris water maze. In both cases, in vivo expression of the vector was demonstrated by three independent methods, at the mRNA, protein, and NGF-like bioactivity levels. An in vivo strategy using adenovirus has been also tested, and the recovery of cholinergic neuron atrophy by transgenic NGF was confirmed (although in this case behavioral assessment was not performed) (Castel-Barthe et al. 1996).

In subsequent work, performed using the NGF secreting clonal immortalized progenitors, we have shown that independent stimulation with NGF of any of the two major cholinergic groups in the basal forebrain (the medial septum or the nucleus basalis) are equally effective in inducing these effects and that the cognitive recovery may be attributable, at least in part, to a cholinergic mechanism (Martínez-Serrano et al. 1996). In the latter study, actual demonstration of increased NGF levels was provided by measuring the content on NGF protein in the transplanted regions, 10 weeks postimplantation.

These experiments demonstrate the reversal of an acute degenerative condition by gene transfer. Given the long-term stability of NGF production by the genetically engineered immortalized neural progenitors, we decided to test these cells in a preventive paradigm. In this study (Martínez-Serrano and Björklund 1998), instead of implanting the cells in aged animals that were already impaired, the NGF producing cells were transplanted into middle-aged rats, still performing well in cognitive tests, in order to provide a continuous supply of NGF to the basal forebrain cholonergic system for the subsequent 9 months (about one third of the life span of these rats). The results show that NGF ex vivo gene transfer initiated early in life may prevent the appearance of cholinergic neuron atrophy and cognitive deficits at advanced age. The transplanted rats reached the age of 25 months with a nonatrophic cholinergic system and a significantly improved spatial learning and memory ability than corresponding control animals that had received nontransduced control cells. Continued expression of the vector could be demonstrated, although this experiment clearly indicated that the retroviral vector underwent a slow inactivation in its transcriptional capacity with time.

In conclusion, the experiments summarized demonstrate that gene-transfer-based supply of a neurotrophic factor is a valid procedure to reverse the symptoms associated with a spontaneously occurring neurodegenerative condition, at both the cellular and the functional levels, and that this strategy may be effective also to prevent or slow down a progressive neurodegenerative process. Experiments in nonhuman primates

(Emerich et al. 1994b; Kordower et al. 1994; Tuszynski et al. 1996) have provided additional support for this approach, However, further experiments in clinically more relevant models will be necessary in order to clarify to what extent the neurotrophic strategy may be feasible to apply clinically. It should be remembered that while aged animals provide a good experimental model of age-related cognitive decline, they do not reproduce any of the specific neuropathological features, such as tangles and placques, which are characteristic for human dementias of the Alzheimer type. For this reason, the availability of accurate Alzheimer models in transgenic mice or rats may be essential for further progress along these lines.

An alternative approach to the neurotrophic factor strategy, based on the restoration of cognitive function though the supply of a deficient neurotransmitter, acetylcholine, by fibroblasts engineered to produce the actylcholine-synthesizing enzyme choline acetyltransferase, has been explored by Winkler et al. (1995). In this case, actylcholine-producing cells implanted into the cortex were shown to improve the spatial learning performance in young animals subjected to a lesion of the cholinergic neurons in the nucleus basalis. From these results, it may be tempting to speculate that combined trophic and neurotransmitter supply could produce additive effects.

Metabolic Disorders

In general, these types of inherited disorders pose a formidable problem for any gene-therapy approach, since they affect all cells of an organ, or an individual, and therapy should therefore be targeted to correct the genetic defect (or its effects) in as many cells as possible. However, these diseases can be genetically detected early in postnatal life, which may help in the design of effective therapies (Snyder and Fisher 1996; Snyder and Wolfe 1996).

One of the first attempts to apply gene transfer to a metabolic disease affecting the brain was carried out in the mouse mucopolysaccharidosis type VII (MPS VII). This condition, which is a model of Sly disease, results from a mutation in β-glucuronidase (GUSB), leading to a defect in the catabolism of glycosaminoglycans that accumulate in lysosomes. This severe neurodegenerative syndrome (among other pathologies) is responsible for mental retardation in human patients. In the first series of experiments, Wolfe et al. (1992) tried to provide these mutant mice with a correct copy of the GUSB enzyme, using peripherally injected (corneal inoculation) herpes-simplex-derived vectors, since they can infect postmitotic neurons. GUSB activity was detected in the brainstem and trigeminal ganglia of injected mice, showing that it was possible to reintroduce the correct metabolic enzyme in the nervous system. Because of the low number of cells transduced, however, phenotypic correction of the disease was not attained. In the next step, Snyder and Wolfe decided to use immortalized neural progenitor cells, which upon engraftment in the early postnatal brain can participate in the last stages of brain development and become structurally integrated into the host tissue. For this purpose, the C17-2 cell line (Snyder et al. 1992), which expresses normal levels of GUSB, was made to overexpress the enzyme by retroviral transduction; in this case, the grafted animals showed widespread correction of the main histological symptoms of the disease (Snyder et al. 1995). Lacorazza et al. (1996) have adopted a similar approach to another common metabolic disease, Tay–Sachs disease, caused by a deficiency in β-hexosaminidase α-subunit. Cells derived from human patients were metabolically corrected after gene transfer of the defective enzyme, and cells grafted into normal embryonic mice were shown to survive well and contributed to increase the levels of this metabolic activity. In the absence of suitable animal models, however, it remains unclear whether this strategy could have any impact on the disease phenotype in a living organism.

Motor Disorders

Parkinson's Disease

The debilitating motor impairments of Parkinson's disease (PD) develop as a consequence of the degeneration of a specific neuronal group in the brain, the dopamine neurons of the substantia nigra (SN), which provide a regulatory dopaminergic innervation of one of the important motor controling regions of the brain, the striatum. Degeneration of the neurons in SN will deplete the striatum of its dopamine neurotransmitter, resulting in the three cardinal symptoms of the disease, hypokinesia (difficulty to initiate and execute movement), rigidity (muscle stiffness), and tremor.

Efforts to implement gene therapy for the treatment of Parkinson's disease have concentrated on two main strategies: (1) the transfer of the gene coding for the enzyme tyrosine hydroxylase (TH), i.e., the enzyme that constitutes the first, rate-limiting step in the synthesis of dopamine from the amino acid tyrosine (see Fig. 3A), with the aim of restoring dopamine production in the critical region in the basal ganglia; and (2) the transfer of genes that encode neurotrophic factors that can stimulate the survival and growth of damaged dopaminergic neurons in the substantia nigra.

Dopamine Replacement

Gene therapy based on the transduction of the tyrosine hydroxylase (TH) gene, ex vivo or in vivo, relies on a simple principle: the restoration of dopamine synthesis with the help of TH enzyme locally in the region in which the production and release of dopamine in Parkinson patients has failed, i.e., above all in the striatum. The first trials with dihydroxyphenylalanine (DOPA)- or dopamine-producing cells were carried out with tumor-derived cell lines. These studies demonstrated that bioactive TH could be transferred to a target cell by the use of retroviral vectors, and induce DOPA and dopamine synthesis in vitro, and when transplanted to the striatum in unilaterally lesioned animals, the DOPA- or dopamine-producing transduced cells were found to improve the rats' motor function, as assessed in the drug-induced rotation test (Horellou et al. 1989; Wolff et al. 1989). In addition, Horellou et al. (1990a,b) could demonstrate, by in vivo microdialysis, that the transplanted cells actually synthesized and released dopamine in the grafted striatum.

The tumor-derived cell lines, however, are of limited use in vivo since they continue to proliferate and form tumors within a few weeks after grafting. In later experiments, therefore, there has been a general switch toward different types of primary cells suitable for transplantation into the brain, such as fibroblasts (Fisher et al. 1991) and astrocytes (Lundberg et al. 1996), or conditionally immortalized neural progenitors (Anton et al. 1994). In all cases, the main secreted product is DOPA, which is due to the fact that the TH-transduced cells do not contain the decarboxylating enzyme DOPAdecarboxylase (DDC) required for the conversion of DOPA into dopamine. To enable DOPA to have any effect at all in vivo, the released DOPA must be converted to dopamine in the surrounding brain tissue.

The in vivo effects of the TH-transduced cells have been examined following transplantation into the dopamine-denervated striatum region of rats with unilateral 6-hydroxydopamine (6 OHDA) lesions. Until now, however, the functional effects have been short lasting and limited to apomorphine-induced rotation, which is a measure of hypersensitivity of dopamine receptors in the denervated striatum. In experiments conducted so far, the greatest problem has been the down-regulation of the TH transgene in the transplanted cells in vivo. In both fibroblasts and astrocytes, the expression of the TH transgene was greatly reduced within the first few weeks after transplantation, and the functional effect declined over time (Fisher et al. 1991; Lundberg et al. 1996). To date, the best results have been achieved with TH-expressing conditionally immortalized neural progenitors in which the initial effect persisted unchanged for 4–6

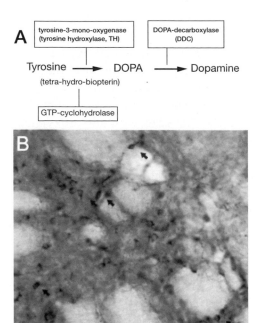

Figure 3 Dopamine replacement: reconstruction of a metabolic pathway. (*A*) TH gene transfer by itself may not be sufficient for the synthesis of dopamine: GTP-cyclohydrolase I might be needed for the synthesis of tetrahydrobiopterin (BH$_4$), an essential cofactor for TH activity, and DOPA decarboxylase might also be needed for the cells to convert DOPA to dopamine. In a case like this, where a complete metabolic pathway needs to be restored in the brain, in vivo gene-transfer systems may require polycistronic vectors, or combinations of different viruses. Both ex vivo and in vivo gene-transfer strategies will face the problem of keeping stable expression of multiple promoters. (*B*) Immortalized neural progenitors are unable to synthesise DOPA after TH transfection unless BH$_4$ is supplied exogeneously, or GTP-cyclohydrolase I is expressed simultaneously. The photomicrograph shows doubly transfected progenitors (immunostained for TH), integrated into the lesioned rat striatum (A. Martínez-Serrano and A. Björklund, unpubl.). This approach may be used to provide the recipient striatum with DOPA which may be converted into functional dopamine in the surrounding host striatum. Promising results along these lines have been reported by Anton et al. (1994) and Bencsics et al. (1996).

months after transplantation (Anton et al. 1994). Recent experiments by Kang et al. (1993) and Bencsics et al. (1996) have indicated that transduction of cells with TH may not be sufficient for efficient dopamine production and that both DDC and the TH cofactor enzyme GTP cyclohydrolase may be required (Fig. 3).

Direct in vivo gene transfer by TH-containing virus vectors designed to directly infect cells in the striatum have also been explored. Adenovirus (Horellou et al. 1994), herpesvirus (Kaplitt et al. 1994), and AAV vectors (During et al. 1994) have all been tested experimentally in unilateral 6OHDA-lesioned rats. However, in all three studies, the resulting TH gene expression was very limited. Functional effects were observed in the apomorphine rotation test, but because these effects were poorly correlated with the extent of vector-induced TH expression in the infected striatum, the functional results may have to be interpreted cautiously (see Isacson 1995).

For future developments of the neurotransmitter replacement strategy, a number of important issues remain to be solved. Most importantly, both the level and duration of TH transgene expression must be greatly improved in order to obtain sustained DOPA or dopamine levels of sufficient magnitude to have a functional impact on large areas of the denervated striatum. Future studies will not only have to be concerned with simple parameters of drug-induced rotation, which may be misleading, but include assays of spontaneous sensorimotor behaviors of relevance for the symtomatology in Parkinson's disease. Finally, it must be clarified whether production of DOPA alone is sufficient to induce symptomatic relief or whether the transduced cells will have to be made to sythesize dopamine, i.e., by concomitant transduction of both the TH and the DDC genes.

Supply of Neurotrophic Factors

Neurotrophic factors are macromolecules that play a part in the development of the peripheral nervous system and probably also have a role as neuronal survival and growth factors in the CNS. In recent years, there has been an intensive search for survival factors for the dopaminergic neurons in the substantia nigra. Among the factors that have been tested in cell cultures, glial-cell-line-derived neurotrophic factor (GDNF) has attracted the most attention due to its high potency and survival effects on dopaminergic neurons both in vitro and in vivo. Several groups, therefore, have tried to develop systems for the supply of transgenic GDNF to degenerating neurons of the substantia nigra.

Both adenoviral and AAV vectors have been tested for their ability to supply GDNF to nigral neurons and to counteract dopamine neuron degeneration induced by intrastriatal 6OHDA (Bilang-Bleuel et al. 1997; Choi-Lundberg et al. 1997; Mandel et al. 1997). In these experiments, researchers have used a partial lesion model that more closely resembles the slow and protracted neurodegeneration seen in PD. Gene transfer was carried out either at the cell body level in the SN (Choi-Lundberg et al. 1997; Mandel et al. 1997) or in the striatum, since GDNF can be retrogradely transported from this location to neuronal cell bodies in the SN (Bilang-Bleuel et al. 1997). In all cases, injection of the vector was carried out before lesioning the cells, to test its effects in preventing neuron death. Significant rescue of nigral neurons was observed in all cases, ranging from 20% to 40%. Expression of the GDNF protein was confirmed, but, as usual, it was found to decrease with time (Choi-Lundberg et al. 1997; Mandel et al. 1997). Also, dopaminergic neuron rescue had some effects reducing drug-induced rotational asymmetry (Bilang-Bleuel et al. 1997). However, since only the nigral neurons were preserved, and not the dopamine innervation in the striatum, it seems possible that this effect was mediated at the level of the SN. Tseng et al. (1997), working with intranigral implants of encapsulated GDNF-producing cells, have obtained results in support of this possibility.

In case of ex vivo gene transfer, the most interesting in vivo results so far have obtained with fibroblasts infected with retrovirus containing the brain-derived neurotrophic factor (BDNF) gene (Levivier et al. 1995). The results showed that implantation of BDNF-secreting fibroblasts, but not control fibroblasts, in the striatum 2 weeks before an intrastriatal 6OHDA injection could afford a significant, approximately 50%, protection of the dopaminergic neurons in the SN. In addition, intrastriatal transplants of BDNF-producing astrocytes have been reported to reduce the behavioral symptoms of experimentally lesioned animals, presumably through stimulation of striatal function (Yoshimoto et al. 1995). These data provide support for the idea that a continuous supply of neurotrophic factors from transduced cells into the area surrounding the lesion can protect the dopaminergic neurons from a subsequent neurotoxic insult.

The relevance of these exploratory animal experiments for the treatment of patients with Parkinson's disease is unclear. The experimental animals used in these studies have Parkinson-like damage to the dopamine system, but they do not model the specific disease process characteristic for Parkinson's disease. Consequently, it is unclear whether the cellular mechanisms responsible for neurotoxic damage are representative of the neurodegenerative process in the Parkinson-diseased brain. As far as the use of gene-therapy methods in neurologic diseases is concerned, Parkinson's disease is probably one of those conditions in which this technique will first be tested. This is based on the fact that the underlying neuropathology is well defined and there are clear biological mechanisms (dopamine production in the striatum, neurodegeneration in the substantia nigra) that offer targets for gene-therapeutic intervention. However, the technique is still very much at the exploratory stage of its development. The vectors and transduction techniques used today are unsatisfactory both with respect to their effectiveness and safety and in terms of the long-term expression in vivo. Research in this field will thus face numerous challenges and exciting possibilities in the future.

Huntington's Disease

Huntington's disease (HD) is a genetic neurodegenerative disease affecting primarily the striatum and the neocortex. The first attempts to apply gene transfer for the amelioration of striatal neuronal cell death in a rat HD model were carried out in O. Isacson's laboratory (Schumacher et al. 1991; Frim et al. 1993a,b). These experiments showed, for the first time, that NGF supplied by ex vivo gene transfer by intrastriatal transplants of transduced fibroblasts could reduce the extent of cell death in rats subjected to excitotoxin-induced lesions of the striatum (a condition reproducing some of the features of HD). Subsequently, similar neuroprotective effects have been obtained with either NGF-producing encapsulated cells (Emerich et al. 1994a) or NGF-secreting immortalized neural progenitors (Martínez-Serrano and Björklund 1996b) implanted into the striatum. In the latter case, local supply of transgenic NGF to the striatum saved about 60–70% of the neurons that would otherwise have died after the lesion. In addition, Emerich et al. (1994a) reported amelioration of pharmacologically induced rotational asymmetry induced by the lesion. In all these experiments, neuronal protection and functional sparing was assessed short term, within a month after the lesion. Whether the protective effects can be long lasting is unclear: In a recent study performed in our own laboratory, we have observed that the NGF-treated animals performed as poorly as the control lesioned animals in tests of drug-induced rotation or skilled paw use, and 6 months after the excitotoxic lesion with quinolinic acid, the NGF-treated animals did no longer show any significant neuronal sparing compared to animals receiving transplants of control cells (A. Martínez-Serrano et al., unpubl.).

In parallel with these experiments, other groups have tested the neuroprotective effects of a different trophic factor, ciliary neurotrophic factor (CNTF) in the same lesion paradigm, using encapsulated cells for the purpose of ex vivo gene transfer. These results have been more satisfactory: Significant sparing of striatal neurons has been obtained in both rats and primates, and in the rat experiment, this effect was accompanied by preservation of motor function (Emerich et al. 1996, 1997). These results suggest that intrastriatal supply of neurotrophic factors may be feasible as a neuro-protective strategy in HD patients, not least since HD is a genetically deter-mined disease that can be diagnosed before symptoms and massive neu-rodegeneration have appeared, so that neuroprotective strategies by means of gene transfer will probably have a great chance for success. Recently, a transgenic mouse that mimics human HD has become available; this mouse may indeed provide a new useful model for preclinical tests of alternative gene-transfer strategies (Mangiarini et al. 1996).

Amyotrophic Lateral Sclerosis (ALS) and Other Motor Neuron Neurodegenerative Conditions

ALS is a fatal progressive motor neuron disease, causing skeletal muscle wasting and ultimately leading to paralysis and death. Several neu-rotrophic factors, including BDNF, CNTF, and GDNF, have prominent survival-promoting effects on motor neurons and may thus be of interest for the development of gene therapy (for review, see McMahon and Priestley 1995). Similar to the metabolic diseases previously discussed, the problem of ALS and other neurodegenerative conditions of motor neurons is how to target a widespread population of affected neurons. Two possible solutions have been tested: either infecting all neurons with a virus-based vector or alternatively, implanting cells that produce a fac-tor that is released to the cerebrospinal fluid (CSF).

The first experiments along these lines (Sentner et al. 1992; Sagot et al. 1995) made use of a genetic model of neurodegeneration, the *pmn/pmn* mouse, which bears a genetic mutation leading to progressive motor neu-ropathy. In both cases, CNTF was the neurotrophic factor of choice, and gene transfer was carried out using ex vivo techniques. Sendtner et al. (1992) made use of a teratoma-derived, tumorigenic mouse cell line that was engineered to produce CNTF and implanted intraperitoneally, where-as in the study of Sagot et al. (1995) baby hamster kidney (BHK) cells were encapsulated and the capsule device implanted intrathecally. In both cases, it was shown that biologically delivered CNTF could prolong life expectancy and improve motor function in the affected mice, and when examining particular motor neuron groups like the facial nucleus, a sub-stantial rescue was found. These experiments thus provided proof of prin-ciple for this gene-transfer approach. Later on, capsules of CNTF-secret-ing cells have been considered for a gene-therapy trial (see protocol in Aebischer et al. 1996b). The first set of results obtained in these patients (Aebischer et al. 1996a) demonstrated that CNTF was actually produced in large amounts, and delivered to the CSF, but surprisingly, and contrary to the results obtained in the mouse model, the patients did not experience any recovery as tested by clinical scores, at least during the 3-month observation period.

Other neurotrophic factors, particularly GDNF, have been shown to have similar neuroprotective effects on motor neurons in vivo (Yan et al. 1995), and this factor has also been applied in an encapsulation ex vivo strategy in the *pmn/pmn* mouse (Sagot et al. 1996). The results showed a delay of the death of motor neurons, but axonal degeneration and the life span of the animals were unaffected.

In vivo gene transfer approaches have also been explored in the *pmn/pmn* mouse model. In particular, adenoviral vectors coding for neurotrophin-3, particularly when combined with adenovirus-CNTF, have been shown to ameliorate cell death in these mice, inducing a similar degree of protection as in the ex vivo experiments described above (Haase et al. 1997). In other studies (Baumgartner and Shine 1997; Gimenez y Ribotta et al. 1997), the efficiency of different neurotrophic factors (BDNF, CNTF, and GDNF) was compared in the model of axonal transection of motor neurons of the facial nerve. In these experiments, the factors were made available to the neurons through an in vivo approach using adenoviruses injected into the facial muscle of neonatal rats before the lesion was made. Comparison of the results obtained with all three neurotrophins indicated that GDNF was somewhat more potent than the others, whereas BDNF and CNTF gave similar results (Baumgartner and Shine 1997). Interestingly, as noted by Baumgartner and Shine (1997), the number of neurons rescued exceeded that of the genetically modified neurons, which suggests that the transduced cells could support the survival of the neighboring neurons through a paracrine mechanism.

Stroke

Research on gene therapy strategies in animal models of transient ischemia and stroke has started to provide interesting results. These studies have explored both extracellular or intracellular mechanisms that could help the affected neurons to survive better after an ischemic insult. A direct comparison, however, between the results of different studies is complicated by the fact that the in vivo models and the methods of evaluation vary greatly between groups. Most of the experiments reported so far have adopted a similar neuroprotective scheme, i.e., gene transfer is carried out before (1–7 days) the ischemic insult is administered to the animal. The genes that have been tried for neuroprotective purposes in these paradigms are of two different kinds: those that encode proteins that act in a paracrine way (like neurotrophic factors) and those acting intra-cellularly when transferred to the neurons affected by the insult. In the first case, NGF has been shown to induce clear neuroprotective effects when transferred ex vivo, using either NGF-secreting fibroblasts in the hippocampus (Pechan et al. 1995) or NGF-secreting immortalized neural progenitors implanted into the striatum, before a transient ischemic insult (Andsberg et al. 1998). In the latter case, around 60% of the striatal neurons that otherwise would have died survived the insult in the NGF-treated animals. Transfer of potentially neuroprotective genes directly to the affected neurons, by infection with either adenovirus or herpesvirus vectors, include the anti-apoptotic protein *bcl-2* (Linnik et al. 1995; Lawrence et al. 1996a), the interleukin-1 receptor antagonist protein (to block the neurotoxic effects of interleukin-1; Betz et al 1995; Yang et al. 1997), the glucose transporter protein (to enhance the energy-rich glucose uptake in otherwise metabolically compromised neurons; Lawrence et al. 1996b), and the anti-apoptotic molecule, NAIP (neuronal apoptosis inhibitory protein; Xu et al. 1997).

Overall, these experiments have provided some preliminary evidence that neuroprotective genes can be effective when they are introduced before the insult. Further work, however, is needed in order to establish the extent to which any of these procedures can be neuroprotective also after the ischemic event has occurred. In support of this, Sapolsky and collaborators have recently demonstrated that there is a time window of a few hours after the insult when *bcl-2* gene transfer is equally neuroprotective as when applied before the insult (Lawrence et al. 1997). Moreover, combinations of different survival factors or anti-cell-death genes should be tested in order to improve the overall rescue effects, and long-term experiments will be needed to evaluate the clinical relevance of these procedures.

CONCLUSION

The use of gene-transfer techniques for therapeutic purposes in the nervous system is still in its infancy. The only procedure that has been tested clinically is the encapsulation technique, where cells engineered to secrete a neurotrophic factor, CNTF, have been enclosed within a semi-permeable membrane and implanted into the intrathecal space in ALS patients (Aebischer et al. 1996a,b). It is easy to see why this approach is early to reach the clinic: The engineered cells remain confined within a capsule so that the device can easily be removed, if necessary, which is a particularly advantageous safety feature of this technique. The clinical application of alternative gene-transfer techniques are clearly associated with much more complex safety issues. For ex vivo gene-transfer techniques, for example, procedures have to be devised to eliminate the risk of tumor formation by the transduced cells, and preferably also to allow elimination or killing of the implanted cells in case something went wrong. For direct injection of viral vectors into the CNS, the potential risks associated with contaminating helper viruses and possible unforeseen recombination with wild-type viruses have to be assessed carefully. Moreover, unless the viral vectors can be targeted to specific disease-affected cell types, the possible deleterious consequences of widespread infection and expression of a therapeutic gene throughout large areas of the CNS, and possibly also other tissues, have to be considered.

Although important safety issues remain to be solved, a more immediate concern is whether any of the currently available gene-transfer procedures are efficient enough, and whether the expression of the transgenes can be maintained at a sufficiently high level for a sufficiently long time to be of therapeutic use in the brains and spinal cords of humans, which are considerably larger than those of rodents that have been targeted so far. In the ex vivo gene-transfer procedures tested, the expression of the transgene in vivo is down-regulated or shut off, usually within a few weeks after transplantation. And in the case of viral vectors for direct in vivo gene transfer, there is a risk that they may induce cytotoxicity and activation of the host's immune system (such as is the case with the adenovirus and herpesvirus vectors currently in use). Moreover, in some cases (such as with AAV vectors), the in vivo delivery systems may have limitations with respect to viral titers and efficiency of in vivo infectivity.

A most promising recent development for gene transfer to the CNS is the introduction of new lentiviral vectors (Verma and Somia 1997; Zufferey et al. 1997). These types of vectors are particularly interesting for gene transfer to the nervous system since they can infect postmitotic neurons with high efficiency and they can be expressed stably at high levels over long times in vivo. The experience with these vectors is still limited, however, and there are particular safety problems with these HIV-1-based viruses that have yet to be solved (see Fox 1997).

After a slow start, the development of gene transfer to the nervous system has been clearly accelerating during recent years. With the introduction of new vector systems, improved versions of existing vector systems, and the development of appropriate safety devices built into the vectors, it should in the future be possible to move more quickly toward the clinic. Much preclinical work, however, remains to be done before any of the experimental gene-transfer procedures, currently explored in various nervous system disease models, can be considered for clinical application. It is important to stress that any attempt to apply an experimental gene-transfer protocol in clinical trials must be based on solid experimental data, and that convincing evidence of efficacy, both anatomically and functionally, in long-term experiments in animal models of relevance for the clinical disease must be available before clinical trials are initiated.

REFERENCES

Aebischer P., Schluep M., Deglon N., Joseph J.M., Hirt L., Heyd B., Goddard M., Hannang J.P., Zurn A.D., Kato A.C., Regli F., and Baetge E.E. 1996a. Intrathecal delivery of CNTF using encapsulated genetically modified xenogeneic cells in amyotrophic lateral sclerosis patients. *Nat. Med.* **2:** 696–699.

Aebischer P., Pochon N.A.M., Heyd B., Deglon N., Joseph J.M., Zurn A.D., Baetge E.E., Hammang J.P., Goddard M., Lysaght M., Kaplan F., Kato A.C., Schluep M., Hirt L., Regli F., Porchet F., and De Tribolet N. 1996b. Gene therapy for amyotrophic lateral sclerosis (ALS) using a polymer encapsulated xenogeneic cell line engineered to secrete hCNTF. *Hum. Gene Ther.* **7:** 851–860.

Andsberg G., Kokaia Z., Björklund A., Lindvall O., and Martínez-Serrano A. 1998. Amelioration of ischemia-induced neuronal death in the rat striatum by NGF-secreting neural stem cells. *Eur. J. Neurosci.* **10:** 2026–2036.

Anton R., Kordower J.H., Maidment N.T., Manaster J.S., Kane D.J., Rabizadeh S., Schueller S.B., Yang J., Rabizadeh S., Edwards R.H., Markham C.H., and Bredesen D.E. 1994. Neural-targeted gene therapy for rodent and primate hemiparkinsonism. *Exp. Neurol.* **127:** 207–218.

Ausubel F.M. et al., eds. 1997. *Current protocols in molecular biology.* Wiley, New York.

Baumgartner B.J. and Shine H.D. 1997. Targeted transduction of CNS neurons with adenoviral vectors carrying neurotrophic factor genes confers neuroprotection that exceed the transduced mechanisms. *J. Neurosci.* **17:** 6504–6511.

Bencsics C., Wachtel S.R., Milstien S., Hatakeyama K., Becker J.B., and Kang U.J. 1996. Double transduction with GTP cyclohydrolase I and tyrosine hydroxylase is necessary for spontaneous synthesis of L-DOPA by primary fibroblasts. *J. Neurosci.* **16:** 4449–4456.

Betz A.L., Yang G.Y., and Davidson B.L. 1995. Attenuation of stroke size in rats using an adenoviral vector to induce overexpression of interleukin-1 receptor antagonist in brain. *J. Cereb. Blood Flow Metab.* **15:** 547–551.

Bilang-Bleuel A., Revah F., Colin P., Locquet I., Robert J.J., Mallet J., and Horellou P. 1997. Intrastriatal injection of an adenoviral vector expressing glial-cell-line-derived neurotrophic factor prevents dopaminergic neuron degeneration and behavioral impairment in a rat model of Parkinson disease. *Proc. Natl. Acad. Sci.* **94:** 8818–8823.

Calos M.P. 1996. The potential of extrachromosomal replicating vectors for gene therapy. *Trends Genet.* **12:** 463–466.

Castel-Barthe M.N., Jazat-Poindessous F.J., Barneoud P., Vigne E., Revah F., Mallet J., and Lamour Y. 1996. Direct intracerebral nerve growth factor gene transfer using a recombinant adenovirus: Effect on basal forebrain cholinergic neurons during aging. *Neurobiol. Dis.* **3:** 76–86.

Chen K.S. and Gage F.H. 1995. Somatic gene transfer of NGF to the aged brain: Behavioral and morphological amelioration. *J. Neurosci.* **15:** 2819–2825.

Choi-Lundberg D.L., Lin Q., Chang Y.N., Chaing Y.L., Hay C.M., Mohajeri H., Davidson B.L and Bohn M.C. 1997. Dopaminergic neurons protected from degeneration by GDNF gene therapy. *Science* **275:** 838–841.

Cooper M.J., Lippa M., Payne J.M., Hatzivasiliou G., Reifenberg E., Fayazi B., Perales J.C., Morrison L.J., Templeton D., Piekarz R.L., and Tan J. 1997. Safety-modified episomal vectors for human gene therapy. *Proc. Natl. Acad. Sci.* **94:** 6450–6455.

Crystal R.G. 1995. Transfer of genes to human: Early lessons and obstacles to success. *Science* **270:** 404–410.

Deglon N., Heyd B., Tan S.A., Jopseph J.M., Zurn A.D., and Aebischer P. 1996. Central nervous system delivery of recombinant ciliary neurotrophic factor by polymer encapsulated differentiated C2C12 myoblasts. *Hum. Gene Ther.* **10:** 2135–2146.

Doerfler W., Schubbert R., Heller H., Kämmer C., Hilger-Eversheim K., Knoblauch M., and Remus R. 1997. Integration of foreign DNA and its consequences in mammalian systems. *Trends Biotechnol.* **15:** 297–301.

Doering L.C. 1997. Towards gene therapy in the nervous system. gene therapy and neurodegeneration (special issue) *Clin. Neurosci.* **3:** 259–260.

Dunnett S.B. and Fibiger H.C. 1993. Role of forebrain cholinergic systems in learning and memory: Relevance to the cognitive deficits of aging and Alzheimer´s dementia. *Prog. Brain Res.* **98:** 413–420.

During M.J. and Leone P. 1996. Adeno-associated virus vectors for gene therapy of neurodegenerative disorders. *Clin. Neurosci.* **3:** 292–300.

During M.J., Naegele J.R., O'Malley K.L., and Geller A.I. 1994. Long-term behavioral recovery in parkinsonian rats by an HSV vector expressing tyrosine hydroxylase. *Science* **266:** 1399–1403.

Emerich D.F., Hammang J.P., Baetge E.E., and Winn S.R. 1994a. Implantation of polymer-encapsulated human nerve growth factor-secreting fibroblasts attenuates the behavioral and neuropathological consequences of quinolinic acid injections into rodent striatum. *Exp. Neurol.* **130:** 141–150.

Emerich D.F., Lindner M.D., Winn S.R., Chen E.Y., Frydel B.R., and Kordower J.H. 1996. Implants of encapsulated human CNTF-producing fibroblasts prevent behavioral deficits and striatal degeneration in a rodent model of Huntington's disease. *J. Neurosci.* **16:** 5168–5181.

Emerich D.F., Winn S.R., Harper J., Hammang J.P., Baetge E.E. and Kordower J.H. 1994b. Implants of polymer-encapsulated human NGF-secreting cells in the nonhuman primate: Rescue and sprouting of degenerating cholinergic basal forebrain neurons. *J. Comp. Neurol.* **349:** 148–164.

Emerich D.F., Winn S.R., Hantraye P.M., Peschanski M., Chen E.Y., Chu Y., McDermott P., Baetge E.E., and Kordower J.H.1997. Protective effect of encapsulated cells producing neurotrophic factor CNTF in a monkey model of Huntington's disease. *Nature* **386:** 395–399.

Felgner P.L. 1997. Nonviral strategies for gene therapy. *Sci. Am.* **276:** 102–106.

Feng M., Jackson W.H., Goldman C.K., Rancourt C., Wang M., Dusing S.K., Siegal G., and Curiel D.T. 1997. Stable in vivo gene transduction via a novel adenoviral/retroviral chimeric vector. *Nat. Biotechnol.* **15:** 866–870.

Finch C.E. 1993. Neuron atrophy during aging: Programmed or sporadic? *Trends Neurosci.* **16:** 104–110.

Fink D.J., Ramakrishnan R., Marconi P., Goins W.F., Holland T.C., and Glorioso J.C. 1996. Advances in the development of herpes simplex virus-based gene transfer vectors for the nervous system. *Clin. Neurosci.* **3:** 284–291.

Fischer W., Björklund A., Chen K.S., and Gage F.H. 1991. NGF improves spatial memory in aged rodents as a function of age. *J. Neurosci.* **11:** 1889–1906.

Fischer, W., Wictorin, K., Björklund, A., Williams, L.R., Varon, S., and Gage, F.H. 1987. Amelioration of cholinergic neuron atrophy and spatial memory impairment in aged rats by nerve growth factor. *Nature* **329:** 65–68.

Fisher L.J. and Ray J. 1994. In vivo and ex vivo gene transfer to the brain. *Curr. Biol.* **4:** 735–741.

Fisher L.J., Jinnah H.A., Kale L.C., Higgins G.A., and Gage F.H. 1991. Survival and function of intrastriatally grafted primary fibroblasts genetically modified to produce L-DOPA. *Neuron* **6:** 371–380.

Fox J.L. 1997. HIV vector challenges gene therapy oversight. *Nat. Biotechnol.* **15:** 832.

Freese A. 1997. Gene therapy for parkinson's disease: Rationale, prospects and limitations (editorial). *Exp. Neurol.* **144:** 1.

Frick K.M., Price D.L., Koliatsos V.E., and Markowska A.L. 1997. The effects of nerve growth factor on spatial recent memory in aged rats persist after discontinuation of treatment. *J. Neurosci.* **17:** 2543–2550.

Friedmann T. 1997. Overcoming the obstacles to gene therapy. *Sci. Am.* **276:** 96–101.

Friedmann T. and Yee J.-K. 1995. Pseudotyped retroviral vectors for studies of human gene therapy. *Nat. Med.* **1:** 275–277.

Frim D.M., Yee W.M., and Isacson O. 1993a. NGF reduces striatal excitotoxic neuronal loss without affecting concurrent neuronal stress. *NeuroReport* **4:** 655–658.

Frim D.M., Uhler T.A., Short M.P., Ezzedine Z.D., Klagsbrun M., Breakefield X.O., and Isacson O. 1993b. Effects of biologically derived NGF, BDNF and bFGF on striatal excitotoxic lesions. *NeuroReport* **4:** 367–370.

Gage F.H. and Fisher L.J. 1991. Intracerebral grafting: A tool for the neurobiologist. *Neuron* **6:** 1–12.

Gage F.H., Ray J., and Fisher L.J. 1995. Isolation, characterization, and use of stem cells from the CNS. *Annu. Rev. Neurosci.* **18:** 159–192.

Gage F.H., Wolff J.A., Rosenber M.B., Xu L., Yee J.K., Shults C., and Friedmann T. 1987. Grafting genetically modified cells to the brain: Possibilities for the future. *Neuroscience* **23:** 795–807.

Gallagher M.G. and Colombo P.J. 1995. Ageing: The cholinergic hypothesis of cognitive decline. *Curr. Opin. Neurobiol.* **5:** 161–168.

Gimenez y Ribotta M., Revah F., Loquet I., Mallet J., and Privat A.1997. Prevention of motorneuron death by adenovirus-mediated neurotrophic factors. *J. Neurosci. Res.* **48:** 281–285.

Haase G., Kennel P., Pettmann B., Vigne E., Akli S., Revah F., Schmalbruch H., and Kahn A. 1997. Gene therapy of murine motor neuron disease using adenoviral vectors for neurotrophic factors. *Nat. Med.* **3:** 429–436.

Harris J.D. and Lemoine N.R. 1996. Strategies for targeted gene therapy. *Trends Genet.* **12:** 400–405.

Ho D.Y. and Sapolski R.M. 1997. Gene therapy for the nervous system. *Sci. Am.* **276:** 116–120.

Horellou P., Guibert B., Leviel V., and Mallet J. 1989. Retroviral transfer of a human tyrosine hydroxylase cDNA in various cell lines: Regulated release of dopamine in mouse anterior pituitary AtT-20 cells. *Proc. Natl. Acad. Sci.* **86:** 7233–7237.

Horellou P., Marlier L., Privat A., and Mallet J. 1990b. Behavioral effect of engineered cells that synthesize L-DOPA or dopamine after grafting into the rat neostriatum. Eur.

J. Neurosci. **2:** 116–119.

Horellou P., Brundin P., Kalen P., Mallet J., and Björklund A. 1990b. In vivo release of DOPA and dopamine from genetically engineered cells grafted to the denervated striatum. *Neuron* **5:** 393–402.

Horellou P., Vigne E., Castel M.N., Barneoud P., Colin P., Perricaudet M., Delaere P., and Mallet J. 1994. Direct intracerebral gene transfer of an adenoviral vector expressing tyrosine hydroxylase in a rat model of Parkinson's disease. *NeuroReport* **6:** 49–53.

Isacson O. 1995. Behavioral effects and gene delivery in a rat model of Parkinson's disease. *Science* **269:** 856–857.

Jones T.J. and Sullenger B.A. 1997. Evaluating and enhancing ribozyme reaction efficiency in mammalian cells. *Nat. Biotechnol.* **15:** 902–905.

Kang U.J., Fisher L.J., Joh T.J., O'Malley K.L, and Gage F.H. 1993. Regulation of dopamine production by genetically modified primary fibroblasts. *J. Neurosci.* **13:** 5203–5211.

Kaplitt M.G., Leone P., Samulski R.J., Xiao X., Pfaff D.W., O'Malley K.L., and During M.J. 1994. Long-term gene expression and phenotypic correction using adeno-associated virus vectors in the mammalian brain. *Nat. Genet.* **8:** 148–153.

Karpati G., Lochmüller H., Nalbantoglu J., and Durham H. 1996. The principles of gene therapy for the nervous system. *Trends Neurosci.* **19:** 49–54.

Kay M.A., Liu D., and Hoogerbrugge P.M. 1997. Gene therapy. *Proc. Natl. Acad. Sci.* **94:** 12744–12746.

Kordower J.H., Winn S.R., Liu Y.T., Muffson E.J., Sladek J.R., Hammang J.P., Baetge E.E., and Emerich D.F. 1994. The aged monkey basal forebrain: Rescue and sprouting of axotomized basal forebrain neurons after grafts of encapsulated cells secreting human nerve growth factor. *Proc. Natl. Acad. Sci.* **91:** 10898–10902.

Lacorazza H.D., Flax J.D., Snyder E.Y., and Jendoubi M. 1996. Expression of human β-hexosaminidase α-subunit gene (the gene defect of Tay-Sachs disease) in mouse brains upon engraftment of transduced progenitor cells. *Nat. Med.* **2:** 424–429.

Lanza R.P., Hayes J.L., and Chick W.L. 1996. Encapsulated cell technology. *Nat. Biotechnol.* **14:** 1107–1111.

Lawrence M.S., Ho D.Y., Sun G.H., Steinberg G.K., and Sapolsky R.M.1996a. Overexpresison of bcl-2 with herpes simplex virus vectors protect CNS neurons against neurological insults in vitro and in vivo. *J. Neurosci.* **16:** 486–496.

Lawrence M.S., McLaughlin J.R., Sun G.H., Ho D.Y., McIntosh L., Kunis D.M., Sapolsky R.M., and Steinberg G.K. 1997. Herpes simplex viral vectors expressing bcl-2 are neuroprotective when delivered after stroke. *J Cereb. Blood Flow Metab.* **17:** 740–744.

Lawrence M.S., Sun G.H., Kunis D.M., Saydam T.C., Dash R., Ho D.Y., Sapolsky R.M., and Steinberg G.K. 1996b. Overexpression of the glucose transporter gene with a herpes simplex viral vector protects striatal neurons against stroke. *J. Cereb. Blood Flow Metab.* **16:** 181–185.

Levivier M., Przedborski S., Bencsics C., and Kang U.J. 1995. Intrastriatal implantation of fibroblasts genetically engineered to produce brain-derived neurotrophic factor prevents degeneration of dopaminergic neurons in a rat model of Parkinson's disease. *J. Neurosci.* **15:** 7810–7820.

Linnik M.D., Zahos P., Geschwind M.D., and Federoff H.J. 1995. Expression of bcl-2 from a defective herpes simplex virus-1 vector limits neuronal death in focal cerebral ischemia. *Stroke* **26:** 1670–1674.

Lundberg C., Horellou P., Mallet J., and Björklund A. 1996. Generation of DOPA-producing astrocytes by retroviral transduction of the human tyrosine hydroxylase gene: In vitro characterization and in vivo effects in the rat Parkinson model. *Exp. Neurol.* **139:** 39–53.

Lundberg C., Martínez-Serrano A., Cattaneo E., McKay R.D.G., and Björklund A. 1997. Survival, integration, and differentiation of neural stem cell lines after transplantation to the adult rat striatum. *Exp. Neurol.* **145:** 342–360.

Mandel R.J., Spratt S.K., Snyder R.O., and Leff S.E. 1997. Midbrain injection of recombinant adeno-associated virus encoding rat glial cell-line derived neurotrophic factor protects nigral neurons in a progressive 6-hydroxydopamine-induced degeneration model of Parkinson's disease in rats. *Proc. Natl. Acad. Sci.* **94:** 14083–14088.

Mangiarini L., Sathasivam K., Seller M., Cozens B., Harper A., Hetherington C., Lawton M., Trottier Y., Lehrach H., Davies S.W., and Bates G.P. 1996. Exon 1 of the HD gene with an expanded CAG repeat is sufficient to cause a progressive neurological phenotype in transgenic mice. *Cell* **87:** 493–506.

Markowska A.L., Koliatsos V.E., Breckler S.J., Price D.L., and Olton D.S. 1994. Human nerve growth factor improves spatial memory in aged but not in young rats. *J. Neurosci.* **14:** 4815–4824.

Martínez-Serrano A. and Björklund A. 1996a. Gene transfer to the mammalian brain using neural stem cells. *Clin. Neurosci.* **3:** 301–309.

———— 1996b. Protection of the neostriatum against excitotoxic damage by neurotrophin-producing, genetically modified neural stem cells. *J. Neurosci.* **16:** 4604–4616.

———— 1997. Immortalized neural progenitor cells for CNS gene transfer and repair. *Trends Neurosci.* **20:** 530–538.

———— 1998. Ex vivo nerve growth factor gene transfer to the basal forebrain in presymptomatic middle-aged rats prevents the development of cholinergic neuron atrophy and cognitive impairment during aging. *Proc. Natl. Acad. Sci.* **95:** 1858–1863.

Martínez-Serrano A., Fischer W., and Björklund A.1995a. Reversal of age-dependent cognitive impairments and cholinergic neuron atrophy by NGF-secreting neural progenitors grafted to the basal forebrain. *Neuron* **15:** 473–484.

Martínez-Serrano A., Fischer W., Söderström W., Ebendal T., and Björklund A. 1996. Long-term functional recovery from age-induced spatial memory impairments by nerve growth factor (NGF) gene transfer to the rat basal forebrain. *Proc. Natl. Acad. Sci.* **93:** 6355–6360.

Martínez-Serrano, A., Lundberg, C., Horellou, P., Fischer, W., Bentlage, C., Campbell, K., McKay, R.D.G., Mallet, J., and Björklund, A. 1995b. CNS derived neural progenitor cells for gene transfer of nerve growth factor to the adult rat brain: Complete rescue of axotomized cholinergic neurons after transplantation into the septum. *J. Neurosci.* **15:** 5668–5680.

McMahon S.B. and Priestley J.V. 1995. Peripheral neuropathies and neurotrophic factors: Animal models and clinical perspectives. *Curr. Opin. Neurobiol.* **5:** 616–624.

Miller N. and Whelan J. 1997. Progress in transcriptionally targeted and regulatable vectors for genetic therapy. *Hum. Gene Ther.* **8:** 803–815.

Naldini L., Blomer U., Gallay P., Ory D., Mulligan R., Gage F.H., Verma I.M., and Trono D. 1996. In vivo gene delivery and stable transduction of nondividing cells by a lentiviral vector. *Science* **272:** 263–267.

Neve R.L. and Geller A.I. 1996. A defective herpes simplex vector system for gene delivery into the brain. *Clin. Neurosci.* **3:** 262–267.

Paillard F. 1997a. Promoter attenuation in gene therapy: Causes and remedies. *Hum. Gene Ther.* **8:** 2009–2010.

———— 1997b. Advantages of non-human adenoviruses versus human adenoviruses.

Hum. Gene Ther. **8:** 2007–2009.

Pechan P.A., Yoshida T., Panahian N., Moskowitz M.A., and Breakefield X.O. 1995. Genetically modified fibroblasts producing NGF protect hippocampal neurons after ischemia in the rat. *NeuroReport* **6:** 669–672.

Raymon H.K., Thode S., and Gage F.H. 1997. Application of ex vivo gene therapy in the treatment of Parkinson's disease. *Exp. Neurol.* **144:** 82–91.

Renfranz P.J., Cunningham M.G., and McKay R.D.G. 1991. Region-specific differentiation of the hippocampal cell line HiB5 upon implantation into the developing mammalian brain. *Cell* **66:** 713–729.

Rinsch C., Régulier E., Déglon N., Dalle B., Beuzard Y., and Aebischer P. 1997. A gene therapy approach to regulated delivery of erythropoietin as a function of oxygen tension. *Hum Gene Ther.* **8:** 1881–1889.

Saez E., No D., West A., and Evans R.M. 1997. Inducible gene expression in mammalian cells and transgenic mice. *Curr Opin. Biotechnol.* **8:** 608–616.

Sagot Y., Tan S.A., Hammang J.P., Aebischer P., and Kato A.C. 1996. GDNF slows loss of motorneurons but not axonal degeneration or premature death of *pmn/pmn* mice. *J. Neurosci.* **16:** 2335–2341.

Sagot Y., Tan S.A., Baetge E., Schmalbruch H., Kato A.C., and Aebischer P. 1995. Polymer encapsulated cell lines genetically engineered to release ciliary neurotrophic factor can slow down progressive motor neuronopathy in the mouse. *Eur. J. Neurosci.* **7:** 1313–1322.

Sah D.W.Y., Ray J., and Gage F.H.1997. Bipotent progenitor cell lines from the human CNS. *Nat. Biotechnol.* **15:** 574–580.

Sakaguchi M. 1997. Eukaryotic protein secretion. *Curr. Opin. Biotechnol.* **8:** 595–601.

Schumacher J.M., Short M.P., Hyman B.T., Breakefield X.O., and Isacson O. 1991. Intracerebral implantation of nerve growth factor-producing fibroblasts protects striatum against neurotoxic levels of excitatory amino acids. *Neuroscience* **45:** 561–570.

Seidah N.G. and Chrétien M. 1997. Eukaryotic protein processing: Endoproteolysis of precursor proteins. *Curr. Opin. Biotechnol.* **8:** 602–607.

Sendtner M., Schmalbruch H., Stockli K.A., Kreutzberg G.W., and Thoenen H. 1992. Ciliary neurotrophic factor prevents degeneration of motor neurons in mouse mutant progressive motor neuronopathy. *Nature* **358:** 505–504.

Snyder E.Y. and Fisher L.J. 1996. Gene therapy in neurology. *Curr. Opin. Pediatr.* **8:** 558–568.

Snyder E.Y. and Macklis J.D. 1996. Multipotent neural progenitor or stem-like cells may be uniquely suited for therapy for some neurodegenerative conditions. *Clin. Neurosci.* **3:** 310–316.

Snyder E.Y. and Wolfe J.H. 1996. Central nervous system cell transplantation: A novel therapy for storage diseases? *Curr. Opin. Neurol.* **9:** 126–136.

Snyder E.Y., Taylor R.M., and Wolfe J.H. 1995. Neural progenitor cell engraftment corrects lysosomal storage throughout the MPS VII mouse brain. *Nature* **374:** 367–370.

Snyder, E.Y., Deitcher, D.L., Walsh, C., Arnold-Aldea, S., Hartwieg, E.A., and Cepko, C.L. 1992. Multipotent neural cell lines can engraft and participate in development of mouse cerebellum. *Cell* **68:** 33–51.

Thompson J.D., Macejak D., Couture L., and Stinchcomb D.T. 1995. Ribozymes in gene therapy. *Nat. Med.* **1:** 277–278.

Tripathy S.K., Black H.B., Goldwasser E., and Leiden J.M. 1996. Immune responses to transgene-encoded proteins limit the stability of gene expression after injecion of replication-defective adenovirus vectors. *Nat. Med.* **2:** 545–550.

Tseng J.L., Baetge E.E., Zurn A.D., and Aebischer P. 1997. GDNF reduces drug-induced rotational behavior after medial forebrain bundle transection by a mechanism not involving striatal dopamine. *J. Neurosci.* **17:** 325–333.

Tuszynski M.H., Roberts J., Senut M.C., U H.S., and Gage F.H. 1996. Gene therapy in the adult primate brain: Intraparenchymal grafts of cells genetically modified to produce nerve growth factor prevent cholinergic neuronal degeneration. *Gene Ther.* **3:** 305–314.

Varley A.W., Geiszler S.M., Gaynor R.B., and Munford R.S. 1997. A two-component expression system that responds to inflammatory stimuli in vivo. *Nat. Biotechnol* **15:** 1002–1006.

Verma I. and Somia N. 1997. Gene therapy—promises, problems and prospects. *Nature* **389:** 239–242.

Whittemore S.R. and White L.A. 1993. Target regulation of neuronal differentiation in a temperature-sensitive cell line derived from medullary raphe. *Brain Res.* **615:** 27–40.

Vile R.G. 1997. A marriage of viral vectors. *Nat. Biotechnol.* **15:** 840–841.

Winkler J., Suhr S.T., Gage F.H., Thal L.J., and Fisher L.J. 1995. Essential role of neocortical acetylcholine in spatial memeory. *Nature* **375:** 484–487.

Wolfe J.H., Deshmane S.L., and Fraser N.W. 1992. Herpesvirus vector gene transfer and expression of beta-glucuronidase in the central nervous system of MPSVII mice. *Nat. Genet.* **1:** 379–384.

Wolff J.A., Fish L.J., Xu L., Jinnah H.A., Langlais P.J., Iuvone P.M., O'Malley K.L., Rosenberg M.B., Shimohama S., Friedmann T., and Gage F.H. 1989. Grafting fibroblasts genetically modified to produce L-dopa in a rat model of Parkinson disease. *Proc. Natl. Acad. Sci.* **86:** 9011–9014.

Wood M.J.A., Charlton H.M., Wood K.J., Kajiwara K., and Byrnes A.P. 1996. Immune responses to adenovirus vectors in the nervous system. *Trends Neurosci.* **19:** 497–501.

Xu D.G., Crocker S.J., Doucet J.P., St-Jean M., Tamai K., Hakim A.M., Ikeda J.E., Liston P., Thompson J.S., Korneluk R.G., MacKenzie A., and Robertson G.S. 1997. Elevation of neuronal expression of NAIP reduces ischemic damage in the rat hippocampus. *Nat. Med.* **3:** 997–1004.

Yan Q., Matheson C., and Lopez O.T. 1995. In vivo neurotrophic effects of GDNF on neonatal and adult facial motor neurons. *Nature* **373:** 341–344.

Yang G.Y., Zhao Y.J., Davidson B.L. and Betz A.L. 1997. Overexpression of interleukin-1 receptor antagonist in the mouse brain reduces ischemic brain injury. *Brain Res.* **751:** 181–188.

Yoshimoto Y. Lin Q., Collier T., Frim D.M., Breakefield X.O. and Bohn M.C. 1995. Astrocytes retrovirally transduced with BDNF elicit behavorial improvement in a rat model of Parkinson's disease. *Brain Res.* **691:** 25–36.

Zufferey R., Nagy D., Mandel R.J., Naldini L, and Trono D. 1997. Multiply attenuated lentiviral vector achieves efficient gene delivery in vivo. *Nat. Biotechnol.* **15:** 871–875.

23

Targeted Gene Repair in Mammalian Cells Using Chimeric RNA/DNA Oligonucleotides

Eric B. Kmiec

Kimeragen, Incorporated, Newtown, Pennsylvania 18940
Kimmel Cancer Center
Thomas Jefferson University
Philadelphia, Pennsylvania 19107

Betsy T. Kren and Clifford J. Steer

Department of Medicine
University of Minnesota Medical School
Minneapolis, Minnesota 55455

BACKGROUND AND EARLY EXPERIMENTAL STUDIES

The potential now exists in many experimental systems to transfer a cloned, modified gene into the genome of the host organism. In the ideal situation, the cloned gene is returned to its homologous location and is inserted at the target locus. The process is a controlled means for repairing DNA damage and ensuring accurate chromosome disjunction during meiosis. The paradigm for understanding this process is derived from detailed biochemical analyses of the RecA protein purified from *Escherichia coli* and genetic studies carried out in bacteriophage and fungi. A compelling picture of the process of homologous pairing and DNA strand exchange has been influential in directing investigators interested in gene-targeting experiments.

A body of information has emerged from these studies on transformation in lower eukaryotes that has been crucial in developing strategies for gene targeting in higher organisms. Unfortunately, the limited biochemical data, as well as the often confusing and sometimes contradictory results from the genetic studies, have not provided a thorough mechanistic foundation for experimentation. For example, it is unclear from the transformation studies that information on the genetic control of integration will be generally applicable to higher eukaryotic systems. The significance of the functionally independent, yet structurally redundant,

RecA-like Rad51, Rad55, Rad57, and Dmc1 genes in yeast is not totally resolved (Story et al. 1993). In addition, the virtual absence of the illegitimate integration events during plasmid transformation commonly observed in many other eukaryotic systems raises certain caveats as to the generality of the recombination system in yeast. Nevertheless, structural homologs of *rad51* and/or *rad52* have been identified in several higher eukaryotes (Shinohara et al. 1992; Bezzubova et al. 1993a,b), providing evidence that fundamentally similar biological principles underlie the mechanism of homologous recombination from bacteria to higher animals and plants. In fact, the rules of gene targeting learned from transformation analysis of lower eukaryotes may, in the end, be widely applicable.

Homologous Recombination in Mammalian Cells

Homologous recombination between a plasmid or vector and the chromosome in higher eukaryotes has been exploited in numerous experimental systems for inactivating or replacing a particular gene (Capecchi 1989). In most organisms, the utility of this process in genetic manipulations is compromised by interference from an alternative illegitimate pathway of recombination irrespective of DNA sequence homology (Roth and Wilson 1988). This process is often viewed as a nuisance by investigators, whose priority is "knocking out" the gene of interest rather than understanding the mechanism of the process. Conversely, the virtual absence of this illegitimate pathway of integration in the more genetically amenable systems of yeast and bacteria has precluded investigation into its molecular mechanism. Therefore, strategies for gene targeting have, for the most part, evolved by the empirical method with only limited guidance from recombination theory or mechanism.

The single most prominent exception to this generalization is the exploitation of the findings from yeast on the mechanism for double-strand break repair. The introduction of double-strand breaks or gaps into targeting vectors is enormously helpful in increasing the accuracy of gene targeting (Kucherlapati et al. 1984; Jasin et al. 1985; Song et al. 1985; Valancius and Smithies 1991). This is based on the recombinogenic potential of open DNA ends in strand invasion that comprises the initial phase of the homologous pairing and strand exchange reaction. Principles evident in the double-strand break repair model (Szostak et al. 1983) have been influential in vector design (Orr-Weaver et al. 1981; Rothstein 1983). These include the introduction of a double-strand break in the cloned sequence of homology to direct insertional integration, or at the border of homology for replacement type integration (Hasty et al. 1992). Because double-strand breaks can be a part of the vector, it now becomes the cell's role to provide the necessary enzymatic machinery to catalyze the specific targeting event. However, the infrequency of homologous targeting in human cells may reflect the low abundance of such machinery. This notion is supported by the observations that similar or identical targeting constructs recombine with varying frequencies in different cell types.

It is likely that the failure to achieve high levels of gene targeting in mammalian cells is related directly to the low frequency of homologous recombination. As described above, efforts to overcome this barrier have focused on the development of genetic enrichment methods. But these methods only eliminate nonhomologous events; they do not improve the frequency of homologous recombination. Experimental evidence suggests that the enzymatic machinery required to catalyze homologous targeting is limited in mammalian cells. For example, Buerstedde and Takeda (1991) demonstrated that gene conversion occurred with high frequency in avian B cells, but not in closely related cells at various stages of B-cell development. Such data led to the hypothesis that the difference in targeting frequencies among cell types is due, in part, to variations in the levels of enzymatic components within these cells. It is thought that gene targeting in mammalian cells is regulated by homologous recombination processes related to DNA repair and that genes known to participate in recombinational repair are likely to be important components of a specific gene-targeting event.

Although DNA-recombinase assays are important in identifying activities, the cellular target for recombination is more complex in eukaryotes. Therefore, a series of biochemical studies was conducted in which the recombination activity of the *Ustilago maydis* Rec2 protein on DNA packaged into chromatin was measured. Such nucleosomal DNA reflects more precisely the status of DNA within the chromosome. Because the environment of the chromosome itself is important for targeting, it was necessary to determine the optimal conditions for recombination. The results of these studies (Kotani et al. 1993, 1994, 1996; Kotani and Kmiec 1994a,b,c) revealed that strand transfer events promoted by Rec2 or RecA proteins were completely blocked when the target DNA was packaged into chromatin or in the form of nucleosomal DNA. Furthermore, the receptive nature of nucleosomal DNA to gene targeting or homologous recombination events relied on the active transcription of the template by RNA polymerase. Finally, the RNA itself became associated with, or part of, the product recombinant molecule.

These three observations led us to consider the role of RNA in recombination. We designed experiments to test the hypothesis that RNA reduced the length of homology that was required for successful joint molecule formation. The joint molecules were measured as representing homologous recombination events promoted by Rec2 or RecA. The results of these experiments clearly demonstrated that, as the size was reduced from 170 to 70 and then to 25 bases of homology between the pairing partners, one of the two partners had to be transcriptionally active. The conclusion from these studies was that gene-targeting events in mammalian cells would be more efficient if the vector contained stretches of RNA (Kotani et al. 1993, 1994, 1996; Kotani and Kmiec 1994a,b,c).

The first attempt to create such a vector was to transcriptionally activate DNA plasmids as one of the pairing partners; but this did not reproducibly improve targeting events in mammalian cells. To create a stable RNA molecule, we designed an oligonucleotide that consisted of both DNA and RNA residues and named it the chimeric oligonucleotide to reflect the heterogeneity of bases in the structure. Direct tests of the recombination capacity of RNA were conducted using these molecules. The results of these studies indicated that oligonucleotides containing RNA residues were able to pair with the DNA target containing short stretches of homology. Oligonucleotides lacking RNA were unable to pair at the same frequency when the target size was reduced. To understand the parameters of these results, the physical properties of the chimeric molecules were initially examined. The importance of purity and melting temperatures was evaluated as well as the association kinetics between the chimeric oligonucleotide and the DNA target. It was also established that the chimeric oligonucleotide could pair with the nucleosomal DNA and produce significant levels of joint molecules in vitro (Kmiec 1996).

As described earlier, the frequency of homologous recombination in mammalian cells is extremely low, even in the presence of enzymes that are capable of catalyzing these events. Attempts to overexpress certain "recombinases" to increase targeted recombination have generally been unsuccessful. In addition, only incremental advances in improving the efficiency of successful mouse knockouts have been seen. Hence, a different strategy was pursued by utilizing the pairing precision of the chimeric oligonucleotide. Rather than attempting to improve the rates of homologous recombination, we developed the concept of a targeted repair using the oligonucleotide as a stable signal to activate the more pronounced and efficient DNA repair processes present in the cell.

It had been widely known that DNA oligonucleotides can be used to introduce site-specific changes in genetic templates, both chromosomal and episomal (Moerschell et al. 1988). In addition, synthetic oligonucleotides can be altered chemically so that on pairing with its target sequence in the DNA, the chemically reactive group is activated and a modification is made in the chromosomal or episomal genome (Glazer et al. 1987). This approach is useful for gene disruption, where the purpose is to disable a malfunctioning gene. Unfortunately, the efficiency of genomic target location and the restrictions for stable base-pairing placed on the targeted site have made it difficult to apply this strategy to a number of genes. In addition, DNA sequence constraints and random mutagenesis reduce the potential of utilizing intact DNA molecules as correction vehicles. Triple-helix-forming oligonucleotides coupled to cross-linking agents have been used with moderate success to alter DNA sequences in cultured cells (Beal and Dervan 1991; Havre and Glazer 1993; Wang et al. 1995). This method of modifying DNA, however, is severely limited because the target DNA sequence must consist of homopyrimidine or homopurine stretches.

The results obtained by Kmiec and coworkers (1993, 1994) enabled us to expand the range of amenable targets and to overcome these limitations. These data suggested that successful target location through homology between the oligonucleotide and genomic DNA was more efficient when RNA replaced a portion of the DNA sequence in the oligonucleotide. This was particularly evident when the chromosomal target sequence was restricted in length to 50 bases or less. A double-stranded, endcapped molecule was envisioned to improve the observed instability of foreign DNA templates in mammalian cells. These facts led us to propose the hypothesis that the use of an RNA/DNA hairpin molecule might be stable enough to localize to the target site. Once bound at the site, the creation of an RNA/DNA hybrid stretch would extend the half-life of the complex, a parameter affecting the rate-limiting step. If the vector was designed to contain a base that creates a mismatch with the mutant base in the gene, then the distorted helix position would be recognized and the mutant base rectified by the cell's inherent DNA repair machinery.

After several rounds of failure and redesign, the chimeric oligonucleotide illustrated in Figure 1 provided some positive results. Although there were a number of distinguishing features of this molecule, the most profound was that nucleotide exchange is directed presumably by the mismatched base-pairing between the chimeric molecule and its target. The ends of this single-stranded molecule are capped by a string of thymidines, permitting the greatest flexibility for the complementary base-pairing. This configuration also prevents the concatemerization of double-stranded ends, which is a common occurrence when linear DNA is transfected into mammalian cells. A single unligated phosphodiester bond allows one end to open and close enabling topological interwinding of the oligonucleotide and its target (Havre and Kmiec 1998). The RNA

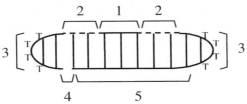

Site	Function
1. DNA mutator region *(1–6)*:	Provides the correction information and template for repair enzymes.
2. RNA bridges *(8–12)*:	Binds more avidly to the target strand and increases half-life of the complex.
3. T-hairpin loops *(3–4)*:	Increases stability of the oligo and protects against concatermerization.
4. Strand break *(1)*:	Permits topological interwinding of the chimera with the DNA helix.
5. RNA/DNA *(30–50)*:	The 2′-O-methyl RNA-DNA base pairing produces an "A" form structure that may activate repair/recombination pathways.

Figure 1 The chimeric oligonucleotide. Key features of the chimeric RNA/DNA molecule designed for genomic site-directed mutagenesis or repair. The number of nucleotides is indicated in parentheses.

stretches provide for increased stability once the complex has been made, thereby lengthening the half-life of the conjoined molecules (Kotani and Kmiec 1994b). Each RNA residue is modified at the 2′ position to protect against enzymatic degradation. Finally, the overall pairing of alternating RNA and DNA regions adopts an "A" form conformation and, due to the intermittent stretches of DNA and RNA bases, the molecule can form stable joint molecules with targeted DNA in vitro (Havre and Kmiec 1998). Each characteristic of the chimeric oligonucleotide may vary in its importance, particularly among cell types, and it is clear that a critical number of molecules must reach the nucleus for the correction process to occur. Furthermore, the delivery of the chimeric molecules into cells is a critical aspect of any potential therapeutic use.

The next step was to test the chimeric oligonucleotide on an episomal target in mammalian cells. We chose an alkaline phosphatase gene containing a point mutation in the coding region. The genetic readout occurs even in the mutated form of the gene, but the translated protein folds improperly and is undetectable using a zinc-based stain. Chimeric oligonucleotides were found to be effective in correcting this mutation in Chinese hamster ovary (CHO) cells, as demonstrated by the appearance of red-staining cells. These results were confirmed both at the genetic level and at the protein level by the detection of a wild-type alkaline phosphatase enzyme (Yoon et al. 1996).

To evaluate other mammalian genes that may be amenable to this targeting event, we chose the β-globin gene in sickle cell anemia. Using a chimeric oligonucleotide, we intended to change the T-A base pair in the β-globin to an A-T base pair, regenerating the wild-type β^A genotype (Fig. 2). This particular mutation was chosen because it is uniform among the effected population. It also provides a convenient restriction poly-morphism cleavage site that enables facile detection of successful con-version. Although it is clear that hematopoietic stem cells would be the optimal cell type for the treatment of sickle cell anemia, we used Epstein-Barr virus (EBV)-transformed B cells as the prototypic system. These cells grow readily in culture but have a number of disadvantages. First, they are basically primary cells, and we have found that they must be transformed routinely to generate cells that are receptive to chimeric oligonucleotide-mediated conversion. Second, it is also possible that the EBV may contribute to some parts of the reaction. The protocol for trans-

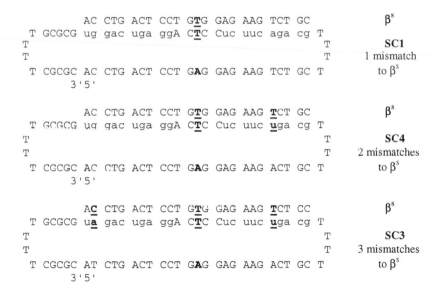

Figure 2 Mismatch sequences of chimeric oligonucleotides to the β^S-globin gene. The site of mutation for the β^S-globin gene for sickle cell disease is an A-to-T exchange within the sixth codon, indicated at *top* in *bold* and *underlined*. Chimeric molecules with one, two, and three mismatches to the homologous 25-bp sequence in the β-globin gene are similarly indicated and designated SC1, SC4, and SC3, respectively. DNA residues are *capitalized*, and the 2′-O-methylated RNA residues are shown in *lowercase*. DNA from β^A contains the sequence CCTGAGG and is cleaved at a *Bsu*36 I restriction site to yield fragments of 228 and 117 bp, whereas β^S genomic DNA with sequence CCTGTGG is resistant to cleavage.

fer of the oligonucleotide into the B cells was performed using standard lipid carriers. After successful transfection, we were able to detect significant levels of correction from the β^S to the β^A genotype. This exchange was detected by restriction fragment length polymorphism (RFLP), by Southern blots, and by direct DNA sequencing (Cole-Strauss et al. 1996). There are, however, indications that PCR may be artifactual in some systems. Therefore, a series of detailed experiments was performed to ensure the specificity of the results. In fact, there was no evidence that the reaction mixture nor the oligonucleotides served as a source for PCR priming. In addition, the chimeric oligonucleotides do not appear to integrate randomly into cells (Cole-Strauss et al. 1997).

Mismatches within the RNA Regions of the Chimeric Oligonucleotides

A modification of the standard SC1 design is one in which one or two bases in the RNA region are mismatched with the complementary DNA base in the target (Fig. 2). Hence, for SC3 and SC4, the DNA center section was identical to SC1 (1 mismatch with the target), but SC4 contained a single mismatch in one RNA region and SC3 contained one mismatch in each RNA region. After transfection of the β^S/β^S B cells and generation of the PCR-amplified fragment, $Bsu36$ I digestion revealed some conversion of $\beta^S{\rightarrow}\beta^A$ in populations only with SC4. No evidence of gene conversion was obtained when SC3 was transfected into the cells using RFLP or DNA sequence analyses. These data suggested that two mismatches outside the center DNA section in the RNA regions reduce the conversion of the β^S locus to levels that are undetectable by our methods.

Experiments with the SC3 and SC4 mismatch chimeric molecules also point to a correlation between the presence of mismatches and the ability of the molecule to alter the targeted base while not promoting random mutagenesis. Sequencing results indicated no apparent alteration at the outlying RNA/DNA mismatches independent of whether the targeted central DNA mismatch had been converted. To further address and confirm this result using a more sensitive technique, correction events mediated by the SC2 chimeric oligonucleotide in a β^A/β^A cell line containing a heterozygous polymorphism at codon 2 (Orkin et al. 1982) were monitored by RFLP analysis at both the sickle globin locus and at the polymorphic site. β^A/β^A cells were transfected with increasing concentrations of SC2 oligonucleotide. Cell extracts, PCR, and enzyme digestion were performed on each sample. When the DNA was cut with $Bsu36$ I, which is diagnostic for the β^S locus, a dose-dependent increase in noncleavable

DNA was noted, indicating a genetic change in the β^S locus. When the same samples were digested with *Bsi*HKA I, which is diagnostic for the codon 2 heterozygous polymorphism, the relative proportion of cut and uncut DNA (50/50) remained unchanged. On the basis of these observations, it is likely that the chimeric oligonucleotides mediate corrections only at mismatches that are present within the DNA targeting region.

Single-cell cloning has been achieved using transfected B cells with the β^S/β^S genotype (A. Cole-Strauss et al., unpubl.). After treating these cells with the appropriate chimeric SC1 molecule, the population number was reduced by limiting dilution and plated at a ratio of 0.3 cells/wellplate. The cells were incubated with irradiated B cells (β^S) as feeders, and at periodic intervals, the genomic DNA was isolated and analyzed by RFLP analyses. In every case, the isolated clones expressing a genotypic change were corrected at both alleles.

Mechanism of Targeted Gene Repair

Although we had envisioned using DNA repair processes to catalyze nucleotide conversion, none of the previous studies were aimed at understanding the molecular mechanism. The success of cellular targeting experiments prompted us to generate an experimental system that could be used to define the reaction parameters of nucleotide conversion. Hence, we developed a mammalian cell-free extract from cells shown to catalyze chimeric-directed nucleotide conversion in tissue culture (A. Cole-Strauss et al., unpubl.). As a genetic readout, we utilized antibiotic resistance in *E. coli* and confirmed the targeted change by DNA sequencing. The system relies on altering the gene encoding tetracycline (tet^r) or kanamycin resistance (kan^r) on a plasmid during incubation with a cell-free extract. The source of this extract can be cells grown in culture or from embryonic stem cells. A chimeric oligonucleotide designed to facilitate the correction in the tet^s or kan^s plasmid is added directly to the reaction mix; no preforming of the template is required. After an appropriate time, the purified plasmid is electroporated into bacteria and DNA from colonies resistant to tetracycline or kanamycin is isolated and sequenced. Using variations of this approach, we have determined that targeted gene repair appears to result from a two-stage reaction: pairing and repairing. Preliminary evidence suggests that the pairing step is mediated through the creation of a molecule containing a double D-loop structure (Fig. 3). The repair step is likely to be facilitated by either the mismatch repair pathway or a variation of this process. The cell-free sys-

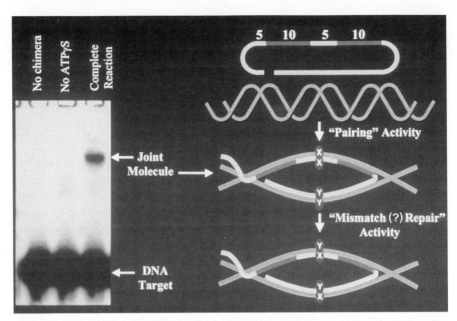

Figure 3 Mechanism of targeted gene repair. A possible two-step process involving "pairing" and "repairing" is shown. As indicated in text, a double D-loop structure is created between a chimeric RNA/DNA oligonucleotide and a double-stranded DNA target. The RNA strand of the chimeric molecule may drive the initial pairing phase while the DNA stretch regulates nucleotide alteration, possibly through mismatch repair.

tem has helped us design new chimeric structures that may be useful for clinically relevant systems and to identify the enzymatic machinery active in the process.

GENE TARGETING OF THE RNA/DNA OLIGONUCLEOTIDES TO LIVER

As a target organ for gene therapy, the liver is especially attractive because of its unique microarchitecture, which results in the hepatocyte plasma membrane being in direct contact with the blood as well as being the site of numerous genetic-based metabolic diseases (Chang and Wu 1994; Strauss 1994; Bowles and Woo 1995; Martinez-Hernandez and Amenta 1995). With the onset of recombinant DNA technology in the 1970s, it was presumed that the ability to treat genetic hepatic metabolic deficiencies would soon become a reality. In fact, although the progress has been remarkable, it has fallen considerably short of the therapeutic benefits that were initially thought to be possible. Numerous problems arose as the techniques for the introduction of the corrective genetic material were transferred from the culture dish to the intact organism. In particular, the lack of persistence in expression of the introduced material as well as the inability to transfer a sufficient quantity of the genetic corrective agent to sustain therapeutic levels in the host organism became apparent. The past decade has seen a dramatic improvement in the ability to both deliver and sustain the expression of the corrective genetic material in liver. Some difficulties, however, remain, including the random integration of retrovirus vectors and genomic rearrangements by adeno-associated viruses, as well as a requirement of actively replicating cells to permit the integration of the viral vectors. In addition, repeated in vivo use of recombinant adenoviral vectors has been shown to result in immunologic activation and transgene extinction (Ye et al. 1996). The development of a replication-deficient adenovirus has allowed recombinant genes to be introduced into nondividing hepatocytes with great efficiency. However, sustained expression of the transgene at therapeutic levels remains suboptimal (Bowles and Woo 1995; Askari et al. 1996). Recently, significant progress has been made in overcoming both the transient expression and the immunogenicity of these virus-based vectors (Connelly et al. 1996; Takahashi et al. 1996; Ilan et al. 1997), but this approach still requires repetitive treatments to maintain therapeutic levels.

An alternative approach, targeted homologous recombination between an identified mutation and the exogenous DNA vector, appears to be the most effective method for long-term genomic correction. The utility of this technique for gene therapy is limited because homologous recombination occurs at a very low frequency in hepatocytes (Thyagarajan et al. 1996) and other somatic cells (Prouty et al. 1993; Arbonés et al. 1994) and is complicated by random insertion in the absence of sequence homology (Capecchi 1989; Zheng and Wilson 1990; Sakagami et al. 1994). Although considerable progress using this approach in cultured cells has occurred, in vivo use of this technique has been limited (Yáñez and Porter 1998). The development of a novel experimental strategy to correct single-nucleotide mutations in genomic DNA that is unaffected by the base composition of the target sequence may, in fact, have represented a milestone in gene therapy (Cole-Strauss ct al. 1996; Yoon et al. 1996). Could a desired nucleotide exchange in hepatocytes be mediated using a chimeric oligonucleotide composed of RNA and DNA residues in a duplex conformation? Although the approach was based on the observation that RNA/DNA hybrids were highly active in homologous pairing reactions in vitro (Kmiec et al. 1994), the ultimate test was whether it could be applied at the bedside.

Targeted Genomic DNA Conversion in Cultured Human Hepatoma Cells

The well-differentiated HuH-7 human hepatoma cell line was used (Nakabayashi et al. 1982) to determine whether this novel strategy for site-directed targeted nucleotide exchange using an RNA/DNA duplex molecule was potentially useful in modifying hepatic genomic DNA. The alkaline phosphatase gene was selected because targeted correction of episomal DNA encoding a missense form of the protein had been successfully demonstrated with a frequency approaching 30% in cultured CHO cells (Yoon et al. 1996). The alkaline phosphatase gene product, although expressed at high levels in osteoblast-derived cells in culture, is expressed at very low levels in hepatocytes (Kiledjian and Kadesch 1990; Matsuura et al. 1990).

The RNA/DNA sequences of the chimeric molecules used for these studies were designed from the transcribed DNA strand that is the complementary strand to the reported cDNA (Kishi et al. 1989; Matsuura et al. 1990). Two blocks of ten 2′-O-methyl RNA residues were constructed to flank a pentameric stretch of DNA, with the center nucleotide targeted for exchange. The chimeric oligonucleotide AP1 was identical in sequence to the wild-type gene, whereas AP2 had a C→T substitution at the central nucleotide of the five DNA residues corresponding to the targeted alkaline phosphatase nucleotide at position 935. AP2mt was identical to AP2 but with a wild-type G in the complementary DNA strand, resulting in a T:G mismatch. Thus, when HuH-7 cells were transfected with the AP2 chimeric molecule, a targeted C→T nucleotide exchange at position 935 of the AP gene sequence would conceivably be introduced. In short, the conversion of both genomic DNA strands would represent a switch in the gene sequence from the wild-type alkaline phosphatase C:G pairing to a mutant A:T pairing.

The HuH-7 cells were transfected with the AP1 and AP2 chimeric oligonucleotides or with a vehicle alone using polyethylenimine (PEI) as the carrier (Boussif et al. 1995). This polycation molecule was chosen for the transfections because of its efficiency and its potential use as an in vivo delivery agent (Boussif et al. 1995; Abdallah et al. 1996; Boletta et al. 1997). HuH-7 cells were transfected in OPTIMEM™ media with the chimeric oligonucleotides at final concentrations of either 150 or 300 nM using the 800-kD PEI carrier at 9 equivalents of PEI nitrogen per chimeric phosphate (Boussif et al. 1995). Vehicle control transfections utilized the same amount of PEI as the chimeric oligonucleotide transfections. After 18 hours of incubation, additional medium containing fetal bovine serum (FBS) was added, thus reducing the chimeric molecule concentrations to 50 and 100 nM, respectively, for the remaining 30 hours of culture. The transfection frequency of the cells with the chimeric molecules was established by parallel transfections using the same concentrations of fluorescein-12-dUTP 3′-end-labeled chimeric oligonucleotides. Transfections were performed as described above; after 24 hours, the cells were fixed and counterstained with DiI to visualize the cytoplasm. The cells were examined using a BioRad MRC1000 confocal microscope, and the number of fluorescently labeled nuclei present in the 1-μm step collection series was used to calculate the transfection efficiency.

Genomic DNA larger than 100–150 base pairs was isolated from the cells 48 hours after transfection and used for PCR amplification of a 160-nucleotide fragment of exon 6 in the human liver AP gene corresponding to nucleotides 886 to 1048 of the liver alkaline phosphatase cDNA (Weiss et al. 1988; Kishi et al. 1989; Matsuura et al. 1990). The PCR amplification products were subcloned and used to transform *E. coli*. The conversion of nucleotide 935$^{G \to A}$ was determined by the hybridization of duplicate colony lifts of the PCR-amplified and cloned region of the alkaline phosphatase exon 6 with ^{32}P-end-labeled 17-mer oligonucleotide probes specific for either the 935^A or 935^G nucleotide sequence. The differential hybridization of the probes spanning the targeted nucleotide distinguished between the converted 935^A and the wild-type 935^G cDNA sequences (Melchior and Von Hippel 1973; Kishi et al. 1989). Autoradiograms of the hybridization patterns obtained using this technique exhibited specificity and low background activity. The overall frequency of conversion of the targeted nucleotide was calculated by dividing the number of clones hybridizing with the 935^A oligonucleotide by the total number of clones hybridizing with both oligonucleotide probes.

The G→A conversion at position 935 was detectable only in HuH-7 cells transfected with the AP2 molecules or AP2mt, although the conversion rate using AP2mt was substantially reduced compared to the AP2 oligonucleotide (Kren et al. 1997). In contrast, neither the AP1 nor the vehicle-transfected cells yielded any clones hybridizing with the 935^A oligonucleotide probe. The G→A conversion rate in cells transfected at a 300-nM concentration with AP2 was 11% and approximately twofold greater than cells transfected at one-half the dose. Independently amplified genomic DNA aliquots from different experiments exhibited similar conversion frequencies, suggesting that the exchange of G→A at position 935 was site-specific. Interestingly, when corrected for transfection efficiency using the number of fluorescently labeled nuclei present in the parallel transfected cultures, the G→A conversion at nucleotide 935 was approximately 40% in AP2-transfected cells. In fact, the percentage of exchange correlated directly with the transfection frequency.

The hybridization results were confirmed by sequence analysis of independent clones hybridizing to either the 935^G or 935^A probe These results indicated that colonies hybridizing to the 935^G oligonucleotide probe exhibited the wild-type alkaline phosphatase sequence, that is, a G at nucleotide 935 of the reported liver cDNA. In contrast, those colonies hybridizing to the 935^A probe exhibited an A at position 935. The complementary DNA strand of these colonies exhibited a conversion of C→T at nucleotide 935. The sequence of the entire 160-nucleotide PCR-amplified region of the alkaline phosphatase exon 6 in the clones analyzed exhibited no alterations other than the indicated changes at nucleotide position 935, suggesting that the cloned and sequenced DNA was derived from the HuH-7 genome rather than non-degraded chimeric oligonucleotides.

Targeted Genomic Nucleotide Conversion in Primary Rat Hepatocytes

Having established the utility of this technology for promoting site-directed targeted nucleotide exchange in genomic DNA in replicating cells of hepatic origin, we examined the efficiency in isolated early G_1 rat hepatocytes. Ultimately, the therapeutic benefit from this approach for the treatment of hepatic diseases requires not only the targeted delivery of these chimeric molecules to the liver, but also the nuclear localization and directed nucleotide exchange in quiescent G_0 hepatocytes. To develope an effective method for delivering these molecules to liver, we made use of the hepatocyte-specific asialoglycoprotein receptor. In fact, previous studies have reported the high specificity and efficiency in using galactose for targeting nucleic acids to the asialoglycoprotein receptor (Fig. 4) (Spanjer and Scherphof 1983; Chowdhury et al. 1993; Steer 1996). Therefore, we chose to modify the PEI because it was an effective carrier of the oligonucleotides in vitro and with its free amino groups was amenable to chemical modification and sugar attachment. Lactosylation of the PEI was carried out by the modification of a previously described method for the conjugation of oligosaccharides to proteins (Gray 1974). Briefly, the PEI in ammonium acetate was incubated with sodium cyanoborohydride and lactose monohydrate for 10 days at 37ºC. The reaction mixture was then dialyzed against distilled water, and the amount of sugar (as galactose) associated with the PEI was determined by the phenol-sulphuric acid method (Dubois et al. 1956). The number of moles of free primary and secondary amines in the dialyzed lactosylated-PEI was determined using a ninhydrin reagent.

There had also been success in transfecting the chimeric oligonucleotides in other cell types in vitro using the cationic lipids DOTAP

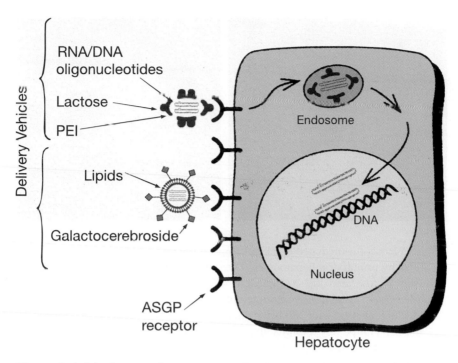

Figure 4 Asialoglycoprotein receptor-mediated targeted delivery of the chimeric oligonucleotide to hepatocytes. The chimeric molecules can be targeted to the hepatocyte through either a liposomal or lactosylated-PEI delivery system. The targeting ligand is a galactose moiety attached as the disaccharide lactose to PEI, or incorporated into liposomal vesicles as galactocerebroside. This sugar binds to the hepatocyte-specific asialoglycoprotein receptor, and the entire liposomal and/or polycationic complex is internalized via the clathrin-coated pit/endosomal pathway. Once inside the cell, the RNA/DNA molecules accumulate within the nucleus and mediate the proposed change in gene sequence. (ASGP) Asialoglycoprotein; (PEI) polyethylenimine.

(Cole-Strauss et al. 1996) and Lipofectin® (Yoon et al. 1996). In addition, liposomes targeted to the asialoglycoprotein receptor were shown to deliver nucleic acids in vivo (Nandi et al. 1986). Therefore, we developed a targeted liposomal delivery system for the chimeric molecules. Negatively charged liposomes with dioleoyl phosphatidylserine (DOPS):dioleoyl phosphatidylcholine (DOPC):galactocerebroside (Gc) (for targeting to the asialoglycoprotein receptor) at a 1:1:0.16 molar ratio were used for encapsulating the chimeric molecules (Bandyopadhyay et al. 1998). Size extrusion of the liposomes to 50 nm was done to promote preferential uptake by the hepatocytes in vivo (Templeton et al. 1997). The

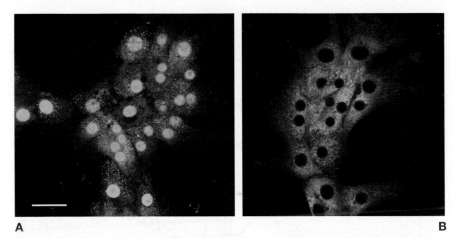

A   B

Figure 5 Receptor-mediated uptake of fluorescein end-labeled chimeric oligonucleotides by primary rat hepatocytes. The confocal micrograph (*A*) demonstrates
that the lactosylated-PEI complexed fluorescein-labeled chimeric oligonucleotide (*green*) is highly concentrated in the nucleus after uptake into the hepatocyte. Coincubation with 100 mM galactose or 10 mM EGTA during transfection
(*B*) almost completely inhibits the endocytosis of the fluorescently labeled
chimeric molecules, as well as any detectable nuclear staining. The cytoplasm
was visualized with DiI (*orange*) and the hepatocytes were examined using a
BioRad MRC1000 confocal microscope. Bar, 20 μm.

specificity of the asialoglycoprotein receptor for uptake of the chimeric
molecules by primary rat hepatocytes was determined by confocal
microscopy of fluorescein-12-dUTP 3´-end-labeled chimeric oligonucleotides. In fact, the fluorescein-labeled lactosylated-PEI:chimeric complexes were delivered to essentially all cells, which also exhibited significant nuclear accumulation after endocytosis (Fig. 5). Cell uptake and
nuclear staining were markedly inhibited by the coincubation of the lactosylated PEI:chimeric complex with 100 mM galactose or 10 mM EGTA.
Similar results were obtained using the targeted negatively charged liposome encapsulated chimeric molecules (Bandyopadhyay et al. 1998).

Having demonstrated the efficacy of these two targeted delivery systems to promote receptor-mediated uptake of these chimeric molecules in
primary rat hepatocytes, we determined the efficiency of site-directed
genomic nucleotide conversion in these nonreplicating cells. We designed
chimeric oligonucleotides that targeted both the transcribed and nontranscribed factor IX genomic DNA strands (Kren et al. 1998). The RNA/DNA
sequence of the molecules was identical in sequence to the wild-type cDNA
(Sarkar et al. 1990), corresponding to nucleotides 706–730 except for the

central nucleotide substitutions in the pentameric stretch of DNA residues. The all-DNA strand was also identical in sequence to the wild-type gene except for the corresponding change in the central nucleotide. The base substitutions corresponded to the targeted nucleotide at Ser-365 and would introduce a missense mutation in the rat genomic sequence resulting in the active site Ser-365→Arg-365, which was characteristic of certain human factor IX mutations (Reiner and Davie 1995).

The cultured hepatocytes were then transfected in the presence of 10% FBS with either transcribed or nontranscribed factor IX chimeric molecules at comparable concentrations using 800-kD PEI as the carrier, and nontranscribed factor IX using 800- or 25-kD lactosylated PEI. Forty-eight hours after transfection, the cells from the chimeric and vehicle control transfections were harvested, and both DNA and RNA were isolated and used for PCR or RT-PCR amplification of a 374-nucleotide fragment of the rat factor IX gene corresponding to nucleotides 433–806 of the cDNA (Sarkar et al. 1990). The PCR products were subcloned, and the A→C targeted nucleotide conversion at Ser-365 was determined by the hybridization of duplicate colony lifts using ^{32}P-labeled 17-mer oligonucleotide probes corresponding to nucleotides 710 through 726 of the cDNA sequence. The probes readily distinguished between the wild-type 365^A and the converted 365^C factor IX sequences. The conversion frequency of the targeted nucleotide was calculated by dividing the number of clones hybridizing with the 365^C oligonucleotide by the total number of clones hybridizing with both oligonucleotide probes.

The results for nontranscribed factor IX indicated that the A→C conversion at Ser-365 was dose-dependent, ranging from 5% to 20%, respectively, at 90 and 270 nM with the lactosylated-PEI. RT-PCR and hybridization analysis of RNA isolated from cultured hepatocytes transfected in parallel with lactosylated-PEIs resulted in A→C conversion frequencies ranging from 12% to 22%. The 800-kD nonlactosylated PEI was 50% less efficient at promoting the A→C conversion; however, similar conversion frequencies were observed using either the transcribed or nontranscribed chimeric molecules. DNA isolated from untreated hepatocytes and PCR-amplified in the presence of increasing doses of the oligonucleotides, as well as cells transfected with vehicle or an unrelated chimeric molecule, yielded no clones hybridizing with the 365^C oligonucleotide probe. In addition, sequence analysis of selected isolated clones indicated that the exchange of A→C at Ser-365 was site-specific and apparently independent of the transcriptional activity of the targeted strand. The nontranscribed factor IX chimeric molecules were also encapsulated in negatively charged targeted liposomes and used to transfect primary rat hepatocytes. This method of introducing the chimeric oligonu-cleotides into the cells resulted in targeted A→C conversion that was dose-dependent and slightly more efficient, ranging from 10% to 25%, at 90- and 270-nM concentrations, respectively. Interestingly, under condi-tions in which RNA/DNA hybrids were active in promoting nucleotide exchange, the corresponding all-DNA duplexes, despite uptake by the nucleus, were essentially inactive (Cole-Strauss et al. 1996; Kotani et al. 1996; Yoon et al. 1996; Kren et al. 1997).

Site-directed Nucleotide Conversion In Vivo

The ability of the factor IX chimeric molecules to promote targeted nucleotide exchange in vitro in the genomic DNA of primary rat hepato-cytes suggested the potential for success in vivo. The targeting strategy developed for the delivery of the oligonucleotides to isolated hepatocytes was associated with an efficient nuclear uptake of the complexes follow-ing receptor-mediated endocytosis, thus overcoming a known barrier to nucleic acid delivery within the cell (Zabner et al. 1995). Additionally, the ability of these targeted delivery systems to function in the presence of serum indicated their potential for effective in vivo delivery of the chimeric molecules to the intact liver. The 25-kD lactosylated-PEI was chosen for the initial intravenous injections because PEI had been shown to be an effective transfecting agent for in vivo delivery to both brain and kidney tissue (Abdallah et al. 1996; Boletta et al. 1997). The chimeric oligonucleotides were prepared in 300–500 μl of 5% dextrose using lac-tosylated-PEI at a ratio of six equivalents of PEI nitrogen per chimeric phosphate (Kren et al. 1998). Male Sprague-Dawley rats (~50 g) were administered the aliquots by tail vein injection either as a single dose of 100 μg or as divided doses of 150 and 200 μg on consecutive days. Liver tissue was removed for DNA and RNA isolation 5 days postinjection. DNA and RNA were isolated for PCR and RT-PCR amplification of the previously defined 374-nucleotide region of the rat factor IX gene. The PCR- and RT-PCR-amplified products were analyzed using the same duplicate filter lift hybridization procedure that was used for the isolated hepatocyte experiments. The animals injected with the vehicle alone yielded no PCR or RT-PCR products that hybridized with the mutant 365^C ^{32}P-labeled 17-mer oligonucleotide probe. In contrast, the animals injected with the lactosylated-PEI:factor IX chimeric oligonucleotides exhibited a dose-related genomic DNA conversion frequency of 15% and 40%, respectively, at 100 μg and 350 μg. Furthermore, RT-PCR of RNA isolated from the high-dose livers indicated A→C conversion frequencies of 26–28% by filter lift hybridization analysis, suggesting a potential bio-logical effect of the genomic DNA conversion.

Sequence analysis was done to confirm the site-specific A→C con-version at Ser-365 in the PCR- and RT-PCR-amplified material from DNA and RNA isolated from the in vitro and in vivo rat experiments. Colonies derived from the transcribed and nontranscribed factor IX trans-fected primary hepatocytes hybridizing to the 365^C oligonucleotide probe converted to a C at Ser-365. The same A→C conversion at Ser-365 was observed in the clones derived from the transfected rat liver that hybridized with the 17-mer 365^C oligonucleotide probe. In contrast, colonies derived from vehicle-transfected hepatocytes or animals hybridizing to the 365^A radiolabeled probe exhibited the wild-type factor IX sequence, that is, an A at Ser-365 of the reported cDNA sequence. For every 365^C clone sequenced, the entire 374-nucleotide PCR-amplified region of the factor IX gene exhibited no alteration other than the indi-cated changes at Ser-365. In addition, the start and end points of the PCR-amplified products from both the primary hepatocytes and the intact liver corresponded exactly to those of the primers used for the amplification process, indicating that the cloned and sequenced DNA was derived from genomic DNA rather than nondegraded chimeric oligonucleotides.

Activated partial thromboplastin times (aPTT) were determined to establish whether the converted RNA produced by the animals treated with the high dose of the factor IX chimeric molecules resulted in altered fac-tor IX coagulant activity. Twenty-one days after the second tail vein injec-tion, blood samples from the vehicle- and oligonucleotide-treated rats were collected, and aPTT using human factor IX-deficient plasma and rat test plasma were determined using an American Bioproducts ST4 coagu-lometer. Factor IX activity of samples was determined from a standard curve constructed from the aPTT pooled plasma from normal male rats of similar age. Factor IX coagulant activities for the oligonucleotide-treated animals were 54% of normal, in contrast to those from the vehicle-inject-ed animals, which were 132% of normal values. To determine the replica-tive stability of the targeted nucleotide correction, the animals received an additional 500 μg of chimeric:lactosylated-PEI complexed material, and 24 hours later underwent a 70% partial hepatectomy.

Rats subjected to 70% surgical resection of the liver undergo com-pensatory regeneration of the liver within 10 days, a process in which 95% of the hepatocytes replicate in two synchronous waves (Higgins and Anderson 1931). Subsequent analysis of the factor IX activity in the par-tially hepatectomized animals by aPTT at 9 and 24 weeks postinjection/hepatectomy indicated that the factor IX coagulant activity in the chimeric oligonucleotide-treated animals decreased to approximately 40%, whereas little change was noted in the vehicle-treated controls (113%). Additional animals received daily injections of 200 µg of the factor IX chimeric complexed with the lactosylated-PEI for 5 days or an equal amount of lactosylated-PEI (B. Kren et al., unpubl.). Seven days after the last injection liver tissue was removed, and a genomic DNA A→C conversion frequency at Ser-365 of almost 50% was determined by PCR amplification and filter lift hybridization analysis. Blood was drawn from the experimental and vehicle-injected animals 3 and 18 weeks after their last tail vein injection, and the factor IX activity was determined by aPTT. The factor IX activity of animals treated with the chimeric oligonucleotides was 54% and 56% of normal values at 3 and 18 weeks, respectively. In contrast, the vehicle-injected controls averaged 112% and 111% of normal values, respectively, at 3 and 18 weeks. Liver tissue was removed from the experimental animals at 18 weeks, and no significant difference was noted between the genomic A→C conversion frequency at Ser-365 and that observed at 1 week. Thus, the site-directed genomic conversion of the factor IX gene in intact liver results in long-term replicatively stable genomic DNA sequence conversion and phenotypic change in the coagulant activity measured by aPTT assay.

The Potential Use for This Technology in Hepatic Gene Therapy

As described above, these chimeric molecules were capable of promoting site-specific nucleotide exchange in intact liver that resulted in a sustained phenotypic change in the factor IX coagulant activity. This introduction of a missense mutation in vitro in primary rat hepatocytes was apparently independent of the genomic strand of DNA, suggesting that the transcriptional activity of the targeted genomic strand does not affect the conversion rate. This result corroborated the work in both the HuH-7 hepatoma and lymphoblastoid cells, indicating that transcription of the β-globin locus was not required for these molecules to effect a missense site-directed nucleotide conversion (Cole-Strauss et al. 1996; Kren et al. 1997). This would suggest that the technology may be applicable to gene loci that exhibit little transcriptional activity in the intact liver. Structural aspects of the targeted nucleotide in the DNA loci may have an effect on conversion efficiency, because repair of mismatched DNA has been shown to be strand-specific and bidirectional and also depends on the nucleosomal position (Fang and Modrich 1993; Kolodner 1995; Shinohara and Ogawa 1995; Klungland and Lindahl 1997; Wellinger and Thoma 1997). However, the ability of chimeric molecules to mediate the site-directed mutagenesis of the factor IX gene in hepatocytes and intact liver suggests that the endogenous DNA repair pathway is present in non-replicating and quiescent cells.

The efficient site-directed mutagenesis observed in this study suggests that the mismatch repair pathways in hepatocytes may be sufficiently active to make this strategy feasible for liver-related disorders resulting from single-point mutations such as hemophilia B (Reiner and Davie 1995) and α_1-antitrypsin deficiency (Teckman et al. 1996). In fact, these chimeric oligonucleotides were able to promote site-directed nucleotide exchange in isolated hepatocytes from the Chapel Hill strain of hemophilia B dogs, correcting the missense mutation responsible for their factor IX deficiency (B. Kren et al., unpubl.). Although some success with virus-based vectors has recently been reported in the Gunn rat model of Crigler-Najjar syndrome type I (Askari et al. 1996; Li et al. 1998), transgene expression has been transient. We have recently determined that RNA/DNA chimeric oligonucleotides can be used to insert a single base in the UDP-glucuronosyltransferase-1 gene (UGT1) in SV40 immortalized and primary Gunn rat hepatocytes as well as in vivo (B. Kren et al., unpubl.).

Using the same strategy outlined above, the chimeric molecule was designed to exist in a duplex conformation with 2'-O-methylated RNA/DNA stems, a 3' GC clamp, and poly-T hairpin loops for chemical and thermal stability as well as resistance to helicases and both RNA and DNA nucleases (Monia et al. 1993). The RNA/DNA sequence of the duplex oligonucleotide was complementary to the targeted region, except that it contained one mismatched nucleotide when aligned with the homologous genomic DNA sequence (Fig. 6). This unpaired nucleotide is apparently recognized by endogenous repair systems (Kolodner 1995; Shinohara and Ogawa 1995), thus resulting in specific modification of the DNA sequence of the targeted gene (Fig. 7). The insertion of a nucleotide into the genomic DNA that corrects the nonsense mutation in UGT1 using this technology appears to be less efficient both in vitro and in vivo than the site-directed targeted nucleotide exchange using these molecules. However, increased UDP-glucuronosyltransferase-1 enzymatic activity and conjugated bilirubin were detected in the Gunn rat animals treated with the correcting chimeric oligonucleotides. In addition, serum bilirubin levels dropped more than 50% relative to pretreatment levels.

The ability of this chimeric oligonucleotide-directed gene repair/mutagenesis strategy to correct frame-shift mutations resulting from a single nucleotide deletion significantly broadens its application for therapeutic use. Because the endogenous genomic nucleotide mismatch repair

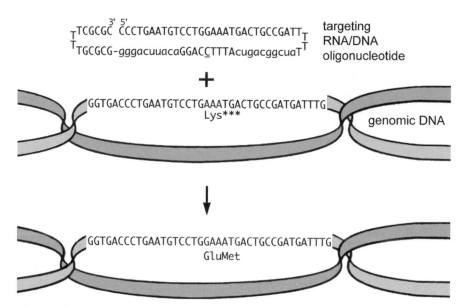

Figure 6 Gene targeting to correct the *UGT1* frameshift mutation in the Gunn rat. G insertion at position 1206 of the UGT1 cDNA using chimeric RNA/DNA oligonucleotides corrects the frameshift mutation in the Gunn rat, removing the premature termination codon and restoring the production of UGT1. It is speculated that the RNA/DNA oligonucleotide aligns with the targeted genomic DNA site and forms the paired intermediate structure, thus allowing the complementary base insertion to C (*underlined*) to be catalyzed by endogenous DNA repair activities. The 2′-*O*-methyl RNA residues in the targeting oligonucleotide are indicated in *lowercase* letters. (UGT1) UDP-glucurosyltransferase type I.

pathways function independently of DNA replication, this technology is compatible with the hepatocyte's quiescent G_0 state in vivo. However, joint chimeric/DNA molecule formation in vitro is increased significantly by the *U. maydis* Rec2 recombinase protein (Kotani et al. 1996), and the cell cycle regulation of a human analog (Yamamoto et al. 1996) suggests that the efficiency of nucleotide conversion by these molecules might be increased in replicating cells. Moreover, proliferating cell nuclear antigen significantly stimulates the "long-patch" base excision repair pathway both in cell-free systems and in cultured cells, suggesting that replicating cells may also exhibit increased activity of specific mismatch repair pathways (Klungland and Lindahl 1997).

The ability to induce site-specific nucleotide exchange without selection may be particularly powerful for ex vivo gene therapy because human hepatocyte transplantation has been shown to be a successful

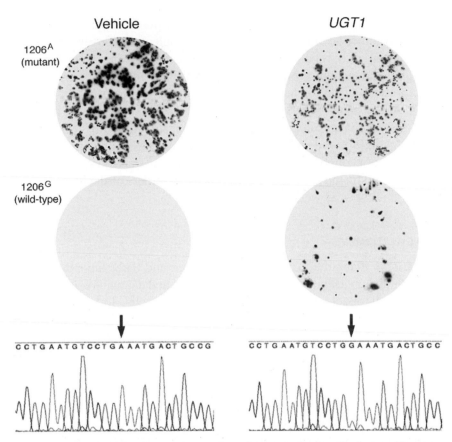

Figure 7 Hybridization patterns and sequence analysis of primary Gunn rat hepatocytes transfected with chimeric oligonucleotides. Hybridization patterns of duplicate filter lifts are shown of the cloned PCR products from cells transfected with vehicle (*left*) or the UGT1 chimeric molecules (*right*). Only cells transfected with the UGT1 oligonucleotides yielded clones that hybridized with the ^{32}P-labeled wild-type 1206[G] 17-mer oligonucleotide probe, while the mutant 1206[A] probe hybridized with both. The nucleotide sequences of plasmid DNA isolated from clones that hybridized with the two different radiolabeled probes are shown. The clones that hybridized with the wild-type 1206[G] oligonucleotide displayed a guanosine insertion at position 1206 (*arrow, right*), whereas the vehicle-transfected 1206[A] mutant hybridizing clones showed an adenosine residue at the targeted insertion site (*arrow, left*). UGT1, UDP-glucurosyltransferase type I.

approach for the treatment of Crigler-Najjar patients (Fox et al. 1998). This ex vivo approach might also permit the incremental repair of genetic lesions greater than 1 nucleotide, thus expanding the use of this technology not only for treating genetic diseases of the liver but other tissues as well. Another advantage of this targeted nucleotide conversion strategy is that, in vivo, regulation of gene expression remains under the control of the endogenous promoter at its native site, which can be critical for the expression of certain hepatic genes such as α_1-antitrypsin (Bulla et al. 1992). As we better understand the cellular repair processes that mediate the observed changes in gene sequence demonstrated by this technology, coupled with further development and improvements in delivery systems, these chimeric oligonucleotides may provide an important advance in gene-directed therapy not only for metabolic disorders of the liver, but for other tissue types as well.

ACKNOWLEDGMENTS

The authors thank the members of the Kmiec and Steer laboratories for their seminal contributions in the development and characterization of the chimeric oligonucleotides and their delivery to cells. In particular, the work of Hidehito Kotani, Allyson Cole-Strauss, Kyonggeun Yoon, Sarah Ye, Yufei Xiang, Howard Gamper, and Paramita Bandyopadhyay deserves special recognition. We also acknowledge the intellectual and experimental efforts of Ramesh Kumar, Richard Metz, MaryAnn Gruda, and Takao Kurihara of Kimeragen, Inc.; William K. Holloman, Cornell University, New York, NY; and Jayanta and Namita Roy Chowdhury, the Albert Einstein College of Medicine, Bronx, NY. Finally, the authors thank Michael Blaese (National Institutes of Health, Bethesda, MD) for his continued enthusiasm and support.

REFERENCES

Abdallah B., Hassan A., Benoist C., Goula D., Behr J.P., and Demeneix B.A. 1996. A powerful nonviral vector for *in vivo* gene transfer into the adult mammalian brain: Polyethylenimine. *Hum. Gene Ther.* **7:** 1947–1954.

Arbonés M.L., Austin H.A., Capon D.J., and Greenburg G. 1994. Gene targeting in normal somatic cells: Inactivation of the interferon-γ receptor in myoblasts. *Nat. Genet.* **6:** 90–97.

Askari F.K., Hitomi Y., Mao M., and Wilson J.M. 1996. Complete correction of hyperbilirubinemia in the Gunn rat model of Crigler-Najjar syndrome type I following transient in vivo adenovirus-mediated expression of human bilirubin UDP-glucuronosyltransferase. *Gene Ther.* **3:** 381–388.

Bandyopadhyay P., Kren B.T., Ma X., and Steer C.J. 1998. Enhanced gene transfer into HuH-7 cells and primary rat hepatocytes using targeted liposomes and polyethylenimine. *BioTechniques* **25:** 282–292.

Beal P.A. and Dervan P.B. 1991. Second structural motif for recognition of DNA by oligonucleotide-directed triple-helix formation. *Science* **251:** 1360–1363.

Bezzubova O.Y., Schmidt H., Ostermann K., Heyer W.-D., and Buerstedde J.-M. 1993a. Identification of a chicken *RAD52* homologue suggests conservation of the *RAD52* recombination pathway throughout the evolution of higher eukaryotes. *Nucleic Acids Res.* **21:** 5945–5949.

Bezzubova O., Shinohara A., Mueller R.G., Ogawa H., and Buerstedde J.-M. 1993b. A chicken RAD51 homologue is expressed at high levels in lymphoid and reproductive organs. *Nucleic Acids Res.* **21:** 1577–1580.

Boletta A., Benigni A., Lutz J., Remuzzi G., Soria M.R., and Monaco L. 1997. Nonviral gene delivery to the rat kidney with polyethylenimine. *Hum. Gene Ther.* **8:** 1243–1251.

Boussif O., Lezoualc'h F., Zanta M.A., Mergny M.D., Scherman D., Demeneix B., and Beher J.-P. 1995. A versatile vector for gene and oligonucleotide transfer into cells in culture and *in vivo:* Polyethylenimine. *Proc. Natl. Acad. Sci.* **92:** 7297–7301.

Bowles N. and Woo S.L.C. 1995. Gene therapy for metabolic disorders. *Adv. Drug Del. Rev.* **17:** 293–302.

Buerstedde J.-M. and Takeda S. 1991. Increased ratio of targeted to random integration after transfection of chicken B cell lines. *Cell* **67:** 179–188.

Bulla G.A., DeSimone V., Cortese R., and Fournier R.E.K. 1992. Extinction of α_1-antitrypsin gene expression in somatic cell hybrids: Evidence for multiple controls. *Genes Dev.* **6:** 316–327.

Capecchi M.R. 1989. Altering the genome by homologous recombination. *Science* **244:** 1288–1292.

Chang A.G.Y. and Wu G.Y. 1994. Gene therapy: Applications to the treatment of gastrointestinal and liver diseases. *Gastroenterology* **106:** 1076–1084.

Chowdhury N.R., Wu C.H., Wu G.Y., Yerneni P.C., Bommineni V.R., and Chowdhury J.R. 1993. Fate of DNA targeted to the liver by asialoglycoprotein receptor-mediated endocytosis *in vivo.* Prolonged persistence in cytoplasmic vesicles after partial hepatectomy. *J. Biol. Chem.* **268:** 11265–11271.

Cole-Strauss A., Nöe A., and Kmiec E.B. 1997. Recombinational repair of genetic mutations. *Antisense Nucleic Acid Drug Dev.* **7:** 211–216.

Cole-Strauss A., Yoon K., Xiang Y., Byrne B.C., Rice M.C., Gryn J., Holloman W.K., and Kmiec E.B. 1996. Correction of the mutation responsible for sickle cell anemia by an RNA-DNA oligonucleotide. *Science* **273:** 1386–1389.

Connelly S., Gardner J.M., Lyons R.M., McClelland A., and Kaleko M. 1996. Sustained expression of therapeutic levels of human factor VIII in mice. *Blood* **87:** 4671–4677.

Dubois M., Gilles K.A., Hamilton J.K, Rebers P.A., and Smith F. 1956. Colorimetric method for determination of sugars and related substances. *Anal. Chem.* **28:** 350–356.

Fang W.-h. and Modrich P. 1993. Human strand-specific mismatch repair occurs by a bidirectional mechanism similar to that of the bacterial reaction. *J. Biol. Chem.* **268:** 11838–11844.

Fox I.J., Chowdhury J.R, Kaufman S.S., Goertzen T.C., Chowdhury N.R., Warkentin P.I., Dorko K., Sauter B.V., and Strom S.C. 1998. Treatment of the Crigler-Najjar syndrome type I with hepatocyte transplantation. *New Engl. J. Med.* **338:** 1422–1426.

Glazer P.M., Sarkar S.N., Chisholm G.E., and Summers W.C. 1987. DNA mismatch repair detected in human cell extracts. *Mol. Cell. Biol.* **7:** 218–224.

Gray G.R. 1974. The direct coupling of oligosaccharides to proteins and derivatized gels *Arch. Biochem. Biophys.* **163:** 426–428.

Hasty P., Rivera-Pérez J., and Bradley A. 1992. The role and fate of DNA ends for homologous recombination in embryonic stem cells. *Mol. Cell. Biol.* **12:** 2464–2474.

Havre P.A. and Glazer P.M. 1993. Targeted mutagenesis of simian virus 40 DNA mediated by a triple helix-forming oligonucleotide. *J. Virol.* **67:** 7324–7331.

Havre P.A. and Kmiec E.B. 1998. RecA-mediated joint molecule formation between O-methylated RNA/DNA hairpins and single-stranded targets. *Mol. Gen. Genet.* **258:** 580–586.

Higgins G.H. and Anderson R.M. 1931. Experimental pathology of the liver. I. Restoration of the liver of the white rat following partial surgical removal. *Arch. Pathol.* **12:** 186–202.

Ilan Y., Prakash R., Davidson A., Jona V., Droguett G., Horwitz M.S., Chowdhury N.R., and Chowdhury J.R. 1997. Oral tolerization to adenoviral antigens permits long-term gene expression using recombinant adenoviral vectors. *J. Clin. Invest.* **99:** 1098–1106.

Jasin M., de Villiers J., Weber F., and Schaffner W. 1985. High frequency of homologous recombination in mammalian cells between endogenous and introduced SV40 genomes. *Cell* **43:** 695–703.

Kiledjian M. and Kadesch T. 1990. Analysis of the human liver/bone/kidney alkaline phosphatase promoter *in vivo* and *in vitro*. *Nucleic Acids Res.* **18:** 957–961.

Kishi F., Matsuura S., and Kajii T. 1989. Nucleotide sequence of the human liver-type alkaline phosphatase cDNA. *Nucleic Acids Res.* **17:** 2129.

Klungland A. and Lindahl, T. 1997. Second pathway for completion of human DNA base excision-repair: Reconstitution with purified proteins and requirement for DNase IV (FEN1). *EMBO J.* **16:** 3341–3348.

Kmiec E.B. 1996. Genomic targeting and genetic conversion in cancer therapy. *Semin. Oncol.* **23:** 188–193.

Kmiec E.B., Cole A., and Holloman W.K. 1994. The *REC2* gene encodes the homologous pairing protein of *Ustilago maydis*. *Mol. Cell. Biol.* **14:** 7163–7172.

Kolodner R.D. 1995. Mismatch repair: mechanisms and relationship to cancer susceptibility. *Trends Biochem. Sci.* **20:** 397–401.

Kotani H. and Kmiec E.B. 1994a. A role for RNA synthesis in homologous pairing events. *Mol. Cell. Biol.* **14:** 6097–6106.

———. 1994b. DNA cruciforms facilitate in vitro strand transfer on nucleosomal templates. *Mol. Gen. Genet.* **243:** 681–690.

———. 1994c. Transcription activates RecA-promoted homologous pairing of nucleosomal DNA. *Mol. Cell. Biol.* **14:** 1949–1955.

Kotani H., Kmiec E.B., and Holloman W.K. 1993. Purification and properties of a cruciform DNA binding protein from *Ustilago maydis*. *Chromosoma* **102:** 348–354.

Kotani H., Sekiguchi J.M., Dutta S., and Kmiec E.B. 1994. Genetic recombination of nucleosomal templates is mediated by transcription. *Mol. Gen. Genet.* **244:** 410–419.

Kotani H., Germann M.W., Andrus A., Vinayak R., Mullah B., and Kmiec E.B. 1996. RNA facilitates RecA-mediated DNA pairing and strand transfer between molecules bearing limited regions of homology. *Mol. Gen. Genet.* **250:** 626–634.

Kren B.T., Bandyopadhyay P., and Steer C.J. 1998. In vivo site-directed mutagenesis of the *factor IX* gene in rat liver and in isolated hepatocytes by chimeric RNA/DNA oligonucleotides. *Nat. Med.* **4:** 285–290.

Kren B.T., Cole-Strauss A., Kmiec E.B., and Steer C.J. 1997. Targeted nucleotide exchange in the alkaline phosphatase gene of HuH-7 cells mediated by a chimeric RNA/DNA oligonucleotide. *Hepatology* **25:** 1462–1468.

Kucherlapati R.S., Eves E.M., Song K.-Y., Morse B.S., and Smithies O. 1984. Homologous recombination between plasmids in mammalian cells can be enhanced by treatment of input DNA. *Proc. Natl. Acad. Sci.* **81:** 3153–3157.

Li Q., Murphree S.S., Willer S.S., Bolli R., and French B.A. 1998. Gene therapy with bilirubin-UDP-glucuronosyltransferase in the Gunn rat model of Crigler-Najjar syndrome type 1. *Hum. Gene Ther.* **9:** 497–505.

Martinez-Hernandez A. and Amenta P.S. 1995. The extracellular matrix in hepatic regeneration. *FASEB J.* **9:** 1401–1410.

Matsuura S., Kishi T., and Kajii T. 1990. Characterization of a 5′-flanking region of the human liver/bone/kidney alkaline phosphatase gene: Two kinds of mRNA from a single gene. *Biochem. Biophys. Res. Commun.* **168:** 993–1000.

Melchior W.B., Jr. and Von Hippel P.H. 1973. Alteration of the relative stability of dA•dT and dG•dC base pairs in DNA. *Proc. Natl. Acad. Sci.* **70:** 298–302.

Moerschell R.P., Tsunasawa S., and Sherman F. 1988. Transformation of yeast with synthetic oligonucleotides. *Proc. Natl. Acad. Sci.* **85:** 524–528.

Monia B.P., Lesnik E.A., Gonzalez C., Lima W.F., McGee D., Guinosso C.J., Kawasaki A.M., Cook P.D., and Freier S.M. 1993. Evaluation of 2′-modified oligonucleotides containing 2′-deoxy gaps as antisense inhibitors of gene expression. *J. Biol. Chem.* **268:** 14514–14522.

Nakabayashi H., Taketa K., Miyano K., Yamane T., and Sato J. 1982. Growth of human hepatoma cell lines with differentiated functions in chemically defined medium. *Cancer Res.* **42:** 3858–3863.

Nandi P.K., Legrand A., and Nicolau C. 1986. Biologically active, recombinant DNA in clathrin-coated vesicles isolated from rat livers after *in vivo* injection of liposome-encapsulated DNA. *J. Biol. Chem.* **261:** 16722–16726.

Orkin S.H., Kazazian H.H., Jr., Antonarakis S.E., Goff S.C., Boehm C.D., Sexton J.P., Waber P.G., and Giardina P.J.V. 1982. Linkage of β-thalassaemia mutations and β-globin gene polymorphisms with DNA polymorphisms in human β-globin gene cluster. *Nature* **296:** 627–631.

Orr-Weaver T.L., Szostak J.W., and Rothstein R.J. 1981. Yeast transformation: A model system for the study of recombination. *Proc. Natl. Acad. Sci.* **78:** 6354–6358.

Prouty S.M., Hanson K.D., Boyle A.L., Brown J.R., Shichiri M., Follansbee M.R., Kang W., and Sedivy J.M. 1993. A cell culture model system for genetic analyses of the cell cycle by targeted homologous recombination. *Oncogene* **8:** 899–907.

Reiner A.P. and Davie E.W. 1995. Introduction to hemostasis and the vitamin K-dependent coagulation factors. In *The metabolic and molecular bases of inherited disease* (ed. C.R. Scriver et al.), pp. 3181–3221. McGraw-Hill, New York.

Roth D. and Wilson J. 1988. Illegitimate recombination in mammalian cells. In *Genetic recombination* (ed. R. Kucherlapati and G.R. Smith), pp. 621–653. American Society for Microbiology, Washington, D.C.

Rothstein R.J. 1983. One-step gene disruption in yeast. *Methods Enzymol.* **101:** 202–211.

Sakagami K., Tokinaga Y., Yoshikura H., and Kobayashi I. 1994. Homology-associated nonhomologous recombination in mammalian gene targeting. *Proc. Natl. Acad. Sci.* **91:** 8527–8531.

Sarkar G., Koeberl D.D., and Sommer S.S. 1990. Direct sequencing of the activation peptide and the catalytic domain of the factor IX gene in six species. *Genomics* **6:** 133–143.

Shinohara A. and Ogawa T. 1995. Homologous recombination and the roles of double-strand breaks. *Trends Biochem. Sci.* **20:** 387–391.

Shinohara A., Ogawa H., and Ogawa T. 1992. Rad51 protein involved in repair and recombination in *S. cerevisiae* is a RecA-like protein. *Cell* **69:** 457–470.

Song K.-Y., Chekuri L., Rauth S., Ehrlich S., and Kucherlapati R. 1985. Effect of double-strand breaks on homologous recombination in mammalian cells and extracts. *Mol. Cell. Biol.* **5:** 3331–3336.

Spanjer H.H. and Scherphof G.L. 1983. Targeting of lactosylceramide-containing liposomes to hepatocytes in vivo. *Biochim. Biophys. Acta* **734:** 40–47.

Steer C.J. 1996. Receptor-mediated endocytosis: Mechanisms, biologic function, and molecular properties. In *Hepatology: A textbook of liver disease* (ed. D. Zakim and T.D. Boyer), pp.149–214. W.B. Saunders, Philiadelphia.

Story R.M., Bishop D.K., Kleckner N., and Steitz T.A. 1993. Structural relationship of bacterial RecA proteins to recombination proteins from bacteriophage T4 and yeast. *Science* **259:** 1892–1896.

Strauss M. 1994. Liver-directed gene therapy: Prospects and problems. *Gene Ther.* **1:** 156–164.

Szostak J.W., Orr-Weaver T.L., Rothstein R.J., and Stahl F.W. 1983. The double-strand-break repair model for recombination. *Cell* **33:** 25–35.

Takahashi M., Ilan Y., Chowdhury N.R., Guida J., Horwitz M., and Chowdhury J.R. 1996. Long term correction of bilirubin-UDP-glucuronosyltransferase deficiency in Gunn rats by administration of a recombinant adenovirus during the neonatal period. *J. Biol. Chem.* **271:** 26536–26542.

Teckman J.H., Qu D., and Perlmutter D.H. 1996. Molecular pathogenesis of liver disease in α_1-antitrypsin deficiency. *Hepatology* **24:** 1504–1516.

Templeton N.S., Lasic D.D., Frederik P.M., Strey H.H., Roberts D.D, and Pavlakis G.N. 1997. Improved DNA:liposome complexes for increased systemic delivery and gene expression. *Nat. Biotechnol.* **15:** 647–652.

Thyagarajan B., Cruise J.L., and Campbell C. 1996. Elevated levels of homologous DNA recombination activity in the regenerating rat liver. *Somat. Cell Mol. Genet.* **22:** 31–39.

Valancius V. and Smithies O. 1991. Double-strand gap repair in a mammalian gene targeting reaction. *Mol. Cell. Biol.* **11:** 4389–4397.

Wang G., Levy D.D., Seidman M.M., and Glazer P.M. 1995. Targeted mutagenesis in mammalian cells mediated by intracellular triple helix formation. *Mol. Cell. Biol.* **15:** 1759–1768.

Weiss M.J., Ray K., Henthorn P.S., Lamb B., Kadesch T., and Harris H. 1988. Structure of the human liver/bone/kidney alkaline phosphatase gene. *J. Biol. Chem.* **263:** 12002–12010.

Wellinger R.E. and Thoma F. 1997. Nucleosome structure and positioning modulate nucleotide excision repair in the non-transcribed strand of an active gene. *EMBO J.* **16:** 5046–5056.

Yamamoto A., Taki T., Yagi H., Habu T., Yoshida K., Yoshimura Y., Yamamoto K., Matsushiro A., Nishimune Y., and Morita T. 1996. Cell cycle-dependent expression of the mouse *Rad51* gene in proliferating cells. *Mol. Gen. Genet.* **251:** 1–12.

Yáñez R.J. and Porter A.C.G. 1998. Therapeutic gene targeting. *Gene Ther.* **5:** 149–159.

Ye X., Robinson M.B., Batshaw M.L., Furth E.E., Smith I., and Wilson J.M. 1996. Prolonged metabolic correction in adult ornithine transcarbamylase-deficient mice with adenoviral vectors. *J. Biol. Chem.* **271:** 3639–3646.

Yoon K., Cole-Strauss A., and Kmiec E.B. 1996. Targeted gene correction of episomal DNA in mammalian cells mediated by a chimeric RNA•DNA oligonucleotide. *Proc. Natl. Acad. Sci.* **93:** 2071–2076.

Zabner J., Fasbender A.J., Moninger T., Poellinger K.A., and Welsh M.J. 1995. Cellular and molecular barriers to gene transfer by a cationic lipid. *J. Biol. Chem.* **270:** 18997–19007.

Zheng H. and Wilson J.H. 1990. Gene targeting in normal and amplified cell lines. *Nature* **344:** 170–173.

24

Human Gene Therapy: Public Policy and Regulatory Issues

Nelson A. Wivel

Department of Molecular and Cellular Engineering
Institute for Human Gene Therapy
University of Pennsylvania
Philadelphia, Pennsylvania 19104

W. French Anderson

Gene Therapy Laboratories
University of Southern California
School of Medicine
Los Angeles, California 90033

Although human gene therapy owes its current status to the development of recombinant DNA technology, several prescient investigators postulated that gene transfer into cells might be possible, even before all of the basic experiments in molecular genetics had been performed (Wolff and Lederberg 1994). In the 1960s, Edward Tatum suggested that viruses could effectively be used for man's benefit by using them to introduce genes into cells, and he actually described a set of conditions that accurately characterizes ex vivo transduction of target cells (Tatum 1966). Joshua Lederberg anticipated that the in vitro culture of germ cells would be possible and that it would be possible to interchange chromosomes and their segments. Furthermore, he predicted that ultimately it might be possible to directly control nucleotide sequences in human chromosomes along with recognition, selection, and integration of the desired genes (Lederberg 1968). After announcing the first successful in vitro synthesis of DNA, Arthur Kornberg predicted that it might be possible to attach a gene to a harmless viral DNA and to use such a virus to cure a patient suffering from a hereditary defect by delivering a gene to the cells (Kornberg 1971). In 1968, one of the authors (W.F.A.) submitted an article to the *New England Journal of Medicine* in which he described how human gene therapy could be accomplished. Although the article was not accepted because it was too speculative, a portion of the text is instructive as it stated that, "In order to insert a correct gene into cells containing a mutation, it will first be necessary to isolate the desired gene from a normal chromosome. Then this gene will probably have to be duplicated to provide many copies. And, finally, it will be necessary to incorporate the correct copy into the genome of the defective cell."

There was a point at which these discussions moved from the theoretical to the actual when, in the late 1960s, Stanfield Rogers conducted the first human genetic engineering experiment when he administered the Shoperabbit papillomavirus to two German girls suffering from argininemia. The basis of this study was derived from the observation that animals infected with this virus had a decreased level of blood arginine and that a number of scientists working with this virus also had reduced blood arginine levels, apparently without any side effects (Rogers and Moore 1963). Although no clinical efficacy was observed, the ethics of the experiment were defended on the premise that this intervention offered a chance to prevent progressive deterioration and that no other treatments were available.

Other investigators were concerned that the Rogers experiment had been premature and that a number of problems should have been addressed before involving human subjects. A principal concern was that this might serve as an example or an impetus for other groups to initiate similar experiments (Friedmann and Roblin 1972)

In the meantime, there was continuing progress in the area of recombinant DNA technology with improvement in techniques of transfection and selection systems for cultured cells. By 1971, Paul Berg was beginning to develop the first recombinant viral vector system by splicing genes into the simian virus, SV40. As a result of his successes, there were some concerns among the members of the scientific community about the safety of such experiments. Because of these concerns, the first Asilomar Conference was held in January 1973. This produced the first detailed discussion of the potential hazards that could be associated with recombinant DNA research. Following the conference, a letter was published in *Science* that emphasized the paucity of knowledge about the consequences of this type of experimentation (Berg et al. 1974). The National Academy of Sciences established a committee to study recombinant DNA technology, and these developments captured the attention of both the press and the public. In February 1975, the second Asilomar Conference was convened, and this provided the forum for an extensive debate about the putative dangers of recombinant DNA experimentation. Despite the differences of opinion, the scientists in attendance agreed to a voluntary moratorium on certain kinds of experiments. On the day following the conference, the first meeting of the National Institutes of

Health (NIH) Recombinant DNA Advisory Committee (RAC) was convened, and its members began the task of creating the *NIH Guidelines for Research Involving Recombinant DNA Molecules*. On June 23, 1976, the *NIH Guidelines* were first published in the *Federal Register*.

THE ORIGINS OF HUMAN GENE THERAPY OVERSIGHT

On June 20, 1980, a letter signed by the General Secretary of the National Council of Churches (Protestant), the General Secretary of the Synagogue of America, and the General Secretary of the United States Catholic Conference was sent to President Jimmy Carter. They expressed concerns about the dangers inherent in genetic engineering in the context of moral, ethical, and religious questions. There was particular emphasis on maintaining the fundamental nature of human life and the dignity and worth of the individual human being.

A few months later, Dr. Martin Cline of the UCLA School of Medicine attempted to perform gene therapy by using the calcium phosphate method to transfect the β-globin gene into autologous bone-marrow cells and transplant such cells into two patients (one in Italy and one in Israel) with thalassemia. This particular protocol did not receive approval by the local safety committees at UCLA, and when knowledge of these experiments became public, there began a series of proceedings that led to censure by the NIH and withdrawal of research funding.

Subsequent to the two aforementioned events, a Presidential Commission was formed to carry out a detailed examination of the ethical issues surrounding molecular genetics and the capacity for intervention into the human genome. By 1982, this commission published a document (*Splicing Life*) summarizing its findings and concluding that there were no fundamentally new ethical issues ingrained in implementing recombinant DNA technology for human use (President's Commission 1982). However, it was strongly emphasized that there needed to be a well-organized system of public scrutiny for any human gene therapy protocols that might be proposed. Since the NIH RAC had been in existence since 1974 and functioning as a review body since 1976, it was correctly pointed out that it had the most significant expertise and experience in evaluating many types of recombinant DNA experiments. Although the original RAC was made up of only scientific members, in 1978, the Secretary of Health Education and Welfare (HEW), Joseph Califano, required that nonscientific or "public" members be added to the committee. In some quarters, this mixed membership (two-thirds scientists and one-third nonscientists) was referred to as the "second-generation" RAC, an appropriate term for a committee considering issues relating to genetics! Both the members of the Presidential Commission and the Director of NIH concurred that it was appropriate to create a "third-generation" RAC that could oversee proposals for human gene therapy.

In the meantime, the U.S. Congress maintained a continuing interest in human genetic engineering, and at one time, there were 16 pieces of proposed legislation for the regulation of recombinant DNA research. But the advent of the RAC served to allay some of the fears of the politicians, and no formal statutes were ever passed by either body of Congress. Following the report of the Presidential Commission, the House Committee on Science and Technology convened its Subcommittee on Investigations and Oversight to review major issues relating to human genetic engineering. Interestingly enough, this subcommittee was chaired by Albert H. Gore, Jr., then a young congressman from the state of Tennessee.

On April 11, 1983 the Chair of the RAC asked his committee if it would wish to study and respond to the report of the Presidential Commission. The response was affirmative with two goals in mind: the creation of a mechanism for reviewing human gene therapy proposals, and acceptance of the responsibility for the review of these proposals at such time when they would be put forward. A RAC subcommittee (Human Gene Therapy Subcommittee [HGTS]) was created for the purpose of developing a document that could be used as a guideline for the review of human gene therapy protocols. The membership of this subcommittee included laboratory scientists, physicians who were clinical researchers, lawyers, and ethicists. By early 1985, this subcommittee had completed its initial draft of a document entitled "Points to Consider in the Design and Submission of Somatic-Cell Human Gene Therapy Protocols."

This "Points to Consider" document was published in the *Federal Register* with the request for public comment, and the document was further refined in response to these comments. A revised version was published in the *Federal Register* for the second time, and further modifications were made. On September 23, 1985 the "Points to Consider" document was presented to the full membership of the RAC and accepted by that body. By February 1986, the Executive Secretary of the RAC sent a letter to all "interested parties" calling for the submission of preclinical data that might pertain to human gene therapy research proposals. On May 12, 1986, W. French Anderson resigned from the RAC HGTS because of his stated intent to submit a gene therapy protocol in the foreseeable future.

There were a number of seminal discussions at the meetings of the HGTS during 1986 and 1987. Included were topics such as retroviral vectors, transgenic animals, and the Food and Drug Administration (FDA) process for the regulation of investigational new drugs. By 1987, French Anderson and his collaborators had developed and presented for review a sizeable tome entitled "Preclinical Data Document." Because of the vast amount of data in this document, it was physically large and was affectionately and scathingly referred to as the "telephone book."

On July 29, 1988, the first request for a clinical trial was formally presented to the HGTS by Steven Rosenberg, French Anderson, and their colleagues. It was not a typical example of a gene therapy protocol, but rather it was a "gene marking" protocol designed to establish that retroviral vectors containing the transgene encoding for neomycin resistance could be given to patients without inducing serious side effects. The Subcommittee formally decided to extend the range of its oversight to both gene transfer and gene therapy protocols. Following an initial discussion, approval was deferred pending the receipt of additional data. A second deferral was issued in September, again based on the request for additional information. On October 3, 1988, the RAC approved this protocol by a majority vote, but the Director of NIH, Dr. James Wyngaarden, did not give his approval, and on October 18, 1988, he requested that the protocol be resubmitted to the Subcommittee with more qualifying information. On December 9, 1988, the HGTS approved the protocol and subsequently the NIH Director gave his approval.

It was at this December meeting that a more precise mode of interaction between the HGTS and the RAC was worked out, and it was agreed that this national review would be a two-step process. An initial review would be conducted by the HGTS and, once full approval was given, the protocol would then be reviewed by the RAC.

On January 30, 1989, the NIH Director publicly announced the approval of the gene marking protocol, and immediately the Foundation on Economic Trends (FET) filed a lawsuit to prevent patients from enrolling. Jeremy Rifkin of the FET successfully advanced the contention that a telephone conference call among RAC members was not equivalent to a public RAC meeting, and thus the vote of approval would have to be repeated in a public venue. After several months of legal discussions, the matter was successfully resolved.

On March 30, 1990, French Anderson and Michael Blaese formally submitted a gene therapy protocol for the study of adenosine deaminase deficiency (ADA), and by June 1, the HGTS gave provisional approval for this trial, pending the receipt of additional data. On July 30 and 31, 1990, a number of seminal events occurred. On July 30, the HGTS approved two protocols, the ADA protocol and a protocol for the study of cancer using tumor-infiltrating lymphocytes to deliver the tumor necrosis factor (TNF) gene to melanoma cells. At the RAC meeting on July 31, both protocols were approved.

On September 14, 1990, Ashanti DeSilva was admitted to the Clinical Center at the NIH, where she was given approximately 10^9 peripheral blood T lymphocytes containing the normal ADA gene. With this event, human gene therapy clinical research became a reality. (For an excellent accounting of this chronology, see the publication of Walters [1991] and the published minutes of the RAC meetings [National Institutes of Health 1989a,b, 1990a,b,c].)

SCHEMA FOR REVIEW OF HUMAN GENE THERAPY PROTOCOLS

There were two principal components that constituted the NIH-directed oversight of human gene therapy protocols. First, there was the local level of review at the individual institutions or medical centers. This involved two separate committees, the Institutional Review Board (IRB) that was the ethics board charged with protecting patients from unnecessary research risks, and the Institutional Biosafety Committee (IBC) whose origins are embedded in the requirements of the *NIH Guidelines for Research Involving Recombinant DNA Molecules* (National Institutes of Health 1997b). One always risks a certain degree of inaccuracy in making generalizations, but in the beginning, the IBCs tended to focus on the scientific aspects of the protocols while the IRBs maintained an emphasis on informed consent. At the outset, both of these review boards faced the challenges that occur with the advent of any new departure in the field of scientific and clinical research, the relative lack of expertise in confronting new paradigms. In many instances, the most knowledgeable people in the field were the principal investigators who were sponsoring the protocols, and these individuals bore the baggage of a direct conflict of interest when debating the merits of a given research proposal. Not infrequently, the local review boards gave a provisional approval and deferred to the HGTS and the RAC before issuing a final approval. For want of a better descriptor, these were the early "growing pains" that were present at the outset of human gene therapy research.

The second component of review occurred at the national level and initially involved both the HGTS and the RAC. A compendium of specific requirements was included in the "Points to Consider" or Appendix M of the *NIH Guidelines*. There have been continuing modifications of the original document, whose origins date back to 1985, but most of these changes reflect an accrual of experience in studying patients undergoing gene transfer. The fundamental questions concerning a trial are little changed from the time they were first formulated. Included in this section is a listing of the principal questions that an investigator must answer before a protocol is submitted to the RAC or the FDA.

Background and Rationale for the Trial

The following information is to be provided about the proposed study:

1. Title, principal investigators, and the participating institutions.
2. State concisely the overall objectives and rationale of the proposed study. Provide information on the specific points that relate to whichever type of research is being proposed.
3. For research in which recombinant DNA is transferred to treat a disease or disorder (e.g., genetic diseases, cancer, and metabolic diseases), the following questions should be answered:

    a. Why is the disease selected for treatment by means of gene therapy a good candidate for such treatment?
    b. Describe the natural history and range of expression of the disease selected for treatment. What objective and/or quantitative measures of disease activity are available? Are the usual effects of the disease predictable enough to allow for meaningful assessment of the results of gene therapy?
    c. Is the protocol designed to prevent all manifestations of the disease, to halt the progression of the disease after symptoms have begun to appear, or to reverse manifestations of the disease in seriously ill victims?
    d. What alternative therapies exist? In what groups of patients are these therapies effective? What are their relative advantages and disadvantages as compared with the proposed gene therapy?

Vector, Target Cell, and Transduction Procedures

1. What is the structure of the cloned DNA that will be used? Describe the gene (genomic or cDNA), the bacterial plasmid or phage vector, and the delivery vector (if any). Provide a complete nucleotide sequence analysis or a detailed restriction enzyme map of the total construct.
2. Identify the intended target cells to be transduced by the proposed vector. If the cells are transduced ex vivo, are the transduced cells to be lethally irradiated?

3. What is the minimal level of gene transfer and/or expression that is estimated to be necessary for the gene transfer protocol to be successful in humans? How was this level determined?
4. Explain in detail all results from animal and cultured cell model experiments that assess the effectiveness of the delivery system in achieving the minimally required level of gene transfer and expression.
5. To what extent is expression only from the desired gene (and not from the surrounding DNA)? To what extent does the insertion modify the expression of other genes?
6. In what percentage of cells does expression from the added DNA occur? Is the product biologically active? What percentage of normal activity results from the inserted gene?
7. Is the gene expressed in cells other than the target cells? If so, to what extent?
8. If a viral vector is being used as a delivery system, what assay is used to detect replication-competent virus? What is the level of sensitivity of the assay for the replication-competent virus?
9. Is there any toxicity related to the vector or the gene delivery system? What animal studies have been conducted to determine if there are pathological or other undesirable consequences of the protocol? Is there evidence for insertion of DNA into cells other than those treated, particularly germ-line cells?

Clinical Protocol

1. What in vitro or in vivo systems were used to determine peclinical efficacy? (A) Cell culture model systems; (B) animal models; (C) relevant human studies.
2. What are the clinical endpoints of the study? Are there objectives and qualitative measurements to assess the natural history of the disease? Will such measurements be used in patient follow-up? How will patients be monitored to assess specific effects of the treatment on the disease? What is the sensitivity of the analyses? How frequently will follow-up studies be conducted? How long will patient follow-up continue?
3. What are the major beneficial and adverse effects that are anticipated? Are there any anticipated immune reactions with the vector or gene product?
4. Selection of patients: What are the number and ages of research subjects who will participate? What are the inclusion and exclusion criteria?

Qualifications of Investigators and Adequacy of Laboratory and Clinical Facilities

1. Provide curriculum vitae for professional personnel.
2. Identify hospital or clinic where studies will be conducted.

Informed Consent Process and Document

1. Indicate where in the submitted materials information is provided about the manner in which the study will be communicated to potential participants, including: (A) personnel involved (and measures taken to avoid conflicts of interest if the researcher has responsibility for the subject's medical care); (B) procedures used to explain the experiment to potential subjects; (C) the length of time for decision making; and (D) any special arrangements in place for pediatric or mentally handicapped subjects.
2. Provide the date that the informed consent was approved by the Institutional Review Board. Include the documentation in the submission.
3. Indicate where the following general consent form elements are found: (A) a description of the study; (B) a listing of possible treatment or research alternatives; (C) an indication that participation is voluntary; (D) a statement of the possible personal and/or societal benefits; (E) a statement of the possible risks, discomforts, and side effects; and (G) an indication of any costs for which the subject will be responsible.
4. Indicate clearly on the document where the following specific requirements of gene transfer research are found: (A) information on any reproductive risks or restrictions; (B) an explanation of the need for long-term follow-up and the arrangements in place for such follow-up; (C) an indication that permission to perform an autopsy will be sought from the family, whatever the cause of the subject's death; (D) a statement explaining the possibility of media interest and steps the institution will take to protect the subject's privacy; and (E) the identity of those groups (agencies, institutions, companies) with access to the subject's records (for further details, see National Institutes of Health 1997b).

Although the basic structure and main requirements of the "Points to Consider" have remained fundamentally unchanged, there have been a number of modifications that reflect natural developments. One of the first involved vectors. In the beginning, the predominant gene delivery system was the retroviral vector, and many of the early groups of investigators used the very same vector backbone with only the transgene and perhaps the promoters being changed. Therefore, there was no continuing need to require the complete vector sequence, but rather the specific request was only for the sequences unique to that particular class of vectors. A similar development occurred when the adenovirus vectors became the dominant system for in vivo gene delivery. Many of the first-generation vectors were deleted in the early (E1–E2) gene region, and the vector backbones were essentially identical. It became readily apparent that there was no necessity to ask investigators to provide the entire 36-kb sequence of the adenovirus genome, but again the required information was limited to the unique sequences associated with the transgene and its promoters and enhancers.

During the first several years of public oversight of gene therapy protocols, the RAC added a number of specific requests with regard to the manner in which the informed consent process was carried out. These changes were aimed at maximizing the objectivity of the process while minimizing any potential conflict of interest that might compromise the efforts of the investigators. The review elements described above reflect the latest iteration of the required information related to informed consent. Throughout its history of protocol review, the RAC debates over the form and substance of informed consent documents were marked by considerable entropy. Despite the unhappiness of some RAC members, there was the underlying and untidy reality that the RAC's role with regard to informed consent was purely advisory. The federal government had already established the Office of Protection from Research Risks to oversee the protection of human subjects, and its mandate was derived from a Code of Federal Regulations (Protection of Human Subjects 1983). Thus, the administrative responsibility for informed consent procedures was not even assigned to NIH, but was assigned to a discrete entity within the Department of Health and Human Services, and the final control over the content of informed consent documents was delegated in autonomous fashion to the individual institutions sponsoring the research. Notwithstanding the fact that the RAC had to operate in a somewhat consultative capacity, there is documented evidence that it had an influential role in shaping both the process and text of informed consent documents.

EVOLUTION OF RAC OVERSIGHT OF GENE THERAPY

In the early stages of human gene therapy research in the United States, there were two complete, but independent, processes for the review and approval of protocols: the NIH oversight process conducted by the NIH

RAC, and the FDA regulatory process that is mandated by federal statute. The NIH reviews are conducted in a completely open public forum, and the FDA reviews are conducted in a closed forum because of legal requirements that information pertaining to new drug development be treated as proprietary and confidential. In retrospect, it is readily apparent that the existence of a public review process did much to allay the fears of the populace concerning this type of genetic intervention in human beings. Even before the first human gene therapy protocol was approved, a poll conducted by Louis Harris and Associates revealed that 52% of the respondents felt that gene therapy was not morally wrong, although 63% admitted that they knew very little about genetic engineering (Office of Technology Assessment 1987).

As has been true for much of recombinant DNA research, many of the earliest concerns and fears about safety in human gene therapy have failed to materialize. It was entirely appropriate to be very conservative at the outset, but the acquisition of substantive experience forces a constant reevaluation of oversight paradigms. As early as 1991, the RAC and its HGTS had to come to grips with the reality that there was a redundancy in this dual review process that was prolonging review without enhancing safety or quality of the protocols. At its meeting on October 7–8, 1991, the RAC agreed to consider disbanding the HGTS and merging its membership with that of the RAC (National Institutes of Health 1991). This was based, in part, on the correct premise that the principal business of the RAC, at this point in its history, was the review of human gene therapy protocols. At the meeting of the RAC on February 10–11, 1992, the HGTS was formally disbanded, all national review of protocols was redesigned to be conducted solely by the RAC, RAC meetings were to be conducted four times a year instead of three, and a 1-year transition period was established during which time the members of the HGTS not already on the RAC would be appointed to the RAC (National Institutes of Health 1992).

During the period from 1990 through most of 1992, the majority of reviewed protocols were for the purposes of studying cancer, and there was a significant degree of repetitiveness in the strategies being proposed. The most common approach involved the use of retrovirus vectors to deliver cytokine genes such as interleukin-2 (IL-2) to autologous tumor cells to modify them in such a way that there would be immune recognition and the induction of cytotoxic T cells that would be tumor-specific.

In December 1992, two investigators (Drs. Ivor Royston and Robert Sobol) requested that the Director of the NIH and the Commissioner of the FDA grant them a compassionate plea exemption so that they could use gene therapy to treat a single patient with a specific type of brain tumor, glioblastoma multiforme. This request represented a significant departure because, if approved, it would bypass all of the oversight procedures utilized by the RAC. This type of request posed less of a problem for the FDA because this agency had in place a mechanism for approving compassionate treatment of single patients in lieu of a full-fledged regulatory review.

At the December 1992 meeting of the RAC, serious concerns were expressed about this alternative approach to approving protocols. It was emphasized that this field of research was quite new and that there was no evidence of efficacy in any of the approved trials that were under way. The investigators who made this request were unable to present any preclinical data; thus, there was no way to evaluate possible risks associated with the trial. There was a general agreement among the members of the RAC that expedited review could be tenable if the protocol in question represented a minor variation of a previously approved protocol. However, the Royston–Sobol protocol did not readily fit the minor variation category, and the RAC did not approve the request. Subsequently, the Director of the NIH and the Commissioner of the FDA approved the request on a compassionate plea basis, which was completely legal for them to do because they were federal government agency heads. Committees such as the RAC are advisory to the NIH Director, but their actions constitute recommendations that can be accepted or rejected by the Director. Interestingly enough, this incident proved to be one of a kind, and there were no recurrences of this type. However, the RAC did create procedures for expedited reviews that were designed to involve ad hoc reviewers who could be called into service on very short notice (National Institutes of Health 1993). At this same time, the RAC also established other categories of protocols that could be exempted from full review.

In 1994, a new government advisory committee was created by the Department of Health and Human Services, the National AIDS Task Force on Drug Development. This task force had a very mixed membership that consisted of government officials, physicians, AIDS clinical researchers, pharmacologists, pharmaceutical company executives, and members of several AIDS activist groups. It was this latter contingent of the task force that demanded a change in the approval procedures for human gene therapy protocols. It was their contention that AIDS patients were being denied possible cures for HIV infection because the RAC and FDA approval processes were totally redundant and unnecessary. It was proposed by the AIDS activists that the RAC be abolished and that the FDA conduct sole reviews of human gene therapy research proposals.

This request was taken under advisement by the Director of the NIH and the Commissioner of the FDA, both of whom were members of the task force, and a compromise counterproposal was forwarded to the task force. Under the provisions of this new plan for review, both the NIH and the FDA would review all new protocols simultaneously. Appropriate staff members of the two government agencies would consult, and if the protocol represented a significant departure in concept or design as compared to previously approved protocols, it was to be fully reviewed by both the RAC and the FDA. If the protocol was not significantly different from previously approved protocols, then it was to receive a single review by the FDA. This new scheme for review was accepted by the AIDS Task Force on Drug Development, and at its meeting on September 12–13, 1994, the RAC voted to approve changes in the *NIH Guidelines* that would accommodate the consolidated review process (National Institutes of Health 1994).

In 1995, the NIH Director, Dr. Harold Varmus, appointed an ad hoc review committee to assess the activities of the RAC, to provide recommendations about its changing role, to define ways to modify its operations, and to determine how it should function to coordinate and facilitate productive gene therapy research. This committee met several times during 1995, and on September 8, 1995, it issued an executive summary of its findings to Dr. Varmus. Among its major recommendations were the following:

1. To avoid duplication of effort and unnecessary delay, the RAC should no longer carry out a case-by-case review of every clinical gene transfer protocol.
2. Review of protocols by the RAC in an open public forum should continue in several areas of concern in which a particular protocol or new technology represents a significant degree of departure from familiar practices. Such departures include, but are not limited to, the use of novel vectors, particularly in cases in which modified human pathogens (such as herpesviruses or lentiviruses) are being evaluated, gene transfer in utero, potential germ-line modification, and gene transfer in normal volunteers.
3. The RAC should continue to provide advice on policy matters revolving around gene therapy and other recombinant DNA issues to the NIH Director and the public.
4. A mechanism should be devised to enable the RAC to continue to be provided with the data needed for monitoring clinical gene transfer protocols, even though it does not continue to review all such protocols. (National Institutes of Health 1995a).

In May 1996, the Director of the NIH announced that he had made a decision with regard to the future of the RAC. It was his proposal that the RAC be disbanded and that it be replaced with a new, smaller group of scientists and ethicists who would meet on an ad hoc basis to advise the Director on pertinent public policy issues affecting human gene therapy research. The Notice of Intent was published in the *Federal Register* (1996) with a request for public comment. Approximately 60 separate comments were received, and the vast majority spoke in favor of retaining the RAC as the public advisory body for all recombinant DNA research activities. In September 1996, the NIH Director announced that he had decided to retain the RAC, but its membership would be reduced from 25 to 15, it would relinquish the approval process for human gene therapy protocols, and it would assist in the organization of regularly scheduled gene therapy policy conferences covering such topics as novel vectors, use of gene transfer for enhancement, and germ-line gene modification. To accommodate these changes in the role of the RAC, the *NIH Guidelines* were amended and the latest version of this document was published in October 1997 (National Institutes of Health 1997b).

In keeping with its new mandate, the RAC began to organize and conduct a series of gene therapy policy conferences starting in 1997, with the following topics being addressed: the use of the techniques of gene transfer for purposes of enhancement; the use of lentiviruses as gene delivery vehicles; the use of herpesvirus vectors in gene transfer; the use of normal controls in human gene transfer studies; in utero gene therapy; and inadvertent germ-line gene modification. Although RAC approval of human gene therapy protocols is no longer required, selected protocols have been discussed to provide a continuing public forum for analysis of new directions in the field. Thus, the "fourth-generation" RAC continues to extend its influence on the field of recombinant DNA research.

RISK-BENEFIT RATIO AND CHOICE OF DISEASE TARGETS IN GENE THERAPY

Prior to the advent of the first human gene therapy protocol, there was consensus among knowledgeable scientists and ethicists that the first disease targets for gene therapy should be restricted to uniformly fatal diseases for which there was no effective treatment. When ADA deficiency was chosen as the initial disease for the study of gene transfer, it met the criteria in a precise way. However, as is often the case with science, there were other therapeutic developments that occurred during the prolonged approval process for the ADA protocol. Partially matched bone-marrow transplants and the development of polyethylene glycol (PEG) conjugated ADA appeared as alternative therapeutic interventions. On balance, there were sufficient shortcomings in both the cellular and drug approaches to allow approval of the ADA gene therapy protocol.

Another major influence on the approval process for gene therapy was derived from the phase I paradigm established by the FDA for cancer trials. Throughout years of drug development, phase I cancer trials largely have been restricted to patients who have failed all standard forms of therapy, and it would have been inconveivable that gene therapy, with all of its inherent unknowns, would have been permitted in patients with anything but the worst prognosis. Although there is no unequivocal evidence for efficacy in any of the cancer trials, which constitute approximately 80% of all of the approved protocols, it has to be noted that phase I trials are primarily safety trials, and that many of the adoptive immunotherapy trials show experimental evidence (in rodents) of being most effective when there is a small tumor burden. Given the considerable difference in phylogeny between mice and humans, it is still not an unreasonable assumption that adoptive immunotherapy ultimately might have some important adjunctive role in the overall treatment of various kinds of cancers, but it will need to be tested in patients who are not in the terminal course of their disease.

A similar and not inappropriate conservatism prevailed in the choice of the first patient cohorts for the cystic fibrosis trials. Clearly, if gene therapy were to introduce unpredicted toxicities, one would not wish to compromise patients with a reasonable disease prognosis. Again, one would would hope to see the time when the effective nature of gene therapy would justify an early intervention in cystic fibrosis, maybe as early as the age of three, before irreversible lung pathology has begun to develop. Perhaps patients eventually could be considered for gene therapy before their lungs are colonized with *Pseudomonas aeruginosa*.

If there is a constant in science and its attendant translational research, it is change. Indeed, changes have occurred in gene therapy research over the last 7 years, and these changes have occurred in the context of a public oversight process. Not only has the review process become less prolonged and cumbersome, but the choice of research subjects has come to reflect an accrued experience with human gene transfer and a willingness to consider the risk–benefit ratio in terms of a history of numerous trials. In 1995, the RAC made a new departure when it approved a trial for the study of ornithine transcarbamylase deficiency (OTC), a rare disorder of urea metabolism (National Institutes of Health, 1995b). In this case, the chosen research subjects were heterozygotes for the condition, the mothers of the affected males who were homozygotes.

While these women are at risk for the consequences of the disease (hyper-ammonemia, seizures, coma), they are more often asymptomatic and may never exhibit positive phenotypes of the disease. This trial received approval from the FDA and, as of January 1998, several patients have been enrolled in the trial and studied without adverse consequences.

In March 1997, the RAC reviewed a protocol in which normal subjects would be exposed to adenovirus vectors, using both the skin and lung as sites of administration (National Institutes of Health 1997a). After considerable discussion and ad hoc presentations by several ethicists, the protocol received a vote of confidence, although approval is no longer required. Thus, in approximately 7.5 years, gene transfer studies have moved from disease states that represent the worst-case scenarios to the use of normal volunteers. It would have been difficult to predict the time course for this evolution of events.

EPILOGUE TO THE FIRST PUBLIC OVERSIGHT OF HUMAN GENE THERAPY

When one looks back at the history of NIH oversight of recombinant DNA experimentation, and human gene therapy in particular, one is necessarily confronted with a whole series of quixotic exceptions to the intended orderliness of the federal bureaucracy. In terms of origin, the RAC was not created by the Congress, and therefore, it has no statutory origins, although its creation persuaded Congress to resist legislation that could have restricted and significantly impeded recombinant DNA research. Creation of the RAC by NIH Director Dr. Donald Frederickson was a prescient act that facilitated important research advances. Since the NIH is not a regulatory agency and has no statutory authority to create regulations, the creation of the *NIH Guidelines* was of critical importance. Perhaps the most attractive feature of said guidelines is the inherent ability to modify them easily and quickly in response to the rapid changes in science. The *NIH Guidelines* have been modified scores of times; having a "living" document has been a great strength.

There are those in the legal community who have called the *NIH Guidelines* de facto regulations because of the penalty clause related to noncompliance, and still other legal scholars have suggested that the *NIH Guidelines* violate the Administrative Procedures Act because of the penalty clause and because of improper lineage in creating the document. Ordinarily, the Secretary of Health and Human Services would order the creation of a set of guidelines, but the NIH Director chose to move promptly in response to the deliberations of the second Asilomar Conference.

An experienced observer and RAC participant has noted that the NIH role in sponsoring the RAC created a conundrum that was laden with ambiguities (Walters 1991). On the one hand, the RAC is an advisory committee of the NIH that was originally charged with fulfilling its role in a critical and independent manner, while, on the other hand, the RAC sponsor is the chief funding agency for biomedical research in the United States. Thus, the NIH put itself in the position of fulfilling two roles that had the potential for discordance; in one context it was promoting biomedical research in the most fundamental way, and in the other context, it was pursuing quasiregulatory oversight on certain aspects of that same research. Superficial analysis would suggest that the two areas of responsibility coexisted without any serious compromises to either, but that is a matter that is better for historians to judge.

In keeping with all of the other aberrations surrounding its existence, the essential structure of human gene therapy protocol review by the RAC has some unique aspects. At the time of its creation, the RAC was fundamentally a safety committee and the rather elaborate containment procedures that became a part of the *NIH Guidelines* were for the purposes of protecting both the investigators and the environment. In the first 7 or 8 years of the committee's existence, the members were involved in setting the appropriate biocontainment levels for various categories of experiments, but they made no judgments as to the quality of the science.

With the advent of gene therapy review, an important new dimension was introduced. For the first time, the RAC and the HGTS were asked to assume certain review responsibilities akin to those assigned to an NIH study section. At once, the issue of quality of science leaped onto the page. It now became fundamentally clear that one could not extricate the matter of scientific quality from matters of research subject and environmental safety. Not surprisingly, members of the RAC frequently engaged in spirited discussions with principal investigators in requesting additional data, additional experiments, and clarifications regarding data already presented. Yet this entire review process could never be analogous to the peer review process enacted by the study sections and the Advisory Councils of the categorical institutes and centers of the NIH. Put in its simplest terms, the RAC was never granted funding authority, and therefore it had no authority to assign priority scores to individual protocols. Human gene therapy protocols were never competed against each other in a precise way, i.e., the assignment of priority scores. However, it would be dismissive to say that the RAC review process was fundamentally flawed. There was an unvarying concern for safety, and the repeated requests for additional data had a salutory effect on the quality of the science. In addition, many of these protocols received limited funding from the NIH, with the majority of financial support coming from the biotechnology industry. Without the entire control of the funding, the NIH was placed in a position of limited leverage.

Although the RAC no longer exercises approval authority for human gene therapy protocols, it still retains its importance as a public forum for discussion of new scientific and ethical issues in this particular field of research. There is an ongoing ethical obligation to uphold the needs of the public. If the past has been instructive in any way, it has clearly indicated the virtue of public discussion. In its essence, public discussion presents the penultimate opportunity for education and the important result is the opportunity for public acceptance of a new mode of molecular therapy. We are beyond the stage when gene therapy per se needs further justification for its existence. Perhaps one of the most daunting threats to gene therapy lies in confusing this technology with all other forms of genetic manipulation, such as human cloning. There is always the danger that the scientifically unsophisticated will lump all of the elements of genetic engineering into one common, mistakenly homogeneous, category. As long as the public has access to open discussion of new issues in gene therapy, and as long as the press reports those discussions, there is reason to be optimistic that the science will eventually go forward and that there will be philosophical support for the attendant policymaking process. Like the unusual history of the RAC, what is eventual progress may be curiously entangled as the process unfolds, but that is the nature of democracies, and it is undoubtedly a tolerable price to pay for scientific advances with humanitarian potential.

REFERENCES

Berg P., Baltimore D., and Boyer H.W., Cohen S.N., Davis R.W., Hogness D.S., Nathans D., Robin R., Watson J.D., Weissman S., and Zinder N.D. 1974. Letter: Potential hazards of recombinant DNA molecules. *Science* **185:** 303.

Friedmann T. and Roblin R. 1972. Gene therapy for human genetic disease? *Science* **175:** 949–955.

Kornberg A. 1971. Remarks announcing the in vitro synthesis of DNA. In *Genes, dreams, and reality* (ed. M. Burnet), p. 71. Basic Books, New York.

Lederberg J. 1968. Tomorrow's babies. *Proc. World Congr. Fertil. Steril.* **6:** 18–23.

National Institutes of Health. 1989a. *Minutes of the Recombinant DNA Advisory Committee*, January 30, pp. 6–14.

—— 1989b. *Minutes of the Recombinant DNA Advisory Committee*, October 6, pp. 25–27.

—— 1990a. *Minutes of the Recombinant DNA Advisory Committee*, February 5, pp. 5–14.

—— 1990b. *Minutes of the Recombinant DNA Advisory Committee*, March 30, pp. 4–7.

—— 1990c. *Minutes of the Recombinant DNA Advisory Committee,* July 31, pp. 3–17.

—— 1991. *Minutes of the Recombinant DNA Advisory Committee,* October 7–8, pp. 25–29.

—— 1992. *Minutes of the Recombinant DNA Advisory Committee,* February 10–11, pp. 37–41.

——1993. *Minutes of the Recombinant DNA Advisory Committee,* December 2–3, pp. 51–53

—— 1994. *Minutes of the Recombinant DNA Advisory Committee,* September 12–13, pp. 43–48.

—— 1995a. *Ad Hoc Review Committee of the Recombinant DNA Advisory Committee. Executive summary of findings and recommendations,* September 8.

—— 1995b. *Minutes of the Recombinant DNA Advisory Committee,* December 4–5, pp. 19–26.

—— 1997a. *Minutes of the Recombinant DNA Advisory Committee,* March 6–7, pp. 4–17.

—— 1997b. *NIH guidelines for research involving recombinant DNA molecules. Federal Register* (62FR59032), October 31, pp. 92–102.

Notice of Intent to Propose Amendments to the NIH Guidelines for Research Involving Recombinant DNA Molecules regarding Enhanced Mechanisms for NIH Oversight of Recombinant DNA Activities. 1996. *Federal Register* (96FR35744), July 8, pp. 35744–35777.

Office of Technology Assessment. 1987. *New developments in biotechnology—background paper: Public perceptions of biotechnology.* U.S. Government Printing Office, Washington, D.C.

President's Commission for the Study of Ethical Problems in Medicine and Biomedical and Behavioral Research. 1982. *Splicing life: The social and ethical issues of genetic engineering with human beings.* U.S. Government Printing Office, Washington, D.C.

Protection of Human Subjects (45 CFR 46). 1983. *Federal Register* (48FR9269), March 4.

Rogers S. and Moore M. 1963. Studies of the mechanisms of action of the Shope rabbit papilloma virus. I. Concerning the nature of the induction of arginase in the infected cells. *J. Exp. Med.* **117:** 521–542.

Tatum E.L. 1966. Molecular biology, nucleic acids and the future of medicine. *Perspect. Biol. Med.* **10:** 19–32.

Walters L. 1991. Human gene therapy: Ethics and public policy. *Hum. Gene Ther.* **2:** 115–122.

Wolff J.A. and Lederberg J. 1994. An early history of gene transfer and therapy. *Hum. Gene Ther.* **5:** 469–480.

25
Ethical Issues in Human Gene Transfer Research

Eric T. Juengst

Center for Biomedical Ethics
School of Medicine
Case Western Reserve University
Cleveland, Ohio 44106-4976

LeRoy Walters

The Kennedy Institute of Ethics
Georgetown University
Washington, DC 20057

The autumn of 1995 celebrated a number of important anniversaries for the ethics of human gene-transfer research. October 1995 marked 30 years since R.D. Hotchkiss introduced the label "genetic engineering" in an essay entitled "Portents for a Genetic Engineering" that appeared in the *Journal of Heredity* (Hotchkiss 1965). December 1995 marked 20 years since the federal Guidelines for Recombinant DNA Research were developed to regulate genetic engineering, after scientists went public with their concerns in the early 1970s upon actually developing the tools to do what Hotchkiss anticipated (Krimsky 1985). November of 1995 was the tenth anniversary of the submission of the first formal protocol to conduct a clinical trial of gene therapy on a human being, and that trial began 5 years later, in September 1990 (Thomson 1994). Today, that first human gene therapy trial has been declared a success, more than 200 other trials are under way, and squadrons of molecular geneticists and biotechnologists now blithely call themselves "genetic engineers," oblivious to the ironic origins of the label.

One of the striking features of this 30-year history is the extent to which the discussion of human genetic engineering has been open to and influenced by concerns over social values and the public's voice. From the beginning, human gene transfer research seems to have been recognized to involve social value commitments that require the approval of the democratic process. Hotchkiss set the tone by concluding his prescient prediction of genetic interventions we are now capable of performing by prescribing that "The best preparation will be an informed and forewarned public, and a thoughtful body scientific. The teachers and the science writers can perform their historic duties by helping our public to recognize and evaluate these possibilities and avoid their abuses. For these things surely are on the way" (Hotchkiss 1965, p. 202).

In the early 1970s, prospective gene therapists such as W. French Anderson made this point the cornerstone of their arguments about how society should proceed, proposing that "This area holds such promise for alleviating human suffering, and yet is so basic to the needs and emotions of all men [*sic!*], that no individual or group of individuals should take it upon themselves to make the decisions. Only the conscience of an informed society as a whole should make these decisions" (Anderson 1972). This point of view found fertile ground in the nation's new interests in controlling the "biological revolution" in biomedicine, and the national public review process established through the Recombinant DNA Advisory Committee (RAC) was explicitly designed to help achieve that goal. In the years since, the ethos that has shaped the professional and public policy discussions of human gene-transfer research has remained grounded in the remarkably populist view that "the public review provided by the RAC assures both policymakers and the general public that they are well informed about developments in gene therapy and thus partners in the progress of this exciting new field" (Capron et. al. 1993). With the development of local Institutional Biosafety Committees, the RAC's *Points to Consider* for the design of human gene-transfer research, and the process of open public discussion and national approval of all human gene-transfer protocols, the first 211 human gene-transfer trials in the United States have had the distinction of being the most thoroughly reviewed experiments in the history of biomedical research.

This populist tradition is important to recall, because it is the backdrop against which new professional ethical and social policy issues in gene-transfer research are and will be framed. There are four possible types of human gene-transfer interventions that can be performed on individuals. Such interventions can be targeted either toward an individual's somatic cells or toward the germ-line cells, and the goal of the intervention can be either the cure of disease or the enhancement of human capabilities. In this chapter, our goal is to review the ethical issues that arise with each of the four possible combinations of target cells and goals: somatic cell gene therapy, germ-line gene therapy, and the use of gene-transfer interventions to attempt to enhance normal human traits in either germ-line or somatic cells. In each case, our focus will be on the intersection of the ethos that guided the birth of gene therapy and the challenges that the field's future will bring, and on the implications of that intersection for professional practice and public policy.

ETHICAL ISSUES IN SOMATIC CELL GENE-TRANSFER RESEARCH

The literature that captures the public discussion of the ethics of human gene therapy displays an interesting metamorphosis between the early 1970s and the 1980s. The first wave of writing was characterized both by an awe in the face of the "New Biology" and by a sense of being ethically disoriented by its prospects. This period's response to these challenges was a sweeping reach of reflections that quickly took investigators to questions of philosophy and theology in search of moral bearings (Ramsey 1972). Much of this early work is still fresh and valuable at those deeper levels. But it was difficult to translate this work into practical policies for scientific research, and not much concrete policy development emerged from it.

In the early 1980s, driven by the news of Martin Cline's experiment (Kolata and Wade 1980) and lured by the prospect of coming clinical trials for somatic cell therapy at the National Institutes of Health (NIH) (Anderson and Fletcher 1980), the discussion rebounded. This second wave of discussion is captured best in the reports on human gene therapy by the President's Commission for Ethical Issues in Medicine and Biomedical and Behavioral Research (President's Commission 1982) and the congressional Office of Technology Assessment (U.S. Congress 1984), and culminates in the publication of the NIH's *Points to Consider* document in 1990 (Subcommittee on Human Gene Therapy 1990). In this literature, the community seems to have gained a moral compass that could allow policy to be developed without having to triangulate against deep questions of individual belief. As a result, commentators could thus write: "From the present vantage point, it may be hard to remember that in 1980, when the Commission began its work, critics (including respected religious groups) argued that 'fundamental dangers' were posed by any alteration of human genes" (Capron 1990).

This new sense of direction was the hybrid result of two sets of national ethical deliberations that occurred over the course of the 1970s: the discussion of biomedical research with human subjects and the debate over the use of recombinant DNA technology (Areen 1985). Neither of these discussions was conducted with human gene therapy in mind, but together they provided the investigators of the 1980s with a repertoire of widely recognized and well-grounded moral and policy considerations that was unavailable to the investigators of the first phase. With the formation of the RAC's Working Group on Human Gene Therapy, these two discussions came together. Collectively, through the RAC's *Points to Consider* document, their moral considerations now form the ethical framework against which protocols for human gene-transfer research are evaluated and the substantive principles for the next phase of deliberations.

The first important accomplishment of this second wave of discussion was to clarify and integrate into public policy the basic conceptual distinctions between the four possible types of human gene transfer introduced above (President's Commission 1982). This led to the recognition that somatic cell gene therapy, if it worked, would simply be an extension of traditional medical efforts to modulate the expression of genes (such as transplantation or gene-product therapy) and did not raise the special problems created by the other possible forms of gene transfer. Because the science required for successful germ-line interventions in humans was also undeveloped, this allowed the public discussion simply to table the most problematic issues until further notice and concentrate on helping to launch the first form of gene-transfer research in a responsible manner.

Framing somatic cell gene therapy as an extension of traditional medical approaches was progress, because it allowed the public review process to assess gene-transfer research protocols through the ethical questions that had already become normative within biomedical research with human subjects: questions that have to do with the anticipated benefits and risks of the intervention (including biosafety risks), the selection of research subjects, and the protection of the rights of research subjects and their proxies to informed consent, free withdrawal, and privacy. Ironically, however, framing somatic cell gene-transfer research as simply another form of innovative medical therapy also sowed the seeds of views that now challenge the need for any special societal oversight of investigators' ability to address these questions.

Risks and Benefits

Concerns about the relative risks and benefits of gene-transfer interventions reflect a commitment to the basic principle of research ethics that "risks to the subjects [must be] reasonable in relation to anticipated benefits, if any, to subjects, and the importance of the knowledge that may be reasonably expected to result." (45 CFR, Sec. 46.111a). Thus, the RAC says in the introduction to its *Points to Consider:*

> In their evaluation of proposals involving the transfer of recombinant DNA into human subjects, the RAC and its subcommittee will consider whether the

design of such experiments offers adequate assurance that their consequences will not go beyond their purpose, which is the same as the traditional purpose of all clinical investigations, namely to protect the health and well-being of individual subjects being treated while at the same time gathering generalizable knowledge (Subcommittee on Human Gene Therapy 1990, p. 96.)

In practice, this concern means that human gene-transfer researchers must have enough prior knowledge to believe that the probable benefits to the subject outweigh both the possible risks of the experiment and the known benefits of any alternative treatments available. Thus, it lies behind the RAC's questions about, and the community's discussion of, the safety and efficacy of somatic cell gene-therapy techniques, the adequacy of preliminary animal studies, and the relative value of emerging therapeutic alternatives (Brenner 1995).

Moreover, one of the striking legacies of the recombinant DNA debate in the public discussion of gene therapy is the acceptance of the need to prepare for the unforeseen as well as predictable risks in designing gene-transfer research. In the public review of gene-transfer protocols, this concern has concentrated primarily on the "biosafety" risks involved in gene transfer. Both the RAC and the scientific community have gone to unprecedented lengths to assess and minimize both the risks of "insertional mutagenesis" involved in the delivery and integration of exogenous DNA into the subject's cells and the risks that vectors may infect germ-line cells as well as their targets, even when the risks seem quite remote. In accepting this emphasis on safety, the community seems to agree with Howard Temin, when he writes that:

> There are often unexpected effects of new technologies.... The use of retrovirus vectors for somatic therapy of human genetic disease can be looked upon as either a novel improvement on present means of drug delivery or as the introduction of potentially dangerous technology. Although scientists and physicians may believe the first characterization is correct, I am convinced that we must act as if the second characterization is correct. We must design vector systems and protocols so that even quite unrealistic fears about safety are allayed (Temin 1990).

The Problem of "Compassionate Use"

Unfortunately, one consequence of the public's preoccupation with the special risks of gene-transfer research has been its relative inattention to the issues involved in defining the benefits of gene-transfer research, and that is where issues are now emerging. Despite the fact that all initial gene-transfer experiments are ostensibly only Phase I investigations of the safety of a particular protocol, some patients and physicians seek to become involved in gene-transfer research out of hope of clinical benefit. In situations in which no alternative treatment exists and death is imminent, the risks of gene transfer pale, and any possibility of benefit, no matter how remote, seems worth attempting. Since little generalizable new knowledge can be expected from these situations, they raise a fundamental professional and public policy question: Are we ready to move beyond gene-transfer research, with its elaborate process of public review and approval, and to treat gene therapy as an "ordinary" innovative medical practice, performed at the discretion of clinicians and their patients? Given that the public review process was established, in part, in response to an attempt at human gene transfer defended on therapeutic grounds (the Martin Cline case), this question cuts to the heart of that process.

Our experience with this issue to date is instructive. In 1992, the NIH was asked to grant an exemption to the regular review process for gene-therapy protocols, on the basis of the clinical need of a particular patient. Borrowing the notion of "compassionate use" from the U.S. Food and Drug Administration (FDA), the investigator argued that a protocol involving the use of gene therapy for brain tumors that had not yet been approved by the RAC should be allowed to be performed on a terminally ill patient. After initially denying the request, the Director of the NIH granted the exemption on compassionate grounds and instructed the RAC to develop guidelines for the expedited review of such requests in the future (Thompson 1993).

In defending her decision to grant an exemption from public review for this intervention, the NIH Director offered an interesting response. She suggested that far from relinquishing public review of gene transfer, the time had come to expand that review beyond the research setting and to establish a public review mechanism that could oversee even the emergency clinical use of gene-transfer interventions (even if that meant bringing them to the public's attention after the fact) (Healy 1993). The RAC took that invitation seriously and developed an oversight process for the compassionate use of gene transfer in "single-patient protocols" that makes such uses the single hardest form of clinical care to perform legitimately in modern medicine. By requiring Institutional Review Board and Institutional Biosafety Committee (IBC) approval, declaring all gene-transfer interventions as "experimental," and advocating the use of the *Points to Consider* as review standards, the RAC essentially argues that emergency clinical uses of gene transfer should meet the same tests as research studies (Recombinant DNA Advisory Committee 1993). In essence, they suggest that no one should perform human gene transfer in a patient in hopes of therapeutic benefit that they would not do in hopes of simply learning something about the safety of the technique, regardless of the desperation of a particular patient's plight. This is a conservative standard: more conservative than that imposed by the FDA for the compassionate use of investigational drugs (Flannery 1986), and even more conservative than the standards used to implant the first artificial heart (Annas 1985).

Selection of Subjects

The development of national policies for research with human subjects in the 1970s was motivated in part by a concern that the burdens of participating in biomedical research were unfairly distributed among particularly vulnerable populations within society, including children and the seriously ill. Children have traditionally been viewed as vulnerable because of their inability to consent to research, and the seriously ill, like the parents of sick children, are capable of neglecting important risk considerations out of clinical desperation, to their own detriment. As a result, whenever it is possible, new biomedical interventions are usually first tested for their safety in healthy adult volunteers in phase I trials before being tested for efficacy in patient-subjects or children (Levine 1986). But both children and the seriously ill have been prominent among the initial subjects of human gene-transfer phase I trials.

Children have been involved in gene-transfer trials when the diseases targeted are fatal in childhood, making their selection as initial subjects unavoidable. Seriously ill adult patients, however, have been typically selected as subjects against another rationale: Like cancer patients in chemotherapy trials, they are recruited because they are the least likely to suffer the consequences of any unanticipated harms from gene-transfer research, and they have the most to gain from any unanticipated benefits. This justification depends heavily on the assumption that, like chemotherapy, gene transfer involves serious potential "toxicities" that would not be appropriate to impose on healthy subjects with nothing to gain. If it were only a matter of distributing potential benefits, the argument would fail: The expectation of benefit from a Phase I study is by definition "unreasonable," and can only become a factor against the backdrop of the risks involved.

In this context, the new issues are the challenges presented by the lower expectations of risk from gene-transfer procedures. Here, the question is, as gene-transfer research becomes safer and our confidence in its techniques grows, when should it shift to the standard model for biomedical research and begin its work by testing the safety of its vectors with normal volunteers? This is the shift of ethical orientation that now confronts the public review process.

Respecting the Rights of Subjects

The need for informed consent by patient-subjects is one of the central tenets of research ethics. Thus, it is no surprise that this issue would be considered in ethical discussions of gene-transfer research, although the challenges in securing voluntary and informed participation do not differ in principle from other areas of biomedical research. Two complicating features of the gene-transfer context are worth noting. First, the proposed interventions in gene-transfer experiments are usually at the cutting edge of research and require that subjects or their proxies have at least a basic understanding of molecular biology. Effective techniques for conveying complex technical information to laypeople do exist and should be employed in the consent process for these research protocols. Second, to the extent that subjects are also patients, those securing their participation should be especially sensitive to the fact that prospective subjects will often be in relatively desperate clinical circumstances, which can exert a powerful influence on the motivations of the subjects, their families, and their physicians. One crucial precaution in this regard is to avoid fostering the "therapeutic misconception" that can be created when terms suggesting clinical benefit are used to describe basic research interventions. Thus, for many protocols, it is preferable to use the phrase "human gene transfer" rather than "gene therapy" to emphasize the remoteness of therapeutic benefit.

The protection of privacy and confidentiality for the subjects in gene-transfer research has also been important. In other cases involving innovative therapy—for example, in the early heart transplants—a virtual media circus has surrounded both the subjects and the research team. Thus, the RAC's *Points to Consider* asks researchers to provide plans for dealing with the public disclosure of their research in responsible ways. A proper approach to subjects will not isolate them from public view, but will attempt to strike a balance between disclosure to an interested public and respect for the subject's privacy.

Protecting the privacy of research subjects, like the job of ensuring that research subjects participate in biomedical studies voluntarily and in an informed way, is in most areas of biomedical research the responsibility of the Institutional Review Boards that oversee research at the local level. Increasingly, concerns have been expressed that the additional public scrutiny of these protections at the national level by the RAC has become redundant and unnecessary. As a result, in 1998, on the basis of the recommendations of an external advisory panel, the NIH Director removed the RAC's authority to review and approve individual protocols for human gene-transfer research, effectively undercutting the public's ability to participate in the decision-making about gene-transfer research and the 30-year tradition of seeing gene therapy as very much the public's business.

In some respects, the national public review process that has overseen the development of human gene-transfer research has been the victim of its own success. With the widespread public acceptance of the idea of gene therapy and the successful review of over 200 pioneering protocols, the RAC and its process are often held up as a model of other scientific policy-making challenges (Wolf 1997). In sum, however, that very success has led to challenges on three fronts: (1) Some argue that the similarity between human gene transfer and other innovative medical therapies has become so strong that the prospects of clinical benefit from gene-transfer interventions now justify decisions to apply gene-transfer techniques to seriously ill patients in the clinical setting without prior public discussion and review; (2) others argue that the risks of gene transfer have now been reduced to the point that it is time to shift into a more traditional biomedical research and development paradigm, and to test new interventions for safety on normal volunteers before recruiting seriously ill patients; and (3) still others argue that both the risks and benefits of somatic cell gene therapy are unremarkable enough to no longer need public discussion at the national level over and above the ordinary system of research review. As the field of human gene-transfer research matures and proliferates, each of these arguments is likely to be pressed further. At the same time, the next phase of our national conversation about the ethics of human gene transfer remains to be addressed.

ETHICAL ISSUES IN HUMAN GERM-LINE GENE TRANSFER

One of the problems that would be created by a premature dismantling of the public review process for human gene-transfer research is that the issues that were set aside during the development of criteria for assessing somatic-cell gene therapy would be left unaddressed at the national level. This would be an important loss, since these issues—the challenges raised by the prospect of germ-line gene-transfer interventions and the uses of gene transfer for enhancement purposes—were the ones that provoked the development of this process in the first place. As the discussion to date already shows, both of these categories raise fundamental social policy issues that require widespread public reflection and debate.

The major difference between somatic cell gene therapy and clinical techniques aimed at germ-line genetic intervention is that the latter would produce clinical changes that could be transmitted to the offspring of the person receiving the intervention. This simple difference is often the only consideration cited in the many official statements that endorse somatic cell gene therapy while proscribing or postponing research aimed at developing human germ-line gene therapy. Behind these official statements, however, lies a longer argument, revolving around four sets of concerns: scientific uncertainties, the need to use resources efficiently, social risks, and conflicting human rights concerns.

Scientific Uncertainties

Even the proponents of germ-line gene therapy agree that human trials under our current state of knowledge would be unacceptable. For gene-therapy techniques to be effective, the genes must be stably integrated, expressed correctly and only in the appropriate tissues, and reliably targeted to the correct location on a chromosome. If the intervention cannot eliminate the parents' risks of transmitting the alleles they carry or can only do so by substituting other genetic risks, its promise remains weak. Critics maintain that, given the complexity of gene regulation and expression during human development, germ-line gene-transfer experiments will always involve too many unpredictable long-term iatrogenic risks to the transformed subjects and their offspring to be justifiable (Council for Responsible Genetics 1993).

Proponents, however, respond that our current ignorance only justifies postponing human trials of germ-line therapy techniques until their promise can be improved. A more optimistic reading turns the argument around in that to the extent that the barriers to effective therapy can be overcome, its' promise should encourage research to continue. Proponents add that by focusing on the obvious barriers to performing clinical trials today, critics of germ-line gene therapy ignore the fact that it will take future research to determine whether or not they are right. So the question remains as to whether current barriers should ultimately dissuade society from contemplating clinical trials in the future (Munson and Davis 1992).

Proponents bolster their technological optimism with an argument from medical utility: that germ-line gene therapy offers the only true cure for many diseases. If illnesses are understood to be, at root, "molecular diseases," then therapeutic interventions at any level above the causal gene can only be symptomatic. From this perspective, all gene therapies involving the simple addition of genes are palliative measures on the road to complete "gene-replacement surgery" in the germ line.

Allocation of Resources

One common criticism of the argument from medical utility is that it betrays a reductionistic attitude toward illness that fails to appreciate approaches that could achieve the same ends more efficiently. Since it must become possible to identify pre-embryos in need of therapy before their transformation, the argument goes, it would be more efficient simply to use the same techniques to identify healthy pre-embryos for implantation (Davis 1992). Many clinical geneticists argue that even our current methods of prenatal screening serve this function. Against these convenient, effective approaches, they conclude, germ-line gene therapy will never be cost-effective enough to merit high enough social priority to pursue.

One scientific rejoinder to this argument is that screening will not help with all cases. Presumably, for example, as more beneficiaries of somatic cell gene therapy survive to reproductive age, there will be more couples whose members are both afflicted with the same recessive disorders (Wivel and Walters 1993). Gene-therapy strategies that affect the germ line may also be the only effective ways of addressing some genetic diseases with origins very early in development. Moreover, by preventing the transmission of disease genes, germ-line gene therapy could obviate the need for screening and somatic cell gene therapy in subsequent generations of a family.

Social Risks

Proponents of germ-line gene-therapy research also point out that screening prevents genetic disease only by preventing the birth of the patients who would suffer from it. This, they point out, is a confusion of therapeutic goals that runs the long-term risk of encouraging coercive eugenic practices and tacitly fostering discrimination against those with genetic disease. By attempting to prevent disease in individuals rather than selecting against individuals according to genotype, germ-line gene therapy would allow us to maintain our commitment to the value of moral equality in the fact of our acknowledged biological diversity (Juengst 1995).

Critics reply to this that, to the contrary, it is germ-line therapy that has the more ominous social implications, by opening the door to genetic enhancement. One line of argument recalls the historical abuses of the eugenic movement to suggest that, to the extent that the line between gene therapy and enhancement would increasingly blur, germ-line interventions would be open to the same questions about the proper vision of human flourishing that eugenics faced. Even those who dispute the dangers of the "slippery slope" in this context take pains to defend the moral significance of the distinction "between uses that may relieve real suffering and those that alter characteristics that have little or nothing to do with disease" (Fletcher 1985, p. 303). Proponents must then argue that appropriate distinctions among these different uses can be confidently drawn (Anderson 1989) and point out that the same eugenic challenges already face those engaged in preimplantation screening or prenatal diagnosis (Fowler et al. 1989).

Human Rights Concerns

Finally, however, some critics argue that the focus of germ-line gene therapy on the embryonic patient has other implications that foreclose its pursuit. If the primary goal of the intervention is to address the health problems of the pre-embryo itself, germ-line gene therapy becomes an extreme case of fetal therapy, and the pre-embryo gains the status of a patient requiring protection. Germ-line gene-therapy experiments would involve research with early human embryos that would have effects on their offspring, effectively placing multiple human generations in the role of unconsenting research subjects (Lappe 1991). If pre-embryos are given the moral status of patients, it will be very hard to justify the risks of clinical research that would be necessary to develop the technique.

This objection to human germ-line gene-therapy research is framed in several ways. For European commentators, it is often interpreted as the right to one's genetic patrimony (Mauron and Thevoz 1991) in that germ-line gene-therapy interventions would violate the rights of subsequent generations to inherit a genetic endowment that has not been intentionally modified. For advocates of people with disabilities, this concern is interpreted in terms of the dangers of society's willingness to accept their differences (Buchanan 1996). Some feminists join this position as well, out of a concern for the impact on women of taking the preembryo too seriously as an object of medical care (Minden 1987).

Proponents can offer several responses to these concerns. First of all, some of these appeals, such as the appeals to the rights of future generations to an unmodified "genetic patrimony," can be criticized simply on scientific and conceptual grounds as incoherent (Juengst 1998a). Beyond that, proponents also argue that germ-line gene therapy is a reproductive health intervention aimed at the parents, not the embryo (Zimmerman 1991). Its goal is to allow the parents to address their reproductive risks and have a healthy baby, in cases in which the parents' own views prohibit preimplantation screening and embryo selection. In taking this position, proponents acknowledge the moral uncertainty over the status of the pre-embryo and defend parental requests for germ-line therapy as falling within the scope of their reproductive rights. Their argument is that, as a professional policy, medicine should continue to accept and respond to a wide range of interpretations of reproductive health needs by prospective parents, including requests for germ-line interventions (Fowler et al. 1989).

Germ-line gene transfer has traditionally been perceived as scientifically remote. However, the development of techniques in reproductive medicine, like in vitro ovum nuclear transfer (Rubenstein et al. 1995), makes the prospect of applying somatic cell gene-transfer techniques to preimplantation embryos a good bit more realistic. Although the application of gene-transfer techniques at that stage would still lack adequate background evidence of safety and efficacy, the relatively unregulated environment in which innovations in reproductive medicine are introduced increases the risks that such experimentation could occur. The danger of muting the public discussion of human gene transfer at the national level, of course, is the risk of creating a climate in which such experimentation could proceed without prior knowledge or review.

ENHANCEMENT USES OF HUMAN GENE-TRANSFER TECHNOLOGIES

The second distinction that has become standard in discussions of the ethics of human gene therapy contrasts the use of human gene-transfer techniques to treat health problems with their use to enhance or improve normal human traits. Whereas the somatic cell germ-line distinction is accused of lacking adequate ethical force (Moseley 1991), the conceptual line between those two classes of intervention is at least clear. The treatment–enhancement distinction, however, often seems in danger of evaporating entirely under its conceptual critiques and to that extent seems to pose the larger risk to our efforts at assessing human gene-transfer technology.

The Treatment–Enhancement Distinction

The treatment–enhancement distinction is usually used to argue that curative or therapeutic uses of genetic engineering fall within (and are protected by) the boundaries of medicine's traditional domain, whereas enhancement uses do not and to that extent are more problematic as a professional medical practice or a legitimate health-care need (Anderson 1989; Baird 1994). There are several interesting rejoinders to this argument. Some argue that medicine has no essential domain of practice, so that a coherent distinction between medical and nonmedical services can never be drawn in the first place (Engelhardt 1986). Others accept the distinction between treating and enhancing but take on the traditional values of medicine, by arguing that privileging treatment over enhancement is itself wrong (Silvers 1994). Others argue that, in any case, for psychological and economic reasons, the line between treatment and enhancement will be impossible to hold in practice (Gardner 1995).

There is another response to the treatment–enhancement distinction, however, that is particularly relevant to the immediate goals of gene-transfer research. This response criticizes the distinction by showing how it dissolves in the case of using human gene-transfer techniques to prevent disease when such interventions involve the enhancement of the body's health maintenance capacities. The argument is that, to the extent that disease prevention is a proper goal of medicine, and the use of gene-transfer techniques to strengthen or enhance human health maintenance capacities will help achieve that goal, then the treatment–enhancement distinction cannot confine or define the limits of the proper medical use of gene-transfer techniques (Walters and Palmer 1996). This argument is bolstered by the fact that the technical prospects for such preventive enhancement interventions already look good, given gene-therapy protocols now under way to treat ill patients in just those ways (Wilson et al. 1992). One gene therapist summarizes this current biomedical work by saying:

> Over the next few years, it appears that the greatest application will be in the treatment of cancer, where a number of genes that have been isolated have the potential to *empower* the immune system to eliminate cancer cells ... Human gene therapy cancer trials have also been initiated for insertion of the tumor necrosis factor (TNF) gene into T-lymphocytes in an effort *to enhance the ability* of T-lymphocytes to kill tumors. Another approach has been to insert the TNF gene into tumor cells in an effort *to induce a more vigorous immune response* against the tumor (Culver 1993 [emphasis added]).

If human gene-therapy protocols like these are acceptable as forms of preventive medicine, the critics ask, how can we claim that we should be "drawing the line" at enhancement?

There have been a number of attempts to articulate the concept of "enhancement" in a way that will allow it to function as a necessary and sufficient criterion for drawing a boundary for medical practice that can include legitimate forms of preventive medicine. Some accounts use quasistatistical concepts of "normality" to argue that any intervention designed to restore or preserve a species-typical level of functioning for an individual should count as treatment, leaving only those that would give individuals capabilities beyond the range of normal human variation to fall outside the pale as enhancement (Daniels 1992). Others attempt to draw the line between interventions addressed to diagnosable pathologies and those aimed at improvements that do not bear on the individual's health (Juengst 1997). All of these efforts, however, face a fundamental practical challenge in that no matter how the line is drawn, most of the gene-transfer interventions that could become problematic as enhancement interventions would not have to cross that line to be developed and approved for clinical use, because they will also have legitimate therapeutic applications.

For example, consider the efforts to develop gene-transfer techniques for promoting the growth of blood vessels. As long as these techniques are developed and assessed as efforts to help treat coronary artery disease, the issues of their potential enhancement applications need not arise. However, once approved for use by the FDA and absent further public oversight or regulation, the same techniques could be used "off-label" to improve the oxygenation of the limb muscles of athletes as a permanent alternative to the various blood-doping schemes now used in competitive sport. Moreover, given the current regulatory vacuum surrounding the private practice of reproductive medicine, there is little to prohibit the application of these techniques to early human embryos as well, in hopes of effecting germ-line transformations.

These realities have pressed those who would use the treatment–enhancement distinction for policy purposes to articulate the moral dangers of genetic enhancement more clearly (Parens 1998). After all, personal improvement is praised in many spheres of human endeavor, and, as purely elective matter, biomedical interventions like cosmetic surgery are well accepted in our society as means to achieving personal improvement goals.

Enhancement as a Form of Cheating

There are two lines of thought that have emerged from this recent work. The first focuses on the idea that biomedical enhancements, unlike achievements, are a form of cheating. This is the view that taking the biomedical shortcut somehow cheats or undercuts the specific social practices that would make the analogous human achievement valuable in the first place. Thus, some argue that it defeats the purpose of the contest for the marathon runner to gain endurance chemically rather than through training, and it misses the point of meditation to gain Nirvana through psychosurgery. In both cases, the value of the improvements lie in the achievements they reward as well as the benefits they bring. The achievements—successful training or disciplined meditation—add value to the improvements because they are understood to be admirable social practices in themselves. Whenever a corporeal intervention is used to bypass an admirable social practice, then, the improvement's social value—the value of a runner's physical endurance or a mystic's visions—is weakened accordingly. If we are to preserve the value of the social practices we count as enhancing, it may be in society's interest to impose a means-based limit on biomedical enhancement efforts (Murray 1984).

Interpreting enhancement interventions as those that short-circuit admirable human practices in an effort to obtain some personal goal has special utility at two levels of ethical analysis. First, for individuals, it highlights the challenge that enhancement poses to their moral integrity. To what extent can they take credit for their accomplishments if they do not achieve them through the socially valued practices that have traditionally produced them? This question is not one about either causation or responsibility. Clearly, they are still the authors of their accomplishments. It would be a mistake for the student whose Ritalin-induced concentration yields a high exam grade from one night's cramming to think that it was literally the Ritalin that took the test and made the grade. The question is whether the student earned the grade, that is, whether the grade is serving its usual function of signaling the disciplined study and active learning that the practice of being a student is supposed to involve. If the grade is not serving that function, then, for that student, it is a hollow accomplishment, without the intrinsic value it would otherwise have.

Moreover, this interpretation of enhancement also has implications for the policies of the social institutions that maintain the practices we value. To the extent that biomedical shortcuts increasingly allow specific accomplishments, like test-taking, to be divorced from the admirable practices they were designed to signal, the social value of those accomplishments will be undermined. Not only will the intrinsic value be diminished for everyone that takes the shortcut, but the resulting disparity between the enhanced and unenhanced will call the fairness of the whole game (be it educational, recreational, or professional) into question. If the extrinsic value of being causally responsible for certain accomplishments is high enough (like professional sports salaries), the intrinsic value of the admirable practices that a particular institution was designed to foster may even start to be called into question. For institutions interested in continuing to foster the social values for which they have traditionally been the guardians, this has two alternative policy implications. Either they must redesign the game (of education, sports, etc.) to find new ways to evaluate excellence in the admirable practices that are not affected by available enhancements, or they must prohibit the use of the enhancing shortcuts. Which route an institution should take depends on the possibility and practicality of taking either, because ethically they are equivalent.

Enhancement as an Abuse of Medicine

Unfortunately, some of the social games we can play (and cheat in) do not turn on participants' achievements at all, but on traits over which individuals have little control, such as stature, shape, and skin color. The social games of stigmatization, discrimination, and exclusion use these traits in the same manner that other practices use achievements: as intrinsically valuable keys to extrinsic goods. Now it is becoming increasingly possible to seek biomedical help in changing these traits to short-circuit these games as well (Parens 1995). Here, the biomedical interventions involved, like skin lighteners or stature increasers, are enhancements because they serve to improve the recipient's social standing, but only by perpetuating the social bias under which they originally labored. Any normal medical tool that cured one patient by making other patients' problems worse would be considered medically perverse, and in this case, the perversion is compounded by the fact that the problem in question is social, not somatic. When enhancement is understood in this way, it warns of still another set of moral concerns.

With this interpretation, what makes the provision of human growth hormone to a short child a morally suspicious enhancement is not the absence of a diagnosable disease or the dictates of medical policy or the species-atypical hormone level that would result: Rather, it is the intent to improve the child's social status by changing the child rather than by changing her social environment. Enhancement interventions are almost always wrong-headed under this account, because the source of the social status they seek to improve is, by definition, the social group and not the individual. Attempting to improve that status in the individual without regard for its social nature amounts to a moral mistake akin to "blaming the victim": It misattributes causality, is ultimately futile, and can have harmful consequences. This is the interpretation of enhancement that seems to be at work when people argue that it inappropriately "medicalizes" a social problem to use Ritalin to induce cooperative behavior in the classroom or that breast augmentation surgery only exacerbates society's sexist vision of beauty (Morgan 1991). In each case, the critics dispute the assumption that the human need in question is one that is created by, and quenchable through, our bodies and assert that both its source and solution really lie in quite a different sphere of human experience.

This interpretation of the enhancement concept is useful to those interested in the ethics of personal improvement because it warns of a number of moral pitfalls beyond the base line considerations that the enhancement–treatment distinction provides. Attempting to improve social status by changing the individual risks being self-defeating (by exacerbating the individual's sense of inadequacy by inflating expectations), futile (if the individual's comparative gains are neutralized by the enhancement's availability to the whole social group), unfair (if the whole group does not have access to the enhancement), or complicitous with unjust social prejudices (by forcing people into a range of variation dictated by biases that favor one group over others). For those faced with decisions about whether to use performance-enhancing drugs in sports or to insert the leadership gene into embryos, this way of understanding enhancement is much more illuminating than attempts to distinguish it from medical treatment.

The medicalization of social problems, as opposed to merely assisting patients to cheat in specific social achievements, also has significant implications for the boundaries of medicine. Whereas stature or speed can at least be measured by medicine, there is a large class of traits, for example, loyalty and competitiveness, leadership and stewardship, aggression and altruism, that cannot even be perceived without a larger frame of reference. These latter are all traits that characterize our interactions with and evaluation by other people: the machinery of social flourishing. In fact, they are traits that are impossible to identify or evaluate without reference to the social context of the individuals who display them: The solitary shipwreck survivor does not display them. Thus, the improvement of these traits has traditionally been the domain of our social engineers (i.e., teachers, ministers, coaches, and counselors) supported by policymakers charged with creating environments in which such traits can be cultivated. When attempts are made to shift the improvement of these traits into the domain of the doctor, by working directly on the substrates for social capacities that our bodies provide, medicine runs up against a basic epistemic boundary. It is not clear how one would identify any optimum or even maximum conditions for social traits through medical means alone. For example, what dosage of human growth hormone will ensure optimum social advantage? This suggests a test for one kind of enhancement boundary. If criteria drawn from other spheres of experience seem like better measures of improvement than medical measures, then the intervention in question should probably count as an enhancement that goes beyond medicine's domain of expertise. Because of its own epistemic limits, the argument goes, biomedicine should restrict its ambitions to the sphere of bodily dynamics, which it knows something about, and leave the sphere of social dynamics in the hands of others (Juengst 1998b).

CONCLUSION

No other biomedical intervention in history has received as much international and interdisciplinary attention as human gene therapy (Fletcher 1990). Three points of striking consensus have emerged from that global discussion.

The first point is that somatic cell gene transfer research does not constitute a major break with medical therapeutics, but this technique still should be regulated through public review processes. The dangers of confusion over the therapeutic benefits of gene-transfer research, and the need for public discussion of the applications of gene-transfer techniques to the germ line and for enhancement purposes, all reinforce the importance of the unique populist tradition that governed the birth of the field. As somatic cell gene-therapy techniques mature into reliable clinical alternatives for different medical domains, this commitment will raise hard questions about how best to achieve public participation in this field.

The second point of consensus is that current concerns about clinical risks and biohazards make human experimentation with germ-line interventions unacceptable for the time being. Beyond that, discussion is flourishing about whether germ-line interventions could ever be ethically acceptable and, if so, under what circumstances.

The third point of consensus is the view that gene therapy, both somatic and germ line, should be evaluated as a clinical tool employed on behalf of the seriously ill, and not as a eugenic public-health tool designed to benefit the population nor as a shortcut to personal social advantage. In contemporary political argot, genetic medicine should continue to be an empowering, not exclusionary, science. It should be about helping living people address their individual health problems, not about protecting the "gene pool" from the ebb and flow of human alleles in populations or the social classifications that we impose our genetic differences.

ACKNOWLEDGMENTS

The research conducted for this chapter was supported in part by National Institutes of Health grant R01 HG-1446-02.

REFERENCES

Anderson, F.W. 1972. Genetic therapy. In *The new genetics and the future of man* (ed. M. Hamilton), pp. 110–125. W.B. Eerdmans. Grand Rapids, Michigan.

———— 1989. Human gene therapy: Why draw a line? *J. Med. Philos.* **14:** 681–693.

Anderson F.W. and Fletcher J. 1980. Gene therapy in human beings: When is it ethical to begin? *N. Engl. J. Med.* **303:** 1293–1297.

Annas G. 1985. The Phoenix heart: What we have to lose. *Hastings Cent. Rep.* **15:** 15–16.

Areen J. 1985. Regulating human gene therapy. *W. Va. Law Rev.* **88:** 153.

Baird P. 1994. Altering human genes: Social, ethical and legal implications. *Perspect. Biol. Med.* **37:** 566–575.

Brenner M. 1995. Human somatic gene therapy: Progress and problems. *J. Intern. Med.* **237:** 229–239.

Buchanan A. 1996. Choosing who will be disabled: Genetic intervention and the morality of inclusion. *Soc. Philos. Policy* **13:** 18–47.

Capron A. 1990. The impact of the report, Splicing Life. *Hum. Gene Ther.* **1:** 69–73.

Capron A., Leventhal B., Post L., Walters L., and Zallen D. 1993. Requests for compassionate use of gene therapy: Memorandum from the subcommittee to the RAC. *Hum. Gene Ther.* **4:** 199–200.

Council for Responsible Genetics. 1993. Position paper on human germ line manipulation (Fall, 1992). *Hum. Gene Ther.* 4: 35–37.

Culver, K. 1993. Current status of human gene therapy research. In *The genetic resource,* vol. 7, pp. 5–10. New England Regional Genetics Group, Newton, Massachusetts.

Daniels N. 1992. Growth hormone therapy for short stature: Can we support the treatment/enhancement distinction? *Growth Genet. Horm.* (suppl. 1) **8:** 46–48.

Davis B. 1992. Germ-line gene therapy: Evolutionary and moral considerations. *Hum. Gene Ther.* **3:** 361–365.

Engelhardt H.T. 1986. *The foundations of bioethics.* Oxford University Press, Oxford, United Kingdom.

Flannery E. 1986. Should it be easier or harder to use unapproved devices? *Hastings Cent. Rep.* **16:** 17–23.

Fletcher J.C. 1985. Ethical issues in and beyond prospective clinical trials of human gene therapy. *J. Med. Philos.* **10:** 293–309.

———— 1990. Evolution of ethical debate about human gene therapy. *Hum. Gene Ther.* **1:** 55–69.

Fowler G., Juengst E., and Zimmerman B. 1989. Germ-line gene therapy and the clinical ethos of medical genetics. *Theor. Med.* **10:** 151–165.

Gardner W. 1995. Can enhancement be prohibited? *J. Med. Philos.* **20:** 65–84.

Healy B. 1993. Remarks for the RAC regarding compassionate use exemption. *Hum. Gene Ther.* **4:** 195–197.

Hotchkiss R. 1965. Portents for a genetic engineering. *J. Hered.* **56:** 197–202.

Juengst E. 1995. "Prevention" and the goals of genetic medicine. *Hum. Gene Ther.* **6:** 1595–1605.

———— 1997. Can enhancement be distinguished from prevention in genetic medicine? *J. Med. Philos.* **22:** 125–142.

———— 1998a. Should we treat the human germ-line as a global human resource? In *Germ-line intervention and our responsibilities to future generations* (ed. E. Agius and S. Busuttil), pp. 85–102. Kluwer, Dordrecht, The Netherlands.

———— 1998b. The meanings of "enhancement" for biomedicine. In *Enhancing human traits: Conceptual complexities and ethical implications* (ed. E. Parens). Georgetown University Press, Washington D.C. (In press.)

Kolata G. and Wade N. 1980. Human gene treatment stirs new debate. *Science* **210:** 407.

Krimsky S. 1985. *Genetic alchemy: The social history of the recombinant DNA controversy.* MIT Press, Boston, Massachusetts.

Lappe M. 1991. Ethical issues in manipulating the human germ line. *J. Med. Philos.* **16:** 621–639.

Levine R. 1986. Ethics and regulation of clinical research. Urban and Schwarzenberg, New York.

Mauron A. and Thevoz J.-M. 1991. Germ-line engineering: A few European voices. *J. Med. Philos.* **16:** 649–666.

Minden S. 1987. Patriarchal designs: The genetic engineering of human embryos. In *Made to order: The myth of reproductive and genetic progress* (ed. P. Spallone and D. Steinberg), pp. 102–109. Pergamon Press, Oxford, United Kingdom.

Morgan K. 1991. Women and the knife: Cosmetic surgery and the colonization of women's bodies. *Hypatia* **6:** 25–53.

Moseley R. 1991. Maintaining the somatic/germ-line distinction: Some ethical drawbacks. *J. Med. Philos.* **16:** 641–649.

Munson R. and Davis L. 1992. Germ-line gene therapy and the medical imperative. *Kennedy Inst. Ethics J.* **2:** 137–158.

Murray T. 1984. Drugs, sports and ethics. In *Feeling good and doing better: Ethics and nontherapeutic drug use* (ed. T. Murray et al.), pp. 107–129. Humana Press, Clifton, New Jersey.

Parens E. 1995. The goodness of fragility: On the prospect of genetic technologies aimed at the enhancement of human capabilities. *Kennedy Inst. Ethics J.* **5 :** 141–153

———. 1998. Is better always good? *Hastings Cent. Rep.* **28:** S1–S17.

President's Commission for the Study of Ethical Problems in Medicine and Biomedical and Behavioral Research. 1982. *Splicing life: A report on the social and ethical issues of genetic engineering with human beings.* U.S. Government Printing Office, Washington, D.C.

Ramsey P. 1972. Genetic therapy: A theologian's response. In *The new genetics and the future of man* (ed. M. Hamilton), pp. 157–179. W.B. Eerdmans, Grand Rapids, Michigan.

Recombinant DNA Advisory Committee. 1993. Procedures to be followed for expedited review of single patient protocols. *Hum. Gene Ther.* **4:** 307.

Rubenstein D.S., Thomasma D.C., Schon E.A., and Zinaman J. 1995. Germ-line gene therapy to cure mitochondrial disease: Protocol and ethics of in vitro ovum nuclear transplantation. *Camb. Q. Healthcare Ethics* **4:** 316–339.

Silvers A. 1994. Defective agents: Equality, difference and the tyranny of the normal. *J. Soc, Philos.* **25:** 154–175.

Subcommittee on Human Gene Therapy (Recombinant DNA Advisory Committee, National Institutes of Health). 1990. Points to consider in the design and submission of protocols for the transfer of recombinant DNA into the genome of human subjects. *Hum. Gene Ther.* **1:** 93–103.

Temin H.M. 1990. Safety considerations in somatic gene therapy of human disease. *Hum. Gene Ther.* **1:** 111–123.

Thompson L. 1993. Gene therapy: Healy approves an unproved treatment. *Science* **259:** 172.

——— 1994. *Correcting the code: Inventing the genetic cure for the human body.* Simon and Schuster, New York.

U.S. Congress (Office of Technology Assessment). 1984. *Human gene therapy.* U.S.

Government Printing Office, Washington, D.C.

Walters L. and Palmer J. 1996. *The ethics of human gene therapy.* Oxford University Press, Oxford, United Kingdom.

Wilson J.M., Grossman M., Raper S.E., Baker Jr. J.R., Newton R.S., and Theone J.G. 1992. Ex vivo gene therapy for familial hypercholesterolemia. *Hum. Gene Ther.* **3:** 179–222.

Wivel N.A. and Walters L. 1993. Germ-line gene modification and disease prevention: Some medical and ethical perspectives. *Science* **262:** 533–538.

Wolf S.M. 1997. Ban cloning? Why NBAC is wrong. *Hastings Cent. Rep.* **27:** 12–15.

Zimmerman B.K. 1991. Human germ-line therapy: The case for its development and use. *J. Med. Philos.* **16:** 593–612.

Index